AIDS

A SECOND OPINION

AIDS
A
SECOND
OPINION

Gary Null, Ph.D.
WITH JAMES FEAST

SEVEN STORIES PRESS
New York | Toronto | London | Sydney

Seven Stories Press
140 Watts Street
New York, NY 10013
http://www.sevenstories.com

In Canada:
Hushion House, 36 Northline Road, Toronto, Ontario M4B 3E2

In the U.K.:
Turnaround Publisher Services Ltd., Unit 3,
Olympia Trading Estate, Coburg Road, Wood Green, London N22 6TZ

In Australia:
Tower Books, 9/19 Rodborough Road, Frenchs Forest NSW 2086

Library of Congress Cataloging-in-Publication Data

Null, Gary.
    AIDS : a second opinion / Gary Null.—1st ed.
       p. cm.
    Includes index.
    ISBN 1-58322-062-3
    1. AIDS (Disease)—Etiology. 2. AIDS (Disease)—
Treatment. 3. AIDS (Disease)—Alternative treatment. I. Title.

RC607.A26 N85 2001
616.97'92—dc21

                                        00-051013

9 8 7 6 5 4 3 2 1

College professors may order examination copies
of Seven Stories Press titles for a free six-month trial period.
To order, visit www.sevenstories.com/textbook,
or fax on school letterhead to (212) 226-1411.

Book design by Cindy LaBreacht
Printed in the U.S.A.

Contents

Introduction

What is truth? said jesting Pilate,
and would not stay for an answer.

—Sir Francis Bacon, "Of Truth"

In late April 1984, there was a historic press conference, which *Time* magazine introduced as "high noon in Bethesda, Md."[1] It was at this conference
that Dr. Robert Gallo from the National Cancer Institute announced his
discovery that a recently identified virus caused Autoimmune Deficiency Syndrome, or AIDS.

Staying with the *Time* account, let's look at the form of this conference.[2]
Was it the sober issuing of a scientific statement given in a sterile government
office? Well, not quite. The meeting was held in a French restaurant. It did not
consist of the reading of a scientific paper but of the kind of contest one would
more likely associate with game shows than medicine.

Dr. Gallo was facing off against a rival, Dr. James Curran, from the Centers for Disease Control (CDC), which was also searching for the cause of
AIDS. (In February 1985, a report by the federal Office of Technology Assessment, the "Public Health Service's Response to AIDS," would indicate that the
bickering between these two agencies was instrumental in derailing and delay-
ing progress in responding to the disease.)[3]

Dr. Curran gave Dr. Gallo 205 anonymous blood samples: some from patients with AIDS, some from healthy donors, some from people sick with other
diseases. Dr. Gallo had previously tested the samples, checking to see if they
contained the virus he was touting as the culprit in AIDS. Presumably, if traces
of the virus were found in and only in the AIDS patients, he would have given
some proof of his contention of this virus's causative role.

As Dr. Curran read off the designations of the samples, "M5... M28" and so forth, Dr. Gallo responded with either "positive" or "negative." At the end, "he had correctly identified nearly all of them."

It seemed, at least to the journalists present, that Dr. Gallo had proven his point. He had proven something else, too. Serious science, affecting millions of people's health, was no longer a thing of the laboratory but of the grandstand.

The article brought out a number of other significant phenomena, whose implications we need to assess. For one, although this was the first official presentation of Dr. Gallo's results—the conclusions would appear in four forthcoming articles in *Science*—"word of the discovery [already] began to leak out" the previous week, through advance copies of the papers underhandedly obtained by *New Scientist* magazine.[4] Also, as word got around, other scientists vied for the spotlight. As *Time* put it, "scientists... bicker[ed] over who first discovered the virus." Dr. Luc Montagnier of the Pasteur Institute in France said a virus he and his group had discovered previously closely resembled the virus Dr. Gallo identified. And, Dr. Gallo and his group were not resting on their laurels but racing ahead to develop an "accurate and inexpensive blood-screening test."

There are both foreshadowings and symbols in this account. The last two items, the claims of the Pasteur Institute and the promises of a blood test, presage events to come. Dr. Montagnier's remarks concerning the originality of Dr. Gallo's discovery foretell a colossal lawsuit brought by the French government three years later. The question would be put as to whether Dr. Gallo's discovery did not so much innocently mirror certain French results—everyone acknowledged the French had already discovered and published a paper on LAV, a retrovirus said to cause AIDS—as use falsified data, concealing how dependent his discovery was on French advances. The French lawsuit, though, turned more upon the French-American dispute over the origin of the blood test (which would be developed later) since no matter how careful governments are of their scientists' prestige, they are usually more concerned about their countries' financial wherewithal. The blood tests, which both the French and Dr. Gallo would quickly develop, would net the U.S. government millions and Dr. Gallo himself a cool $800,000. By government mandate, these tests would have to be used by U.S. blood banks, which had proved shockingly negligent in screening blood even as cases of AIDS by transfusion mounted. This is what the French lawsuit would center on: Who deserved the monetary rewards that poured in from selling blood tests?

Also appearing in the article are a number of symbolic counters that suggest unsettling facts about modern science, not only in the grandstanding we have mentioned but in location of the meeting and the fact that the press conference preceded scientific publication.

A key point brought forward by Randy Shilts (a journalist whose work we will examine later) is that the timing was dependent on political agendas. The announcement had to make the Reagan administration look good, so it was delayed so that higher-ups, such as Margaret Heckler, chief of the Department of Health and Human Services (HHS), could be present. Yet plans to perfectly orchestrate the event were foiled and the press conference held precipitously when, according to Shilts, a reporter for *New Scientist*, rushed into print, even though he said he would hold the news till after the papers were published. Meanwhile, an AIDS researcher in San Francisco inadvertently blurted out news of the discovery to a radio interviewer. Thus, Dr. Gallo's presentation was hastily scheduled to head off further reporting on the issue, which would steal the thunder from his own staging.

Of course, it may be thought—though it is unseemly—that a scientific statement has to be slated in accordance with the whims of the media; this hardly affects the discovery itself. True enough, but other aspects of the media fanfare do tend to strike more deeply.

Consider for a moment what science has traditionally been about and contrast with the way the media conceives and publicizes the enterprise.

The media imagines that every scientific breakthrough is analogous to Archimedes' discovery of principles of volume displacement. The Greek was sitting in a tub at the public baths when the idea hit him. He ran out naked into the streets near the Acropolis yelling "Eureka" (I found it). The law was immediately accepted and has not been disputed since.

Perhaps very simple laws of physical nature were discovered that way, self-evident once pronounced and accepted by one and all; but the findings of modern research are not so easily and cheaply won. What happens, in the best case, is that a team of scientists publish the results of an experiment. Then this experiment is collegially tested by other interested workers in the field. On the one hand, the experiment is carefully reduplicated in other laboratories to see if it was replicable. On the other, different scientists question the interpretation the original team made of the initial experiment. Even if the results the first group found were repeatable, this does not mean the interpretation arrived at is logically or empirically sound.

This description should alert us to the fact that whatever the ultimate validity of Dr. Gallo's research, he threw out conventional scientific decorum at his press conference in favor of sideshow antics. Of course there was validation of a sort, but not the sort where peer testing goes on in a straightforward, sober manner. When Dr. Curran presents his samples so that Dr. Gallo can pass judgment, we have the kind of testing we expect from a mind reader, like the Great Wazoo, who tells a naive, dumbfounded matron what color hankie she has in her purse.

But what about the interpretation of this information? Here, the weakness of Dr. Gallo's claims are even more patent. He has found a virus that is discovered in AIDS patients. Is it a necessary, logical step to say this causes the disease? Obviously not. Logically, its presence may be a byproduct of another cause, even of another as yet unidentified virus. What would you think of a policeman who arrested a man for murder because he kept finding him at the scene of the crime? Of course, it looks suspicious, but it is not proof.

(To be fairer, Dr. Gallo and his coresearchers' original publications presented the causal role of HIV as hypothetical, if strongly implied, while it was *Time*—with its report that "a breakthrough has finally been made in the understanding the deadly AIDS epidemic... [so that] we now know the face of the enemy"—and other media outlets that jumped the gun in drawing conclusions.[5] However, as we will show later, scientists, even Dr. Gallo himself, soon followed suit, taking the correlation as proof of the definitive causal role of the virus—as if they were letting the media determine the nuances of a scientific debate.)

Moreover, there is one further symbol lurking in the article: the bistro. It was "a small French restaurant, with hanging baskets and beamed ceiling." Why pick such a place for the announcement of a significant discovery? In our society, French restaurants are viewed as the epitome of chic and moneyed dining.

If we temporarily narrow our evaluation of American medicine, simply to look at the question of inequality in recruitment and treatment, we may say that it is a sad fact that the health care system has been geared to the elite. Since early in the century, "opportunities for nonwealthy and nonwhite medical students [who had been prevalent before this time] almost vanished."[6] Although the situation may have improved a little in the past few years, since the beginning of the twentieth century, the increasing cost of medical schools, combined with the decreasing number of such schools and their ever more stringent entrance requirements, has closed them to most minority and poorer applicants. Moreover, given that physicians tend to be drawn from the upper middle class, medical research has been inclined to center on diseases of the well-off. According to complaints voiced by dissidents at the American Medical Association's (AMA) conventions in the 1970s, "Medicine has focused on the diseases of the rich and the established—cancer, heart diseases, stroke—and ignored the diseases of the poor, such as malnutrition and still high infant mortality."[7]

This is the dirty little secret behind our government's and media's criminal neglect of the AIDS crisis in its developmental years, as so ably chronicled by the aforementioned Randy Shilts in *And the Band Played On*. U.S. medicine has long had two modalities, one for treating white, middle class heterosexuals and one reserved for others, the poor and minorities. So, when the cry went out that a disease was ravaging gay males and IV drug users, it was met by blanket silence and indifference from President Reagan, government bureaucrats, and most scientists and journalists, who felt they could safely ignore this situation, which did not affect people with clout.

Eventually, political uproar (along with the principled stance of lone researchers) forced the government and media to turn attention to the disease; yet it might be conjectured that giving the news briefing in a tony restaurant was a way of reasserting that the handling of it would stay in the elite establishment's control.

One thing has been left out of this *Time* story, by the way, which is what had suddenly accelerated the snail's pace of research into the disease that had gone on during the first years of its spread. What overthrew the establishment's expectations that if they ignored the disease it would go away was an unexpected boomerang effect. The excluded set about getting included. For gay males who saw themselves endangered yet ignored this sometimes took the route, as it did in San Francisco, of deepening political activism, which forced elected officials, beholden to gay votes, to put the city at the forefront of government funding of prevention, treatment, and research. Or, as in New York City, brought about the creation of a totally alternative system of health care centered around the Gay Men's Health Crisis (GMHC), drawing its funding from private donors and its dedicated workers from volunteers.

(As an aside, we may note the species of irony the disease sometimes posed for gay males. Belonging to the homosexual minority differs from belonging, for example, to such groups as racial minorities in that it is quite feasible to "pass" as a straight in everyday life, thus avoiding discrimination. Paul Popham, for instance, the president of GMHC, was at first wary of the word "gay" in the organization's name because he was "passing" in his straight job. So disclosing that one had AIDS could be a double blow to a man who was passing, both revealing he had a disease that quickly led to death and tearing down his straight facade, since his cover would be blown. Paradoxically, it may well be that AIDS's potential for destroying masks, and thus indicating both the pretense and undependability of these masks, played a part in galvanizing "closeted" gay males like Popham to embrace their gay identity more fully and become strong fighters in the battle against the disease.)

I don't know how many readers would have drawn all these implications from the *Time* magazine article. Certainly, anyone reading it today, even if he or she overlooked many of the symbols and foreshadowings, would note the outstanding positive and negative prongs; that is, the *coming together* and *splitting apart* that have occurred since that day.

By the latter, the *splitting apart*, I mean that the seeming consensus that existed at that moment—in which everyone agreed that the HIV virus, then called HTLV, was at the bottom of the disease, and that its activities and a sound blood test to detect it were just around the corner— no longer exists. The first part of this book will measure how such dogma, though still held by the orthodox, has come to be shaken at its core.

To preview, the assertion at the Bethesda conference that HIV is the cause of AIDS will be one of the first doctrines we evaluate; and we will see that its lack of scientific credibility is becoming increasingly evident, no matter how much the medical establishment, the government health ministries, and media declare it infallibly true.

The dogma laid down since 1984 is that HIV infection leads to AIDS and so on to death. If the causative role of virus HIV is doubted, then it will immediately follow that the purported cures, which attack the virus, including toxic drugs such as AZT, become insupportable. Further, the social campaigns in response to the disease may be invalidated. Take the recommendation that we give clean needles to drug addicts because stopping infection with HIV is more important than discouraging addiction to dangerous drugs. If HIV is not the main culprit, however, this recommendation falls to the ground.

The first half of the book will run through many of the championed ideas of the establishment (e.g., that AIDS originated in and spreads like wildfire through Africa, that it can be heterosexually transmitted, and that the number of AIDS patients has increased astronomically in the United States over the last decade) and show that, brought to the bar of objective science, they are found wanting.

This will be balanced against the *coming together*, that is, the example of scientists and activists pitching in to work on solutions to the AIDS crisis.
In the *Time* article, coming together was suggested by the way teams of scientific researchers on both sides of the Atlantic had striven semi-collaboratively to understand the disease. In my book coming together is represented by the lab scientists who have come up with new, more convincing ideas of what causes AIDS, as well as the clinicians who have supplied ways of treating the disease that outdistance the often pernicious effects of the drugs prescribed by the establishment.

This is the part of the book that I am most excited about because I feel it is here that the solutions reside for combating AIDS. What moved me to write this book was not an urge for Juvenalian satire, lashing the rogues and clowns of modern establishment science, though there is a place for that. I wrote it because I see hope for AIDS patients.

Though I will begin by casting a cold eye on the inflated claims of much recent research, Part Two will offer alternative positions on the origin of AIDS. If any of these alternative views are correct, then treatments deriving from them will be more successful and will tend to have happier results than those deriving from a faulty analysis. This is where the hope comes in.

The last portion of my book will present three alternative ways of treatment (originating not only from a different understanding of AIDS but from a different perspective on health and disease) that have had phenomenal results in keeping AIDS patients well and thriving. It is the desire to communicate to the concerned these valuable treatments—ones that offer hope but are not getting the attention they deserve from the unadventurous media—that has prompted me to say that in looking at AIDS, we need to rethink all our findings and all our options. We definitely need a second opinion.

From Where I Speak

I have chosen to write this book mainly in first person, just as Charles Dickens wrote *David Copperfield*. Like it was for Dickens, this is more of a rhetorical than a factual convenience. Of course, I have written large portions by myself, but my cowriter, Feast, has also contributed a share, and there are parts each of us has rewritten for the other, all woven into a seamless whole. I also use "we" when I want to convey the collaborative nature of the project.

But there is another sense of collective authorship in this book that is more profound. Although I have directed studies as a lab researcher, the scientific discussion I will present will not be based so much on my own findings as on those garnered from hundreds of hours of interviews, both on and off the air, and study of the works of reputable doctors and researchers— including some top names in their respective fields, who dissent from the dominant AIDS paradigm. This work, then, is a tapestry of voices, often imbricating generous quotations from their works and always sipping inspiration from their fearless, imaginative, and compassionate work.

As a longtime journalist, I can contribute not only my own direct research on the disease but a degree of media savvy. Thus I will open this first critical chapter of my book, before I turn to the disputed scientific claims involved in the debate, with scrutiny of the effect of the media on the scientific method.

(Let me add that Feast has a doctorate in English from New York University and did three years of postgraduate work on the sociology of science under Dr. Stanley Aronowitz, from whose works we have both learned much. So, for Feast's part, he can help in the close reading of various texts as well as alert us to questions about the overall legitimacy of science that transcend the perspective of this one illness.)

Sir Francis Bacon's Watchword

"Knowledge is power" was his cry. Sir Francis Bacon, an Elizabethan lawyer, diplomat, and thinker, is credited with being one of the first people in the West to urge thinkers to turn away from speculating about God and metaphysics and toward scientific queries seeking to understand the nature of the world. These studies, he felt, should be conducted by doing experiments on materials, not solely by using pens to scratch on paper. In vain did he urge Queen Elizabeth and her successor, King James, to devote some of the royal treasury to fostering science, because he felt the more people knew, the more they could enhance human life.

There are two reasons to invoke Sir Francis Bacon here. The first is this. One of his contributions, which is of more contemporary relevance than many of his other thoughts, since most of his ideas (such as the need for government funding of science) lost force by being so thoroughly incorporated into current practice, was his study of what made people unprepared to think objectively. Finding that his ideas were commonly greeted with scorn and incomprehension, he went to work to see why so few people were willing to look rationally at his proposals, and he ended by cataloging four types of blocks. At the time, he was showing how these blocks, which he called idols, prevented people from setting aside religious speculations on nature so that they could adopt a scientific view. The curious situation I would like to reveal, in re-presenting his four idols, is that in our age these generic obstacles to freer thought have become central building blocks of modern science, at least establishment science, which is hindered by undue conformism.

The second reason to mention this antique writer is that the intent and scope of his book *The Advancement of Learning* so closely match my own. There, Bacon argued that for current work in science, little of it though there was, to go forward, it would help if there were a general summation of current knowledge. In *Advancement*, he proceeds through all current disciplines assessively; that is, he rates them as to whether they adequately encompass the field and whether the knowledge they do contain is sound. In his view, it was too often the case that ideas that had been around a long time were believed not because of their truth but because of their pedigree.

Bacon states that it is a general but erroneous prejudice that "of former opinions and sects, after variety and examination, the best hath still prevailed and suppressed the rest," which is to say, that whatever idea is current must be the closest to the truth.[8] So, according to this misconception, there is no reason to go back over the ideas that were discussed before orthodoxy was installed. "If a man should begin the labour of a new search, he were but like to light [on] somewhat formerly rejected, and by rejection [correctly] brought into oblivion."[9] Bacon conceives, on the contrary, that piously held concepts are often the most in need of questioning.

Though *AIDS: A Second Opinion* does not have the vast scope of Sir Frances Bacon's endeavor, the intent is the same. What I purpose is to examine accepted knowledge on AIDS as well as current work by dissenters, to provide a thorough anatomy of the contemporary state of knowledge. This will not simply be a guide through material but a judicious weighing of claims, which will sometimes allow me to say that to the best of our current evidence some point should be accepted, but will just as often make me say that the case must remain open until more information is gathered.

Let me begin, before settling down to a hard look at the scientific claims of the orthodox by doffing my media critic hat to glance over journalism's influence on the way science is being done today. (When I talk here and elsewhere of the media, I am aiming at the mainstream, always acknowledging that small alternative outlets, such as gay papers and alternative health bulletins, as well as a very few exceptions among the larger media, such as the *San Francisco Chronicle* and *Spin*, have swum against the tide in being open-minded in their reporting on the disease.)

I start by noting the evidence of a certain herd mentality in accepted opinion on the disease, and of the scapegoating of dissenters.

PART ONE

EVIDENCE AGAINST THE CURRENT AIDS HYPOTHESES

The Media and Medical Establishment Support Their Beliefs

Media Blitz

To the classically trained scientist, it is galling that Dr. Gallo and his fellow celebrity scientists involved in the AIDS debate seem to publicize their findings in the media before subjecting them to the normal processes of scientific review and discussion.

Philip Johnson has commented that AIDS research has to be distinguished from ordinary science. "Its basic premise was established at a press conference, and never subjected to the kind of critical scrutiny that ordinarily protects the scientific community from endorsing a catastrophic error."[1]

The point is made more strongly by another author:

By announcing his theory to the media without providing any evidence… Dr. Gallo had violated rules of scientific process. Researchers must first publish evidence for a theory [hypothesis] in a medical or scientific journal documenting the research or experiments that were used to construct it. The hypothesis is then debated by other experts and attempts are made to duplicate the original experiments and confirm the original findings. Any new theory must stand up to this debate and scrutiny before it can be considered a viable hypothesis, and any new hypothesis must be confirmed by successful experiments before it can be considered a fact.[2]

Three researchers writing in the *San Diego Union-Tribune* concur, saying the media's decision to blame AIDS on HIV in 1984 could not have been scientific, "for at that time no scientific papers had been published, and the normal critical procedures of the scientific community had not been allowed to operate.[3]

Just because AIDS appeared as a new plague, which would have terrifying repercussions, did not have to mean—but, as it turned out, did mean—that people would be so in need of assurance that preliminary findings would be accepted as received wisdom; a party line would immediately develop; and a pack mentality would reign among respectable scientists, doctors, and researchers.

The first step in the media-driven side of the AIDS discussion was the abandonment of probability. "Overnight, the word probable ceased to exist and it became dogma, literally engraved in stone," says investigative journalist John Lauritsen.[4]

Steven Epstein did a content analysis of seven major science journals from 1984 to 1986, looking at their citations of the seminal 1984 article by Dr. Gallo et al., where it was hypothesized that HIV is responsible for AIDS. He divided the references between those that say the Dr. Gallo article made a qualified claim about HIV's causal role—which in truth the article did, stating the evidence was not conclusive—and those that say the article's claim is unqualified; that is, that there is no doubt about the virus's causal role. (Note that here and later I use the term "HIV" for convenience, although the virus did not yet go by that name.) Epstein's findings are shocking.

> Only 3 percent of the 1984 articles that cited Dr. Gallo (2 out of 59) did so in conjunction with unqualified claims about the virus being the cause. Fifty-eight percent of them cited Dr. Gallo to support qualified claims. [The rest did not touch on this subject and cited him for other reasons.] By the following year, the percent citing Dr. Gallo to make unqualified claims had jumped to 25 percent (26 out of 106). And by 1986, 62 percent (49 out of 79) of the articles cited Dr. Gallo to make unqualified claims, and only 22 percent did so in conjunction with qualified claims.[5]

Remember, this is the same article, that in fact made qualified claims about HIV's role, but in the minds of Dr. Gallo's peers (even in Dr. Gallo's own later citations), it gradually becomes more overweening in its claims.

After the silent liquidation of qualifications to the claim comes the establishment of dogma, partly kept in place because alternatives are not represented. Mark Garish Conlan compared the prevalence of this mind-set to that of Soviet Communism. "People believe that HIV causes AIDS for the same reason... folks who lived in the Soviet Union thought Communism was wonderful." They "weren't given a frame of reference," and any dissenting opinions were kept out of the media. Soviet truth, like that produced by the dominant media, gains acceptance not because it is so plausible but because it is repeated ad infinitum, like a mantra.[6]

The average scientist, unless directly involved in AIDS research, does not have the time or energy to carefully study all the controversies in scientific literature. This is certainly excusable and understandable. However, what is culpable is moving from an acknowledgment of one's limited knowledge to an unquestioning acceptance of the current opinion simply because "everyone believes it." This is herd mentality.

I interviewed Charles Thomas about this in 1996 and he spoke about the equation HIV = AIDS = Death. "It is exactly this view," he points out, "that is held by anyone of the top twenty or one hundred of our most prestigious scientists!" However, he said, this view is not based on knowledge but ignorance. "If you put one of these guys on the spot and ask them: 'Have you ever looked into it?' He says, 'No, no, I haven't. Of course, I haven't—I haven't had time.'" This is an easy out, Thomas notes, since if they actually did look into it they would have to be responsible and say yes or no, regarding this or that. It's far easier to run with the pack.[7]

Of course, not any idea will be readily accepted as scientific truth. It will have to be one that jibes with the current paradigm. Before turning to that topic, however, let us add one last caution about modern science. We have tried to suggest thus far that many scientific speculations have been illegitimately promoted to truths by the media, and gullible or overworked scientists have gone along with them. Mr. or Ms. Average Reader may say, "This is impossible. Modern science is not subject to the type of quackery that ruled in the nineteenth century, when people believed in phrenology and other pseudosciences. Our age is too skeptical and enlightened."

David Rasnick has usefully pointed out that pseudoscience, like voodoo economics, is still very much with us. He recalls the hoopla around cold fusion, which became a fad among reputable scientists. Not too many years ago, "two scientists in Utah claimed to have produced nuclear fusion on a laboratory bench (in essence a controlled hydrogen bomb)." The media trumpeted the discovery before other physicists could replicate the experiment. "The same was true of the 'AIDS virus.' HIV is to medicine what cold fusion was to physics." He continues wryly:

> Both the 'AIDS virus' and cold fusion were unleashed upon the world during hastily called press conferences, and received intense international coverage. Previously unknown scientists—though respected in their fields, or course—were, regrettably, transformed during prime time into television celebrities. Sadly, none of those gentlemen was prepared for what lay ahead. The perpetrators of cold fusion and HIV are either embroiled in litigation or have other legal problems, and have generally become quite embittered by all the scrutiny they brought upon themselves. [8]

Dr. Bernard Forscher, formerly managing editor of the *Proceedings of the National Academy of Sciences*, was so disheartened by the way AIDS research is being played to the media and not to scientific peers that he said, "It is a hoax that has become a scam."[9]

AIDS Hypotheses Fit Doctors' Predilections

We have already seen a number of reasons that may motivate a scientist to jump on the HIV = AIDS = Death bandwagon, such as a willingness to listen to what others say when one doesn't have the time to study the situation for oneself. There are also many positive motivations, such as the desire for a cure and the urge to offer a certain type of hope that comes from identifying the culprit. All this is only human.

A particular theory will also be appealing to medical researchers if it accords with widely held presuppositions, both social and scientific.

Social presuppositions are those that are congruent with one's worldview. Dr. Joe Sonnabend mentions that the equation HIV = AIDS, which was connected to the idea that HIV was sexually transmitted, was welcomed by some. "People who wanted to promote family values liked it, because it was nice to say extramarital sex could be lethal. Others liked it because they could promote celibacy before marriage."[10] It proved convenient for politicians who wanted to dismiss the social and economic problems of Africa, which many would blame for the poor health of the continent's people. "With a killer virus theory, it was easy for politicians to say our economic policies, and the fact that many people live in squalor, have nothing to do with this disease" on the continent. David Mertz, a philosopher of science at the University of Massachusetts, Amherst, locates a number of other groups attracted to the equation. On the one hand, there would be "gay civil rights campaigners who wanted to remake AIDS as an equal opportunity killer, while on the other, there would be the neoconservatives and members of the religious right, who wanted an agent to concretize the wrath of God that they fantasized as visited upon gays."[11]

Turning to more scientific influences, we might recall the pioneer work done by Thomas Kuhn, who put forth the influential thought that science is guided by large-scale paradigms. These are overarching conceptualizations, such as the once reigning idea that the earth is the center of the solar system. In a given era, all research uses the paradigm in setting the terms of experimentation, while any findings that would upset or contradict the foundation are set aside as anomalies that deserve further study but that hardly detract from the explanatory value or truth of the central conceptualizations. Only when the anomalies pile up to the sky do some try to construct a new, replacement paradigm.[12]

Indeed, if we wanted to spend a minute dealing with the psychological considerations related to scientists' tendency to blindly adhere to a paradigm long past the time when it seems to hold all the answers, we might do worse than to turn to Sir Francis Bacon's fourth fetish of the mind, which he calls the Idol of the Theater, referring to ideas brought in "from the playbooks of philosophical systems."[13] In a passage strongly redolent of the controversy to come on AIDS, he finds a particularly obnoxious feature of this idol worship to be that thinkers may come to rely on the few, crucial discoveries that established the paradigm—such as the Dr. Gallo articles that identified AIDS with HIV infection—and go on to discount newer evidence. Science created in this way will be "based on too narrow a foundation of experiment… and decide on the authority of too few cases… [such cases] neither duly ascertained nor diligently examined or weighed, [while the rest of the science based on these initial findings is left] to meditation and the agitation of wits."[14]

In the circumstances under discussion, the HIV/AIDS thesis proved attractive because it fell in with the underlying forms of thought, agreeing with both long- and short-term trends in medical theory.

Peter Duesberg, a research scientist and, as we will explore more fully later, one of the strongest opponents of the idea that HIV plays a causal role in the AIDS disease, thinks that the HIV thesis was quickly accepted by the scientific community because of scientists' respect for germ theory, which earlier in the century played a yeoman role in public health. He points out, "The enthusiastic acceptance of the virus-AIDS hypothesis… was grounded on the universal admiration and respect for the germ theory."[15] Germ theory, which held that illnesses arose when people were infected with malignant microorganisms, was responsible for some of medicine's greatest victories in quashing diseases and "celebrated its last triumph in the 1950s with the elimination of the polio epidemic." Remember that polio is caused by a virus just as AIDS, it is hypothesized, is caused.

In his book *The AIDS Mirage,* Hiram Caton makes the same point about the popularity of this theory, but he adds it is not only doctors but the public who embrace it. The idea that HIV is at the bottom of things "enjoys plausibility with physicians as well as the public who have been inoculated with the germ theory."[16] The public sees the scientists' job as tracking down and slaying these insidious germs. Scientists, of course, are more than willing to be seen as heroic hunters. In Caton's words, the search "activates the 'Tally Ho!' pose of medicine, featuring gallant doctors in pursuit of low and cunning pests, whose carcasses will be triumphantly exhibited to the cheering multitude and to the Nobel committee."[17]

Furthermore, just like the protocol for a tourist's African safari, the procedure for hunting a virus is also well known. Duesberg explains,

> If you claim a virus or a microbe [is causing a disease], everybody knows what to do. A company starts making vaccines or test kits or drugs. The scientists crank out papers on viruses and everybody can count on a vaccine coming out and a solution.[18]

The emergence of a new paradigm, by the way, does not necessarily mean the wholesale rejection of the previous one. It could be, as when Einsteinian physics supplanted the Newtonian version, that previous concepts are not discounted but merely situated. Those in the Einstein camp who developed the new paradigm did not prove Newton was wrong, but showed that his ideas were more specialized than previously thought. Einsteinian theory brought in Newtonism as applicable to a subfield, while the theory of the new physicists could be used across the board.

Later I will discuss the analogous situation in relation to the new paradigm of wellness and illness that is emerging among alternative health practitioners, a paradigm that does not toss aside germ theory but imagines it as part of a different picture. The necessity for this new explanation can be partially attributed to germ theory's having hit a wall as far as being able to account for dominant diseases such as cancer.

For decades, scientists have tried to prove that cancer was caused by a microbe, and their lack of success had been spectacularly expensive. So far, most evidence seems to indicate that these researchers have been barking up the wrong tree. If AIDS were caused by a virus, though, this would be a vindication, indicating the enduring relevance of germ theory. In *Reappraising AIDS*, Philip Johnson comments the HIV hypothesis would be supremely attractive to "the virus hunters who had failed to find a cure for cancer [and now] could justify and continue their very expensive laboratories."[19]

I bring up expense at this point to suggest that a reigning theatrical idol, such as the germ theory, holds pride of place not only because medical researchers' egos and prestige are invested in it, but because this prestige translates into dollars. The money temporarily gives them room to maneuver, as long as they seem to be producing results. "By the time AIDS rolled around, in 1981, the virus hunters dominated the medical establishment and controlled the lion's share of biomedical funding, prestige, and awards," says Bryan J. Ellison from the University of California, Berkeley. "These people were set to blame AIDS on a virus. They couldn't think of anything else."[20]

It is because this hypothesis fit in with so many already entertained ideas, not because of the proof that was offered in experiments, that the HIV= AIDS idea so quickly swept through the public and scientific circles. The result was, as Lamar Graham put it in 1993, that "over the past decade, HIV has infected every facet of American life. In addition to whatever it does inside the human body, it now exerts great power over the collective psyche."[21]

Dissenters

In Bacon's view, one reason to avoid giving homage to the idols of the theater is that they are jealous gods. Their acolytes are not forgiving of heretics or of those who burn sacrifices at other shrines. Such intolerance "render sciences dogmatic and magisterial." The epitome of such learning for Bacon was the school of Aristotle, which, "after having by hostile confutations destroyed all the rest... has laid down the law on all points."[22]

If Elizabethan Scholastic Aristotelians were not welcoming to fresh concepts—Bacon's own ideas were ill-received and ignored by the leading thinkers of the era—neither is the current climate of AIDS science a pleasant one for dissenters who refuse to bow down to the current idol.

The case of Peter Duesberg, the prominent researcher already quoted, is perhaps the most spectacular. Duesberg, professor of biochemistry and molecular biology at the University of California, Berkeley, was one of the first to attempt to refute many of the central claims of the HIV = AIDS equation. (Nobel Prize-winning scientist Walter Gilbert also argued against the equation early on.)

The reaction toward such free thinkers was not Popperian. (The important Austrian philosopher of science Karl Popper argued that the test of a good theory was how well it could withstand objections. The more criticisms it could forthrightly overcome, the better it had proved. If those from the mainstream AIDS camp were followers of Popper, they would be happy to field criticism, since if their hypotheses were correct, it would give them a chance to strengthen them.[23])

However, the non-Popperian reaction of the orthodox was lamentable. "Instead of urging them [the dissenters] on in an attempt to help mankind, they were ridiculed and their funding was stopped."[24] A. Liversidge details the story:

> The Berkeley professor of retrovirology [Duesberg] who so rashly took on this role [of questioning the HIV = AIDS equation] was and is one of the most prominent figures in retrovirology, blessed at the time [of his first attacks] with one of the richest federal grants ($350,000 a year) in science to pursue research avenues wherever his mind led him. Today [1995]... Duesberg is virtually without grants, graduate students or influence, prevented from replying to his critics in leading journals and routinely ignored... in the mainstream press. The Nobel he was expected to win for his earlier work has gone to others.[25]

John Heilbron, professor emeritus of history and history of science and former vice chancellor of the Berkeley campus, sees this persecution as more reminiscent of the medieval Catholic Church than what he would have expected of modern science. "The establishment," he argues, "sought to suppress Duesberg by the methods of the priests of old: censorship, ostracism, excommunication, and refusal of sacraments—in this case, invitations to meetings, outlets for publication, and money for research."[26]

Robert Gallo in an interview where he was speaking about Duesberg's dissenting opinion on the HIV/AIDS equation, quickly went beyond temperate, considerate discussion of the issues to characterizing Duesberg's ideas as "baloney," using, on top of this, some well-chosen four-letter words, which modesty prevents me from repeating. He explains, "No thinking scientist involved in the problem knows anything else but that there is one single cause of AIDS, period. I said it in… early 1984, publicly in the spring and there is no need to say it again."[27] Hardly a dispassionate (or modest) presentation. This is the type of rhetorical overkill that has replaced collegial discussion among the AIDS priesthood.

One can well imagine that if this is the treatment meted out to a prominent figure in the field, others will think twice before objecting to the current dogma. Frank Buianouckas feels many scientists do not voice any objections they may have to the current AIDS opinions because "they may be afraid of [losing] their grants." He points to the case just mentioned:

> Look what has happened to one of the great scientific geniuses, Peter Duesberg. What has he gained from his courageous stand? He doesn't have a laboratory anymore. [The chilling effect is obvious.] Lesser scientists see that and will keep their mouths shut.[28]

Neville Hodgkinson is a medical reporter, not a scientist, but when he began presenting the views of scientists who dissented from the accepted opinions, he felt he'd entered a hurricane. When his reports appeared, "There was a tremendous weight of criticism for questioning the HIV theory. Usually people justified their attacks on our coverage on the basis that we were weakening the public health efforts to fight the deadly peril posed by HIV." This was not more than he expected, but what surprised him was "mainstream science journals which are supposed to adopt a reasonably objective view… they too behaved extremely irresponsibly and seemed not to want to look at this evidence that could challenge the HIV theory."[29]

Let me mention a few other cases where dissenting voices were effectively silenced. Others could be brought forward, but at this juncture they are simply noted to indicate that Duesberg and Hodgkinson are not isolated cases but part of a pattern. For example, Charles Geshekter, section chair of the Pacific Division's American Association for the Advancement of Science, was thwarted when he tried to organize a symposium of distinguished scientists to discuss the controversy surrounding the HIV = AIDS hypothesis.[30]

Second, in June 1991, forty scientists sent a letter to the five leading scientific publications in the English-speaking world (*Nature*, *Science*, *The Journal of the American Medical Association*, *The New England Journal of Medicine* , and *Lancet*) to say the HIV = AIDS hypothesis was faulty. None published the letter.[31]

Third and last, in 1987, I personally sponsored a conference featuring a hundred AIDS survivors who beat the odds using alternative therapies. Although press releases were issued on three separate occasions, not a single member of the media attended.

Popper saw open-mindedness in entertaining objections as the hallmark of science, since an honest refutation of objections would fortify theories that met the challenge. It appears that AIDS science is not conducted in that spirit. "The research agenda was set in concrete, and skeptics were treated as enemies to be ignored or punished. As a result, the self-correcting processes of science have broken down."[32]

The
Establishment
View

While media and many in the scientific and medical establishment have often steamrollered the public by asserting their views dogmatically while shutting out dissenters, we still need to consider the validity of their views. We have argued so far that many of the leading theses of the HIV = AIDS camp have not been subjected to rigorous experimental verification. However, lest we follow in the footsteps of those against whom we are raising objections, let us keep clear that the seemingly unscientific practices of some of these scientists do not disprove the truth of what they are saying. A scientist may have made an earth-shaking, revolutionary discovery and still treat his or her colleagues shabbily and give the results to the media before submitting them to peer review. What we have said so far about trendy journalism, herd mentality, the pillorying of dissenters, and so on does not bear on the truth of what Dr. Gallo and his supporters claim. It neither disproves nor proves their positions.

So, the truth, not the trappings, of the orthodox view must be our next subject of discussion. In this chapter, then, I turn to the views of dissenting scientists who dispute the findings of the AIDS orthodoxy. What we want to look at is whether establishment findings are replicable and, most central for this examination, properly interpreted. We will begin by reviewing a fair number of issues concisely, the particular point here being to indicate the breadth of dissent. We will look at, in turn, (1) the question of whether AIDS is a new disease; whether HIV causes AIDS, a point that includes (2) the cofactor idea, (3) the question of whether any retrovirus, such as HIV, could produce an AIDS disease, (4) questions about the strength of HIV, (5) the question of latency, (6) the problem of T cells' condition when an AIDS patient dies, and (7) how increasingly Byzantine explanations of HIV's mechanism of action have failed. Then, remaining with brief considerations but turning to less strictly scientific issues, we will note (8) the problem of a discrepancy between money spent and results obtained, (9) review ethical complexities of the crisis, and (10) look at aversion programs. Following this, we will take up eight other issues in more depth. These will be (1) the disputes over the scientific work that Dr. Gallo did for his original 1984 linking of HIV to AIDS, (2) the question of what evidence there is that HIV plays a causal role in the genesis of AIDS, (3) an important discussion of the validity of the HIV tests, (4) an examination of the thoughts of some scientists who believe there is no actual HIV virus in existence, (5) the problematically shifting definition of AIDS, (6) the inaccuracy of predictions that AIDS would spread through the general population, (7) the doubts being raised about the sexual transmission of AIDS, and, finally, (8) the whole thorny question of whether AIDS originated in Africa and whether it is widespread there.

The Disease Called AIDS

AIDS is unprecedented as an illness. With a traditional, life-threatening disease, one gets infected and sick, gets progressively worse, and then dies. With AIDS one gets the virus, which weakens the immune system, and death occurs from some other disease. It is not always the same disease, but from one or more of a list of possible killers. As Susan Sontag pithily puts it, AIDS "takes its identity from the presence of some among a long, and lengthening, roster of symptoms... symptoms which "mean" that what the patient has is this illness."[1]

One can imagine that this will cause problems with diagnosis. Two patients die of tuberculosis. One is found to have HIV infection and the other does not. So, one dies of TB and one of TB secondary to AIDS. It makes things complicated. TB in both cases brings death. Where does HIV influence the death?

It is perhaps not surprising that some researchers question whether AIDS itself is new. In a 1995 interview, researcher John Lauritsen, who, as we will see later, did much to uncover problems with the testing of the AIDS drug AZT, said, "In a nutshell, the idea of AIDS is a phony construct."[2]

In the book The AIDS Cult, he elaborates on this position, arguing, "The question must be raised whether AIDS even exists as a coherent disease entity... People are undeniably sick, but are they sick with a new disease requiring its own name?"[3] He sees AIDS as a sort of catch-all term that "spuriously links 29 (at last count) old and extremely heterogeneous AIDS indicator diseases together with a presumption of HIV infection." Those who believe in AIDS as a new syndrome, he says, would have a case if, indeed, it were "a serious disease of acquired immune deficiency without preexisting or induced immune deficiency." The problem is that "in all verifiable cases, demonstrable immune-suppressive disease and/or treatment have always preceded " the onset of AIDS. In other words, he finds that "AIDS" patients always have had diseases, such as syphilis, that weaken immune function before they came down with the syndrome, leading him to believe AIDS is not a new condition, but merely a later stage of an immune system deterioration that began with the first illnesses.

Lynne McTaggard, author of *What Doctors Don't Tell You: The Truth About the Dangers of Modern Medicine*, said in an interview that AIDS is a new medical condition, but one caused by the treatment, not HIV.

You take a totally healthy person, you give them a completely unscientific test... [and say] they have HIV. And you give them a set of totally toxic chemotherapy type drugs which are known to have a range of side effects including death, then watch the progressive destruction of the person's immune system. This person, sure enough... dies.[4]

This wouldn't explain those who got AIDS before drugs were available but does underscore how powerful and potentially toxic anti-retroviral drugs are.

Does HIV Cause AIDS?

THE COFACTOR CONCEPT

Let us preliminarily accept the fact that AIDS is a new medical condition that has recently struck various populations around the world. The first step in dealing with this situation is trying to find out what causes the syndrome. The 1984 Bethesda conference was touted as a breakthrough, because Dr. Gallo had identified a retrovirus as responsible for the syndrome.

If the HIV virus were found in every AIDS patient, this alone does not necessarily prove it plays a causal role, though it is strongly suggestive. We will examine scientific positions that dispute the point that HIV is behind the disease constellation called AIDS.

HIV could be identified as a cofactor, an agent that is necessary but not sufficient to cause the syndrome. Prominent AIDS researcher, Prof. Root-Bernstein, in *The Scientist*, writes, "We also [at one time] thought... HIV alone is sufficient to cause AIDS. But such researchers as Dr. Luc Montagnier, Shyh-Ching Lo, Joseph Sonnabend, and many others—including me—now believe that cofactors are necessary and, therefore, that HIV by itself cannot cause AIDS."[5]

The presence of a retrovirus in the blood of people with AIDS does not automatically prove that this retrovirus causes AIDS or was caused by AIDS. Nobel Prize winner Dr. Kary Mullis, taking a Popperian position, comments, "As applied, the HIV theory is unfalsifiable and useless as a medical hypothesis."[6] "I am not an agnostic," claims Dr. Fabio Franchi, a specialist in infectious diseases and preventive medicine. "I am well convinced HIV is harmless."[7] The list of those who agree with these scientists includes immunologist Edward S. Golub, president of the Pacific Center for Ethics and Applied Biology,[8] Dr. Beverly Griffin, director and professor of virology at the Royal Postgraduate Medical School in London,[9] and Prof. Harry Rubin, professor of molecular and cell biology, University of California at Berkeley.[10]

Dr. Luc Montagnier led the team at the Pasteur Institute who discovered HIV (called LAV) before Dr. Gallo. At the Eighth International Conference of AIDS in Amsterdam, he stated clearly: "I think we should put the same weight now on the cofactors [as we have on HIV]."[11]

The case for labeling HIV as partially responsible for AIDS is summed up by Prof. Root-Bernstein. He notes, "All AIDS patients do have multiple, well-established causes of immunosuppression prior to, concomitant with, subsequent to, and sometimes in the absence of, HIV infection."[12] If HIV is always found with other debilitating conditions—he lists some of these as "chronic or repeated infectious diseases caused by immunosuppressive microorganisms; [use of] recreational and addictive drugs; anesthetics; antibiotics... and malnutrition"—then it seems reasonable to conclude these cofactors are necessary for HIV to lead to disease.[13]

Dr. Richard C. Strohman, professor emeritus of the Department of Molecular and Cell Biology at UC Berkeley, is studying the question of whether HIV alone is responsible for the immune suppressive syndrome. He found that those who concluded that HIV is causative of disease were using "the same [poor] argument that we had for genetic predictability."[14] Here he was referring to researchers who published results that the media used for stories such as "Scientists Search for Gene That Causes Criminality." These researchers contend that a particular "gene is necessary but not sufficient" to cause its possessor to manifest criminal behavior unless combined with other factors, such as environmental or emotional deprivation. In other words, the proposed genetic cause of certain behavior traits is not the cause alone. Dr. Strohman goes on, "It's the same with HIV. If 'other conditions' must be combined with it [for AIDS to appear], full causality can be imputed to those other conditions and HIV may well play no role at all."[15]

The outspoken criticism of Peter Duesberg prompted Dr. Strohman—a scientist who wants to judge for himself, not from hearsay—to investigate. Duesberg, in response to the harshness of his treatment from traditional scientists, responded in *California Monthly*:

> "[Some in] the establishment consider HIV... [not to be] sufficient to cause AIDS. A co-factor may be involved. Duesberg has discovered this co-factor: it is the establishment itself. "HIV," he says, "causes AIDS only under the influence of the National Institutes of Health, the CDC, their corresponding agencies abroad, and interested drug companies."[16]

RETROVIRUSES AND THE IMMUNE SYSTEM

A virus is a simple organism that contains genes and sometimes a few chemicals—the retrovirus always has chemicals—within a membrane that can be double-layered. According to a common view—though not the only one found in science—genes are chemical strands of either DNA or RNA that hold in their chemical configuration the information needed by the virus to replicate and survive. The virus is so simple that it cannot reproduce, nourish itself, or carry out any function on its own. To survive, it has to get into a more complex living cell to use that cell's machinery. It is a parasite, though—perhaps surprisingly—not always detrimental to its host. After all, in the long run, it is self-defeating for a virus to kill its host, since it will die along with it. Though many viruses do kill the host, many more are benign.

In this context, we might bring up the much cited case of the myxomatosis virus's Australian career. This animal virus was first introduced into the continent in 1950 as a way to cut down the overpopulation of rabbits. It was so effective that in five years nearly every rabbit in Australia had died. Then, a curious thing happened. Before the last rabbit was gone, a mutated nonlethal virus strain appeared and ended up replacing the fatal version. "So, evolution, driven by the natural selection of variant viruses, was gradually tailoring the virus population to make it more compatible with the Australian rabbit."[17] This example should help us see that viruses will mutate to preserve the host. Contrary to popular perception, killer viruses are the exception rather than the rule.

The more benign viruses move from cell to cell in the host, reproducing in moderate amounts, leaving genetic material behind as they travel, but not damaging or eliminating their host. In the best case, they leave behind genetic material that may be useful to the host, and, if nothing else, can account for increased genetic diversity.

The retrovirus is a type of virus only lately discovered. Until 1970, it was believed that the DNA in a cell's nucleus was, like a bank vault with the bank's funds, the sole possessor of the genetic code. When it was necessary to direct cell operation, RNA would be manufactured from a bit of the DNA code and proceed to carry out the DNA orders, just as tellers may be given funds from the vault to carry out transactions. If new funds or new code elements were needed for new transactions, they would need to be taken anew from the central depository.

This seemed to be an adequate explanation of DNA procedure, except that it couldn't account for the actions of some viruses. We said viruses were so simple they could contain either RNA or DNA, not both. While a DNA-carrying virus could be easily understood—it entered a host cell and used viral DNA to form RNA out of materials purloined from the host cell—the operation of RNA-carrying viruses seemed inexplicable.

In 1970, the biochemists Howard Temin and David Baltimore, overthrew accepted thinking by showing that certain RNAs could themselves manufacture

DNA. RNA did this by using an enzyme called reverse transcriptase (RT). (This is one of the chemicals that retroviruses must carry in their packages.) At first, it was thought that only retroviruses could make this backward transformation. Temin and Baltimore deservedly won a Nobel Prize for their findings.

Now we can see how an RNA-carrying virus operates. Once it enters a likely host, it employs RT and the host's machinery to manufacture its own DNA, which then takes over operation of the host cell. This is the defining characteristic of the retro-, that is, backward-acting, virus.

For future reference, we might note that it is the retrovirus's use of RT that makes it comparably easier to identify, once it is in a host cell, than a DNA virus. After all, the DNA of the virus injected into the cell will often head straight into the nucleus, where it begins issuing instructions just as would the cell itself. But the injected RNA has to produce its own DNA first, operating in a reverse direction from that taken by the host cell, hence using chemicals RT that the cell itself doesn't possess.[18] This makes the retrovirus easier to identify and possibly to combat. This reverse action of the new virus type, which made it seem rather suspicious, has to be combined with another dramatic trait: it is self-effacing. HIV excepted, retroviruses as a group have proven to be non-threatening, leaving hosts unaffected.

In my interview with Christine Johnson, she told me at one time it was supposed retroviruses caused cancer. What put the scientists on this track was their knowledge that when retroviruses infect a cell, "they actually sometimes cause that cell to grow faster than it's already growing, [but] they never killed the cells."[19] Eventually, this explanation did not prove feasible and cancer is no longer blamed on retroviruses. As Johnson put it, similar knowledge about the retrovirus's proclivities led Duesberg to reject them as causes for AIDS. Johnson said, "He thought that HIV could not possibly cause AIDS because retroviruses were never known to kill cells."

Dr. Harvey Bialy, editor of the science journal *Bio/Technology*, rounds out this point by mentioning, "HIV is an ordinary retrovirus. There is nothing about this virus that is unique." It does not differ substantially from other innocent retroviruses. In his considered opinion, "It contains no gene different enough from the genes of other retroviruses to be a possible AIDS gene."[20] As he explains, "HIV uses all of its genetic information when it first infects." It doesn't hoard any for later use, so "there is no conceivable reason HIV should cause AIDS ten years after infection, rather than early on when it is unchecked by the immune system."[21] His conclusion is wry and on target. "Contrary to widespread belief, HIV is not mysterious. It is neither insidious nor clever. These are obviously the properties of the Virologist, not the virus."[22]

On top of that, where HIV retroviruses have no unusual characteristics that seem to fit them for the production of such an unusual disease as AIDS, the immune system appears to respond to HIV in a typical fashion. Initially, a newly infected person will experience mild flulike symptoms. Then the immune system will attack the virus and reduce its numbers to insignificant amounts. "If HIV destroys the immune system," argue Thomas et al., "it must do so years later, after the immune system has already destroyed the virus."[23]

The previous quote should not be taken to imply—and let me stress this because, I confess, it has sometimes appeared as an implication in statements by dissenters who are not as careful with words as they need to be—that a virus will only strike once and then either conquer or be conquered by the immune system. There are many viruses that, once driven back by an antiviral response, take up residence in the nuclei of infected cells, waiting their chance to sally forth for fresh attacks. Genital herpes, for instance, has this mode of operation. As Andrew Scott explains in his book *Pirates of the Cell,* after a human's genital herpes infection is put down by a strong immune reaction, the virus "retreats" into nerve tissue. There it can remain inactive for years, until, "when deactivation does occur the virus appears to multiply briefly in the nerve cells, and then travels back… to the initial site of infection. There it establishes another bout of acute infection in the same type of cells as were infected in the first place."[24]

Although herpes' resurfacing may seem to present a cut-and-dried case of a virus causing a disease and is viewed as such under the established paradigm, which links disease occurrence to a single factor, we have to ask, "Why did the herpes virus get stirred up again?" A broader view of what causes illness, one we will discuss more fully below, might like to say that presence of the virus in latent form is not enough to explain the new outbreak of illness, but needs to be supplemented by noting the causal role of a weakened internal bodily environment, which the virus has some way of monitoring, so it can strike at an appropriate moment.

Duesberg explains this by saying the virus, whether the first time its particles get in the body or when it is ready to reemerge from a latency, will only make someone sick in a set sequence. A primed or hardy immune system would quickly wipe out the invading particles. If a successful incursion is found, it takes this route: (1) a weak immune system, (2) infection with virus, and (3) a victory for the virus that is not strongly opposed by the body's defenses. However, according to Duesberg, the HIV virus seems to reverse the normal procedure. Here we have (1) a strong immune system, (2) infection with virus, (3) the virus goes to the trouble of weakening the immune system, and (4) it can then begin spreading its infection. Duesberg expresses it like this: "It is backwards thinking to say that a virus weakens an organism. An organism must already be weak before it is receptive to a viral takeover."[25]

Let me warn that these arguments are rather cautionary than clinching. As we saw in the case of pre-1970 biology, it was once unthinkable that RNA could itself direct the creation of DNA, its supposed master, just as it could be unthinkable for a teller to take control of the vault. It turned out, though, that the unthinkable was possible. We don't want to fall into the same trap, saying that because other retroviruses act one way, HIV must also follow this line. We are only emphasizing that once the actions of other retroviruses are examined,

HIV seems less of a strong candidate as a cause of disease than it might have first appeared.

THE STRENGTH OF HIV

Another difficulty with the thesis that HIV, after its latency period, breaks out, infects the body, and weakens the immune system by killing T cells so the body can no longer withstand disease assaults, is that HIV is not found in many cells of AIDS disease sufferers. Peter Duesberg, writing with Bryan J. Ellison, notes, "It is very difficult to understand how HIV would be able to devastate the immune system while never infecting more than a tiny fraction of its [T] cells." The fraction affected is so small, he claims, they could easily be replaced. "Such losses could be sustained indefinitely without affecting the immune system, because the body constantly produces new T cells at far higher rates."[26]

A couple of scientific articles provide us numbers. They explain that "HIV knits itself into the genetic material of approximately 1 in 500 CD4 cells." Of these, it replicates in no more than 1 in 10,000. "The regenerative capacity of the immune system is sufficient to replace CD4 cells at the rate of 3 percent every 48 hours." Ergo, CD4 cells are replaced at a far faster rate than that at which HIV might be damaging them.[27]

Dr. Paul Philpott, from the Graduate Department of Biology at Florida A&M University, summarizes: "Compared to other germs, HIV has no biological activity. You hardly ever find it in people who are sick." He continues, "When other viruses cause a problem you find millions of them per milliliter of blood. I think it's safe to conclude that HIV is basically a harmless germ that poses no threat to anyone."[28]

LATENCY PERIOD

Lastly, Duesberg calls attention to the peculiarity of the retrovirus's long period of hibernation. "The virus-AIDS hypothesis," he argues, "offers no explanation for… why the mean latency between infection and disease is five years, whereas antiviral immunity is established in a few weeks."[29]

Of course, we have already seen that some viruses, such as genital herpes, are characterized by extended periods of hidden inactivity; but I believe what is really disturbing to Duesberg is the elasticity of HIV's latency. The five-year latency period Duesberg mentioned in 1987 quickly expanded. As Michael Verney Elliot explains, the incubation period for AIDS was embarrassingly inconsistent.

> First of all it was considered to be a year, then three years, then five years, then it was stretched to ten years. It grew roughly one year for every year the virus was studied. Today [1996] it stands at between one and fifteen and even twenty-five years.[30]

One can readily see that, as scientists posit an ever longer latency period, when studying the sick it becomes an increasingly vexing problem to decide whether a sufferer from an opportunistic disease associated with AIDS really has the immune suppressive syndrome or whether the associated disease should rather be attributed to natural causes associated with aging.

T-CELLS THAT DIE ARE NOT INFECTED WITH HIV

There is yet another facet to the question of HIV and T cells. We've noted that the retrovirus is not found in many T cells, but beyond that, as Duesberg pinpoints, "The T-cells are disappearing but they are not infected by HIV. One in 1,000 at the most is infected." So if one believes that HIV is what is behind the elimination of the cells, then one has the unprecedented belief that "a cell that is not infected is dying from a virus." In a talk he gave on alumni day at his college, Duesberg put this situation in simple terms.

> Viruses… have to get into the cell and then they mess up the machinery… They cannot send a signal, "Okay, I'm staying here, I'm too busy with something else, but you're going to die over there." Viruses cannot work that way.[31]

UNSATISFACTORY EXPLANATIONS OF HIV'S ACTION

Given the unusual character of HIV, whose activities are so out of keeping with those of other retroviruses that it was bound to call forth some skepticism, it became increasingly necessary for defenders of HIV's causal role to provide a plausible sketch of its mechanism of action in bringing down the immune system. The theories presented so far have not been noticeably successful in dispelling doubts about the retrovirus's responsible role. Yet unless a believable account is produced, attempts to help people with AIDS are bound to be unintentionally misguided, since—if HIV is responsible—without knowing the virus's means of working, any counteractions will be striking out in the dark.

As Philip Johnson explains in the journal *Reappraising AIDS*:

> There is no agreement about how HIV causes the damage to the immune system… nor is there even agreement as to whether HIV is supposed to be suppressing the immune system, or over-stimulating it, or somehow doing both at the same time. In the absence of a basic understanding of what the virus does and how it does it, there is no way to know whether stimulating the immune system to produce more antibodies will be beneficial or harmful.[32]

Wrestling with such already mentioned findings as the lack of presence of virus in destroyed T cells, Dr. Gallo, giving up his early attempts to find a way HIV could act directly, now believes there is an indirect way HIV sets off a course of programmed cell destruction, which can continue even when it has left the scene. At a 1993 conference, Dr. Gallo said, "The molecular mimicry in which HIV imitates components of the immune system sets events into motion that may be able to proceed in the absence of further whole virus."[33]

Another set of theories, such as those put forward by Dr. Harris, focus on explaining HIV's elastic latency period. They label HIV a lentivirus, that is, a slow-acting viral entity whose leisurely rate of cell growth may cause inactive periods of various lengths.

Lentiviruses have been discussed for decades. It has been conjectured that lentiviruses affect a few animals, though with quite mild effects. The problem with using the lentivirus as the basis of one's explanation is that they "have never been isolated nor proven to do what is alleged of them." As a report in *Continuum* by Stefan Lanka puts it, "Fiddling around with them has been going on for 20 years—to no effect."[34] However, in January 1995 this theory was put to rest when it was shown that HIV is not a slow actor at all but replicates at a rate of 1,000 million particles a day, every day, for an average of ten years.

Another train of thought had it that HIV, after first being beaten back by immune response, was able to hide out in the lymph nodes for years, like robbers retreating to their mountain dens when the Texas Rangers were patrolling the mesas. When the immune system becomes unwary, a second invasion is launched. At this point, opinions differ on how HIV proceeds.

> Either the virus kills T4 cells by any one of various mechanisms but none in particular (according to the Old Guard), or T8 cells attack HIV-infected T4 cells while massive amounts of virus are produced and cleared (according to Ho and Shaw's "new view"). Eventually the surviving T4 population wears down trying to replace its fallen brethren, and AIDS results.[35]

Again, the ingenuity of the theory has not been matched by the ingenuity of the virus. The study of the lymph nodes of HIV-positive people "reveals the concentrations of HIV found in the lymph nodes are entirely unimpressive and are in fact, still minuscule."[36]

It might be said that while the cries of doubters, who can't believe AIDS is created by HIV if no plausible mechanism for its operation can be adduced, grow louder, the establishment's attempts to find such a mechanism seem to grow more outlandish.

Money and Investigation Brought No Answer

In a moment, we will go over a number of other medical issues in more detail, but it might be worthwhile to keep in mind that the questions around AIDS turn on more than simply the rightness or wrongness of scientific hypotheses. Ethical, financial, and therapeutic questions are implicated as well.

No one reasonable would regret money devoted to curing a disease. Yet, by the same token, in a world of limited resources, every dollar spent on research that is traveling down a blind alley is one dollar taken away from subsidizing the research that might be taking the royal road to a solution.

Quite a few doctors and medical researchers have pointed to the perverse equation in AIDS study whereby it seems more and more money is spent for less and less enlightenment.

As a pamphlet from the AIDS awareness group HEAL states, a lot of money and time are involved in this research. The group estimates that since 1984, more than 100,000 papers have been published on HIV. "Though more research money has been spent on HIV than on the combined total of all other viruses ever studied in medical history, there is still no scientific evidence... that HIV is the cause of AIDS or that AIDS has a viral cause."[37]

This thought is amplified by David Rasnick, a scientist with twenty years' experience, who feels the amount of money spent fruitlessly in trying to prove HIV causes AIDS has convinced him of the opposite premise. He says, "The only reason why I'm absolutely convinced... is because HIV/AIDS is the most studied thing in all of science. Absolutely the most studied thing in all of science."[38] Yet, as he sees it, all this study and the billions spent on it have not produced a shred of evidence in favor of the Dr. Gallo hypothesis.

When the HIV causal hypothesis was first advanced, many leads looked promising that today seem played out. This is the opinion of the researchers Thomas, Mullis, and Johnson as summarized by Duesberg in 1992. They argue, "The present stalemate contrasts dramatically with the confidence expressed in 1984." At that time, Gallo said HIV was directly eliminating healthy cells. The government foresaw that a vaccine would be available in, at most, two years. Ten years later, after spending billions of dollars, "no vaccine is in sight, and the certainty about how the virus destroys the immune system has dissolved in confusion."[39]

On the bright side, one could say we've learned scads about HIV as a virus, but that is little consolation for those who expected knowledge about treatment of the disease. This was the point made by Charles Thomas when I interviewed him in May 1996:

> In a limited sense, you can't say that all these papers are wrong... we know a good deal more about HIV than we do about any other virus in the whole world, now. So in the narrower sense, the molecular biology of HIV might be a perfectly valid undertaking. But what is not correct is to say that the outcome of this research is in any way going to benefit to people who are classified as AIDS victims.[40]

Prof. Root-Bernstein put it epigrammatically: "By the end of the century we will know everything there is to know about HIV and nothing about AIDS."[41]

According to Dr. Bernard Fields, it was the restricted scope of the undertaking, set off by the Dr. Gallo conference in Bethesda, that has condemned the enterprise in advance. Fields, professor of molecular biology at Harvard and a senior member of the HIV/AIDS research establishment, directs our attention to this limitation in an article in *Nature*, where he states, "The nation's $1.3 billion AIDS research program is on the wrong track." This is because the program has been centered on "drug development and vaccine programs that have little scientific rationale." Real scientific progress comes from more basic research into how the body and its immune system operates, since "we still have too many serious gaps in our fundamental knowledge to know how to prevent and treat AIDS."[42]

Kary Mullis, in fact, believes the money has ended up producing more rather than less confusion about HIV. Mullis invented a technique that is used in virtually every HIV study, but looking at the money spent without result, he sardonically notes, "The mystery of that damn virus… has been generated by the $2 billion a year they spend on it. You take any other virus, and spend $2 billion, and you can make up some great mysteries about it, too."[43]

Ethics

The most distressing aspect of the highly debatable proposition that HIV equals AIDS is that after it has been trumpeted by the media and establishment doctors as an unquestionable truth, it appears as a death sentence to those who find themselves HIV positive. Celia Farber outlines the climate of hysteria this creates. "They call HIV the AIDS virus, and it's this generation's symbol of terror. It has come to rule us, our lives, our relationships."[44]

Aside from the fact, which will be taken up in detail later, that tests for HIV's presence in a person's blood are often unreliable, it has not yet been proven that those who are HIV positive will get AIDS. In consequence, untold suffering is spread to those positives who implicitly trust the counsels of the establishment.

Alfred Hassig examines the ethical implications, arguing in an essay in a German magazine, "Never in the long history of medicine has a collective death sentence been passed as is the case today with AIDS." He feels basic human rights are violated when doctors discourage HIV-positive patients with overly pessimistic prognoses. Further, he believes, "Many HIV carriers have ended their presumed, hopeless psychological situation by committing suicide." In this context, Duesberg calls on doctors to remember their pledge, "It is the duty of every medical doctor to preserve life at any cost—the Hippocratic Oath—and not death-curse people based on any test so they are so frightened they kill themselves."[45]

Programs to Prevent AIDS

We've already said that, lacking a decent theory of how HIV acts to cause AIDS, there is less chance of developing a workable cure given that scientists will be proceeding by a hit or miss method. Let's flesh out that thought. In the second and third parts of this book I will consider this at greater length, so here I am only suggesting the dimensions of the problem, which can be illustrated by two examples.

The needle exchange program, where drug users are encouraged to turn in used hypodermics for sterile ones, assumes HIV can be spread by using infected needles, and since those behind this program believe HIV causes AIDS, for them making sure those needles are sterile is the priority. Yet what if, as others argue, AIDS is the result of an immune system devastated by such things as drug addiction. Then, as Duesberg and Ellison argue, "Sterile needles may limit the transmission of hepatitis and other infectious diseases, but they do not guard against the immunosuppressive effects of heroin, cocaine, and overuse of antibiotics."[46]

Duesberg also notes how foolhardy it is to give HIV-positive people a toxic drug to eliminate the HIV virus if, indeed, the virus is not at the root of AIDS. "The HIV-AIDS hypothesis has become a threat to public health," he feels, because it accounts for the administration of "cytotoxic DNA chain terminators originally designed to kill growing human cells for chemotherapy like AZT, that are now prescribed as anti-HIV drugs."[47] Again the argument is that if it is a weak immune system that should be given more responsibility for AIDS than the presence of HIV, then taking immune-system distorting drugs because of HIV infection is doing more to help than hinder the disease.

Both of these criticisms of current treatments for AIDS will be given manifold support in the later sections of our study.

Critique of Established AIDS Hypotheses

I t is time to move to another level and bring a more searching analysis to some of the claims of the establishment. Many, more than you might suspect, have already been wrecked by history, as was the idea, put forth in 1984, that within two years a vaccine would be on the market. Just as exploded, we will see, have been ideas such as that AIDS would spread to the general population outside the initial target groups and that the way HIV damages the immune system would be quickly worked out. Others are affected by multiple problems of logic or inability to generate any proof in scientific experiment.

By this point, it may seem that a number of the more extravagant claims made by the dissidents, such as that all the money spent on HIV research might just as well have been thrown down a rabbit hole or that some respected scientists seem driven more by a lust for fame than a desire to improve our understanding of human biology, are to be reckoned as argumentative hyperboles rather than serious claims. All I will say is this: See if you hold this opinion, after looking in more depth at what will be brought out in this chapter, when we try to bring the extravagant claims of the establishment to the bench of justice.

Dr. Robert Gallo

In 1987, Dr. Robert Gallo's name was again prominently featured in the news. It involved another press conference, although this time he was not personally present, taking place on March 31 at the White House, where President Reagan and French Premier Jacques Chirac met to sign a truce. The men signed a pact by which scientists in each country shared the credit for discovering the HIV virus and, subsequently, a blood test to detect its presence.

The disputants were Dr. Gallo of the National Cancer Institute for the United States and Dr. Luc Montagnier at the Pasteur Institute for France. This was a purely diplomatic agreement, which did not try to adjudicate the claims of each scientist but simply allowed both to share the glory and the substantial amount of money, $5 million a year, accruing as royalties to the HIV tests each had developed. As the *New York Times* put it, the agreement left "it to historians to sort out the credit."[1]

Rather than rendering judgment on who discovered what, the agreement contained a chronology, which indicated that Dr. Montagnier's group had isolated the virus well before Dr. Gallo's team, which independently isolated a similar virus. Dr. Gallo's team claims credit, however, for being the first to say the virus caused AIDS. Before continuing to sift the epidemiological and scientific evidence concerning AIDS and HIV, let's spend a few moments looking into the underpinnings of this scientific joust.

We need to be clear about the reasons for this historical excursion. It is not to cast aspersions on Dr. Gallo, although he comes out looking far from heroic. Remember, our overall purpose is to evaluate the HIV-causes-AIDS equation and related claims. Mudslinging will not settle anything; although, as we noted earlier, Dr. Gallo and others in the establishment have been known to stoop to it.

Ad hominem arguments do not work in legitimate science. Suppose Dr. Gallo, or Peter Duesberg for that matter, was found with his hand in the poor box of a local church. Sad as this might be as a comment on either's morals, it would have little bearing on the results of any scientific experiment either man carried out.

If we find out instead that Gallo's lab occasionally played fast and loose with the truth, the conclusions we want to draw relate to what this says about the state of modern science as a whole, jockeying and rivalry between competing scientific agencies and the idols of the cave.

Scientific research is expensive, so it demands funding. To get that funding it helps to have a high profile. To get that high profile, it's good to be willing to put on media sideshows as Dr. Gallo did in 1984. And to put on these sideshows, it is helpful to have breakthrough discoveries to publicize.

Don't get me wrong. I'm all in favor of breakthroughs. However, let us bear in mind the sage advice of Dr. Bernard Fields, quoted a few pages back. He argues that the basic research that is needed to actually grasp the functioning of the human body is not being done because it is not flashy or productive of fast results. Instead, as we will see, quick fixes are the order of the day and a lab like Dr. Gallo's will cut corners and act in ways that most would consider wrong to achieve a breakthrough, in media terms.

What I am going to argue, then, is not so much that Dr. Gallo's group was dishonest, but, a much more damning statement, that it is representative of many other high-profile medical research centers. The situation is parallel to that of those corporations that put the achievement of high quarterly results at the top of their agenda in lieu of growing the company with plans for long-term stability and enhancement. In the one case, the economy suffers, in the other, knowledge.

We know a lot about what Dr. Gallo's lab did toward discovering the HIV virus, since it has been the subject of a number of government investigations, most particularly a two-year inquiry by the Inspector General's Office of the Department of Health and Human Services, a three-year investigation by the Federal Office of Research Integrity, and one conducted by Rep. John Dingell as head of the House subcommittee that oversees the NIH. All these groups, and others we will mention, found rather discouraging facts about both the history of Dr. Gallo's discovery of HIV virus cells and the subsequent patent application process for a blood test to determine if a person had the virus.

We also want to look at Dr. Gallo the man: his early career (including his fall from grace and reinstatement), his promotion of infighting and backbiting, and his one-sided obsession with retroviruses. This look will give us some insight into the operation of idols of the cave. According to Bacon, such figures "are the idols of the individual man. For everyone… has a cave or den of his own, which refracts or colors the light of nature."[2] Which is to say, each person's temperament and experience cloud his or her perceptions.

DR. GALLO'S "CAVE"

Dr. Gallo's background was in cancer research. He was inspired to work on cancer by the death of his younger sister, who was carried off by leukemia when Dr. Gallo was fourteen; but he was dissuaded from working directly with patients because of how troubled he was by his first assignment for the National Cancer Institute on a ward of children dying of leukemia.

Knowing that some animal cancers were caused by viruses, Dr. Gallo directed his research in that direction throughout the 1970s, trying to see if any viruses could be responsible for human cancers. He was responsible for a number of breakthroughs, being one of the number who isolated the first retroviruses, and also having to his credit the cultivation of a cell line for growing human viruses.

(Since viruses do not really live outside of cells, it can be seen that they will be harder to cultivate in the laboratory than, say, bacteria. Independent bacteria will grow by themselves in a proper medium, but viruses will only grow inside host cells. Moreover, if the virus to be grown is a harmful one, it will continually kill the hosts that are sponsoring it. Thus, finding a way to cultivate a line of living cells that provide viable homes for a type of virus and can successfully be replenished was central to viral research.)

Dr. Gallo's star was eclipsed, however, in 1976. "It appeared that he had discovered a new virus, and proudly, Dr. Gallo announced that to the world. When it turned out that an animal germ had contaminated his cell line, and there was no new virus, Dr. Gallo's reputation plummeted."[3] However, he recouped his standing a couple of years later when he did find a human virus that played a role in a rare form of leukemia. He called it human T-cell leukemia virus (HTLV).

Bacon's idols of the cave are rooted in both positive and negative impressions that have marked the individual. Positive experiences shape the direction of one's interest. "Men become attached to certain particular sciences and speculation, either because they fancy themselves the authors and inventors thereof, or because they have bestowed the greatest pains upon them and become the most habituated to them."[4]

In April 1983, when Dr. Gallo first became interested in AIDS, before he had done any research into the topic, Shilts reports him saying, "I believe a retrovirus is involved and we are going to prove or disprove it within a year."[5] By August, not only was he claiming that he had been justified in thinking the disease was caused by a retrovirus, but that HTLV, the same virus he had previously identified as causing cancer, was behind it. To other doctors, this seemed wrongheaded in that HIV and cancer were at opposite ends of a continuum in how they damaged the body, the first killing cells, the second making them overactive.[6] Later, Dr. Gallo discarded this particular idea, but he would maintain at Bethesda and after that HIV, then called HTLV-III, was still part of the HTLV family. Later, retreating even further, he changed the meaning of the acronym, so that the "L" no longer meant "leukemia" but "lymphoma."

The value of Bacon's intuitions are certainly evident here. Dr. Gallo had recouped his reputation by discovering HTLV. "Men become attached," remember, to things that "they fancy themselves the authors… thereof." Once Dr. Gallo found one retrovirus, he began finding them everywhere. This is not to say, "Here is proof that he was wrong," but it does indicate how extrascientific values influence the direction of beliefs. This is why the Elizabethan counseled

Let every student of nature take this as a rule: that whatever his mind seizes and dwells upon with peculiar satisfaction is to be held in suspicion, and that so much the more care is to be taken in dealing with such questions to keep the understanding even and clear.[7]

I've talked of how negative experiences also register forcefully in the individual. Savor this passage. Back when Dr. Gallo was favoring the hypothesis that the first HTLV caused AIDS, he was disturbed by reports of Pasteur Institute discoveries. According to Shilts, "Privately, he spread the word that the French isolates were not human isolates at all, but contaminants from other viruses kicking around their labs."[8] Sincere as his belief may have been, we can see that in making this supposition about the French, he is reviving the accusation made against him at the time of his flubbed viral isolation experiment.

So, we see that unless screened out, personal history enters into scientific practice and may urge one to support theories, such as that supposing HIV causes AIDS, that do not merit such backing on more objective grounds.

There is also a role for basic character traits to play in determining scientific progress. Later, when we talk about the culture of laboratories, we will bring up some of Dr. Gallo's, such as irascibility, in order to illustrate that such characteristics, though off-putting, are advantageous for bureaucratic warfare. For now, let's turn to the specifics of one particular campaign in this battle.

THE VIRUS USED IN DR. GALLO'S LAB

There is no question that Dr. Gallo's earlier work in developing cell cultivation techniques and in finding a legitimate disease-causing virus influenced the French researchers at the Pasteur Institute to look for a virus as the basis of AIDS.

It is also not disputed that the scientists at the Pasteur Institute: Francoise Barre, Jean-Claude Chermann and Luc Montagnier, were the first to isolate the AIDS virus.

When the group sent a paper to *Science* in April 1983 describing their early results—isolation of the virus from one patient with swollen lymph glands—Dr. Gallo, according to journalist Seth Roberts, "reviewed the paper for *Science* before publication, [and] added a sentence to the introduction. It stated that the French virus "appears to be a member of the human T-cell leukemia virus (HTLV) family [i.e., the family of the virus he discovered]. The main text of the paper said no such thing."[9] To give Dr. Gallo his due, however, what Roberts is calling the introduction was, in fact, an abstract, a condensed report of the findings, which *Science* includes with each of its articles but that Montagnier had neglected to provide. Moreover, we might add that Dr. Montagnier requested that Dr. Gallo be the reviewer.[10] In any case, Dr. Gallo did egregiously add a sentence connecting the French discovery with his research.

Dr. Gallo's lab asked for and was courteously given samples of the virus the French had isolated. Once the French virus arrived, Dr. Gallo's lab got it to grow.

With the knowledge thus gained, according to one source, the Dr. Gallo researchers were able to isolate other examples of the same type of virus from AIDS patients. But the French virus grew best—and it was the French virus that Dr. Gallo's lab used for their research. Dr. Gallo's lab notes, obtained by the *Chicago Tribune*, show that the French virus was renamed a couple of times, apparently to hide the fact that it was being used.[11]

The bone of contention here is whether Dr. Gallo's lab, which admits it used the Pasteur Institue's virus as an example, then isolated its own retroviruses from AIDS patients' blood and used them to conduct its experiments. Dr. Gallo's group does not say it was first, but does claim to have autonomously discovered the virus in the blood of people with AIDS.

Its findings about HIV's causative role, if true, would be valid no matter what sample were used; but Dr. Gallo's avowal that he had independently found the retrovirus would seem less believable if to do so he used the same blood sample the French had used. Moreover, it is a matter of pride. It certainly wouldn't look "good" if the American lab couldn't even isolate its own samples and had to rely on French handouts, especially since, as noted, Dr. Gallo had been a pioneer in perfecting ways to grow retroviruses.

But since that's precisely what they did have to do, use the French sample, given that they wanted to quickly make a media splash and culturing a viable sample was very time-consuming. So, they pasted their own labels on Pasteur Institute (IP) blood and pretended they had done the work. This, at least, was the conclusion of the three-year investigation of the Federal Office of Research Integrity. According to the *New York Times* report, the office "found that Dr. Robert Gallo had intentionally misled colleagues to gain credit for himself [through pretending not to use French samples] and diminish credit due to his French competitors."[12]

But putting all the blame on Dr. Gallo would be like saying Stalin alone was responsible for the rise of Russian state Communism. There had to be collaborators all along the line, and there were. Looking at this reveals a shockingly pervasive infidelity to the goals of science on the part of the powers that be.

A separate investigation, conducted by a subcommittee of the House of Representatives, discussing the actions within Dr. Gallo's lab (called the LTCB), said: "A substantial body of circumstantial evidence… shows that at the very inception of their seminal experiments, the LTCB *scientists* knew or had reason to know that the virus they were working with and claimed as their own was the IP virus."[13]

Not only did these scientists disregard this knowledge, a sin of omission, they actively kept it from the public's attention, a sin of commission. The subcommittee report continues, "The LTCB *scientists' own actions* [showed] there was something to hide, that the LTCB scientists knew there was something to hide, and that they made every effort to do exactly that."[14]

The report unearthed that Dr. Popovich, who had oversight over all of the LTCB's experiments with the IP virus, and actually performed many of them, attempted to give proper credit to the IP scientists. "In the first draft of his seminal paper"—this was one of the papers that the famous Bethesda press conference was heralding—"Popovich explicitly acknowledged he had used LAV." LAV was the virus provided by the French. "But Dr. Gallo, chastising Dr. Popovich, extensively revised the paper, removing all references to LAV and making it seem they had been using their own sample all along."[15]

This seems to exonerate Popovich, while indicating a surprising lack of ethics on the part of Dr. Gallo. But, think more deeply. Does a real scientist knowingly cover up the truth because his boss tells him he should? Popovich is equally damned by this finding.

The corruption does not just extend to the scientific community. There was one responsible scientist found in the story, Gerald Myers at the Los Alamos National Laboratory in northern New Mexico, who specializes in the genetic analysis of viruses. According to the *Chicago Tribune*, "Myers compared the genetic codes of the French and American AIDS viruses," which turned out to be exactly the same. This would be impossible if there were two individual viruses involved, albeit of the same type. He "determined they were not independent discoveries but had undoubtedly come from the same patient."[16]

He shot off a letters to senior officials at the National Institutes of Health. "Myers' memo… was promptly buried in the NIH's files where it remained until it was accidentally discovered late last year by investigators for Rep. John Dingell."[17] The report of Dingell's staff on the situation was the most damning of all. According to the *Tribune*:

> The interment of the Myers' memo represents a single example, albeit a particularly egregious one, of what the report describes as a "continuing coverup" by successive administration of the role played by American scientists in the discovery of the AIDS virus…
>
> While the Department of Health and Human Services… "did its best to cover up the wrong-doing," the report states, "the failure of the entire scientific establishment to take any meaningful action left the disposition of scientific truth to bureaucrats and lawyers, with neither the expertise nor the will essential to the task."[18]

The pressure on the bureaucrats to make Dr. Gallo look like the Columbus of AIDS, one can imagine, did not come from scientific but political quarters. I noted already that the Bethesda press conference was to be juggled to fit a government department head's schedule. On this score, Shilts reports on Dr. Gallo's finding of HIV:

> Rather than be heralded as an accomplishment of the National Cancer Institute or the National Institutes of Health, credit for the breakthrough was to go to the Reagan administration. The announcement would counter liberal criticism that the government had dragged its feet on AIDS funding.[19]

Shilts amply documents that the Reagan government had constantly backpedaled on AIDS funding, generally proposing minuscule amounts, amounts that were upped by the legislature. Thus, as Shilts suggests, the Reagan government would use a scientific finding as a way to grossly distort the truth in picturing itself as seriously dedicated to facing the disease. Highly ironic, isn't it, that a small lie (that Dr. Gallo's lab developed its own retrovirus samples) is used to further a bigger one (that the government gave a damn about an illness that attacked marginal groups)?

THE PATENT DISPUTE

The same pattern of falsification on Dr. Gallo's part and a helpful complicity on the part of the government agency, which should have been weighing evidence carefully, is seen in the case of the patent application he made for his HIV-detecting blood test.

On the surface, this is a mirror image of the previous one, where there was a misrepresentation of which AIDS virus was used in Dr. Gallo's laboratory. Here the French scientists at the IP developed a way to test if a person has traces of HIV in his or her blood. A few months later Dr. Gallo's team developed their own test. There is no evidence that Dr. Gallo copied the French test in developing his group's own, which the Americans did find independently; but there is strong evidence that Dr. Gallo went out of his way to conceal the prior existence of the French test from the patent examiners. They, as we will see, apparently wore some fairly large blinders to ignore what was going on in the scientific world and even in other parts of their own office.

As in the previous instance, here a government investigation brought to light numerous irregularities. A two-year inquiry into the patent application process by the Inspector General's Office of the Department of Health and Human Services noted that Dr. Gallo didn't tell the Patent Office that scientists at the IP already had made an AIDS blood test of their own. This is "despite a legal obligation to disclose all information material to his claim of inventorship." According to a summary of the report, "The patent examiner who granted HHS [Health and Human Services] the 1985 patent on Dr. Gallo's AIDS test told investigators she would not have done so had she known that French scientists already had developed such a test."[20]

But is the Patent Office innocent in this matter? The report states that "HHS officials accepted uncritically everything they were told by Dr. Gallo and his colleagues." Where they should have been inspecting the claims of Dr. Gallo—this is the job of the office, after all—they accepted whatever they were told. "incorporating the LTCB scientists' information unqualifiedly... into official reports of the Department." When evidence appeared indicating that some of this information was erroneous, "the evidence was ignored, discarded, and/or suppressed."[21]

Not only was the agency less than perspicacious in relation to what was going on elsewhere in the scientific world, they were even unaware of what went on in their own office. The French had already applied for an American patent for their blood test—four months earlier than Dr. Gallo! But while "the Dr. Gallo et al. patent was issued in record time... the IP [French] patent application. . . had not been touched."[22]

After studying these reports, one's confidence in how science is conducted in these United States must be shaken.

POLITICKING AND SCIENCE

Dr. Gallo is known for his cantankerousness.

When the Centers for Disease Control, which tracks the number of AIDS patients, decided in August 1983 to set up its own retrovirology laboratory, it hired Dr. V. S. Kalyanaraman from Dr. Gallo's group. Enraged by this "poaching,"

Dr. Gallo said he would not give the CDC any HIV antibodies or other samples to work with. He had already accused the CDC of trying to undercut him by not giving him top-quality samples of the blood from people with AIDS. He carried through on this threat, withholding samples and thus setting back the CDC's own investigations.[23]

A second incident occurred at an April 4, 1984, meeting in Paris between Dr. Gallo and high representatives of the CDC and the IP, who were discussing Dr. Gallo's forthcoming Bethesda announcement. Informally, Dr. Gallo agreed he would share credit for the discovery—by this time, trailing the other two groups, the CDC had itself isolated a retrovirus from AIDS sufferers' blood. As Shilts tells the story, Dr. Gallo was not above some behind-the-scenes trickery. He took the French representative, Jean-Claude Chermann, aside and proposed they bifurcate the credit, cutting the CDC out altogether, thus breaking a promise he had just sealed with his word.[24]

It goes without saying that scientific progress is not going to be fostered if teams working on the same problem waste energy trying to sabotage each other, in terms of both credit and professional courtesies. Yet, looking at the various evidence in this chapter, it is possible to draw the perverse conclusion: For all the chicanery and skullduggery uncovered, the scientific ideal itself is unsullied. In other words, Dr. Gallo's lab never tampered with the science involved in its project. Its less than kosher goings-on revolved around trying to harvest the credit for certain discoveries. But then, on the obverse side, it might seem scientists are as much or more exercised by problems of reaping credit as of, for example, curing diseases.

If one were willing to fall in with the ideas of Steve Woolgar and Bruno Latour, though, as expounded in *Laboratory Life*, it might seem these twin goals are not to be separated. After studying the workings of a neuroendocrinology lab for two years, they conclude:

> It is at best misleading to argue that scientists are engaged, on the one hand, in the rational production of hard science, and, on the other, political calculation of assets and investments... Their political ability is invested in the heart of doing science. The better politicians and strategists they are, the better the science they produce.[25]

How do the writers arrive at this conclusion? Simplifying, we can say that a scientist's chief concern is credibility. His or her credibility is measured in the fact that "people phone him, his abstracts are accepted, others show interest in his work, he is believed more easily and listened to with greater attention, he is offered better positions," and so on.[26]

Credibility is gained by producing usable results, such as, for the neuroendocrinologist, isolating and characterizing chemicals found in the brain. Such a finding gains one credit, not because it is a contribution to the world of knowledge but because it can be used by other scientists to build their credibility.

Thus, if one isolated Chemical X, then other scientists could use this characterized chemical in their own research. A discovery would be adjudged a "breakthrough" among scientists if it could be seen by them as providing material for many others to gain credit.

Popper can be brought in here also. Remember, he saw as the hallmark of a good theory that it had withstood many attempts to invalidate it. In the Woolgar/Latour view, a theory would only be creditable if it would support other's strivings for credit, which would mean it would have to be replicable. Moreover, when a seemingly exciting breakthrough appeared, some scientists would try to prove it wrong, because doing so would be one route to building their own credibility.

The authors argue that certain types of politicking (if not all)—such as waiting for an area to get "hot" before beginning research in it, trying to get the most mileage possible from earlier credits (as Dr. Gallo did with his HTLV virus), claiming priority of place for one's own discovery of a phenomenon, and, perhaps, hoarding samples or not quickly revealing newly unearthed knowledge of nature as one races ahead in exploiting these samples or information for further credit—make for a feverish, sometimes impolite atmosphere, but one that also makes for signal achievements, since credibility is only achieved by authentic science.

At this point, as long as for the sake of this discussion we are accepting the causal role of HIV, the Woolgar/Latour thesis should make us look more charitably on Dr. Gallo's behavior. Contra Bacon's type of argument, they would say that certain of Dr. Gallo's splenic traits, such as prickly irascibility, would hinder his ability to be a dispassionate experimenter; Woolgar/Latour might say that his adverse traits aided his politicking ability, which is needed as an essential, not peripheral, skill in doing modern science.

However, there are multiple problems with the HIV = AIDS thesis, and so the infighting and sniping that are so much a part of AIDS medicine do not seem to be producing the good science the analysts expect. Contrary to the picture in *Laboratory Life*, the system of checks and balances seems inoperative if those who make highly dubious propositions are awash in credit.

We've already seen that the media may bear some responsibility for this. Once a scientific dispute gets picked up by mainstream journalists, it seems to be less responsive to purely scientific controls. However, I don't think that is the central issue. The problem brought to light by the AIDS controversy, which seems to vitiate the Woolgar/Latour model, is the situation that arises when alternate paradigms that view the fundamentals of science differently are in conflict. The AIDS controversy pits a mono-causal theory of disease against one that sees illness as arising from a more complex framework.

This topic can only be taken up fully after we have done more to establish the weaknesses of orthodox science. So, let's hold an examination of this thought in abeyance until we have undertaken some pivotal discussions of the causal role of HIV, the validity of the HIV-detecting blood tests, and about whether the existence of the HIV retrovirus has ever been proven. Then we can swing back to this discussion of competing paradigms.

Correlation and Classification Dilemmas

Previously, we wondered whether, given the nature of retroviruses and the immune system, HIV could really be behind AIDS. We looked at the nature of retroviruses, the posited reasons for HIV's long incubation, and the status of T cells when a person is dying of AIDS, among other topics. We now seek to address the following postulates: Does HIV exist? If HIV does exist, does it have any strong connection to AIDS? Does HIV's presence prove that it causes AIDS? How does one explain those who die of AIDS but are not HIV positive? What about those who are HIV positive but never get AIDS?

We will put off the first question, picking it up when we are a little more familiar with retroviruses and scientific technique. The second question is simple enough, while its ramifications are about as unsettling to the orthodox as any could be.

Earlier I brought up a telling metaphor: the thought of a man arrested for murder simply because he is found at the site of the violence. HIV has been found in many AIDS patients, while not found in healthy humans. This proves that HIV has some, but not necessarily a causative, relationship to the disease. It could arise as a byproduct of immunosuppression created by some other means.

In the French film *Moon in the Gutter* (based on a work by the American writer David Goodis), the hero's sister is raped and murdered. The police have neither leads nor hope of solving the crime. The despairing brother, finding his own investigation equally futile, begins going every night to stand on the corner where her body was found. Should he be arrested for the crime? In other words, guilt is not the only reason one may be in the vicinity. This holds also for disease. The presence of particular accompanying conditions in a sick person does not de facto explain why those conditions appear.

CORRELATION-ONLY EVIDENCE

HIV has been linked with AIDS by correlation; that is, it is said HIV is present in patients who have AIDS, therefore it causes the disease One looks in vain through the literature for alternative proofs of HIV's causal abilities.

Correlation is the only arrow these establishment experts have in their quiver. When confronted with the question of what other arguments they have to prove that HIV is the cause of AIDS, they fall back on talking around the subject, ignoring the question, or handing out patently sophistic answers.

In his reporting on the case, Serge Lang in *Yale Scientific*, notes that leading dissenters Duesberg and Mullis have both asked Dr. Montagnier, Dr. Gallo, and others for papers showing "any scientific justification that HIV causes a disease. They asked for such papers but none was forthcoming."[27] Mullis took the bull by the horns and spoke to Dr. Montagnier directly. In Lang's words,

> In San Diego, after Dr. Montagnier had given a talk on AIDS. Mullis noticed that Dr. Montagnier hadn't said one word about why we ought to think HIV is the cause of AIDS. After the talk Mullis asked Dr. Montagnier directly for a scientific reference, and Dr. Montagnier admitted that none existed.[28]

More indirectly Duesberg met the same stonewalling.

Duesberg wrote a letter dated 11 February 1993 to Harold Jaffe, Director of the HIV/AIDS Division at the CDC. In that letter, Duesberg asked: "Exactly which papers are now considered proof or, if there is no proof, the best support for the HIV-AIDS hypothesis?" Not a single specific paper was mentioned in Jaffe's reply. Jaffe only gave what he viewed as epidemiological evidence.[29]

In an article by Mullis and others in *Reason*, they make clear that correlation is a very thin reed on which to support the causal link between the disease and the retrovirus, especially in light of the absence of an animal model, that is, an example of an animal being infected by a like virus and coming down with a like disease. All that those behind HIV as a cause have to prove their hypothesis is correlation, but, as the *Reason* article puts it, "Very sick persons with damaged immune systems carry many microbes, and it is impossible with correlation studies alone to prove that a particular microbe is the effective cause of the syndrome rather than a mere 'passenger.'"[30]

In an earlier newspaper article by the same authors, a vast discrepancy is highlighted between the cottage industry churning out articles on AIDS—many on minuscule and secondary issues (in Woolgar/Latour terms, examples of scientists building on each others' credibility claims), all of which accept the holy writ that HIV is the villain in the AIDS drama—and the paucity of attempts to come to grips with the outstanding problem of what proof there is that HIV is really culpable. "Although more than 75,000 scientific papers have been published on AIDS," they say, "no paper has seriously considered all relevant evidence and attempted to prove that HIV causes AIDS."[31] This certainly says little for the willingness of the orthodox majority to take on difficult tasks, let alone for their ambition, in that "if such a paper were possible to write, it would have been written, and been the most widely cited scientific publication of this century."[32] Which is also to say that this absence strongly hints such a paper cannot be written, since the causative role of HIV is nonexistent.

To summarize and restress the importance of this point, let's listen to Root-Bernstein:

> We must be absolutely certain that HIV is not an epiphenomenon of AIDS before we assert that it is the primary cause. The fact that it is an extremely frequent finding in AIDS patients is not logically com-

pelling. It is only suggestive. Other infections, such as cyto-megalovirus, are also nearly universal among AIDS patients. If both are correlated with AIDS, which is the cause? Or are both viruses reactivated by previous and perhaps more diverse causes of immune suppression? How do we know what is cause and what is effect?[33]

HIV WITHOUT AIDS

One reason to doubt whether HIV causes AIDS is the fact that many people who have been diagnosed as HIV positive never become sick. The editors of *Rethinking AIDS* point out that approximately 75 percent of American hemophiliacs have had HIV for over seven years and yet only 2 percent annually develop AIDS indicator diseases.[34] According to accepted predictions about the incubation period, about 50 percent should have developed AIDS.

Moreover, the deaths of the HIV infected should have gone in the same direction, increasing as the patients became older. Instead there has been a wildly skewed distribution. As Duesberg and Rasnick report, "Although HIV is widespread in American/European hemophiliacs, the mortality of hemophiliacs has decreased (until 1987, when most started receiving AZT)," while that for male homosexuals increased, and that of intravenous drug users stayed about the same since HIV was diagnosed. "If HIV were the cause of AIDS, the mortality of all infected groups should have followed the same curve."[35]

However, let me caution that the Duesberg/Rasnick report is not beyond dispute. Another dissident, Vladimir Koliadin, argues that the high rate of death of infected hemophiliacs as reported in British studies did not follow the trajectory the two scientists presented. "By 1992, mortality in HIV positive hemophiliacs was about 10 times (severe hemophilia) and 20 times (moderate or mild hemophilia) higher as compared to the pre-1985 overall mortality rate in these groups of patients."[36] He discounts the influence of AZT in causing these increases, especially in that he dates the increase in mortality to 1985, two years before the antiviral was introduced! As Koliadin writes forcefully,

> Rapid growth in mortality [for this group] began not 'exactly in 1987' (as you [that is, Duesberg] insist) but in 1984-1985—when HIV-antibody tests were introduced. There are no increases in the slope of the curve in 1984-1985 and 1987-88, at least in severe hemophilia in [the] UK and US at large.[37]

Thus, we need to take this particular assertion of Duesberg and Rasnick with a grain of salt.

To return to our theme, there are also incidences of HIV+ homosexual men who have remained healthy for over a decade.[38] Further, in experiments, chimps repeatedly inoculated with HIV, in an attempt to create an animal model, never develop the syndrome.[39]

Let's dwell on those experiments a minute. There has been much talk in the mass media about the existence of an animal virus in monkeys that in some ways parallels the human one in type and disease-causing effect. However, a close examination of the scientific literature reveals that no such parallel virus/disease exists, and, furthermore, trying to give AIDS to animals has proven fruitless.

As a number of scientists point out in *Reason*, "The absence of an animal model [that is, cases of animals with a disease similar to AIDS] is not cured by tall tales about the so-called simian immunodeficiency virus [SIV]." The monkey disease called SIV does not resemble AIDS, they claim, since, unlike AIDS, "SIV disease follows the primary infection closely, and does not occur after a latency period of ten years." Thus, it is like other short-term viruses. Moreover, they point out, this virus only sickens laboratory animals, who are often weakened by being taken from their natural environment. They go on, "The same retrovirus is found in wild populations causing no ill effects."[40]

Since the parallels between SIV and AIDS are limited, scientists have tried to infect animals directly. As a writer in the *Miami Herald* mentions, "Scientists have stuck all kinds of mice and rats and monkeys and chimpanzees [with HIV], and none of them get anything resembling human AIDS."[41]

Dr. Paul Philpott draws our attention to how desperate researchers have become to produce AIDS in animals through HIV infection. One experiment followed infected chimps for more than ten years. "When none of the 150 or so HIV-positive chimps developed AIDS after ten years, that should have falsified the official view," Philpott notes. "Instead, the primate experiments were continued into their eleventh year." Eventually, one chimp grew sick with some symptoms resembling AIDS, although these conditions might as well have been attributed to age. The chimps in the experiment were about fifteen years old, many growing close to the end of their natural life expectancy. So, "at some point they are bound to start losing body mass and mental capacity, and become susceptible to opportunistic infections such as pneumonia, just like aging humans." Philpott wonders if their deaths by natural causes, then may be willfully attributed to the effect of the original HIV infection.[12] There seems to be no other way than this (flawed one) to infect animals with AIDS.

When I interviewed Dr. Kary Mullis, he brought up an article in *Nature*, which reports on another group that has had HIV for a long period but no death or disease. Further, it shows the length scientists will go to ignore what would seem the obvious suggestions of their own findings. You would think that if HIV-positive people were not getting AIDS as predicted, this would shake the scientists' belief in the HIV = AIDS equation, but instead they weaseled out of this deduction. He told me in May 1996:

This article named about ten or fifteen doctors [who] studied the prostitutes in some little eastern African country above Liberia. They had gone there five years before and found 75 percent of the prostitutes were HIV positive. They predicted if they came back in five years, half of them would be dead, all right? So they came back in five years, and there're no bodies to count... no dead prostitutes to do autopsies on. They're still HIV positive, according to their test...

And the conclusion of the paper was that—when that should have told them HIV doesn't seem to be hurting these people—the conclusion was these... people have a special strain of HIV which, number one, does not cause any disease; number two, it protects you from strains rampant throughout Africa....

I mean even a sixth-grader, I think, looking at the logic there would say wait a minute these guys—the emperor has no clothes here.[43]

AIDS WITHOUT HIV

Being HIV positive without developing AIDS, in itself, does not discount causation; it may only suggest a longer than expected incubation period. What is more damaging to the case for HIV is the reverse scenario, people who have what looks like AIDS while being HIV negative. There are ample documented cases of this.

As we will see, these cases reach back to the first reports of AIDS; though, understandably, the establishment is not eager to publicize them. They are quite prevalent in Africa. Moreover, the finding of such cases throws into doubt one of the prevalent practices in AIDS autopsies. A person dies of AIDS, judging by the symptoms, and so doctors list him as HIV positive, presuming he or she has the retrovirus, without testing for HIV presence. This procedure would be respectable if 100 percent of those dying of AIDS had been found to have HIV. Since this is not the case, this use of presumptive diagnosis seems like another way to fudge statistics and give HIV a prominent role in AIDS that it has not rightfully earned.

According to Peter Duesberg, numerous cases of AIDS without HIV have always existed, but these deviations from the theory have been camouflaged. "These cases had all the AIDS-defining illnesses, the characteristic drop in the number of CD4 cells [and so on], but no HIV or antibodies to HIV."[44] The Centers for Disease Control in Atlanta eventually called an emergency meeting to discuss these cases. The upshot was that the disease (AIDS without HIV) was renamed "idiopathic CD4 lymphocytopenia." Thus, the tricky question of how these patients got AIDS without HIV was solved, not by rethinking the situation but by terminological sleight of hand. They no longer had AIDS if they were HIV negative, no matter how closely this disease matched AIDS in every other respect.[45]

Researcher Michael Bomgardener elaborates on the persistence of these cases. "The correlation between HIV and AIDS at the very early stage at best was about 80%. So from the very beginning we had AIDS without HIV."[46]

Duesberg cited reports up to 1990 that showed approximately five thousand cases of AIDS without HIV in groups considered to be at risk of developing the syndrome. He adds, "There may be many more as in the U.S. only 50% of all cases of AIDS are confirmed HIV positive, the rest are presumptive diagnoses." These last are those cases we noted in which it is presumed the patient is HIV positive without actually doing any tests to prove this.[47]

Moreover, some of the best-known HIV researchers have turned in studies in which they confirm the existence of what would seem to be AIDS except for the lack of HIV retrovirus. For example, in 1991, Jacobs et al. (1991) reported that at the New York Hospital-Cornell Medical Center, during a three-month period, they diagnosed PCP, an infection commonly linked to AIDS, in five adults who had low T4 counts, another typical trait of AIDS patients. Of the patients, one was not tested for HIV, but the other four were and came up negative. Similar studies are cited in the notes.[48]

AIDS minus HIV is especially prevalent in Africa, where being HIV positive is not even a factor in diagnosis. According to Professor Geshekter: "The definition of AIDS in Africa is decisively and fundamentally different than the definition of AIDS that is used in industrialized countries." In Africa, he explains, if a patient has a persistent cough, a high fever, chronic diarrhea, a 10 percent weight loss in the last two months, he or she has AIDS. These identical symptoms can be caused by any number of diseases endemic to African countries.[49] The HIV test is generally not even given and when it is, it often registers negative.[50] A recent *Lancet* report from Japanese doctors working in Ghana, for example, showed that out of a group of 227 diagnosed "AIDS" patients who had the four telltale signs, as well as other "AIDS-related conditions," 59 percent showed no trace of HIV in their blood.[51]

The fact, then, that African statistics of AIDS patients being HIV positive is another case of largely presumptive diagnoses should make one wonder about the status of the disease on that continent, something we will look at in more detail when we focus on Africa. It should make one wonder whether HIV plays a significant role in the country. Nevertheless, Western scientists claim that 85 percent of the African population is positive for HIV antibodies.

As can be readily imagined, these cases of AIDS without HIV have caused a number of students of the AIDS phenomenon to reassess the HIV = AIDS thesis. Charles Thomas explains his disillusionment with establishment views of AIDS causation in this manner:

> The main reason I think that HIV has very little to do with AIDS is that there are so many AIDS cases that have been published that they can find no evidence of HIV at all. No antibody—or no culture—no nothing. Yet these same people are coming down with AIDS diseases. Well the response to this… [by the establishment] is that something else is causing the AIDS in these cases and my response is, if something else can be causing AIDS in cases where there is no HIV then why couldn't that something else be causing AIDS where there is HIV.[52]

Root-Bernstein sees it this way, "If non-HIV immuno-suppressive agents can cause AIDS in HIV-free people, they can also cause AIDS in HIV-infected people."[53] In other words, if we have AIDS without the retrovirus, it would seem to indicate HIV is a frequent accompanier but not necessary companion of the disease, hence not its instigator.

Testing for AIDS

Adjunct to the problem of the possible misclassfication of sick people due to presumptive diagnosis is the even more startling miscategorization that may occur when a person receives a false positive reading from a blood test, which is assumed to indicate a person is infected with HIV. Whatever the causal role of HIV, the tests now used are not highly accurate in determining whether a person actually has HIV in his or her blood, because, for example, the tests often come up positive in reaction to conditions other than HIV infection.

Public unawareness of the unreliability of the tests is particularly pernicious because of how seriously people take the results. Writing from the orthodox viewpoint, Dr. Schoub puts this way:

> The awesome consequences and implications of a positive result make it a unique test amongst all other clinical laboratory tests. Not only is it a confirmation of a terminal illness, but it also categorizes the individual as a source of lethal, transmittable virus, who will remain infectious for the rest of his or her life and, in addition, affirms a status which will permanently be a severe social and economic hardship for the rest of that person's life.[54]

Unfortunately, as noted, these tests are not like pregnancy tests, which are usually accurate. That this is not widely known can have dire consequences.

Let's look first at the question of assessing the distribution and number of AIDS cases. What if we find, for example—and this is not an imaginary case—that a man is given the test and comes up positive. He takes another test and comes up negative, in fact, he does so on every other such test. If this frequently happens, as we will show it does, then many people who are diagnosed as HIV carriers based on one test and who never take a second test, may not actually be positive. Thus, when such a person dies of TB, he or she may be falsely labeled an AIDS statistic. If such inaccurate diagnoses are widespread, and I believe they are, then this is more evidence that the actual existence of AIDS is being overreported.

Further, to turn to more emotionally wrought issues, this inaccuracy level means that many victims of a faulty test erroneously think they are HIV positive. For a person that believes with the establishment that HIV = AIDS = DEATH, a false diagnosis will lead to tragedy.

You may think of the case I cite in the short AIDS chapter in my book *Get Healthy Now*. I bring up the long-running Broadway play *Rent*. I am not interested in its theatrical value, such as it is, but in its espousal of popular misconceptions of AIDS. In the play, a woman is diagnosed as HIV positive. She goes home, writes a note to her boyfriend, and kills herself. She doesn't know that most doctors recommend at least two blood tests because of the recognized un-reliability of the tests now used.

Because of the widespread lack of information around this issue, then, I will devote special attention to explaining how these tests operate, in what ways they are seen to be inadequate (in that, for example, they can't distinguish between HIV and malaria), what futile attempts have been made by test advocates to make the tests more accurate, and what part these tests have played in stoking irrational fears.

ELISA AND WESTERN BLOT TESTS

If you asked the man or woman on the street—I'll designate this person Mr. Well Informed—what the HIV blood test does, he would tell you, "It checks for HIV in your blood."

Suppose you went on, "Well, how does it do that?"

Mr. Well Informed would answer, "It looks in there to see if there are any viral particles."

You might go further, asking, "What good is knowing that?"

Well Informed says, "If you are HIV positive, you will get AIDS."

In point of fact, none of this is strictly true. How many of those who are checking their blood via a test read the materials accompanying it? One ELISA test has a leaflet that explains:

> The test for the existence of antibodies against AIDS-associated virus is not diagnostic for AIDS and AIDS-like diseases… Positive test results do not prove that someone has an AIDS or pre-AIDS disease status nor that he will acquire it.[55]

We have already argued that HIV infection does not necessarily lead to AIDS, as the leaflet states, but look at the first line closely. The test is for antibodies, entities created by the body to battle against viral infection, not for HIV itself.

Well Informed might ask, "Why don't they just test for the retrovirus itself?"

Such a question would be overlooking the unfortunate fact that, according to Stefan Lanka, there are "photographs of virus-like particles in cell cultures, but [there are] none of isolated viruses, let alone of a structure within the human body having the shape ascribed to HIV."[56] There are no photographs of HIV in an isolated state simply because it has never been possible to isolate HIV according to accepted methods. Suffice it to say that a blood test that would identify HIV in the body requires a clear picture of HIV, which could only be obtained through isolation.

Since this is not available, second best will have to do. The ELISA test (the most commonly used) identifies an antibody that would be produced only if HIV were present. The logic is that the antibody, whose purpose is to counter the virus it fights against, was produced in response to an antigen of HIV. To be defined as HIV positive rests on the assumption that the HIV antibody was created but was not successful in eliminating the retrovirus.

The possibly flawed character of this set of assumptions has not escaped the critics. Duesberg nails the point home. "Antibodies," he says, "are always indications of immunity, not disease. Why, then, are HIV positive results interpreted to mean that full-blown AIDS is imminent?"[57]

His reasoning has to be taken with a grain of salt. Duesberg underlined how HIV is a virus that anomalously is more potent after antibodies have waged war on it. He continues, "There is no microbe anywhere in microbiology that causes disease only after it is neutralized by antibodies. That is the very reason why vaccinations work so well." In a vaccine, one receives a mild dose of the disease, which creates antibodies. Then when the person comes in contact with the full-blown disease, the body does not have to produce antibodies de novo because they are already formed and circulating in the blood. With most viruses, the antibodies' presence indicates that the virus is being ushered out. "With one exception. When you have made antibodies to HIV, then all of a sudden everyone tells you, 'Now my friend, you're in trouble.'"[58]

Duesberg is being excessive here, since, as I've already explained, a number of viruses, like genital herpes, do survive an initial rebuff by antibodies, hiding out in nerve cells. (Although, like most viruses that temporarily retreat, subsequent attacks of the herpes are measurably less virulent than the original onslaught.) In fact, an earlier hypothesis, since shelved, about AIDS' latency period saw it as following a herpes-like strategy.

Barry Schoub, the director of the National Institute of Virology in Johannesburg, and a scourge of the dissidents, argued this point in 1994. A person could be infected by two strains of HIV, he said. One would be fast-acting and obvious, the other stealthy and slowly replicating (making it harder for the body's immune system to recognize), hiding in blood cells where it replicated without harming its host, escaping the policing of the immune system. When HIV enters, Schoub wrote, "the more virulent variants probably undergo several cycles of replication but are overcome by an intact, functioning immune system."[59] Thus the presence of numerous antibodies at this stage indicates that, yes, an immune counterattack against the virus has been successful; but no, this is no a positive result, since the second strain of the virus is hiding out and working on a sapping strategy. This theory had to be discarded when it was found out that HIV is more prevalent in the blood of patients during the latency period than had been thought.

I bring up this theory to indicate that, contra Duesberg, it is possible to construct plausible accounts of disease action in which presence of antibodies would not indicate the total eradication of the virus.

The ELISA test relies on antibody detection rather than finding traces of the virus itself because HIV has not been isolated.

To be fair, I should note that there are blood tests for other common viruses, such as measles or mumps, that detect antibodies rather than the virus itself. Antibodies are detected because they are easier and cheaper to detect, not because the virus could not be isolated.

Let's look at how the ELISA test works to identify antibodies to HIV. The other main diagnostic tool is the Western Blot test. This test is used less frequently because its interpretation is less straightforward and it is more expensive. Like ELISA, it is also based on detection of HIV antibodies.

Dr. Lanka explains how the ELISA test detects antibodies:

> This is routinely done by layering proteins ostensibly from the virus in the wells of a plastic rack and adding blood serum to be tested to each. If antibodies are present, they bind to the proteins, and when this happens sophisticated staining procedures can make this visible.[60]

To be more specific, the viral proteins are from HIV antigens. Antigens are markers that appear on the outside of all cells. Each type of cell has specific markers, which, in the human body, for example, makes it possible for circulating cells, like those of blood, to identify and be identified by them for exchange of material or information. Viral particles also have these outer markings.

The HIV proteins are taken from viruses or, more often, parts of viruses that have been cultured in the laboratory. Certain proteins tend to be characteristic of retroviruses and it is conjectured that these proteins are from a retrovirus. Traditionally, to obtain viral materials, the cells in which they have been growing have to be burst open. "After opening up the cells a messy mixture remains, made up of the viruses *and* the unwanted debris from the host cells."[61] Presumably, scavenging through that the scientists find the desired antigenic proteins. These proteins were first used "as is" and then, with the creation of a second generation of tests, were created through further processing using genetic engineering.

The Western Blot blood test utilizes a similar connection between HIV antigen and antibody but it is more finely calibrated. Individual proteins from the HIV antigen are put in bands on nitrocelluose paper. The alluded-to controversy in interpretation of this test hinges on how many proteins a patient must have to be called HIV positive, since any single one of the proteins on the paper, taken by itself, could be derived from a source other than HIV.

After studying this description, you might ask, "If HIV has never been isolated, how do scientists figure out what proteins are in HIV antigens?" Ah, there's the rub. However, before looking into that, we should finish describing how the Western Blot test works.

(On the lighter side, I might mention the curious way this test got its name. Dr. Edward Southern developed a blot test for DNA identification and named it after himself. A similar test for RNA was waggishly called the Northern Blot. The Western Blot follows by taking the next available point on the compass.[62])

To get back to the crux of the difficulties with this test's modus operandi, we need to examine both the broader question of whether a method would ever be practical which sought to work with the antibody of a virus that had never been isolated, and the narrower one of whether Dr. Gallo's attempt to finesse this seeming impossibility and obtain these antibodies was a valid one.

Dr. Valendar Turner broaches the broader question via analogy. To understand his comparison, we need to merely touch on some points about particle isolation that we will study more fully later. "Isolation," as the word implies, has traditionally meant separating the virus from everything else so that one has obtained in a solution a band of pure HIV, for example. But this has not been done for the retrovirus, rather HIV or segments of it have been identified within a heterogeneous living culture, where it has appeared with other entities. The same can be said for the HIV antibody, which has also not been isolated. Here is the analogy.

> Imagine this experiment... someone hands you a test tube containing milks obtained from half a dozen different animals. In other words, a mixture of several different proteins but you don't know from which animals. Now in place of a mixture of antibodies from AIDS patients you obtain a second test tube containing a number of different acids. You add the mixture of acids to the mixture of milks and produce curdles. Now you claim you isolated [the proteins of] a cow. Or a goat. And not just any cow or goat. A completely new species of cow or goat. One never seen before...
>
> [So, to translate the analogy back to the real situation,] AIDS... is accompanied by antibodies [ones never before identified, so we don't know [which] virus or microorganism they were created to fight against] which are interpreted as proof that AIDS-diagnosed patients are infected with HIV. Then the antibodies are used to prove that HIV is the cause of AIDS. In other words, AIDS proves it's HIV proves it's AIDS.[63]

To take the leading point from this somewhat baroque analogy: If one has never isolated HIV in the first place, how can one claim that antibodies found in blood are ones attuned to that retrovirus? It's a rather flawed logic that would hold one can prove the existence of one unidentified entity by its correlation with another unidentified one, like proving goblins exist because ghosts say they do.

Dr. Paul Philpott broaches a second defective assumption underlying these tests. Assuming for the sake of argument that one had found an antibody that will identify HIV, one must also assume "the proteins... be exclusive to HIV, and the antibodies that react with them must react with no other proteins."[64]

As we saw, in the test antibodies are to bind with proteins on the wall of a plastic rack. For this test to be valid, the proteins must be ones found only in HIV. We saw also in discussing the Western Blot that such proteins may not be exclusive to HIV. Why else would there be a controversy over how many must be present for HIV to be indicated? Philpott adds that Dr. Papadopulos-

Eleopulos has studied all the proteins used in the test to bind the antibodies and given a list of cellular proteins from which these proteins may have been derived as readily as they might have from HIV. She has also drawn up" a list of non-HIV entities that cause reactions with 'HIV' antibody tests, [and] absolutely falsify the specific antibody ideal for HIV."[65]

But if these antibodies are so suspiciously cross-reactive—in a moment we will say more about this problem—where in the world did they come from?

Recall that Dr. Gallo's first triumph was discovering the virus that he argued was responsible for a form of leukemia, and that he called HTLV-I. When it came time to work on the HIV blood test, as Lanka reports, "since no DNA from HIV existed to hybridise with the prepared DNA, Dr. Gallo and Dr. Montagnier [who developed the first ELISA tests] simply used stretches of DNA from what they said was specific to HTLV-I." Their arguable proposition was that HTLV-I was a close enough cousin of HIV that any antibody formed to it would be the same as one formed to HIV. "The DNA [from HTLV-I] detected in this way was replicated and certain stretches of it cloned and declared to be the DNA of HTLV-III (later to be called HIV)."[66]

Moreover, even if we grant that the two viruses are enough alike to call up the same antibody, laboratories have great trouble extracting pure viral proteins from what we earlier characterized as a "messy mixture," containing "unwanted debris from the host cells" with the viral particles. In this case, it is admitted that the proteins, presumed to be from the retrovirus, obtained for the first generation of ELISA tests were contaminated by cellular proteins. "Because no proteins which are viral and free from contaminants have ever been obtained, one cannot be sure what the antibodies are that bind to the proteins," Lanka argues.[67] In other words, there is no way of knowing whether these antibodies are binding to viral or cellular protein. (To be even-handed, there has been less evidence of contamination with second- generation tests, which rely on gene splicing.)

In sum, as we inquire systematically into the subject, it appears the makers of the test keep moving further out on a limb. We learn that the test does not actually identify HIV particles but antibodies to them. But these antibodies were not actually created as reactants to HIV proteins themselves but to proteins of another virus, which supposedly closely resembling HIV. Still, the antibodies found were not actually reacting only to these supposedly analogous viral proteins but also to inevitable contaminants.

Let me emphasize that piling on unwarranted assumptions is not the normal method employed for detection of a virus. As Dr. Duesberg says acidly, "If a virus is to be claimed for a disease, you want to see the virus, not an antibody against the virus." When you are considering whether a patient has such diseases as polio, hepatitis or measles, he goes on, "you can find the virus, you don't have to mess around with antibodies."[68]

The multiple assumptions made in deriving this test have caused some scientists, such as leading AIDS researcher Dr. Eleni Papadopulos-Eleopulos (hereafter Eleopulos) and her group in Perth, Australia, to conclude, that the "AIDS test detects antibodies against proteins produced in the procedure itself."[69] In other words, intelligent doubt may reasonably claim the antibodies found in the test are not ones the testee had in his or her blood to combat HIV but ones that are reacting against other substances mixed in the protein samples. In Eleopulos' opinion, the idea that the antibody found in the test was formed to react against only HIV protein is an ill-founded hope rather than a rational assumption.

However, even if we were willing to believe the antibodies reacting in the AIDS test were specifically formed in reaction to the presence of HIV, this itself would not end the problems. The plain fact is that human antibodies are not so individualistic that they will react only against one invader, or tiny group of invaders. "It is often believed that an antigen [the particle or cell marker] and the antibody it elicits are 'soul-mates' and only react with each other," Christine Johnson explains. "In reality, antigens (and antibodies) are not so selective and often cross-react with antibodies (and antigens) that they don't belong with."[70]

Dr. Turner describes antibodies in more humorously seamy terms. "The problem is that antibodies indulge in casual and indiscriminate relationships. They are in fact promiscuous. Antibodies meant for one agent may react with another agent, a perfect stranger."[71]

If one entertains the hypothesis that the antibodies found in the test are reacting to HIV proteins, then, like other antibodies, they will also react to other, similar non-HIV proteins. How else to explain that the two standard tests (ELISA and Western Blot) so often contradict each other? Turner points us to a study of 1.2 million applicants for U.S. military service. Of twelve thousand who had first-time positive ELISAs, only two thousand were ultimately shown to be also WB positive and, thus, HIV infected. "That left 10,000 positive ELISAs which must have reacted for reasons other than 'HIV antibodies,' a fitting testimonial to the problem caused by cross-reacting antibodies."[72]

As it is described by orthodoxy, HIV itself illustrates this principle of promiscuity. Viruses are limited in what type of cells they will assault, identifying them by their external markers. HIV is primed to attack cells with the CD4 marker, but that means it does not only attack T cells, which are cells that are part of the immune system and which prominently display these antigens, but also cells in the brain and other bodily systems that also have CD4 markers on their surfaces.

So, with this information about the non-one-to- one correspondence of antibodies to particular viruses firmly fixed in mind, it should come as no surprise, later in this chapter, that the HIV blood test will register as positive for quite a number of viruses besides HIV.

VERIFICATION OF TESTS

Imagine you had created an intelligence test to be administered to schoolchildren. In order to be assured this instrument is worth using, you would have to check it for validity and reliability. Reliability is a measure of consistency. If a student's ability has not changed between a test and a retest, then he or she should get about the same score on each test. The blood test equivalent would be specificity. If a patient tests positive on one test, he should not test negative on a retest. This, as we are seeing, is one of the weaknesses of the two major HIV tests, since they often contradict each other. Moreover, each one is unreliable, liable to record a positive reading one time and a negative reading another.

The validity rating of an intelligence test would tell us if the test measured what it set out to measure. If the test is meant to evaluate how well a student learned chapter 12 in a history book, then the test should be representative of what is in that chapter. It shouldn't take all the questions from the first or last half of the chapter, nor should it draw them from chapters 9 or 13. We saw that questions around analogous issues arose around the Western Blot test. Even establishment scientists couldn't agree on whether what some were saying was evidence of HIV (the finding of certain proteins in the blood) was, indeed, truly evidence. There was no consensus, in relation to this test, of whether certain readings of it were measuring for occurrence of HIV.

The scientific equivalent of having a test that is valid is having a test that meets the gold standard, which is one where the test assuredly measures what it is said to measure. For the HIV blood tests, this would mean, "the reaction [a positive or negative test result] must be correlated with the presence or absence of the HIV in the body."[73] Just as a test that does not reasonably cover the subject it is meant to assess would be considered invalid, so, too, would a scientific test that does not meet the gold standard. Obvious problems with meeting this standard apply as much to the ELISA as to the Western Blot, for while the latter's reliability is weakened by disputes over which grouping of proteins adds up to positive assurance that HIV is present, the former's is weakened by questions of whether proteins used to attract antibodies are truly HIV-derivative. However, in relation to the gold standard, these points may almost be called quibbles in comparison to the main weakness of both tests, which goes back to the inability of scientists to isolate HIV.

The gold standard operates in this manner. If a person tests positive on the blood test, say, by finding antibodies or viral proteins, then a separate test is run to check for the retrovirus itself. This would obviously be an impossible standard for these blood tests to meet, for since HIV has never been isolated, there is no independent way to verify whether it exists in anybody's body.

This point has not been lost on the scientists at Perth, who have stated, "The HIV antibody tests, both ELISA and Western Blot... have yet to be verified against the only suitable gold standard, viral isolation."[74] Since at the current state of research there is no likelihood this isolation will be achieved, "the

HIV antibody tests are useful prognostic markers [but]... their use as diagnostic and epidemiological tools for HIV infection is questionable."[75]

But then, what, if any, standards was Dr. Gallo able to bring to bear when he tried to validate the original blood tests he was developing? In other words, if a blood test came up as positive, seeming to show a person was HIV positive, did he have any way to double-check that this person, indeed, was infected with the retrovirus?

As Christine Johnson reports, he resorted to circular reasoning. "To begin with, Dr. Gallo simply assumed that his AIDS patients were all infected with HIV. After all, HIV caused AIDS, and therefore they must be infected." Meanwhile, he also took for granted that random blood donors used as controls were not infected. "Since they were healthy and didn't have AIDS, they must not be infected." Thus, the validity was proven by assuming that every AIDS patient tested should come out positive—although we have seen some AIDS patients are negative—and every non-AIDS sufferer was negative. Moreover, just to stack the deck even further, it was decided "that any negative test results in AIDS patients must be false negatives, and any positive results in the random blood donors must be false positives." Johnson's conclusions about the one-sided nature of these tests are scathing.

> Perhaps Dr. Gallo had been to one of those positive thinking seminars where you are instructed to keep affirming that something is true, even if it isn't, in the hope that your belief will make it so. The reality is that he didn't know if his [test] was accurate, he didn't know if AIDS patients were really infected, he didn't know if the random blood donors were not infected... There was an awful lot of assuming going on in Dr. Gallo's study, and in subsequent years, no one else made any different assumptions when they were attempting to verify the accuracy of their HIV antibody test kits.[76]
>
> Dr. Eleopulos also objects to this procedure. "One cannot simultaneously use the presence of AIDS as proof of HIV infection, and conversely, the presence of a positive HIV test as proof that HIV is the cause of AIDS, as presently is the case."[77] In other words, anybody with AIDS who tests positive for HIV confirms that HIV causes the disease, and anyone with AIDS who tests negative, is a false negative, (i.e., the test wasn't done right). This is a type of science that is very convenient for producing any type of results that you want, though its relation to the truth does not seem as harmonious.

ARE TESTS BEING USED ACCORDING TO STANDARD PRINCIPLES?

To be fair to Dr. Gallo, some of the points we have just raised about his cavalier dismissal of test results that do not accord with his expectations are not as damning as they may appear on the surface. The throwing out of contradictory results is not unusual in epidemiological studies, even though his usage seems unwarrantedly extreme. To explain why this practice is accepted to a degree, before looking at the problems that have cropped up when the test is used in the field, let us note some general principles of disease testing. At first these principles will appear counter-intuitive, but they are generally accepted for reasons we will explain.

Common sense would say that if a person tests positive for any disease, he or she has that disease. Such an attitude would make sense if tests never lie. To put this positively, there is no blood test for any disease that does not record false positives, many of them often; because, for one thing, there is no blood assay that is so specific that only one thing (the culprit microbe) can possibly trigger it to read positive.

With this in mind, as Johnson alerts us, "It is a classical principle of diagnostic test interpretation that when a person has no logical reason to have the disease being tested for, [a positive test result] is most likely a false positive."[78] (This is why there is some justification for Dr. Gallo's classification procedures.) Dr. Griner states in the *Annals of Internal Medicine* hat "When the likelihood of disease is low [for a particular patient]… an unexpectedly positive result is not particularly helpful in confirming the disease."[79]

If we apply this to AIDS, we see that researchers, including Dr. Gallo, are abiding by accepted, reasonable principles and discounting the value of a positive registered by, say, a virgin with no history of drug use or blood transfusions. Dr. Langedijk states that in both ELISA and Western Blot blood tests, "almost all reactions… in low-risk population, represent false-positive results."[80] Where these researchers are crossing the line, though, is in applying this principle outside of the field for which it was developed. The disregard of seemingly false results is a rule for testing. It means the virgin should be assured she or he probably does not have HIV. and should take another test. This rule was not established to appraise whether a particular virus should be correlated with a particular medical condition.

A second trouble spot in relation to this rule is that the media, with no grasp of accepted scientific procedures (such as that of disregarding unlikely positives) but with a great nose for sensational stories, are flooding the public with cases of anomalous transmission of HIV, where the plain fact is that these are almost certainly cases of false positives. As Johnson notes,

> People in no risk groups, practicing no risk behaviour, who have absolutely no reason to be HIV infected, have blamed their "HIV infections" on such diverse and absurd sources as dentists, oral polio vaccine, and surgeons somehow picking up HIV from one patient and transmitting it to another during sequential surgical procedures.[81]

At the root of the media's confusion may well be the scientific confusion that reigns in the orthodox camp insofar as it is claiming to have linked AIDS to a virus, which has never been isolated. Thus, orthodoxy has to try and settle the infection status of a person who has anomalously registered positive without being able to rely on the gold standard. If the gold standard could be used, then, as with other infections, it could definitively settle the case one way or another. According to Philpott's exposition of Elena Eleopulos's views, "All viral tests... cross react with entities other than their intended targets."[82] When such a reaction registers a false positive for a person who has the test for another virus, this is not alarming, she explains, because the gold standard can be applied to investigate whether the patient's plasma shows evidence of the checked-for virus. "Virus isolations are rarely achieved in symptom-free people who test positive," she goes on. If the positive tester has no virus in his or her blood, "the only sensible interpretation for positive results in healthy people is that these people have experienced, sometime in the past, an infection that is no longer active (and is thus inconsequential), or they were exposed to cross-reacting proteins."[83]

However, without this way of establishing unambiguity whether the virus is present, it is possible to dream up all types of fantastic scenarios in which healthy, risk-avoiding people could have become infected even if a false positive is produced—a positivethat cannot be countered by the absolutely reliable information that would be provided by the ability to isolate the virus via the gold standard.

RELIABILITY

Once the HIV tests got under way in earnest, there appeared, if possible, even more reasons to be dubious about them than there were after looking at how they work and were verified in the first place.

Perhaps the most vexing difficulty is the one we already discussed when we mentioned that there is no antibody that will react against only one type of virus. If a particular antibody were created by the body to combat HIV infection, this antibody, if it were like all other known ones, would be attuned to the antigens of a number of similar viruses. Thus, if the HIV antibody is found in the blood, there is no way to know whether that antibody appeared to drive out HIV or some other virus. Say the anti-HIV antibodies were found in a patient due to the presence of one of these other viruses. If a person were given the HIV blood test, he or she would have a "false positive" result for this disease. In other words, the patient tested positive but did not have HIV.

When I interviewed Professor Charles Geshekter, he explained that the most common HIV tests (ELISA and Western Blot) are known to frequently produce false positive results, because the tests cannot distinguish between HIV antibodies and microbes that are symptomatic of malaria, leprosy, or tuberculosis. (Anita Allen points out, further, "Pregnancy is one condition which leads to false positives.) [84]

I asked him to give me an example of how the false positive might show up. He told me:

If someone recently had a flu shot or even a hepatitis B inoculation, and then was given an HIV-positive blood test, he would have to be very, very dubious of the results because there would be a high rate of false positives, meaning that HIV tests in some of these areas are, on the face of it, utterly unreliable. All but useless in fact.[85]

He mentioned leprosy, malaria, and tuberculosis. Only the last would be found, and even that not in great frequency, in the United States. However, another, more frequently encountered disease on these shores, which also fools the HIV test, is hepatitis. Christine Johnson reports that there have been numerous documented cases of false positives appearing when tests are given to people who have been exposed to hepatitis B or vaccinated for hepatitis B. She explains, "This disease is common among AIDS risk groups in the United States and Europe... Many gay men and almost all hemophiliacs have been infected."[86]

Just as the vaccination against hepatitis will give a false positive, the innocuous flu shot will create the same mistaken result. The *New York Times* reported, "Some blood donors who received the Beijing flu shot were testing positive to HIV... Upon further blood testing, the 'positive' HIV tests were determined to be 'false-positive.'"[87] A more scholarly examination appeared in the *American Journal of Epidemiology*, which noted that in December 1991, "U.S. blood centers reported an unusual increase in donations that tested falsely reactive" for HIV antibodies and other infections. This unusual increase of false positive, it is conjectured, comes from the fact that this was during flu season. "Many of these donations were from people who had recently received the 1991-1992 influenza vaccine, raising the possibility that this vaccine had somehow specifically caused the problem of multiple false reactivity."[88]

What about the three diseases Geshekter highlighted? It is the frequency of false positives with these that probably has created enormously inflated AIDS statistics for Africa and Asia, a point we will deal with in more detail later. Here it is enough to underline that these results point to further inadequacies of the blood test.

Malaria, for example, has frequently been shown to render false positives. According to Christine Johnson, this may account for the fact that "HIV appears to be transmitted predominantly via heterosexual intercourse in Africa and Asia"; whereas in the United States and Europe, when it is spread sexually, it seems to move between homosexual males. The reason would be that "mosquitoes do not practice gender discrimination and anti-malaria antibodies are common in men and women alike in malaria-endemic countries." Since the blood test cannot distinguish between HIV and malaria, it is likely that it is malaria rather than HIV antibodies that are being passed around by sexual contact. In fact, she says, a map published by the *Los Angeles Times*, purporting to show the regions in Asia and Africa where HIV infection was most prevalent should be compared with the World Health Organization's (WHO) map of worldwide malaria prevalence zones. Johnson asserts, "It is interesting to note that the malaria belt and the AIDS belt coincide almost exactly."[89]

Leprosy is another disease that seems to account for many false positives. In a book-length study of the problems of modern science, Hodgkinson, the reporter we met earlier, mentions a revealing experiment carried out by Dr. Max Essex, a strong believer in establishment views of AIDS and professor of the Harvard School of Public Health, who worked with scientists from the University of Kinshasa and the Health Ministry in Zaire on leprosy patients. "The results, published in the *Journal of Infectious Diseases*, were remarkable." Although Essex himself did not draw the obvious conclusions, they are ones that undercut both the validity of the ELISA test and general testing patterns applied to Africa.

As Hodgkinson describes the experiment's results, "Out of serum samples from 57 leprosy patients, 41 tested positive using one Elisa kit, 39 using another, and 37 (65 percent) by both." However, since this was a well-funded study, the group decided to test further. "When the sera were tested with western blot (WB) and radioimmunoprecipitation analysis, however… only two of the leprosy patients were confirmed as positive. Even those two could be considered false positives."[90]

Tuberculosis, of all the diseases mentioned so far that may be cross reacting to produce positive blood-test readings, is the most prevalent disease globally. Indeed, the progress of this disease has been leapfrogging ahead, partly, some say, because too much money is being spent on AIDS and too little on combating this real scourge.

An essay in *Continuum* points out, "More humans—about 3 million—died from tuberculosis (TB) in 1995 than in any other year in history," going by statistics in a WHO report. "One third of the world's population… are considered to be infected with mycobacterium tuberculosis, the causative agent." This report elaborated that between 9 and 11 million people have an active infection—95 percent of them in Asia, Africa, and Latin America.[91]

The WHO report links TB's rise to a rise in AIDS, which supposedly weakens people's immune systems so that they can be overwhelmed by the respiratory illness. However, Hodgkinson sees this as more of a "tragic error" than evidence of a rising epidemic of AIDS. This error has been created because "people with active TB infection are at greatly increased risk of testing positive because of M. tuberculosis, not HIV."[92]

Joan Shenton, who studied the health care in Haiti, tells a story that exhibits how devastating the confusion of TB and AIDS can be. "In one Haitian hospice run by nuns I visited," she said, she saw a woman's ward, "full of youngish women and they didn't look terribly ill and they were sitting despondently on their beds." It was known that they had tuberculosis when they came in. However, the nuns were given money to test them for HIV. They found 90 percent of them to be positive according to the test, not knowing that it was probably cross reaction with the TB bacillus that was giving the positive result. The nun then told Shenton, "So, as we know, they're going to die anyway, they don't get any medication for TB."[93] This was depressing end result of the test's unreliability.[94]

While we will look further at the misdiagnosis of Africa later, at this point we should underline that the frequent occurrence of cross reactions with diseases endemic to the continent may be causing a wildly exaggerated sense of the prevalence of HIV infection there. Prof. John Papadimitriou, a researcher in the Perth group, states, "There are signs that in Africa, at least, the realisation is growing that the problem of false positives could be enormous, and that it might explain why the predicted devastation of that continent by AIDS has not occurred."[95]

Labius Mutanda, a Uganda scientist, commented in *The Blue Sheet* that "ELISA and Western Blot assays may not always be able to reliably ascertain HIV infections in many African individuals."[96] In his experience blood samples would be positive when tested with the kits of one manufacturer, whereas the same samples would be negative with the test kits from another company.

As we will see, this is not the main problem—since the majority of Africans who are counted as HIV positive are never blood tested at all—but it makes a contribution to the skewed and unreliable statistics that are produced for Africa.

When I interviewed Hodgkinson in May 1996, I asked him to sum up the conclusions we should draw after noting how many infections and vaccines— and this list of cross- reactors is not exhaustive—can obtain a false positive from the HIV blood tests. He told me:

> There's such a long list of conditions that can cross-react that it's really quite a scandal actually… But the common factor [in accounting for false HIV readings] seems to be not… a short-term illness like, you know, some simple viral infection, but when your immune system is chronically activated. It might be a series of flu jabs over a period of time, or it might be a chronic infection that isn't clearing up or re-peated reinfection.[97]

Johnson, in fact, believes that it is the proneness of common infections in the high-risk groups that accounts for their tendency to register positive on HIV blood tests, regardless of actual infectivity. "People in the main AIDS risk groups," she explains, including gay males, hemophiliacs, IV drug users, and blood transfusion recipients, "all have something in common—and that is they have been exposed to a multitude of different microbes and foreign proteins." Obviously, the more exposure you have to microbes and foreign proteins, "the more likely you are to have some kind of antibody somewhere floating around in your bloodstream, that will cross react on these HIV antibodies tests." That's why, in her surmise, people in AIDS risk groups tend to have a high rate of positive test results. Not because, as the common wisdom would have it, they are

HIV positive, but because through shooting drugs, having multiple sex partners, using a lot of other people's blood because of needing transfusions or in other ways, "they've picked up a lot of different infections that way. So their bodies have a lot of different antibodies."[98]

If this is a correct reading of the situation, it might even suggest that the non-HIV infections, which, we already noted, invariably accompany HIV infection and may be misidentified as HIV, are actually more responsible for AIDS than HIV. As Hodgkinson stressed when I interviewed him, the real basis of faith in the blood test was not that there were scientific reasons to have faith in it, but "that people who tested positive often went on to become ill, and healthy people, who rarely tested positive, they stayed well." However, we should factor in the possibility of the test reading falsely positive because it cross-reacted with another infection, which also may be bringing about illness. In Hodgkinson's words,

> And so, the mistake was made; that because there was an association between testing positive and falling sick, it was concluded that this was evidence HIV was to blame. But, actually, the association there came about because your people who tested positive were more at risk, in that the test picked up immune-system activation from this great variety of causes.[99]

FALSE POSITIVES

With all the stimulants besides HIV that can make ELISA and Western Blot tests register positive, it's no wonder a rash of positive results have later been found to be false alarms; that is, "false positives." According to data on the ELISA test, for example, in Russia in 1990, "out of 20.2 million such tests performed, there were 20,000 false positives and only 112 'confirmed' positive results."[100]

More evidence of this high rate of mistaken diagnoses comes from the figures of the Centers for Disease Control on tests of heterosexuals with no risk factors. We noted that it is general policy for health workers to disregard positive results of people that are not in risk groups. Looking at the blood tests results from 1988 to 1989, we find that the CDC estimated that the HIV infection rate was 2 in 10,000 for the population classified as heterosexual without specific identified risk. However, what was actually measured by the blood tests among 302,005 such heterosexuals voluntarily testing for HIV at counseling centers? CDC's estimate would give us 62 positives. The actual HIV positive results from this sample: 6,954 (2.3 percent)! M. P. Wright draws out the conclusion, "Accepting the CDC's estimate that the true infection rate was .0002... then only 60 of the 302,005 testing should have been positive. More than 99% of the positive results—about 6,000—were likely to have been false."[101] Not a very good track record for a test on which rest so many hopes and fears.

Moreover, even when a positive ELISA test is confirmed by the Western Blot, this does not add up to a reliable positive result.

Dr. Valendar Turner, also from the Perth group, explains the problem:

> HIV researchers perform an HIV antibody test in a number of individuals and then repeat it half a dozen times using a slightly different technique or a different brand of test. But they're all the same test.[102]

In other words, as Turner points out, given that all the blood tests are known to cross-react with a number of other infections, resulting in a positive reading whenever these infections are present—regardless of the presence or absence of HIV—a number of positives from a slew of tests may only indicate they are measuring the same cross-reactions. As he puts it, "the way you get your cross-reacting… antibodies is to give your immune system a few belts." The more assaults your immune system suffers, the more likely a false positive from cross-reaction. "But we know," he goes on, "that in places like Africa [and for other high-risk groups like drug addicts] this kind of thing [exposure to other diseases and infections] is happening all the time." The end result is "the very people you're testing for HIV are those with the greatest chance of having cross-reacting or non-specifically induced antibodies."[103]

DO FEAR AND GREED PLAY A ROLE?

We've probably said more than enough about the scientific aspects of these problematic blood tests; but we have yet to go into the disastrous consequences the overvaluation of the tests have, consequences such as the one mentioned earlier where we brought up a case (fictionalized but with precedents in reality) where a woman killed herself when she found out she was HIV positive according to one test. Moreover, we mentioned that selling and administering blood tests is very profitable; that's why the French and U.S. governments were wrangling over it. It's also profitable to individual practitioners. Your run-of-the-mill doctor, like anyone, is going to be less inclined to question a practice that is lining his pockets.

The media, as we said, is a sucker for easy answers. If you go by what you hear from it, you will think these blood tests are foolproof, giving your doctor infallible power to tell if you are HIV-infected or not. Since the substantial doubts aired here are not usually mentioned by the daily press or establishment outlets, the hapless, uninformed patient who goes to be tested will be desolated by a positive reading. The results are predictably grim. As journalist Celia Farber sums up the situation, "No single diagnostic test in the history of modern medicine has had such a momentous impact on the lives of the individuals who rely on it." Cases have been seen where people react in self-destructive ways upon receiving positive results, including "lapsing into severe chronic depression and anxiety, quitting, or losing their jobs… getting divorced, having abortions, taking their lives and sometimes even other people's lives," because of one inconclusive blood evaluation.[104]

When I interviewed James Whitehead in May 1996, he told me, "If a person was to come back so-called HIV antibody positive, this has a profound, psychological effect." This is so strong that, over a period of years, "the psychological effect by itself… can kill a person. Just the psychological affect alone."[105]

Professor Duesberg, bete noir of AIDS orthodoxy, gave me a scenario that can be produced by blind faith in the reputable tests:

> Imagine that you are a relatively healthy person who has recently received a flu shot. You are entering a new relationship and decide to get an HIV test to show your partner that you are not infected. When you test positive for HIV, you think, how is this possible? You haven't had unsafe sex… You think you are HIV positive but in reality you are not. You simply have an antibody cross reaction. The flu vaccine cross-reacts with the HIV antibody so you test positively… A likely result of the diagnosis is that you will be frightened into believing you have AIDS and coerced into taking 250 mg of AZT, DDI or DDC every four to six hours. After being on a very toxic medication, you will start to manifest symptoms of AIDS and your life will be ruined.[106]

Understandably, given the general acceptance of what is pronounced by the mass media, not only the public but even physicians are uninformed about the possible shortfalls of the test. As Kary Mullis said sarcastically when I interviewed him in May 1996, "I'll bet you there's scarcely fifty physicians in this country that know what a Western Blot really is. They know when to order it. They know how to get a kickback on it, probably!" All they don't know is how unreliable the test is and how to advise their patients soundly on test results. Why worry when the test is so lucrative? They "make some money from having to take the blood or whatever and they make money off prescribing AZT. They get money every time they turn around," Mullis pointed out.[107]

Much earlier, I mentioned that the reason there was a rush to identify the virus was, strangely enough, not only because this was a step toward curing the disease but so a blood test could be developed and sold. Whatever the motives of the researchers, the drug companies were certainly clamoring for such a result, and they pushed even harder once Dr. Gallo proclaimed his breakthrough to get these tests ready and on the market. As Mullis explained, one reason for this urgency in getting the test on-line was the money involved. Remember, the test didn't actually help anyone suffering from AIDS, it merely identified those who would likely be getting it, and so, perhaps, warned them to take preventive measures, if any could be found. However, what these preventive measures were was unclear, so the health value of the tests alone would not seem to have called for such a frantic search, which seemed to outstrip any search for a cure.

Once they developed the blood test, those who were going to distribute it, according to Mullis, must have been thinking, "We're going to make a lot of money off this! Everybody in the world is going to have to take this test to see whose going to die of the plague!"[108] Since Mullis shares my conviction that the test is extremely fallible, he concluded, "There's all kinds of people all the way down the line getting paid for doing something that is absolutely insane."[109]

Unfortunately people on each side of the doctor's desk are accepting the reliability of the blood test as another unshakable pillar of AIDS dogma, upholding the idea that if a patient tests positive, he or she has an HIV infection. In light of what we have brought out, we might say two emotions are forming a spiral around the blood test. Patients, from their side, are so afraid of the disease, their fear possibly linked to guilt about sex, that they are panicking and not looking critically at the test. Their doctors, knowing only what they hear in the mass media, are making good money out of giving the tests and so are unlikely to question the source of an easy cash flow. Fear and greed are working in tandem to dull the critical edge people should bring to living their life and preserving their health.

The Existence of HIV

We have already wondered about whether HIV could cause AIDS, given, for example, comparable retroviruses' qualities; but there is a more fundamental question to ask about HIV's status, one you may have pondered as we discussed blood tests.

Consider what we said about the gold standard. It involved independently checking whether the virus or other agent a blood test had located was actually present. It could not be carried out for HIV, since the retrovirus has never been isolated.

Just as you may be scratching your head over this, it has also caused a number of scientists to step back and ask the obvious question. "If a number of prominent scientists, such as Duesberg, have argued that AIDS is not caused by HIV but by other factors, and since HIV has never actually been isolated, then is it possible there is no such thing as HIV?"

While Duesberg was the first prominent scientist to assail early AIDS orthodoxy with a 1987 paper, the second major blow against the establishment came with the article "Reappraisal of AIDS: Is the Oxidation Induced by the Risk Factors the Primary Cause?" published in 1988 in the journal *Medical Hypotheses.* by Dr. Eleopulos and members of her Australian group of researchers. Her group's opinion differed from the ideas of Duesberg in that it rejected the existence of HIV altogether.[110]

We need to roll back our narrative to the picture we opened with: Dr. Gallo disarmingly charming the media in a French restaurant with his "breakthrough discovery," so-called, of the virus that allegedly caused AIDS. If Dr. Gallo hadn't properly isolated the retrovirus, then what had he done to achieve his results? As we will see, he had carried out some but not all of the scientifically recommended procedures for isolating a virus. In fact, we will have to go back even further to the Pasteur Institute studies and see what they did, again deficiently, to "isolate" the virus they found and originally labeled as C-type retrovirus LAV. And before we do any of that, we need to see what viral isolation is all about.

ISOLATING A NEW VIRUS

Dr. Paul Philpott in a series of articles in *Reappraising AIDS*, which I will draw upon, has presented a highly readable account of how standard isolation is carried out in a laboratory and where the isolation of HIV fails to adhere to these standards.[111]

Imagine this scenario. A patient is sick and it is suspected that he has either a known or an unknown virus. To determine if either suspicion is true, a virus has to be found in his blood (or other tissue). To find it, the clinician begins by filling a test tube with a sucrose density gradient. This is a solution of ordinary sugar arranged in such a way that the mixture is thinner at the top, gradually growing thicker toward the bottom. A drop of the blood or other fluid from the patient is put on top of the solution. Then the test tube is centrifuged, spun for a few hours at high speeds. Due to the force of the rotation, particles from the sample penetrate into the solution until their buoyancy stops them from going further. It is already known that retrovirus particles will stop at a density of 1.16 gm/ml. They are said to band at this density.

(Note, the term "retrovirus particles" is often chosen by scientists over "retrovirus," though they mean the same thing, in order to stress that viruses are not cells. As noted, they have no cell machinery such as a nucleus.)

The second step is to extract this band and photograph it with an electron microscope. If the retrovirus is one that has been previously identified, and thus whose density, appearance, and size are already known, its existence in the patient will either be confirmed or denied by the pictures. But what if the retrovirus is one that has not been previously isolated?

If it cannot be identified, the next step will involve subjecting the band of particles to further manipulations. However, to even begin these further steps, the scientist has to be sure that he or she has nothing but the retroviral particles in the extract. As Philpott explains, "There is only one way to confirm this: a photograph with an electron microscope that contains *nothing but* identical virus-looking objects."[112]

But what if the photograph reveals two or more objects in the band? "If the micrograph reveals contaminating entities, that means the sample contained some material that had the same density as the virus-looking objects. In that case, scientists would have to add additional steps to the isolation procedure."[113] These steps would involve separating entities with the same density based on other differentiating traits. He also comments that these later steps are rarely needed, but, as we will see, they were needed when the HIV isolation was attempted. There was contamination, but the scientists moved on to the next steps as if they had already obtained a pristine band of retroviral particles.

If the scientist is searching for a new virus, then these early steps have only given him or her a particle that has the density usually found in retroviruses; but, since other non-retroviral microbiological entities have the same density, there is still a distance to go before there is proof that what has been glimpsed is an unknown virus.

Two important sets of procedure remain to be carried out. One, which is not conclusive, but informational, is to break up the particles and see what they are composed of, what chains of RNA, for example. This alone, though, does not prove a particle is viral.

The only definitive proof will come if it is shown that the particles behave like retroviruses, that is, by reproducing exact copies of themselves when placed in the appropriate medium. To see if the particles isolated are proper viruses, they have to be placed in an appropriate cell culture. Real viruses will then infect the host cells and rapidly multiply. To see if this has happened, after infection a drop from the cell culture is centrifuged. Identical particles should now be seen at the same sucrose density and there should be more of them, indicating that the particle has both replicated and proliferated. If these are found, the scientist has uncovered a new retrovirus.

This is a major discovery and the scientist who has made it will gain great prestige and a mantle of authority. Both Dr. Montagnier and Dr. Gallo tried to don this mantle without having accomplished all that was requisite for wearing it.

DR. LUC MONTAGNIER'S INITIAL DISCOVERY REDUX

In *Impure Science*, Steven Epstein ably chronicles the Pasteur Institute's early research into HIV. The group had noticed that people with AIDS generally had an abnormally low number of T cells. They hypothesized that a virus was killing these cells and, further, that they would find more viral cells if they caught the disease at an early stage. A third hypothesis was that lymphadenopathy syndrome, a disease causing swollen lymph glands, was a precursor to AIDS, and so would have a large number of the culprit viral cells. In Epstein's words:

> Dr. Montagnier's team extracted T cells from the tissue [assuming they already were contaminated with the retrovirus] and put them in an incubator with nutrients, hoping to grow a virus. When tests showed the presence of reverse transcriptase activity, the enzyme that is the distinctive marker of retroviruses, they knew they had found something... With the aid of electron microscopy, the Pasteur group also succeeded in photographing viral particles.[114]

It would seem at this point that real investigation, involving centrifuging and so on, should have begun. Instead here is where it ended. What is described in this quote became the new (illegitimate) basis of recognizing a new retrovirus.

The findings, which would be published in *Science*, and were taken to prove the existence of a retrovirus, rested on three points: (1) the presence of reverse transcriptase (RT) activity, with a characteristic budding associated with this RT—RT, you'll recall, is used by retroviruses to transmute RNA into DNA— (2) the glimpse of particles of the suspected retrovirus, and (3) the claim that T cells were being eliminated by the retrovirus.

Solely the second claim is consistent with accepted procedures for isolation. Yet this should not have been done before the first step, when these blood samples would have been centrifuged in order to obtain a banding. Instead, the HIV particles were being picked out from the cell culture debris. At least, though, this claim could be considered a preliminary finding that would lead to the carrying out of a strict isolation procedure. According to traditional parameters, claims one and three did not even have that much going for them. At most, they might have been mobilized as supplementary arguments, which would have been illuminating but would have done nothing to prove the existence of a new entity.

Before looking at some of the criticism these findings have elicited, let's recall what we said earlier about scientific work and Popper's ideas on testing a theorem. When we find that a number of other scientists have gone at Dr. Montagnier's (and Dr. Gallo's) findings hammer and tongs, we should not take this as evidence of a cranky disposition on the part of the questioners. No, such criticism is paramount to the development of intelligent scientific theorems, which ideally will prove themselves unshakable in the face of rigorous critique. Nor should we follow the gullible media in thinking that publication in a prestigious journal like *Science* represents a blanket endorsement by the editors of what the authors published. Acceptance by the journal connotes only that the research is credible and deserving of further review, which will either fortify or puncture the positions laid out by the published authors. Further, to be fair, some of the points raised by the scientists who claimed to have discovered HIV would look worse in the light of developing knowledge about retroviruses than they did according to the wisdom of the time.

Let's turn to the criticisms, bringing in for convenience here also complaints against Dr. Gallo when these complaints were aimed at the same type of claims made by the Pasteur Institute.

BUDDING To start, we will take up one of the less significant items in Dr. Montagnier and his colleagues' argument, the subclaim of claim one, that in the T cell sample were observed cells that were budding in a way indicative of RT. Dr. Montagnier and his colleagues wrote in 1984, "Several characteristics indicate that LAV or LAV related viruses belong to the retrovirus family. Budding particles at the plasma membrane have been observed in electron miscroscopy."[115]

However, according to Eleopulos, Dr. Montagnier has admitted: "We published images of budding which are characteristic of retroviruses. Having said that, on the morphology alone one could not say it was truly a retrovirus.... With the first budding pictures it could be a type C virus."[116] Thus, if one went by the photos alone, positive identification of the particles as retroviruses was impossible, since the budding could be evidence of a more typically found DNA virus.

REVERSE TRANSCRIPTASE Basic claim one was that reverse transcriptase, which, it was believed, cannot be carried on by T cells themselves, was observed in the T-cell culture of the man infected with lymphadenopathy syndrome. It was conjectured that this RT signaled the presence of a retrovirus. However, there is a disquieting problem with this conclusion.

Given that banding had not been carried out and these retrovirus particles were located in a mixed culture, how were the scientists so dead sure RT activity was due to a retrovirus? Is it possible anything other than a virus might be causing it?

Indeed, it is. This is the question of specification. It is has been shown that "not all particles with RT activity and visual properties of [that is, resembling] retroviruses are viruses."[117] Eleopulos asserts further, "This is a fact acknowledged even by Dr. Gallo well before the AIDS era." She elaborates:

> RT is not truly specific of retroviruses. Non-infected cells as well as bacteria or viruses other than retroviruses have RT. According to some of the best known retrovirologists including its discoverers, as well as Nobel Laureate and Director of the US National Institutes of Health, Harold Varmus, reverse transcriptases are present in all cells including bacteria. Indeed RT activity has been reported in many of the cell lines from which 'HIV' is isolated.[118]

The important nuance, in her eyes, is that some scientists are not carefully distinguishing between "characteristic of" and "exclusive to." "We agree," she says, "that RT activity is characteristic of the retrovirus. However, 'specificity' does not have the same meaning as 'characteristic.' Hair is characteristic of human beings but not every animal with hair is human."

She sums up.

> It is true that Dr. Montagnier et al reported RT activity at the density of 1.16 g/ml but since: (a) Barre-Sinoussi [Dr. Montagnier's collaborator in the first "isolation" of HIV] and Chermann accept that cells and cellular fragments also have RT activity; (b) at the 1.16 g/ml band no particles with the morphological characteristics or retrovirus were seen... it follows that the evidence for the existence of "HIV" by detecting RT activity at that density was not only not specific but non-existent.[119]

KILLING OF T CELLS One typical pattern of virus operation is as follows. The virus is inoperatively floating through the host's bloodstream. It hooks onto a compatible cell, and the DNA or RNA inside the viral envelope is injected into the cell. This matter travels to the cell's nucleus, where it commandeers the host's machinery and begins manufacturing more viruses. These viruses fill up the cell to the point when it is ruptured and dies. The new viruses now float through the blood looking for new victims.

Imagining this process was taking place, Dr. Montagnier and colleagues noticed a decrease of T cells in their sample, which they believed could be attributed to the ruthless activity of the AIDS virus. They counted the number of T cells in this manner. "The number of T4 (CD4+) cells in the culture were measured by counting the number of cells able to bind [to] a monoclonal antibody claimed specific for the CD4 protein."[120] In plainer language, the scientists were not tallying numbers of T cells directly, but the number of T cell/antibody dyads, imagining that any existent T cells would naturally bind with these antibodies.

But this is no foolproof method. "At the beginning of the AIDS era there was ample evidence that treatment of cell cultures with PHA and other oxidising agents leads to decreased binding of the CD4."[121] In other words, it is possible the same number of T cells were in the culture but they were no longer binding to the antibody because of their exposure to agents used in preparing the culture. "In this type of culture, T-cells lose their CD4 marker and acquire other markers, including CD8, while the total number of T-cells remains constant."[122]

Eleopulos remarks acidly, "It is... surprising that Dr. Montagnier and his colleagues did not have controls, that is, cultures of T4 cells originating from patients who were not at risk of AIDS but who nonetheless were sick and to which they added PHA and IL-2."[123] If such controls had been used, it could have been determined if the seeming disappearance of T cells was due to the effect of the preparation or to their being destroyed by some retrovirus, since the healthy T cells in the control would also seem to disappear as they suffered the effect of the preparation.

To recap, we see that Dr. Montagnier's group's claim to have identified a retrovirus could be assailed in a number of ways, especially in that reverse transcriptase budding and activity were not exclusive to retroviruses, and that it is not definitively known whether T cells were being killed or merely having their detectability altered by the culture's preparation.

DR. GALLO'S RETROVIRUS SEARCH

Dr. Eleopulos and her group in Perth have centered their work on unraveling some of the flaws and fallacies surrounding the HIV blood test and in studying the claims for the existence of a retrovirus in the blood of pre-AIDS patients.

When Neville Hodgkinson, the journalist we met earlier, interviewed her for his book on AIDS, she recalled how unfriendly scientific circles were to the original retroviral hypothesis of the French team. "*Nature* turned down the Dr. Montagnier team's paper that was later held to have been the first to demonstrate the presence of HIV." It was felt at that point that it was unlikely that AIDS was caused by a retrovirus.[124]

Later, one of these papers was published in *Science*, where, you may recall, Dr. Gallo was one of the reviewers. When this was mentioned earlier, it was only to bring up how he threw in an unsolicited comment about his own research. What was not mentioned then was that though he later relied on their findings, at the time he thought the French discovery of this retrovirus was bogus. In fact, quite ironically, he brought up some of the same complaints against the French that are now being turned against him.

"'No one has ever been able to work with their particles," Dr. Gallo wrote in *Lancet* , taking the French to task. He went on, "Because of the lack of permanent production and characterisation it is hard to say they [the retroviruses they detected] are really 'isolated' in the sense that virologists use this term." Moreover, he had doubts that the microscopic pictures accompanying the *Science* piece were genuinely of a human retrovirus.[125]

Eventually, as we know, Dr. Gallo turned a corner on this. Like a sinner who has long resisted the evangelists, but then goes through an overnight conversion, Dr. Gallo, following a similar path in viewing the French retrovirus, suddenly made an about-face. According to Dr. Eleopulos, this turnabout could be put down to the pressure he was under. She feels:

> With the National Cancer Institute suddenly coming under acute pressure from the Reagan administration… to do something about AIDS, he [Dr. Gallo] seems to have allowed his competitiveness to overcome his judgment, leaping on to the bandwagon of the French "Aids virus"—but naming it as "HTLV-3"—once it appeared to have started to roll.[126]

She believes Dr. Gallo entered "what he later called the 'passionate' stage," forcefully arguing for the existence of an AIDS-causing virus, "when his own candidate as the 'AIDS virus,' HTLV-1, proved a non-starter with the scientific community."[127]

To tell a little more of the back story, though, remember that Dr. Gallo was constrained to switch his allegiance to the French virus, because the first candidate for AIDS-causing retrovirus Dr. Gallo identified (HTLV-1), the one he tried to foist on Dr. Montagnier in his abstract for their papers, did not hold up. Dr. Montagnier's team said it was not the same retrovirus they had found. Meanwhile, in further trials, Dr. Gallo himself was not finding it with regularity. That's why he eventually traded in his virus for the one found by the Pasteur Institute.[128]

Some of the same objections might be raised to Dr. Gallo's "isolation" of his version of this virus, as were brought against Dr. Montagnier's group, such as questions about the equation of RT activity with the presence of retroviruses. However, let's cover two other topics about the research leading up to his 1984 *Science* paper in which he tentatively claimed HIV was responsible for AIDS: (1) the antibodies used to identify HIV, and (2) the percentage of cases of HIV infection among the patients whose blood was studied.

HIV ANTIBODIES We saw the importance of antibodies when we observed how the Pasteur Institute used antibodies to T cells to count the number of T cells in a culture. In Dr. Gallo's work, antibodies to HIV were used to determine whether a specific blood sample had traces of the retrovirus. He obtained the antibodies by first infecting the rabbits with "AIDS" and then drawing out the antibodies they produced.

Aside from the dubiousness of expecting the processes in rabbit and human blood to run parallel, there is a peculiar circularity in the procedure—his procedures seemed plagued with circularity—as was brought out in an interview Christine Johnson did with Dr. Eleopulos. The Perth doctor notes:

> Dr. Gallo claims he had a serum from rabbits that contained antibodies specific to HIV. Just imagine for a moment the scene in Dr. Gallo's laboratory. They've cultured H9 cells... and when they came to determine which proteins in their cultures originate from a presumed virus they reach up on the shelf and, lo and behold, they pull down a bottle labeled specific antibodies to HIV. How did they manage to get those antibodies? This was the first paper they wrote but they already had a bottle containing rabbit antibodies specific to a virus they were currently attempting to isolate for the very first time.
>
> They say they prepared rabbit antibodies by repeatedly infecting rabbits with HIV. But if they were preparing antibodies to HIV they would have had to inject rabbits with pure HIV which again means they must have already isolated what they were now attempting to do for the first time. It doesn't make sense...
>
> To even begin to talk about specific antibodies to specific HIV proteins first you have to prove the particles are constituents of a retroviral-like particle that is able to replicate. And the only way to do that is to isolate the particles and do everything else I've described. You need the virus BEFORE you go looking for proteins and antibodies.[129]

The questionable method used here, Eleopulos hazarded, since the details of how Dr. Gallo obtained the HIV to give to the rabbits were not published, is that they took banded specimens of proteins that were said to be viral particles. These would then be the type of particles Dr. Montagnier claimed were evidence of a retrovirus, such as those that showed budding. The same weaknesses we saw in his claim that these particles were necessarily that of a retrovirus could then be appended to Dr. Gallo's research.

THE NUMBER OF CASES OF HIV PRESENT IN AIDS PATIENTS Another weakness in the argument of the 1984 *Science* paper, which purported to have found the HIV retrovirus, was how infrequently HIV was present in the studied AIDS patients who were supposedly made sick by this virus. As Eleopulos states: Dr. Gallo

> only managed to isolate it from 26 out of 72 AIDS patients. That's only 36 percent. Only 88% of 49 AIDS patients had antibodies. And if the virus was present in only 36% of patients, why did 88% have antibodies? I mean there were more patients with antibodies without virus than there were patients with virus?[130]

THE STICKING POINT

According to the Perth group, all of these caveats, both about Dr. Montagnier's and Dr. Gallo's processes of pinpointing HIV, are small potatoes compared to the central difficulty: that the accepted practice of retroviral isolation was not carried out. Other objections about the attempts to nail down HIV, such as questions about the presence of RT, pale into insignificance compared to the point that viral isolation was never completed. HIV was located among the debris in a cell culture rather than in a purified band. Eleopulos clarifies:

> A culture contains a myriad of things and thus by definition is not evidence for isolation of an object. The only way possible to claim that one had made a culture of a virus, is to have had proof for the existence of the virus before making a culture. For example, observing something in the ocean which looks like a fish (even if it is a fish), is not equivalent to having the fish in your frypan separate from everything else that occurs in the ocean.[131]

Let's dwell on this.

However, before swallowing everything I say, note that the following discussion is based on the Perth group's views on isolation. As we will see later, not all scientists consider their understanding of isolation as the correct one.

ISOLATION

The French and the American teams both stopped short before centrifuging blood and carrying out the accepted procedures of viral isolation. Instead they worked with objects in nonpurified cell cultures. We saw that according to standard operating procedure, since a homogeneous sample of particles had never been obtained, it was requisite that the scientists engage in further first-step isolation procedures, paring down what was in their sample till they ended with a single type of particle in a band. Then they would be assured they were studying a single entity before moving to further steps to determine if this entity was a new retrovirus.

This was not done. (We will look at possible reasons for this elision in a moment.) In both labs, later steps in the isolation process were taken without this crucial one having been completed. It could be imagined the scientists harped so loudly on inessential factors, such as the presence of RT, because they were (subconsciously) covering up for the absence of the decisive proof. Their "isolation" of HIV amounted to the replacement of orthodoxy with a new, less rigorous, more error-prone way of finding viruses.

Eleopulos puts the case quite baldly:

> There's an old, logical, reliable, commonsense method of proving the existence of a retrovirus. It's based on nothing more than the definition of a retrovirus as a particle having a particle size, shape, appearance and constituents and the ability to replicate. But for some unknown reason this method has been abandoned in the HIV era. Don't ask me why but it has. In its place we have a disparate collection of data including particles not photographed in density gradients and some evidence for reverse transcription... Neither of these are proof that a retrovirus exists in the cultures.[132]

In an interview with H. Christie, Dr. Valendar Turner has echoed her point by insisting that all the secondary points Dr. Gallo has brought forward as evidence of his having isolated HIV, from the presence of RT to the resemblance of certain particles to other retroviruses, do not add up to true isolation. As he says, "The only scientific proof that a particle is a virus is purification and analysis followed by experiments to prove particles make more particles exactly the same."[133]

In the previous block quote, Dr. Eleopulos brings out the ad-hoc procedure for "isolating" retroviruses that the French and Americans have scared up and used in place of accepted methods. In another place, she drives home the fact that customary usage has not been followed.

> For isolation of retroviruses the stage of purification IS obligatory. One CANNOT ISOLATE retroviruses WITHOUT PURIFYING. By definition, isolation means "to place apart or alone" (Concise Oxford Dictionary) and purify means "to clear of foreign elements" (Concise Oxford Dictionary). Thus, unless the contaminants are removed from around the "HIV" particles... the "HIV" particles are NOT ISOLATED.[134]

The lack of isolation throws the whole of AIDS science into question, according to the Perth group. Eleopulos hammers on the point that the various facts used as evidence for the existence of HIV are all nonspecific. That is, RT activity can indicate retroviruses are present, but it can just as well indicate other conditions when retroviruses are absent. "What we see, the phenomena collectively known as HIV, are non-specific. No one has ever proved its existence as a virus. We [the Perth group] don't believe it exists."[135]

This is not to say that the orthodox camp have remained without polemical resources in this battle. In fact, Duesberg, who, as we saw, believes in the existence of HIV although he dissents from other accepted views, has argued that the Perth group is erroneously interpreting Koch's postulates. As David Rasnick points out, "Koch's postulates deal with infectious agents. An infectious agent must be separated from all other infectious agents... If a biological specimen contains only a single infectious agent, regardless of whatever else may be contaminating the specimen, then the specimen satisfies Koch's rule of isolation."[136] Thus, in the view of Duesberg and Rasnick, the Perth group is being overly punctilious, and their bruiting about the violation of the postulates is not as damning as they think.

Here we see that two groups of scientists are arguing about the proper interpretation of Koch's postulates. Later we will see that some have called for a complete junking of these postulates for viral identification. In both cases, the AIDS controversy is disturbing the bedrock beliefs of viral biology.

Accepting in order to follow their argument a bit further that the rules of viral isolation as the Perth group interprets them are in effect—unlike Duesberg, most scientists involved in work on HIV ignore the whole problem—we can see that, as Eleopulos makes clear, the situation holds both irony and tragedy.

> In spite of AIDS being regarded as one of the gravest conditions ever to afflict the human race, no has deemed it necessary to use a proven method to establish the existence of the putative cause of this dread disease. Instead everybody's opted for a set of non-specific criteria and appear to imagine that if you put all these together they must somehow metamorphose into the right answer.[137]

CONTAMINATION

According to the Perth interpretation, if the culture being used for further experiments contained not a single type of particle but a mix of particles, then further work with the culture (if this work assumed it had an isolated entity) would be fruitless. Until more was done to isolate the entity, one would be working with "contaminated" specimens.

Philpott, citing Eleopulos, explains that HIV cultures were contaminated in just this disabling way.

> Electron microscope pictures, micrographs, of samples Dr. Gallo calls "HIV isolates"—and of all "HIV isolates" produced by Luc Montagnier of France, or since by other scientists—show some objects that

look like viruses (the "HIV"), plus lots of other things, including things that clearly aren't viruses.[138]

What Eleopulos lists as found in the samples, besides particles that resemble retroviruses, were microvesicles (which are unstable but noninfectious organelles that bud from cells) and endogenous retroviruses. These are retroviruses normally found inside the body, not brought in from outside. They are also noninfectious.

We might add at this juncture that the very appearance of endogenous retroviruses in the samples plays havoc with many of the theses of those who claim to have isolated HIV. Given that "these endogenous retroviruses cannot be distinguished from exogenous retroviruses either morphologically or chemically," then many of the processes observed, such as budding and RT activity, and taken to be markers of the presence of an invasive retrovirus could just as well be signs of an endogenous retrovirus. In fact, a number of technical scientific points (which it is not necessary to go into), such as the fact that HIV does not affect as many types of cells as theory predicted it would, even make the idea that HIV originates in an aberrant retrovirus within the body (not invading from outside) quite attractive. This possibility has led Eleopulos to suggest, "It is more probably that 'HIV' (if proven to exist) is an endogenous retrovirus rather than an exogenous retrovirus."[139]

To go back to the discussion of contamination, we might begin by anticipating an objection. You may ask: How could such an important and obvious fact, as that the cultures from which HIV particles were obtained were filled with other non-retrovirus particles, be overlooked? Two points come to mind.

Most obviously, as we have noted, the Rasnick/Duesberg interpretation of Koch's postulates holds that contamination is not important as long as the contaminants are non-infectious agents.

Secondly, for those who want a completely isolated specimen, Eleopulos offer the following explanation for scientists' general acceptance of the samples of HIV offered:

> Perhaps it's because pictures are so powerful. There are pictures containing particles which look like a virus and there's reverse transcriptase in the same cultures as the particles. It is possible mentally to connect particles, reverse transcription and proteins and the antibodies that react with the proteins and make this into evidence for the existence of a retrovirus. Especially for a retrovirologist.[140]

Another part of the answer may be that not until March 1997 (in a study done by Hans Gelderblom of Berlin's Robert Koch Institute that was printed in *Virology*) were electron microscope pictures of HIV particles banded at the accepted density published so that they could be studied by other researchers. Recall that centrifuging forces particles to separate at different densities. It should be added that there is a typical band at which isolated retroviruses tend to cluster. Gelderblom had made some progress toward separating HIV according to the first step of traditional isolation. But there were still problems. He did not achieve isolation, that is, at the band there were still a mixture of particles, not the single type that would prove isolation. Further, the fact that he had gone this far proves nothing about the reality of HIV since, (1) this may be simply a heterogeneous mix of materials, none of which is a retrovirus, (2) even if he obtained a single type of particle in the band, it may not be a retrovirus. Further steps would have to be taken to prove it was indeed a viral entity.

Earlier pictures of the retrovirus were not taken from centrifuged bands. Before this time, HIV may have been "isolated" (that is, in the limited way accomplished by Gelderblom) through banding, but photos of banded materials had not been in print. It was only in March 1997 that it came out that "the electronmicrographs disclose major contaminants in 'pure HIV'" in banded gradients.[141]

According to a summary that appeared in *Continuum*, the paper accompanying the photos "describes the contamination as an excess of vesicles particles of cellular proteins, that may contain DNA or RNA."[142] An article in *Virology*, done by a U.S. research team from the AIDS Vaccine Programme in Maryland to comment on the Koch Institute photos, stated, "It is unknown how these cellular proteins associate with the virus." It went on to warn, "The presence of microvesicles in purified retroviruses has practical implications." One is that all tests that have been done on HIV heretofore are nonspecific, that is, because of the contamination, they cannot be said to reflect results that can be definitively pinpointed as due to the HIV, since it may well be that the results, such as RT activity, are due to one of the contaminants. [143]

Eleopulos brought out another problem with the pictures, noting that even the particles identified as HIV in the gradient did not resemble other known retroviruses. She uses a nautical metaphor to bring out the unprecedentedness of this finding. First, though, she stressed, "Retroviruses are not esoteric, nuclear or cosmological notions whose postulated existence can only be inferred by indirect observations. They are particles which can be seen, albeit not with the naked eye."[144] That Gelderblom's group's noted that the particles they identify as HIV do not have the morphology of other retroviruses makes their contention so "totally unsubstantiated [it] defies belief," she claims. Further:

> The 1.16 g/ml band can be likened to a fishing net. The difference is that the band traps objects according to their density, not their size. Imagine a fisherman who sees in the ocean many different objects some of which may be fish. He throws the net, waits, and upon retrieval of the net performs a thorough examination of its contents and shows that it contains many sea creatures but nothing that looks like a fish. Yet strange as it may seem, he claims to have caught fish. In fact, he claims that the net has nothing else but pure fish.[145]

When we looked at the way Montagnier's and Gallo's groups had "isolated" HIV, we saw that the fundamental problem, displacing all others in terms of seriousness, was that known and valid isolation procedures had not been embarked on. The groups had relied on, among other factors, pictures of hard-to-identify particles taken in cultures. However, when the researchers at the Koch Institute did take the first step toward fulfilling Koch postulates on isolation by attempting to band the retrovirus, instead of solving the problem by obtaining purified HIV, they encountered new difficulties. Where Gelderblom's group expected answers, they found new questions, such as these:

> If one of these particles really is a retrovirus experts call HIV, what are all the others? If the HIV particles originate from the tissues of AIDS patients, where do all the others come from?... If the HIV particles cause AIDS why doesn't one or several of the other particles also cause AIDS?[146]

THE DISADVANTAGES OF TECHNICAL ADVANCES

By idols of the tribe, Bacon referred to judgment-warping effects that stem from humanity's biased cognitive and sensory apparatus. "For it is a false assertion that the sense of man is the measure of all thing."[147] On the one hand, as Sir Francis Bacon sees it, there are general tendencies of consciousness, such as the willingness to be overly touched by first impressions, that make for perpetual mistakes in comprehension. On the other hand, and more to our point, the senses themselves know well how to play us false. "But by far the greatest hindrance and aberration in the human understanding proceeds from the dullness, incompetency, and deceptions of the senses."[148] By this he meant that the senses make us prone to such errors as the following. "Things that strike the sense outweigh things which do not immediately strike it, though they be more important."[149] Although Bacon was referring to our unmediated vision—which is more interested in the most prominent features of a scene, so "of things invisible there is little or no observation"—this thought can be extended without violence to the idea recently expressed by Eleopulos to the effect that "pictures are so powerful."[150] She meant that micrographs of viral-like particles looked very convincing to scientists, pressing them to believe in the reality of HIV in lieu of more acceptable scientific evidence.

Even closer to home is a further observation Bacon produces on this topic. "The sense by itself is a thing infirm and erring; neither can instruments for enlarging or sharpening the senses do much."[151] In his day, scientific instruments were in primitive condition and so he puts little trust in them. In our time of Big Science, though, this apparatus, ranging from radar telescopes to chromatographs, has outdistanced and supplanted the eye and ear as the proper devices for making scientific observations. So, it seems fitting, in our updated application of Bacon's themes, that we take the idols of the tribe as applicable as much to scientific instruments as to the five raw senses. Using this extension, we will be able to catch typical errors that arise from a naive faith in technology, which among scientists seems to have replaced the naive faith in God that reigned in Newton's day.

In this section, we will see how a representative tribal (technology-fetishizing) idol enters into the situation and, like a pied piper, leads scientists' astray, encouraging them to draw conclusions about the existence of HIV that were not merited by the evidence.

It is worth considering how the discoverers of the retrovirus and their followers could move ahead so blithely in tagging and studying what they took to be HIV without considering the controversies around what proper viral isolation is.

When I interviewed Michael Verney Elliot in May 1996, he made the suggestive point that it is the advancement in science itself, in the realm of creating more extraordinary tools, that paradoxically led AIDS researchers to violate accepted canons for identifying a new retrovirus.

He began with some history. In 1976, when "the definition of isolation of a virus was established by the Pasteur Institute," it simply formalizing the Koch postulates, postulates we described above.[152] (Note: Some, like Rasnick, do not see the Institute's codification of the rules as holding much validity.) As we mentioned, once the banding step was completed and one had a homogeneous set of particles at one gradient, one later step was to break the suspected retrovirus into its constituent proteins. Partly, though, this was a matter of the available technology. Even if one had wanted to, in 1976 it would have been impossible to go into a culture soup (one with varied types of particles) and fish out a likely-looking snippet of DNA or RNA in order to study whether it had the characteristics of a retrovirus.

Remember that a retrovirus is an extremely simple entity composed of an outer shell surrounding two strands of RNA. It also may not be out of place to remember some elementary biology. RNA contains a set of genes. A gene is one section of nucleotides that serves as a code for composing a protein. The whole operation of DNA and RNA is to activate specific genes in order to build proteins that are then used in various ways to power the cells.

As noted, circa 1976, to look at the genes in a suspected retrovirus entailed first isolating the virus in a gradient band. There was no other way to do it. However, that necessity was swept aside with the advent of more high-powered laboratory instruments. Elliot had this to say:

Since the advent of very, very super technology which can actually detect one molecule of something in a billion cells it is now considered sufficient to show a constellation of proteins without showing that it aggregates into a complete viral particle—you can see this is just suggestive of a virus.[153]

In other words, in studying viruses with the new, more penetrating technology, bits and pieces of genes or the proteins made from those genes found in a culture can be investigated without having to take the trouble of deriving these materials from a preisolated virus. (Lest you get the wrong idea, let me be clear that we are dealing with a major trend of HIV research: working with scraps, but there are, nonetheless, scientists—to be discussed later—who have been working with complete HIV chromosomes, that is, the virus's complete set of genes.

The group who has worked with fragments have, in Elliot's eyes, given birth to a rather dubious set of scientific practices, which involves engaging in full-scale research on viruses that have neither been isolated, which would be bad enough, nor even been photographed as whole entities. He cites the massive research into the hepatitis B, research, for example, that has been carried on full tilt in spite of the fact that "we still can't culture hepatitis B in a petri dish." He goes on:

> The point I'm trying to make is that technology has zoomed into space age wizardry but the method of interpreting the data of that technology is still practically in the stone age. I've started to question whether HIV really exists or whether it's a laboratory artifact… I believe what they may have found is an odd protein here and there or the stretch of DNA or RNA, but do they all add up to a retrovirus or any other kind of virus?[154]

This whole discussion may conjure up for knowledgeable readers one of the arguments made against nuclear weapons. So far the validity of this argument has not been put to the test—if it had we wouldn't be able to sit around and discuss it—but it went like this.

The more atomic weapons we in the United States and our allies and enemies manufacture, the more we will be tempted to use them. Put crudely: There is an inherent tendency in mankind to play with its toys. By analogy, once we have the ability to extract individual gene patches from a soup of cellular debris, there will be a tendency to find a use for this ability.

We can further refine this insight by seeing it less as an overall human characteristic than as an aspect of the reign of experts. No one has brought this out with more devastating wit than Zygmunt Bauman in his *Modernity and Ambivalence.* There he writes that as a field, such as the creation of medical imaging devices, becomes more specialized and focused, "the chance grows that new skills (meaning new technical capacities) will be invented, which will at first have no clear application."[155] So, for instance, new techniques are developed that have the ability to view molecules and patches of RNA; but there is no particular need for them, at the moment. The most likely result of this situation, Bauman argues, is:

> Their presence [that of the new techniques], however, will bring into relief areas of the life-world previously unnoticed; will redefine previously neutral or easily coped-with elements of life routine as… factors not-

adequately defined, opaque, ambiguous, insufficiently controlled… as problems that need to be "dealt with."[156]

In our example, this would mean that the ability to view bits of RNA from retroviruses was technically developed and then it was "fortuitously" discovered that there was a great advantage to identifying viruses based on a few strands of genetic material. The old way of first needing to isolate the virus in a banded gradient was now (retroactively) found to be time- consuming and inadequate. This is what Elliot means by unevolved "stone age" interpretations of the data produced by "super technology." The interpretations are naturally so, since they arose as justifications of the use of highly expensive scientific techniques, not from a need for such techniques communicated by the scientists to the developers of the technology. This is why Bauman characterizes such development in this way. "Skills seeking application masquerade as problems needing solution."[157]

But let's ponder whether this newfound ability to study small portions of a retrovirus's RNA is as useful as it seems.

THE VIRTUES AND DEFICITS OF WORKING WITH FRAGMENTS

We saw that Dr. Gallo, Dr. Montagnier, and others, had not gone the distance in isolating a virus but worked with likely looking particles taken from a cell culture containing a heterogeneous mix of elements—but now, on closer examination, we will see that they and those scientists who worked with fragments have gone even farther from established methodology. According to Koch's rules, at an advanced stage of trying to isolate a virus the virologist should learn the chemical structure of the entity by breaking down whole strands of DNA or RNA; but this group of scientists began with selected fragments. To put it more graphically, they did not pick out a retrovirus particle that, even if elbowing in among a crowd of other microscopic entities, was at least discrete. They were dealing with jagged, broken-off pieces of the retrovirus.

"Most analyses of so-called 'HIV' genetic material," according to Hodgkinson, "are based on small segments of the purported virus genome, typically covering between 2% to 30% of it, since the longer sequences are so rarely found."[158]

But, it may be asked, if the scientist has only this small segment of the total chromosome to work with, then how can he or she get an idea of the larger structure? Dr. Lanka answers:

It may be hard to believe that all maps purporting to represent a whole retrovirus including HIV, are always compilations, many bits and pieces cobbled together by their authors to the best of their beliefs. They are collages. No complete retrovirus nor its RNA in its entirety has ever been proved to exist.[159]

Perhaps the danger of this approach is already evident. We have a soup of different entities in a patient's blood culture. In this soup are identified a number of pieces of supposed HIV RNA or DNA (as we know, the original retrovirus would not have DNA but would manufacture its own after it had infected a cell) or the proteins they code for. Whatever of these pieces contain proteins or gene material thought typical of retrovirus are then brought together in a thought experiment and said to be the components of one virus.

What we have is a kind of paleontology of living entities that is like the dinosaur variety that tries to reconstruct, say, a two hundred-foot-long reptile, from two of his buck teeth. Here, one is constructing the complex chromosome of an entity from a few genes or proteins.

But the obvious difficulty, given such facts as we have already noticed, such as that not only retroviruses are capable of budding and RT activity, is that pieces of the retrovirus (if it exists) may be mixed up, in the imaginative creation of the whole virus, with pieces of other non-retroviral material.

Moreover, although the pieces being used are ones, as I noted, "thought typical of retroviruses," the idea that some such gene sequences are found typically in viruses is more of a fond hope than a solidly proven fact. As Eleopulos points out, it is not possible "by sequencing a piece of DNA to say that it is truly a retrovirus unless prior proof exists that these sequences are present in a retroviral particle and nowhere else. There is nothing specific about the structure of retroviruses."[160]

Indeed, the most radical rethinkers of orthodoxy, such as Eleopulos, would say there is no retroviral material in the HIV construct at all. In Hodgkinson's summary of her views, she holds, "the 'HIV' found in the mitogenically stimulated laboratory cultures may be no more than a package synthesised as a result of activation and recombination of some of these many, naturally present gene sequences."[161]

And the problems extend to proteins that have been located with these attempts at piecemeal reconstruction. Three proteins that have been associated with HIV are p24 antigens p41 and gp 120. (Remember an antigen is a chemical marker that is found on the surface of a cell or particle.)

P24, as noted, is HIV antigen; p41 and gp 120 are said to be components of the studs that protrude from the HIV shell and are used to fasten the virus to a likely-looking host. However, taking the presence of these proteins in a culture as evidence for the presence of HIV is highly speculative. Eleopulos states, "The p24 protein is not sufficient for diagnosing 'HIV' infection because it is not specific. Indeed, no other 'HIV' protein not even p41 has been reported to react more often with sera from healthy (at no risk of AIDS) individuals."[162] Which is to say that in testing for the presence of these proteins, it is very common to find them in healthy, uninfected individuals, which would indicate either that these proteins are not particularly distinctive of HIV or that tests used to identify them are unreliable (and so, then, would be attempts to specify them as elements of HIV).

It has been pointed out that the use of gp 120 as an indicator of the presence of HIV is also problematic. Where p41 is said to reside in the base of the HIV studs, the protruding studs themselves are said to be composed of glycoprotein 120. It should be remembered that HIV has a life cycle, going from floating in the blood stream to attaching to a cell, invading it and replicating. If one had a specimen of HIV taken from its "cruising" stage, one might possibly find gp 120 as part of the stud. If one were examining HIV inside infected cells, one could not find gp 120, since the stud remains outside the cell.

However, and this will lead us to the next section, all the intact HIV particles so far seen have been without knobs. "Given the fact that the p120 protein is said to be present only in the knobs, [and that] no cell-free 'HIV' particles possessing knobs have been reported so far, it follows neither the particles in the culture supernatant nor the 'pure' virus will have gp 120249 In other words, there is an anomaly in the evidence where gp 120 is used as evidence for HIV in situations where that HIV does not have the knob structure that would include gp 120."[163]

A QUESTION OF VIRAL KNOBS

I hope it has not been misleading to focus on AIDS researchers who rely on fragments of HIV particle material for their imaginative constructions, since, as I've tried to keep clear, that this was not meant to imply that no specimens of intact HIV (at least intact except for their spikes) have ever been found or published in photographs. This is not true. All I have been getting at is that much research in the field has been carried out by concentrating on segments rather than wholes.

Let's amplify the concluding point of our last section, since it does undercut the value of some of the specimens of HIV particles that have been found. After all, if one is using the blood of infected individuals, one would expect to find examples of HIV in both the first stage of its cycle, when it is studded with knobs, and at later stages. The fact that stage one examples are never found may be one reason, besides the impetus of available technological means, that led scientists to rely so excessively on fragments of the retrovirus.

To repeat the plain fact, "Nobody has published EM [electronmiscroscope photographs] of cell-free particles having the dimension of retroviral particles and also knobs, spikes, that is gp 120."[164] But then, to take the implications as far as they will go, this also can be taken as proof of Eleopulos's thesis that HIV does not really exist. Here is how she reasons:

> Dr. Gallo and all other retrovirologists... agree that retrovirus particles
> are almost spherical in shape, have a diameter of 100-120 nanometres
> and are covered with knobs. The particles the two groups [Dr. Gallo
> and those from the Koch Institute] claim are HIV are not spherical,

[have] diameters exceeding twice that permitted for a retrovirus. And none of them appear to have knobs.

All AIDS experts agree that the knobs are absolutely essential for the HIV particle to lock on to a cell. As the first step in infecting that cell. So, no locking on, no infection.… If HIV particles do not have knobs how is HIV able to replicate? And if it can't replicate, HIV is not an infectious particle.[165]

CLONING

At this point, let's look at those scientists who have been able to extract complete specimens of what they believe is HIV DNA or RNA, which, to begin, we can correlate to what was said previously about canonical isolation procedures.

From one angle, it seems that while these positive thinking scientists have redefined the first step of what it means to isolate a virus, casting aside the older requirement of obtaining a homogeneous band of particles before proceeding further, they have nonetheless complied, to an extent, with certain later steps of the procedure. We saw that part of the accepted practices for viral isolation was to take the retrovirus that had been banded out and put it back into a congenial culture of cells. Then, as a final step in proving one had a new retrovirus, it would be observed whether this added particle-infected cells and replicated itself.

These later steps are followed insofar as scientists such as Fisher (as reported by Duesberg) have taken full-length HIV DNA from infected cells and "produce[d] particles that contain reverse transcriptase, HIV specific antigens… and produce infectious virus."[166] The last point is the clincher; it accommodates traditional procedure in finding that the virus reproduces and can infect other cells.

There is another technique used in this work with complete sets of DNA, one that has recently been greatly perfected, especially because it is one so centrally significant to bioengineering, which is one of the most profitable medical fields and, ipso facto, one in which research is most heavily funded. This is the cloning of genes and protein.

Dr. Duesberg, who, you may recall, rejects the Perth group interpretation of Koch's postulates and has considered setting aside these postulates in favor of new principles, sees cloning as a central means toward establishing the reality of HIV. He has weighed replacing earlier methods of virus isolation with cloning, as the following quotation makes clear.

HIV has been isolated by the most rigorous method science has to offer. An infectious DNA of 9.15 kilo bases has been cloned from the cells of HIV-antibody-positive persons… This DNA "isolates" HIV from all cellular molecules, even from viral proteins and RNA. Having cloned infectious DNA of HIV is as much isolation of HIV as one can possibly get."[167]

Here is where I have to draw a line; which is to say, as much as he holds this argument, what the AIDS dissidents have to say on isolation falls to the ground. It's not that in any sense he has won the argument, but that he has taken a stand that puts him outside the discussion, like a prizefighter who ends a bout by exiting the ring. For on this issue the dissidents in their arguments assume that their opponents in the orthodox party hold, as most of them still do, to the old standards of viral isolation; so these opponents are violating their own principles if they continue to consider HIV isolated before they have carried out the accepted method of establishing this.

Duesberg has had the forthrightness to face the most fundamental question involved, namely: What should be the standards of viral isolation?

Let me interject before going further, that in making this argument I am drawing from an article by Duesberg written in 1996 for *Continuum*. In this piece, in an exceedingly lucid manner, he entertains the idea of substituting new substantiating methods for those of Koch. However, it is not clear whether he is actually calling for a replacement to be made or whether he is merely presenting certain merits that would come with this replacement. It is tempting to use his discussion to focus on this argument simply because he lays out the new position so clearly. However, I would be doing him a disservice if I said this were actually his final thought on the matter. In fact, Anita Allen has written me that "the molecular cloning argument is dead in the water," and, she claims, "Duesberg knows it." [168]

Joan Shenton has called to my attention to the fact that in 1990-1991, Duesberg did speak strenuously in favor of the value of Koch's postulates and, in particular, castigated defenders of the HIV = AIDS thesis for their failures to prove themselves against those postulates. This fascinating story can be followed further in her book *Positively False—Exposing the Myths around HIV and AIDS*. Here, let me merely quote her to the effect that

> Duesberg had long maintained that molecular biology and hence retrovirology has spiraled downward into a chasm of reductionism. The more scientists saw under the microscope, the more mistakes they made as they tried to attach what they were seeing to specific diseases, often blaming them on infectious agents.
>
> Koch was important, said Duesberg, because he laid down criteria for infectivity which needed to be met. These criteria were not met for HIV.[169]

Duesberg's analysis of where HIV failed to satisfy the criteria runs parallel to the ones I include in this book. So, as we progress, remember that these thoughts on cloning are not Duesberg's final word, although he does give a precise articulation of ideas that others hold more rigidly.

To return to our thread, we must ask what methods of viral isolation should be acceptable. He answers by saying the older, widely agreed-upon practices of this isolation may need to be abandoned in favor of new ones that are more dependent on new technology, since "molecular cloning of infectious HIV DNA exceeds [in a positive way] the criteria of the old Pasteur rules."[170]

(I don't want to give the impression that this is the sole grounds on which Duesberg depends for his belief in the existence of HIV. He also bases his opinion on the presence of HIV antibodies in HIV-infected people and the usual absence of these antibodies in those without HIV, among other reasons. But among his reasons, it is only cloning that he brings up in considering whether some of Koch's postulates may be outdated.)

One may note that this position is even more paradoxical when it is held, as it is, by more orthodox scientists. In these cases one is treated to the spectacle of an establishment figure, in a scrambling effort to defend a certain conservative position, dropping some of the most hallowed standards of the profession, as if a military commander allowed the castle to be captured so that he could better guard the walls surrounding it. However, as I have striven to be open-minded about all scientific questions, so I will grant that the idea that the cherished "Pasteur rules" are no longer viable certainly is one way for the orthodox to combat the dissidents who claim AIDS scientists have failed to properly isolate HIV. Still, Duesberg's position in this piece would be more plausible if he made a systematic comparison of the older and the newer methods of isolation rather than baldly asserted the new method is preferable.

This said, let me add that a number of scientists have located weaknesses in Duesberg's opinions about cloning.

The obvious point is that simply duplicating a particle proves nothing about what that particle is. "Since any piece of nucleic acid can be cloned and sequenced, cloning and sequencing of a piece of nucleic acid cannot be used to prove the existence of a retrovirus or of its genome," Eleopulos writes.[171] Instead, she argues, any value cloning may contribute to the study of retroviruses would lie further down the road, after the virus has been isolated by traditional procedures and its RNA shown to be "a unique molecular entity belonging to particles with morphological, physical and replicative characteristics of retrovirus particles."[172]

Thus, she crosses swords with Duesberg (as he expresses himself in this piece) by contending the cloning method is not sufficient to establish the existence of HIV.

> Since the propriety of naming an RNA "HIV RNA" is contingent upon prior isolation of a particle proven to be a retrovirus, on this basis alone, "the most rigorous method," i.e., "molecular cloning of infectious HIV DNA," cannot prove HIV isolation.[173]
>
> …We agree that analysis of the proteins of the virus demands mass production and purification. It is necessary to do that. In this respect they [orthodox AIDS researchers] have not just partially failed [since they haven't purified, i.e., isolated the retrovirus], but TOTALLY FAILED.[174]

Eleopulos argues against cloning a strand of DNA as an adequate replacement for the older method of viral isolation.

This is not the place to engage in a full-scale tug-of-war over the validity of Koch's rules for isolation. That is why I have purposely narrowed my discussion to Duesberg's ideas, although other scientists have advanced other thoughts on how to overhaul the accepted rules for isolating a virus.

Epstein lays out this chronology, "In the early 1980s, many researchers, including even Dr. Gallo, seemed to have Koch's postulates at least generally in mind as they sought to isolate the virus," but when Duesberg needled them about not fulfilling these postulates (Shenton's point), "researchers argued that Koch's postulates were rarely satisfied in practice."[175] Dr. Gallo himself said that Robert Koch "has been taken too literally and too seriously for too long."[176]

Eventually, not only did Duesberg entertain the possibility that science needed new rules for viral isolation, but others took this route. Notably, Dr. Warren Winklestein from Berkeley's School of Public Health proposed a five-part criteria for establishing the existence of a virus. What radically separates his postulates from the older version is that although his, perhaps, could prove the existence of a virus, they do not do so by identifying or manipulating the particles. One of his suggested typical criterion, for example, is that the disease rate will be higher in those exposed to the virus than in those not exposed Another is that physical responses to exposure to the virus should follow a gradient, for example, as revealed in the slow decline of a patient's T cells.[177] Thus, Winklestein would change the rules more radically than Duesberg by moving outside the laboratory altogether and shifting the burden of proof to epidemiological evidence. If one accepted his principles, "isolation" would no longer seem to be a very fit name for what he is doing to prove the existence of a new infectious agent.

Looking any closer at the various attempts to find substitutes for Koch's postulates would take us too far afield; however, at the conclusion of this discussion on whether HIV exists, I will say more about the broader implications of this debate. For now let us conclude by considering the difference between Duesberg's and Koch's approaches. In the older method, one starts with a band in sucrose that, whatever else it is, shows a grouping of distinctive particles that share density and shape. In the new method, one has taken one chromosome strand from cell debris and assumed it is representative of a class of entities. On the surface, deriving scientific conclusions from the second procedure seems far riskier.

Limitations of Retrovirology

We noted that Duesberg shared with orthodoxy the view that HIV is existent, though he had become an apostate in many other areas. One might ask: Why has he stubbornly clung to this particular element of acceptable belief, although he stolidly aligned himself with the dissenters on so many other issues? Could it possibly have to do with his position within the scientific world?

Disputants on both sides include many scientists. However, Duesberg, along with Dr. Gallo and Dr. Montagnier and other leading adherents of AIDS dogma, and as distinguished from Eleopulos and other dissenters, is a retrovirologist. He was a pioneer in the field and achieved recognition and prominence within it. It could be teased out, then, that since he owes his success and reputation to what has happened in that area, he is likely to remain loyal to it.

Let us imagine a scenario—to conduct a little thought experiment of our own—in which new findings strongly suggest AIDS is caused by a bacteria. Imagine further that two camps formed, one vociferously assailing and the other supporting the new theory. One side, let us project, was made up of bacteriologists and the other of retrovirologists. Can anyone doubt who would be on each side?

In our world, one's prestige and even ability to function as a scientist depends on funding. There are few gentlemen scholars anymore who, like Baron von Frankenstein, can afford to equip their own laboratories. When a cure for a disease is being sought, government or corporate funding will flow into the pockets of scientists who work in the area of medicine that appears to promise relief and so if bacteriologists, say, manage to begin convincing people that AIDS was due to a new strain of bacteria, the funding floodgates would open into their domain, while being closed to retrovirologists.

This is not meant cynically. It stands to reason that a person will grant more importance to what she or he is familiar with. Scientists will magnify the significance of events and entities in their own field while diminishing the value of developments in neighbors' areas of expertise. The retrovirus, which a bacteriologist will probably consider, from afar, a rather simple and dull creature —after all, it's not really alive like a bacterium—will be viewed by a retrovirologist, from close up, as capable of amazing feats of longevity, violence, and penetration.

In this section, we will look at the deformations that come from total immersion in one's specialized discipline—ending, however, for once, on a more positive note, in that I will call up, in contrast, two representative instances of how the AIDS situation has been the proving ground for pioneering moves to free medicine from its elitist and overspecialized shackles.

To our first topic, when Duesberg, Dr. Gallo and the others first made names for themselves in the early 1970s, retrovirology was a "hot" area, a trendy and exciting field in which, for the first time, with reverse transcriptase, RNA took on a role no one had believed it could perform. Yet as the decade wore on, retrovirology lost some of its sheen when it was discovered that RT was not as eccentric or dangerous as it first appeared. It lost even more luster when further breakthroughs were not as dazzling as anticipated.

Some writers have argued that both these experiences affected the mindset of retrovirologists who shifted into AIDS research in the early 1980s, that is, they were influenced both by the then-current eclipse of their field and by memories of glory days.

Stefan Lanka believes that once retroviruses declined in interest for the rest of the scientific world, especially when all those identified at that time were benign, those who studied them sometimes went off the deep end.

> Until AIDS was invented, retrovirologists were a minority sect who were happy to accept each other's flights of fancy without being too critical. They could fiddle around to their hearts' content, safe in the knowledge that "retroviruses were the least dangerous of all viruses." Well meaning and credulous colleagues, as well as aspiring virologists, journalists, and through them, laymen were mesmerized by incomprehensible jargon.[178]

Fiddling around is not necessarily a bad thing. Sometimes it is the first step toward an imaginative leap that can lead to a discovery; but there was one definitely bad practice in the retrovirology camp that a number of dissenters have noted. This was the refusal to give up on early contentions that had brought fame and fanfare to the retrovirus, but that later proved to be less than scientifically accurate, such as the belief that reverse transcriptase was, like a budding Hollywood starlet, under an exclusive contract to the retrovirus, and that these viruses, whichstood previous biological dogma on its head, must be diabolical.

Writing in *Continuum*, K. Krafield argues that the earliest supposition about reverse transcriptase activity was that it was confined to the posited action of assumed-to-exist retroviruses. The excitement created by the existence of RT action made retrovirology a cutting-edge discipline. But the idea that retroviruses were the unique sites of RT activity was soon challenged. "It later became accepted that reverse transcription is not an exceptional marvel at all, but a normal process in cells, no 'retrovirus' needed to explain it."[179] It was seen, for example, that RT activity took place in such mundane locales as in the chromosomes of yeast, where "it was shown that reverse transcription reflects a repair mechanism of damage to cellular genetic material."[180]

Yet, was it likely that the study of retroviruses, once they were seen to be not as special as they first appeared, would be sidelined? Krafield notes that by the time of the deflation of the virus, "Great retrovirologists had arisen, known worldwide, like Robert Gallo, Luc Montagnier and Duesberg."[181] With so much prestige and funding arising from top scientists' work in the field, it is unlikely the study of retroviruses would be abandoned, let alone would their existence be doubted. "Few then questioned whether these marvelous viruses actually existed."[182]

For here's another catch. Because RT activity at first seemed so out of the ordinary, it seemed natural to assume it had to be carried out by a novel being, such as the retrovirus; but while the operation of the RT enzyme itself was well documented, the existence of retroviruses was not. "Retroviruses as a source of reverse transcription had still not been proven to exist as a biological entity, thus to exist at all."[183]

All the problems that we saw plaguing the isolation of HIV were first met here, right at the beginning. For it was not, as you might assume, that sudden, unexpected difficulties arose around the isolation and identification of HIV, which was a particularly pesky retrovirus. These same difficulties had been beset attempts at isolation of all human retroviruses.

Dr. Lanka puts it like this:

> No retrovirus was ever shown to exist for example, by being able to isolate and characterise it, and to demonstrate its transmissibility. These failures (obviously not for want of trying) should have sufficed to kill off the whole concept. It may be hard to believe that all maps purporting to represent a whole retrovirus, including HIV, are always compilations.[184]

I quoted the very last sentence before, but the emphasis was then on HIV, so you wouldn't have known that Lanka meant to say the difficulties in isolating particles has been endemic in retrovirological research.

Even before HIV, some retrovirologists had considered dropping Koch's postulates and substituting new and unacceptable (to many) methods of identifying viral particles. Lanka explains, "The rules demonstrating the existence of HIV (and retroviruses in general)," which is to say commonly accepted procedures of specimen isolation and use of that isolated specimen to infect cultures, "were never adhered to… nor were they ever validated."[185] As we saw, retrovirologists would bandy about terms such as "isolation" and "the most rigorous method," but these terms no longer meant what they would mean to other scientists. "Terms were used in retrovirology," according to Lanka, "rather as [they were used] in *Alice in Wonderland*—'It means what I say it means.'"[186]

Dr. Lanka has personalized this whole tragic situation by writing about Peter Duesberg, who, we saw, straddles orthodoxy and dissidence. Lanka acknowledges that "Duesberg's enormous services to mankind are beyond dispute… [for] nearly 10 years now, [he] has steadfastly and at great personal cost, been the anchor of sanity and decency in a world driven mad by the simpleminded HIV theory."[187]

Yet his notable courage has understandably not extended so far as to allow him to rethink some of the basic dogmas of the field he has helped found. How could he do so? This would be, like Sigmund Freud, for example, late in life after his reputation had been made, suddenly saying he was wrong and that there was no such thing as a subconscious.

Let's look in more detail at Lanka's argument, since it will loop back to points we raised earlier about the manner in which new technology may bias thought.

Retroviruses were postulated as being that species of micro-organism which caused reverse transcriptase to occur, which, as a working hypothesis at the time in the early 1970s, was entirely reasonable. The mistake was to elevate the hypothesis to a dogma. Early gene detection techniques lent some credence to the existence of an entity that could be transmitted from one cell to another, which was unfortunate, because this, too, turned out to be wrong.

Errors of this kind occur whenever technology makes available for general use a new experimental procedure, which propels a whole army of researchers into mass producing experimental data, heedless of what the biological significance, if any, of their work might be.[188]

This is more substantiation to our claim that technology often drives research. Here it is the capacities of new apparatus to provide electromicrographs of ultra-minute goings-on within cells as well to allow them to analyze the makeup of these entities that encouraged scientists to begin studying fragments and strands of protein and DNA found in cells. But while the technology could depict and analyze these fragments, it could not tell scientists what they were or what they meant. Here, in interpreting the data, Lanka argues, it was the prejudices of the retrovirologists that had a field day. When one interpretation proved invalid, they immediately supplied another that was even more jerry-built, without even questioning the soundness of the foundations.

Even worse, is the [retrovirologists'] habit of making countless ad hoc adjustments to the original theory, which completely distort the original hypothesis. Correct science demands that there should be a radical rethink when this happens. If there isn't, as in this case, a fundamentally flawed concept goes haywire ending in disaster.

This lack of intellectual rigour has, in a contemporary metaphor, debased molecular biology to a virtual science, leading to the deplorable state of having a disease (AIDS) with a virtual definition, due to a virtual pathogen (HIV).[189]

The personal cost to Duesberg was that he would be forced to fight a battle with holds barred, which is to say, he would have to halt his research and questions within narrow lines, in that he was trying to preserve the basic edifice of retrovirology while renovating some of its floors.

Duesberg went along with mainstream AIDS researchers, limiting his objections to the relatively minor aspect of whether HIV could cause AIDS or not, whereas he really ought to have smelt a rat regarding the whole concept of retroviruses, given his earlier, courageous stance in admitting the mistaken role of retroviruses in causing cancer long before anyone else did.[190]

This brings up a further topic, the role of retroviruses in causing cancer. Once the pristine uniqueness of retroviruses was tarnished by further research, such as that showing the benignity of RT behavior in normal cell repair, it would seem retrovirologists would need to find new lands to conquer or risk being spurned by fellow researchers and, more importantly, funders. This is when Gallo leapt into the breach as he began to study retroviruses as a cause of cancer.

As Eleopulos tells the story, "He was one of the many virologists caught up in President Nixon's decade of war against cancer."[191] As we mentioned, he was spurred by his sister's tragic death from leukemia (and thus professional and deeply personal motives intertwine) to study this disease and in the mid-1970s claimed he had "isolated" HL23V, which was said to be the retrovirus responsible for human leukemia. His proof, as my quotes around isolation will have alerted careful readers, indicate that this did not involve accepted procedures of isolation, but those less exacting ones being cooked up by retrovirologists.

Eleopulos's version of what happened next is:

> Not long afterwards others [retrovirologists] claimed to have found the same antibodies in many people who did not have leukemia. However, a few years after that these same antibodies were shown to occur naturally and be directed against many substances that had nothing to do with retroviruses. Then it was realised that HL23V was a big mistake. There was no HL23V retrovirus. So the Dr. Gallo data turned out to be an embarrassment and HL23V is now extinct.[192]

However, this fiasco was of some value, at least to retrovirologists, in that it helped them perfect ways of "isolating" viruses and working with antibodies to identify particles. Though these methods were not strictly kosher scientifically, they did provide an alternative, seemingly rational approach to working with viral materials when older methods failed to obtain any results.

And this is no small matter, for to be convincing one has to have means of establishing credibility, if not with previously consensually agreed upon methods, then with newly created ones that seem to serve the same purpose. Strategically, the ability to tie these new methods to new technologies is a strong plus in that it intimates that these means of proof positing are replacing the old—not as a measure of desperation because the old tests aren't working, but as an advance allowed by breakthroughs in the creation of assaying apparatus.

We have seen, then, that in the dissenters' opinion, the AIDS crisis has provided something of a salvation to the retrovirologists, whose early promise to help in the study of diseases had shown no very significant results (aside from Dr. Gallo's discovery of a retrovirus believed responsible for a rare leukemia). The identification of HIV was a godsend for retrovirologists in that it put them back into the center of medical science. The dissenters argue further that regardless of the reality of HIV, the retrovirologists' "years in the wilderness" gave them time to cultivate some bad habits, such as a disregard for older scientific standards, which they had trouble meeting. It is this that predisposed them to such actions as throwing over the Koch principles as so much useless baggage.

However, though understandable within their terms of argument, the dissenters may give a false impression when they gang up on retrovirologists. They may suggest that this discipline is particularly egregious in its sins of disciplinary inbreeding. I would like to argue—and will say more about this at the end of Part One—that medicine as it is practiced under the older paradigm is endemically crabbed, that is, weakened by narrowness and elitism.

The later parts of this book will deal with the emergence of a new paradigm in alternative medicine, which, let me emphasize, will not only be innovative theoretically, for example, in how it looks at disease, but in practice, particularly in how it emphasizes empowerment of patients and cooperation, rather than paternalism and the coerced or frightened passivity of patients. Here let me preview a few thoughts on cooperation, since in the alternative paradigm, mutuality provides a valuable antidote to the rampant specialization.

Let me begin by saying that I don't damn the retrovirologists' urge to magnify their domain, distorting reality to do so; for this tendency is part of human nature—an idol of the theater in Bacon's usage. However, as the Elizabethan counseled, it has to be resisted and can be resisted for the good of science.

However, when Bacon fashioned his warnings against such idols, he had in mind subjective resistance, that is, he held that an astute individual could resist the blandishments of these idols by a strong-minded self-control. Here I break with this master. Just as I have suggested that, it is easy to flow with the pack, so I would argue that, against accepted positions, it is most productive to forge a new trail with the pack. One becomes a rebel en masse.

Let me try to remove the paradoxical sting from this assertion, first by a word about my own practice, then with some thoughts about AIDS and community, offering a different version of science from one conducive to disciplinary solipsism.

I've said this book is the product of a group consciousness, both in that it was written as a flowering interaction between Null (health) and Feast (rhetoric), and as it contains the distilled wisdom I have gained from interviews and reading of my colleagues in the struggle to put AIDS on a new footing. Anyone who knows how I go about working on particular health problems will know further that my preferred form is the study group. In these groups, I do not sit at the head of the class and pontificate, but engage in dialogue with group members so that together, and taking the advice of alternative health practitioners, we can build a mutually inspired protocol. Moreover, I work to recruit so that each study group will contain people from different backgrounds, believing the heterogeneity of such groups adds to the overall insight and impact.

For just as I decry a one-sided individualism, I am against the one-sidedness of single entrenched disciplines, especially when they act as exclusive clubs that restrict entrance to those who act and look the same. In the first pages of this book, I mentioned that one of the shames of American medicine is how it has worked to shut out all but the elite from its ranks. In *Get Healthy Now*, for example, I document how hard it is for a woman to enter the closed male ranks of surgeons. Although one woman I interviewed, for example, was qualified and hired by a hospital, her male co-surgeons did all they could to make her life a misery and drive her out of that medical specialty, which they desired to maintain as a male preserve.

I also said earlier that this situation has seen some opening in recent years, but not so much in a greater amount of higher medical personnel being drawn from nonelite groups, although this is a small part of the change. What is most significant in changing medicine care is the way the average citizen is more forcibly taking part in her or his own health care. Here's a dark irony. One of the reasons for the great surge forward of people taking more responsibility for their physical well-being is AIDS.

The limitations that have arisen in medicine and biology due to the infatuation of enclosed disciplines and elite control are being counterbalanced by the input from the grassroots. By grassroots I mean patients, those who are their advocates, and alternative health care workers. This is a major breakthrough, and, I daresay, medicine will only progress to the degree it becomes a democratic enterprise.

Let me present two examples of how the rules are changing in medicine because of this new input: one quite specific and limited to one hospital; and one broader, as it concerns networks of patients and physicians who have intervened to change the way drug trials are carried out as well as created community-based research.

Randy Shilts tells the moving story of Cliff Morrison, "a gay clinical nurse specialist," who in late 1983 set up and took charge of the AIDS ward (5B) at San Francisco General Hospital. He was able to take over the responsibility of running a ward, a task not usually delegated to a nurse, "because the most important hospital administrators all seemed embarrassed by the ward."[193] Morrison

> disliked the hierarchical doctor-nurse-patient model that dominated hospitals... [he felt] patients should have a louder voice in their own care, which only made sense, Morrison noted, because they usually knew more about the intricacies of their often-experimental medications than their doctors... Community groups... had free reign in 5B.... Morrison also rejected the idea of visiting hours as a concept designed for the convenience of the nurses not the patients, and he instituted policies to permit visitors to stay overnight if they wished.[194]

Need I say that Morrison's ward became a model in its own right, as a place providing a compassionate and humane oasis for people with AIDS and their loved ones?

Research into drugs is a place where the larger effects of the democratizing tendencies that have developed in response to AIDS can be viewed. It would be no exaggeration to say that the face of science was changed by activist patients and alternative practitioners.

As Steven Epstein notes, according to the accepted wisdom, to correctly test new drugs, it is necessary to set up studies that put control groups on placebos. Thus, while one group is taking the experimental medication, a second group is unknowingly swallowing sugar pills. Of course, when such experiments were done on AIDS patients, this course had extreme drawbacks. "In blunt terms, in order to be successful the study required that a sufficient number of patients die." This was because "only by pointing to deaths in the placebo group could researchers establish that those receiving the active treatment did comparatively better."[195] No matter how scientifically necessary this seemed, it was hard to sell to subjects, who would enter the experiment not knowing whether they were getting any help at all from a new drug or, in fact, getting useless counterfeit pills. In consequence, in such programs, noncompliance (refusing to follow the rules set down by experimenters) became rife. In one study, "some patients were seeking to lessen their risk of getting the placebo by pooling their pills with other research subjects"; in another, "patients had learned to open up the capsules and taste the contents to distinguish the bitter-tasting AZT from the sweet-tasting placebo."[196] What the episode of sharing pills makes explicit is that noncompliance is often accomplished by a group. Epstein argues brilliantly that, whereas doctors running studies tend to view noncompliance as "purely an individual effort, a truer understanding of the trend sees it as a community effort… [that] represents more than simply a questioning of medical orthodoxy, but [tends to] also involve the setting up of alternative clinics, the support of unique therapies, and the democratization of medical knowledge."[197]

The mention of alternative clinics leads us easily to the second part of this topic, the establishment of community-based research. Not only were people with AIDS dissatisfied with the way drug trials were conducted, they were upset by how few AIDS drugs were in the pipeline. "Instead of waiting for NIAID to test drugs in its lengthy, cumbersome clinical trials at academic centers, primary-care physicians and people with AIDS decided to go about designing such trials themselves."[198] In such trials, the testing of the drug would be combined with normal treatment and would avoid the bureaucracy and cost of trials carried out by the government. The research that was done in this way in New York City was highly democratic.

> From the start, people with AIDS and HIV infection participated in decision making about what trials CRI [the coordinating body sponsoring community initiatives] should conduct and even how they should be organized, "[setting] policies on placebo use and [insisting] that trials… be effectively open to women and minorities, not only gay men.[199]

This way of doing science proved its "credibility and viability" in the testing of aerosolized pentamidine, a treatment for pneumonia, which was eventually approved by the FDA, "the first time in its history that the agency approved a drug based solely on data from community-based research."[200]

This long aside has, perhaps, seemed to take us a distance away from the more "objective" question of whether HIV exists as a physical reality. But there's more to it than that. If, as I have already suggested, the dissenting view of AIDS tends to support a different medical paradigm from the one valued by the orthodox, then part of this paradigm is that it looks at health, sickness, and the healing process from a broader perspective. This viewpoint demands that when treating a question, such as a controversy over the existence of a virus, one combines an examination of the scientific data with a grappling with larger issues, such as the sociology of how scientists are affected by the corporate culture of their discipline.

This section of my critique seemed just the place to raise these concerns, because it pushed the controversy closest to the breaking point. I will explain what I mean by that in more detail at the end of the chapter, where I will comment on paradigms and the Woolgar/Latour model of scientific research.

Retrovirology So Far

I want to begin with a discussion of ideas about scientists we bandied about before. Only toward the end will the relevance of this topic to our general theme, the existence of AIDS, become clear.

In *Laboratory Life*, we saw that the authors depicted the best scientists as consummate politicians as well as extraordinary researchers. If we went a little further in paging through Latour's oeuvre to study his work *The Pasteurization of France*, we would see that there he applies his concept of what a scientist does both retroactively and frenetically. Just as Pasteur is head and shoulders in scientific stature over the more run-of-the-mill scientists viewed in *Laboratory Life*, so he towers over them in his degree of political hyperactivity. Pasteur's portrait might cause us to rethink my early assertion that the opening image of Dr. Gallo, flashbulbs popping as he spoke at Bethesda, was something new in science.

One of the most pervasive images of the scientist in the public's subconscious is that of the Dr. Frankenstein we mentioned. Shorn of his criminal activities, such as grave robbing, he is a rather idealized figure in that he is working on his experiments out of pure love of the game. He is not out to win points with colleagues, who have laughed at him, or to please grant givers. He is pouring his own fortune into his laboratory. Although he does hanker after fame, he is not looking for material success and is driven ultimately by a Promethean urge to penetrate the secrets of life.

Such scientists have long ceased to exist if they ever did. Latour, in his depiction of Pasteur, renders a 3-D portrait of the contemporary practitioner.

Pasteur was not a Renaissance man; but one could say that he was tripartite. As a professional, he was equal parts scientist, networker and, publicist. He did make groundbreaking discoveries; and once these were made, he began beating the drum to make his new practices acceptable to the public, while intensively cajoling, flattering, and threatening political supporters to form them into a broad supporting coalition.[201]

Regardless of the fact that he was making genuine scientific break-throughs—while, I believe, those who have claimed to pinpoint HIV as the cause of AIDS were not making such advances—both he and they followed the same public relations strategy. The initial insight is quickly set in stone and all doubts are forbidden. Reservations or doubt, though they may be of value for a scientist, are anathema for advertising.

I would like to go at Latour's idea in two steps. First I will question his basic position, which holds that good science can be created in the midst of a three-ring circus, as it were. Then, retracting my ground a little, I will ask whether it is possible that Latour's portrait could apply to scientists in noncritical moments of research, but not at moments when paradigms are in flux.

The first point, that Latour's type of opinion should be rejected altogether, is held by many dissenters. They argue that as mainstream scientists are involved, to one degree or another, in creating a heroic media image along with doing hard science, it is up to those of us who are more concerned with the search for truth than reputation to make clear that it is not the amount of fanfare a discovery manages to sustain but its ability to be proven in light of the most stringent possible tests and to sustain itself against practical and interpretative criticism that will prove it right.

Dr. Lanka brings out what he sees as the heavy cost of being swayed by belief in a theory that has found great success and popularity in the media in lieu of its proving its mettle in the laboratory. He is speaking of how the fight against AIDS has hinged on a belief in the existence of HIV, which belief stands on a very shaky collection of unstable premises.

> The AIDS-myth with its exploitation of the fear of an alleged deadly sex-plague has been blindly launched into the realm of seemingly scientifically-based biological fact through the pseudo-rational "HIV" detection technique [that is, the claim to have isolated HIV through new, previously unapproved methods]. But the two terms [HIV, AIDS] according to the rules of construction depend on each other. The clarification of the question whether "HIV" exists, with the most secure method of identification available (and this would be only the isolation of complete "HIV-viruses") is a sine qua non for dismantling the mass-delusional trance called "AIDS."[202]

In other words, the edifice of belief constructed around AIDS, which Lanka sees as containing as much disabling falsehood as verity, has as its core element an abiding faith in the never yet substantiated retrovirus HIV. In his opinion, if people saw how checkered and inconclusive is the history of scientists trying to isolate this will-of-the-wisp, they would begin to revamp all their ideas about the disease.

> That "HIV" has never been identified as secure biological matter is of the greatest importance and must immediately be told to every stigmatized person. No HIV—no false diagnosis AIDS—no death sentence—no false treatment—no unnecessary suffering—no needless dying, but new chances for people who for complex reasons got seriously ill, amongst them being labeled as "AIDS" cases and "HIV" positives at all, and then falling victim to medical shortsightedness based on laboratory-technical constructs.[203]

However, let me play devil's advocate a moment and accept that Latour's view has some validity, that is, I am accepting that backbiting, currying favor, chasing the limelight, and other such unsavory traits may play a part in advancing science—particularly with the Latour/Woolgar thesis added—that all the individual scientist's ploys to increase his or her credibility will only pan out if the science being advanced through them is good. I still maintain this situation is of limited application, to a certain paradigm.

A few words about the scientific paradigm as the concept was developed by Thomas Kuhn in *The Structure of Scientific Revolutions*. His fundamental insight, which everyone cites, is that an older paradigm, such as the Ptolemaic idea that the sun revolves around the earth, becomes weak and has to be abandoned when too many anomalies begin to crop up. Too many astronomical events clash with the worldview and scientists begin to doubt its validity. A further point made by Kuhn—this is the one not so often cited—is that the abandonment of an insufficient paradigm is not usually an orderly process whereby the old guard gracefully concedes they were wrong and jettisons their view. No, the common chronology is that the group who has profited from the old paradigm holds onto it as long as possible, no matter how many anomalies are found. The new paradigm comes to power when older members of the discipline have died out.[204]

Kuhn poses the paradigm shift as based on both scientific and social grounds. While he identifies the social component as generational changes, other research links the support of alternative paradigms to different social groups. Everett Mendelsohn states, "The activities that scientists carry out are socially structured activities and are then related to social interests."[205] These interests extend from the type we saw attributed to the retrovirologists—defense and aggrandizement of their scientific specialty—to "broader direct social influences and interests, such as religion, social class, or political position."[206] When two paradigms are in opposition, scientists and others cling to either one depending on these interests. In relation to recent scientific controversies in the

United States, such as those over cancer treatment, Ronald Giere says, "The most likely source of the increasing controversy is a divergence in relevant values between those who were in fact making the decisions [such as establishment scientists] and sizable groups in the population at large [such as people with AIDS]."[207] Each backs one of the conflicting paradigms.

I will temporarily hold back on discussion of the content of the paradigm I see emergent in the alternative health movement; but what I said in the previous section about the democratizing forces that have become important in the AIDS controversy suggests where supporters of the new perspective lie and also gives a clue to the limitation of Latour's ideas. His presentation of the politicking, grandstanding scientists is one that fits within the bounds of an old boy's network, such as that often found among AIDS orthodoxy. It seems less applicable as a model of the relations within the more democratic, multicultural organizations we have witnessed in such phenomena as community-based treatment. There science is advanced through more cooperative action.

This takes us to our last point, returning us to the topic of HIV's existence. In discussing evidence for and against HIV, I highlighted Duesberg's attempt to overturn Koch's postulates. I wanted to stress that the debate over AIDS has a marked tendency to move from argument over fine details to a probing of the deepest fundamentals of science. When such a tendency surfaces, I propose, this registers that an issue stands abridge the basic question of the validity of the dominant paradigm of science. Even beyond its presentation of colossal health risks and scientific puzzles, AIDS is an extremely sensitive topic, since many of the difficulties it raises seem to suggest, as part of their solution, the overturning of the existent philosophy of biomedicine.

Let me stop there for the moment. When we reach the end of this critique, I will make concrete points about where the AIDS controversy has exposed key weaknesses of the established paradigm. First, let's turn to the thorny topic of language.

What Is AIDS?

It was not ill said by Plato, "That he is to be held
as a god who knows well how to define and to
divide."—Sir Francis Bacon, *The New Organon*

E ven if the existence of the retrovirus HIV was disproven, this would not
lessen the number of people getting sick with AIDS, but it would redi-
rect research in a new and hopefully more productive direction. However, is it also possible to lessen the numbers affected? That is, is it possible that
if we look closely at the statistics of those who are counted as having AIDS, we
might find they are not as awesome as we have been led to think?

The Definition of AIDS

We've already noted that AIDS tests regularly render false positives. Moreover,
we've seen that, especially in Africa, presumptive diagnoses are rife, cases in
which a person who dies of a disease that can be associated with AIDS is presumed to have the syndrome and classified as such without administering any
tests of HIV positivity or other factors that would make the categorization reasonable. In this section, though, the focus is on how different agencies arrive at
the figures of how many are suffering from AIDS. Why is it that there are sudden boosts and occasionally even dips in these figures?

In the 1980s and early 1990s, one of the periodically alarming moments in
the AIDS drama came whenever the CDC released new calculations of how
many people had the disease and the number was seen to have dramatically and
unexpectedly (given previous projections) increased. The media would ballyhoo

these frightening numbers as signs that the unstoppable AIDS plague is scything more than thought possible and so, it would be reported, the situation is growing even more dire.

In this chapter, we will try to erase some of these startling numbers, not because I in any way want to lessen the seriousness of the crisis, but to show that this is yet another area in which sensationalism seems to have shunted aside sane science.

On the surface, it might seem foolhardy to think one could merely look at reports on paper, without reexamining patients, and arrive at different conclusions than the ones presented by those who made the reports. Yet, this attempt seems less delusive as soon as it is borne in mind that there have been continual problems in deciding just who has AIDS.

To repeat some basics: People don't die of AIDS directly; their immune systems are weakened enough so that they are brought low by another disease. Think what this may mean in specific borderline cases. There are many diseases in which the immune system weakens. It's the trademark of almost every devastating illness. If we think of all who have died of illness, we can say it's hardly feasible to test this whole population, post mortem, for the presence of HIV. The only people who are likely to be tested are those who went down from diseases like PCP, which are strongly correlated with AIDS.

Fine, but what if, as happens regularly, it's suddenly found that a new disease, one that was not thought to be related to AIDS, is found to be so related? What would one do, in terms of classification, about all those patients who died from this disease over the last few years? Short of exhuming them, it would be impossible to determine whether they died of what it said on their original death certificates or from AIDS. I only bring up this hypothetical case to show that the questions involved in deciding whether a person has AIDS are far messier than those that come up around other diseases.

In the previous chapter, we descended to the level of the cell and particle to take note of how some evidence for the existence of the HIV retrovirus was based on a perversion of or, more charitably, a rewriting of the rules for viral isolation. In this chapter, we will ascend to the study of large masses of people and notice how the fact that there are sudden increases in the numbers of AIDS patients seems to depend on a rewriting of the definition of AIDS rather than doctors finding more cases.

We will begin (1) with a nod to Francis Bacon, by looking at this confusing definition, seeing that lack of a firm agreement on what AIDS is makes it all the harder to accurately calculate the number of affected. Then we will see (2) how, when the number of AIDS patients suddenly show an unexpected spurt, it may be attributable to a manipulation of this unstable definition, or to other paper juggling. Then we will look further into (3) the implications of these paper amplifications to see how they affect other parts of the AIDS doctrine.

MUDDY DEFINITION

Sir Francis Bacon names the fourth obstacle to objective perception the idols of the marketplace. These are "idols which have crept into the understanding through the alliances of words and names. For men believe that reason governs words; but it is also true that words react [in a distorting manner] upon the understanding."[1] This can happen in manifold ways, but for us Bacon's most pertinent example is this: There "are names of things which exist, but yet confused and ill defined, and hastily irregularly derived from realities."[2] He takes the word "humid" as an example of this, showing that, chameleon-like, it alters its meaning according to the situation in which it is found. The great danger is that a reasoner will imagine that because "humid" is the same five-letter word every time it is employed, the idea behind it has the same inflexibility. This thinker will then blithely ignore the gyrations of "humid"'s meaning as it moves from context to context.

My extrapolation here is that researchers' inability to settle on a single, set definition of AIDS is behind many problems in examining the syndrome, including settling on how many people actually have the disease.

Let's begin by asking why a single definition has not been agreed on. Serge Lang sees the lack of a scientific consensus on precisely what constitutes AIDS as due to multiple factors, including lack of institutional agreement, and the theoretical difficulty (already noted) that arises when two people can die in exactly the same way from the same disease, yet be classified as having died from different illnesses.

One problem is that different institutions have alternative ideas of how AIDS should be characterized. Lang states, "There does not even exist a single proper definition of AIDS on which discourse or statistics can reliably be based.... Some statistics from some sources are based on [the] CDC definition, while others are not."[3] He asks, "What good are statistics obtained or reported under such circumstances?" since what one source would classify as AIDS another would not.[4] Thus, for example, if two institutions with contrasting definitions of the disease studied the same patients, they would end with different reckonings on how many had AIDS.

Such lack of congruity among definitions is only to be expected, Lang argues, because diseases that are said to arise once AIDS has weakened the immune system are so disparate. About 60 percent are ones, like tuberculosis, that are connected to immunodeficiency. The other 40 percent, including cancers, such as cervical cancer, are not connected to immune deficiency. This "schizophrenic" definition has prompted John Lauritsen to lash out that the definition "is absurd because the diseases have nothing in common. Although the central idea of 'AIDS' is immune deficiency," he states, "some of the AIDS-indicator diseases—like the cancers, wasting, and dementia—have nothing whatever to do with immune deficiency."[5]

What unites these diseases is HIV. "The CDC calls these diseases AIDS only when antibodies against HIV are confirmed or presumed to be present. If a person tests HIV negative, then the diseases are given another name."[6] AIDS, said to be caused by a retrovirus that can only survive by leeching off host cells, is also the illness that gains most of its hallmarks by taking them from other diseases. As Lauritsen notes, "Other diseases, such as mumps, measles, polio, chicken pox, rabies [and so on]... can readily be described and diagnosed. Not 'AIDS,' which is defined entirely in terms of other, old diseases."[7]

However, the inability of AIDS to stand on its own would not necessarily hinder scientists from arriving at agreement on how to define it. The perplexity is that there is contention over what diseases should be associated with AIDS.

If we look just at the CDC definition as an instance of this, we see that it has kept expanding. Prof. Root-Bernstein has traced the dramatic changes that took place in the earliest period. "Beginning in 1984... the CDC revised its definition by adding to the list of diseases diagnostic for AIDS any lymphoma (cancer of the lymph system) limited to the brain."[8] It also made other adjustments (this was before HIV's role was identified). Once HIV was considered the cause of the disease, there was "a second revision in June 1985. To the previous set of fourteen diseases predictive of cellular immune suppression, the CDC added seven more diseases."[9]

It would seem sensible to reclassify patients once it was believed that HIV causes AIDS. I have no quarrel with the logic there. However, the significant point (to be brought out fully below) is that sudden explosions in the number of AIDS patients are frequently due to changes on paper, that is, to the altering of the list of associated diseases, rather than to (as the media suggests) the sudden uncovering of a new multitude of sick people.

I should add that though it makes sense logically to add "presence of HIV" to the markers of the disease, once one believes HIV is the virus base of the disease, it is hardly justifiable scientifically, since the causal role of HIV has never been definitively established. If proof of the HIV/AIDS connection is still being sought, then it creates further confusion by saying, in effect, "A patient's having HIV is to be taken as a distinctive sign of AIDS, although what role HIV plays in the disease, if any, is not yet known." (To repeat, it could be, for example, that HIV is an accompaniment of a deteriorating immune system, which has broken down for other reasons.) As Thomas and others argue, it is prejudicing the study of the illness by precipitously enrolling HIV as part of AIDS's definition before HIV's function has been acceptably analyzed. They say, "That the CDC does not understand the need to define the syndrome independently of the hypothesized cause is further evidence that their experts never did the epidemiology work impartially in the first place."[10]

But let's turn to how the kaleidoscopically changing definition of AIDS (along with other purely intellectual exercises) has seemed to ratchet up the number of AIDS sufferers.

CHANGING DEFINITIONS AND SUSPECT STATISTICAL AND RECORD KEEPING PROCESSES

After rehearsing the fact that "reporters, health officials, university professors and medical doctors advance the impression that each year more Americans become HIV positive and develop AIDS than the year before," Philpott makes the following startling claim. The impression these experts are giving is false.

> The fraction of Americans who are HIV-positive has never increased since testing began in 1985. And since 1993, the number of new AIDS diagnoses reported each year has been fewer than the year before, for all groups, including women, blacks, heterosexuals, and young adults.[11]

The reader must already suspect that the plausibility of Philpott's claim rests on the way increasing numbers have been appearing as the result of redefinitions and other paper manipulations.

In 1993, a new definition of people with AIDS was instated by the CDC so that it now included "any person who had developed a significant loss of… T-helper lymphocytes." As Prof. Root-Bernstein explains, a healthy person ought to have thousands of these cells per cubic millimeter of blood. When the T cell count falls below two hundred per cubic millimeter of blood, then, by the new definition, the person has AIDS.[12] As Celia Farber reports, this was a stunning departure from previous definitions, in that formerly a person wasn't said to have AIDS "until his or her first AIDS-defining illness. But by the new definition, anybody who is HIV positive and has a CD4 count below 200 has AIDS, regardless of symptoms."[13] The CDC argued that the new definition was needed because, due to antiviral drugs and other improved treatments, "people were going longer without developing infections. So the definition was broadened in an attempt to include all people whose health was threatened because their T-cell count had dropped."[14]

As could be expected, "implementation of the new definition artificially boosted the number of AIDS cases reported in the national AIDS statistics at least 2-2.5 times," that is by 200 to 250 percent, V. Koliadin states.[15] "In 1993, medical status of about 36,000 asymptomatic HIV-positive Americans was changed on purely formal basis—they became 'AIDS-patients' only because [of] the new definition of AIDS."[16] These positives had simply been moved from one pigeonhole to another, which had a drastic effect on the appearance of overall AIDS statistics. A HEAL pamphlet brings this out by saying: "The new definition caused the number of AIDS cases to double overnight. Prior to this change, AIDS cases had actually leveled off it all risk groups."[17]

Writing in *Continuum*, Molly Ratcliffe gives this example of the effect of a another rewriting of the classification that occurred in the same year. "In the beginning of 1993 the CDC changed the definition of AIDS to include two conditions specific to women."[18]

These new additions to the list were the female-specific disorders: P. I. D. (Pelvic Inflammatory Disease) and Invasive Cervical Cancer. "Interestingly, in the year from March 1993 to March 1994, just after the AIDS definition was changed, there was a 44% increase in the number of women diagnosed with AIDS in the UK," Ratcliffe notes.[19] This is a sterling example of how a sudden boost of AIDS cases in a population segment, which will be reported as frightening evidence that AIDS is now attacking a new fraction of the populace, is, in fact, due to a revision of a classifying schema.

The third of the three changes made in 1993 was to add pulmonary tuberculosis to the list of AIDS-associated diseases. Christine Johnson belittled this addition as one that is as apt to confuse as clarify who has AIDS. This third change "means that anybody with TB can have AIDS. And you ought to know that TB all by itself causes a depletion of T cells, and when you have TB it can look exactly like you have AIDS—but you don't, you have TB. They are virtually indistinguishable."[20] She went on eloquently, "Now, as we go along, we change the AIDS definition, I think three times now [by 1996]. And every time it includes more people who are less sick."[21]

Indeed, John Lauritsen has neatly laid out how the upward curve of AIDS patients has paralleled the lengthening definition. In 1984, he mentions, AIDS diagnoses depended on a person having Kaposi's sarcoma or PCP along with a weakened immune system. "When the CDC expanded the list of AIDS-indicator diseases in 1987, it caused an immediate boost of new AIDS cases by approximately one-third."[22] If the CDC had stuck with the older criteria, "the incidence of new AIDS cases [would have] peaked in 1988, but with the net cast ever wider, incidence continued to rise."[23] As we saw, with the addition of three new defining conditions in 1993, the count was again bolstered.

And it is not only medical writers like Lauritsen that have been concerned with the unreliability of the advancing statistics, but also actuaries, whose job is to calculate risks of death and disease for insurance companies. A story in *Continuum* reports that the Society of Actuaries, "one of the most conservative and respected professional associations in the US," with an international membership, set up a task force "to examine concerns that Americans—and others—are being misled over the nature and extent of the AIDS epidemic. They say the increases are artificial, arising from drawing an ever-widening list of conditions into the AIDS net."[24] According to Robert Maver, a member of the task force, his group was particularly vexed by the 1993 inclusion of T cell count among defining characteristics of AIDS in that "people who test HIV-positive and whose immune system cells fall below a certain level can now be called AIDS cases, even though they may feel fine—and even though many different lifestyle factors can affect the T-cell numbers."[25]

His opinion echoes that of David Mertz, writing in *Rethinking AIDS*. Maver argues that when the newly added associated diseases "are subtracted from the total, it turns out that the epidemic had leveled off... in 1993, the opposite of what the official statistics seemed to show."[26] While, in Mertz's view, AIDS numbers from 1993

> will continue to decline at roughly the same rate, year by year, that they increased prior to 1994... the fact is that AIDS is not going to be fundamentally different from every other new disease in human history in following a bell curve of initial incidence [and then later decline].[27]

Indeed, Mertz's predictions have been roughly correct in that AIDS cases have been steadily declining. We will talk more about flawed predictions in the next section, but it's worth mentioning that, according to the *HIV/AIDS Surveillance Report*, the decline has continued unabated. In the period from July 1996 to June 1997, there were 64, 597 cases reported in the United States In the following period, from 1997 to 1998, there were 54,140 cases, and then in 1998 to 1999 47,083 cases were found—a large diminishment each year.[28]

It might be apropos to comment that these redefinitions don't always have it one way. There are times, though these are rare, when associated diseases are removed from the list of AIDS when they conflict with other parts of the definition. We noted in the introduction to this section that in 1985 being HIV positive was added to the defining characteristics of AIDS. However, this played havoc with the list of accompanying diseases, since not all of them were attended by presence of the retrovirus. Duesberg

> shows that the number and nature of AIDS diseases have been changed to save the [HIV] doctrine, when too many people died of Kaposi's sarcoma (once considered the typical AIDS disease) without carrying the required antibodies, the CDC dropped it from the official AIDS list.[29]

Definitions are not the only way figures are artificially or mistakenly inflated. Philpott, who has made a particular study of the topic, points out how cumulative numbers can mask declines in cases. He points out that according to one government agency fewer new U.S. cases were reported in 1995 (73,380) than had been in 1994 (78,863), a 7 percent drop, while 1994 itself was lower in cases than 1993. However, you wouldn't know that from glancing at the media where "HIV/AIDS alarmists able to claim that AIDS is growing."[30]

The sleight of hand here involves cumulative figures. "Although the number of new cases has dropped each year since 1993, the cumulative number keeps growing."[31] In other words, though it's true that the number of new cases in 1995 is less than those of the previous years, alarmists report a 16 percent increase in the total number of cases, the number arrived at by adding the year's 73,380 to the total. With such arithmetic, if only five new people were found to have AIDS, it could still be contended that the epidemic was burgeoning.

Philpott goes on to point to another problematic manner of reporting AIDS figures, which is to stress how particular groups "are grabbing a bigger share of the AIDS pie" without bothering to mention "the AIDS pie is shrinking." He illustrates:

> For example, the 13,764 new cases among women in 1995 was 19% of the 73,380 annual total, compared with the 14,081 new cases among women in 1994, which was 18% of the 78,863 total for that year. So while 2% fewer women were diagnosed with AIDS in 1995 than in 1994, the fraction of women among total new cases was one percentage point higher.[32]

He has seen this biased mathematics used on the statistics for children and teens. In 1995, infant cases were down to 800 reported from 1,034 the preceding year; the teenage numbers were 405, slightly down from 412 in 1994. However, by using percentages to show that these risk groups are increasing their weight in the overall batch, it is made to seem they are under attack by the virus in an extraordinary way. It follows, "Most Americans are under the [false] impression that AIDS is overrunning maternity wards and high schools."[33]

Another means of "cooking" the numbers is by presumptive diagnosis. As we've mentioned, in this type of procedure, a person who has one of the diseases that are commonly found in HIV positive people is labeled as an AIDS case, without having the necessary tests performed to actually establish if the person is infected with HIV, which is presumed to accompany AIDS. As Mullis et al. note, in the CDC's recording of AIDS cases, "frequently, no HIV testing has been done, a 'presumptive diagnosis' is made and it is included in the tabulations as an AIDS case. The CDC admits to 40,000 such cases, but... Duesberg argues that the number must be much higher."[34] The just referred to HEAL pamphlet highlights the absurdity of this policy by using an illustrative case. "Even though the only difference between 'pneumonia' and 'AIDS' is a positive HIV antibody test, the test is not necessary for a diagnosis of AIDS," when these presumptive diagnoses are made.[35]

Lest I make this discussion too one sided, it should be injected that those who feel presumptive diagnoses are justified can argue rightly that a number of the AIDS-associated diseases that are taken to signify the existence of AIDS are not everyday ones. As Epstein observes, some of these diseases, such as PCP or toxoplasmosis, are rare, while others, such as herpes or candidiasis, which are more regularly found, are only considered as AIDS indicators if they occur in parts of the body where they are not usually located.[36] However, this is by no means true of all AIDS-associated diseases, particularly newer ones added to the list, such as cervical cancer.

Also highlighting the problematic nature of presumptive diagnosis is what has been found when presumptive diagnoses are followed up, although such following up is rare. This happens when, after a presumptive diagnosis has already been conveyed to the epidemiological authorities, subsequent blood tests reverse the findings. As Michael Elliot stated in an interview with me, Duesberg's estimates of these mistaken categorizations were based on "examples in the medical literature where individuals were diagnosed with AIDS on the basis of disease conditions or immune deficiencies and then the diagnosis was reversed when HIV was not found."[37] Such cases make clear that relying on presumptive diagnoses is likely to prove another means of artificially boosting figures of AIDS deaths.

Again let me stress that what is being teased out here is not that the CDC doesn't have every right to change its disease definitions as emerging situations warrant, but that such alterations are giving the public a highly misleading picture of how large the epidemic is, to the point that declines are seen as increases and bigger slices are cut from a smaller pie.

It is possible to say that these violent discrepancies between the dismaying statistics that appear in the press and the more hopeful ones—those found if the numbers are looked at with an enlightened knowledge of how they often crest due to redefinitions and other paper changes—can be totally blamed on the bad faith of the media who lust for frightening, hence audience-attracting, numbers.

But then, why didn't the CDC and others producing statistics for public consumption articulate in their press releases that in the United States there have been no dramatic increases after 1993? It is possible to suppose that one small factor in play was a desire to cement the orthodox case. We saw that Kaposi's sarcoma was dropped from the list when it contradicted the postulation that HIV causes the disease. Let's examine, in a speculative vein, other reasons that may have led reporting agencies to go with misleadingly high figures.

IMPLICATIONS OF ALTERATION

Let's recognize first off that, although there were early bruited triumphs in the fight against the syndrome, such as the discoveries of Dr. Gallo and Dr. Montagnier, in the long run AIDS has been a blow in the face to medical science that has not been able to find a vaccine or anything approaching a cure. (Later I will argue that this blow has not only rattled the public's confidence in medicine, but damaged biology's guiding paradigm.) Not only has no cure been forthcoming but the disease is one from which, it seems, there is no escape. According to the redefinition that made HIV infection tantamount to having the disease, and recognized that HIV infection cannot be eliminated:

> Once a person is diagnosed, he or she will have AIDS forever after, regardless of any improvement in state of health and regardless of whether death results from a non-AIDS associated disease (for example, heart disease or diabetes).[38]

In Prof. Root-Bernstein's pithy explanation, "This is another way in which the definition of AIDS is a medical novelty. This makes AIDS the first disease that no one can survive, by definition."[39]

However, regardless of this drawback of this particular definition, all the redefinitions that have been formulated and that have had the effect of increasing the numbers have had certain positive sides, at least for doctors engaged in studying and trying to cure the disease.

For one, as Mark Conlan remarked to me, by adding people at an earlier stage of the disease (those who are infected with HIV) it makes it appear that the campaign against AIDS is succeeding insofar as less people, percent-wise, are dying. "The reason people got AIDS and died very quickly early on and now it takes longer is that they've changed the rules so much."[40] He points out that as the CDC extended the definition so that, for example, healthy people infected with HIV were included, it naturally lowered the death rate, since this new group may be ten years or more away from full-blown AIDS.

Thinking more along the lines invoked previously, where I observed that from one perspective separate scientific disciplines are like warring fiefdoms, each battling for the lion's share of prestige, funding, and Nobels, Conlan argued that part of the reason that numbers released to the press are artificially kept high is so that the AIDS researchers can maintain their turf.

> Because you have this vast infrastructure that needs new cases, that needs to justify itself to stay in business, you have this constant redefinition—constant reclassification, constant rewriting of the rules, all so you can continue with the idea that there is an AIDS epidemic and you need to put billions of dollars into it every year.[41]

Duesberg voices the opinion that the production of upthrusting numbers is a method of winning constituencies. You'll remember that in my discussion of Latour's book on Pasteur, I alluded to his theory that a scientist must be a networker, garnering adherents to his belief not only by presenting convincing arguments to back up elegant experiments, but by pandering to their hopes and fears. For Duesberg, the fact that parasitic AIDS is dependent on other diseases to devastate a patient leaves researchers studying it open to appeal to those who have all AIDS-associated diseases and opportunistic infections.

> But the coup to rename dozens of unrelated diseases with the common name AIDS, proved to be the most effective weapon of the AIDS establishment in winning unsuspecting followers from all constituencies. By making AIDS a synonym for Kaposi's sarcoma and candidiasis and dementia and diarrhea and lymphoma and lymphadenopathy, the road was paved for a common cause.[42]

The thoughts Duesberg and Conlan entertain on what has motivated health agencies to make the numbers impressive are offered as suppositions. I would not preclude the less Machiavellian possibility that high numbers come about and are trumpeted simply because of a nonconspiratorial confluence between agencies' attempting to rethink the definition as more is known about the disease, and the media, which has endless thirst for scare heads.

The more meaty point I hope to establish is that the panic over rising numbers of AIDS cases is largely a "virtual" catastrophe, to use the term introduced earlier. In other words, the numbers rush upward not when someone uncovers an unsuspected part of the population that is dying, but when people are shifted from one category to another.

Playing fast and loose with the statistics has caused more than one student of the problem to take a cynical view of the agencies that have concocted the numbers. David Rasnick envisaged a scenario whereby AIDS would be eliminated simply by more tampering with the figures. His imaginative picture begins with the thought that

> AIDS is really a disease of definitions. Since the definition of AIDS is now so inflated as to comprise a staggering litany of diseases and conditions (29 as of last count), watch for the pillars of government science to attempt to extricate themselves from their embarrassing situation by shrinking the definition of AIDS, thereby reducing—by means of this accounting trick—the number of reported new cases in the future.[43]

Although he was overly pessimistic about the government agencies' ability to face reality—since, as our numbers from 1997, 1998 and 1999 indicated, the agencies are now admitting (without having changed the definition) that, in the United States if not the world, the AIDS caseload is shrinking—his broader thesis, which is that changes in the numbers have quite often simply accompanied redefinitions, is sound.

Since the CDC and other agencies have come to accept a declining U.S. AIDS rate, this makes some of their predictions, over the years, even less palatable. Even if, in many cases, these predictions have been superseded by events, especially as lower numbers keep appearing, it is still a quite valuable exercise to look over some recent excesses in orcer to learn more about the ways and means that an interplay of media and medicine have continued to rain down dispiriting consequences.

Prevalence of AIDS in Different Populations

By quoting from the *HIV/AIDS Surveillance Report*, for one, we have shown that in the United States—we will look at African figures later—the alarming prophecies about the geometric increase of AIDS in the population, like the predictions of millennialists who expect the end of the world next Tuesday, were never fulfilled. An October 7, 1998, press release from the U.S. Department of Health and Human Services stated that AIDS had fallen out of the top ten leading causes of death. "Age adjusted death rates from HIV infection in the U.S.," it was reported, "declined an unprecedented 47 percent from 1996 to 1997, and HIV infection fell from 8th to 14th among leading causes of death in the U.S. over the same time."[44] The following year, AIDS fell out of the top fifteen causes of death.

This is more evidence, if we need it, that AIDS is not the fearsome epidemic it was long taken to be. However, it's time to move to another topic within the question of the spread of the disease, that of which groups the illness would target.

One of the more terrifying postulations of the early students of AIDS was that although the original outbreak was confined to certain groups within society, such as IV drug users, it would soon blossom out and begin attacking the whole population without discrimination. This is a separate point from that of the number of sick individuals as a whole. After all, the fact that AIDS cases here are trending downward could simply testify to better therapies or more health-conscious living in regard to sex and drugs in the general population. By itself a decrease does not indicate that the belief that the disease would pass on to all groups in society was wrong. It may be AIDS is spreading more broadly, if less deeply.

In this section, we will put that prediction to rest, providing data to indicate that in the United States AIDS has stayed largely within the confines of the first groups within which it surfaced. Such restriction obliquely comments on the science of the diseases. Viruses, after all, tend to be unselective, aside from being more likely to attack those in a rundown or weakened state than those in the pink of condition. So, if it is shown that AIDS has not been moving equably through all fractions of the social world, it would certainly shift if not weaken the claim that AIDS is caused by a virus. There could still be plausible extenuating explanations of why the virus is expressed in certain groups and remains latent and unexpressed (if it is contacted at all) in others, but the argument that HIV is the sole cause of the problem would have lost some of its force.

However, there is another point raised in this discussion, for we will see that it is not simply that no one is getting AIDS because no one is becoming infected with HIV. On the contrary, while some groups are not finding HIV transmitted to them, there are other groups that are indeed getting HIV infection, such as the spouses of AIDS patients, but who are not suffering any ill effects and never coming down with the disease. So when we say that AIDS is not spreading as expected, this means both that HIV is not entering certain groups and that HIV, while entering other susceptible groups, is not causing AIDS.

We'll start our examination lightly by simply (1) noting sectors of the more general population to which AIDS should have moved early on if it were branching out, since these included groups who were sexually active or already HIV infected, but to which AIDS did not move; and by (2) looking in passing at the figures for the UK, which long ago showed up the weakness of the claim that AIDS would spread far and wide. This may be also be the place (3) to note another underreported fact about the spread of the disease, which is that in the United States people of color seem more susceptible than other groups, a point that may be connected to racial differences or, equally likely, impoverished living conditions. From there we will be (4) turning to the overall invalidity of the idea that AIDS will engulf the whole of the population. Once we have satisfactorily shown the invalidity of this last belief, we will be (5) drawing out the implications of the distorted view that for a long time has been fed the American people.

GROUPS LEFT UNTOUCHED

One group that should have been quickly enrolled in the AIDS statistics, if HIV infection indeed causes the disease, is that of health care workers ministering to people with AIDS patients. Remember such accidental infections are expected in the medical profession. As Peter Duesberg told me when we talked in 1995, "Normally these people are at a high risk for contracting an infectious disease. For example, fifteen hundred cases of hepatitis are reported each year from accidental needle sticks."[45] Accidents occur, for example, when nurses get jabbed by needles laced with infected blood and then develop the disease. Something similar happens to those who work with AIDS patients. They get jabbed with HIV-tainted needles. But there the parallel ends. Few seem to become infected and almost none of them gets sick.

The discrepancy between what would have been expected to happen if AIDS were caused by HIV and what has happened is brought out by Dr. Root-Bernstein:

> The CDC figures show that, of several thousand reported cases of needle sticks, cuts, and other contaminations [from AIDS infected blood], only 5 percent have developed HIV seropositivity; and of these, only one individual lacking other identified risk factors has thus far developed AIDS. Since identification or risk requires personal testimony, we can never be sure that a lack of identified risk does not mean simply a lack of candor on the part of the patient [who may not want to admit, for example, that he or she takes illegal drugs], so that this one case may or may not be significant.[46]

The amount of infection in this group is, to say the least, underwhelming, especially when we consider that "the U.S. employs some five million health care workers, treating a cumulative total of over 100,000 AIDS patients for almost 10 years; thousands of American scientists also work on HIV."[47]

However, going against Prof. Root-Bernstein and Duesberg, Koliadin counterargues that, given Prof. Root-Bernstein's estimate of thousands of risky punctures per year, the actual infection rate is not out of whack. (Koliadin will appear frequently in these pages as one dissenter from the mainstream who nonetheless feels it is imperative to give as much credence as possible to orthodox arguments in order to develop plausible dissident positions.) According to the HIV = AIDS viewpoint, which has looked at this possible discrepancy as seriously as have the dissenters, this lack of HIV transmission can be attributed to the retrovirus's low infectivity. As Bell et al. put it in discussing this issue, "The average risk of HIV transmission after percutaneous exposure to HIV-infected blood is approximately 0.3%"[48] We cannot, then, at this point give too much weight to this particular argument.

A second group we need to look at is one that became HIV positive but never came down with the syndrome. This is the set of women who were artificially inseminated with HIV-contaminated semen before the dangers of HIV were recognized. Though long-term sufferers from HIV infection, as Prof. Root-Bernstein reported in 1990, "All are currently healthy, as are their children and spouses."[49] Again this puts the lie to the belief that being infected with HIV inevitably leads to the AIDS disease.

What about the case of infected hemophiliacs? In the early 1980s, when AIDS was first cropping up, it was learned that the majority of them, who frequently need blood transfusions, were infected with HIV, because before the HIV retrovirus was identified, there had been no screening blood donors for this agent. Yet, like health care workers, this group did not produce AIDS patients in anywhere the expected numbers. As reported in the *Journal of International Health Research* in 1994:

> Out of 15,000 American hemophiliacs who have been infected with HIV for over ten years, with a ten-year incubation period for HIV, less than 3,000 have developed AIDS: This implies a mean incubation period of 25—30 years and is equivalent to saying that for most hemophiliacs, HIV does not cause AIDS, since the incubation period exceeds their life expectancy.[50]

The same journal, discussing the opinion of science editor Harvey Bialy, also addresses the question of the low rate of teenage infection with HIV. Here what is noted is not that teenagers don't develop AIDS once infected, but that they are not infected at all, a finding that runs counter not only to what would be expected if HIV were a highly contagious, sexually transmitted virus but to what has been blared out by the megaphones of the mass media, which dun into the public that teenagers have to stop having sex or they will quickly become HIV positive.

Bialy's point is that "it would be predicted that the most sexually active group would have the most HIV and AIDS; however, fewer than 200 cases of AIDS per year are reported by the Centers for Disease Control (CDC) in teenagers."[51]

Moreover, to sift these statistics further, Robert Maver, the actuary we met earlier, who has been working on a task force to determine the risks of AIDS, has determined that if we set aside from these two hundred cases those of teenagers involved in homosexual or I. V. drug activity, the number of cases of AIDS is reduced to five. His conclusion is that, for those not practicing behavior associated with infection, "The chance of ending up as an AIDS case... is less than the chance of being struck by lightning."[52]

In Germany the numbers are even lower for this group. Dr. Jager, one of the leading German AIDS authorities, and president of the Curatorium for Immunodeficiency, Munich, has noted that from 1981 to 1996, "there has not been a single case of male or female HIV infection in the age group of 14—20."[53]

It's disheartening to balance these figures against the battle cry of American educators, who are dead set on linking teenagers to AIDS regardless of the actual figures. As H. Kremer comments, this is "proof that the advocates of HIV/AIDS are by now unwilling to separate fact from fiction, even for the sake of the patients entrusted to their care."[54]

INFLAMMATORY PREDICTIONS IN THE UNITED KINGDOM

Since we are on the topic of how distorted a picture of AIDS spread has been forthcoming from the authorities, an undertheme of our study being the irresponsibility of those who disseminate public health information, we might use Britain to bring this further into focus, since, like the United States, it has been awash with what now seem outlandish predictions of how AIDS would be flooding all groups in the general population, predictions that happily have been proven ill founded.

Writing in the London paper the *Daily Mail*, T. Gallagher provided a litany of exaggerated prognoses made in the U.K. and the eventual real figures that undercut them. We can arrange them in a bulleted series.

➤ In 1988, a committee chaired by Sir David Cox forsaw "up to 17,000 people in England and Wales would have died" of AIDS by later 1992. The deaths to the end of October 1992 were 4,933.
➤ In 1988, the government saw AIDS deaths as rising to about 13,000 per annum. In 1991, the government scaled back its prediction to 6,500. In December 1993 the year's toll was 446.
➤ In 1989, the Royal College of Nursing stated that by 1995 one million Britons would have AIDS. By year end 1993 the number was 8,252 cases. The statistics that most concern us at this point are those dealing with erroneous predictions of AIDS spreading into the broader population.
➤ It was trumpeted that "while at-risk groups would initially be worst hit by the disease, it would quickly spread through the ranks of heterosexuals." Yet, by 1993, there were only 63 AIDS cases "in which no special risk has been identified."
➤ In 1991 the Public Health Laboratory Service "predicted 3,000 heterosexual AIDS cases by 1997." In 1993 the PHLS lowered the figure to 1,000.[55]

Writing in 1997 on the British data, Philpott notes, "AIDS there is 99.9% confined to the official risk groups."[56] He cites an article from the *Sunday Times* (London), which highlighted that of the 12,565 people in Britain diagnosed with AIDS since 1982 "a mere 161 (0.1%) have been heterosexual [who were] not exposed to a high-risk category, such as [being] drug abusers or bisexual men."[57]

While the British Health Services claim these astonishingly low numbers—astonishing if one believes in the ease of sexual transmission of the disease—are due to the Services splendid public education campaign, an editorial in the British magazine *Continuum* says such a flippant boast "does not explain why figures remain so high in the gay community which received [the same Health Services] message."[58]

Hodgkinson brings us back home by indicating the parallel in the lack of spread of AIDS to heterosexuals without high-risk behavior in the UK and the United States.

> AIDS cases outside the original "risk groups"—that is, where no special risk has been identified—are almost nonexistent… a cumulative total of sixty-three in the UK in cases; and sixty-six out of 37,600 in New York City, [both using figures available as of November 1995] where from the start the health department has put considerable effort into investigating the background to cases.[59]

As we have said, if AIDS is not being engendered outside of the first at-risk groups, then, at the very least, there are gaps in the current understanding of the disease. Let's look more closely at what is portended by AIDS's confinement to these population segments.

HIGHER INCIDENCE OF MINORITIES IN AIDS STATISTICS

One at-risk group in the United States that, until recently, was seldom highlighted is minorities. Those among them that are in *at-risk* groups seem to get a disproportionate share of the disease, although not of news coverage. (My italicized emphasis is to underscore that we are not talking here of the disease creeping out of the at-risk groups, but of its variation within this limited pool.) This early failure to highlight the particular susceptibility of certain minorities to AIDS—keep in mind there are many diseases, such as sickle cell anemia, which tend to target specific racial or ethnic groups—may be a product of the media's general disregard of issues that do not affect the white middle class.

This failure is now being redressed to a limited degree, but the susceptibility of minorities (within at-risk groups) still needs more exposure. Starting with a glance at the worldwide problem—though taking this first set of statistics with a grain of salt, in that we have said (and will go on to establish more fully) that the numbers of AIDS cases in Africa was vastly inflated—we learn that "63% of AIDS patients world-wide are people with dark skin-colouring." A more detailed breakdown was provided by the World Health Organization up to July 1992. It said, from the world's 501,272 AIDS patients, "152,463 lived in Africa and, apart from a few exceptions, were of dark skin-colouring; in the USA there were 105,329, in Latin America and the Caribbean—58,000: thus a total of 316,533."[60]

Of more moment for us are statistics for the United States as of July 1992 from the CDC.

> In the USA, 29.1% of the AIDS patients are "Blacks" and a further 16.4% of afflicted patients are "Hispanics" 5… 3% of the USA citizens affected by AIDS were "Whites," meaning that 47% were dark-skinned.
>
> As a reminder: the share of "Blacks" in the USA population comes to 12%, that of "Hispanics" is 6%,… and that of "Whites" 81%. In view of their share of the population as a whole, the infection rates of dark-skinned or "Hispanic" people are strikingly high… Male "Blacks" and "Hispanics" are thus infected with AIDS three and a half times more frequently than "Whites"; females fifteen or ten times more frequently, and children nineteen or sixteen times more frequently.[61]

As emphasized, these comparable racial statistics refer to those already in high-risk groups, such as among gay males or IV drug users. "In the USA, people of dark skin colour and "Hispanics" represent 79% of victims who are said to have infected themselves with the agent causing AIDS through the joint use of syringes."[62]

A more localized study of New York City found that 82 percent of whites and 68 percent of black drug users shared syringes. "Yet 56% of 'white' iv. d. u. [iv drug using] persons are HIV-infected as opposed to 62% of 'black' iv. d. u. persons."[63]

W. Geisler remarks the mainstream media commentators were less worried by such reports of AIDS among minorities and IV drug users than they were by the fear that AIDS would quickly go mainstream. "Those infected by the new, deadly disease, namely homosexual men, iv. d. u. persons, and 'Blacks' were shown as being abnormal and have now become a risk to the 'normal' people."[64] Thus prejudiced whites were given new reasons to fear the underclass minorities. "The press speculated," for instance,
about the spread of AIDS, that "'normal' 'white' men involved with 'coloureds,' possibly prostitutes using intravenous drugs, could also be regarded as a source of infection."[65]

Such a reaction is not only repulsive in its continuation of negative stereotypes, but objectively wrong in that, as I am about to argue, such an overflow of the disease outside of at-risk groups was not to occur.

AIDS SPARES GENERAL POPULATION

Let's begin somewhat obliquely. It's no secret that the countries of Western Europe have much fuller and more thorough social services, from education to health to unemployment, in fact, all across the board, than are offered to the citizens by the penny-pinching United States, where the government is only generous when it comes to doling out funds to business. Moreover, when even these measly funds are trimmed, as when President Clinton slashed welfare programs, there is hardly a whimper of protest as contrasted to the strikes and protests that, in the same period, have met and turned back downsizing of social programs attempted in France, among other Western European countries.

It has been said that the willingness of the public to take to the streets to defend such programs has to do with the way they are couched. In Western Europe, programs such as universal health care are given to the whole citizenry. Thus when they are endangered by a governor's knife, the people rise up to preserve what they take to be their rights. In fact, the few U.S. programs that do apply to everyone—Social Security being the most notable—are hotly fought for when they appear endangered. But the majority of U.S. programs, such as Aid for Dependent Children, Medicaid, or Food Stamps, are means tested and only granted to a qualified minority. If the government decides to take away these benefits (not rights), this only upsets the numerical minority receiving them, and the removal is ignored by the majority who have no stake in them.

By analogy, it might be felt that those affected by any disease, whether as patients, doctors, or those studying it, will tend to gather more support the more broadly that disease is said to strike out at the whole population. Although this is neither just or scientifically justified, in the last decades of the twentieth century in the United States, in which the funds offered to health care are constantly snipped away at by either government or HMOs, the possibility that AIDS might be limited to large but discrete communities would tend to circumscribe the attention and resources devoted to the problem (especially if this tendency had not been countered, as it was, by strong community activism). Thus, there were pressures within the affected patients and researchers to believe in the potential of AIDS to become widespread, irrespective of contrary evidence.

However, if we take this from more narrowly scientific grounds, ignoring the allocation situation, whether the disease is pandemic or restricted in scope is a crucial piece of the puzzle in knowing how the disease operates. Knowing what type it is reveals a lot about it, and, on these grounds we have to know the truth. — In the United States and Western Europe (although not everywhere), evidence has it that AIDS is largely restricted to those who are in officially categorized risk groups, namely, gays, drug injectors and their heterosexual partners, hemophiliacs, and transfusion recipients, with minorities accentuated within these at-risk sectors. What is this evidence?

One of the ways to ascertain the prevalence of a disease that can be detected in the blood is by looking at large sample groups, such as Army recruits. Philpott calls our attention to the Centers for Disease Control's HIV Serosurveillance Report, which shows "HIV seropositivity for each year from 1985 through 1993 for blood donors, military recruits, and Job Corps applicants."[66] In his opinion, of these three sets of statistics, the one that would be most representative of people who do not belong to an official AIDS risk group is blood donors, since people at risk are discouraged from giving blood. As Philpott records, "In 1993... only one first time donor in 7,000 (or 1.75 donors in 10,000) was found to be positive for HIV."[67] Extrapolating from that figure would give 44,000 HIV positives out of a "possible 250 million risk-free Americans."[68]

This helps confirm our suspicions, which run smartly counter to what had been predicted, that AIDS has not spread broadly. When I interviewed author Philip Johnson, he generously noted that the early expectation that AIDS would break down all barriers to immerse all of society was not illogical.

> When AIDS was first declared an epidemic, government agencies confidently predicted that before long HIV would spread from the initial risk groups... to the population at large. After all, that is what normally happens when people have not yet developed an immunity to a new virus.[69]

It turned out that this confidence was misplaced, and, as Duesberg says, "Predictions based on the HIV theory have failed spectacularly."[70] As Kary Mullis and his cowriters put it in the article "What Causes AIDS?" the disease has not spread indiscriminately in the United States and Europe. "Rather, it remains almost entirely confined to the original risk groups, mainly sexually promiscuous gay men and drug abusers."[71]

This opinion was backed up by a 1992 report issued by the U.S. National Research Council, which stated "The convergence of evidence shows that the HIV/AIDS epidemic is settling into spatially and socially isolated groups and possibly becoming endemic within them."[72]

More support for this, which can be added to what was established earlier, was provided by a study of American sexual behavior, *Sex in America: A Definitive Survey* . The book concludes, "AIDS is, and is likely to remain, confined to exactly the risk groups where it began: gay men and intravenous drug users and their sexual partners."[73]

Charles Geshekter, discussing this book, says that the scientists who compiled the survey, "warned that 'it is better to tell the truth than to behave like scaremongers, telling the country that a disaster will soon strike us all, no matter what the data say.'"[74]

Geshekter goes on to mention that from 1981 to 1995 there were 476,899 AIDS cases in the United States, "but over 90% of the victims suffered from hepatitis, malnutrition, drug abuse, previous infections that required antibiotics, or had a sexually transmitted disease before any exposure to HIV."[74a] He affirms that this same pattern is seen in HIV positivity where the number in 1995 who were not from at-risk groups but were infected totals about "50,000 and 60,000. That's about 1/500 of 1% of the American population."[75]

As did the authors of *Sex in America*, many others have decried the widespread tendency among AIDS educators to fly in the face of the facts about AIDS prevalence. Already in June 1992, Celia Farber wrote in *Spin*, the first mass-market magazine to report on the dissenters' opinions, "The vast campaign to convince the heterosexual community that 'AIDS does not discriminate' is turning out to be politically correct but factually bankrupt AIDS-speak."[76] Farber quotes Dr. Gordon Stewart, who, speaking even more vehemently, states, "Nobody wants to look at the facts about this disease.... The fact is, this whole heterosexual AIDS thing is a hoax."[77]

Three years later the same universalist slogan was being pronounced by ill-informed AIDS education spokespeople and was still causing the more completely informed to despair. As one writer put it,

> Stories about HIV and AIDS appear everywhere. Phrases like "HIV-equals-AIDS" and "AIDS doesn't discriminate" are accepted as true without a second thought. Yet during the past year, researchers have admitted to glaring inconsistencies and errors in what they have been telling Americans about AIDS.
>
> . . . It's time to tell the truth: most Americans are at little or no risk for AIDS.[78]

The following year I interviewed Barry Craven, who said that politicians and AIDS establishment spokespeople were finally admitting AIDS had not spread broadly, but were ignoring the obvious conclusion that the risk-free group was far less susceptible. Instead they were saying its advance was stopped dead by the health counterattack. "The politicians have been able to claim credit for avoiding the AIDS epidemic by saying we've thrown so much money after it. And it's worked."[79] This ignores the glaring counterevidence that was brought out in the discussion of UK statistics, that is that AIDS is still spreading in at-risk groups who should have been equally helped by the thrown money. Because of this, Craven argues, "They say it [AIDS] hasn't spread to the general population, which is true, but it hasn't spread not because they've given lots of money to it but [because] it wasn't going to spread."[80]

AIDS CONFINED TO AT-RISK GROUPS: IMPLICATIONS

Accepting AIDS's confinement to certain groups, what can we say further about how the disease has been mischaracterized, how the fear generated by this characterization has helped it draw funding, and how, because of the fallible definition, much of the money was then misspent?

One term that was bandied about unmercifully in the early days of AIDS was that of "plague." But AIDS is not a plague, not only in that its spread has been locked inside certain at-risk groups but because of its limited killing power. Speaking about this issue in the *New York Native*, John Lauritsen points out that in the ten years from 1978 to 1988 there were "35,188 deaths in the United States from AIDS, out of a population of 250 million."[81] Tragic as this is, it is hardly of epidemic proportions. In the influenza epidemic of 1918, by contrast, "more people than that died in a day. In the bubonic plague, one-third of the population of the entire world died within the space of five years."[82]

Not only is AIDS not of epidemic proportions, but it is hardly as devastating as many other mammoth diseases now in play, Lauritsen avers. He notes that cigarette smoking-associated diseases, such as lung cancer and emphysema, a thousand people a day, so that "as many people die from cigarette smoking in five weeks as have died from AIDS in a decade."[83]

More relevant to our line of argumentation is that, unlike the traditional plague, AIDS is too discriminating in choosing its victims. "AIDS also fails to qualify as an epidemic because it does not affect most of the population. AIDS is compartmentalized."[84]

This aspect of the disease has drawn Duesberg, who you'll remember believes HIV is a benign accompanier of AIDS, to feel that HIV is not a new retrovirus. It has been around, unrecognized for a long time. He says,

> You can deduce that from the distribution of HIV in the population over time. If a microbe is new, then it increases exponentially. In epidemiology, that's called Farr's law. But if a microbe is long established in a population, then the distribution of the number of people carrying that virus over time doesn't change because it's equilibrated. It has found its equilibrium in the population. That is the case with HIV.[85]

Although I don't necessarily agree with Duesberg's position, it does present another way of looking at the virus's limitations. One might suppose that the virus—we're not talking about the disease itself—was staying within high-risk populations because there was something about these populations that proved friendly to HIV's existence. Although Duesberg agrees with the idea of a welcoming host, he adds the nuance that HIV long ago found these ideal groups and has settled into them. For him, the testing and finding of those who are HIV positive means the geography of HIV's presence is being mapped but what is being clarified are not new cases but those that have long played host.

In any case, the belief that AIDS would become plague-like in its plowing through heterosexuals and homosexuals alike, which at outset was at least plausible, went on being bruited in the face of contrary evidence. Some argue that this willed ignorance was motivated by the AIDS establishment's fear that funds would dry up if the disease was less "well-rounded" than it had originally appeared. For instance, Charles Thomas and others writing in *Reason*, refer to the lack of attention paid to the aforementioned 1992 National Research Council report, which downplayed the possibility of AIDS spreading widely. According to Thomas et al., "This factual picture [in the report] is so different from what the theory predicts, and so threatening to funding, that the AIDS agencies have virtually ignored [it]... and have continued to preach the fiction that 'AIDS does not discriminate.'"[86]

Author P. Plumely goes on to argue that coaxing the public and funding agencies into believing that AIDS will soon broaden its scope into all sectors of the population is not only savvy in terms of maximizing funding, although it is that, but valuable in other ways. He enumerates three major benefits of espousing the idea that AIDS will soon affect everyone. First, it is because "no one wants to appear complacent."[87] I take it that he means here that if AIDS does spread broadly, it will have proven better for scientists to have positioned themselves as quick off the mark in anticipating this change rather than as laggards who react rather than proact to a developing situation. This is a case of acting in advance of the trend in order, especially, to present an attractive image whether or not this pose is conducive to accomplishment. Second, to hark to our main theme, he notes, "the worse the epidemic is projected to be, the more money will be available for public health work."[88] Third, also with their minds on possible outcomes, scientists realized that it was the safest course to predict a large number of future patients, since "when the numbers turn out to be lower, the officials can take credit for having done a good job of AIDS education."[89] We've already heard Craven's reporting that this third point of Plumley's was well taken. Once AIDS's lack of generalization across the population had to be acknowledged, scientists and politicians began touting this as evidence of the benefits of proper educational efforts rather than of a limitation inherent in the disease itself.

Further, in looking at reasons tied to funding that would make AIDS's widespreadedness an attractive belief, we can't discount legitimate fears of homophobia and other prejudices among the patients. It was felt that if AIDS were stigmatized as a disease restricted to homosexuals and druggies, then this would probably make it a low priority among politicians in charge of funding research, at least for those who felt no ties to these communities.

In both the U.K. and the United States, these fears have been said to motivate the stress on what was taken to be the coming wide penetration of the disease. According to an article in *Continuum*, Jamie Taylor of the London-based Gay Men Fighting AIDS (GMFA), stated that his group's promotion of the idea that AIDS would soon strike heterosexuals was based on fears of support dwindling if the disease were felt to be restricted to male homosexuals. He said, "The type of money we needed to combat the epidemic in the gay community would only be forthcoming if the disease appeared likely to affect Middle England."[90]

Farber reports that Dr. Sonnabend, one of the earliest dissidents, whose positions were fueled by his early work in New York City with AIDS patients, found the same attitude in AmFAR's then public relation director, Terry Beirn. AmFAR is a key fund-raiser in New York city for people with AIDS. According to Sonnabend, Beirn (who has since died of AIDS)

> knew this heterosexual AIDS thing was a hoax, but he said I have to do it to raise money. And certainly, you could argue that unless those heterosexual male politicians in Washington thought that sex could kill,

they weren't going to release any money. But my response to that was, if you raise money on a false premise, that money's going to be put to no good. And in fact, that's exactly what happened. The money was raised to protect heterosexual men from a disease they're not going to get anyway.[91]

The last point Sonnabend made shows the double bind into which AIDS activists could fall. On the one hand, if they told the truth, that AIDS was likely to be confined (outside of IV drug users) to homosexuals, they would likely find many government agencies turning a deaf ear to their requests that money be spent on researching the disease. On the other, if they lied to up their funding, much of this funding was likely to be spent on irrelevancies.

We can empathize with the fears of the patient community, but we also need to lament the waste when money is wrung out from funding agencies under false pretenses. The waste is easy to document. According to research by David Mertz "only 5% of all U.S. AIDS educational materials are directed at gay men."[92]

> I will even confess [he goes on] that not every safe-sex pamphlet and billboard not specifically targeting gay men thereby automatically excludes them. But the overall pattern is clear: a sizable majority of safe-sex material is… targeted to young, white heterosexuals. Injecting drug use receives similar short shift in these materials.
>
> When, occasionally, the actual demographics of AIDS faintly tugs at the consciousness of safe-sex pamphleteers, gay men and IVDU's might receive a passing footnote for their specificity of risk. The tone here is generally one where in a pamphlet warning of the dangers of unsafe-sex, one might read a parenthetical allusion to the fact that gay men are particularly high risk, or that sharing needles should also be avoided.[93]

Taking to task the CDC officials, Maggie Gallagher notes that by 1987, they already knew that AIDS would probably be confined to already affected at-risk groups; yet they "embarked on a deliberate PR campaign to mislead the American people into thinking that AIDS was spreading inexorably into the mainstream."[94]

According to a investigation by epidemiologist James G. Kahn, "Each dollar spent on high-risk populations prevents 50 to 70 times as many new infections as the same money spread out among low risk groups;" but flying in the face of such facts "of the almost $600 million the federal government spends on AIDS prevention, probably less than 10 percent is spent on high risk groups."[95]

With these facts in hand, Gallagher complains that the CDC 1987 AIDS education program, titled "Respect Yourself, Protect Yourself," was framed to warn heterosexuals. "According to *The Wall Street Journal*, 'A current focus of the campaign is to discourage premarital sex among heterosexuals.'"[96]

As the *Journal*'s writers so trenchantly put it, ""The emphasis on the broad reach of the disease has virtually ensured that precious funds won't go where they are most needed." They specify, for instance, that the federal AIDS-prevention budget of that time (1996) "includes no specific allocation for programs for homosexual and bisexual men."[97]

Thus, on the educational front the mistaken notion that AIDS was perpetually on the verge of spreading widely through all classes of the population ended up accruing more funds than would probably have been available if other notions prevailed, but, on the downside, saw to it that these monies were often spent ineffectually.

On the scientific front, believing this wrongheaded opinion would certainly direct research away from a deeper understanding of the disease, since, if it is confined to certain segments of the population, this should be a sign that these segments have a special susceptibility (understanding that susceptibility would be one avenue by which to approach AIDS). But as long as the plain facts of restricted spread were ignored, it would be an avenue not traveled.

Perhaps the most exemplary lesson of this section is that it indicates once again how, in the direction of research, public relations efforts with an eye toward maximizing resources have come to take precedence over less glossy but ultimately more purposeful scientific goals.

Sexual Transmission of AIDS

We've just explained that AIDS has not spread into the general population but stayed within certain high-risk groups. We've also made note that if physicians and scientists found that a disease were confined to certain groups, this would be one clue in helping them to understand the illness. Of course, the ideal situation would be if the disease was confined to a single population group or, saving that, ones that are closely related. A difficulty with AIDS is that two of the major groups in the United States that the illness strikes, male homosexuals and IV drug users, don't seem to share much in common.

If we think merely of the first group for a moment, and wonder why gays have been afflicted while straights generally have not, we might say that homosexual and heterosexuals tend to practice different types of sexual intercourse, so perhaps something in the difference between anal and vaginal intercourse is at the root of disease susceptibility. This might seem a rather obvious conclusion, although one that has been blatantly overlooked by all the crusaders who have taken the AIDS crisis as an opportunity to decry any type of sex. Nonetheless, this is simply a hypothesis and it may well be that some other difference between heterosexuals and homosexuals is responsible for the disparity in the groups' contacting the disease.

Nonetheless, there is a great deal of evidence that (1) vaginal sex is not a route to transmission. By looking at such subsets of the population as prostitutes, discordant couples (where one partner has AIDS and the other doesn't), and spouses of infected hemophiliacs, we can document that AIDS is not generally spread by way of vaginal intercourse as well as suggest that where it has been spread by heterosexuals, this may be due to anal sex or drug use rather than traditional intercourse. As a complement, then, we will examine (2) the conclusions of researchers who have suggested anal intercourse is a prime means of spreading infection. However, there are admittedly puzzling elements that are not accounted for in this view. For one, the sperm of AIDS sufferers has been found to contain low levels of HIV. We will then be called on to (3) entertain the speculation of some researchers who believe AIDS is not sexually transmitted at all. Some of them, who espouse the "fast lane" theory, argue that only those gays who were engaging in heavy drug use and other activities that strain the immune system besides engaging in promiscuous sexual activity, became infected. Whatever the conclusions eventually reached in this ongoing debate, one thing that remains the same in this controversy, as in so many others we have touched upon, is that (4) popularized presentations adeptly ignore both the established facts (such as the lack of heterosexual transmission) and areas of controversy (such as whether anal sex is or is not a route of transmission) so that they can present hysterical screeds that cast more darkness than light on the subject.

AIDS IS NOT TRANSMITTED BY VAGINAL INTERCOURSE

We can begin this discussion by looking at a number of populations in which, if vaginal transmission were a direct route for AIDS infection, the disease would have been expected to appear but in which it has not surfaced. From there, we can move forward to more general remarks about heterosexual transmission.

One group whose lack of infection seems to indicate the nonviability of heterosexual transmission is discordant couples—in which one person has AIDS but the partner remain uninfected. According to Peter Duesberg, "There are thousands of those," although to make his point more dramatically, he likes to point to celebrities who are in the public eye. "One of the most famous examples is Arthur Ashe. He had HIV for ten years and died of AIDS but his wife and daughter are both HIV negative. They don't have AIDS and they don't have HIV."[98] Another example is Magic Johnson, "who is also HIV positive and his wife is negative."[99]

Another place where we might have expected large numbers of AIDS cases, if heterosexual transmission was rife, is among heterosexual prostitutes. Here again the statistics discount such a transmission route.

The first point to make is that the low rate of infected prostitutes was totally unexpected by early researchers into AIDS who confidently predicted that the disease, which began among gays, would use prostitutes as a conduit by which it would spread to heterosexuals. Hodgkinson, in summarizing an article done for the *Wall Street Journal* by Root-Bernstein, notes, "A single, HIV-infected female prostitute might, it was thought, infect dozens of heterosexual men, and equal numbers of women through those men."[100] Eventually, though, it was found few prostitutes in the United States, became HIV positive. Between 5% and 10% of female prostitutes are HIV-infected in major U.S. cities such as Los Angeles."[101] Those figures refer to prostitutes as a whole; but for New York City, the numbers have been broken out to look at them according to different classes.

> Forty to 50 percent of streetwalkers (a very low caste of prostitute) who have used IV drugs over the past decade are HIV seropositive. (Whether these streetwalkers had other immunosuppressive risks such as non-IV drug use, unprotected anal intercourse, multiple sexually transmitted diseases, and/or anemia and malnutrition that may have predisposed them to HIV and other infections has never been studied.) Among call girls in New York City (a higher caste of prostitute), no seropositivity was found among those who were drug free. These figures were constant between 1984 and 1989.[102]

As this citation makes clear, those prostitutes who did become infected "almost all were intravenous drug abusers. Cases of sexually acquired HIV were almost unknown among prostitutes who did not take drugs."[103]

Incidentally, the fact that AIDS contractors in this group were drugs users is not necessarily support for Duesberg's thesis that abuse of drugs, not HIV, is the main culprit in bringing on the AIDS disease. As Koliadin puts it, the mainstream might explain the correlation (among prostitutes) of drug abuse and coming down with the disease by pointing out:

> Drug-abuse might be interpreted as a marker of a longer time spent in the profession, hence (i) longer living with HIV positivity, (ii) lower general health status (which itself accelerates progression from asymptomatic "HIV-infection" to AIDS).[104]

Still, no matter what the reason, the correlation still stands. A 1997 article in *Reappraising AIDS* finds drug use as one of the commonest transmission factors in all cases of female HIV positives. "There is virtually no HIV to be found among American females who deny having injected drugs."[105] While only one in ten thousand women who give blood for the first time tests positive; this has to be contrasted to the high positivity found in drug users. In New York City 38.6 percent of the female drug injectors have HIV.[106]

In line with this, a 1993 report from the CDC found that out of 96 percent of AIDS patients in the United States, being those on which full data was available, only "10.1% of female AIDS patients (3,773 of 37,314)" deny being in high-risk groups."[107] Those in high-risk groups "included 32% who were IV drug users, 20% [with] a history of prostitution, and 44% a history of crack cocaine use."[108]

Moreover, to return to prostitution, there have only been a handful of cases discovered in which those prostitutes who were HIV infected passed on the virus to a customer, and in these few cases the client was an IV drug user. The conclusion reached by Prof. Root-Bernstein was this: "Every major review of female prostitution by the medical authorities of Western nations had concluded that drug-free female prostitutes were not susceptible to HIV and would not be the means of infecting the general population."[109]

The acceptance of this fact, by the way, is reflected in medical statistics. "Men and women who are prostitutes do not represent a 'risk group' with respect to AIDS and are not listed in the hierarchy of 'risk groups.'"[110]

According to Root-Bernstein, these facts alone lead him to believe "vaginal intercourse and oral forms of sex... (by far the most common forms practiced by the prostitutes interviewed)... are not high risk activities for either the acquisition or transmission of HIV and AIDS."[111] Although that conclusion may be a little broad when based solely on the evidence of prostitutes' infection rates, a second point he urges does seem reasonable. He insists, "Female prostitutes have not been and cannot be vectors for transmitting HIV or AIDS to a healthy, drug-free heterosexual population."[112] (This is not to say that sex with prostitutes is safe since one is liable to pick up other STDs from them, which can lead to a disabling of the immune system.)

The next group it might be useful to observe are the partners of hemophiliacs. Root-Bernstein stated in the aforementioned *Wall Street Journal* article that it is generally believed that between 1981 and 1984, when there were no blood tests to screen out HIV in blood transfusions, about "90%, or some 15,000, of the hemophiliacs in the U.S. were infected with HIV."[113] As we mentioned briefly before, here is another group that if sexual transmission were the norm should have become pipelines through which AIDS gushed into the wider society, outside of the early high-risk groups. But "hemophiliacs have not become vectors for spreading AIDS into the heterosexual population."[114] As of January 1992, there were only 104 known cases where HIV was passed to another person by an infected hemophiliac, and, as in the case of the clients of prostitutes who went positive, "most of the affected individuals have documented assaults on their immune systems beyond HIV exposure."[115] Further, as I mentioned in a magazine piece, given the large number of HIV positive hemophiliacs, "there isn't one study that shows that the wives of hemophiliacs get AIDS from their partners."[116]

We will focus later in this chapter on some common misperceptions about the existence of AIDS in Africa, we might insert at this point a few words about the situation there, especially because it is said by some establishment figures that while AIDS in the West, as far as it is spread sexually, follows homosexual routes of transmission, in Africa it is primarily moving along heterosexual lines. Although this seems to be true, even there, according to Geisler's examination of the statistics of the World Health Organization for 1994, the chances of getting infected with HIV through heterosexual intercourse are not high. "These figures show that on average, HIV will be transmitted from an infected man to a woman, even at the most unfavorable calculation, once during 700 [incidents] of sexual intercourse."[117] When I interviewed Michael Elliot in spring 1996, he further noted that in a 5-year study of 15- to 19-year-old pregnant women in Uganda, "supposed to be one of the real hot spots for AIDS in Africa, the figure for HIV infectivity in heterosexuals was coming down."[118]

We will have more to say about African AIDS in a moment, but if we return to our main focus now, heterosexual transmission in the West, we can conclude that this has turned out to be unexpectedly rare. Just as in Africa, it is not easy to pass HIV by heterosexual intercourse ."Several studies have calculated that it takes between 500 and 1,000 unprotected sexual encounters before an HIV-negative partner seroconverts to HIV-positive," Farber reports.[119] To put this in mathematical terms, we can cite a scientific paper, "Heterosexual transmission of HIV," which sees the risk from heterosexual encounters as negligible. "We estimate," it reads, "that infectivity for male-to-female transmission is low, approximately 0.0009 [percent] per contact, and that infectivity for female-to-male is even lower."[120]

Agreeing to this assessment but trying to make the figures more salient, writing in the academic *Journal of Sex Research*, Symon has put this in terms of comparative risks, beginning by noting that for heterosexual intercourse, "the risk (in the industrialized world, at least) is not of great magnitude, either in comparison to the major modes of transmission, or to other voluntary assumed risks of daily life (such as driving)."[121]

According to Brody's synopsis of Symon's report:

> Having vaginal intercourse (without condoms) with 5,400 American partners (of unknown HIV status, but not known to be in a high-risk group) would result in a lost life expectancy (LLE; an epidemiological value used to quantify and compare risks) comparable to 12 months [lost] for being 12 pounds overweight.[122]
>
> Again, bear in mind, this should not be taken as an indication that unprotected sex is without risk since the danger of picking up STDs remains.

Going further, we might note that when heterosexual transmission does seem to take place, it is generally from the male to the female, not the other way around. "In the United States, for instance, in a period of over 10 years [1984-1994] there have been only 11 documented cases of female to male transmission out of a total of 30,943 positive men."[123] The low transmission rate is the same if we narrow the locus of our study to the Big Apple, one of the epicenters of the disease. According to Hodgkinson, "In New York City, out of 21,421 men with AIDS as of 2 March 1990… the number said to have acquired the disease [from] sex with women at risk was only eight."[124] Looking at some of the annual statistics, according to Farber, would give us results like these. "Adult cases of AIDS in New York City in 1990 as a result of female-to-male transmission totaled one. In 1991, there were none."[125]

As M. Sushinsky summarized, "Even if all of the 3,328 men and 5,545 women with AIDS [in the U.S. in 1998] who report heterosexual contact as their sole risk are making accurate reports, it is important to notice that these numbers would contribute only in the small measure to the total health risks facing heterosexuals."[126]

When I interviewed Hodgkinson, he pinpointed this limited spread of AIDS among heterosexuals as one of the key determinants pressing scientists to rethink AIDS theory. In his words, "One of the factors that strengthened the case for looking again at the HIV theory was that over the years, it became clear despite continued protestations to the contrary that AIDS was not spreading outside the original risk groups."[127] A normal STD would have eventually been found equally distributed between both sexes; "but it hasn't been like that in the countries most affected by AIDS, it has stayed confined to the original risk groups."[128]

EVIDENCE AIDS IS SPREAD BY ANAL INTERCOURSE

To a degree, the exaggerated predictions about heterosexual spread of AIDS stemmed from an erroneous calculation that, as with most other sexually transmitted diseases, the immunodeficiency syndrome would spread with equal facility through all types of sexual intercourse. However, a second reason for these mistaken generalizations about HIV transmission arose from HIV-positive individuals who either gave false reports or were not questioned in enough depth.

In the 1990 book *AIDS and SEX*, editors Bruce Voeller, June Machover Reinisch, and Michael Gottlieb brought out one such case of insufficiently meticulous research.

> Government researchers published data indicating that United States armed forces personnel infected with HIV-1 had caught the virus from prostitutes, triggering calls for increasing campaigns against prostitution. When infected soldiers were interviewed by nonmilitary researchers whom they trusted, it became clear that nearly all had been infected through intravenous drug use or homosexual contact, acts for which they could expelled from the armed services, which prevented them from being candid with the original military researchers. In each of these flawed published studies, researchers, journal editors, and peer reviewers failed to correct mistakes that should have been recognised.[129]

A second cause of misdiagnosis is this: In discussing AIDS, we are using heterosexual intercourse as shorthand for vaginal sex and homosexual intercourse as a signifier for anal sex. In both instances, though, this is a simplification, since both groups can engage in oral sex and heterosexual couples may experiment with anal sex although vaginal sex is the statistical norm. Many of those that believe HIV is spread sexually see it as passed by anal sex, regardless of the sexual orientation of the couple. For them, we might say, if there were a heterosexual community that exclusively practiced anal sex, its members would be as at risk of getting AIDS as any male homosexual group. Mutatis mutandis, a gay community that only practices oral sex would be classified as low risk. Although we are unlikely to find such communities, there are a good percentage of heterosexuals who practice anal sex; and it is argued by some students of the topic that it this percentage who account for what heterosexual spread of the disease we do have.

This is backed up by a comprehensive study published in the *Archives of Sexual Behavior* that indicated that "researchers investigating AIDS often fail to ask about anal intercourse," which is "often dismissed when heterosexual intercourse is reported as the primary sexual activity." The piece goes on to assert, "Without clarifying these factors [both the heterosexuals' use of anal sex and, mentioned elsewhere, IV drug use], it is impossible to assert the existence of purely vaginal transmission of HIV."[130]

An article reviewing the work of medical statistician Stuart Brody stresses the possibilities of misinformation inherent in this procedure. "HIV researchers who publish these papers do not seem to be serious about accurately accounting for anal intercourse," Brody says. "The very studies that claim to document vaginal transmission show that coitus frequency does not correlate with seroconversion, but that frequency of receptive anal intercourse does."[131]

This failure of questioners to ask HIV-positive women or heterosexual men about rectal intercourse has unconsciously skewed the figures of those infected by heterosexual intercourse upward. Brody feels, "A total liar rate of 5% is more than adequate to account for all the cases of HIV transmission and AIDS which are classified as heterosexual."[132] That is to say, he discounts vaginal transmission altogether, preferring to believe "that the rare seroconversion observed in HIV [vaginal] coitus transmission studies represents participants who have not been forthright in their self appraisal of risk exposure, or who have not been asked the specific questions that would have identified their risks."[133] As an example of a positive who might be disingenuous about his answers, he mentions a man who has been raped in prison. Men like this, he says, "might not consider themselves to have ever had sex with a man," but nonetheless, it would have been the anal sex, not other instances of vaginal sex, by which the virus was transmitted.[134]

"I expect," Brody states, if the reporting of behavior of HIV positives was accurate, "the true HIV rate for Americans," who engage only in vaginal sex, would be "far less than one in 7,500."[135] Putting this in more picturesque terms, he opines, "a risk-free American who has a single act of unprotected coitus with a random risk-free partner is about as likely to become HIV-positive as of being struck multiple times by lightning in one year, or winning several state lotteries."[136]

Statisticians would call the vaginal passing of the virus "secondary transmission," which occurs when "a person not in a primary risk group acquires AIDS from someone in such a group," such as when a male IV drug user passes HIV to a non-drug-using female through vaginal sex. Secondary transmission of any type constitutes "only 3% of all AIDS cases ever reported in the U.S."[137]

To return to the article in *Archives*, we find that, as it notes, "heterosexual anal intercourse is common." It gives as representative figures that about 18 percent of North American college women have had anal intercourse; while in Denmark, 27 to 36 percent of women between ages twenty and thirty have experienced anal intercourse. It is held to be significant that these numbers "are close to the 20% male to female transmission rate reported among monogamous heterosexual couples with one partner being an HIV-positive male."[138]

The possibility that anal intercourse is the way HIV infection is passed among heterosexuals is supported by "a European study [that] has shown the relative risk to women of becoming HIV-positive through sex with an HIV-positive partner is increased more than five-fold in anal intercourse, compared to vaginal intercourse."[139] The data from this investigation indicated 46 percent of women who had anal intercourse, with HIV-positive partners became HIV positive.

Dr. Bruce Voeller mentions, by the way, that in the notorious story of Kimberly Bergalis, in which she claimed to have contracted AIDS during a single dentist's visit to the infected Dr. David Acer, Bergalis "was found to have venereal warts on the anus upon being autopsied. These can be contracted only by anal intercourse. This, in spite of the fact that she testified before Congress that she was a virgin! Perhaps vaginally she was."[140]

To get back to the *Archives* study, it adduces a reason why anal intercourse may be a more likely route of transmission. This has nothing to do with sexual proclivities, but with possible health hazards. The article argues:

> There is no question that anal intercourse and unsterile IV activity are the main vectors of HIV transmission, and are the behaviors associated with those people in the primary risk-groups. During both of these activities, the protective skin barrier is compromised, something that does not happen during normal vaginal intercourse. The rectal tissue is more permeable than that of the vagina, and HIV transmission in those having anal intercourse has been directly related to the amount of rectal trauma experienced.[141]

Those that agree with this reading of the available data would be inclined to narrow the range of the susceptible even more, so it would not be engaging in anal sex itself that put one at risk but being the receptive rather than the active partner. This is because the rectal trauma that was said to facilitate transmission would only be (possibly) suffered by the receptive consort.

Let's follow Hodgkinson's analysis of material that supports this viewpoint. He notes that by 1987 researchers reported in the *Lancet* that anal sex put one in danger, but "the risk was almost exclusively confined to receptive partners."[142]

Another set of studies tracked a large number of HIV-negative homosexual men, noting both their sexual patterns and retesting them to see if any became positive over six-month periods.

> Of men who did not engage in receptive anal intercourse within six months before the start of the study and during the six-month follow-up, only 0.5 percent (three out of 646) seroconverted to HIV. By contrast of men who engaged in receptive anal sex with two or more partners during each of the six- month periods, 10.6 percent (fifty-eight out of 548) seroconverted. No seroconversions were seen in 220 homosexual men who did not practise anal intercourse at all during the twelve months.[143]

It was also discovered that HIV risks for the receptive partners would increase depending on their number of partners. Those who had five or more partners would increase their risk eighteen-fold. On the other hand, those men that "reduced or stopped receptive anal intercourse decreased their risk of becoming seropositive."[144]

According to this side of the argument—and remember, we said there is other research that would undercut this position—a key means of risk enhancement is anal intercourse. The unexpected paucity of cases of heterosexual transmission in the West, this side believes, can be explained by the fact that vaginal intercourse is not a likely or feasible means of spreading HIV. That said, let us look at counterarguments, both of those who say, hands down, HIV is not sexually transmitted, and those who agree HIV can be spread through anal sex, but feel this can only take place for an individual who combines recipient intercourse with other, unrelated risk-posing factors such as infection with other STDs and drug use.

EVIDENCE AIDS IS NOT SPREAD BY ANAL INTERCOURSE

If the HIV virus does move from body to body through intercourse, then it stands to reason that it would be transmitted through sperm. But here's a factor that those who vouch for the sexual transmission of AIDS are hard put to explain. The amount of HIV in semen of positives is extremely low.

"In a study of [the sperm of] 25 antibody positive men, only one single provirus of HIV could be found in over one million cells in only one of the men. No evidence of HIV was found in the other 24."[145] In a study done in 1992 of ninety-five men with HIV, only nine of the patients showed evidence of the virus in their sperm.[146] Finally, a study from 1991, which appeared in the medical journal *Fertility and Sterility*, ended with the same conclusions. "Semen samples of 25 HIV-positive men were studied, and it was discovered that only four showed any trace of HIV."[147]

On this front, the evidence for sexual transmission seems to go the other way from what would have been expected if HIV were being moved through sperm. Geisler states, "So far there has not been any evidence, for example molecular biological evidence, which proves that HIV has been transmitted [through semen, which in]... infected men contains only tiny quantities of HIV."[148]

A second difficulty facing the anal sex transmission theory brings us back to the same set of facts that were brought up in strong support of it. According to the theory, one fact that was biasing some to believe there was heterosexual HIV spread was that many heterosexual couples engage in anal intercourse, which is not reported to interviewers. This, according to them, is the reason HIV positivity is sometimes found among heterosexuals. But, the opponents ask, if heterosexuals are often engaging in anal intercourse, which, they agree, is undoubtedly true, why is the extent of HIV positivity among them so low.

In *Continuum*, F. Cline talked to a number of doctors who make this argument. According to Voeller, the U.S. female population between fifteen and sixty-four in 1987 was about 81.6 million. If, following common estimates, about 10 percent of women frequently engage in anal sex, then this would put 8 million at risk. Voeller states, "This startling number of women at risk through anal intercourse probably exceeds the number of homosexual men at similar risk.... What then accounts for the vast difference in the number of AIDS cases between the two groups?"[149] V. L. Koliadin has the same question. He says it's true anal intercourse is frequently performed in heterosexual coupling and the "heterosexual population is many times larger than homosexual. Nevertheless, AIDS is not associated with heterosexual anal sex."[150]

It could also be said that if anal sex alone is all that is necessary for contacting the virus, then even among the gay population, the figures are not as high as would have been expected. As Sean Currin told me, "The percentage of the American gay population that supposedly has AIDS has always been and always will be probably about 2 percent... It has never increased."[151] To put this in other terms, as John Lauritsen does, "All across America are tens of millions of males who have had sex with each other, and who remain healthy."[152] He acknowledges that 58 percent of U.S. cases of AIDS are gay men, but "as a whole they are not at risk for developing AIDS."[153]

Let's keep Lauritsen's paradoxical thought in the back of our minds for a moment: Male homosexuals are the most high-risk group for the disease, yet AIDS is not spread by sex. Before looking at the implications of this notion, it's worth noting that a number of scientists have drawn the obvious conclusion from the enumerated data that there is no proof AIDS spreads sexually. "I don't think AIDS has anything to do with sex at all," Mark Conlan told me.[154] Jody Wells writes, similarly, that there is zero proof of sexual transmission. "The fact that not a single clinical study exists which supports the idea that HIV is spread sexually is important because such a study would not be too difficult to engineer if indeed [there was] evidence the HIV virus existed in the patient."[155]

This returns us to an earlier inference. Previously, we acknowledged that in the West, gay males were a high-risk group for contracting AIDS, and then we hypothesized that what may have distinguished them from other groups (and hence would have been a transmission route for the disease) was their penchant for anal sex. However, this hypothesis has been disturbed by some of the facts mentioned in the previous few paragraphs. They led up to the Lauritsen paradox, which held that, indeed gay males were high risk, not because of their sexual habits, but rather, presumably, due to some other shared characteristic.

At this point we are back to square one. However, there have been some attempts to wrestle with the Lauritsen paradox. Let's go back to what Conlan said. He continued his conversation with me in these terms:

> I'm willing to acknowledge the possibility that in people who don't do recreational drugs there's a possibility that people have gotten multiple cases of the sexually transmitted diseases—the ones that really are sexually transmitted—and have taken multiple antibiotic treatments and trashed their immune system in the same way that recreational drugs would do, but aside from that I don't see a connection between AID and sex at all.[156]

Carefully examined, his statement lays out two alternate routes through which gay men may have become liable to the disease. One is through overuse of recreational drugs, and the other is through overdosing on antibiotics after experiencing multiple STDs.

Let's dub this the "fast lane hypothesis." According to Dr. Steward, who supports this idea—and it is one that has been broached a number of times since AIDS was discovered—"a general disease burden on the body arising from risk behaviours, rather than a specific infection, is the real cause of the epidemic of immune deficiency, with accompanying risk of testing HIV-positive and developing AIDS."[157]

To circle back, though, there is a part played by sex in this theory: It is surmised that having a great number of partners can increase risk by making the engager more liable to other sexually transmitted diseases, and it exposes the passive partner to a greater possibility of rectal damage.

Let's keep one thing clear before looking further into this idea. In our puritanical society, it is very easy to find a hypothesis like this attractive simply because it implicitly condemns promiscuity. This is not what we are about, though. The question raised here is simply whether promiscuity in anal intercourse can induce AIDS. To answer this realistically, it is imperative to avoid moral posturing. We are not talking about the rightness or wrongness of having multiple sex partners in any nonmedical sense. If, to posit another unlikely scenario, a gay male was promiscuous to the point of having twice the partners that any of those in the following descriptions had, but only engaged in protected oral sex, then his chances of getting AIDS (due to this sexual factor) would be nil. The point being: It is not promiscuity per se that is under discussion so much as the wear and tear caused by indulgence in certain sexual practices.

To move to one of the most eloquent defenders of the fast-lane causation thesis, let's listen to Michael Callen, himself HIV positive and confessedly once a member of the fast lane. In his endorsement of this thesis, he told Neville Hodgkinson that the Centers for Disease Control study of the first hundred men in the United States to come down with AIDS, the median number of partners over each lifetime was 1,160. "That number indicates a very specific type of gay person with a very specific lifestyle.... In plain English, only excessively promiscuous gay men like me are developing AIDS."[158]

From anecdotal evidence, Callen concludes that the number of partners found in this early study would hold true for most gay males who have been hit by the disease, particularly because those who have had so many encounters have contracted a corresponding number of sexually transmitted infections. He further argues that, just as the stigma of being gay (which appears strongly in some contexts, such as when one is in the military) has caused underreporting of homosexuality in some circumstances, so the stigma that may be attached to what is viewed as sexual excess will cause some people with AIDS to reduce the number of sexual partners they have had when they talk to doctors. Callen feels it is this that accounts for spurious tales of getting AIDS from one encounter and other mythological tales. "You will continually hear rumors of monogamous or sexually celibate gay men (whatever that means) who have developed AIDS," he told me. "Let me assure you, this simply is not true. There are none. I reiterate: Each AIDS victim has a long history of many sexually transmitted infections," which he claims can be proven simply by giving blood tests to see whether they have antibodies to diseases found among the promiscuous males.[159]

Going on to examine the question more sociologically, he argues that promiscuity is not so much the end result of a particular man's psychology and taste for a wide range of pleasures as the consequence of prevailing political attitudes. After the Stonewall riots in New York City in 1969, when gay clubgoers fought back against a police shakedown, the homosexual community began speaking out about police harassment and systematic discrimination and standing up for their rights. This, Callen avers, led to a certain elevation of promiscuity as an expression of defiance. But Callen explains this better than I could, so let's look at what he has to say. We will take a little time over this passage, which will generously repay scrutiny. He begins by lamenting his own carelessness, then moves on to characterize early AIDS sufferers.

"You try having three thousand men up your butt by the age of twenty-six and NOT get sick.... And I was a baby! I knew the first wave of people with AIDS: they were the founders of what we called the ten thousand club; they had 10,000 or more different sexual partners." Callen said [in Hodgkinson's words] that among homosexual men in general, only a small minority had "pigged out" on sex in this way. But he insisted it was within this "brotherhood of lust" that most of the homosexual AIDS cases were to be found.[160]

Then he goes to link this "brotherhood" to the new feelings of empowerment brought on by post-Stonewall activism. Hodgkinson paraphrases his thought.

The gay liberation years of the seventies had brought unprecedented opportunities for gay men to have sex with each other, as well as unprecedented peer pressure to take advantage of these newly won freedoms, and he and others had subscribed to the idea that the more sex a gay man had, the more liberated he became. Although the lesbian and gay political agenda was much broader than sex, he was only interested in the part that dealt with pleasure. "Never before had so many men had so much sex with so many different partners."[161]

This is an important insight because it reveals how a sexual practice is coded as part of the relationship to the larger community and to that community's connections to the society. One doesn't have to be a Foucault or other writer who has connected sexuality to regimes of power to make the further extrapolation that because promiscuity was tied to gay liberation, crusades by conservatives to police gay sexuality—centered on moral outrage—were bound to fall flat; campaigns for safer, less promiscuous sex within the gay press and community tended to be effective simply because they recognized the context. As Epstein makes clear, the later outreach, done by the community itself, perceived the roots of the promiscuity and labored to change practices by revising the image of what would be politically correct (in a non-pejorative way). As much as safe sex was persuasively explained to the gay community as a way to be responsible and to move the community forward in its demand for equal participation, it struck a sounding note and was accepted, becoming a new guide to behavior.

We might remark, in passing, that this nicely backs up the educational ideas of Brazilian pedagogue Paulo Freire, who often centered his discussion on the attempts of outsiders to come into an enclave and try to "modernize" it. In *Education for Critical Consciousness*, he castigates, for example, the intellectual agronomist who comes to the countryside to educate the peasantry. This reformer inevitably fails because he does not grasp that education is a dialogue, not a one-way street. To succeed, he or she must be as willing to learn as to teach, taking as much from the farmers as he offers. Key is finding the right vocabulary that takes from both sides of the dialogue. "The verbal expression of one of the Subjects must be perceptible within a frame of reference that is meaningful to the other Subject."[162] In our example, the frame of reference is the way aggressive pursuit of sex partners is intimately tied (in a indirect way) to an expression of gay pride. For those seeking to change sexual practices within the male homosexual public, only the ones speaking from within this frame would be able to reach those to whom they wished to communicate.

But let's get back to Callen's thoughts on why having multiple sex partners led to disease proclivities. For one, "In Callen's view, it wasn't just a matter of having too much sex."[163] Rather, for the passive party, having sex "involved repeated exposures to secretions from many different partners."[164] Add to that the "hazard from the cumulative effects of drugs used habitually to drive this sexual merry-go-round," and that they "had concentrated among themselves just about every sexually transmitted microbe available. They were a diseases group, repeatedly infecting one another."[165]

We see, then, that on this fuller explanation, which leads us back to Conlan's remarks, the fast-lane thesis does not rest solely on sexual behavior but also on a behavior linked to use of drugs. Moreover, beyond what was said about rectal damage—which, it would seem, could occur as frequently in a monogamous relationship, given the same amount of intercourse—is the danger of contracting STDs, a danger that increases with the number of contacts.

This last point can be backed up by more analytic research. "A Denver study found that an average bathhouse patron risked a 33 percent chance of walking out with syphilis or gonorrhea."[166] Another study showed that 94 percent of sexually active male homosexuals had been infected at one point with CMV (a sexually transmitted disease), while "7-14 percent tested at a venereal disease clinic were actively infected, with the virus present in their semen."[167]

Koliadin argues, moreover, that because of the low rate of infection, which he estimates as one or two cases for every thousand sexual encounters based on evidence from the transmission between discordant couples, promiscuity should not be a central factor in promoting the disease. (Although we mentioned that certain people with extremely high rates of promiscuity, such as members of the "10,000 Club," did have AIDS, in general those classified as promiscuous do not have anywhere near such a large number of encounters.) The fact that there does seem to be a connection between promiscuity and HIV infection indicates that another accompaniment of promiscuity, other than number of sexual contacts, must be involved. Koliadin notes that "Promiscuity is directly associated with antibiotics treatment and/or permanent prophylaxis of STDs," so a study of these factors is indicated as a way to pin down whether they are associated with spread of the retrovirus.[168]

It's worth noting that, whereas the first theory we investigated that heterosexual transmission, when it occurred, was probably due to anal intercourse, this second theory argues that heterosexual transmission may occur in cases where heterosexuals have histories of great promiscuity.

The one case in which the US researchers presumed female-to-male transmission of HIV, from a wife to her husband, was in any case unique and fully consistent with Stewart's [fast lane] thesis. The couple participated in a swinging singles club, and over the five years before entry into the study, the wife had over 600 male partners, including over 2,000 contacts with a bisexual man, an unidentified number of contacts with an intravenous drug user, and over 1,000 contacts with a person she knew to be HIV positive. As part of their swinging activity, the woman would often have sex with another partner while her husband first watched and then [the husband would have] intercourse with her himself."[169]

The thought might be introduced now, though it is somewhat tangential to the fast-lane thesis, that a number of writers, such as A. E. Albert, have it that the spread of STDs and HIV because of promiscuity can hardly be partially attributed, as it is by some, to the fact that sex is "unprotected." Albert's notion is that condoms are unreliable. "The FDA found leakage of HIV-sized particles through 33% of condoms tested, and the overall failure rate of condoms has been estimated to be between 1 and 12.9%."[170]

Let's think of some of the implications of the material on sexual transmission developed thus far. For one thing, it should be acknowledged that the fast-lane thesis, which I have presented as one response to weakness in the direct transmission theory, does not really solve the problems of that earlier model. If there is little or no retrovirus in the semen (assuming for this argument that HIV infection is at least one crucial component of the disease), then even drivers in the fast lane won't be in danger of contacting the virus through promiscuity. However, I have presented this later theory as a derivative merely as a way to organize this chapter. In reality, the fast-lane view originated independently of concerns over sexual transmission. Many who hold versions of this hypothesis, such as Duesberg, do not believe HIV is a central factor in AIDS and so would not be bothered by the low rate of the retrovirus in sperm.

This is not to say, on the other hand, that the fast-lane view is without flaw. Not enough research and reasoning have been done for anyone to say the fast-lane theory wins hands down over other ideas. It can also be said that the theory that AIDS can be produced by sexual transmission through anal intercourse, a theory that we saw had certain strengths (such as taking into account studies that showed passive partners are more likely to contact the disease as well as considering the part rectal trauma may play in generation of the illness), is not totally compelling either. The only certainty we have found in this section is that AIDS is not transmitted by vaginal intercourse. This is about the only subject in this whole controversy of which the "straight" media has taken cognizance, though, in doing so, got it wrong.

MISINFORMATION FROM THE MEDIA

We have seen that in the mass media what is sensational takes precedence over what is scientifically accurate. Going further, we might say that what is sensational is what plays upon the prejudices and fears of the media's viewers/readers. When we looked at statistics projecting the future spread of AIDS, for example, we saw that the most extravagant figures were seized on without checking up on such things as whether rising estimates were based on paper manipulations. Media naiveté in this instance, it might be conjectured, was founded in its pandering to the fears of its consumers who were titillated by the dread of a new plague.

By this same method of interpretation, we can imagine (and this is what happened) that any material hinting that AIDS can be passed sexually would also be sensationally pictured, because it plays upon many long-held prejudices of the solid middle class. Many in the middle class believe that teenagers having sex and general promiscuity are both bad and should be stopped in any way possible. If it were discovered, then, that a dreaded new disease was spread sexually, this would provide an opening for even more calls for the stigmatization of the sexually active and for abstinence as a strategy to preserve health. We might even say that there would be pressure on the media to back up this prejudice by enlarging upon ideas of heterosexual transmission and downplaying material that expressed disbelief in such transmission.

If we may digress a moment and step back from the data to look at the general nonpermissive attitude that seems to have been ruling the airwaves for the last twenty years, we might argue that this puritanism is not an endemic characteristic of the middle class but a historically mediated attitude.

At least this is the opinion of Edward Luttwak, who in *Turbo Capitalism* has drawn some interesting conclusions about the connections between the general economic conditions in this country and the rise of "prohibitionism." The two key economic trends that have had the most impact on people's lives in the United States in the last two decades of the twentieth century are shrinking wages and increasing instability of employment.

To take the second point first, "The decisive fact is that, except for particular specialties, the U.S. labor market is chronically oversupplied in all sorts of high-skill categories.... in addition to the unskilled, whose utter uselessness in a 'high-tech' economy is by now well established."[171] In the current lean and mean corporate environment, where an oversupply of labor exists, companies prefer to hire temps (who do not have to be given any benefits) over full-timers, to the point of firing full-timers and rehiring the same people back as casuals. Luttwak cites as a representative example a woman who had worked for Pacific Bell as a full-timer for twenty-two years, was fired at the time of a "downsizing," and then rehired as a temp without benefits or job security.[172]

The second major long-term trend is the falling back of wages. For those in white-collar jobs, "the more highly skilled minority has done very well, the less skilled majority has suffered an actual loss of earnings"; for the lower paid service workers, whose jobs have replaced those of workers in production, Luttwak found that "the new service economy pays less well than the old-fashioned industries in which people still make things"; and for more upscale middle managers, there is "a downward slide [in wages] of the entire population from which middle managers are drawn in the U.S. economy."[173]

For us, though, the key point the economist has to make is that instability of tenure in jobs and tightening belts creates embitterment and resentment. Given that these feelings cannot be expressed in politics, with neither party much interested in the plights of the average citizen, Luttwak surmises that these soured emotions give birth to "decency" crusades. Without judging the value of such calls for prohibition, he lists the confluence of demands for the ban on smoking, rich foods, types of speech, flirtation, pornography, drugs, and topless bathing. No doubt many of these, except the last, should be banned; but Luttwak's point is not to judge the merits of any of the demands but to underscore the climate of banning, which he sees as rooted in unexpressed, unreflected-on economic grievance.

> It is no coincidence that prohibitions seem to multiply when the middle classes are especially insecure. In the past… [a scapegoated minority] would have paid the price for the declining relative incomes of most Americans, and the insecurity of many more in the brave new economy of unlimited compassion and endless structural change, which offers splendid opportunities for financial acrobats [like well-heeled stock market traders], and which keeps so many Americans awake at night fearfully wondering what the morrow may bring…. The unexpressed fears of insecure breadwinners [who in our society cannot persecute minorities] can find their outlet only in the prohibition of all that can be prohibited.[174]

Actually, this digression is meant to partially excuse the media, for, as I've said repeatedly, my point is not to tar this or that adversary, bad as some of them may be, but to make clear how even their destructive consequences have some logical rationale. In knowing the true grounds of different beliefs (even knowing more about these grounds than the holders of the beliefs may comprehend or admit to knowing), pernicious beliefs can be combated all the more strenuously.

In this instance, it appears (if one holds with Luttwak) that a promotion by the media of the view that AIDS can be spread heterosexually will often go on, in the face of evidence to the contrary, in order to pander to the public's lust for antisex messages.

Let's look at a couple of instances of how the media willfully plays up shocking elements of this part of the AIDS story, even when these elements are known to be unrepresentative or altogether wrong. It's known, for example, that even if AIDS were spread heterosexually, then transmission must parallel the cases of homosexual transmission in which it is believed that it takes five hundred to a thousand encounters to receive the virus. However, as Farber reports in *Spin*, there are cases reported where a person claims to have gotten AIDS after one encounter, and these are the ones dwelt on lovingly by the mass media. "It is noteworthy that those cases are so rare that we know each victim by name," Farber states.[175] She continues:

> Alison Gertz, for instance, the young, affluent heterosexual woman who said she caught AIDS from a man she slept with once, has graced magazine covers across the country. Suddenly, the media were making a tremendous deal out of what is in reality a very rare example, while insisting that the cases like it were the wave of the future.[176]

In 1992, to cite another example, the press went to town with a statistic that indicated HIV was to be found in one out of every 250 Americans, this being used to suggest a widespread infection through both homosexual and heterosexual channels. However, this was another example of the old mathematicians' adage that untruths can be divided into "lies, damn lies, and statistics." A cooler head in the media pointed out, with little effect on the general hysteria, that "if you do the math," going by a population of 250 million Americans, then "if 1 in 250 has HIV, that's 1 million people—less than the previously predicted figure of 1.5 and 2 million."[177]

Reports by prestigious agencies are splashed across front pages when they provide ammunition for calls for antisex campaigns, but buried when they indicate the opposite. When the London paper the *Sunday Telegraph* (one of the media outlets that does not run with the pack) featured a story by Ambrose Evans-Pritchard on a meeting of research scientists in Washington in December 1993, at which the participants downplayed the spread of the disease except through high-risk groups, this was barely noticed by the bulk of the mass media. The Washington meeting featured the unveiling of a study from Dr. Geraldine McQuillan of the National Center for Health Statistics. In it, according to Hodgkinson, were the findings of the first "large random screening survey for HIV-positivity in the US. Of nearly 8,000 people tested, twenty-nine tested positive, including just one white woman."[178] The obvious conclusion, made in the study, was "HIV was confined to specific pockets of the population," such as HIV drug users, particularly minorities with less access to health care and other amenities than had users from the majority group.

However, this story was not given much coverage elsewhere and so could hardly deflect the funding juggernaut, with dire consequences for dealing with the disease.

> ...[after the study was ignored] the dogma of heterosexual AIDS ... [went on possessing] the American establishment from top to bottom.... So, instead of targeting funding on those at risk, the money is scattered around the country indiscriminately. Programmes for sexu-

ally transmitted diseases in the inner cities, an effective way of controlling the spread of HIV, have been gutted to help pay for AIDS
awareness campaigns for Mormons in Utah and Baptists in the Bible
Belt... The heterosexual AIDS scare is one of the great scandals of
modern journalism and modern science.[179]

That's a note that was already struck in our study when we brought up how
AIDS education money has been centered on educating heterosexuals on how
to avoid AIDS (which they weren't going to get) while only a trickle has been
earmarked for IV drug users or homosexuals (who needed the education most).
Something else wrong with this spread of misinformation is set forward by
Symon, who points to the psychological and emotional toll.

> Natural selection has provided us with only a few ways of experiencing
> intense pleasure, and surely, for many people, sexual intercourse is at
> or near the top of this short list; hence, anything that diminishes sex-
> ual pleasure constitutes a significant cost. Yet many writers, both pop-
> ular and professional, appear to assume that sex is the one area of life
> where we should strive to reduce risk no matter what the cost.[180]

Thus, where the media and political pundits are often talking about the irresponsibility of people in spreading the disease, it seems to me it is equally just to
turn the tables and ask about the irresponsibility of the media in following what
it takes to be its readers' and viewers' proclivities rather than reporting the facts.
And if the United States media has generally distorted the situation when the hot
button of sexuality is pressed, imagine how they will begin to warp perceptions
when the even more explosive topic of race lights up on the control panel.

AIDS and Race in Africa

In 1974, Walter Abish published the novel *Alphabetical Africa*, which, if it was
not always as playful or iconoclastic as it purported to be, did include many
scathing passages where, metaphorically, Westerners have viewed Africa as they
wanted it to be (a foil to their own self-conception) rather than as it really was.
Take the following description that appears when the hero arrives in Tanzania:

> Life in Tanzania is predicated on the colored maps of Africa that hang
> in the palace courtesy of National Geographic. On the maps Tanzania
> is colored a bright orange.... The maps are the key to our future pros-
> perity. The maps keep everyone employed [,] said the Queen [a Euro-
> pean usurper of the throne]. Slipping into a pair of alligator boots she
> then proceeds to take me on a tour in her helicopter. From an altitude
> of five thousand feet everything below, including the city, is a bright
> orange. Astonishing. I thought it was the effect of the sun. No dear
> boy. She patted my knee. Each day one hundred thousand Tanzanians

carrying ladders, buckets of orange paint and brushes, are driven and also flown to different parts of the country. They paint everything in sight.[181]

It's absurd, yet suggestive. A country is to be remade so that it matches the pictures in National Geographic. An imported Western expert is in charge of the recoloring of the bush. Doesn't it seem like an allegory of what took place in relation to AIDS in Africa? Scientists in the West decided on slim evidence that AIDS must have originated there and, they extrapolated, would prove a scourge. Then they sent researchers with unreliable tests to scout out whether their suppositions were correct. Although, as the dissidents argue, evidence about the existence of AIDS was slight and contradictory, the AIDS heralds, with exaggerated statistics, poor testing, a heavy dose of racism, and other factors, managed to find what they wanted to find, just as Queen Quat is intent on making Tanzania look as she presumes *National Geographic* wants it to look.

We can turn from literature to the criticism of literature to make another necessary point. The book *Blank Darkness: Africanist Discourse in French* by Christopher Miller examines how Europeans (not just the French, despite the title) have persistently presented Africa as a country of inferiors, who, being a lower order of humans than Europeans, certainly should not object if Europeans take over their countries and economies, since these Westerners are better fitted to run them. Even the casting of blacks into slavery could be justified by these ideas, since it was claimed the slaves would be given Christianity along with their chains and thus have their souls saved, souls that would have been lost if they continued in primitive idolatry.

The belief that Africans, at least those outside the influence of Islam in the north, were all idolaters—though from the Middle Ages to the nineteenth century this was based largely on fantasy rather than anthropology—was central to the thesis of the continent's people's inferiority. This idolatry was presented as so casual as to be almost accidental. That is, as one sixteenth-century European "authority" had it:

> [Africans] are all idolaters without knowledge of God or of his Law. And this poor nation is so blind that the first thing encountered in the morning, be it bird, serpent, or other animal, domestic or wild, they take it for the whole day… [as their] God.[182]

Refusal to believe in Christianity would not be seen as such a great fault now, as it was then, but the traits connected to this refusal, we will find, are still being attributed to Africans when racist science is on the march. Another European "expert" quoted by Miller draws this connection: "The cause of idolatry is that man, through sin, left behind the contemplation of divine, invisible, and intellectual nature and sank wholly into the sense."[183] This immersion in the senses had to be coupled with the fact that the gods were incredibly lax in restraining their worshipers. Miller states, "Instead of a God of authority, repression, and

all-defining constancy, there is [as the Europeans portrayed the African reli-
gion] a god of released tension, wish fulfillment and malleability," who stands
benignly to the side as his people indulge in polygamy, concubinage and even
cannibalism.[184]

There is almost a melancholy quaintness to some of the racist utterances of
yesteryear, and yet, as we will see, parts of these submerged doctrines still resur-
face in the discussion of AIDS. Long-exposed racist concepts—such as that
African are children of the senses, who feel free to indulge in constant orgies,
thus spreading AIDS unconsciously—find renewed vigor in some of the popular concepts of AIDS. Of course, scientists have not stooped so low as to make
mention of such vile remainders of racial theories, yet below, I will (1) look at
how some of these myths have acted as an unadmitted or unconscious background to some discussion of the supposed epidemic of AIDS on the continent
as well as, especially, to the conjecture that AIDS originated there. From there,
we will break with the myths. Beginning with (2) an examination of some of the
glaring factual contradictions in the twin theories of AIDS's African origin and
its pandemic proportions there, I will move on to show how (3) bad science and
(4) politics have been influential in helping promulgate the mythical views. I
will end by showing some of interests that seem to have spurred the creation of
myths as well as briefly highlighting some of the tragic aftermath that has resulted when policies have been guided by these spurious insights.

Let it not be thought, though, that I have nothing positive to say here. No,
each time a theory based on myth is exposed, I will try to add a countervailing
theory, one based on facts, as an alternate hypothesis. For instance, after showing some of the fallacies involved in the idea that AIDS is of African birth, I will
bring forward hard data that suggest, on the contrary, that AIDS originated in
the United States or Europe and was later transported to the dark continent.

RACISM AND AIDS SCIENCE

Neoconservative liberal bashers make much of the fact that on certain college
campuses there seems to be an exaggerated concern for politically correct
speech that makes it impossible for people to offer even the most innocent joke
for fear it might offend a minority or other stigmatized group.

I am not going to pass judgment on this issue one way or another but rather
refer to the broader issue of historical forces. We live in a civilization that for
hundreds of years portrayed blacks as the despicable inferiors. The United
States itself got much of the wealth needed for its economic takeoff from the
labor power of black slaves. Once they were freed—in a civil war, by the way,
begun not due to any tender feelings about the injustice of slavery but to keep
the Union from splitting apart—African-Americans were not welcomed into
the bosom of the nation. In the South, where the majority resided, Jim Crow
laws were quickly put in place, establishing an unequal system with superior

whites only facilities, from schools to graveyards, that made sure blacks didn't rise much above their former slave status. It was only with the civil rights struggles of the 1960s that southern blacks, largely due to their own nonviolent protests, were able to abolish these laws and gain a fuller measure of equality.

Mr. Liberal Basher might be willing to agree with our capsule history but he would then insist, "All that injustice was long, long in the past. It doesn't concern us anymore."

Let's take another step in the argument. For hundreds of years, racist doctrine, which was not only part of popular culture but taught in universities as a scientific discipline, pronounced that blacks were not as high on the racial pyramid as the whites because, as we learned from Miller, they were sensualists whose appetites always overbore their rationality. This was especially shown by the blacks' ungovernable sex drive. "Now," our liberal basher laughs, "such ideas are left well behind us, thrown on the ash heap with other sick fairy tales, such as those that said Orientals were all coolies or mandarins." But is he right?

Our next step. We examine a scientific hypothesis about the epidemiology of AIDS. Researchers are arguing that its spread will be greatest in Africa. Not everything is said in so many words, but it is implied that part of the basis of the theory is the fact that Africans are such sensualists that, with all the horny men and women over there, it is quite likely the disease will spread like wildfire.

Don't you see what a terrible pitfall for science is on this road? Imagine that there were piles of good evidence that AIDS was affecting more people and spreading much faster in Africa than elsewhere. We don't have such evidence but imagine we did. Even if we had this backup, I would argue that it would have to be treated even more skeptically than normal scientific evidence, because the centuries of stereotypes, lies, and slanted reporting have established a proclivity toward a certain exploded view.

We might put it like this by use of a simile. It is as if gently sloping land had deep channels cut in it from long ago and now the work was to cut new channels, which were presently quite shallow. It would be more a hope than a justified expectation to imagine that current runoff would avoid the old pathways, unless we constantly monitored and blocked them.

This is why I would take Mr. Liberal Basher to task. He refuses to realize that no matter how free of prejudice he is personally, racism is encrusted in our words and institutions. He is being feckless in saying we shouldn't have to worry about words.

I don't mean that racism is inherited in our genes, but that it often underlies a confluence of established institutions and interlinked attitudes. Take the common U.S. prejudice that power and wealth are the badges that proclaim an individual's worth. This can easily lead to racist thinking.

Why? The plain fact is that power and wealth were long ago distributed so that (with token exceptions) the majority groups that obtained them then continued to hold the lion's portion. This is not to say no outsiders can ever penetrate the precincts of privilege, but that those in the upper groves will, on the main, be the ones born with silver, if not platinum, spoons in their baby food. The media, of course, tend to let this pass without comment, and so men like Donald Trump and Bill Gates, who have risen to great wealth are praised as self-made men, regardless of the fact that they came from rich families—Trump the son of a multimillionaire—and so overlooking that it is so much easier to be self-made when, like Gates or Trump, who both started their careers with a few inherited millions in the war chest.

Now add this ignorance, ignorance of the fact that those touted self-made heroes started from a bed of roses, to the aforementioned idea that those without wealth, are losers and you have a formula for racist thought. Our Liberal Basher wonders why so many WASPS are rich and so many minorities are poor, drawing damaging conclusions about the minority's capacities while overlooking the way wealth is drawn to already established wealth.

This is hardly the place to begin a psychology of prejudice, but it is necessary to insist both on how subtle influences can shape unstudied facts into the premises of racism and how such influences still play roles, even leading roles, in the background to our thought. A knowledge of this background has led R. Harrison-Chirimuuta to say that from the earliest days of AIDS study, "Western scientists have attributed its origin to black people, first in Haiti and then in Africa, yet the scientific literature that claims to prove an African origin is full of inconsistencies and sheer racist nonsense."[185]

This same author in an essay in the volume *Western Medicine as Contested Knowledge* laid out in more detail the chronology of this evolving belief whose one continuity was the thought that black people must somehow be responsible for the problem.

> When in 1982, Haitians were diagnosed with AIDS, the as yet unidentified infectious agent was immediately assumed to come from that country. A Haitian origin soon fell into disfavour, at least in part because no evidence could be found that AIDS existed in Haiti prior to its appearance in the United States, and was soon superseded by the African hypothesis, with or without the rider that the virus had stopped over in Haiti on route to the U.S. The African origin of AIDS was immediately accepted in the West with minimal critical assessment, and there was significant debate only on the issues of whether the reservoir of HIV was in a remote tribe of Africans or in African monkeys, and at what point it is spread from Africa to the West.[186]

He continues by elucidating the nexus between readiness to believe in the black source and an unacknowledged racial bias:

> As Europe began to come to terms with the Nazi holocaust, racism lost much of its intellectual respectability, a process hastened by independence in the colonies and the black civil rights movement in the United

States. But whilst the expression of racism may be less acceptable, the underlying racist beliefs, particularly those pertaining to Africa and Africans, remain integrated into the European and American World view. Thus within the scientific literature about AIDS and Africa all the racist themes can be found underpinning arguments for which scientific evidence is contradictory or absent: Africans are primitive peoples living in isolated tribes cut off from civilization, so could have harboured diseases for centuries before they spread to the rest of the world; they are evolutionarily closer to monkeys, thus could more readily acquire monkey diseases, perhaps by having sexual relations with monkeys or at least involving them in their sexual practices; they are sexually unrestrained, and a sexually transmitted disease would therefore spread more rapidly amongst them than any other people… and their objections to being attributed with its source are harmful to themselves and do not need to be taken seriously.[187]

In the last extended quote, Harrison-Chirimuuta has raised a point already familiar from our discussion of Miller's work, namely, the myth that Africans are sexually unrestrained, as well as adding a point we will consider later in this chapter, which is that African health workers are seldom consulted on what's going on in their continent. It turns out that African doctors and other health workers, who one might think should be given special credence, since they are literally at "the scene of the crime," are the ones whose opinions seem to be given least weight by Western researchers.

Before tackling that topic, let's look a little further at this racist idea of African hypersexuality. Harrison-Chirimuuta links the willingness of Western scientists to give immediate credence to the idea that there has been a speedy spread of HIV through middle and southern Africa with stereotypes about Africans' sensuality. "Africans were said to indulge in anal intercourse and intercourse during menstruation, and were far more promiscuous than other people."[188] This belief in African overindulgence has become a caricature, he argues. "If claims that a third of sexually active adults in some central African countries are infected with AIDS are true, life in these countries must be one endless orgy."[189]

Guided by prejudice, Westerners were quick to jump to conclusions about the African origin and ready transmission of the disease. "Given the racist stereotyping of black people as dirty, disease carrying and sexually promiscuous, it was virtually inevitable that black people, on the first sighting of the disease amongst them, would be attributed with its source."[190] So as soon as a few Haitians were seen to harbor the disease, "the gay plague changed overnight to the Haitian disease."[191] Even though, as mentioned, attempts to lays the disease at the door of the Caribbean natives didn't pan out, Harrison-Chirimuuta expands, "the idea of black people as the source of AIDS was too attractive to abandon. Attention shifted to the African continent itself."[192] His conclusion: "Racism, not science, motivated the search for the origins of AIDS."[193]

Charles Geshekter, a professor of African studies, by the way, has noted that not only is the idea of African indulgence a myth without any foundation in reality, but if we compare the incidence of sexual promiscuity in the southern part of Africa with, say, its incidence in the United States, our country comes across as very much the loser—if the winner is the society where a restrained morality is practiced. According to Geshekter, the myth of African immorality can be dated back quite a ways: "Early European travelers returned from Africa bringing tales of black men allegedly performing carnal athletic feats with black women who were themselves sexually insatiable."[194] He goes on:

> The reality, however, is very different from our perceptions: Widespread modesty codes for women, whose sexuality is considered a gift to be used for procreation, make many African societies seem chaste compared to the West. The Somalis, Aars, Oromos and Amharas of northeast Africa think that public displays of sexual feelings demean a woman's "gift," so that sexual contacts are restricted to ceremonial touching or dancing. Beginning sexual relationships are geared to the beginnings of making a family. The notion of "boyfriends" and "girlfriends," virtually universal in the West, has no parallel in most traditional African cultures.[195]

We might note in passing that another myth dredged up by Harrison-Chirimuuta, that Africans have much contact with monkeys and so might have caught AIDS from a simian virus, is worth about as much credence as the myth of Africans' unrestrained sexuality. Geshekter notes, "Most Africans… have little contact with monkeys, and amongst those who regularly hunt monkeys, for example the pygmies of the equatorial rain forests, AIDS is notable for its absence."[196]

Moreover, the claim that the disease originated by jumping species from ape to man could be used as a textbook example of bad science. First ideas that "SIV [a virus that is found in monkeys] was similar to a virus isolated from west African prostitutes were disproved when the SIV was found to be a laboratory contaminant."[197]

Then it was hazarded that "SIV mutated into HIV in the last few decades, the conclusion based on estimates of the rate of mutation of these viruses and their degree of genetic dissimilarity."[198] Even those uninitiated in scientific mysteries can see this is a flight of fancy. One might as well compare the virus that causes the common cold with HIV, note how much they differ, then count up the number of mutations it would take for the former to mutate into the latter, taking that as evidence that the cold was the origin of AIDS. Besides, Harrison-Chirimuuta writes, "Even if such an improbable event [the mutations] did occur, given the existing colonial ties and trading links between Africa and Europe, the virus would have caused an epidemic in Europe at the same time as—or before—the epidemic in the United States."[199] Yet taking only the Western countries, evidence points to the United States preceding Europe as a center of infection.

It's worth making the further point that the enterprise of installing AIDS in Africa has even been given a sentimental patina in a spurious tracing of the first European victim struck down by the African plague. According to the Harrison-Chirimuutas (the husband and wife acting as coauthors of the piece, who will be hereafter designated the "Chirimuutas"), "One of the most cited papers in support of an African origin for AIDS, 'AIDS in a Danish surgeon (Zaire, 1976),' described the case of a surgeon [Dr. Rask] who had supposedly caught HIV from her patients whilst working in a hospital in Equater province."[200] One reason this paper is so relied upon is that it was incorporated into the best-selling book by Randy Shilts, *And the Band Played On*, about the AIDS epidemic, where the story was told in a crowd-pleasing way, with the use of inflated rhetoric. After noting that Shilts labels Rask as "Danish surgeon in Zaire, first Westerner documented to have died of AIDS," the Chirimuutas looked at what was really known about the Rask death.

> A British researcher, Dr. Janie Grote, wrote to Bygbjerg, the author of the paper about the Danish surgeon, inquiring further about Dr. Rask, and received the following reply:
>
> "We have found a very small serum sample, which was taken shortly before the patient died in 1977, and we know already that it was HIV-I ELISA antibody negative. Thus, we cannot prove that the patient had AIDS, although the clinical picture was very suggestive."
>
> So this famous AIDS case was seronegative. This information has not, of course, been presented in the scientific literature, and Bygbjerg's article continues to lead the citation index of cases 'proving' an African origin for AIDS.[201]

We need to say a little more about the Shilts book—which does include valuable, hard-hitting information that we have already drawn upon. In his brief sections on Africa, he falls flat. In fact, in these parts the vulgarization, sensationalism, and, yes, fictionalization of the AIDS crisis in Africa reaches a kind of giddy height. Although Shilts is powerfully compassionate when treating gay males and drug users in the United States , when he moves to Africa it's time to trot out the stereotypes, not only of the pathetic blacks, but of the great white virus hunters, who appear on the scene to selflessly battle for their well-being.

My collaborator has drawn my attention to how Rask is depicted as a headstrong heroine, who decides to work in Africa despite the fact that a "surgeon's work had risks that doctors in the developed world could not imagine," such as lack of adequate supplies and greater danger of infection from patients, which places Zaire "on the frontiers of the world's harshest medical realities."[202] Rask is up to the task, however, because of her strong Anglo Saxon racial heritage ! She grew up in the northern part of her native land: "Far from the kings in Copenhagen, these hardy northern people had nurtured their collective heritage," which made them "direct and decisive, independent and plainspoken."[203]

It is such qualities that the Africans apparently lack, at least so it appears from their reaction to the Eboli disease, which Rask had earlier witnessed in her area. "Frightened African health officials swallowed their pride and called the [Western] World Health Organization."[204] It's no wonder they couldn't handle it on their own, given the irrationality of your common African. "Natives were infuriated when the Americans [brought in to clean up the Africans' mess] banned the traditional burials of the victims since the ritual bathing of the bodies was clearly spreading the disease further."[205]

Let's stand back a minute. I'm not saying that anything here didn't happen—that the WHO wasn't called in or that the natives didn't become recalcitrant—only that details are selected to build on stereotypes. Shilts does not mention Africans doing anything positive to help the WHO, only the fact that they are willing to block efforts. He even tries to get in the promiscuity angle by adding about the first carrier of the disease, "The man apparently had picked up the disease sexually."[206] Not only that, as if to rivet the idea that one's relation to disease is linked to racial heritage, Shilts cites an unnamed historian to vaunt the following far-fetched idea.

> One historian has suggested that humans, who first evolved in Africa eons ago, migrated north to Asia and Europe simply to get to climates that were less hospitable to the deadly microbes the tropic so efficiently bred.[207]

Is it too much to read this as hinting that the enterprising Asians and Europeans moved to climes where they could become "direct and decisive, independent and plainspoken," while their more shiftless cousins roosted in the same spot?

That last point may be an exaggerated interpretation, but it was no exaggeration to use the term "fictionalization" for the book. Although *Band* is about real events, using the literary precedent of 1960s New Journalism, which borrowed fictional techniques to jazz up reporting, it goes into the minds of the leading characters as if Shilts were an omniscient narrator privy to everything they thought. Whatever value such liberty may have in making the book more readable, one notable effect it does have is helping the self-aggrandizement of Western medicine. It's very reassuring for readers to be able to see into the minds of doctors to learn how selfless and dedicated they are. For instance, when Dr. Bygbjerg, the physician who wrote the cited paper on Rask, finds out about her death—they were close friends—Shilts reveals his torment this way: "Bygbjerg would never forgive himself for taking her away from the cottage north of the fjord."[208] He had brought her the city for tests, and she died in hospital.

> Bygbjerg decided he would devote his life to studying tropical medicine... he wanted to know what microscopic marauder had come from the African jungles to so ruthlessly rob the life of his best friend, a woman who had been so intensely devoted to helping others."[209]

If the reader can get past the purple prose, perhaps it will be evident that Shilts is implicitly drawing a contrast with the passage three pages earlier in his book about how Africans react to death. While Africans are irrationally demanding to wash their dead, irrationally demanding physical contact with the corpses, a Westerner, by contrast, has a purely intellectual connection to his deceased best friend. Rather than allowing himself idle tears, he pledges himself to the pursuit of truth in her name.

By no means am I suggesting Shilts is a racist, but rather that he is thoughtlessly picking up the stereotypes about Africa that are all too ready to hand in popular and serious Western culture.

There is one last thing to say here. Although our main concern in this part of the study is Africa, lest we think that an implicit racism can only be found in AIDS thinking that centers on that continent, it will be enlightening to look for a moment at how this racism can also extend to the coverage of blacks in the United States. To just take one example witness an editorial that appeared on May 12, 1992, in the *New York Times* titled "The AIDS Plot Against Blacks." The piece began, "Bizarre as it may seem to most people, many black Americans believe that AIDS and the health measures used against it are part of a conspiracy to wipe out the black race."[210]

Note the politic language, absolutely de rigueur nowadays to put racist ideas into circulation in our touchy environment. The writer doesn't say the notion is bizarre, but that it may seem bizarre, nor does the author even say everyone will find it bizarre, only most; yet the beliefs being discussed are made to seem as nonsensical as if they had been rejected outright. Imagine if one were to say this: "Irrational as it may seem to most, Duesberg believes there is a *conspiracy* to prove HIV causes AIDS." Note the italicized word "conspiracy." Using this word in connection to an opinion is another not so subtle indicator that such opinion is held by a kook, since conspiracy connotes the delusive belief that world events are secretly controlled by a cabal of behind-the-scenes operators.

Let's go a little further into this article, but this time I will highlight such indicator words that are thrown into seemingly objective discussion in order to slant opinion in a way selected by the author. You can work out how these words work on prejudice for yourself.

A survey of black church members in 1990 found that an astonishing 35 percent believed AIDS was a form of genocide. [A poll] found that one black in ten believes that the AIDS virus was "deliberately created in a laboratory in order to infect black people."… Worse yet, the treatments and preventatives against AIDS have become suspect. Some blacks believe that AZT… is a plot to poison them.[211]

Nina Ostrum has written some eloquent words on the overall racism found in this editorial; however, for me, the key point is the note hit on AZT. As we will see in a later part of this book, there is widespread dissent about AZT being an adequate prophylactic among health workers of all races. The *Times* article, then, does double duty for the establishment, on the one hand serving its racial prejudices by indicating how weak-minded African-Americans are, in that they prefer dubious emotional responses to the AIDS crisis rather than proper scientific ones; and on the other hand serving the establishment's scientific prejudices, by suggesting that dissenting views of the utility of AZT are confined to groups that are scientifically marginal.

However, for all these brave words, there are flaws in the AZT thesis just as there are flaws in the connections generally made between AIDS and Africa. Let's turn to some of the latter.

WEAKNESS OF AFRICA THESIS

The orthodoxy's opinions on how Africa and AIDS are connected might be said to involve even more than the usual effrontery. While establishment ideas about the existence of HIV and the fact that HIV causes AIDS, for example, have a surface plausibility, notions about the African genesis and spread of the disease seem to involve a glaring contradiction, or at least conflict, in their base premises. Think about this. Orthodoxy holds to the following two concepts: (1) AIDS originated in Africa from whence it spread to the West, and (2) African AIDS, which affects mainly heterosexuals, is fundamentally different from AIDS as it appears in the West. Am I the only one who smells fish when he hears these two ideas proclaimed in tandem?

Diseases are found in two different regions of the world, which have significant differences, yet it is confidently asserted they are the same disease. That perhaps one could swallow. But what about the point that in one region is the original of the two diseases, and that found in another region is the copy? Diseases do not, like passports, get stamped at customs and transform themselves as they travel across national borders. Even if we accept the idea that HIV is very apt to mutate into new forms, how can it be explained that every HIV strain that has been carried from Africa to, say, the United States undergoes the same type of transformation, moving from a disease that is heterosexually transmitted to one that had been homosexually passed? Such an idea beggars belief.

Dr. Eleopulos lays out this point firmly:

> If AIDS in Africa is the same condition with the same cause as anywhere else in the world then AIDS in Africa and AIDS in the West should be identical. This is not the case and what is called AIDS in Africa is almost unrecognizably different from AIDS in the West, so much so that if African patients suddenly switched continents, very few Africans would remain AIDS cases. This is due to the existence of multiple AIDS definitions.[212]

There is a tortured way of dealing with this discrepancy, the one chosen by orthodoxy, which holds a mutant strain started the disease in the United States and Europe. There is also a commonsense way to handle it: Understand that these are different diseases.

Before grappling with this problem, let's note some of the less startling contradictions that have to be appended to the African AIDS theses, ones that have moved many health workers to doubt the underpinnings of these concepts.

For one thing, we may measure the exaggeration in the statistics generally reported from the continent, which have often proved as fallacious as those that used to be issued about heterosexual transmission in the West, which was expected to rocket skyward as soon as HIV was passed beyond the original high-risk groups. African statistics have been overblown both in terms of overall projections (of spread and lethalness) for the future and in current assessments, which as we saw in a previous section were often based on presumptive diagnoses and unreliable tests for HIV positivity.

However, even if, for the sake of this argument, we accept the large number of cases of HIV infection recorded by orthodoxy in Africa, we find that having HIV does not translate to getting AIDS as readily as it does in the West. Here, I suppose, is another firm difference between African and Western HIV, that is, in Africa one doesn't die of infection, or at least not at the same high rate as in the West. This is not a difference that is made much of, especially in comparison to the difference in sexual transmission, since it goes against the grain of the hyped-up fears of an African epidemic. According to Farber, once HIV was detected in Africa, "the media was pouring out reports of a continent on the brink of virtual extinction."[213] She goes on that in the eyes of mainstream reporters, "There was no question: AIDS was bulldozing Africa, taking out entire villages—men, women, and children, changing the demographics of central Africa forever."[214]

Yet these forebodings were based on the idea that HIV would be as quickly lethal in Africa as it was in the West. (If AIDS truly originated on the continent, it should even be more lethal there, since the cases would have had more time to pass the incubation period.) Farber notes, "According to official sources such as [those from] the World Health Organization, 7 million central Africans are infected with HIV."[215] In comparative terms, worldwide 69 percent of HIV positive people are in Africa, while only 16 percent are in the United States . "However, in terms of actual reported AIDS cases, 44% come from the U.S., whereas only 30% come from Africa."[216] This adds up to 230,179 people with AIDS in the United States and 151,455 in Africa as of 1993.

The Chirimuutas note the same findings about infection not leading to the disease that come from field studies of a particular country, Kenya. They mention "findings which emerged from a study conducted in Kenya by 12 researchers including Robert Biggar and Robert Gallo."[217] Researchers came up with high figures for HIV infection: "Overall, 21% of the population studied were seropositive, but in one ethnic group, the Turkana, 50% were positive. Still, there were no cases of AIDS."[218] Other studies have yielded the same results: rampant infection, but almost no illness. It is an understatement by the Chirimuutas when they conclude, "It is unlikely that AIDS illnesses of the recognized varieties are occurring at a rate commensurate with the prevalence of antibody against HTLV-III," the antibody's presence, as we have seen, being the recognized way to establish the existence of HIV in a patient.[219]

A broader-scale report, which covered 1984 to 1995, was published by the World Health Organization's *Weekly Epidemiological Reports*. It showed that "97% of those who test HIV-positive" in Africa don't have AIDS."[220] Although this could be discounted as simply evidence that many HIV-positive people are still in the latency period, it should be added that those with AIDS in Africa don't die at the same rate as those in the West. "It is interesting to note that only 13% of Kenyans with AIDS have died, whereas in Europe and America the figure is around 50%."[221]

Everything said so far in this section about HIV positivity, however, has been accepting of the legitimacy of the establishment's measures based on blood tests, ones that we already know are extremely fallible. When more accurate tests have been conducted, it has often been seen that HIV infection in Africa has been grossly overstated due to reliance on untrustworthy testing. A study done in Germany and published in the *British Medical Journal* "found that about 10% of serum samples were positive with the ELISA test [the one commonly used], but only 0.07% were positive on more specific testing."[222] That is to say, one out of 143 positive ELISA tests was accurate. If such figures were to be projected, one would find that "estimates of the incidence of AIDS virus infection in central Africa were exaggerated by a factor of about 150."[223] To find the real size of the epidemic one would have to take current orthodox figures and divide by 150.

This study was focused on older blood samples, ones collected from 1976 to 1984, both to consider if HIV was as widespread as claimed and, as importantly, as a way to partially test the hypothesis that AIDS originated in Africa. If it did, then there should have been early evidence of seropositivity in the countries from which it supposedly sprang. The Germans were wide-ranging in their sampling and examined blood from 6,015 people, including 789 from Senegal, Liberia (935), Ivory Coast (1,195), Burkina Faso (299), Nigeria (536), Gabon (1,649), Zaire (15), Uganda (164), and Kenya (433). The researchers discovered that "only four samples contained antibodies. Three of these were from Senegal."[224] Their conclusion was "fewer than one in 1,000 subjects were seropositive for AIDS at the time of sampling before 1985 [which proved lower than the positivity found in German blood samples for the same time period] and do not support the hypothesis of the disease originating in Africa."[225]

FAULTY SCIENCE

QUESTIONING THE ORIGIN OF AIDS VIRUS As we move to a fuller look at where science has been weak in its study of AIDS in Africa, let's remain with the topic of the origin of the disease for a moment. We have already canvassed the mythological component of the belief, which includes such ideas as that Africans engage in peculiar animistic rituals where they smear themselves with monkey blood, thus making them liable to the diseases of the apes; but we need to say a little more about what facts were marshaled as the scientific foundation for this concept. Farber lays out:

> This theory was based on a few reports that a virus similar to HIV had been found in African blood samples dating as far back as the 1950s. A virus said to be "closely related to" HIV was isolated in the African Green Monkey, and before long the theory evolved that HIV had somehow crossed species, jumping from monkeys to humans through some unidentified mode of transmission. This idea was bandied about in leading scientific journals during the mid to late 80s by AIDS researchers, who also claimed that AIDS was spread more efficiently in Africa due to extreme sexual promiscuity, blood drinking rituals, and children playing with dead monkeys.[226]

Another hypothesis disagreed with the cross-species' attribution and held "AIDS was an 'old disease of Africa' that had reached the West via recent intercontinental travel."[227]

These thinly supported or purely speculative theories matched up, as we saw, with preconceptions about Africans' notable lubricity in order to account for the quick spread of the disease once originated. I keep mentioning these claims in tandem—that of origin and quick spread—because although it may seem they are disconnected points, there is a submerged logical operator that would use the numbers as a justification for the origin thesis. It would make a kind of sense to say that in the place where the infection is the most widely seen, there must be its origin. Isn't it logical to imagine the most cases would appear wherever the disease has had the most time to spread? I am not saying this link wouldn't be a good one, but that the alleged instance of skyrocketing infections in Africa is insisted on so stridently because it acts as an underlying proof for the origin idea, which definitely needs such help, with so many facts standing against it.

So, what are these facts? One that we've already rehearsed at length is the simple yet shocking point that the ELISA test, which is given to determine the presence of AIDS antibodies, cannot distinguish these antibodies from evidence of malaria, leprosy, and other diseases endemic to Africa but unknown in the West. To get further information on what is wrong with how HIV infection is tested for, let's refer to the opinion of an African journalist.

As an aside, let me repeat that African scientists, medical workers, and journalists have largely been ignored as the Western establishment set up an orthodox consensus. This makes good sense in accord with Western racism, for, after all, if Africans are out-and-out sensualists who spend more time on bodily pleasures than intellectual matters, their opinions on scientific material are of little importance. Indeed, intelligent, scientifically rigorous studies of the AIDS crisis by Africans, of which I have seen many, undercut the mythology of African lack of rationality and thus must be ignored by the establishment in order for it to keep upholding its racist mythology—which needs to be kept in place as an underlayer to the whole dream that Africa is the cauldron where AIDS was originally brewed.

However, ignoring African writings makes bad sense if one is striving for an honest evaluation of what is going on, since in that case it always helps to go straight to the root of the matter and consult with people "in the field." By that last expression I don't refer to Western scientists who, acting as surveyors, jet in for a few months' visit, but to those African health workers who live on the continent and so are most conversant with disease conditions prevalent there. I have taken special care to listen to their voices and to those who have interviewed them.

One outstanding African writer on the AIDS is the Kenyan columnist, Philip Ochieng. In January 1987, he wrote in *New African* about the questionable nature of tests done to evaluate rates of HIV infection on the continent. His concern is not the scientific validity of the screening instruments but the way high figures are extrapolated from small and unrepresentaive evidence.

> The methods used to determine the spread of AIDS in the West and in Africa are vastly different. In the West the emphasis is on actual cases reported and projections based on lifestyles. In Africa, a few people, mostly prostitutes, are rounded up and subjected to screening. Whenever AIDS antibodies are discovered, the medical boffins juggle their calculators and come up with figures that suggest that half the entire nation is suffering from AIDS! If the same system of calculation was applied to California, the results might well indicate that every citizen there had not only contracted AIDS but had died from it![228]

From there Ochieng goes on to enunciate his own provocative ideas about the origin of AIDS—provocative for the West, that is, where they are never discussed, though are widely entertained in Africa. He holds that AIDS may not be emigrating from but to Africa. "Why is it that we had never heard of AIDS or anything approaching it until after it appeared in the US and Europe?"[229] He goes on to question, "Why is it that those areas of Africa most in contact with the Europeans and Americans over the past 15 years are the worst affected by AIDS? What is the only logical conclusion to draw?"[230]

Even Dr. Jonathan Mann, director of WHO's AIDS program, though not willing to positively locate the origin anywhere, sees the proposition that AIDS began in Africa—a "cheap hypothesis,"—as what it is: one of the "thinly disguised prejudices about race, religion, sex, social class, and nationality" that have beset study of the illness."[231] However, if its place of origin cannot be shown, there is evidence that for Africa, it was an import.

This evidence is based on investigating which people on the continent were first found sick. If, as the orthodox theory goes, the illness "had originated in some obscure rural area but appeared in the cities because of greater urban sexual promiscuity," then as the Chirimuutas note, it should have shown up among sexually active citizens who would have had contact with the lower-class peasants who first got the disease. These would be lower-class female prostitutes and their equally lower-class clients. On the contrary, "AIDS first appeared amongst educated Africans with a history of foreign travel. As a study in Kenya revealed, this group frequented 'high class' prostitutes whose clientele also included foreign tourists and businessmen."[232]

To take another example, the Chirimuutas suggest we look at studies of African Kaposi's sarcoma, which was at first considered the main precursor to AIDS. When doctors tried to pinpoint who was suffering from this disease, "they found a disease concentrated in the wealthy elite."[233]

More evidence of the foreign origin of AIDS—since it is a possibility not countenanced by most Western doctors, it has not been given a great deal of attention—can be taken from a South African study that was published in the *New England Journal of Medicine* in May 1985 and in *South African Medical Journal* the following month. Using blood from 139 Kenyans, 357 black South Africans and Namibians, 39 white South African homosexuals, and 79 primates (baboons and vervets), tests were given for HTLV-III [HIV] antibodies using indirect immunofluorescence assay, which is held to be more accurate than the ELISA or Western Blot tests . The results were reported in this way: "The only positive subjects were in the group comprising [white] male homosexuals. The majority of these positive subjects had either recently been to the United States or had had sexual contact with other homosexuals who had visited the United States."[234]

None of this is enough to definitely substantiate that AIDS originated in the West, although it should give one pause, especially when it has been shown with, I think, good evidence that there is no proof or even good reason to hypothesize AIDS comes from Africa. All the evidence on that thesis points the other way.

If we suppose for the sake of argument that a new virus does exist, though, it might be asked: Why would it originate in the West? Can we just reverse the arguments, seeing that sexual promiscuity is a more pronounced characteristic in Western nations than it is in Africa?

No, that would be to jump from one stereotype (oversexed Africans) to its mirror image (oversexed Americans and Europeans). There is really no answer to this yet; although, since we have reviewed a great deal of what Richard

Chirimuuta (writing alone or with his wife) has to say about the falsity of making Africa AIDS's birthplace, a subject to which he has devoted great thought and study, it might be worthwhile to examine his ideas on this subject. As he sees it, the AIDS virus may have been produced accidentally in a scientific laboratory. Bear in mind, he is mainly sighting the potential for errant biotechnology to cause diseases such as AIDS, not identifying this technology as the culprit. He argues:

> One possibility that has been given scant attention in the vast scientific literature about HIV and AIDS is an artificial origin of a mutant virus. This would seem rather surprising, as the risks of mutant viruses emerging from laboratories has been widely debated for many years.[235]

To back up his account of a general concern over the issue, he notes that in the early 1970s several molecular biologists published the book *Biohazards in Biological Research*, while in February 1975 an international conference of molecular biologists was held in California where they agreed on a policy of self-regulation. In 1976 the National Institutes of Health "released guidelines for research on recombination DNA molecules," and in 1977 "the Federal Interagency Committee on Recombinant DNA Research issued an interim report," suggesting laws that might be enacted to regulate this field, suggestions that were put into law the following year.[236]

Not only should scientists have been aware of this research as a possible production site for new viruses, but, Chirimuuta notes, they should have been aware that monkeys that might have had SIV were often experimental subjects.

> Hundreds of thousands of African green monkeys and other species have been exported from Africa to research laboratories in Europe and America, where they have been subjected to experimental infections and their tissues used in cell culture.[237]

He asks which one would prefer if faced with these two hypotheses about the origin of the disease: (1) a lab worker in a Western country was bitten by a monkey who had been given an artificially altered virus, or (2) a monkey with a mysteriously naturally altered virus infected an African in the bush? Each hypothesis is equally unlikely, Chirimuuta says, but the choice of which to believe is symptomatic of the prejudice of Western scientists. "That the latter hypothesis and not the former has been pursued suggests that factors other than science have been guiding the activities of AIDS researchers."[238]

The thoughts closing this discussion are merely suppositions about how scientific endeavors may have erred and played a part in unleashing a virus. However, we no longer need to depend on suppositions in the next section, where we will document some of the bad science that has been used to create the magnified projections of HIV infection and the labeling of any cause of mortality under the sun as an AIDS death.

AIDS SCIENCE EVALUATED BY AFRICAN DOCTORS

As promised, as we examine for the first time or revisit some of the problems with accounting for the amount or actual existence of AIDS and HIV in Africa, we will focus particularly on the opinions of African physicians, health officials, and workers themselves. From them, we will learn especially about the questionable statistics and the tendency of establishment doctors to carelessly attribute many deaths to AIDS without thorough examination of the cases. We will also note the idea, often raised on the continent, that AIDS in Africa is not a new disease, but a new "label" placed on old conditions that have been around for centuries, although recently exacerbated.

Let's begin with a telling anecdote. There is no doubt that the media's drumbeat has often created panic and fear in Africa, just as it has over here. Geisler reports:

> In Uganda, people in some areas (Rakai district) are aware of who is bringing illness to them, and flee from those physicians who approach (allegedly for just doing HIV-tests) them. In Kampala, the Weekly Topic newspaper accused CIA Agents, disguised as scientists and journalists, of spreading lies about the origins of the virus. Shortly afterwards, unknown persons ransacked the house of Wilson Carswell, the leading Aids-research scientist in Uganda [and a member of the British armed forces] and destroyed his computer together with all the AIDS files.[239]

Is this evidence of more African irrationality, pace Shilts, or could it be that the "spreading of lies," although not done by CIA agents, has been of outrageous proportion, enough to sometimes justifiably enrage Africans?

For one thing, as we have already seen when we noted the thoughts of Kenyan columnist Philip Ochieng on how HIV infection statistics were extrapolated to the whole population using prostitutes as the sole people sampled, the method of deriving statistics of overall infectivity leaves much to be desired. All that needs to be added now is that Ochieng is not the only African who scoffs at the oversized estimates of a forthcoming or already present AIDS epidemic. A story in *New African* recounts, "A growing number of African physicians," and the list includes Dr. Mark Mattah (Midland Center for Neurology in England), Dr. Sam Okware (former director of AIDS research in Uganda), and Dr. P. A. K. Addy (director of clinical microbiology in Kumasi, Ghana), "say they think the panic over the heterosexual transmission of AIDS may be a hoax."[240] When Dr. Felix Konotey-Ahulu, a Ghanaian physician at London's Cromwell Hospital, went to several African countries after hearing that AIDS was devastating these nations, he gave his report in the prestigious *Lancet*, where he stated he had found no evidence of an out-of-control epidemic. He wrote, "If tens of thousands are dying from AIDS (and Africans do not cremate their dead), where are the graves?"[241]

Neville Hodgkinson, who was sent by the*Sunday Times* of London to Africa, concentrated on talking to African doctors and officials about the extent of the crisis. He says,

> I couldn't find any evidence of an AIDS epidemic in Africa. I visited Kenya, I visited Zimbabwe, I visited Zambia, and I visited Tanzania. Okay, I was only there about a maximum of ten days in each of these countries, but, I talked to people at a high level within those days that I was there. I found that when you question people more closely, you find that many of them had grave concerns about the picture the world was being given. They felt because of the scare of the large numbers of people testing positive, that and AIDS epidemic was forthcoming. But, they didn't actually have an AIDS epidemic upon them in most instances. The health minister in Zimbabwe told me, that the biggest health threat in his country was an epidemic of what he called HIV-itis. He said it was like a fear created by the HIV story, which was jeopardizing people's health in a major way."[242]

To Hodgkinson's mind, this HIV-itis does not only strike people who become deathly afraid of getting ill, but health workers, who in anticipation of an epidemic are all too ready to either find results of HIV infection that are not there or boost them to help secure funding for the increase of disease that is just around the corner. He says, "AIDS workers genuinely believed HIV presented a huge threat which would destroy millions of lives tomorrow.... Because they believe there is an epidemic, they think they're doing the right thing in multiplying HIV and AIDS statistics out of proportion to the facts."[243] Thus, they posit huge increases in deaths in the near future based on small and skewed statistical bases.

Part of the reason for the overestimation of what terror the disease would unleash on the content—an overestimation registered, for example, in the already noted differences in infectivity and death in Africa and the West, with Africa being higher in infectivity but much lower in deaths from AIDS—is that the ELISA tests tend to cross-react with typical African infections such as malaria. Although we previously took an in-depth look at the tests, there is one nuance that has not yet been noted and that suggests we revisit the topic briefly. This additional nicety is that African doctors usually have not been acquainted with the problematic nature of the tests, which has often caused them to hide their doubts about results.

Before noting some of their chagrined reactions as they became more aware of the limited value of these tests, let's remind ourselves of some of the shortfalls of the tests.

For one, the ELISAs cross-react with both leprosy bacteria and TB germs. A report in the *Journal of Infectious Diseases* states: "The test detects the presence of antibodies to antigens associated with infection by *Mycobacterioum leprae* [leprosy]."[244] Since this leprosy bacteria shares a number of outer shell markers with tuberculosis, the presence of TB in the blood will also cause a cross-reaction. "People with active TB infection are therefore liable to be at greatly increased risk of testing positive because of *M. tuberculosis*, not HIV."[245] As we've mentioned, there is a high incidence of TB on the continent. "Of the 660 million people in sub-Saharan Africa, an estimated 2-3 million have active TB, with an annual mortality of 790,000."[246] We can see that the mistaken identification of TB as HIV infection will jack up the numbers of those who are considered to have been killed by AIDS, particularly in that "30-50 percent of African 'AIDS' deaths involve TB."[247]

In our previous discussion, we went through a gamut of other diseases that can "fool" the tests and render false positives. As is clear, then, "the more infectious agents that a person has been exposed to, the greater the likelihood that such cross-reacting antibodies will be present."[248] It should also be easy to recognize that multiple infection is the norm in Africa. In fact, we can draw on an orthodoxy to confirm this. An article in *Immunology Today*, whose authors take at face value the reliability of the HIV blood tests, nonetheless lay stress on this factor. "The average African host is exposed to a huge number of infectious diseases from early childhood onwards. These include various bacterial, viral, and parasitic infections," they write.[249] "Noteworthy is the wide prevalence of helminth infections, malaria and tuberculosis in most parts of Africa: especially in sub-Saharan Africa, and in East and West Africa."[250] They also mark the prevalence of STDs. Their argument is that reason different patterns and rates of transmission of AIDS found in Africa in comparison to those found in the West are due to Africans being more susceptible—because of their multiple latent or active infections. The dissidents' interpretation is simpler. The average sub-Saharan African, being more liable to certain infections than the average Westerner, is more likely to be falsely registered as HIV positive because of a flawed exam.

It's time to turn to how Africans themselves, earlier on misled by testimonials to the value of the blood tests, have begun to see their weaknesses. In a moment, we will see how Hodgkinson alerted Africans to articles questioning the tests; but if we started with his journey we would be seriously distorting the picture. We would be in danger of falling into another variation on the "Ignorant Native Meets Apostle of Enlightenment from the West" scenario, which has already been found so obnoxiously present in the orthodox literature. Although Hodgkinson did inform many Africans in some areas about unknown dissident views, in other areas African researchers took the lead in finding holes in the tests.

Hodgkinson summarizes an African study in Zaire from mid-1994 in these terms:

> Scientists had been setting out to investigate if there was a link between HIV positivity and vulnerability to leprosy. They tested a large number of people at a leprosy area and found that something like 70 per-

cent tested HIV positive.... But when they did more detailed examination of the blood of these patients, they found that the real reason for the positivity wasn't HIV at all. It was that the microbacterium responsible for leprosy, which is in the same family as the one responsible for TB, was actually causing the proteins to be put into the bloodstream that causes these people to test positive.[251]

When I interviewed him in 1996, Hodgkinson said that based on his own extensive meetings with African health personnel, not only were these women and men finding these tests unreliable, but even those who hadn't personally gone into the subject were skeptical. "When one questioned them about their personal experience," he said, "you found that the doubts were there."[252]

He sees a learning curve, the type that I believe occurs generally with the introduction of medical "breakthroughs." First a new drug or process is trumpeted with much fanfare. Temporarily, it is embraced by the public as unlocking a new means of healing or helping us understand the body better. Then, as unanticipated problems arise, people begin to see through the hype and make a more realistically gauged assessment. It's a cycle that should be familiar from the marketing of a new or "improved" commodity. I'm not suggesting, by the way, that this indicates gullibility on the part of the public, since when the new item comes on the scene, there's not been time for any independent evaluation, and the only knowledge available about the innovation is purveyed by those whose purpose is to "flog it," in the older sense of that term.

This is more or less the sequence Hodgkinson outlines for the early acceptance of and later disenchantment with the HIV-locating blood tests.

> Tests based on supposed antibodies to HIV were first developed in the mid-eighties, the scientists responsible went over to Africa because they needed to create some kind of theory about where this bug had come from and Africa seemed like a good bet to test these kits. And sure enough they found that very large numbers of people—40 to 50 percent of some populations—were testing positive with the HIV kit. So the idea was put around in the second half of the eighties that HIV had been affecting Africa for some years and that huge numbers, millions, were in the pipeline of death...
>
> As time passed, it became clear in fact from AIDS statistics that this wasn't really happening.[253]

Hodgkinson traveled in late 1993 to Africa, where he was greeted both by those who were wavering about the validity of the tests (and other orthodox opinions about AIDS's relation to Africa) and those who, at personal cost, rejected the orthodox view. Of the first type, he mentioned that once he had shown them an article in *Bio/Technology*, that tended to discredit the tests, "they said, 'Well, in fact, although we didn't know about this, we're not surprised to hear this because we've often noticed that different HIV test kits give widely different results for the number of people testing positive, and we've had our doubts.'"[254]

The second group in Africa had unflinchingly opposed the establishment view. "Some had even gone so far as to argue that in areas that were supposed to be epicenters of AIDS in Africa, there was no AIDS! None whatsoever! It was a case of mistaken identity arising from the fear induced by this unvalidated and invalid test."[255] "The personal cost" of espousing such views, even much milder ones, was that with the burgeoning international AIDS scare, more and more money was being earmarked simply for this disease, and any health worker who wasn't working with AIDS "victims" could expect funding to become a trickle or dry up altogether. As Hodgkinson puts it in describing this group of African dissenters:

> The people that I'd spoken to were respectable workers, and they had everything to lose in fact by speaking out like this. There was one couple who'd started a charity making money for supposed orphans in northern Tanzania, on the border with Uganda. They risked their money, and in fact they lost a lot of funding as a result of speaking out in exactly the way I described, concluding there were no AIDS orphans. In fact, they were the children of poor people that needed help, but they often would describe these children as AIDS orphans when it was know that charitable money was available for AIDS.[256]

We might observe, by the way, that the misspending of health dollars, based on a misapprehension of the extent or locus of the disease, is not dissimilar here from what we witnessed in relation to sexual transmission. In the United States where homosexual transmission was the norm and heterosexual spread a minute aberration, the refusal to see this on the part of health commissioners meant that the bulk of AIDS education funding would be wasted on warning those who had little chance of getting the disease. In Africa, the analogous case involves targeting money toward an anticipated disease crisis rather than the one already present. "It's a tragedy," Hodgkinson states, "because it's [AIDS funding is] diverting desperately needed health resources away from the real areas of need, like fighting TB, like fighting malaria and leprosy."[257] We will have more to say about this in a moment.

But, to return to our theme, we might say AIDS in Africa is being overcounted at both ends. To calculate the number of infected people, a fallible AIDS test is used; while to calculate how many people are dying of the immune deficiency disease, a large number of deaths that seem to have been due to other causes are attributed to the illness. Remember this fundamental point: People do not die from AIDS but from associated diseases that take over once AIDS has weakened the immune system. Under such circumstances, a death from tuberculosis is not necessarily caused by tuberculosis but perhaps by AIDS, which prepared the way.

The contention of dissenters is that any death from TB, for example, is being added to the AIDS casualty list. More extensively, any disease that has even the remotest connection to AIDS is said to have been caused by immune deficiency. As the aforementioned Dr. Konotey-Ahulu complains, "Today, because of AIDS, it seems that Africans are not allowed to die from these conditions [from which they used to die before the AIDS era] any longer."[258] In a more jocular vein, a Ugandan doorman told a joke that's making the rounds in his country, which is, "A man got killed by a car while crossing the street—it was listed as an AIDS suicide."[259]

Moreover, the institutional system of available hospitals adds to the confusion in that each facility sees itself as treating a designated disease. Thus, as Kary Mullis explained to me, "In Africa, the only kind of place you can get any medical treatment is at WHO hospital for AIDS, right? So anything you come in there for—you cut your knee with a machete—you're an AIDS patient in some little book somewhere. I mean presumably you have HIV, too."[260]

The underlying problem is not simply that a "TB death" is reclassified as a "TB death led to by AIDS-induced immune system collapse," for such mislabeling could happen as well in the United States. What is particularly galling to African dissidents is that, just as African AIDS is said to have a different means of transmission from Western AIDS, the disease is defined according to a wider and laxer criterion than is used in the West. This classification casts a longer shadow over the health of the populace than does the Western categorization. Moreover, the problem goes beyond the fact that the grab-bag assortment of diseases linked to AIDS makes diagnoses inaccurate; the classifying system itself is of dubious merit insofar as it claims for AIDS a symptomology that has long been common to unrelated tropical diseases.

How do the classifications of Western and African AIDS differ? "In the West, AIDS consists of a person having one or more of about 27 relatively rare diseases," Eleopulos explains.[261] You may remember our earlier note that although there is a tendency in the West to keep enlarging the number of opportunistic diseases tied to AIDS, the ones originally tied to immune deficiency (such as PCP) and many of those added later had been relatively rare conditions. "In Africa, AIDS, as defined by the World Health Organization's 1986-1987 Bangui African AIDS definition," Eleopulos continues, "is no more than a collage of common symptoms and signs such as cough, fever and diarrhea, and a few diseases, some of which have been endemic in Africa for generations."[262]

Specifically, "The clinical definition of AIDS in Africa was established at a WHO meeting in the city of Bangui in 1987, and came to be known as the Bangui definition of AIDS."[263] By this criterion, a person can be said to be under attack from AIDS if he or she has a month of fever, a month of diarrhea, and then a dry cough, as well as a weight loss of 10 percent. Vomiting, abdominal pain, and a number of opportunistic diseases are also given as common symptoms. Celia Farber elaborates, "CD4 cell counts are not part of the definition, as testing is too expensive and therefore entirely unrealistic. Pulmonary tuber-culosis (TB of the lung, the most common form of the disease), in the presence of HIV, is called AIDS."[264]

Farber submits, "Some doctors and researchers, both African and West-ern… [say] the WHO definition for African AIDS is so nebulous and broad that it is virtually meaningless."[265]

By "broad" she means, as suggested, the definition does not center on new conditions of disease found to be existing. Rather, the Bangui lineup of symp-toms merely replicates the ones that had already been found in other, older dis-eases. She develops this thought: "Slim disease, as AIDS is commonly known in Africa, has become a kind of mop-up term for every disease involving diarrhea, vomiting, fever, or a cough, which are unfortunately also the primary symptoms of several tropical diseases."[266]

The same point was made in a front-page story in the *Times*, of London en-titled "AIDS in Africa: 'The Plague That Never Was," which argued, "The symptoms… used to identify AIDS are identical to those associated with com-mon African conditions such as malaria, tuberculosis, parasitic infections, and the effects of malnutrition and unsanitary water, all of which have troubled the continent for decades."[267]

A few pages back we took one remark of Dr. Konotey-Ahulu out of con-text, quoting his thought that Africans "are not allowed to die from these con-ditions any longer." By this time, the meaning of his epigram should be clear. It seems to him, a doctor who had a long record of service in Ghana, that all the old diseases for which he treated his patients have suddenly been renamed AIDS, although the change of nomenclature is the only thing that separates them from the diseases they were before AIDS was heard of.

Perhaps a little background attesting to the thoroughness of his research into the problem would be relevant. In his report in *Lancet*, he enumerated:

> In February and March of this year [1987] I made a six-week tour of twenty-six cities and towns in sixteen sub-Saharan countries, including those most afflicted by AIDS, did ward rounds with doctors and nurses, met ministers of health, directors of medical services, and re-search workers (native and expatriate)…
>
> [Presenting his findings that after rating "AIDS in Africa on an ar-bitrary scale from grade I (not much of a problem) to grade V (a cata-strophe)," only in five countries was AIDS to be denominated as much as of grade II situation, he moves to the topic of throwing so many dis-eases into the AIDS basket.]
>
> Before the days of AIDS in Ghana there was a death a day… on my ward alone of thirty-four beds…. They died from one or another of the following: cerebrovascular accidents from malignant hyper-tension, hepatoma, ruptured amoebic abscess, haematemesis, chronic

renal failure, sickle-cell crisis, septicaemia, perforated typhoid gut, hepatic coma, haemoptysis from tuberculosis, brain tumour, Hodgkin's disease… Today [to repeat], because of AIDS, it seems that Africans are not allowed to die from these conditions any longer.[268]

You must have already gathered that the pandemic nature of malaria, TB, and other illnesses on the continent, due to lack of medical care and poor living conditions, results in generally poorer health conditions than in the West, which, among other things, will manifest in a general public with weak immune systems that will be in more danger of infectious invasions. When I interviewed Joan Shenton, she put it like this: "People [in Africa] are dying of pathogenic assault, which is [caused by] filthy water, bacteria, tuberculosis, parasites."[269] Commonly, she says, one will find "children having malaria eight times before they're eight, children having enormous bouts of terrible diarrhea eleven times in their first year."[270] It is not surprising, then, that there is high mortality, especially for children. "These are the conditions that are leading to death," she told me, then noted further, "so people are called AIDS [patients] if they're showing these symptoms."[271]

In a moment, we will delve a little further into the uncanny coincidence of unhealthy living conditions and the incidence of AIDS in Africa, but a couple of final points should be made on the scientific aspects of the AIDS scare. These have to do with what would need to be done by scientists if, rather than relying on all-encompassing definitions, fallible blood tests, and skewed statistical projections, they wanted to objectively assess whether there was evidence of an African AIDS epidemic. This would mean taking account of existent disease patterns and studying the history of health and health expectations on the continent.

The first point seems shockingly obvious, although it has been systematically overlooked in discussions of the subject. It is an unfortunate fact that in southern and middle Africa, governments or health organizations have never had the resources to collect statistics on disease or mortality. Without health registers, there is no way to know how bad, in relative terms, a disease outbreak is. Philippe Krynen puts it in this form: "The hypothesis of a new virus needs to be checked against the pre-existing mortality pattern. But there are no such data."[272] In Africa, he explains, "births and deaths have only been registered in the memory of the people. It is a very poor source [from which to try] to calculate the specific mortality per age group and per disease one hundred years ago."[273] Let's imagine, it's true, that in a small southern African country there are now a thousand deaths per annum from TB complications, arising, it is said, from AIDS. This number might seem startling, but it holds less meaning than it might appear to as long as we do not possess information on previous rates of death. It may be that five years ago 999 people died annually in that country of tuberculosis.

Moreover, the West's sudden interest in Africa cannot make up for the years of neglect expressed in the West's lagging memory of previous incidence of disease there, coupled with underappreciation of sociocultural factors. Krynen explains that in Africa the sudden onset of a new and for-a-time unidentified disease is not unprecedented.

> If it was the first epidemic of this type in the region, it could be of some meaning. But it is not the first time that many young adults have died here. In the 1940s, sudden and mysterious diseases (later qualified as syphilis), caused such havoc that the elderly went to the border of Burundi to look for women to repopulate their villages which are said to be, now, devastated by AIDS.[274]

He also calls attention to the evolution of rising expectations in sub Saharan Africa, which in the later twentieth century has brought Africans to, if not achieve, at least be aware that better health is possible. This perception, Krynen feels, is what is behind the average African's sense of crisis. He says, "There is a rampant, very subjective feeling that something has gone wrong with the health of the adults. Several decades ago, at a time when life expectancy was low, such a feeling may not have taken off."[275]

However, with upped aspirations, there is a palpable unease. "But since colonization, people have seen hospitals, medicines, immunizations gradually shortening the usual course of their pathologies," Krynen writes, "and the present generation has forgotten the heavy toll that life had to pay when natural immunity… was the only potion for longevity."[276]

This last comment calls our attention once again to the need to wrestle with the bigger picture in Africa, including the continent's history, in order to reach an adequate understanding of what is going on.

We might say it is a hallmark of the bad science associated with orthodox study of AIDS to unduly isolate one factor, without taking the trouble to situate that factor in the wider universe of impinging facts. Notwithstanding that the experimental method itself involves radical separation of elements of study, this type of careful disaffirmation and separating out is only proper in certain fields of science, aiming at certain results. Outside its natural time and space, such isolation becomes fetishized, that is, put forward to the detriment of truth seeking. We saw it in the establishment's insistence that there must be a single cause that can be held to be the guilty party behind the AIDS illness, and now we are seeing it in the belief that it is this outlandish disease that is causing the health devastation in Africa. As I hope to show in the next section, it is not a micro-cause (a retrovirus) but a macro-cause (the collapse of the infrastructure) that is producing in Africa conditions that resemble (but are not the same as) a plague. In both cases, though, it often seems that what was wanted was not the true culprit but a patsy. In the former case, as we saw, some have argued that the rush to label HIV the sole cause of AIDS aimed to produce a sense of closure and dampen public fears, regardless of the scantiness of the evidence. What I mean by the second contention will be shown in the next two sections in which I lay out some ideas about the culpability of the African political economy for a generalized health breakdown and give some suggestions why the West is reluctant to admit to these conditions, instead preferring to charge them to the HIV scapegoat.

SOCIOCULTURAL CONDITIONS AND POOR HEALTH
OF SUB-SAHARAN AFRICA

Manuel Castells in an ambitious three-volume text has produced one of the most responsible works of futurology that appeared at the end of the twentieth century. By calling it "responsible" I mean that though he occasionally shares the breathlessness these works are known for, he counters these passages with sober assessments of the downside of present trends. In the first volume, *The Rise of the Network Society*, one of the most negative of all the developments he charts for the coming millennium is the absence of subSaharan Africa from the emerging world economy. The results will be Africa's continuing turmoil, which, he believes, will be ignored by the West since, unlike developments in the Middle East or Eastern Europe, for example, the region's chaos does not impinge on the priorities of the industrialized world.

Castells's characterization of the history and unhappy conditions in the area, some of the most concise and precise I have seen, have to be quoted in depth to give us background on why general health there seems in steep decline.

First, a capsule history.

> When the oil shock of 1979 and the rise of interest rates provoked their financial bankruptcy, most African economies came under the direct control of policies of international financial institutions that imposed liberalization measures supposedly aimed at generating trade and investment. The fragile African economies did not resist the shock... The competitive position of African countries in international trade was not brilliant in the 1960s, but deteriorated dramatically after the intervention of international financial institutions.... While liberalization policies were not able to attract investment or improve competitiveness, what they did was to destroy large sectors of agricultural production for local markets, and, in some cases, subsistence agriculture. As a result, African countries were left without defense against the impact of bad harvests. When drought stuck in Central and Eastern Africa, widespread famine followed... aggravated by civil wars and banditry induced by the heritage of surrogate military confrontation between the superpowers. Because states, as intermediaries between their countries and meager resources transferred from abroad, were the main source of income, control of the state became a matter of survival. And because tribal and ethnic networks were the safest bet for people's support, the fight to control the state... was organized around ethnic cleavages, reviving centuries-old hatred and prejudice.[277]

Bringing the situation up to the present, Castells notes that Africa has been cast to the sidelines of the newly integrated circuits of the world economy.

> Overall, the systematic logic of the new global economy does not have much of a role for the majority of the African population in the newest international division of labor. Most primary commodities are useless or low priced, markets are too narrow, investment too risky, labor not skilled enough, communication and telecommunication infrastructure clearly inadequate, politics too unpredictable, and government bureaucracies inefficiently corrupt... Under such conditions, the only real concern of the "North" (particularly of Western Europe) was the fear of being invaded by millions of uprooted peasants and workers unable to survive in their own countries.... Yet what can be said of the experience of the transition of Africa into the new global economy is that [the land's position has become one of] structural irrelevance from the system's point of view.[278]

The way it worked was like this (taking Senegal as an example): In the late 1970s, many villages went into debt. Landlords and the state, guided by pressure from Western experts, demanded the land be put into a cash crop, notably peanuts. Even peasant groups who did not have outstanding IOUs were tempted to conversion because of high prices being paid for the food. Then, when the price collapsed on the international market, the peasants had neither money nor food. The choice was usually to go into the city to swell the unemployed, starve on the land, or join the military for plunder.

It is the devastation wrought by these bloodcurdling social conditions that many researchers target as playing a causal role in the plummeting health conditions in the region, which the orthodox blame on AIDS. Charles Geshekter at California State University, Chico, says bluntly, "It is the political economy of underdevelopment, not sexual intercourse, that is killing Africans."[279] He elaborates by stressing such factors as "poor harvests, rural poverty, migratory labor systems, urban crowding, ecological degradation and the sadistic violence of civil wars," which have disrupted such minimal health care facilities and sanitation as had existed. "When essential services for water, power and transport break down, public sanitation deteriorates and the risks of cholera and dysentery increase." He repeats his conclusion, "African poverty, not some extraordinary sexual behavior, is the best predictor of AIDS-defining diseases."[280]

Like the fact about lack of comparable mortality statistics making it hard to gauge how bad the disease is, the thought that degenerating living situations would be playing a large part in new disease formations on the continent would seem a rather obvious conclusion to draw, but it is not one put forward by the establishment. Some have seen this obtuseness on their side as resulting from the same mind-set that has often marred Westerners' opinions on Africa: Conclusions are being drawn without consulting people on the ground. According to Duesberg: "Most of the people who talk about AIDS in Africa are living here in Atlanta at the CDC or somewhere. Or the World Health Organization in Geneva. They're not bothering to go to Africa."[281] As our guide Sir Francis Bacon puts this point more generally, "In nature, the more you remove your-self from particulars, the greater peril of error you do incur."[282]

This contention is echoed by other dissidents, whose opinion, reported in a London Sunday*Times* article, is that "*those who have worked with Africans* know that what is considered AIDS over there is nothing more than a new term for old diseases."[283] Philippe and Evelyn Krynen, who led a medical relief organi-zation in Tanzania, met people said to have the immune deficiency disease. Ac-cording to the story, "Contrary to their expectations, they learned that: 'There is no AIDS. It is something that has been invented. There are no epidemiolog-ical grounds for it; it doesn't exist for us.'"[284]

Celia Farber, whose trip to Africa was already mentioned, also has written of how going to the African countries deeply influenced her point of view. "We saw many sick, dying, and even dead people during this trip, but the task of trying to decipher just what they were truly dying of struck me as impossible. Often, their doctors—didn't even know."[285]

These last set of quotes, contrasting Dueberg's opinion on why Western doctors may be exaggerating and/or overlooking aspects of the African reality not visible from afar with the opinions of those who had extensively worked in or visited Africa, highlights the division between those who know and those who dream they know.

But, as we've said, there is also a basic difference in how researchers assess the causatives factors behind disease. Why should those establishment doctors, "chill-ing" in Geneva and Atlanta, waste time on field trips if they remain convinced the disease is caused by a virus, which can be as easily studied with flown-in African blood in one's air-conditioned office as at the site? From this perspective, the poor social conditions in Africa could be discounted in diagnosing what seemed to be increases in TB and other diseases that were being associated with AIDS. For dissenters like Krynen, though, such a view is part of the problem. "The western look at health, after peering into so many microscopes, has become myopic. It does not see that, in Africa, it is no so much the tiny microbe which has to be tracked down as the conditions favouring diseases' proliferation."[286]

SELF-INTERESTS AND AIDS IN AFRICA

"AIDS is a perception. The more you look for it, the more you see it," said Dr. Kassi Manlan, the Cote d'Ivorie's director general of Health and Social Ser-vices.[287] He might have added, the more you want to see it, the more you will find. In other words, macabre as it may seem, there are some groups that are motivated to make as much of the crisis as possible. This would include, most obviously, those whose livelihood depends on treating the disease, but it would also serve such interests as the media, who increase readers or viewers by touting catastrophic figures, those who want to shift responsibility for spreading the disease away from particular high-risk groups, and even those who cynically hope to test AIDS drugs on African guinea pigs. For all these groups the projection of astronomical figures of African AIDS cases can be wielded as a weapon of convenience. Let's look at some of the groups who are strangely allied in their willingness to give easy credence to the thought that all Africa is or will be devastated by this new immune system deficiency.

Earlier, I called a number of African doctors "dissidents." I did not mean to imply that these health workers and doctors are fiery foes of orthodoxy who rail against the West's stereotyping, nor that they represent a majority. In fact, many, perhaps the majority of African health workers are establishment oriented and accept orthodoxy's views on causation, transmission, and, with less fervency, African origins of AIDS and Africa's approaching devastation. However, these conservatives can be separated from their Western counterparts insofar as they have different reasons for agreeing with the dogma. Just as their Western brethren, they may well be sincere advocates of these opinions, but the comparative financial advantages to them for holding this view will be much greater. This is not to say that an African doctor, for instance, can pull down a bigger salary than a CDC retrovirologist, but that given the region's unreservedly bleak economic situation, AIDS funding from abroad can be quite lucrative compared to other sources of income. Farber reports, "Many believe that the statistics [of the number of AIDS sufferers in Africa] have been inflated because AIDS generates far more money in the third world from Western organizations than any other infectious disease."[288]

When she went on an investigative trip to the continent, she saw some truth in what many believed, at least insofar as AIDS produced funding. "This was clear to us when we were there: Where there was AIDS, there was money— a brand new clinic, a new Mercedes parked outside, modern testing facilities, high-paying jobs, international conferences."[289] Her next point is this: She was forewarned by an African doctor in London, who told her that she would have trouble learning the facts because it pays for many health workers to conceal them. He said, "You will never get these doctors to tell you the truth. When they get sent to these AIDS conferences around the world, the per diem they receive is equal to what they earn in a whole year at home."[290]

Farber learned, according to Hodgkinson, that to obtain information from establishment officials, it was wise to in turn conceal something of her own intentions. "In Uganda, obstacles were overcome when she said she wanted to tell her young American readers what a terrible toll AIDS was taking."[291] She found that African officials had sound economic reasons for wanting such a message spread, for as she wrote:

While the public seems resentful of the constant repetition that AIDS is wiping out Africa, the officials generally nod in somber agreement when the most dire statistics are cited. And to what end? AIDS generates far more money than any other disease in Africa. In Uganda, for example, WHO allotted $6 million for a single year, 1992-1993, whereas all other infectious diseases combined—barring TB and AIDS—received a mere $57,000. One of those diseases, malaria, is still the leading killer of people worldwide, and several drug-resistant strains of malaria have been emerging in recent years.[292]

Similar reflections on the way AIDS funding can appear as a gravy train to health care workers have been made, and as cuttingly, by Joan Shenton. Her emphasis, however, is on the tragic waste of resources entailed when more threatening diseases—again the case cited is malaria—have their financial allotments cut so as to swell the AIDS-dedicated coffers. In one African country, she writes, "I saw a gleaming of Toyota vans with young sex counselors in them with bags of condoms in the back racing around the countryside telling people not to make love to anyone new."[293] Meanwhile, malaria control, "which used to have a lot of money," could hardly afford such vans.

Now, to return to Farber's claim that she wanted to inform readers about the terrible ongoing tragedy of AIDS, we can say that what made her story quite plausible to officials is their previous experience with Western journalism, whose interest is in eye-catching, exciting coverage rather than sober assessments. As Gestekher characterizes newspapers in particular, "The disaster voyeurism of tabloid journalism enables them to use AIDS to sell more newspapers than any other disease in history. It is a sensational disease—with its elements of sex, blood and death it has proved irresistible to editors across the world."[294] This is a tendency that has reached a nadir in stories, as mentioned by Shilts, such as the one appearing in the supermarket tabloid *The Globe* in 1983, which traced the African origin of AIDS to King Tut! The article claimed his mummy spread the disease "following the tour of Tut's treasures to the United States in the late 1970s."[295] In a moment, we will turn to a more sinister piece of this ilk.

It's worth emphasizing first that, as Gestekher outlines, most journalists seem to prefer statistical speculation over hard news. So, when the Global Burden of Disease Study appeared, showing mortality and morbidity data for Africa that indicated AIDS should be downplayed as a killer compared to other, old and familiar tropical diseases, it was brushed aside by journalists intent on preserving the illusion that AIDS was mowing down lives like a steamroller. "The media's appetite for scary scenarios and its disdain for alternative perspectives enables it to treat Africa in apocalyptic terms... Oblivious to data [such as that from the Global Burden report] journalists reflexively maintain that 'AIDS is by far the most serious threat to life in Africa.'"[296]

Most revealing and depressing in this light was an article by a white South African in which abstract, unsavory speculations about money to be made once the continent was cleared of obtrusive natives is combined with acceptance of a worst-case scenario of AIDS's depopulation. The article reads in part:

What if current projections are correct, and more than half the population of countries like Uganda, Kenya, Zaire and much of equatorial West Africa are wiped out before the turn of the century? Who will fill the vacuum? Immense tracts of African real estate will become available, literally for the taking.... A curious sidelight in East Africa is the attitude of the local Asian and European residents. Some whites in Nairobi refer rather cynically to Kenya as a good place to buy property for the future. But it is the Asian community which takes a rather more serious view of the matter. Some Asians in Nairobi are already talking of a depopulated East Africa being ideal for the settlement of some of continental India's overpopulation.

It could be that a major Indian migration would turn Africa into the continent of the future. With the economic gusto and enterprise which Indians have shown in dozens of countries around the globe, they could be the best influence Africa has seen for generations. Certainly, they could change the economic face of a continent at present slipping into somnolent chaos.[297]

Our brief historical overview of Africa, which detailed how bankrupt countries were disastrously restructured and looted by multinationals and indigenous power brokers, shows how poor the above author's economic knowledge is, since he or she seems to attribute the poleaxed economies of these countries to natives' business incapacity. More disturbing, though, is the point Konotey-Ahulu highlights. "The most frightening thing about this article is not that it may be true [it is nonsense] but that powerful interest groups in the West and South Africa wish it to be true."[298] That is, such groups revel in high death-rate predictions as opening up new investment opportunities.

In mulling over the topic of sensational journalism, we might extract a further lesson concerning the dialectic between causing and damping panic. If you think about this a little, you'll see it is not all one way. Not every potentially panic-inducing event is worked for all it is worth. This only seems to happen when the fear that is likely to be provoked either will issue in further demands for social control (of the type Luttwak described) or will help cement stereotypes. An example of the former would be hysteria driven up around heterosexual transmission of HIV, which was guided toward creating sexual repression and demands for celibacy. (This is not to say that safe or reduced sex may not be called for in particular instances.) An example of the latter use of hysteria is the case at hand, in which fears of AIDS depopulating Africa are stoked because they help accord with the stereotype of rampant African sexuality, as well as (we will see below) letting the West off the hook for its role in causing deteriorating health conditions in that part of the Third World.

By contrast, measures are taken to reduce panic when the fear might rebound against a sacrosanct component of the system. Take the case of blood transfusions. By January 1983, as Shilts reports, scientists at the CDC thought there was strong enough evidence that AIDS could be transmitted by way of blood transfusions to call for screening of all donated blood. As Dr. Bruce Evatt stated at a consultative meeting with the major blood banks, before 1982 hemophiliacs, who routinely needed blood transfusions, had never gotten AIDS. "In only the past year, however, 6 out of just 100 hemophiliacs in Ohio were dead of AIDS... and nearly 10 percent already were sick with something having to do with AIDS."[299] The blood bank representatives refused to see these cases as statistically significant or proven cases (since the diseases or deaths might have been caused by other behaviors rather than by transfusions). At bottom, though, their arguments were "almost solely on fiscal grounds... the blood industry represented big money, with annual receipts of a billion dollars." As Shilts lays out in his hard-hitting account, "The cost of testing for hepatitis antibodies [whose presence had been correlated with AIDS infection], Kellner from the New York Blood Center suggested, would be $100 million annually for the entire nation. That was simply too much."[300]

By July of that year, many more cases of transfusion-linked AIDS had appeared, and Dr. Harry Haverkos of the CDC had prepared a formal report showing the probable connection. This time there was a strong reaction. No, it was not that the blood bankers would now seriously consider or institute tests; rather the government and media rushed forward to assure the public, which now was becoming restive in light of rumors about the unreliability of the country's blood supply. "Local public health officials demonstrated their interest by minimizing the threat of transfusion AIDS. In Los Angeles, for example, the announcement that three infants had died of AIDS probably contracted through transfusions brought heated denials from hysteria wary officials."[301] The federal government also got into the act, with the Health and Human Services Secretary Margaret Heckler denying the relevance of the L.A. cases. As a photo opportunity, she also went to the Red Cross (one of the groups most adamantly opposed to any expensive changes in blood bank policy) and donated blood, saying, "The blood supply is safe both for the hemophiliac... and average citizen."[302] The media tried to throw a wet blanket over fears. "Seeing themselves as the bastions of common sense, science writers and reporters covering the epidemic... wrote curb-the-panic stories and avoided asking the blood bankers tough questions."[303]

Here a panic would have run against the current the media is set on making flow, that of a near religious faith in the beneficence of powerful players and the undilutedly positive qualities of products for sale. It would seem that when a panic would injure the interests of any powerful institution, it has to be starved not fueled.

Broadsides such as that of Geshekter, where he tars all productions of the media as fear arousing, are too lacking in discrimination, no matter how applicable they may be in specific circumstances. Before imagining that a trend or event will cause panic, the type of panic it would bolster must be gauged against the archetypes and positions journalists are pledged to uphold.

To move to the next reason that might motivate people to want to lend ear to the idea of AIDS overwhelming Africa, we might begin by referring to a typical strategy visible in the view of Mr. Real Estate Speculator. This is the way one disadvantaged group is pitted against another to the benefit of neither. In Speculator's report, Africans were balanced against East Asians, the former group indicted for supine lack of entrepreneurial spirit, the latter for cold-hearted greed in wanting to take advantage of growing misery.

Some have seen a similar playing off of oppressed sectors in the way a number of gay male spokesmen seemed to exult in the attention brought to Africa in the AIDS discussion, since it took homosexuals out of the media glare. Here we can listen more sympathetically than we did to Mr. Speculator, who was obviously from neither of the groups about which he opined; the gay spokesmen speak from having been under the scourge of discrimination, which they understandably wish to elude.

In talking about this tendency of one already stigmatized group to attempt to escape unmerited blame by shifting the public's onus to another such group, the Chirimuutas bring up a book by Dennis Altman's and the statements of Peter Tatchell, a gay activist, who wrote in 7 Days, "Contrary to the mythology that AIDS is a gay plague, the overwhelming majority of people with AIDS are heterosexuals, including as many women as men and almost as many children as adults. Most of these AIDS cases are in central Africa."[304] Tatchell then cites the large number of cases in varied African countries, which the Chirimuutas decry. "This attempt to deflect anti-homosexual fire onto central Africans shows little respect for either truth or logic."[305] They cite counter numbers, upholding many of the points we have already made about the falsification and exaggeration of African AIDS statistics, and conclude, "Although Altman and many other homosexuals quite rightly rejected attempts to blame them for the AIDS epidemic, they uncritically accepted and propagated suggestions that AIDS had originated in black people and was primarily a black and not a homosexual problem."[306]

What we have seen so far is that, while no one is happy about all these deaths, for some they can be functionalized. In other words, they can be put to use in solving a problem. If the media needs supercharged stories, ones featuring the usual lurid combination of sex and death, they can paint a picture of the specter of death, transmitted by prostitutes and truck drivers, stalking sub-Saharan Africa. If U.S. gay males are under attack, as they were, by right wing ideologues, such as Patrick Buchanan, who say gays' connection to AIDS means they should be interred or restricted in other coercive ways—they can argue that the long figures in Africa prove responsibility for the disease rests abroad. In neither case does this entail that the actors do not believe African AIDS is as devastating as has been portrayed, but that their needs make it all to easy too believe, despite strings of contrary evidence.

The seemingly large number of African AIDS victims could also be turned to account by the pharmaceutical companies that needed to test drugs that might offer a cure or palliative to the disease. Remember, testing AIDS drugs posed a number of problems that were not necessarily faced in tracking other diseases. There was no reliable animal model, as there was, for example, with smallpox, whereby variants of the human disease appear in animals. Moreover, ethically sensitive laws were in place that restricted human trials, such as those once used to test yellow fever vaccine. They had been given to prisoners, who would be offered parole if they survived.

With these hindrances, is it any surprise that pharmaceutical companies would be pleased to find a large number of cases of the disease in Africa, where governments were lenient in overseeing research and poverty, and lack of medical care could be expected to induce many sufferers to enroll in studies? In Geshekter's words, "The serious consequences of claiming that millions of Africans are threatened by infectious AIDS makes it politically acceptable to use the continent as a laboratory for vaccine trials and the distribution of toxic drugs of disputed effectiveness like ddi and AZT."[307]

Geshekter's words do not present purely speculative fears. Drug trials have already been called for and carried out on the continent. The *New York Times* reported on February 19, 1988, an announcement by Dr. Anthony Fauci, director of the National Institute of Allergy and Infectious Diseases, that the United States "is planning to conduct large scale human experiments in Africa to test potential AIDS vaccines."[308]

Silvia Federici, in an article titled "White Doctors in Africa (on the trail of AIDS or in service of the empire?)," wonders why neither the press nor the public questioned the ethics of blandly calling for such experiments in light of the fact that "we have repeatedly been told that there is no safe method… for conducting such tests on humans."[309] She believes the lack of concern is partially based on the reporting of record numbers of cases of infections and disease in Africa.

> [The] public conscience has been appeased by the official explanation, which tells us that Africa provides a "natural" lab because AIDS in Africa is… "spreading like wildfire." From this we are expected to deduce that the tests are in line with the best utilitarian tradition, that times of extreme hardship demand the sacrifice of a few for the benefit of the majority.
>
> [Moreover, as long as there is no citizen's outcry in the United States, acquiescence is to be expected in Africa.] For the sake of a little

hard currency, some African governments can be blackmailed into delivering their citizens over to unscrutinized medical experimentation, in the same way that some are now turning their countries into dumping grounds for chemical and radioactive waste, in exchange for dollars or pounds.[310]

Indeed, even before Fauci's announcement, drug trials were under way in Africa. In its March 19, 1987, issue, *Nature* reported the testing of a live AIDS vaccine in Zaire. Zimbabwean scientist Davis Gazi stated, "The so-called AIDS vaccine is already in use in Zaire although European and American doctors refused to experiment with it on their own people."[311]

(As an aside, it's worth noting that people of color in the United States, people with less clout than other social groups, also seem most liable to be cast as guinea pigs in drug-use dramas. At one point, the government recommended that HIV-positive pregnant women be given AZT to shield their newborns from the disease, although evidence of such shielding's usefulness was dubious. "In the only published federally-sponsored study, [it was found] more than 74 percent of the woman given AZT were black or Hispanic; when the 'adolescent mothers' were separated out statistically, African-American teenagers comprised more than 90% of the 'treated' group."[312]

To conclude this discussion of motivation, let's hark back to its beginning where I said, without providing details, that the thesis that AIDS is expanding at exponential rates in Africa can be taken as something of a patsy, or perhaps I should have labeled it a fall guy. That's the little guy in the crime syndicate who takes the blame for a crime so that the heat will be off Mr. Big. In our specific case, Mr. Big would be the seriously eroded conditions of life in Africa, which had gone downhill since the 1960s, and which to a great degree, following Castells's argument, can be laid at the door of Western interference, although conditions in the African nations themselves also contributed. Hodgkinson quotes Farber, who lays out the logic of this displacing of the responsibility. "Blame falls squarely at the feet of the AIDS patient, while the socioeconomic setting is ignored—a scenario that is unsettingly convenient for the establishment."[313]

Pointing the finger at one-on-one sexual behavior as the culprit behind all the turmoil also concurs with the long-term Western urge to see everything in terms of the responsibility of the self-reliant individual. As the more recent tendency, already highlighted, is to look on curbs of pleasurable activity as a promising solution to general feelings of malaise and insecurity. So, for instance, in looking at the sad state of health of the population of Uganda, which had been "subject to two decades of turmoil, war, decay, and the unparalleled dictatorship and wreckage of General Idi Amin," these factors were ignored in favor of "the politically correct, simplistic and unproven party line that... sex is the primary cause," and skyrocketing statistics of infection and death are due to unconscionable promiscuity.[314]

It might even be ventured that dire as an African AIDS plague may be—which, if the West were intent on helping solve it, would call for massive educational campaigns and a great upgrading of heath care—it is far less imposing than accepting that the near universal breakdown of people's well-being is due to three decades of socioeconomic restructuring, whose solution would have to be sought in a national redistribution of wealth as well as a humanizing of international relations. Instead, between two unpalatable options, the one that would demand less of the West, both in terms of taking blame and righting wrongs, is chosen. "Numerous interests… are served by the idea that an epidemic of a new sexually transmitted disease is ripping through Africa. It distracts attention from rising death rates caused by economic exploitation and decline following periods of unsustainable population growth."[315]

Medicine and Dream Time

In *The New Organon*, Bacon lays out the path a reality-testing, experiment-based science should take, not presuming to walk this route himself, but acting as a surveyor who marks out the road's curves. Still, he knows that humans need strong encouragement to undertake anything innovative and, moreover, that "by far the greatest obstacle to the progress of science and to the undertaking of new tasks and provinces therein is found in this—that men despair and think things impossible."[1] So he ends with a presentation of what has been achieved by experiments, ones fitting in with the logic of his new science but undertaken haphazardly over the years. The reason for appending these is so he can combat overpowering despair. "I am now therefore to speak touching hope.... [And I find that] the strongest means of inspiring hope will be to bring men to particulars," he says.[2] The presentation of these tonic particulars, which he calls his first vintage, should rally his readers to delve into the new way of acquiring sound knowledge.

So, analogously, to induce readers to continue on to the positive portions of this text—after having crossed a rather depressing patch in which the liabilities and shabby shifts of orthodox science have been laid bare—I want to continue my thoughts on how the AIDS crisis throws light on the emergence of a new paradigm.

I've said that our examination of AIDS has shown biologists grappling with fundamental issues in the doing of science, such as whether or not to hold on to Koch's postulates, and that this indicates that the AIDS issue strikes bedrock, calling mainstream biology and medicine to account, while questioning whether they are as viable as ways that would approach health alternatively and naturally. And I have interweaved discussions of mainstream science and mainstream media. Now we can bring those two points, on the challenge and on the media/science connection, together.

In terms of dealing with epidemics, until the appearance of AIDS medicine basked in the memories of a glorious past, when it had eliminated many infectious diseases that had burdened humankind for centuries, while looking forward to a splendid future, when the few remaining stamping grounds of such diseases would be cleared of these predators on health.

In truth, many of these memories were illusionary. Thomas McKeown's close study of the relaxation of the grip of diseases on early-nineteenth-century England shows:

> The reason for their disappearance as major causes of death were, in order of importance, first, improvements in the standard of living (eg, nutrition, housing), second, improvements in the control of the environment (eg, water supply and other sanitary devices) and only third medical care.[3]

The coming of AIDS, meanwhile, shot down the hopes that humankind could look forward to a rosy future without large-scale outbreak of infectious diseases. I contend that this shattering of the illusion of the medicine is not only a blow against the amour propre of medicine, but weakens the whole edifice of our business civilization. Under the reign of a commercial social structure, each major component, from particular enterprises to education to entertainment, not only must put forward its own self-serving justification but must help rationalize the system by downplaying its sins and magnifying its virtues.

Projections of the healthy future provided by medicine always entailed a blatant disregard of the vast inequality between the rich and the poor in U.S. society. New cures for cancer, genetic engineering, ways to overcome infertility, and so on are imagined. Whenever they are discussed, it is tacitly assumed that everyone will have an equal right to them and share in their benefits. Thus, the unequal distribution of health care that now reigns— without a national health insurance or a just allotment of society's wealth, which precludes the majority from sharing any of the boons of medical innovation—is carefully kept out of discussion of the future triumphs of medicine. To move into this future world, where none of our untidy power inequities exist, is to give a boost to ideology and produce a *dream time.*

While each area in American life does double duty by justifying itself and the whole system at once, many areas backstop each other. Biological science creates rationales for pouring research money into treatments whose expensiveness means that, at present, few but the rich can afford them, while pretending, in the great bye and bye, that they will prove advantageous for everyone. It is seconded by the media, who jolly the public along with thoughts about upcoming medical miracles. I call this *the politics of underlining*. Each statement made by agency A is more solid because it seems to take on added stress from B's, C's, and D's statements.

The shattering of medicine's mythological future has exposed not only the bankruptcy of its manner of doing health care, but the insufficiency in the hierarchical way U.S. society is arranged. As the orthodox medical paradigm collapses, it brings down, at least by exposing them, other unjust and indecent setups in the socioeconomic order. And through its dust is visible something that has been around since the 1960s but often hidden, an alternative paradigm of health.

In *The Turning Point*, Fritjof Capra argues that a new worldview is on the rise, which, as opposed to the older mechanistic view (which saw each individual as an isolated speck, self-reliant and unperturbed by others), emphasizes "the essential interrelatedness and interdependence of all phenomena."[4] One element of this novel perspective is that in place of the older theory of disease, by which a single cause is to be held responsible for any physical ill health, this theory sees illness as the result of

> patterns of disorders that may become manifest at various levels of the organism, as well as in the various interactions between the organism and the larger system in which it is embedded.... Accordingly the systems view of health can be applied to different systems levels, with the corresponding levels of health mutually interconnected. In particular we can discern three interdependent levels of health—individual, social, and ecological. What is unhealthy for the individual is also unhealthy for the society and for the embedding ecosystem.[5]

One of the striking corollaries of this view is that one can only improve individual health by improving society!

You should be able to cast your mind back to our earlier examples of community-based research and the San Francisco hospital ward run by the nurse to see the truth of this last combined claim. In both cases, advances were being made (in treatment in one, in drug testing in the other), through the confluence of two factors of the new paradigm. Simultaneously, there was the use of a holistic outlook on science, and a democratic, cooperative relationship among all parties, from doctors to people with AIDS to activists, who were working with the new perspective.

Showing the clear superiority of this paradigm will be one of my objects in the rest of this book.

PART TWO

EXPLORING ALTERNATIVE CAUSES OF AIDS

O ur discussion of the emerging paradigm will suggest to the intelligent reader the new directions in research being pursued by dissidents. If, as just laid out, orthodoxy sees disease as the result of a single intrusive invader—in this case, a retrovirus—dissidents, coming from the enlarged perspective, understand AIDS as the resultant of multiple impinging factors, including (possibly) drug use, stress, the combined effect of infections by a number of microbes, and nutritional deficiency.

Acceptance of the new point of view is spurred by two factors: the plausibility of the perspective in terms of conformity to the evidence and the miserable performance of the orthodox view in terms of explanation and devising of a successful treatment. On the latter point, that of orthodox treatment, let us remember that even if orthodox successes are cast in the best possible light, and AZT is taken seriously, the most that has been achieved is the slowing down of the disease, though in no way has it been staved off. One is still doomed by HIV infection, but doomsday is pushed back a little (but not much) further.

As can be imagined, such paltry results after so much time and money spent in the vain search for either a cure or, at least, a vaccine has ended in the mass production of disillusionment. As was mentioned earlier, disenchantment has even crept into the camps of scientists, such as that of Montagnier, who were once the doughtiest defenders of the HIV = AIDS thesis and all that thesis carries with it.

Let's pause a moment to chart the disappointment experienced by even those who first rode the bandwagon.

In December 1994, Richard Horton, the North American editor of *Lancet*, assessed what advances had been made that year in dealing with AIDS. The progress he saw came not in any improvements in understanding or treatment of the disease but in the showing up of the insufficiency of current activity.

> It's been a year when there's been a painful reassessment of all the as-sumptions that have been the foundation of AIDS science during the past decade. People have been forced to admit uncertainties they were unwilling to admit before.
>
> That's what's so momentous about 1994. One, we are reversing the entire clinical trials program. Two, we are reversing the point of view that we know the cause of immunodeficiency. Three, we're reversing the entire vaccine strategy. Four, we are reversing the notion that you need to focus on HIV. Most people are arguing strongly that you need to focus on the immune system, not the virus.[1]

In other words, orthodoxy's chief contribution so far has been to explore all the dead ends, thus revealing clearly what paths are not worth following any further.

In such a situation, we have to get back to basics. Dr. Dean Black says we have to ask: "If HIV is not the cause, what is the cause?"[2] Or, as Richard Strohman puts it, "We need research into possible causes such as drug use and behavior, not a bankrupt hypothesis."[3]

Here in Part Two, we will see what this new research, implicitly or explic-itly guided by the progressive paradigm, has to say about the cause of AIDS. First we consider, taking a point raised earlier, whether AIDS is not a peculiar compilation of already known diseases, working in a diabolical synergy, rather than a new entity. Next, we will question present standards of diagnosing im-munodeficiency. Finally, we will look at a range of factors believed to cause the syndrome.

We can also call Sir Francis Bacon to our side again. When praising Bacon, most writers are careful to restrict their comments to the brunt of what he says. That is, they indicate that his orientation in general was a godsend to justifying scientific experiment and moves toward the pragmatic study of nature. His own feeble efforts to carry out his own programs, however, are more of an embar-rassment than a help to his defenders.

One of the "experimental" methods he thought would be most promising for science was the study of prerogative instances. These are examples of nat-ural processes, the study of which would be particularly apt for illuminating natural laws, especially because of the untowardness of the phenomena. He mentions, for example, the strange cases where animals seem able to draw on powers of reasoning, as when a thirsty crow, seeing water at the bottom of a tree trunk, plunked in pebbles till he raised the water to a level at which he could drink. This, Bacon avers, that in certain exceptional cases, "brutes too have the power of sllyogizing."[4]

The fantastical nature of this example suggests that his method is not likely to be rehabilitated as a scientific protocol, but it still can have virtue as a rhetorical tool. By this I mean that often in studying our topic, we come upon spectacular instances that stand as representative patterns of the failures of orthodoxy or the strengths of the dissident viewpoint. These instances do not increase the objective argument in favor of adopting the dissident side, but they can play a valuable pedagogical role by helping the student of this subject to fix in her or his mind the outlines of the case I am making.

So, as I move from section to section, I will often point my discussion by calling your attention to a prominent instance, classifying it according to Bacon's categories, and using it to dramatically underscore a moment in our argument.

Is the AIDS
Syndrome New?

I've harped on the fact that AIDS is sui generis in being a killer disease from which one doesn't die. The syndrome weakens the immune system to the point at which it can't fight off the onslaught of a secondary complication. I've also brought up the difficulty this poses when it comes to deciding the cause of death for various patients. What if the secondary infection that ultimately eliminated the sufferer was one that could prove fatal even without AIDS? Then how does the conscientious doctor decide if a person died from a disease or a disease allowed access by AIDS?

Don't forget—to complicate the decision—as Rene Dubos established in the 1960s, various effects of malnutrition and vitamin deficiency weaken the immune system, thus making the human body more liable to disease attack.[1] Thus the equation infection by TB (or other deleterious agent) + weakened immune system = disease was known long before AIDS came on the scene.

Those who are inclined to doubt whether AIDS is a new disease on the scene point to these difficulties in deciding what disease has struck down a sufferer. If one tries to work with the CDC definition of the disease, recognizing, as we've shown, that this definition has changed over the years, one will find numerous pitfalls.

We've noted that many diseases were considered accompaniments of AIDS especially because they were rare. For one, saying a disease was rare before AIDS, means that it used to occur without AIDS, which, granted that it is then still so occurring, jeopardizes any attempt to characterize all present occurrences of the disease as supplements to AIDS. Moreover, some of these diseases, though rare, were becoming more and more prevalent in the decades before AIDS. Such facts have caused Prof. Root-Bernstein, associate professor of physiology at Michigan State University, to write in *Lancet*, "There are hundreds of cases of opportunistic infections satisfying the CDC criteria for AIDS that have been around for decades. AIDS therefore should not be viewed as a new syndrome."[2] Let's make mention of some of these diseases.

Let's take as an example Kaposi's syndrome, a vascular tumor. Although Kaposi's syndrome came to prominence in relation to AIDS, as Prof. Root-Bernstein spells out, it can be traced back to the beginning of the century. "[3] Cases of Kaposi's syndrome were found by Katner and Pankey (1987) between 1902 and 1978 that satisfy the definition of AIDS." He also mentions "several hundred [examples] of KS dating back to Kaposi's first paper in 1872 which satisfy the CDC surveillance definition of AIDS."[4]

Another disease that has been around for a while is progressive multifocal leukoencephalopathy (PML). The CDC considers its presence in otherwise healthy individuals indicative of AIDS. The disease first gained recognition in 1958. "By 1970, 50 cases of PML were identified."[5] Six years later, there were almost fifty more cases "of which at least 5 resembled AIDS, characterized by deficiencies in cellular immunity due to unknown causes."[6] Doing the arithmetic, since nowadays PML added to immune deficiency equals AIDS, "10% of all PML cases would have qualified as AIDS cases prior to 1981."[7]

Another disease the CDC calls indicative of AIDS is extrapulmonary or disseminated cryptococcosis. We can label this a Baconian instance in the sense that it is a striking exposure of the difficulties entailed by the establishment definition of the disease.

Taking Bacon's terminology, we call it a migratory instance. This is one in which "the nature in question is in the process of being produced when it did not previously exist."[8] Specifically, cryptococcosis is a disease that had started dramatically rising before AIDS appeared; thus, the rationale that this disease should be linked to AIDS's depletion of the body's resources because it was formerly so rare is especially difficult to maintain. The first clinically described case of this disease was recorded in 1914.[9] By 1955, there were reports of three hundred cases of cryptococcal meningitis.[10] "Between 1949 and 1953, 151 fatal cases were reported in the U.S.—indicating that the published cases were small compared to actual cases."[11] From 25 to 50 percent of those with the disease had immune deficiency not caused by any other underlying disease.[12] Although not every case involved immune deficiency and so would not fit current parameters, according to R. Duperval et al., "one may conclude that probably 30 cases a year (until 1963) could be defined as AIDS" under current definitions, although the cases long predated the existence of the immune deficiency syndrome.[13]

As noted, most devastating for the hypothesis that now links the occurrence of this disseminated cryptococcosis with AIDS is the fact that "reports agree that prior to the recognition of AIDS, cryptococcosis increased at least tenfold during the '60s and '70s."[14] Here is a disease that was on the rise, with or without its connection to AIDS. The implications of this increase will be investigated below.

Disseminated CMV infections is the next health problem to consider. The CDC states that CMV is diagnostic of AIDS if it is found in an organ other than the liver, spleen, or lymph nodes.[15] However, cases fitting this criterion were found long before AIDS, the first documented one in 1925. "Between 1925 and 1962, 40 more cases of CMV were reported in adults of which 15 fit the surveillance definition of AIDS."[16]

Root-Bernstein generalizes about all the disease occurrences mentioned to say they are not anomalies but fit the same mold. Not only the ones mentioned, but other AIDS-related illnesses—including PCP, chronic and oesophageal candidiasis, disseminated tuberculosis and atypical mycobacterial infections—"show the same pattern of AIDS-like cases reported for as long as the disease has been described in the medical literature with increasing numbers of such cases during the two decades prior to 1980."[17]

Putting this together, we might say that if we were to look for the reason for the increase of specific, once rare diseases, two hypotheses quickly come to mind: (1) a new retrovirus is weakening the immune system so that the body is more welcoming to these diseases, and (2) for other reasons, there has been an increase of the illness independent of any new disease. Although arguments can be made for either position, what we saw about cryptococcosis lends weight to the second hypothesis.

Root-Bernstein, among others, favors the idea that AIDS is simply a misnomer for a number of diseases that are on the rise. He also suggests that the increase in these so-called opportunistic infections can have other explanations besides the hypothesis I offered, such as "that diagnostic abilities have continuously increased," that is, new techniques make it easier to identify diseases that once might have been missed or miscategorized.[18]

So far, when we have talked about the opportunistic infections that are said to accompany AIDS as diseases in their own right, ones that have been known to cause a deterioration of the immune system and thus would seem to fall under some CDC definitions of AIDS, we have not considered one of the later qualifications of the AIDS definition, which is that the sufferer must be infected with HIV.

It's true that the patients mentioned who had CMV infection and other illnesses before AIDS appeared did not have HIV infection, but this is not to say such infection did not also predate the immunodeficiency syndrome. "Increasing numbers of reports maintain that both HIV-1 and HIV-2 were present in Western Europe and the Americas in the 1950s and '60s," according to Root-Bernstein.[19] And, if one is willing to accept this fact, then the whole question of the origin of the disease as it is now posed becomes moot, since it is not a new disease.

What is new is only its movement from a rare to a frequent occurrence. Prof. Root-Bernstein makes one hypothesis, "The transmission of endemic HIV, non-HIV immunosuppressive agents, or some combination of these with HIV, increased dramatically in the past two decades." It may be possible that this increase, he argues, is to be causally related to another change in health parameters that took place in the same time period: "huge increases in the incidence of multiple immunosuppressive risk factors associated with groups at high risk for AIDS."[20]

How one settles this question—that of whether AIDS is a recent intrusion into the human sphere or a collection of diseases that has been around a long time but only now is spreading—will determine how one studies the disease in every facet. Perhaps it would be appropriate to recall Dubos's predictions, made in the 1960s, for the course of medicine in the West. This writer, not only a renowned microbiologist but respected on all sides as a philosophical thinker on questions of health, believed that better sanitation, nutrition, and vaccines had all but eliminated the danger of old-style epidemics in which a person came down with the disease that he or she caught from outside. This situation, nonetheless, does not represent much of an advance in health, since "the morbidity rate of infections have not increased significantly, and in some cases have actually increased."[21] This is due to the greater prevalence of a new type of disease.

> The microbial diseases most common in our communities today arise from the activities of microorganisms [and viruses] that are ubiquitous in the environment, persist in the body without causing any obvious harm under ordinary circumstances, and exert pathological effects only when the infected person is under conditions of physiological stress. [22]

The idea that AIDS is simply a set of long resident but only now threatening diseases, brought to life by the increased stressors in the postmodern environment, sits well with Dubos's consideration of what types of diseases are most prominent in the West.

How to Accurately Assess Immunodeficiency

Another point the establishment might have in its favor in its argument that AIDS is a new disease would be any new symptom that was occasioned by the condition. That is, if it could be shown that AIDS caused a symptom that was previously unknown, this would militate toward adoption of the thesis that AIDS is novel, rather than being simply the coming into prominence of a number of diseases that have had a long history but had previously not been widespread.

Although this is not precisely the case with the fall in levels of CD4 in the blood, which is taken as a leading indicator of AIDS progression, it is only in relation to this syndrome that the reading of these levels has been considered significant. In the mind of the general public, the sudden concern with CD4 has helped legitimize the idea that AIDS is new. This, in turn, would seem to demonstrate that AIDS is the product of a new virus.

A statement from the *Journal of the American Medical Association* indicates how central the establishment takes CD4 count. "The scientific literature still recognizes that the CD4 cell count is the best single predictor of disease progression to AIDS." Although, the statement continues, there are a number of other indicators that should be taken into account: "however, CD4 cell count will still determine when to begin primary prophylaxis for AIDS-associated opportunistic infections and can be used widely be clinicians in practice."[1]

We need to see that it may not be AIDS but the ability to count CD4 cells that is new in this situation, and the very fact that this scientific skill at totaling up types of blood cells is a recent development makes its use questionable in that there has not yet been time to determine what healthy levels of CD4 are.

But before taking on these topics, we should clarify exactly what CD4 count is. Technically speaking, "CD4" refers to an antigen (cell marker) that is found on a number of different blood cells; but in taking blood counts, CD4 cells are taken to be helper T cells. ("T" refers to the fact that these blood cells are manufactured in the thymus.) These cells act to defend the body against foreign intruders. Once a helper T has been "informed" by a macrophage (another blood cell) that an intruder is on the grounds, the helper begins releasing chemicals that call into play other blood cells, both those that kill viruses (and other invaders) and those that store a memory of the intruder, so that if further down the line the same intruder reappears, "the immune system can respond quicker and more effectively to the reappearance of the antigen [on the unwanted cell]."[2] Once the obnoxious items are removed, it is the helper T cells' responsibility to shut down the killer and memory cells. Thus the helper cell is the central regulator in marshaling antibodies to repel attack.

It would seem that if an assailant wanted to disable the body's immune system, striking at the helper T cells would be a good place to start, since without them, even if an invader were identified, the repelling forces would not be called out, nor, if they had been activated by other means, would they be able to stop even when their job was finished. Indeed, according to orthodoxy, eliminating T helper cells is HIV's number-one priority, so judging the T helper cell count, that is, the CD4 count, can give an accurate estimate of what stage the disease has reached.

Scientific technology has only recently identified, differentiated, and quantified T and B cells and assigned them the function of immunity. Since the ability to tag and count these cells has only been available a short time, and because enough resources have not been devoted to studying the general relationship between the helper T cells and the human body in all states, sick and healthy, I was surprised to learn, after an extensive review of all the literature on T cells and immune deficiency, that there are few T cell studies in healthy individuals. Worse, none of these studies establish a workable benchmark; there is no scientific standard for "healthy" T cell counts and subset ratios. To put this another way, the orthodox are telling people with AIDS that their CD4 counts are dangerously low, while it has not yet been figured out what a healthy person's CD4 count generally is.

What little evidence there is on the CD4 counts of HIV-negative persons has offered little comfort to the orthodox. Dr. James Mosley of the Transfusion Safety Study Group, for example, presented a paper at the International Conference on AIDS in 1993 on just this subject. He tested over two thousand blood samples that were collected in 1985 to track HIV infections. Where AIDS orthodoxy has it that a cell count down to two hundred is a signal for alarm, since a person whose count has fallen so low is considered to be very sick, Mosley found many healthy people with two-hundred counts, and found, on top of that, that a three-hundred count could be considered the low end of a normal reading. Mosley told *Spin*, "No one… looked at uninfected persons in the lower range of normal. We observed these over a period of up to six years and they had no health problems that could be related to an immune deficiency. Everyone assumes that a low CD4 count necessarily means immune deficiency. It does not."[3]

Furthermore, while Mosley's findings indicate a low count can be found in the healthy, it has also been noted that CD4 count is highly variable. For one thing, the amount of these helper cells in the blood can change over twenty-four hours. "One estimate has it that a CD4 count can fluctuate as much as 35 to 74 percent in a single day."[4] Other factors that can cause a variation are age (children have much higher counts), sex (women have higher counts than men), and smoking (smokers have higher counts).[5]

Emotions also have an impact on CD4 levels. In fact, they might be called instances of companionship in Bacon's terminology, since here two phenomena, emotion and helper cell count, seem to run in tandem. According to Bacon, these instances occur when there is "a body or concrete substance in which the nature inquired into constantly attends."[6] He gives as an example flame and heat, which are always conjoined. For us, a less remarked example of companionship is that of a lowering T cell count and a situation of emotional stress. In reviewing the literature on the topic, Michael Ellner and Tom DiFerdinando argue that declining T cell counts and T cell ratios "reflect… the physiological profile of someone who is severely stressed."[7] They link stress to "in vivo hydrocortisone studies from the late 70's, cortisol being a major stress-related hormone," showing that release of this hormone lowers CD4 counts.[8]

Although blamed on HIV, it has been demonstrated by Dr. Fauci, no less, that cortisol in the bloodstream drives CD4 cells into tissue and inverts the CD4/CD8 ratio. Thus, the chronic feelings of helplessness and hopelessness associated with HIV and T cell testing can produce the identical blood picture as that claimed to be the hallmark of "HIV disease!" What this blood picture can therefore reflect is the physiological profile of someone who is severely stressed.

Other factors that one has to be aware of in trying to assess the accuracy of a blood count include discrepancies in laboratory counts and the skewed results that may be produced by coinfection.

First off, the laboratories that count blood are far from precise. One study given at the International Conference on AIDS in 1993 compared CD4 counts from the same blood, different laboratories. "Nine out of 24 patients studied had such different counts from the various labs that the results would lead to different treatment recommendations. Three out of those nine met the revised CDC/AIDS definition according to some lab results, but did not have AIDS according to others."[9]

A second difficulty in interpreting results stems from the fact that if AIDS does lower CD4 counts, there are still some conditions that may accompany

AIDS and, acting in a contrary way, will boost the count. A study in the *Journal of the American Medical Association* reported on Brazilian HIV-positive people with CD4 counts over one thousand who died of AIDS. "They were all co-infected with another virus which causes a proliferation of CD4 cells."[10] For them, counting blood cells told little of the AIDS story.

However, theirs is not an exceptional case. With all these factors other than manifestations of the disease that can cause a difference in CD4 count, it is no surprise that the correlation between T cell count and AIDS is not as self evident as it has often been made out to be. Thus we find that there are many cases where a low helper T cell number has not translated into disease, conversely, some of those with a high count have become deathly ill.

According to a spokesperson for HEAL, an organization dedicated to working with people with AIDS, they have seen, if anything, conditions associated with T cell enumeration opposite of what the establishment predicts. "We have met 1000's of otherwise healthy people, both HIV+ and HIV-, with extremely low CD4 counts who are not developing disease long associated with immunodeficiency." What is most unexpected is that "many of the people coming to HEAL meetings felt their best with low T-cell counts and many have told us they got sick and felt terrible and worse, when their counts were in the 'healthy' ranges." He draws the obvious conclusion that "whatever T-cell counts mean, they don't mean anything in relation to the progression of illness."[11]

Coming from less anecdotal evidence, an article in *Science* arrived at the same conclusions. "CD4 counts are far from being ironclad predictors of progression to AIDS," the essay notes. "HIV-infected people with very low CD4 counts sometimes remain healthy; conversely, ailing people sometimes have comparatively high levels of CD4s."[12]

In reaching this opinion, the author of the *Science* article was probably drawing on earlier studies, such as Flemming's of 1994, and 1996 reviews of AIDS trials. In looking at sixteen major trials of drugs to combat AIDS, Flemming did find, in line with orthodoxy's predictions, that CD4 counts were favorable in cases where people with AIDS were responding well to treatment. On the other hand, out of line with orthodoxy, "the CD4 cell count was [also] significantly favorable in 6 of the 8 trials in which treatment did not improve progression." For those who survived through the course of the trials, CD4 count was favorable in only two out of four cases, while "even worse, [CD4 count] was significantly favorable in 6 of 7 trials in which treatment had no effect on survival. Three additional trials, including the Concorde Trial showed an inverse relation between survival and improved CD4 cell counts."[13]

Some of the researchers who hold that CD4 counts are irrelevant to understanding the progression of AIDS do not believe HIV plays a central role in causing the disease; but there is also a dispute among those who believe that HIV's weakening of the immune system is a key factor in breaking down the body. These biologists, in articles such as those by Gianneti et al., Langhoff and Haseltine, and others, have developed the thesis that helper cells are not the immediate target of HIV and so studies of their amounts in the blood are unilluminating. As mentioned a moment ago, T helper cells spring into action when the invading virus or other intruder has first been identified. This identification is carried out by another immune system cell, the macrophage. The biological heretics see HIV as going after the macrophages, not helper cells. "AIDS is not caused by depletion of CD4 cells," they argue. "It is caused by HIV infection, dysfunction, and the ultimate destruction of the cells that present [that is, identify] the intracellular microbes (virus, bacteria, fungi, and protozoa)…. When infected, these antigen-presenting cells can no longer appropriately activate CD4 and CD8."[14] Thus, from this perspective, any counts of CD4s are rather oblique indicators of the destruction carried out by HIV on other immune system entities.

Even Dr. Gallo, a kingpin of the establishment view, does not see CD4 count as a particularly appropriate way to measure AIDS, having said, "[The faith in] CD4 as a sole surrogate marker is dying."[15]

We've just brought out that there is a difference of opinion in orthodoxy over the mechanism of HIV's attack on the body, with some saying the elimination of CD4 cells is not germane to an understanding of the retrovirus's action. However, such controversies over the suitability of this criterion were consciously ignored by the FDA when, in 1992, it set up guidelines for the approval of AIDS drugs based largely on their improvement of T cell counts. These guidelines made "it easier to approve AIDS drugs rapidly based on only data from surrogate markers. The marker it has relied on most heavily to date is the one that's received the most scientific attention: the number of CD4 cells."[16] At the time of the regulation's adoption, Commissioner David Kessler acknowledged, "We're certainly pushing the limits of the envelope here…. We're approving treatments on less data than ever before."[17]

Although there is certainly a need to streamline the drug approval process, relying on an upping of the helper cell count as a measure of a drug's viability can lead to mistaken findings. For instance, an interesting Italian study into the effects of AZT (reported in *AIDS*, October 1996) looked at eighteen people who began taking AZT under the mistaken impression that they were HIV positive. On taking the drug, "the patients experienced short-term increases of up to 30% in the counts of their circulating T-cells, consistent with the short-term increases seen in diagnosed people."[18] Since it turned out that those taking the drug actually were not infected with HIV, their raised immune-cell levels could not be due to any counteraction of the drug against the virus, but would seem to have been caused by the body's reaction to the upsetting drug. In consequence, if this drug were approved according to its effect on CD4 count, it would be obtaining a false bill of effectivity, given that it will raise cell count irregardless of the presence of HIV.

The topic of how anti-AIDS drugs are approved will come into its own when we turn to the subject of accepted treatments, but for now it is enough to note the interrelatedness of all components of the AIDS discussion. How drugs are approved depends on how they affect the disease, but understanding this implies making judgments on how AIDS impacts on the immune system, since we must know this to know if a drug is counteracting its effects. To know this one has to have an idea of the illness's cause or causes. If, for example, the disease is brought on by a virus, this will demand that scientists look one place. If, however, it is brought on by immune collapse due to drug abuse, then scientists must look somewhere else to see how it is registering its effects. Of course, each of these factors can be read the other way. Thus, for example, noticing a certain number of abnormal bodily conditions coincident with the onset of the disease will suggest that the cause is of a certain type. As we will see, it is this type of consideration that has led many to drop the retroviral cause theory of AIDS.

AIDS as Stress Response

O rthodoxy, following an exhausted mind-set, believes AIDS the product of a single entity. As the National Institute of Allergy and Infectious Disease has it, "AIDS is caused by a virus called HIV. HIV attacks and weakens the immune system, causing vulnerability to certain types of cancer and to various opportunistic infections."[1]

Would it be stretching an analogy to call this diagnosis a counterpart of the individualism that is so prevalent in other realms of U.S. scholarship and popular thought? According to this broad conception, it is the creative activity of the unfettered individual that is responsible for both success in business (and life) and improvement in the citizenry's general well-being by way of inventions and innovations. The fact that the climate of the United States supposedly fosters this individualism is held to account for the country's distinction and post-World War II preeminence.

Whatever truth these arguments may have (and in our earlier discussion of the distortions behind a Trump or a Gates myth, we saw how dubious some of this theory of individualism's claims were), such arguments are tailor-made to exonerate the system from charges of inequality and to protect privilege. In the 1990s, spouting doctrines of individualism, legislators rolled back such measures as affirmative action, which help minorities as a group—because it is claimed, people should be only rewarded for individual effort. Although certain minorities may be at the bottom of the heap at present, their rise up the economic ladder, it is claimed, should take place through the efforts of exceptional members of the minority, not by attempts at blanket improvements in a group's life chances. Left out of the discussion, naturally, is that wide adoption of this stance should ensure those now in the elite that they are not threatened by too many demanding a share of their pie.

In parallel, the single cause theory of disease may serve in another way to shore up the system against charges of inequity by holding to an individualist position. We have seen already that Dubos and others have proven that the gaining control of the great diseases of the nineteenth century, such as TB, smallpox, typhoid, and dysentery, had nothing to do with advances in medicine, which the control preceded, and everything to do with changes in public health practices. As Dubos sees it, harsh living makes people susceptible to microbial attack. "Famine, crowding, wars, and social upheavals can increase the prevalence of microbial disease through… creating conditions which decrease general resistance."[2] He takes the eradication of TB as an example to prove that changes in these conditions led to alteration of disease patterns.

> The tuberculosis epidemic which prevailed throughout the industrialized countries of the Western world during the nineteenth century owed part at least of its severity to the long working hours, the poor nutrition, and the low living standards prevailing among the laboring classes…. As living standards improved, tuberculosis mortality immediately began to decrease.[3]

Contrary to physicians' mythology, it was not isolating and axing nefarious microbes and viruses that took the upper hand from the most dreaded diseases of the nineteenth century, but the results of reformers' valiant and successful crusades (in which they were aided by members of the working class) to do away with slum overcrowding, better sanitation, shorten work hours, and *make other ameliorations that would render the conditions of life (not individual lives per se) more healthy.*

So why were Pasteur, Koch, and other doctors able to appropriate so much glory to themselves, who even now the popular imagination sees as behind the cure of these diseases? Howard Waitzkin argues that the rise of germ theory can be tied to the eclipse of political reform. "After the revolutionary struggles of the late 1840s suffered defeat, Western European governments heightened their conservative and often repressive social policies."[4] Social reform was driven out of fashion. "With the onset of political reaction, Virchow [a pioneer of social medicine] and his colleagues turned to relatively uncontroversial research in *laboratories*."[5] When, later in the century, Koch and others, making real scientific gains, shifted attention from living conditions to microbial agents as the causes behind illness, this was a boon to reactionaries insofar as it meant (if this way of looking at things gained ascendancy) that doctors would stop poking their noses into the way things such as the disadvantaged's substandard housing, food, and circumstances could be blamed for illness. Thus, it was to the ultimate benefit of the ruling orders that a belief in one microbe/one disease would become accepted. The encomiums to Pasteur, echoed, we've seen, in other terms by postmodern "radicals" such as Latour, were not only due to an appreciation of his scientific achievements.

The parallel we can draw to the way the establishment seems dead set on running with a monofactorial causal theory of AIDS is not exact. People getting AIDS in the United States are not half starved mill workers putting in twelve-hour days. However, the dissident theories of the disease's cause that we are about to investigate share with the ideas of the nineteenth-century reformers the concept that *illness arises from a matrix of living conditions*. Some of these circumstances people with AIDS are personally responsible for (such as ingestion of recreational drugs), but others can be laid at the door of the organization of work (which produces exceptional stress in so many jobs), advertising (which encourages the consumption of unnourishing foods), and medicine itself (which blithely overprescribes antibiotics while discounting their negative side effects on the immune system). Saying that such factors—that is, *an ensemble of them working together*—are to be blamed for AIDS makes people uncomfortable, to both those who don't want to change their lifestyle and those who work in support of institutions that will not abandon unhealthy strategies of profit taking or power increasing.

This standpoint unsettles individualism from two ends. On the one hand, it demands that individuals take command of their own health, a deepening of personal responsibility beyond what is expected by the norm, where health is left to the medical practitioner. But, on the other, it calls for attention to collective structures, such as the arrangements of medicine, whose reform must be on the agenda if citizens expect an upgrading of their mental and physical well-being.

However, meditating more deeply, we can see that in a sense, these poles are not as separated as they appear, since even the increases in individual responsibility turn upon transpersonal considerations. You'll remember we quoted Mark Conlan to the effect that the adoption in the 1970s of a fast-track lifestyle by homosexual males, with promiscuity and drug ingestion as prominent factors, was the result of a flawed appreciation of gay liberation. This cohort of fast-trackers saw their behavior as a way to act upon the gains in gay pride manifested in a social movement symbolically started at Stonewall. I call this "flawed" in that, as Shilts and others have shown, gays had still only scratched the surface in their fight against prejudice and discrimination, so the pretense that one could now as a gay male become hedonistic and ignore the political struggle for acceptance and equality was misguided.

Let me caution that this crude formulation, which may seem to indulge in stereotyping, will be given a slightly different and more compassionate interpretation when we take up the discussion, armed with further knowledge, in the evaluation of Schmidt's theories of hysteria and AIDS later in this section.

However, the main proposition is not that this or that behavioral choice was wrong or right, but that individual choice is dependent on assessment of the larger framework of one's reference group. To repeat the extremely significant observation made earlier: A change to safer sexual and drug-using practices by gay males only came about when these changes were linked by gay activists to progress for the homosexual community. Important alterations of behavior were rooted in each person's appreciation of what this change would mean for the larger group.

To return to the health issue: If we wanted to put the difference between the orthodox and dissident approach to causation in a word, we could contrast "virus" and "stress." Where a virus is an isolatable, recognizable single-minded culprit, "stress" refers to fixed pressure on the physical system that comes as the product of multiple impingements.

As noted at the end of the previous section, we can look at the symptoms of AIDS to get some idea of what is causing it. An article by A. Hassig et al. notes, "It has long been established that the chemistry of the body of most people with AIDS… does not look like any typical response to a single viral agent, i.e., HIV. In terms of chemical interactions, AIDS is far closer to resembling a 'stress' response."[6]

Hassig and others have noted that the existence of stress is signaled by an increase in chemical "messengers" called cytokines, who act to guide the immune system.[7] Cytokines include interleukin-1, interleukin-6, and TNF-alpha. When these appear in overabundance, disturbed health results. Too much interleukin-1 is documented as causing fevers; while excess TNF-alpha causes loss of appetite and wasting.[8,9] Though there is no consensus on how too many cytokines affect the body, some have said that their appearance leads to an increase of macrophages, which, not being called up to deal with actual bodily invaders, rattle the normal functioning of the immune system. "This overproduction of cytokines can cause a persistent suppression of immunity that impairs the body's immune response."[10]

In the bulk of this section we will examine the four factors that are the major causes of oxidative stress and overproduction of inflammatory cytokines, all of which, in differing combinations and to varying degrees, are found to occur in people with AIDS and many of those considered to be living at risk.[11] These are:

1. Chemical toxicity/drugs: both recreational and orthodox drug use as well as environmental sources of toxicity.[12]
2. Psychological stresses: for example, fear, anxiety, and sustained-stress.[13]
3. Microbial activity: multiple and concurrent infections, including viral, bacterial, fungal, and protozoal.[14]
4. Nutritional deficiencies: a diet deficient in antioxidants coupled with other nutritional deficiencies.[15]

The Role of Illicit Drugs

Johnny hands Sick Boy his [Johnny's] works.

—Ye git a shot [of heroin], but only if ye use this gear. Wir playin trust games the day, he smiled, but he wisnae jokin.

Sick Boy shakes his heid.—Ah dinnae share needles or syringes. Ah've goat ma ain [own] works here.

—Now that's no very social.... Whit d'ye think ay that? Ur you tryin tae insinuate that the White Swan [that is, the speaker]... has blood infected by the human immuno-deficiency virus? Ma finer feelin's ur hurt.

—Irvine Welsh

Trainspotting

Our thesis is that a combination of stresses weaken the immune system as a prelude to AIDS. This opinion arises from an understanding of disease as the result of an interaction of internal and external pressures. Those without this understanding, which we saw is basic to a new paradigm, will sometimes take up, only to dismiss, one or another of the cofactors we are going to discuss.

An example of this would be the discussion of poppers by health authorities first concerned with explaining AIDS. When AIDS was first reported, as Shilts documents, various scientists wondered if the sickness could be caused by use of amyl nitrite (contained in "poppers"). Some went so far as to hazard one contaminated batch was at the bottom of the whole outbreak.[1] This formulation was abandoned when it was shown that men who had never sniffed poppers had gotten AIDS. Such an abrupt reversal reveals the limitations of the monocausal view. Since these scientists believed in one illness/one provoker of illness, a single case of AIDS absent poppers was enough to make them drop amyl nitrite as a possible suspect. According to the new paradigm, though, disease is caused by a set of factors, none of which is essential, but many of which have to be present in tandem for the immune system to weaken enough for the patient to be liable to microbial or viral inroads. From this perspective, the question is not whether amyl nitrite by itself can give birth to AIDS, but whether it is one of the contributors.

As we will see when we look at how early AIDS sufferers were questioned about their practices, the questioners' persistent ignoring of the possibility of multifactorial causation has hampered both investigation of the etiology and treatment of the immunodeficiency syndrome. It should cause us to take gingerly claims by the establishment that high-risk behavior, rather than HIV alone, has been "repeatedly contradicted." The CDC bases such an allegation on "compelling evidence… from cohort studies of high risk groups in which all individuals with AIDS-related conditions are HIV-antibody positive, while matched, HIV-antibody negative controls do not develop AIDS or immuno-suppression, despite engaging in high risk behaviors."[2] Although these cohort studies may possibly provide evidence for HIV as sole cause of the disease, the fact that, as will be presented below, orthodox studies of AIDS have repeatedly ignored certain key elements of risky behavior to focus on a circumscribed set of them—looking, for example, at use of poppers but not of equally toxic drugs—means that all such claims should be taken under advisement.

So let's start this part of our study by (1) looking at poppers. Then we will (2) notice the role of other drugs, both licit and illicit. Next we'll (3) get back to symptoms, finding out what effects various drugs can have on human architecture. We'll also (4) look also at the demographics of the AIDS crisis, noting at this juncture both how the inception of AIDS coincides with an epidemic in drug use and how early demographics used to determine the types of people who came down with the disease were skewed insofar as they ignored certain important risk behaviors. Finally, we will (5) examine how treatment would change if the cofactorial causative role of drugs was taken seriously, ending with (6) a brief note on the United States's ambiguous attitude toward addiction.

Amyl Nitrite and Other "Poppers"

Poppers were originally developed for emergency use when a person suffered heart pain (agina pectoris). As John Lauritsen reports, "The original poppers were little glass ampules enclosed in mesh, which were 'popped' under the nose and inhaled. Manufactured by Burroughs-Wellcome, of AZT notoriety, they contained pharmaceutical amyl nitrite."[3] Until 1960, amyl nitrite could be sold only by prescription, but in that year this rule was dropped by the FDA. Once it was legalized, "a few gay men, primarily those into S&M, began using amyl nitrite as a 'recreational' drug."[4] In 1969, the FDA put it back on the restricted list, but it could be easily obtained through underground channels, since its distribution was so profitable.

According to Neville Hodgkinson,

> In 1978, in just one city, an estimated total of $50M a year was being made from sales of more than 100,000 bottles a week.... By 1980, sales had reached hundreds of millions of dollars and the National Institute on Drug Abuse reported that more than 5 million people were using poppers more than once a week. Gay men [and no longer just those involved in sadomaschocism] made up virtually the entire market, and many gay organisations depended on a slice of the industry's profits for their existence. Gay publishers carried large ads for the products.[5]

Hodgkinson's assertion that gay males made up the majority of the consumers of this drug is widely supported. "Some US studies have reported that up to 95% of gay men use poppers either frequently or occasionally." Moreover, those who take the drug seem more likely to come down with AIDS. In the UK, Project Sigma did a statistical survey that found 72 percent of HIV positive gay men "used poppers either frequently or occasionally in the previous year," this in comparison to 42 percent who were HIV negative (or unaware of their status).[6]

Studying the historical trends in drug use, Newell, et al., (1985) came to the conclusion that nitrite, while unheard of previous to 1960, "became nearly synonymous with gay life in various cities by the mid-1970s, approaching an estimated rate of 80% incidence in the male homosexual community as a whole."[7]

Amyl nitrite (and the associated butyl nitrite) became the drug of choice since "as vasodilators and muscle relaxants in order to facilitate anal intercourse and to increase sexual response."[8]

Despite the flip-flopping of the FDA over whether amyl nitrite should be widely available, it is known to be extremely toxic. Aside from causing immediate violent reactions on inhalation (including sudden death), the repeated use of the drug can lead to immune suppression and various cancers, including Kaposi's sarcoma.

A summation of the evidence in *Continuum* notes that breathing in the drug "induces profound though transient vasodilation, with blurred vision, headache, flushing, a 'warm feeling,' and a widespread throbbing sensation."[9] Following this, one experiences reflex vasodilation and a sped-up heartbeat. Taking of the drug may be accompanied by nausea, burning in the nose, fainting, eye irritation, cough, bronchitis, coma, and respiratory distress.

Descending to the cellular level, we find that the drug interferes with the red blood cells' ability to carry oxygen, which "without antidotal treatment...

may be fatal."[10] The morbidity of this condition depends on such factors as the frequency of drug use, a person's genetically given ability to handle oxygen depletion and a person's past history of heart disease or anemia, which would compound the problem.

Even if the results are not so drastic, oxidative stress, where the body's ability to uptake oxygen is lessened, has been implicated in causing cancer. "Several scientists point to the oxidative... effect [and cancer inducing stimulus] of volatile nitrites on the tissue of blood vessel walls, one of the primary sites of the effect of these drugs."[11] Kaposi's syndrome is a cancer of these same blood vessels.

The drug's cancer-causing abilities do not end there.

> Amyl, butyl and isobutyl nitrites are extensively documented as being both mutagenic, and carcinogenic. This means that poppers have the ability to mutate the genes of existing cells which can serve as faulty templates for future cell lines and that poppers can combine with other substances to form cancer causing compounds. Nitrites convert most antibiotics into carcinogens in both animal and test-tube studies.[12]

Along with stressing red blood cells, poppers have a baleful effect on the immune system. It has been found that poppers "particularly deplete natural killer cells."[13] In probing their effect on experimental animals, it was found that dosed animals had the "abnormally low helper T-Suppressor T ratio of lymphocytes typical of AIDS patients."[14]

It will be remembered that low T cell count has been taken by the establishment as a signboard proclaiming the active presence of HIV. Yet this and other animal studies see a like T cell drop brought on by the ingestion of poppers. To drive this point home, we might mention another animal study that showed "a depletion in T4 helper cells leading to a reversed T-cell ratio."[15] Add to that two more experiments that indicate "the development of T-lymphocytes and antibody responsiveness decrease about 90% in mice that have been exposed to nitrites at similar levels to human inhalation doses."[16] Finally, one investigation showed popper inhalation by animals caused damage to the thymus gland, which is where the T cells are manufactured in the first place.[17]

It's worth noting that the destructive effects of amyl nitrite on the body are not something recently cooked up by dissidents who were rushing to uncover factors other than HIV that may be blamed for the breakdown of the immune system in AIDS. In 1980, before AIDS, Thomas Haley, one of America's leading toxicologists, published a review of the literature on the hazardousness of amyl nitrite. He noted the immediate effects of inhalation that we have already come across as well as pointing out, "Fatalities have occurred in workers exposed to organic nitrites after strenuous exercise 1 to 2 days after cessation of exposure."[18] This came about because of the overrelaxing of "the vascular bed and pooling and trapping of blood in the veins of the lower extremities," which led to constricted arteries and death.[19]

Other studies published in the same year showed the cancer-causing properties of the drug. Subsequent ones, still in the prehistory of AIDS, indicated the drug's tampering with the body's immune function.[20]

On the seamier side, "Poppers have been used successfully to commit suicide (by drinking) and murder. (The victim was gagged with a sock soaked with poppers)."[21]

Let's say a little more about popper use and Kaposi's sarcoma (KS). It has already been indicated that amyl nitrite (a known carcinogen) causes stress to the walls of blood vessels, where Kaposi's sarcoma manifests itself. Aside from this link on the physical level, researchers point to the correlation between popper users and AIDS patients who have KS.

Lauritsen notes, "In AIDS cases, KS is found almost entirely among gay men who used poppers.... A six-fold decrease in the incidence of KS over the past five years parallels a sharp decline in the use of poppers."[22]

There is also a temporal coincidence between the appearance of poppers on the scene and then the coming of KS-associated AIDS. Poppers were one of the few new factors that altered in the gay male lifestyle in the late 1970s. Eleopulos sees this as connected to the later onset of AIDS. She argues that by "the early 1980s nitrite use became ubiquitous in California and New York," the locales where most AIDS-linked KS sufferers would be found. Further, the latency period for those whose KS is connected to organ transplantation "is similar to that between homosexual exposure to nitrite and the appearance of the disease."[23]

This final point about latency was echoed in a historical review by Drs. Guy Newell and Margaret Spitz of the Department of Cancer Prevention and Control, University of Texas, Houston. They found the use of poppers as recreational drugs " first became widespread between 1970 and 1974, preceding the arrival of AIDS epidemic by about seven to ten years, which was consistent with a carcinogenic effect."[24]

Despite what was known about the dangerousness of amyl nitrite, whose immune-suppressing effects were evident, and despite the fact that it was common paraphernalia of fast-track gay life, its possible role in causing AIDS was quickly dismissed by the CDC and other health-watch groups. This was because of the reason mentioned at the head of this chapter, by themselves poppers didn't cause AIDS, as proven by those who had gotten AIDS but had never resorted to that drug—and because "the CDC conducted a brief mice study in 1982-1983, and claimed to find 'no evidence of immunotoxicity'" from poppers. The CDC stopped its investigation there, although, as Lauritsen notes, "These [positive] results are contradicted by several other studies, which *did* find that the inhalation of nitrite fumes causes immune suppression in mice."[25]

Moreover, not only did the agency call off its own watchdogs, who were probing a possible AIDS/poppers tie-in, but it began trumpeting its flimsily supported view. In 1983 it published the pamphlet "What Gay and Bisexual Men Should Know about AIDS," where the lack of connection between poppers and the disease was underlined. Although the CDC was obviously making this statement because of scientific concerns (and to calm public fears), its message was certainly welcome to drug manufacturers. "The CDC's mice study was cited in a press release sent out by Joseph F. Miller, President of Great Lakes Products, the world's largest manufacturer of poppers."[26] According to Miller, in a story run by most of the gay press, "Dr. Jim Curran of CDC's AIDS Branch had given him a guided tour of the CDC and assured him there was no relationship between poppers and AIDS."[27] A later denial by Dr. Curran of these precise implications—he said that poppers shouldn't be given a blanket pardon, since they might play a role in generating the opportunistic infections that gain ground when AIDS has weakened the immune system—was overlooked by the same papers that had bannered Miller's self-promo.

Perhaps some of the same pressure from advertisers appeared here as, Shilts documented, came from gay bathhouses. In gay papers, where these clubs were frequent advertisers, there was editorial reluctance to chide them for encouraging unhealthy sexual practices because of the possibility of losing their lucrative ad copy. One may guess that papers taking advertising dollars from poppers' pharmaceutical source were in no hurry to dig up the unflattering history of animal experiments that did see immune damage stemming from use of the drug. In any case, as Lauritsen reports, "Great Lakes Products followed through with a series of ads in the *Advocate*, entitled "Blueprint for Health," which gave the impression that poppers, like vitamins, fresh air, exercise and sunshine, were an ingredient in the healthy lifestyle."[28]

As noted, the CDC was not 100 percent behind poppers, as Dr. Curran's reluctance to endorse them wholeheartedly shows. In 1995, when I interviewed Tom Bethel, he said that the CDC by this time had quietly backtracked and was putting together a task force to investigate the relationship between Kaposi's sarcoma and amyl nitrite intake. But, Bethel emphasized, "Hardly a word of this has come out in print, needless to say."[29] By this time, of course, the use of poppers had already been abandoned by most in the gay community, who had become savvier at getting information and had learned not to rely on effusions from drug companies.

I might add that the CDC is not the only government agency with a checkered history on this matter. I noted the flip-flopping of the FDA on whether to restrict amyl nitrite to prescription use. What I didn't mention was the brain-dead excuse the agency used for the period when it left the drug unregulated. As Lauritsen puts it in eminently sarcastic style:

> The FDA refused to recognize the fact that poppers are drugs… claiming that poppers are "room odorizers," since they are labelled as such. The FDA has traditionally been concerned with labelling, and would certainly take action if snake oil were labelled as an "AIDS remedy," or if cocaine were labelled as a "nasal decongestant." Why should they accept the cynically ridiculous claim that poppers are room odorizers?[30]

Thus we see that when a scientific possibility—such as the involvement of poppers in AIDS—runs against the dogmatically held monocausal paradigm, a single mouse study is considered sufficient to shoot it down. We saw also how eager the pharmaceutical industry is to get word of such a study, which it can then wave like an all-clear flag after an air raid, telling everyone they have nothing to fear.

Other Illicit Drugs

It would be wearying to go into as much detail about every drug that may be a cofactor in producing AIDS as we have about poppers, so in the following discussion I will just mention a few studies associating three other groups of drugs or chemicals with immune breakdown and AIDS.

The first group, the one most to be associated with poppers would be other recreational drugs. A number of studies have sustained my belief that there is more statistical correlation between drugs and AIDS than between HIV and the same illness. As a HEAL publication testifies, "Over 80% of all AIDS cases in the West have a history of recreational drug use, yet only approximately 50% of all AIDS cases are confirmed HIV antibody positive by the CDC."[31]

With evidence as strong as this, one would think it would be up to those who felt drugs were *not* a cofactor to have to prove their position, rather than, as the case stands, the other way around.

Heroin is notable among this group of recreational drugs, since it has long been known that inveterate users are apt to come down with pneumonia, a disease now commonly associated with AIDS. Dr. Polly Thomas of the New York City Health Department has noted the similarities between the types of mortality induced by different afflictions, saying, in effect, a heroin addict who dies with pneumonia and HIV would be considered an AIDS victim; while the same addict without HIV would be said to have died from the drugs tout court. "And yet there would be no difference in the clinical profiles: of the 'AIDS-pneumonia' case or the 'drugs-pneumonia' case."[32]

Having mentioned that both illicit IV drugs and poppers wreak havoc on the immune system, it's time to score another point off the establishment view. AIDS in the West has been largely linked to two groups: gay males and IV drug users. Wouldn't it seem to be an elementary step in studying the disease to find out everything these two groups have in common? Orthodoxy went as far as saying sharing needles and anal sex both meant the partners in each practice could be infected with each other's blood. But is it sane to stop there and disregard another, even more telling parallel, which is that both groups used immuno-destructive drugs? The fact that a very large percentage of infected gay males were heavy into poppers is often left unmentioned, when it should be central to our discussion. That IV drug users and fast-track gay males were both overindulging in recreational drugs seems such an obvious way to tie them together that the myopia of the establishment in not following up on this beggars belief.

Those studies that have inquired into AIDS patients' use of cocaine, poppers, and speed "find that 93% to 100% admit using these drugs."[33] Peter Duesberg surveyed the research into this area and tallied the results, as given in the following table, which indicates what percentage of the patients queried in the given study admitted to drug use.

FIRST AUTHOR	PUBLICATION	DATE	PERCENT DRUG USERS
Marmor	*Lancet*	May 15, 1982	100%
Jaffe	*An. Int. Med.*	August 1983	96%
Havarkos	*STD*	Oct/Dec 1995	97%
Kaslow	*JAMA*	261(23), 1989	96%
Archibald	*Epidem*	3(203), 1992	100%
Duesberg	*Genetica*	February 1995	93%

In another examination on this topic—few have been done, regardless of the importance of the subject—"a controlled study among 300 HIV-positive intravenous drug abusers," it was found that "the risk of developing AIDS over sixteen months was three times higher in those who persisted than in those who stopped injecting drugs."[34]

These studies indicate that when drug use is examined in relation to the immunodeficiency syndrome, the correlation between the two tells an illuminating tale.

We will have to look further at this, which I will do momentarily, restricting myself to one point at present. We have talked of how Bacon saw prerogative instances as crucial to scientific endeavor because they draw attention to areas from which much could be learned. One such area is certainly this confluence of drug taking between two disparate groups who are falling prey to the same disease. In Bacon's terms, this would be a bordering instance. "They are those which exhibit species of bodies that seem to be composed of two species or to be rudiments between one species and another."[35] If, to attend to the latter specification, we think of gay males and IV drug users as two "species," then use of toxic drugs would be the "rudiment" that these seemingly disparate series share. Concentration on this bordering instance will be "of excellent use in indicating the composition and structure of things."[36]

Let's bring in another type of drugs. It has also been theorized that gays in the fast lane were weakening their immune abilities not only by ingesting hazardous chemicals but by "overdosing," as it were, on medicine. That sector of gay males who frequented bathhouses and other venues that provided an arena for promiscuous encounters were, as we've noted, highly likely to be infected with STDs. To fight off the effects of varied venereal diseases, many would begin swallowing penicillin pills like they were jelly beans. Byran Ellison says,

"Tetracycline was used [by this group] to prevent even the slightest risk of infection from a venereal disease."[37] Taking these pills several times a week for years "leads to the kind of immune overload, the kind of breakdown, that we're seeing in AIDS patients," Ellison concludes.[38]

As with poppers, it is not a recent discovery that antibiotics can wound the immune system. Although people imagine that these drugs merely weed out bad bacteria while leaving the rest of the body untouched, they, in fact, normally also kill off good bacteria, such as those that reside in the human gut and are necessary for food processing. J. Valnet adds, "Antibiotics can have detrimental side effects on the blood from slight anaemia or a simple fall in the number of white blood cells, to the total disappearance of the white cells."[39] Moreover, extended use can result in diarrhea with impaired food absorption and weight loss as well as lessened resistance to infection, "persistent fevers, skin rashes, weakness and prostration and diseases that affect the nervous system, all of which are considered to be primary symptoms of AIDS."[40]

A more direct example of the implication of medicines in the creation of AIDS-related diseases comes from the experience of kidney-transplant patients in the 1970s. They were given "large doses of immuno-suppressive drugs. . . in order to protect the newly transplanted kidney from destruction by the patient's own immune system," A. Cantwell reports. "In transplant cases, the incidence of KS skyrocketed to 400-500 times greater than normal. When the drugs were discontinued, the KS tumors improved or disappeared."[41]

Lastly, a few words about environmental toxins. Just as people in the early nineteenth century began to find that much of their ill health stemmed from imperfections in their living conditions such as polluted water, unsanitary waste disposal, and lack of ventilation in places of work; so I believe we will find in the early twenty-first century that many of our health problems are influenced by the highly toxic environment in which we dwell. Not only do we breathe in smog and drink water cut with traces of heavy metals or other toxins, but our foods are laced with pesticides. Overall the impact of all these adds to the strain on the detoxifying centers of the body and helps contribute to immune dysfunctions. Some dissidents argue that these environmental pollutants play a role in making the body susceptible to AIDS. As Richard Strohman puts it, "The presence of a viral genome [i.e., HIV] in many cells of the human body, by no stretch of the imagination" is enough, working alone, to cause the AIDS syndrome. "You will still need to have, for the most part, a compelling environment for that genome to be expressed."[42] He goes on to say that an appropriate disease-preparing environment is one that includes "the presence of carcinogens, and other polluting agents that infest the industrialized ecosystem."[43]

Having noted some specific effects of these drugs, let's turn to their general abilities to bring on immune system deterioration, which, we will see, are similar to those the orthodox adduce as due to HIV.

Deleterious Effects of Drugs

Any strong drug will have a particular effect on an individual. We know, for example, the immune-suppressant drugs given to those undergoing kidney transplants can, combined with other factors, push a body toward Kaposi's sarcoma. However, another thing to keep in mind is that the taking of any powerful drug (and that includes all those we have seen associated with AIDS) beyond its specific effects, causes generalized stress to the immune system, whose worst consequences may not manifest till years after the drug has been encountered. Some of these general stresses may be connected to the onset of AIDS, although their implications have not been fully researched.

Paul Philpott, who spoke so eloquently on the isolation problem, points out that the immune system is made up of two components, the specific one (which labels and disposes of intruding agents, and is run out of the thymus, hence includes T cells) and the nonspecific immune system, which more generally keeps the interior milieu healthy. Philpott finds it highly significant that in rodent studies, mice who are missing working thymus glands certainly get sick but they do not come down with the type of illnesses that are connected to AIDS. "The twenty-nine or so AIDS diseases fall into the… categories [of]: cancer, wasting, dementia, and opportunistic infections. These nude mice don't develop any of these things. And they have no functioning *specific* immune system."[44] He concludes that those focusing on HIV's hindrance of the specific immune system are looking in the wrong place. For the type of diseases associated with AIDS, it is destruction of the nonspecific immune system that is indicated. If establishment researchers looked there, they might notice that "drugs destroy not just specific immunity but… the nonspecific immune function which is responsible for controlling cancer and controlling opportunistic infection."[45] He wonders why more research has not been done by immunology into the intricacies of this system and how it might be battered by diseases.

Another symptom in the AIDS picture that has been given much less attention than T cell elimination is atrophy of liver cells, which is commonly caused by mycotoxins. These toxins can be produced by injecting or inhaling certain narcotics. Geisler points out, "Aflatoxin [one type of mycotoxin] was found in the urine of 32% of heroin users in Amsterdam, of 13% in Merseyide (England)."[46] It is also found "in users of amyl nitrite (aka, poppers), including AIDS patients. In one study, people with a history of popper intake exhibited an alteration in T cells such as is typically brought on by aflatoxin poisoning."[47] As we have seen, poppers were connected to many early AIDS cases and so their influence and the culpability of mycotoxins also cry out for further study.

Although one can die of, say, a heroin overdose or one "pipe" of crack, it is not news that the usual disease route for those taking such serious drugs is a gradual sapping of immune system strength, so a longtime user eventually passes a critical point and his or her immune system is no longer up to waging the con-

stant low-level conflict against infectious agents that the functioning immune system can take care of handily. Thus, in common parlance, the diseases ascribed to drug use have a more or less long latency period. If we ponder this and connect AIDS to a period of drug abuse, we can see that it gives a us a new way to interpret the latency period tied to the immune deficiency disease. This period of quiescence between infection and expression may not be because HIV spends a period vegetating in the cells, but because HIV, along with other infections, is omnipresent in human subjects but only proliferates when the host body has been weakened by lengthy periods of stress. Duesberg puts the thought this way: "The very existence of a 'latent period' strongly suggests that years of health abuse are required for such fatal conditions." He goes on, "Among most AIDS patients in the United States and Europe, one extremely common health risk has been identified: the long-term use of hard drugs."[48]

This leads us to the cumulative burden thesis. Those that hold with this etiology believe HIV does have a part to play in causing AIDS, but see the virus as a submerged factor that does not come to prominence unless the immune system is enfeebled. We've noted this type of explanation before, but let's hear what a few researchers have to say on how this pertains to drug-taking's bearing on immune duress. Hodginkson theorizes that the typical road from infection to outbreak may be that a person is "infected previously, and the virus [is then] brought out of latency by some subsequent immune challenge." What brings on the disease, according to this postulation, is not infection but "the cumulative burdens on the immune system that permit activation of the HIV genes."[49] One of these burdens is prolonged drug abuse, which gradually adds up to an immune system stretched too thin.

When I interviewed Andrew Shoft, he hammered away on this note, saying:

> I do not believe HIV can cause AIDS in itself. It is simply not that potent a virus. I believe that the real killer… [is] an unhealthy lifestyle through perhaps drugs on the streets, drugs through the hospitals, anything that is immunosuppressant—those are the cofactors that are the cause of AIDS. I mean HIV if it's carried around on the bullet of a gun, and you get shot—you could say HIV killed you, but I think it was the bullet. HIV was there, but I don't believe it alone can kill you.[50]

As we know, a key advocate of the cumulative burden theory is Peter Duesberg. I don't want to repeat any more of his comments but simply record how his reception has changed as more and more weight is accorded to the idea that drugs have an influence on the etiology of AIDS or AIDS-related diseases. On May 23 and 24, 1994, the National Institute on Drug Abuse (NIDA) sponsored a meeting, filled with establishment figures, to consider the relationship of poppers to KS. The consensus by this time was that this connection must be studied, for it probably offered part of an explanation for the occurrence of the disease. Robert Gallo, doyen of the establishment scientists, who attended, was

moved to amend his earlier belief that HIV is poweful enough to be the sole instigator of AIDS. As Lauritsen paraphrased, Dr. Gallo said, "It is now necessary to consider co-factors. No longer is HIV believed to cause KS by itself; at most it may aggravate KS after it has been caused by something else."[51] However, for those who believe actions speak louder than words, perhaps Dr. Gallo's body language was as significant as his opinion. "Dr. Robert Gallo started to leave, got to the door, hesitated, then came back and sat next to Duesberg. Within a few minutes he had his arm around him."[52] It's hard not to take this as an implicit recognition of the resilience of the idea that drugs play a cofactorial role in AIDS, an idea dismissed before the evidence in its favor had been fully investigated.

Demographics of AIDS Cases, with or without Drug Use

So far I've concentrated on the common nature of immune destruction that can be alternately attributed to AIDS or toxic drugs, in the course of which, especially in relation to amyl nitrite, I threw in a few statistics about the confluence patterns between people with AIDS and drug takers. It's time to scrutinize the statistics behind these patterns more carefully.

One of the most impressive changes in the last ten years concerning the populations who have been coming down with AIDS may be culled from the discussion thus far. I mentioned that gay males have drastically reduced, to the point of near total abandonment, the use of poppers. It might have been conjectured that the percentage of gay males among the stricken would consequently drop. This is what has happened. This group has contracted as contributors to the statistics while the IV drug users group has expanded.

R. M. Selik, writing in 1990, remarks, "Currently, 32% of all US AIDS cases and 33% of all European AIDS cases are accounted for by intravenous (IV) drug users." He goes on, "Those who use drugs intravenously also have a greater tendency to develop tuberculosis than the other so-called 'risk groups.' Similarly, crack (cocaine) users have a greater tendency to develop PCP."[53]

By 1995, as Gina Kolata reported in the *New York Times*, this trend was even more evident.

> According to CDC statistics, people who are now diagnosed with the [AIDS] syndrome differ from the original population in most ways, but one. They are drug users.
>
> Of those newly diagnosed with AIDS in 1993, who were probably mostly infected in the early to mid-1980s, about half were gay men and a little more than a quarter were intravenous drug users. Fewer than ten percent were heterosexuals.... Now, in his analysis of national data for new HIV infections in 1994, [CDC epidemiologist], Dr. Homberg finds

a very different pattern, which is continuing this year. Only a quarter of the most recent infections are in gay men. About half of the new infections are among drug users who shared needles, and about a quarter are heterosexually transmitted.[54]

For more evidence along these lines, I refer you to the 1991 review article, "Aids, Drugs of Abuse and the Immune System: A Complex Immunotoxicological Network," which highlights the neat fit between populations affected by AIDS and those known to indulge in illicit pharmaceuticals. The article concludes, "Nearly three quarters of all American AIDS patients were men aged twenty to forty-four, and NIDA surveys showed that men in that same age group accounted for a similar proportion of 'hard' drug-taking."[55] The argument thus far is that the alteration in the populations affected by the disease is bringing out more clearly that AIDS is more drug-using-related than anything else. We might further the case by noting that an explosion of drug use in the United States can be connected, with the necessary lead time, with the appearance of AIDS. Having established that, we can demonstrate the above-analyzed link between drug taking and AIDS-getting with more specificity

Tim Haring made the point to me that, at the time of the earliest AIDS infection, drug use was peaking. He put it like this, "I would say that we've got an epidemic of drug use in this country. That has been documented by some of the agencies looking at drug use."[56]

As I researched further, I saw that Haring's term "epidemic" was not hyperbole. It was rooted in such facts as these:

HEROIN

Heroin-related arrests tripled during the 1970s, which matched the number of those who died from heroin overdoses. From 1976 to 1985, addicts entering hospitals doubled and then doubled again. In 1985, 580 IV drug users died in hospitals, a death rate that increased to 2,483 by 1990. From 1992 to 1993, heroin-related hospital emergencies increased 44 percent, according to the Department of Health and Human Services.

COCAINE

"More than five million Americans had tried the drug by 1974, but eleven years later this figure had jumped to twenty-two million. Currently [1996], about eight million Americans are regular users of cocaine." By 1985 crack, the smokable form of the drug, was rampant among poor young adults of minority groups. Duesberg noted in 1996, "A recent national household survey of drug use found that one million Americans between eighteen and twenty-five years of age had used crack during the previous year." The DEA impounded 500 kg of coke in 1980, "9,000 in 1983, 80,000 in 1989 and 100,000 in 1990—a total increase of 20,000 percent in one decade." In that time, the number of those who went into hospitals after cocaine overdose also rocketed upward from slightly more than 3,000 in 1981 to over 120,000 in 1993—a 4,000 percent increase. Cocaine-related deaths increased tenfold from 1980 to 1990.[57]

AMPHETAMINES

In 1981, the police confiscated 2 million doses of amphetamines; but by 1989 they netted 97 million doses.

AMYL NITRITES (POPPERS)

The National Institute on Drug Abuse estimated that by 1980 5 million Americans were inhaling the drug at least once a week. "By 1978 the once-tiny poppers industry was already grossing $50 million in annual profits, a figure that continues to climb."[58]

ANTIBIOTICS

Journalist Bob Lederer notes, "Studies have shown… from 1960 to 1980, the period when AIDS first appeared, there was a massive rise, a tripling, a quadrupling, in the prevalence of sexually transmitted diseases in the gay male community." While these diseases are no threat to the immune system, Lederer stresses that when these STDs were discovered, a second immune weakener also came into play. "Antibiotics [were] often dumped into people's bodies to treat them [the STDs] indiscriminately, [which] can cause the kind of immune compromise we see in AIDS."[59]

Peter Duesberg returns us to the subject of the concurrence between PWAs (persons with AIDS) and those who indulged in hard drugs by connecting the dots linking the two epidemiological trends. "Naturally, one might expect major health problems in the wake of this drug explosion," he writes. "If the timing of the AIDS epidemic—following on the heels of the drug epidemic—was no coincidence, then one should also find the spread of AIDS following the spread of drug use."[60]

Let's follow his argument closely here, since both his concision and acuteness are unsurpassed on this issue.

> Not only did the drug-use epidemic take off shortly before AIDS appeared, but it hit hardest among precisely the same risk groups. The parallels are astounding. Both AIDS and drug use, for example, are concentrated in younger men. Between 1983 and 1987 the death rate among American men ages twenty-five to forty-four increased by about ten thousand deaths per year, the same as the average number of AIDS deaths per year in that time. But also during the 1980s death from drug overdoses doubled in men of exactly the same ages, while deaths from blood poisoning—an indirect consequence of injecting drugs—quadrupled.
>
> Ninety percent of all AIDS cases occur in men. But nine of every ten people arrested for possession of hard drugs are also male. Even the age distributions coincide perfectly. Men between the ages of twenty and

forty-four make up 72 percent of AIDS cases, just as they make up 75 percent of people arrested or treated for use of hard drugs.[61]

He goes on to subdivide the AIDS risk groups, using as his baseline the 130,000 cases as of late 1993 found in IV drug users, this being one-third of the total. Obviously, it undercounts the total implication of drugs in the disease since poppers or even crack will not be counted in this accounting, since they are not injected.

> Consider how that number breaks down. This figure includes three quarters of all heterosexual AIDS cases and more than two-thirds of all female AIDS cases. More than two thirds of all babies with AIDS are born to mothers who inject drugs. Even 10 percent of the hemophiliac AIDS cases inject drugs.[62]

Thomas sheds more light on Duesberg's note about babies by noting the CDC reports between 1982 and 1991, where there were 1,328 AIDS-sickened babies recorded in the United States. Of these, 1,260 (95 percent) came from "mothers at risk," that is, largely, drug-abusing ones. Thomas adds, "Babies born to drug abusing mothers who do not have HIV also die from the same diseases [as HIV infected babies], but they are not considered AIDS victims."[63] Here again, as in the similarity we noted earlier between heroin addicts with AIDS and pneumonia and those with only pneumonia, we see that the drugs alone can harry the immune system as completely as does the AIDS disease.

It might seem astounding that this close correspondence between drug taking and the disease has been overlooked by the establishment, even early on in the epidemic before HIV became the cause of choice. Of course, some may say it's possible to mitigate the blindness of the first students of the problem insofar as early on gay males were so prominently affected. But to make this allowance is to allow these researchers off the hook for irrationally ignoring the fact that the gay males touched by AIDS were perennially part of the drug culture. It is to believe in what Michael Verney Elliot labeled as the myth of the "one-screw wonder."

When I interviewed him, he punctured that myth. "We've all heard the story of what I call the one-screw wonders. The gay man who came down from the farm in Nebraska and bright city lights, got laid once, and then went back home and died of AIDS." This has become part of urban folklore, but, as he emphasized, "You just can't find them. The ones who became ill invariably had a fast track lifestyle and usually did drugs."[64]

When I pressed him on the drug-AIDS match, he elaborated:

"You see drugs are now such an inherent part of the gay culture that nobody thinks there's anything wrong with them." This is not to say, drugs are something inherent in all gay lifestyles, but to encourage us to further consider the idea raised earlier by Conlan, who tied promiscuity into a post-Stonewall dubiously grounded belief—held by those in the fast lane—that a militant stance toward homosexual rights also entailed weekends at the baths. In Elliot's eyes, there was also a sense that gay liberation included the right to take whatever drugs made you feel good. Fast-trackers "didn't want to know about the fact that drugs might contribute to their immune suppression because that was considered an infringement of gay civil rights."[65]

He emphasizes that this cultivation of a drug lifestyle was not only confined to those in the fast lane, but historically sited.

> Well you see I came up in the mid-fifties and there were no drugs then. We didn't need an anesthetic to do what we did. We did it because we liked it. Nowadays it seems you can't go out and go to a disco unless you're primed on and psyched out on drugs. If you meet somebody you have to have drugs to cope with that encounter, and then if you go back somewhere with them you have to have drugs to cope with whatever they're going to do. So that the whole gay culture, or what they call the gay culture today, is permeated with drugs."[66]

Allowing for some exaggeration, since there has been a notable new restraint in the community on drug use, which is no longer an inevitable component of even a faster lifestyle, we can say that Elliot has accurately depicted the scene in the 1970s and 1980s.

According to Duesberg, the flourishing of a new homosexual partying lifestyle that developed in the 1970s, when triumphs in civil rights allowed relatively greater freedom of expression, was dolefully accompanied by heavy drug use. Among this subset, Duesberg argues, there was the equation of multiple sex partners + STDs + antibiotics + recreational drugs. "These men accumulated hundreds or even thousands of sexual contacts within just a few years." Naturally, then, venereal diseases and other infections "spread like wildfire," while "heavy doses of antibiotics were taken by many each night before sex, just to prevent unsightly sores or acne."[67] The linchpin holding this all together is, as Elliot also said, was a consciously calibrated use of drugs.

> Such extreme sexual activity cannot be done on a cup of coffee alone or even on natural testosterone. The fast-track lifestyle required liberal drug use—stimulants to get going, poppers to allow anal intercourse, downers to unwind afterward. Several drugs, combined with alcohol and marijuana, became par for the course of an evening, a routine that would go on for years.[68]

Lauritsen lets us see some of the finer details of the lifestyle.

> Gay bathhouses and clubs became drug dispensaries. Some of the gay discotheques had designer drugs manufactured especially for them. There were rituals where thousands of men in a gay discotheque would take a half a dozen different "recreational" drugs in the course of an evening. At one time everyone would take "Special K," then later in

the evening they would all take MDA or Ecstasy, and so on. All night long some of them would snort poppers which they didn't even consider to be a drug.[69]

This would be a good place to put in a number of caveats so that we don't follow the media and tar everyone with the same bush. We need to keep in mind Lauritsen's strong words, "Gay men as a whole are not at risk for AIDS. It is only a small and very particular subset of gay men that are at risk," namely, those who played with recreational drugs.[70]

I should also comment that if fast-track gays' drug abuse takes the central role in weakening their immune systems, displacing anal sex as a key factor, then those who want to rant about how the immorality of gay sexuality was a prelude to the disease outbreak will have lost their thunder.

Also, importantly, we should not let our argument about the unhealthy aspects of the fast-track lifestyle, which we have linked to beliefs about gay liberation, be connected to the typically heard neoconservative laments about the excesses of the 1960s and 1970s, which are seen as devolving from political radicalism. Such complaints do not apply here, since those gay males who were following traditional progressive thought in this same period were aiming to build on the gains of Stonewall. And they were doing so by community building, trying to breach the political system, working to educate the public about gay issues, and raising consciousness—all activities, by the way, as we will later detail, that developed in the fight against AIDS. Those in the fast track who connected the pursuit of pleasure with gay rights were making a wayward interpretation of what liberation was all about, one that deflected activism into a line that was in keeping with the consumerist ethos of U.S. commercial society rather than one aligned with a liberal agenda. In other words—to expand on this slightly—according to the hypothesis that drug use was key to immune-system dampening for the fast-trackers, it was not self-conscious espousal of a gay agenda or gay male sexual practices that was causing the problem (both things which the establishment finds distasteful) but rather the one component of fast-trackers' mores that aped the drug-reliant majoritarian society's way of life—though the favored pharmaceuticals on the main stem were different, ranging from diet pills to tranquilizers to alcohol.

To return to the argument: We started this excursion into a discussion of the fast track by saying that the establishment's overlooking the role of drugs in preparing the body for AIDS seemed inexcusable. Let's see why this obvious etiological factor was disregarded not with a view toward measuring guilt, but only so we can better see how research went astray.

Duesberg quickly laid his finger on two substantial problems in the CDC's methods of preparing data on those who have AIDS. His first point was that their "statistics incorporate only self reported drug injection, for they cannot confirm such illegal habits in people who will not admit to them."[71] Epidemiologist Tim Haring backs him on this by emphasizing that early analysts of

AIDS patients used a "flawed" fact-gathering technique. They would simply ask PWAs, "Do you use drugs?" Haring deduces, "When you go out and ask someone whether or not they're doing an illegal act, they're going to be very reluctant to tell you if they are and how much of it they're doing."[72]

The second, more startling problem that Duesberg enumerates is this: "The CDC does not ask AIDS patients about nonintravenous drug use."[73] But this means snorted cocaine and poppers, the two most popular drugs among the fast-track smart set, are not even being tallied! No wonder the connection went unnoticed.

In his book on AIDS, Snead went into more detail, underlining that the CDC in 1985 was classifying people with AIDS who "had used needles for self injection of drugs not prescribed by a physician" as almost the only drug users counted in its classification. Researchers Krieger and Caceres, after studying the CDC, reported, "Only heterosexual drug users wound up in that [CDC] classification, and that the use of substances such as barbiturates and amphetamines, hallucinogens such as LSD and PCP, cocaine, nitrites and marijuana were only studied in limited, special groups," not including gay males. They contrasted this perspective to findings such as those of Dr. Haverkos, who had gauged in his study of eighty-seven homosexual or bisexual AIDS patients "that 97%... had used recreational drugs at least once, 75% made weekly uses of such substances."[74] Yet all of these patients, if tabulated by the CDC, would have been considered non-drug-users.

Breaking down Haverkos's study, which was done between September 1981 and October 1982, we find that 97 percent of the men had used poppers, "93 percent marijuana, 68 percent amphetamines, 66 percent cocaine, 65 percent LSD, 59 percent quaaludes, 48 percent ethyl chloride, and 32 percent barbiturates."[75] This carefully done study was executed under the CDC's auspices, but "the CDC neither published the study nor cleared it for publication elsewhere, though Haverkos released copies privately."[76]

Having in mind the discrepancy between those who would make it into the CDC's list of people with AIDS who were drug users and those who actually used drugs, in 1996 Hodgkinson argued that the agency was even grossly mistaken about the people it thought it was counting as drug users. He said the CDC "under-represented by more than half the role of intravenous drug use."[77] This came about from their taking the 17 percent of AIDS patients who they identified as IV drug abusers only from among heterosexuals, and ignoring any homosexuals who may also have shot drugs. "Subsequent information suggested that about 25 percent of the gay men with AIDS at the time were also IV drug abusers," he notes.[78] This means a further 19 percent should have been put into the sum of those who were IV drug users in the total. And, note, Hodgkinson is only trying to work back into the statistics uncounted IV drug users, which leaves to one side those who did not inject their favored substances.

In an article quoted by Hodgkinson, Lauritsen pointed out that there was a hidden rationale in not looking into this latter group, that is, those who snorted or swallowed their drugs. This rationale was derived from an implicitly believed theory of transmission. The agency ignored any but IV-administered drugs because

> from the beginning, the hypothesis had been that the only way drugs caused AIDS was when needles were shared, thereby transmitting an AIDS-causing microbe from one person to another. The way the CDC was collecting and presenting its epidemiological evidence was already artificially excluding hypotheses of AIDS causation in which drug abuse played a central role.[79]

Let's talk about needle sharing for a moment. As in other central givens of the orthodox view, such as the equation HIV = AIDS, a tremendous weight is placed on a flimsy bough, as if an osprey had landed on a sapling. For, astounding as it may seem, the foundation stone of this argument on transmission has never been studied. It has never been shown that IV-drug-user PWAs have been passing needles.

Hodgkinson, continuing his discussion of Lauritsen's work, observes, "there has never even been a study to determine whether all, or even most, intravenous drug abusers with 'AIDS' ever did share needles, although such research would be simple and inexpensive."[80] Lauritsen reports that though "public health officials believed that such research existed... none has even been able to provide a reference."[81] In his own conversations with IV users with AIDS, he found that when he asked them about their needle use, the majority responded, "Share needles? Are you crazy? What would I do that for?"[82] (Our epigraph from a novel about Scottish IV drug culture echoed this point.)

What research we do have on needle sharing seems to prove the opposite of what the orthodox contend. Geisler cites a number of studies that show, for example, as did a 1985 examination, "HIV-infected persons did not share syringes as much as the non-infected."[83] Work done on IV users in New York, published in 1987, tracking people between 1979 and 1984 found that "only 27.3% of those who had always used syringes which were also used by other people, were HIV-infected. On the other hand, 39.1% of iv. d. u. persons who had never shared needles, had HIV-antibodies."[84]

If we look abroad, we might note that in Italy, where 80 percent of AIDS cases are of IV drug users, "needle-sharing was almost unknown as anyone could walk into a drug store and buy one."[85]

Meanwhile, in Canada, it has been found that those in needle exchange programs do worse than those who stay out of them. A Canadian study of 1,599 intravenous drug users, carried out between 1988 and 1995, "is showing that IV drug users participating in Needle Exchange Programs (NEPs) in Montreal have higher seroconversion rates, and an increased HIV risk, compared to IV drug users who do not participate in these programs."[86] Vancouver, which has the largest NEP in North America, has also seen a continuing rise in AIDS among those participating in the program.[87]

This Canadian information segues into the last topic that has to be considered in this discussion of drugs. We want to pry into the shortfalls of treatments that must arise insofar as they are based on an insufficient appreciation of the role of hard drugs (not just injected ones) in weakening the immune system, regardless of any blood exchange encountered in using them. This will be contrasted with the positive results of programs that have acknowledged these factors affecting the disease, as well as with programs and research that should be developed to further elucidate and alleviate the results of this connection.

Treatments, Valid and Invalid

As the Canadian example has suggested, if it is not the few drops of foreign blood that may enter an addict sharing a needle, but rather the payload of drugs in the syringe that is causing the immune dysfunction preparatory to AIDS disease, then needle exchange programs (where addicts turn in used for new needles with which to inject themselves) are implicitly condoning the very behavior, taking the drugs, that is most likely to lead to ill health.

Hodgkinson makes this point: The wrong messages are being sent, not only to IV drug users but to gay males. "Drug users have not been told to quit using drugs, only that they must stop 'sharing needles.' Gay men have been told that they must restrict their sexual activities, but not… to stop using cocaine, heroin, quaaludes… poppers," and all the other items from the illicit pharmacology.[88]

Duesberg and Rasnick go so far as to hold that by passing over the possibility that drug use may add to the burden on the immune system, the establishment ends up encouraging drug use, since it seems to give the green light to many hazardous activities. "By ignoring, obscuring and even directly refuting… the possibility that nitrites, cocaine and heroin could cause diseases, medical orthodoxy misinforms a vulnerable and trusting public about the consequences of recreational drug use."[89]

After all, every rule of conduct not only proscribes one type of behavior but at the same time tacitly encourages behavior that is not under ban. If the Catholic church, for example, says sex for pleasure is sinful, then, in this particular formulation, without stating it directly, it is stating that sex for procreation is legitimate. Analogically, if the authorities say that to avoid danger you must use condoms and clean needles, then they are also saying, in Duesberg and Rasnick's words, "Clean needles for unsterile street drugs and condoms protect against all medical consequences of drug use."[90]

Let me add at this juncture—lest it seem I'm supporting the idea (which the previous quotations imply) that there is always a seamless one-way, sender-receiver relationship between a culpable medical establishment and a gullible public—that not all segments of the public are taken in by orthodox propaganda. For instance, as Harlon Dalton records, although he doesn't approve of it, when in 1988 the New York City government put in place a needle exchange program, the African-American community was outraged. The New York City Council, led by the Black and Hispanic Caucus, voted to immediately terminate the program. As Enoch Williams, caucus chair, stated, "The city is sending the wrong message when it distributes free needles to drug addicts while we are trying to convince our children to say no to drugs."[91] Community leaders, such as the Reverend Reginald Williams of the Addicts Rehabilitation Center in East Harlem, also came out against the program. The reverend spoke out, "There will never be a needle-exchange program here. I think the communities and neighborhoods would rise up in opposition."[92]

So, we need to recognize that ill-founded directives proposed by the establishment are not necessarily heeded by their intended audience. This, though, doesn't indicate that any proposal for change will be rejected, but that it must be conceived of and promulgated after and through dialogue with the affected community. As we saw earlier in a crucial discussion, it was not because of externally introduced guides for behavior extended to the gay community from "on high" that motivated its members to drop popper use (to a degree) and introduce safer sexual behavior, rather, it was a largely internally generated change in the male gay community's self-image that brought it to take up these changes as a way to protect and foster that community. The conclusion to draw is not that external messages sent from authorities are never of value, but that they have to be coordinated with the ideas of the group addressed. Dalton concludes on this issue, "To reach the gay community, public health officials had to learn to view the epidemic from the perspective of the gay community."[93] He goes on, "Consequently, today even the most control-minded public health officials take care to involve the gay community in decision-making."[94]

A worthy public-health program that works with a target community on the drug/AIDS issue would steer people away from drugs, recognizing, as studies show, "that regardless of HIV-positivity, the risk of developing AIDS could be drastically lowered by health-promoting changes in behaviour."[95] By cutting out the drugs, whether taken by injection or through other channels, former users find they "so strengthened the immune system that AIDS was either prevented or its onset considerably postponed compared with those who kept exposing their bodies to the same risks."[96]

Such a program would make the deleterious contribution of hard drugs to AIDS further known to the wider public through such campaigns as the one staged to educate about the cancer risks of smoking. Rasnick and Duesberg note that the war against cigarettes presents a good model in that it has racked up a notable success. As they highlight, "Based on education that smoking causes lung cancer, emphysema and heart disease, smoking has dropped in the U.S. from 42% of the adult population in 1965 to 25% in 1995.[97] They feel that similarly mounted and prosecuted publicity about the ties between drugs and immune disease could reverse trends in AIDS just as such campaigns have decreased smoking.

Many feel these same guidelines, in publicity aimed at the general public, can also work on a smaller scale to improve the health of AIDS patients who had been drug users. A sterling example of how a behavioral change, interconnected with other health-restoring actions, can put such PWAs on the road to enhanced vitality, is shown by the program run by Dr. Maurizio Luca Moreti of the Florida-based Inter-American Medical and Health Association. He and his staff work with former IV drug users. Not only did they keep them "clean" but got them on healthy diets and made other reforms in their lifestyle. The result: "Among 139 individuals who had been using heroin daily for an average of more than five years, all were still free of AIDS symptoms after an average of more than four years since they first tested positive."[98] Hodgkinson compares this to general U.S. statistics that show that "32 percent of HIV-positive addicts develop AIDS within two years and more than 50 percent within four years."[99]

To repeat, a different assessment of the factors that make for AIDS—in which frequent use of drugs, even antibiotics, is seen as so harrowing to the immune system that it induces it to fall prey to life-threatening infections—leads us to put more trust in alternative preventive regimes that do not rely so completely on discouraging the exchange of blood. Since the deleterious effects of various drugs on the immune system are known but have not been systematically studied in relation to AIDS, Duesberg and Rasnick suggest that for AIDS research, "Drug toxicity could be tested experimentally in animals, and in human cells in tissue culture. In addition, drug toxicity could be tested epidemiologically in humans who are addicted to recreational drugs or are prescribed AZT."[100]

They further suggest, in an article in *Continuum*, that proceeding from what we already know about strong drugs' ability to do damage—and among strong drugs they include some used in treatments—a campaign to stamp out AIDS would work if

1. AZT and all other [toxic] anti-HIV drugs were banned.
2. Illicit recreational drug use was terminated.
3. AIDS patients were treated for their specific diseases with proved medications, e.g., tuberculosis with antibiotics, Kaposi's sarcoma with conventional cancer therapy, and weight loss with good nutrition.[101]

Although I would moderate some of Rasnick and Duesberg's optimism over the values of publicity in curtailing drug use, and the ability of some of the "conventional" therapies to handle disease, the program outlined does offer the strongest options available for reducing the AIDS menace on this front.

Addictive Behavior

By way of creating a bridge to the next chapter, in which I will bring up the provocative theories of Casper Schmidt—who holds that AIDS can only be completely understood by integrating into one's view the part played by psychic elements in bringing about disease—let's pause briefly to think about why the possibility that hard drugs plays a part in causing AIDS is a note not sounded by the establishment. In the case of antibiotics, it stands to reason that the medical regime that relies on them as a cure-all is not willing to impugn their integrity. That is understandable. But what about illicit IV drugs, whose existence and pernicious effects are widely lamented in other contexts? Why is it that on the AIDS issue, such drugs are presented as benign, at least in comparison to the HIV that may have been left in the syringe by a previous user?

We've already seen the obvious answer. It's important for orthodoxy to boost the idea that HIV has the central causative place in AIDS etiology; hence any other factor, no matter how unhealthy in itself, must be downplayed. However, let's look also at a less overt reason, the United States's ambiguous flirtation with addictive substances.

As my collaborator argues, TV commercials often present addicted behavior as acceptable. We've all seen the TV ad in which a guy has inadvertently run out of his favorite junk food—let's say his last bag of Fatty Puffs—so he runs through his house in a frenzy, ransacking drawers and closets, trying to track down one remaining bowlful. The behavior is imprinted on viewers as comically benign. (Disturbingly, such a plot is even drummed into the heads of children, as in the "Trix Are for Kids" ad, where a cuddly rabbit is always trying to cadge some of this breakfast cereal.)

Such ads are naked emperors. In other words, they unconsciously reveal aspects of the culture that the producers of the ads would prefer to conceal.

Two connotations of such ads are most striking. For one, if a shortage of the junk food, as advertised, gives rise to addictive withdrawal symptoms, this certainly indicates that the food is consumed compulsively; and is taken like any addictive substance: not because it is nutritious or even because it tastes good, but because it produces a temporary mood upgrade due to the food's high sugar or salt content.

Secondly, inasmuch as the behavior of the character in the commercial, albeit exaggerated, is modeled as acceptable and viable within our culture, one must conclude that even if individual addictive substances are condemned, addiction itself is within our norms, as far as they are reflected by the media. We can conjecture that this tolerance is due to the media's business-friendly attitudes. After all, many products from which big money is made, from junk food to nuclear power, are counterindicated for the nation's or individual's health, yet they are allowed by regulators because their health effects are not immediately obvious while the profits they bring to their purveyors are. Many such products, such as cigarettes or alcohol, are known to be addictive; yet, though regulated, they are still given wide license to be sold. This accounts for the media's schizophrenic attitude toward addictive substances, frowning on some, smiling on others. No matter how dysfunctional for a person, addiction cannot be delegitimized by the media because it is functional for the economy.

Psychological Assessment: Prof. Casper G. Schmidt

O ur discussion of the paradoxes of national consciousness ties in with the more daring proposals about the zeitgeist advanced by Casper Schmidt, concerning the interaction of the collective mental states of minorities and the majority in a given society and how they may issue in disease states.

Some of what Schmidt talks about is hardly debatable. This would include his examination of scapegoating that has been so visible during the crisis, with gays or drug users said by some to be dying as punishment for their sins. Another example, which was highlighted by an inflammatory *New York Times* editorial, is the condemnation of African-Americans as irrationalists because they seemed reluctant to embrace AIDS orthodoxy. However, some of his more heterodox points (which I will investigate shortly), as where he contends that certain disenfranchised groups are (to a degree) psychologically making themselves sick, will certainly raise a few eyebrows. You may question such a contention, saying, "Why would a person want to get sick? This sounds like another case of blaming the victim. Not only does it appear Schmidt believes PWAs have engaged in risky behavior, but he suggests they have a subconscious desire to end up in the hospital. This," you might continue, "is crazy. Gary, now you've really gone off the deep end. No one wants to die of AIDS."

Acknowledging I may be getting a little ahead of myself—in that I am making remarks about Schmidt's ideas before I have outlined his thought in any depth— I think it would be sensible to preface my discussion of this thought with a couple of preliminary considerations that will help you approach him in a receptive manner. Let me start by saying that I give his ideas such prominence in my book not because I endorse his conclusions lock, stock, and barrel, but because Schmidt's underlying way of conceiving of illness includes two valuable biases, which are centerpieces of the new vision of healing I am trying to create.

The first of these is his understanding that "psychosomatic" factors have to be treated seriously, as effective in promoting disease or health. I use the word "psychosomatic" advisedly because I want to reject its older meaning, whereby it indicated a disease that was "all in the head," that is, not a real illness at all.

Instead, I am taking it in the way used by Laurence Foss and Kenneth Rothenberg in *The Second Medical Revolution*, where the term denotes the empirically demonstrable ability of the mind to affect bodily health or disease. The authors argue that this connection is recognized in certain forms even in traditional medicine, where, for example, experiments on rats indicate that the animal's immune responses can be conditioned to follow pathways they would not normally take.[1] Here it is a simple case of Pavlovian training. However, Foss and Rothenberg extend the concept of mind/body influence further, stating that for humans, that is, for beings who can be self-aware, the mind can assert conscious control over autonomic processes. They index Dr. Simonton's work with cancer patients, whom he not only treats physically but whom he asks to do visualizations (biofeedback) as part of the healing process. As much as this second component of the cure is effective,

> the clinical use of biofeedback implies volitional control of interior bodily processes. Like behavioral conditioning techniques (those used on rats) it seeks to regulate heretofore inaccessible… bodily functions. Unlike animal behavioral conditioning techniques, however, it implies a self-conscious autocatalytic element. The therapist (or patient) recruits the patient's intellect for medical purposes.[2]

All of which is to say that attitude and mental perspective play a role in well being, and a comprehension of this has to be tapped into in examining AIDS and other physical breakdowns. This belief is implicit to Schmidt's theory.

The second major thesis on which he grounds his views is one we have already treated extensively, which is that stereotypes and stigmatization enter knowledge production at all levels. Just as a racially prejudiced white, lower-middle-class home owner may, because of his prejudices, fight against blacks moving into his neighborhood, so, as we've exposed, may prejudiced upper-middle-class scientists assert that AIDS originated in Africa, where medical care is undeveloped and people still cling to superstitions, such as wanting to wash their dead even if they were infected with Ebola virus.

Let me pause, here, though, since we are on the topic of prejudice, to criticize certain formulations made earlier in the book. Although I was not consciously being sexist, I was drawing on slanted medical reports, which, inasmuch as I did not call attention to the slant, may have appeared to agree to that slant.

In an important essay in *Health/PAC Bulletin*, Kathryn Anastos and Carola Marte have written on "Women—The Missing Persons in the AIDS Epidemic." They acknowledge that in the first days of AIDS spread in the United States, the vast majority of the affected were male. However, as the disease has spread more widely, more and more women have been stricken, and yet this has not been recognized in the research. Women, for example, have usually been left out when subjects are gathered for trials of AIDS drugs. Anastos and Marte note that in the "original studies for AZT in 282 patients [this] included only 13 women."[3] Moreover, female sex-specific developments of the disease are generally ignored. This is something I should have noted when I decried the artificial expansion of the AIDS definition. Some of this expansion, such as the inclusion of cervical cancer in the defining criteria, may not have been nugatory but may have been part of a long overdue recognition that women have a different symptomatology in connection to the disease.

To continue, Anastos and Marte found that when AIDS researchers did look at women, they often honed in on female stereotypes rather than on the effects of the disease on all groups of women. "Until recently one could gain epidemiological information on women and AIDS mainly from perinatal studies and, to a lesser extent, from studies of prostitution."[4] My own discussion of women has often been narrowed to these categories, given that that's where studies have been done. Still, in reporting them, I should have made clear that such studies need to be more broadly gauged. Moreover, such studies are often proceeding with a warped outlook. Ask yourself why researchers often concentrate on transmission of AIDS from prostitutes. As Anastos and Marte explain:

> There are a number of studies of HIV infection in prostitutes, presumably because this affects heterosexual transmission to men. In contrast there has been no discussion in the professional literature of how women's lack of empowerment affects heterosexual transmission to women. Prostitutes are frequently seen as the guilty parties in the infection of women whose husbands or steady partners are clients and major support of the sex industry. This shifts the responsibility away from the man who engages in risk-taking sexual encounters.[5]

We can see, then, that research into the disease rates of prostitutes is likely to be guided by sexist assumptions about how these women are or will cause the spread of the disease to "innocent" men. My self-criticism here should teach us that it is easy to pick up data such as this without sifting it for concealed ideology. This is yet another proof that although some may be immune to HIV, no one seems immune to the invalid stereotypes circulating in our society.

Now, what Schmidt has done is unite the two points I have just raised: that mental states can affect the occurrence of disease and that wrongful stereotypes pervade discourse about the AIDS crisis. Putting them together, he argues that the weight of debasing stereotypes has played a part in generating the disease.

Let's look at how he makes this connection, charting (1) how he was first led to his belief, (2) how he sees AIDS as similar to mass hysteria, drawing with it emotions and rituals associated with such outbreaks, (3) what he sees as the biochemistry of this psychosomatic event, (4) how doctors have often fed this hysteria, and finally (5) what he gives as his prescription for overcoming this aspect of the disease.

The Origin of Prof. Schmidt's Analysis

When Psychoanalyst Casper G. Schmidt began exploring AIDS, he felt nonplused. "The more I read, the more puzzling and the more profoundly disturbing it became. Because it made no sense."[6] What was rattling him was that here was a disease consistently described as a sexually transmitted disease that would soon be at epidemic proportions but that disobeyed the rules governing previous epidemics. "The further I read about this, the less sense it made," he says. "Infectious epidemics do not follow fault lines that are cultural or sociological, which is what we find in this epidemic. You do not find any infectious epidemic in history that selectivity infects a group designated on a basis of cultural factors."[7]

Previous epidemics have followed such pathways as were sketched out for AIDS, such as the transmission route from prostitutes to heterosexuals or from bisexuals into the heterosexual realm. These routes were plotted in early projections of the disease's likely trajectories, but the disease never actually took these roads.

So, Schmidt asked himself what type of disease is confined to marginalized groups. He answered, mass hysteria. "When you have people low down on the totem pole of power: women, prisoners, children, homeless people... people who do not have an avenue to express their rage, they turn this inward and become ill with epidemic hysteria." he then concludes: "AIDS is a clear case of epidemic hysteria."[8]

Epidemic Hysteria

So what is this type of hysteria?I will abbreviate Schmidt's definition from his essay in the book *The AIDS Cult* and flesh it out—since all Schmidt's examples are chosen from the AIDS epidemic itself—by drawing from my collaborator Feast's essay on mass hysteria in *Magill's Survey of Social Science*, an encyclopedia of sociological concepts.

"Epidemic hysteria," Schmidt writes somewhat opaquely, "is (a) a form of collective mental disorder... (b) through the mechanism of a contagion... (c) this susceptibility is defined by the interface between individual character structure (drives) and distal group tensions."[9]

Feast provides this example:

In March 1954 in Seattle, Washington, the papers began reporting that in a nearby small town, residents had found their car windshields scratched and pitted. Police suspected vandalism. However rumors flew that this damage was linked to radioactive fallout from recent nuclear bomb detonations in the Pacific. On April 15 and 16, nearly 250 Seattle residents called the police to report their windshields had been af-fected... the mayor appealed to the president for emergency aid. Yet, in a few days, the fear blew over.... Experts who studied the situation

. . . [said] the windshield pitting was due to average wear and tear, and that this deterioration was something, like slight scratches on an in-dividual's eyeglasses, that one would not notice unless his or her atten-tion was called to it.[10]

This case shows a disorder (in the sense that there was no objective reality to a sudden pitting of the cars' glass) which is contagious, spreading from the small town to Seattle, and based in group fears about atomic tests, "the generalized atmosphere of tension that ushered in the nuclear age." It does not yet show how this hysteria can affect health, nor the mechanism through which a fearful atmosphere coalesces into a particularized, imaginary threat.

In Schmidt's definition, the hysteria "can be a wish, a fantasy or a delusion and... gives rise... to psychosomatic symptoms to resolve[d] conflicts through... unconscious means, since... direct, conscious expression... is impos-sible due to group defenses"[11]

An example embodying these further elements would be that of the "June bug," analyzed by Alan Kerchkhoff and Kurt Back in a book of that name. In the mid-1960s in a textile mill in the South, "a few people became sick and rumor ascribed their illnesses to bugs that had arrived in a shipment of cloth."[12] More and more workers began dropping from the disease, but no insect was ever found. The authors ascribe the mass sickness, in part, to hostility of the workers toward management, which fear of losing their jobs inhibited them from expressing openly.

Here we see that an unexpressed conflict surfaces in a mysterious illness. It's worth observing that this instance is especially typical of contemporary manifestations in that mass-production facilities are sites of most current out-breaks. Since in the global market nowadays, more and more of these factories are located in the Third World and staffed by women, it is among them that mass hysteria is found to be most frequent. The student of the phenomon "W.H. Poon argues that Malay females [as an illustration of this pattern] be-come involved in mass psychogenic illnesses because they are angered by their working conditions, yet do not dare voice their protests for fear of the social disapproval in their milieu."[13]

Schmidt himself refers to the job site when developing another significant perception. He states, "One thing I find fascinating and crucial about epidemic hysteria is that it follows cultural folk lines."[14] He uses the case of mass hysteria in Singapore, a multiracial city with large Chinese, Muslim, Indian, and local populations, to bring out this theme. In one reported incident, a psychosomatic illness affected the Malay workers who were predominant on the morning shift. They said they were being bitten by mysterious bugs; but "workers on the afternoon and evening shifts, working in exactly the same places but belonging to different cultural entities, don't show these symptoms and don't become ill from the same epidemic hysteria."[15]

As Schmidt sees it, the trait just described, whereby hysteria is a phenomenon confined to particular culturally homogenous groups, suggests that this hysteria is like AIDS in its trajectories, since the immunosuppressive disease also infects along cultural lines—centering, at least in the West, on certain stigmatized groups and not moving readily to the general population.

Schmidt's comparison, however, is not very strict in that hysterical contagion is usually centered around one locale, like a factory, and is short-lived. Let's see if Schmidt can find any other parallels.

. Like the Seattle incident or that of the June bug, AIDS comes from far away. "As in other epidemics, that poison which 'causes' the disorder is assumed to originate from outside the country."[16] Our section on Africa established that he is right on that score. U.S. scientists obsessively tried to put AIDS's origin in the jungles. But what of the inexpressible conflict that festers until it finds expression, such as the conflict between workers and foremen seen in the industrial outbreaks?

He begins on this matter by saying, "The unconscious social conflict which AIDS is called upon to resolve is twofold."[17] At the basis of the crisis, Schmidt finds a dialectic between two groups in society, just as the Malay women's outbreak could be rooted in the conflict between labor and management. On the one side, there are "unconscious drug-taking and homosexual fantasies within the general population, brought dangerously close to the surface because of the permissiveness of the 60s and 70s."[18] To restate, he holds that many strands of the majoritarian culture, those who are heterosexual and don't take drugs, are highly attracted to these pursuits, at least in fantasy. For some of them, this produces a guilt over anticipated but not acted-upon desires. Meanwhile, there is also "guilt over sexual and addictive excesses within... these two subcultures" themselves, that is, within the gay and drug-using groups.[19]

During the libidinal 1960s and 1970s, this guilt could be lived with, as long as license seemed to be the order of the day, but things changed when the political structure became increasingly conservative. The New Right "delivered a harangue to shame those" who had been partaking in pleasures of the flesh."[20] In the late 1970s, "you had the rise of the Moral Majority, of Christian Voice, of Religious Roundtable, a whole bunch of conservative organizations designed to 'protect' traditional family values," Schmidt details.[21] "What these people essentially said was that homosexuals were causing the dissolution of the American family. If only they could kill them or make them disappear everything would be fine."[22] This was paralleled by the War on Drugs that went on through the two Reagan and one Bush administrations. From this front, too, the message was sent that the planet would be better off without all these drug abusers. In the end, "this created a wave of reactive depression, caused by repression of the retaliatory fury that the shamed person feels."[23]

The last sentence is a mouthful but can be unpacked as follows. There is a certain type of depression that is built up from internalized anger induced by shame. A character in Patricia Highsmith's novel *The Talented Mr. Ripley*, says, roughly, "No one, no matter what he has done, thinks of himself as a bad person"; yet there are situations in which even the most hardened sinner might blanch. Suppose, to revert to an earlier example, the young Dr. Gallo (or Bruno Latour) was caught by a priest with his hand in the church poor box. Dr. Gallo might be angry at being caught but, as much as he still subscribed to society's ethics, he would probably not express his chagrin by knocking down the father. He would, however, feel ashamed of himself for letting down his church and family. He would have turned his anger inward. Analogically, according to Schmidt, the IV drug user or promiscuous gay male, caught with his sleeve rolled up or pants down by the castigating eye of the New Right zealots, suddenly fills with repressed anger, the type that will eventually lead to an explosive hysterical outburst.

It would seem that in Schmidt's view—for he doesn't trace out his analogy fully on this point—a slow motion hysteria sweeps through the "pariah" communities after the first few cases of a new disease are reported in the media, gradually leading to an increase in this disease, with new sufferers mimicking those already affected.

At this point, let's remember that, as noted, there are ways stress, here expressed in depression and shame, has physiological effects that can undermine health. This is the line pursued by Schmidt, whose argument is that AIDS is produced by an immune breakdown caused by the overwhelming stress brought on by these internalized feelings. In a moment we will look at the physical side, but first we will draw out some corollaries of his basic posture, which unfortunately are not always given in the most coherent manner.

According to Schmidt's theory, AIDS will attack those who have bottled up their rage. Just as it has been repeatedly shown that the type A personality, the person who is hard-driving, aggressive, and in-your-face, is the most prone to heart disease, Schmidt suggests that it is those who are most suffused with shame and inner turmoil who will be taken by AIDS. He writes, "Each and every person who becomes a part of the syndrome is part of a scapegoat group and has had some previous predisposition based on child abuse, child neglect, or child abandonment."[24]

A person's coming from a damaged childhood, though, by itself is not enough to make him or her fall ill. If such a person lived in a caring community, the person's wounds might be salved. However, here the macrostructure (the socioeconomic conditions) intersects the microstructure (the frail individual). Schmidt says, "When individuals don't feel they can influence political life, if you have a conservative swing, with Reagan coming into the White House… that is the kind of thing which creates inescapable stress, giving rise to feelings of helplessness."[25] This stress unseats the healthy reactions of the immune system and leads toward a disease state.

The flip side, of course, is that a person with a combative spirit—a gay male, for instance, who expresses his outrage over the pillorying of homosexuals carried out by right-wing zealots—will not fall ill in this way because his anger is externalized. Schmidt points to ACT UP, an activist group that fights government and scientific bureaucracies over their biased, underfunded, and short-sighted views on AIDS issues, as an example of the healthy, proactive response to an anti-gay public climate. "That's where ACT UP comes in, because there, people are not helpless!" he writes. "They can release their feelings and express them. Which is why most of the people in ACT UP don't easily die from AIDS…. Expressing their rage is what saves them."[26]

Let's think a little further about the character of those who succumb to AIDS after spending years overwhelmed by repressed feelings; this will help us see the complexity of Schmidt's thought. It is not simply a question of biochemistry singling out certain people, those with less ability to physically sustain tension. As we saw, Schmidt talks of those who "had some previous predisposition based on child abuse, child neglect, or child abandonment" as being most susceptible. This is not because their immune systems were weakened by early negligence on the part of their caregivers, although this may be part of the problem, but because they have been socialized into victim roles.

To bring out how this happens, Schmidt refers to experiments with animals. In a series of investigations in the 1960s, scientists "found that if you put animals in a situation they can't escape from, and give them electric shocks; they try to escape for a certain time, but when they realize they can't escape, they lie down and tolerate the shocks, whimpering and moaning." After a while, this reaction to pain becomes second nature. "If you open the gates of their cage, they behave as if they're still prisoners. They develop learned helplessness."[27]

It is not as if these animals can never recover from this behavior. "All you need to do is to drag them out of the cage, and then they learn very quickly."[28] However, if they are resubjected to the pain, after having lived without it for a space, they will not adopt typical animal responses—for example, fight or flight—but return to passivity. Schmidt's analogy is that a child of an unloving or abusive home generally reacts to the situation by some form of passive-aggressive behavior, since he or she does not have the strength or resources to run away from home or physically challenge the abuser.

Moreover, the child will often internalize the messages sent by the authoritarian care provider. "For example, a child is severely abused by his father repeatedly. Each time he starts crying after being beaten... his father... orders him to stop crying. This child... learns that the expression of grief or rage against authority figures is totally unacceptable."[29] After the child grows up, he may have learned more positive ways of dealing with his emotions; but, when taxed by a particularly tense situation, will return to passive, inhibited reactions. "Every time he comes into a situation where he would like to express any of these feelings, he can't do it because of these built-in memories and the profound threat from the memories of being abused, beaten and told not to do certain things."[30]

A point of which we need to be aware is that such attitudes are not necessarily taken up begrudgingly. Although these attitudes are ultimately disease creating and make a person feel powerless, at the same time they offer the comfort of returning to old patterns, as well as the psychic benefits that can accrue for anyone playing the role of victim or martyr in front of sympathetic friends. Moreover, if we can again refer to the macro level, the media and many other establishment agencies lavish attention on victims, making it seem that playing such a role, though it is an unhappy one, is one that is highly approved by the public, whose hearts go out to the miserable. (Bear in mind, too, that encouraging the disenfranchised to adopt the victim role is self-serving for the establishment, who would rather deal with the prayers of victims, who are satisfied with crumbs, than the demands of activists, such as those in ACT UP, who seek to advance their own agenda.)

Schmidt, who is gay himself, contends in a touching passage, that in the gay male community, not only have many AIDS or HIV positive men adopted the victim role in a way that can become alienating, but that this role, backed to the hilt by media presentations of its glamour, has split the group, by making those who are HIV negative sometimes feel guilt ridden.

> After ten years, HIV has become the new identity of gay men... We came out of the closet introducing ourselves proudly to the world with "I am gay"—we now introduce ourselves proudly to the world with "I am HIV."
>
> The reality for many homosexual men unable to identify as "I am HIV" was equally devastating. Since homosexuality is not inherited and many of us experienced abandonment by our physical families, we had been building our own gay families over many years. Suddenly HIV+ challenged those ties, leaving the touched on one side and the untouched on the other—identified ironically by blood testing, the very fluid that does not bind us to our biological families. Much like in those [abandoning genetic] families, friends now started isolating themselves on the grounds of a reality that the "not HIVs" could not relate to...
>
> The effect of the negative HIV test on many homosexual men is devastating. Depressed, having to cope with survivor guilt and feeling left

out, many dive into the night with the unspoken desire of becoming HIV [positive].[31]

You may, as I've said, find the last characterization as counterintuitive. (I will return to this in a minute.) However, aside from the pathos of this quotation, what needs to be noticed particularly is that what seem like purely individual decisions are shot through with social meanings. To put this in another way, we each have roles to play in a drama scripted by society. Schmidt says acerbically, "If you're 'HIV+,' your part is to get sick and die; if you are a doctor your role is to test for an antibody, make healthy people sick [by prescribing toxic drugs] and sick people die, and then blame an alleged virus."[32] This one may accept, but Schmidt argues that even many dissidents fit dysfunctional roles. "If you're a gay AIDS activist, your role is to insure that unproven treatments [like AZT] get into everyone's body… if you're an AIDS organization your role is to deliver 'HIV+s' to the pharmaceutical ovens and silence anyone who questions the insanity."[33] And lastly, "If you're not in any of these groups, your role is to wear a red ribbon, a latex condom and act like you care."[34]

Schmidt's language is wildly intemperate here. After all, the Nazis, who he is comparing to drug companies with their "ovens," had an explicit goal of exterminating certain social groups. Such goals hardly motivate the drug companies, which are concerned with bottom line. I could have clipped out that particular passage, but I kept it to remind readers of my own position on Schmidt's work.

There are places where, in his excitement, he makes insupportable statements that do not deserve belief. Yet, his underlying contentions still have merit. Here he is uncovering an important means of social control in the feedback structure that aligns personality with prepared social roles. Here is where, for all his bumbling with words, he is onto something.

In nuce, he is saying that many adults who belong to disenfranchised communities have learned passivity during deprived childhoods, and that in conditions of stress targeted at their communities they will tend to fall back on these patterns, urged on by a media system that flags these roles as valued, ones treasured in soap operas and pop songs as well as in some depictions of AIDS victims. "There does seem to be a glorification of HIV/AIDS, even as a magnet of compassion and a celebrity life," Schmidt urges. This deification of the victim gives "a meaning and purpose for lives that may have lacked such qualities before being infected and pitied, and even [giving them]… heroic proportions as kinds of martyrs."[35]

But the last important point is still to be made. Given that the stress arises in a hostile political climate, the AIDS role-playing game "serves to keep us all from looking at what's truly going on in the world. The rampant death and subsequent social cleansing artificially absolve everyone of the repressed tension generated by the pre-existing social conditions."[36] To say that again, instead of exposing the lunacies of right-wing politics—the claim that homosexuals are destroying the American family, for example—gays became obssessed with medical issues. (Note that I am restricting my attention at this point to the gay male population that was hit by the illness because this is what Schmidt focuses on.)

As Schmidt sees it, a handful of militants, such as those in ACT UP but also in non-health-oriented progressive groups, are fighting to change the reactionary environment that is creating the larger society's homophobia and the sexual minority's feelings of guilt and depression. Meanwhile, the majority within the gay minority under attack are accepting AIDS as defined by the establishment and not battling back forcefully against the false characterizations supplied by the New Right. Such demonizing characterizations of them and other disempowered groups (such as welfare mothers, "wilding" young blacks, and illegal immigrants) are a prelude to and accompaniment of rollbacks of social spending. Since the mid-1980s, while more and more attention was given to AIDS, the Right and their business associates—under the banner of such "voodoo" economic theories as "Trickle Down" (whereby, it was claimed, money given to the elite wealthy will eventually rain down on the poor) and a gerrymandering of the income tax, which greatly lessened the burden on the over-a-hundred-thousand-a-year set (while leaving the taxes to be paid by the middle and lower classes the same or higher)—society's wealth has been redistributed upwards. As economist Hazel Henderson describes, "Each round of changes, or 'reforms,' in the tax codes is skewed even further in favor of the investor and corporations by massive lobbying, and even the tax cuts manage to favor the already affluent."[37]

Furthermore, the AIDS crisis is not only causing many to avert their eyes from other disasters in the polity. As we saw earlier, it is also used to shift attention from the socioeconomic realities of Africa (looted and mismanaged by Western extractive industries and the so-called Green Revolution, which may have increased agricultural yield, but did so at the cost of converting from subsistence to export-only monoculture, leaving the peasant more vulnerable to droughts and famine) to its medical aspects.

Combining this characterization of AIDS as distraction (though, obviously, he recognizes the disease is a mammoth problem) with his thoughts on how the straight public does not want to confront the issues of the marginal (such as drug takers and homosexuals) since doing so may force the public to face up to the undeniable attraction of these life modes, Schmidt argues that most from the mainstream will avoid confronting the underlying tensions of U.S. society at all costs. The costs, though, are to be borne by people with AIDS, who the representative members of the mainstream would like to sweep beneath the carpet, or, more literally, under the soil.

The War on AIDS is a huge success socially, politically and economically because most people really don't care if gay men and IV-drug users are being murdered. Their homophobia and own guilt and shame dull their basic humanity while contributing to their stupidity. But as terrible as this gay extermination is, it is only the tip of the genocide. Most people really don't care if 30 million of the poorest people on earth [those in Africa] starve to death either. The Big Lie (HIV) simplifies things. These horrific crimes are more easily ignored when these people are sentimentally written off as victims of a mythical AIDS pandemic.[38]

This is "what's truly going on in the world," which Schmidt decries and which he says AIDS is making people forget. (And the New Right's rollback of social programs, by the way, not alternate lifestyles, is the real threat to the American family, since divorce and marital breakdown accompany economic dislocation.) Schmidt is not—to be clear—saying that AIDS shouldn't be a top priority, but only that it has to be one priority among others.

It might be apropos at this juncture to add another wrinkle to his argument. While affirming that AIDS is a terrifying reality that has to be attended to, Schmidt says it is at the same time something of a Trojan horse that has diverted energy from other threats to the gay minority. And, not only has this worked to turn interest away from certain New Right agendas, but, Schmidt argues, some generally adopted responses to AIDS agenda have incorporated conservative goals. "AIDS has succeeded in shifting the left-wing discourse of sexuality away from one of liberation, freedom, and resistance, to one of responsibility, danger, and obligation—concepts more at home with a right-wing scheme of social control, xenophobia and authoritarianism" than with any liberal mindset.[39] He qualifies his assertion by noting that safe sex is not necessarily unerotic, although it is hard to "let oneself go" when every move is hedged around with restrictions. Nor is liberated sex without its own bedevilments. "I would be the last to argue that sexual liberation has always been liberatory, sexual freedom always free, or sexual resistance always contrary to broad forms of domination," he comments sensibly. "But at least there was a time, before this last decade of AIDS, when the discursive apparatus of liberation, freedom and resistance had not been blanched out of sex."[40] He is echoing an earlier thought mentioned in this book, that sex is one of the few inexpensive, interactive pleasures people have in this world; adding to it that being ably to freely exhibit and reciprocate through one's sexuality has always been connected to the creation of a democratic and liberated community.

However, it's time to take up another rather dubious statement of the psychotherapist, to which I promised to return. In the last block quote, Schmidt talks of those who "dive into the night with the unspoken desire of becoming HIV+." Is this more rhetorical overkill or can we give any credence to the ideas of a death wish?

First off, remember, that when he is making this claim about some people's behavior, Schmidt is only referring to a small group. "It is unlikely that anyone with a strong sense of self-worth would deliberately seek to seroconvert," he says. "But for a gay man who has internalized the negative judgments placed on him by family and society, it may well be a lot simpler to think about [having]

AIDS than about being gay."[41] We can broaden this by saying that when the pressure is too great, the most psychically fragile will opt for suicide, no matter to what group they belong. Gays who succumb only differ in having the choice of doing so by getting the immune deficiency disease. "AIDS itself may be horrible, but contracting AIDS is a relatively easy—and pleasurable form of suicide."[42] Schmidt's recipe, "diving into the night," presumably to visit a bathhouse or pick up a one-night stand at a club, is also a particularly inept way of killing oneself, since, as we've argued, simply having multiple sex partners has little to do with getting the disease. (It is rather the drug use and possible acquiring of STDs associated with having many partners that cause problems.)

Schmidt adds that mainstream society has not only offered the AIDS stricken a role—"if you're 'HIV+,' your part is to get sick and die"—but elevated into cult figures gay or bisexual males, like James Dean, who lived hard and went down in (lavender) flames. The establishment media (and the part of the gay subculture that takes its cues from this media) elevates images of those on the fast track who are "living fast, doing it all, not worrying about the consequences, forgetting the future." Simultaneously, there are few images of those who grow gray gracefully. "Giving in to the forces of life, growing gaily older (for those who know of Quentin Crisp), in this society seems to some more difficult and less attractive than dying young and rather glorified."[43] This, of course, adds to the attraction of self-destruction for those who are already finding it hard to bear up in the neoconservative atmosphere.

I've implied that Schmidt's argument is decidedly complex, and there is another fuller historical plane on which it has to be approached. Not only does Schmidt link the AIDS hysteria to the dark climate created among disadvantaged communities by the rise of the New Right, but he also argues that it draws on older historical stereotypes of victim and accuser.

Let's return to the misplaced Nazi analogy. Although Schmidt's particular analogy with ovens misfired, the basic idea that the Nazis were carrying out, that a society will be better off if outsider groups that it contains are eliminated, is a time-tested one. That is, no matter how wrong it may be, every nonhomogeneous society goes about labeling scapegoats on whom it vents its rage during bad times. In the United States of the early nineteenth century, these were blacks, Irish, and other lower-class groups. When things were sour, nunneries would be torn down and the neighborhoods of the undesirables raided. (In the New York City draft riot during the Civil War, whites, rioting over the institution of the draft, even burned down an orphanage for black children, killing as many of the youngsters as they could catch.) In the second half of the century, the scapegoats were the Chinese, Jews… and Blacks. When there was an economic downturn in the 1890s, for example, the workmen of San Francisco were roused to go and burn down Chinatown, based on the harebrained idea that the Chinese were responsible for the depression. (See, it's not only modern-day Republicans that thrive on harebrained economic schemes.)

Gays, too, have often felt the wrath of the majority. "Gay men have long served as repositories for a complex of group fears and fantasies about sex, sickness and death," Schmidt says. He quotes with approval psychohistorian Lloyd deMause, who talks of "society's periodic killing off of its own id—representatives, its youth, who represent itself in the life-phase when it was most sexual and most aggressive."[44] This does not accord exactly with my own capsule history—the Chinese burned out in San Francisco were not particularly young—but deMause does have a valid point. Those who are scapegoated, though it is often for economic reasons, are usually viewed by the majority as indulging in one type of illicit sexuality or another. The Chinese were said to rely on prostitutes rather than wives, while the Nazis thought of the Jews as oversexed. The way such characterizations has been applied to gays by right wingers needs no elaboration.

We can trace the scapegoating phenomenon back further in time, and tie the scapegoat/rampant sexuality dyad to another linkage, that of rampant sexuality and disease. Schmidt asserts, "The fantasy complex of AIDS (i.e., the triad of promiscuity, poison blood, poison sperm) is not new. It is identical to the one in vogue during the Middle Ages to explain the origins of leprosy."[45] At that time, he continues, leprosy, which is not a sexually transmitted disease, "was thought to follow from the 'filth of lechery,'" namely, from unbridled lovemaking, as well as from "the conception of a child in menstrual tyme," since the blood a woman discharges during her period was viewed as supernaturally tainted. The third means of transmission of the disease was through "poisonous sperm: John of Gaddesden warns that a man who sleeps with a woman who has previously had intercourse with a leper will contract the disease," provided "the woman still retains the seminal fluid of the leper."[46]

The erroneous etiology of leprosy is not given to indicate, more generally, that diseases cannot be contracted through sex or blood exchange. Rather, it shows that this triangle of hypersexuality, deadly blood, and semen has a long pedigree in a mythology set up to link scapegoats to a spread of disease brought on themselves by their bawdiness but also endangering the more innocent portions of society. Sound familiar?

These long-range comparisons have been used to bring out the historical continuity among societies (at least those inequitably divided between those with and without power) who choose to discharge tensions by blaming scapegoats. Schmidt is saying that these arrangements—scapegoats are located among the disenfranchised and need to be punished because they are (supposedly) indulging in rampant sexuality which is leading to disease—are ready templates, waiting in the collective unconscious, and available when a new disease strikes or when social tensions grow unmanageable.

In Bacon's terms, the use of this scapegoating procedure might be named instances of power. This is a category, that, at first glance, seems ill-suited to the purposes for which his whole classificatory scheme had been established. It will be recalled that he argued that the identification of these instances was, like the locating of submerged shoals along a coastline, mandatory for the proper plotting of what routes science should travel. By orienting itself according to what has been peremptorily revealed by these examples, science would be able to carry out a discovery process on nature in the most fruitful way. However, the anomaly of the instance of power is that it deals with cultural, not natural entities. These instances are "such works of art, which appear to be the very summits and crowning points of human industry." Yet, he argues, it is just such consummations that may reveal the greatest amount about nature, since "the method of creating and constructing such miracles of art is in most cases plain, whereas in the miracles of nature it is generally obscure."[47] That is, if the inventions of humans are modeled on the organic productions of nature (sometimes in a near, sometimes in a more distant way), then, since we can understand the workings of the human models, we may be able to move toward a comprehension of the more complex figurations of nature. This is the argument advanced nowadays by those who say that studying the design of computers will help us see into the planning of the human brain.

The comparison I am making between Bacon and Schmidt may seem less sure than Bacon's note of the similarities between cultural and natural products, for the connection here is not between the social action of scapegoating and some similar act carried out, for example, in ant colonies. Rather, in this slightly altered sense, it is between two human products, including one that pretends to merely objectively trap and mirror natural events. So, in this version, the first process, the instance of power, is the use of scapegoating as a human means to help a group unify its members while masking the real source of trouble in the organization, as in a pogrom or a lynching. What this process illuminates is the second process, the same one disguised in medical terms as that of an outsider group, with dirty practices, that needs to be policed heavily, since it is endangering the health of the mainstream.

Of course, both with leprosy and AIDS, the diseases are real, but these mythological premises take a hand in how each disease is viewed. So—to enlarge on only one connection—within the scapegoat mytheme (a mytheme is a single element of a larger myth), it is necessary for the disease to be contagious; ergo, one must find something, such as a retrovirus, that can transmit it. If, contrarily, the disease were brought on by each person's abuse of his or her immune system, and hence was not able to be spread, the whole system would fall down. It is essential for the scapegoat myth that the perverse outsiders are endangering the majority. A barely conscious belief in this myth, Schmidt would argue, while not actually helping the scientists to find any transmittable agent, would pressure them to look in this direction in trying to figure out the background of the disease. In terms of this argument, then, Schmidt is not interested in whether HIV exists or not, but in suggesting that biological scientists' initial direction of research may be dependent as much on a society's mythology as on contemporary trends in medicine.

So far, in examining wider historical genealogies, we have talked about how a society's majority in times of trouble has manipulated the scapegoats it has created to let off anger—blinding itself, perhaps, to inoperative programs in its overall organization, that it does not want to face, though they are really causing the trouble. It may also be asked if there are historical and/or mythological parallels to the way gay males, Schmidt's primary concern, have reacted to the AIDS crisis.

Going back even further, Schmidt also sees homologies between current AIDS rituals for gay men and the practices of tribal societies, both of which are used to get younger members of the group to accept the collective's myths. In these early societies, a young man, before he becomes a full-fledged member of the group, must pass through a trial during which he endures various ordeals and learns of the tribe's inner secrets. According to Schmidt, "The gay health crisis has reached a stage in which seroconversion has become institutionalized as the most important rite of passage in the life of a gay man."[48]

He sees a whole list of similarities between these two sets of practices, which are widely different in type of society but akin in use of mumbo-jumbo and methods of gaining allegiance of the young. (At this point, we are, for purposes of the argument, emphasizing the manipulative side of primitive ritual, though a fuller treatment would have to bring out the positive fulfillments also involved in these ceremonies.)

First off, Schmidt remembers that initiations take place on sacred ground, far removed and set off from the land used for normal, domestic activities, where the initiate will meet masked elders in fearsome costumes. So it is in AIDS science, "the HIV testing ceremony, highly ritualized, demanding a visit to a special, rather fearsome place—a sacred place, the Test Site."[49] There the man or woman going for a test is met by "priests and acolytes in attendance: the various physicians, psychiatrists, social workers, peer counselors and AIDS workers hovering or bustling about, many wearing their white robes of office."[50]

Next, the tribesmen endure tattooing or some other painful experience. In the AIDS camp, "the ceremony involves a literal blood sacrifice, drawn with a hypodermic needle by a nurse or paramedic. Some initiates faint."[51] While the young man copes with the pain, he is being bombarded by the tribal secrets. The AIDS test, similarly, "is accompanied by highly structured readings from sacred texts—AIDS Education and Safe Sex scripts."[52]

Very commonly, after the first part of the ritual, the initiate is sent off into the wilderness to accomplish some task, perhaps, to kill his first lion or, as in Southwestern Native American tribes, to have a vision of the tribe's sacred animal. In AIDS science, there is also "a time of trial—the stressful period of several days or weeks involved in waiting for one's results. During this time, one's thoughts are concentrated almost continuously on HIV as the shared object of fear and devotion," just as the Native American boy is fixing his consciousness on the tribal totem.

which is tricky but can be fended off with a technological tool, the blood anti-body test. For those who hold the terrorist view, this blood test is a panacea. Their idea of it is "based on the fantasy that everyone with HIV can be detected." From this vantage, "HIV testing is… a means of primary prevention, rather than a means of access to health care providers, and is justified on behalf of an imagined uninfected 'general public,' with little or no concern with infected individuals."[63]

For the missionaries, HIV is more like an evil spirit that falls upon those who have sinned. It thrives on "immorality, bestiality, unnatural acts and ungodly practices," coming to sweep the damned down to hell, like the devils who appear in the last act of Faust. Here at least the concern is the people with AIDS, not the general public, but the missionaries' cure does not center on saving their skins but their souls. What will keep miscreant homosexuals, drug takers, and their companions in vice safe is to swear off their ungodly practices and return to the fold, where exists "the supposedly 'traditional' values of Judeo-Christian morality and its attendant institutions, above all, marriage and the 'family.'"[64] From either way of looking at the case, the sick are to be condemned first, whether they are then to be pitied like stray lambs or blasted for bringing their pollution on others. Such attitudes create great gloom around those who are infected or ill.

Where Watney gives an overview of the bile (terrorist model) or treacle (missionary version) that the media barons and health ministers are spewing out concerning the problem, Schmidt is concerned more with examining how such verbal and visual poison impacts the individual. He draws out attention, for example, to the new life regimen an at-risk-but-HIV-negative young person now has to face since AIDS has come over the horizon. He or she certainly can't feel abandon in the throes of sex, the abandon that links eroticism to mystical consciousness, since it means a giving up of the self and merging with the other. How can one let go when the fear of contamination is omnipresent, so even in carnality the self has to be ever wary? Schmidt puts it like this: "It may be harmful for the healthy emotional life of young humans to be constantly reminded of an intimate, inevitable 'deadly threat' that will catch you when you don't stay vigilant."[65] The young gay male, for example, who must feel himself at maximum risk, faces harsh enjoinders from both sides. "While fearful of blame from straight parents for being gay, the younger generations now fear blame from their older [cautious, gay] peers for not playing it sexually safe enough."[66]

Once one becomes HIV positive, though, things get much worse, both socially and psychologically. In terms of individual psychology, Schmidt argues, "an 'HIV' diagnosis has directly immunosuppressive effects in its own right."[67] He quotes the *Deconstructing AIDS* report from the HEAL trust to the effect that "the diagnosis is associated with higher levels of psychological and social distress and leads to T-cell abnormalities and increases in latent viral activity."[68] Mention is made in that report to the work of Swiss immunologist Alfred Hassig.

He has analyzed how stress hormones, including adrenaline, noradrenalin, and cortisol, can tamper with the body's oxidative/anti-oxidant balance, "depleting the effectiveness of immune system cells."[69]

Socially, being labeled positive results in stigmatization. Rather than relying on Schmidt here, let's turn to the novel, *Trainspotting*, which we quoted at the head of this section. It describes the life of a group of down-and-out heroin addicts in Edinburgh, the stories narrated by the characters in Scottish dialect. It is a good place from which to absorb what information and misinformation about AIDS exist in the drug community. As we've seen, much less is reported about this society than about gays. Though the addicts are disrespected by society, it is only when they are identified as HIV positive that they are really shunned. One user describes what happened to Julie.

> She wis a really good punter She hud a bairn [baby] whin she wis HIV, but the bairn wis all-clear... The hoespital sent Julie hame in an ambulance wi the bairn, wi tow guys dressed in sortay radioactive-proof suits—helmets, the lot. This wis back in 1985. It had the predictable effect. The neighbors saw this, freaked, and burnt her oot the hoose. Once ye git tagged HIV, that's you fucked.... Harassment followed harrasment. Eventually, she hud a nervous breakdoon and, wi her damaged immune system, wis easy prey fir the onset ay AIDS.[70]

The link between psychological stress and immune disintegration couldn't be any more clear than it is here. It's not surprising that a novelist, who has to be aware of the real texture of the lives he or she is depicting, proves more knowledgeable about psychosomatic disease factors than most medical researchers.

Standing between those who are negative or positive for the HIV virus, each with his or her own dreads, is another category, that of the worried well. "Even though they have repeatedly tested 'HIV' negative and have been repeatedly assured they are not at risk for 'AIDS' these people present a whole range of pseudo HIV disease symptoms," including dramatic weight loss, low grade fevers and low T cell counts.[71] The existence of this class of people is indicative of how deeply the fear of AIDS has sunk in, causing absurd fears. An apprehensive person, for example, coming back from a session where he received a clean bill of health for his blood test, may immediately worry that the test was taken during the window of opportunity, a few days after infection, during which presence of the infection is not registered.

One reason this disease is so frightening and will cause such near paranoid reactions as the one just envisioned is that infection with the retrovirus is set forward as carrying an inevitable death sentence. For any individual who is apt to take on the passive victim role, the discovery that he or she is HIV positive may lead to a withdrawal into hopelessness, since according to the orthodox, there is no escaping the extinguishing of mortality. Notwithstanding the cases of HIV positive people who never get AIDS, which we have already documented, the common wisdom is that HIV infection is fatal. This is maintained even in the light of protests by such eminent figures as Luc Montagnier, who has stated, "AIDS does not inevitably lead to death, especially if you suppress the co-factors that support the disease."[72]

Significantly, Dr. Montagnier made this statement because he, along with Schmidt, believes depression is adding to the disease-creating strain on the bodies of those who find themselves HIV positive. As Dr. Montagnier continues, "It's very important to tell this [that the virus is not 100 percent fatal] to people who are infected. Psychological factors are critical in supporting immune function. If you suppress this psychological support by telling someone he is condemned to die, your words alone will have condemned him."[73] Schmidt drives the point home: "It is a terrible thing psychologically to present someone with the expectation that, from the moment of his 'HIV positive' diagnosis, he is one of the living dead, awaiting only the formalities of cremation and memorial service."[74] Whereas with cancer and heart disease, at least in less advanced stages, every patient feels he or she has a fighting chance, with AIDS it seems there is little one can do before the brief candle of life is snuffed. Having come down with the disease or seen one's T count fall dangerously low, there is little one can do, according to the mainstream view, except begin preparing a will and writing an epitaph. As Schmidt says, given this pessimism, "it is no surprise that the first response to the AIDS epidemic was to teach people with HIV/AIDS how to die well; what it was actually doing was teaching them how to die while well."[75]

What has been said so far may not have convinced you that Schmidt's argument is correct: that unprecedented levels of stress felt by a number of marginal groups led to widespread immune system crackup among them. Even so, it must be admitted that Schmidt is convincing in demonstrating how the whole orthodox AIDS package (with its hair-raising equations) has brought an unsettling amount of anguish into the lives of those at risk.

To repeat, my brief here is not to prove that Schmidt is right but to call attention to a number of his fresher insights, such as those about how AIDS, as reacted to by both at-risk and mainstream groups, borrows imagery and patterns from mythological and religious storerooms. One difficulty with Schmidt, though, is that, like his opposition in the AIDS establishment, he is too enthralled with the bugaboo of monocausal etiology, often seeming to imagine that mass hysteria can account for everything in the crisis. I believe his argument would be stronger if he presented the emotional turmoil created by first homophobia and then the AIDS disease, as key cofactors in leading many to a disease state, rather than taking this turmoil to hold sway by itself in determining the illness.

Although his idea of cause is onesided, he does try to work out how the unhappiness and depression he is concerned with translate biologically into an assault on the immune system. We'll look at that next.

The Biochemistry of Mass Hysteria

Just as I have certain reservations about Schmidt's concepts, so other detractors have voiced concern about his formulations. L. Cooper and M. Walker, writing in *Continuum*, go at Schmidt for overlooking too much in his attempt to account for the existence of AIDS. They begin by agreeing that psychological, even mythological themes play a role in the generation of the illness. "It may be the case that some gay men are tempted to take up AIDS as a tragic mask and act out a death wish choreographed by medical science."[76] Yet, they go on, "in reality the hysteria thesis ignores the complexity of the scientific and lay conflicts in 'AIDS.' It disguises the multiple interaction of the political, industrial, ideological and psychological" factors that contribute to causing the immune deficiency. Thus, Schmidt's narrow focus imitates the limitations of the orthodox outlook. They end, "Belief in a vacuous 'theory of hysteria' is itself as hysterical as the unquestioned acceptance of 'HIV' as the cause of 'AIDS.'"[77]

Although no champion of Schmidt, I think it has to be pointed out that he does try to take account, at least, of the interaction of the psychological with the physical, though he may have given short shrift to other substantial factors involved in AIDS's genesis.

We mentioned Dr. Hassig's work on the way stress hormones can degrade the immune system. This is the tack Schmidt takes in his explanation of how the psychic distress placed on stigmatized minorities, first by the New Right and then by the AIDS miasma, ends by overtaxing human immunity. "When you are placed under this kind of threat, the characteristic response is to have an increase in endorphins."[78] The word "endorphin" abbreviates "endogenous morphine," meaning a morphine produced in the body. (The hard drug morphine, like all hard drugs, is quite similar to a substance found naturally in the body. If a drug were totally foreign to endogenous matter, it would hardly be able to interact with the nervous system, whether stimulating or depressing it.) When a person is suffering stress or pain, endorphins are released and block pain signals, thus producing relief. However, a sustained production of endorphins

> paralyzes the inside of the t-cells, what is called vimentin filaments. It's the vimentin filaments that actually handle or manage the presentation of antigens on the cell surface [which serve as guides, directing other elements of the immune response]. If you have too large an amount of endorphins in your body, that paralyzes these filaments. That explains why before you have a decrease in t-cells you have a paralysis already of the ability to present antigens....
>
> [Moreover] the second thing that happens biochemically when you're under this kind of chronic threat is that you have an increase in cortisol levels.[79]

Cortisol, a steroid (one of a set of fat-soluble compounds with physiological effects), is produced to help the body fight against infection in various ways, particularly infections of the intestines, stimulating the production of a special class of T cells (as we will explore further below), among other operations. Whereas intestinal infections result in the loss of potassium, cortisone tends to conserve this mineral.[80]

We can look to independent studies, ones that have no involvement with psychological issues but were conducted purely to examine the biochemistry of people with AIDS, for support of Schmidt's hypotheses. As we will see, they have shown that cortisol levels are elevated in those who have the disease, and have indicated the mechanism of operation whereby imbalance in the presence of cortisol (versus other steroids) impairs immune function.

A team headed by J. M. Vilette published a piece in the Journal of *Clinical Endocrinology and Metabolism* reporting a study his team carried out to investigate steroids in people with acquired immunodeficiency disease. Their subject sample consisted of thirteen males who were HIV positive, seven asymptomatic, and six PWAs, along with a control group of ten healthy males. It was learned that values "were significantly higher for cortisol and lower for DHEA, DHEA-S, and ACTH in all HIV-infected patients."[81] The article argues that the lower and higher measurements of the different steroids are linked in that "excess cortisol would lower adrenal androgen secretion by shifting adrenal steroid biosynthesis toward glucocorticoids and decreasing pituitary ACTH secretion."[82] In simpler terms, the higher level of cortisol acts to change the more evenhanded production of steroids carried out by the adrenal and pituitary glands, which manufacture these steroids.

Whereas Schmidt attributes this overproduction of cortisol to be the result of long-term stress, Villette's team speculates that what is causing it "might be a stimulating substance secreted primarily by infected immune cells"; but all agree that abnormally high levels of cortisol appear in those who are HIV positive and in PWAs.[83]

A study led by T. Wisniewski has focused on how harmful the buildup of cortisol in the body can be, in combination with the accompanying slackening off of DHEA production. "Steroids can also influence immune function; glucocorticoids [such as cortisol] cause immunoincompetence while dehydroepiandrosterone (DHEA) enhances immune function."[84] Centering their investigation on the relation between these two compounds and CD4 levels, Wisniewski's group noted that DHEA and CD4 levels descend in tandem. "The data exhibit a positive relationship between the immune status of patients with HIV-related illness and DHEA, leading to the hypothesis that DHEA deficiency may worsen immune status."[85] As we saw, a glandular regulatory mechanism ties cortisol increase to DHEA decrease, so Wisniewski's findings add weight to Schmidt's views insofar as indicating that too much cortisol weakens immunity, although the research doesn't touch on whether anxiety affects steroid levels.

In this context, before returning to Schmidt's ideas, we might delve a little deeper into how CD4 cells are affected by amounts of steroids in the body. We can begin by noting that experiments by Mosmann and Coffman in 1986 "demonstrated that two cell groups can be differentiated amongst the CD-4 helper cells," which they designated as Th-1 and Th-2 cells.[86] The Th-1 cells secrete IL-2, IL-12, and IFN in order to stimulate cellular immune reactions; while the Th-2 cells secrete IL-4, IL-6, and IL-10 to increase the activity of the humoral immune reactions. Clerici and Shearer established that the HIV positives were more likely to progress to AIDS if they had a "Th-2 cytokine profile," that is, if the majority of their CD4 cells were of the Th-2 variety. Those whose CD4s were mainly of the Th-1 type resisted the move to AIDS.[87]

All this relates to our discussion of steroids, since, according to research carried out by the Daynes group, "The decisive factor for the type of peripheral regulation of the lymphocyte cytokine production is the production of steroid hormones," which means that the type of CD4 cells produced is dependent on the proportions of different types of steroids available. "It was found that the cortisol to DHEA ratio for cellular and humoral immune reaction is balanced for the Th-1 cytokine profile, whilst the Th-2 profile of the lymphocytic cytokine has an excess of cortisol."[88] The important point to take away from that last sentence is that Th-2 CD4 cells will be the ones manufactured more heavily when there is an overabundance of cortisol.[89]

Scientist F. De Fries, drawing on the work of Hassig, says more about the negative effects of an overabundance of Th-2 helper cells. These cells activate "the defense through antibodies of B cells (immunoglobulins) via the inflammatory substance interleukin 4."[90] If such an inflammatory response persists, as it will if it is being produced by continuous stress, the first signs of imbalance will be depressed activities of the Th-1 system, including lessened production of helper T-4 cells and the impeding of "the signal (interferon gamma), with which T-4 cells, as Th-1 cells, induce the phagocytic cells (macrophages) to isolate and destroy pathogens in their interior."[91] Also, one finds that natural killer cells, "cytolytic, 'cytoxic' cells, which can destroy estranged and degenerate, cancerous cells," are now operating subpar. After a while, if Th-2 overproduction keeps up, there will be damage to the thymus and the Th-1 cellular immunity system will continually weaken until "various viral pathogens, which we normally isolate and destroy with T cells, macrophages, killer cells and cytologic cells, can then increase."[92] De Fries remarks, by the way, that a number of causes can produce this asymmetrical state. He puts in this group not only the psychological stress with which Schmidt is concerned, but oxidative stress, a topic we will handle later.

Putting together what we have learned, we see that excess cortisol is correlated both with a drop in overall number of CD4 cells as well as a shift toward the production of Th-2 CD4 types, both of which are bad for the immune system. The Daynes group remarks further that the extra cortisol production is "stress-induced," which links this part of the biochemical research to Schmidt's thesis.

(High levels of cortisol, by the way, are indicated not only in people with AIDS but in cancer patients. "Cancer patients show an enormous increase in cortisol and a lowering of their DHEA levels, the same as in AIDS.[93])

The two processes we have mentioned in this perhaps overly technical discussion—the disruptions caused by high levels of cortisol and endorphins—markedly weaken immune function. As Schmidt puts it in reference to the second chemical, "Instead of expressing their needs outwardly, they turn inward and over-secrete cortisol, which paralyzes their immune system."[94]

This, then, in answer to some of his critics' qualms, is how Schmidt correlates at least two planes of analysis: the biochemical and the psychological.

Doctors' Misinformation

We have already dealt extensively with how negative stereotypes of people with AIDS and pessimistic views of the course of the disease have been major engines to create anxiousness in those affected already or at risk. We have even brought up how the sobersided pronunciamentos of doctors on the HIV equation have contributed to the depressing atmosphere.

Still, as we move on to an examination of Schmidt's own way of treating HIV positives and people with AIDS, it might be worthwhile to pause briefly to note some of the peculiar practices of doctors and lab technicians that have engaged in prescribing or prejudging that is in line with the flawed assumptions of orthodoxy rather than the Hippocratic oath's injunction that a doctor do nothing that would harm the life of his or her patient.

The first assumption is that those who belong to risk groups are to be treated differently than others when it comes to checking for the presence of HIV. As reported in *Continuum*, "It has been confirmed that 'HIV antibody' test labs use 'risk factor' information before deciding on results."[95] The report flags a statement given out by Britain's Public Health Laboratory Service (PHLS) to the effect that "since clinical testing began, all clinical specimens have been accompanied by information necessary to make an accurate diagnosis"[96] This is to say, if the sample came from, say, an IV drug user, then it would be put in a special category and treated differently. We can look at the situation in the United States to see what this different procedure is. The same article explains that as early as 1985, the CDC "recommended blood samples from people suspected to be at high risk of 'HIV infection' be heated at 56 degrees C for 30 minutes before testing." This, it was claimed, would protect lab workers from risk of contamination, but "heating blood samples led to a large number of otherwise negative samples testing positive."[97] This is what one calls a self-fulfilling prophecy.

The second assumption is AIDS = Death. If doctors were not convinced of that, how explain the report presented at Eleventh International Conference on AIDS in Vancouver that physician-assisted suicides for people with AIDS were common, but generally covered up. "A survey in 1996 of 118 physicians who treat people with AIDS found that 53 percent had helped their terminal-stage patients die. A similar survey in 1990 found that only 28 percent said they had."[98] This should not be taken to indicate that physicians themselves see suicide as a valid option, since there is evidence enough that people with AIDS are forward in entertaining this option. "A British researcher also reported that between 10 and 20 percent of the patients with AIDS in British Columbia seek their doctor's help in dying."[99] This same researcher found that in Canada as a whole 83 percent of the people with HIV and AIDS had looked on self-murder as a reasonable future option.

Treating the Psychological Aspects of Immune Disease

From what has been said so far, you can probably surmise what the general outlines of Schmidt's treatment would be. One could conjecture that he would try to combat the despair that can overpower a person with AIDS or HIV positivity who is met by the negative images produced by the establishment. Also, he would act soothingly to uncover and heal some of the still festering wounds bestowed in the person's childhood. Lastly, recalling his comments on the general healthiness of ACT UP members, he might prescribe therapeutic political activism as a way to turn the person's repressed anger outward.

Indeed, as to the first note, Schmidt urges, "An important element of recovery is hope. One must expect and believe that he or she will live."[100] He cites the testimony of medical hypnotherapist Michael Ellner, Nick Siano, author of *No Time to Wait*, and Nathaniel S. Lehrman, former clinical director of Kingsboro Psychiatric Center. All agree with immunologist, Dr. Alfred Hassig, who says, "The sentences of death accompanying the medical diagnosis of AIDS should be abolished," and instead, people with AIDS should be nurtured with hope.[101]

Hope should grow alongside justified anger, the type found in Shilts's and Larry Kramer's activities—although Schmidt might not approve of these models, since both hold to the idea that HIV is the sole cause of the immune disease—the former castigating government backpedaling in a time of crisis, the latter raging against New York City's do-nothing Koch administration. For Schmidt:

> Our survival depends on not accepting the role of victim. If people [such as Fundamentalist bigots] direct death wishes as us, we should direct death wishes right back again at them. No one should be allowed to at-

tack us with impunity. At the same time we need to retain a sense of cool: an appropriate balance of self-preservation, anger, and a sense of humor. Aside from the fact that our lives are at stake, current events really are pretty ridiculous, aren't they?[102]

Asking him for more specific prescriptions—for after I had read his books I wanted to learn more about his actual treatments—I found first that, given that HIV does not always lead to death, a positive person should avoid at all costs current drug treatments, all of which "have toxicity and adverse side effects which of themselves can cause severe illness." Taking them at an early stage of HIV infection could only be justified if there were a "100% fatal outcome" of positivity.[103]

Turning to more proactive directions, Schmidt says, "The Group-Fantasy Origin of AIDS," "The Group-Fantasy Origin of AIDS," "There are the two things which need to be addressed in psychotherapy with anyone at risk for getting AIDS—the suicide syndrome and the somatization of feelings."[104] By "somatization," he means the tendency to choke back emotions that one was taught to repress in an emotionally cramped or abusive childhood, and that, as we saw, add to the self-hatred and the tension of continually bottled-up feelings. "These things have to be addressed vigorously. Once they're addressed, the person overcomes the difficulties in expressing feelings, stops somatizing, and the immune system has a chance to recover."[105]

Detailing the results of his own practice, which, based on intensive one-to-one work with individuals, has been small in number of people with AIDS or HIV positives served, he brings up a patient who "was abandoned three times during early childhood before the age of five."[106] The parents moved to the United States from Cuba, leaving him behind, while other children were kept with the mother and father. This favoritism was because his parents, quite frankly, didn't like him. "He was identified as the one child who was awful, difficult and bad, and scapegoated relentlessly throughout childhood." To escape this household, but not his feelings of being a bad boy, "at the age of 14 he ran away from home to live for seven years on the streets of New York as a hustler and as a drug addict. Then he came to see me in psychotherapy."[107]

He had accepted his parents' evaluation of his worthlessness, expressing his suppressed rage against them by turning it against himself in thoughts of doing away with himself. Schmidt helped him face his suicidal thoughts, which were expressed, for example, in the fact that "at all stages of illness he refused to be hospitalized, even when he was acutely suicidal." In response, Schmidt says, "I would extend treatment hours… and at some points see him for 27 hours in one week while he was ranting and raving and being actively suicidal."[108] Eventually he helped him come to grip with his problems and break out of the pattern of suicidal thoughts, which was the first step toward recovery.

Schmidt generalizes that the necessary first step in therapy is convincing the client that he as a counselor does not forbid him from expressing his feelings and pain, as a parental figure did in the past, but encourages him or her to be forthcoming. Once the client has learned to trust Schmidt, "this usually results in him or her overcoming these inhibitions and expressing these feelings fully. It's quite wonderful to see this." He mentioned one especially dramatic episode. "I knew of one patient who cried for three to five hours a day for a solid nine months. After that, his t cells started rising and have been rising ever since."[109]

When queried about his other results and he pointed to a patient he began working with in October 1992. "He was having a gentle slow decline in the number of t cells.... Then, as soon as we started psychotherapy, the t cells rebounded. They've since gone up forty-seven percent in the last five months, which I think is significant."[110]

In concluding this chapter, I would like to preview as much as summarize. Often philosophers of science distinguish between a strong and a weak version of a theory, and that is something I am tempted to do in presenting this doctrine of "mass hysteria," which is really a two-part theory elegantly cobbled together from strands taken from individual and collective psychology. The strong version of Schmidt's theories, the one he himself holds, would be that under the increased social strain that falls collectively on "fringe" groups, many fall ill from AIDS individually because an early history of abuse has predisposed the person to internalize feelings. A weak version of this doctrine, which I would find easier to endorse, is that individuals' immune systems are taxed by the United States's growing homophobic climate and the mass hysteria around AIDS invoked and furthered by the establishment. This hysteria, along with other factors, leads to immune breakdown.

After reviewing, in the next two chapters, microbial involvement as a cofactor in disease etiology and considering the overall importance of a multifactorial causation theory, I will come back to a number of questions that Schmidt has answered, but not fully enough. He talks about New Right ascendancy, but explains neither why it gained this ascendancy (which I can only explore briefly) nor how the early AIDS campaigners—especially but not only those in the homosexual community—combined fighting against the illness with battling the darkness spread by neoconservatives. When we look at these early efforts, we will see that Schmidt's defective and misleading view of the gay male community, which he sees as detrimentally fixated on AIDS and HIV positivity, was through this "fixation" achieving advances in community building and activism. This activism, at least in Cindy Patton's argument, is what partially motivated the laggardly Reagan administration to finally take action. Patton's argument, in other words, is that:

> Safe sex was viewed by the early AIDS activists, not merely as a practice
> to be imposed on the reluctant, but as a form of political resistance and
> community building.... It was this libertory subtext that seemed most to

raise the ire of the New Right, and it was the first premise of safe sex organizing to be lost when professionals unveiled their plans for safer sex education. [111]

We will see that Schmidt's complaints about weaknesses in the gay community, although he does praise ACT UP, miss both early strengths and the way in which deformations from outside the community hindered its full empowerment as professionals and others implicitly fighting against the self-assertion and autonomy of the affected, entered the AIDS environment. We might add, as a bridge to our next discussion, that, as Patton shows, early intra-community AIDS activists were highly effective in changing sexual and drug-taking behavior, *relying on an ad-hoc multifactorial etiology* to do so, since much of this work was done before the putative discovery of HIV and before testing. Patton notes, "Men understood and made major effective changes without the benefits of antibody testing."[112]

Before making the case for multifactorial causation and looking at the benefits produced by holding such a view in terms of understanding and working against the disease (as well as the light it sheds on the issue of whether AIDS calls for the adoption of a new medical paradigm), we need to focus on one other group of immune-system weakeners, one we have repeatedly mentioned in passing but not specifically targeted, sexually transmitted diseases and other infections.

Microbial Activity and Sexually Transmitted Disease (STD)

When we talked about the dangers to which fast-track gay males were exposing themselves, as when we took up the issue of poppers, there was always some allusion to the high risk and contraction of sexually transmitted diseases in this population. It was also mentioned that the group was overprescribed or perhaps self-prescribing antibiotics, which, like an STD, was doing no favors to the immune system. So far, though, we haven't concentrated on the specifics of how STDs and other microbial invasions can weaken or rout the immune system. We will do so now.

Syphilis

Though across the board heterosexuals and homosexuals are equally likely to take hard drugs, we showed that one drug, poppers, was taken almost exclusively by fast-track gays (aside from those who took it for medical reasons). In the same way, though STDs are equally widespread over the sexual spectrum, syphilis, according to studies by the Los Angeles Health Department Epidemiological Studies, is most frequently found in male homosexuals. "If you are gay in the US, statistically you are 14 times more likely to have had syphilis than if you are heterosexual."[1] Although whether there is a causal link between syphilis and AIDS is unknown, findings by the Los Angeles Health Department set forth in 1989 and based on independent epidemiological studies, which concluded that "a history of syphilis [is] the most common predictor to the development of AIDS in the absence of all other factors," is certainly suggestive, hinting at some compact between the two diseases.[2]

We do know that syphilis jeopardizes immunity, while "it is of special interest that almost all of the symptoms of AIDS are also documented symptoms of secondary and tertiary syphilis. This has led some researchers to claim that AIDS is a combination of chemical immunosuppression and epidemic syphilis."[3]

Herpes Viruses

To see in a more blow-by-blow fashion how an STD disables immune function, we might consult the work of Nina Ostrum, reporter and managing editor for the now defunct *New York Native*, who has written on findings on the Human Herpes Virus VI (HH-VI) made at Columbia University and other institutions.

HH-VI, Ostrum explained when I interviewed her, has two varieties: A and B. Type B is common and benign, while "type A is found in people with AIDS and other immune-compromised conditions." She maintained, "Scientific paper after scientific paper show what this virus does to people and how it destroys their immune system."[4] Its action is to destroy such vital parts of the immune reaction organization as T cells, B cells, and natural killer cells. So, in her definition, "It's the real heavyweight virus."[5]

Moreover, this is not a virus with whom those with AIDS have a passing acquaintance. As Dr. Marjorie Siebert states, "Herpes is found in almost every case of full-blown AIDS. The herpes virus… has immune- suppressive properties of its own and increases the replication of the HIV virus."[6]

Her contention is supported by research done by Konstance Knox and Donald R. Carrigan, demonstrating "100 percent of HIV-infected patients studied (ten out of ten) had active Human Herpes Virus 6 Variant A infections in their lymph nodes early in the course of their disease."[7] These scientists who have focused their studies on understanding HH-IV adopt the supposition, shared by others, that the long period between infection with HIV and the breakout of AIDS, accompanied by an upping of the amount of HIV retrovirus in the affected, can be accounted for by the hypothesis that the immune system has to be partially deranged already for HIV to rear its head. Since in 75 percent of the patients in their trial had herpes in an early stage of AIDS, they surmise, "The A variant of HHV-6 is capable of breaking HIV latency, with the potential for helping to catalyze the progression of HIV infection to AIDS."[8]

Synergy of STDs

We have seen that herpes and syphilis are found in conjunction with AIDS and that, even without being connected to the immune deficiency syndrome, these sexually transmitted viral infections undercut the immune system. The same can be said for what could be called, in talking of other STDs, all "the usual suspects."

Although it is not charged that all these other agents prepare the way for HIV to emerge from its latency—as we've seen, many dissidents don't feel HIV's role is crucial to AIDS, anyway—these STDS are said to promote the transmission and destructiveness of the disease. Siebert mentions, "Epstein Barr virus… increases the penetration of white blood cells by the HIV virus.… [Moreover] when herpes lesions or lesions of syphilis, molluscum or papilloma virus are present, HIV virus spreads much more quickly," from person to person.[9]

As Philpott notes, many of these STDs have been indictable from the very beginning as accessories to the disease: In Gallo's original findings of HIV in association with the disease, he actually could isolate the retrovirus in only half the blood specimens, yet they all had other infections. "Some of them seemed to have HIV infections but all of them had infections with all sorts of other germs," that is, from other viral infections.[10]

As we've noted, the gay male subgroup, particularly those who ran in the fast lane, had a high propensity to be, at one time or another, hit with STDs as well as other infectious agents. In his book *Rethinking AIDS* , Root-Bernstein has dug into the problem. The most immunosuppressive of those he found are cytomegalovirus, herpes simplex viruses, Epstein-Barr virus, hepatitis B virus, Mycobacteria species, Mycoplasma species, and candida.[11] Dr. Christopher Calapai goes on that one or a combination of these infections always accompany AIDS and should be zeroed in on by scientists who are now too exclusively focused on HIV. "I think we need to focus on dealing with these viruses and infections as well," he says.[12]

Every one of these agents is a bane to the immune system. Cytomegalovirus, herpes, and Epstein-Barr, for example, "infect macrophages and lymphocytes, and… cause a decrease in CD4 cells and an increase in CD8 cells leading to a reversed CD4/CD8 ratio, a situation characteristic of people with AIDS," Root-Bernstein explains. Besides reversing the CD4/CD8 ratio, hepatitis can also cause severe liver damage impairing its ability to cleanse the body from accumulated drug toxicities.

If we are looking for the culprits that helped AIDS break and enter, then it would certainly seem more justifiable to grill sexually transmitted diseases, whose potency and previous trespasses are already down on the charge sheet, than pull in HIV, which may be no more than an innocent bystander and is at most a first offender.

Effects of Parasites and Other Infections

As the list Root-Bernstein prepared suggests, it is not only infections that are passed along by sexual intercourse that seem to have a hand in the AIDS disease. Other infections may well be active in softening up the immune defenses. These frontline sappers include, in the words of Snead, "yeast or candida infections, toxicities, commercial drugs such as lithium, malnutrition, kala-azar and other parasitic diseases, [as well as] measles." He finds them to be candidates for involvement, because each "can cause T-cell changes similar to those we see in AIDS."[13]

R. Watson in an article in *Parasitology* lays a finger on another component that should be gauged when we look into sources of immune suppression. Like poppers, "parasites readily accompany a fast track lifestyle," he points out. "Cocaine and alcohol, for instance, have been shown to reduce resistance to intestinal parasites and intestinal lymphocyte numbers."[14]

When I interviewed Dr. Joan Priestley for a chapter in *The Pain, Profit, and Politics of AIDS*, she said that in her work with PWAs, "I parasite test everyone who comes in through the door, whether or not they have any known risk factors, whether or not they have any diarrhea or other gastrointestinal problems."[15] She recognizes that immune-debilitating effects of varied parasites are contributors to AIDS.

Is There Support for a Multifactorial Etiology?

From one way of looking at it, I am doing exactly what I have accused the establishment of doing: fishing for a connection between AIDS and a physical manifestation that is found accompanying it, and then proclaiming that the two are causally married. That is, I am basing my argument on a correlation, while hinting that there is a causal relationship underlying it. However to upbraid me for following steps I disallow others would be to make a false analogy. Where the orthodox hold a strong version of a cause-effect doctrine—acting alone, HIV causes AIDS—I hold a weak version. I argue that a number of factors in varied combinations, their multiplicity accounting for the different ways AIDS mixes with related diseases, tear down the immune system, issuing in AIDS.

It's not yet the time—this will come shortly—to expound on why multifactorial explanations are generally preferable to monocausalist ones; but I can say that, from the point of view of the orthodox, my method here may be compared to Bacon's search for instances of union. "They are those which mingle and unite natures supposed to be heterogeneous and marked and set down as such in the received divisions."[16] "The received divisions" would be the pockets devised by the AIDS establishment, one of which holds AIDS and HIV, another of which holds syphilis and other STDs, for example, which are said to be unimportant contingencies in connection to the immune dysfunctional illness.

Let's take one of Bacon's examples. He remarks that it is generally believed that there are three separate genuses of fire, whose natures are different and which cannot intermingle. The fire of the sun has a fructifying and generating power, since it helps crops to grow, while earthly fire is generally malign, for "the heat of fire wastes and destroys."[17] However, this too hasty categorization slips the following fact: "When a branch of vine is brought within a house where a fire is constantly kept up, the grapes ripen on it a whole month sooner than they do out of doors… so the ripening of fruit… may be brought about by fire."[18]

The above instance shows that the divisions artificially imposed by science in Bacon's day were shortsighted in that different elements of nature were placed in watertight compartments.

In this discussion of microbial activity, similar claims are being made about the lack of perspicuity in the orthodox's decision to dismiss the implication of viral and other infections in the bringing to birth of AIDS. The statistics provided on, for example, the constant companionship of syphilis and AIDS indicate that here are two entities "which mingle and unite natures [that are] supposed to be heterogeneous." We will plump for the general validity of a re-classification that acknowledges that many factors should be causally associated with AIDS.

Multifactorial Etiology: A Closer Look

In a number of spots earlier in this inquiry, it was necessary to discuss ahead of time other cofactors that may be linked to AIDS's onset. We talked, for example, of passive anal intercourse as potentially playing a role in the illness, because it naturally fit in with an investigation of heterosexual sexual transmission. Although it would hardly do to rehash what has already been said, it would be useful to rehearse some of this earlier material before moving on to general conclusions about the dissidents' views of what causes AIDS. Along with a partial review, this would also be the place to mention a number of other possible cofactors, such as the elimination of microflora, hormones and blood transfusions, that have not yet been looked into but merit brief consideration.

The Loss of Microflora Due to Antibiotics

In an article on the *Rethinking AIDS* Web site, V. Koliadin has set forth a well-argued, radical hypothesis concerning the role of the destruction of the body's microflora in the AIDS disease process. It may be remembered that we previously brought up the way antibiotics upset the delicate balance of these flora in the intestinal tract. As Root-Bernstein lays out the theme of the overreliance on these drugs, "Many gay men, particularly promiscuous ones, also tend to abuse antibiotics, apparently as a prophylactic or remedial measure against repeated sexually transmitted diseases."[1] One costly result is the depletion of T cells. This may be because these antibiotics strip the body of zinc, which is needed for cloning these cells. "Gay men typically have unusually low zinc and sele-nium serum levels and abnormally high copper levels as compared with hetero-sexual men, women and lesbians."[2]

Root-Bernstein is particularly attentive to the role of zinc loss in disease generation, but Koliadin sees the effect of antibiotics in eliminating microflora as much more significant. To understand his perspective, we must look at how he reenvisions what sectors of the body are essential for fighting off infectious attack. He says:

> Contrary to widely shared belief, the immune system (T-cell mediated immunity) plays no significant role in protection from the bacterial and parasitic opportunistic infections observed in AIDS patients.
>
> The main factor which prevents these infections in healthy individuals is normal resident (friendly) microflora—the harmless and frequently use-ful microorganism which populates various body compartments: mucous membranes of the inner tract (guts, lungs, respiratory tract, etc.) as well as skin. These microflora occupy ecological niches in organisms and pre-vent these areas from colonization by other species of microorganisms, including species pathogenic for the host macroorgansim.
>
> The main factor which triggers bacterial parasitic opportunistic in-fections in HIV-positives and AIDS patients is destruction of their nor-mal microflora… by unusually prolonged exposure to broad-spectrum antibacterial and antiparasitic antibiotics.[3]

It seems we could view Koliadin's thesis as adding another step to the causal framework the dissidents have been developing in this section. Their ideas about the effectiveness of viral and other agents in weakening immunity would now become steps two and three, in a new enumeration for which Koliadin has supplied step one in the following:

1. Antibiotics kill off indigenous microflora.
2. STDs and other infections, ones already latent, move into the now empty nooks and crannies in the body.
3. These infections enfeeble the immune system, allowing opportunistic diseases to enter.

As Koliadin urges in explanation of the second point, normally, harmful viruses and other infectious factors are present in small quantities in the body but microflora keep their numbers low. When the body is healthy, the unwanted microorganisms "cannot proliferate effectively because they lose ecological competition with normal resident microflora." Furthermore, these indigenous microflora do double duty. Not only do they outcompete alien microscopic entities but they share vital work in many body functions such as digestion and production of vitamins. "Depletion of the microflora results in depression of these functions and leads to worsening general health and decline of non-specific resistance of the organism."[4] The result of these depredations is the appearance of opportunistic infections "as well as long-term diarrhea, malabsorption, [and] weight loss…. In cases of severe enough opportunistic infection, T-immuno-deficiency developed as natural stress-reaction to the opportunistic disease."[5]

If we assume, for heuristic purposes, that Koliadin's position is reasonable, then we can say that what the establishment has done in relation to the opportunistic infections that arise during an AIDS case is at cross-purposes to solving the problem. Orthodoxy proceeds from an understanding of the disease that is opposite to that of Koliadin, holding that the opportunistic infections, "spotted by the CDC in 1979-1981," were to be explained as the product of "acquired T-immunodeficiency as the primary biological cause."[6] That is, the various health agencies held then, and still continue to hold, that the infections arose because of immune deficiencies. Koliadin's idea, alternatively, is that T cell breakdown and other immune-system lapses happen because of the infections, themselves made viable by the disappearance of microflora, which has been instigated by the omni-destructive effects of antibiotics that mow down both benign and malign bacteria.

Going by their own lights, CDC researchers prescribed what Koliadin might label as "the hair of the dog," more antibiotics to fight infections ultimately derivative from the effect of antibiotics. To begin the cycle, in 1981 cases of opportunistic infection "were announced as first signs of new and deadly immunodeficiency," later designated AIDS." The treatment recommendation, still adhered to, was this: "intensive permanent prophylaxis by broad-spectrum antibiotics…. Thus," Koliadin goes on to comment, "instead of cure, they received just the disease-causing factor; this made their opportunistic infections recurrent, severe, and seemingly incurable."[7]

To present the pattern of establishment treatment module in the way that Koliadin perceives it, which can be counterpoised to our earlier sketch of the disease's progress, we have:

1. Antibiotics are administered to deal with a patient's syphilis or other infection.
2. The antibiotics kill off indigenous microflora.
3. Latent infections have a field day, taking over places left available by the departed flora.
4. Doctors prescribe more antibiotics to handle the infections.
5. The infections are temporarily staved off by the drugs, but any remaining symbiotic microflora and those trying to reestablish themselves (which they normally do after they have been momentarily thinned by antibiotic attack) are also cut back, further weakening

the overall operation of the body and paving the way for the return of the original or a new malign infection.

This deadly cycle was also recognized by G. T. Stewart, who said in *Genetica* that for fast-track drivers, "the unregulated use of antimicrobial drugs… inhibits the competitive flora of the intestine, opening it to bacterial and fungal super-infections." Stewart lists *Pneumocystis carinii, Candida albicans*, cryptosporidia, and organisms causing chronic diarrhea as the mainstays that move into emptied niches.[8]

This type of procedure—in which a prescribed cure turns out to be welcome to the illness, which after the course of treatment ends up stronger than ever—is seen recurrently in the AIDS picture, at least as the dissidents view it. We've seen it before in the discussion of needle sharing, where it was strongly argued by dissidents that hard drugs, not blood trading, are the primary factors promoting AIDS among IV users; so setting up needle exchange programs that facilitate drug taking ends up making it easier for the main instigator of the illness, the input of hard drugs into the body, to have greater influence.

Sexual Practices

I'd like to return to the touchy subject of the influence of different sexual practices on susceptibility to immune dysfunctional disease. This is touchy because it can easily lead to accusations from the politically astute that such a discussion on how, for example, unprotected anal intercourse may have a role in provoking disease is merely a veiled attack on male homosexuality itself.

Rather than dismiss this type of accusation out of hand, after reviewing some of the earlier discussion of sex and introducing a few additional points, I will give some thought to how to fend off such complaints by a close examination of what I am and am not saying.

However, as a foretaste of that discussion, I might say a word about such charges as that of I. Young, who says this about the fast track. "Another risk factor for the gay men who developed AIDS was promiscuity. Most… had dozens of cases of venereal disease in the decade before they developed AIDS. Each time they were treated with stronger and stronger doses of antibiotics," with predictable effects on immunity.[9]

We say that in this case, as will be found analogously in others that will be studied, what is being indicted is not promiscuity alone but a ladder of causes. In this sense: Antibiotics cause death of microflora (and without them opportunistic infections run rampant).

Antibiotics were taken by gay males in response to venereal disease. Venereal disease was picked up via promiscuous behavior. Ergo, the whole ladder deserves blame, not the first rung. Think of it this way. If a man were terribly promiscuous but insisted he and his companions wear condoms, STDs could probably be avoided and the chain of events leading to immune weakening would be snipped. Or, if a man who had caught a venereal disease refused antibiotics and cured himself using natural herbal techniques (assuming for the moment these would be effective), again the promiscuity would not greatly damage his immune system. The focus of homophobes on promiscuity as the sole causative factor making men sick, then, like a belief that HIV is the sole culprit in bringing about AIDS, testifies to a rudimentary understanding of the complexity of causation, which cannot stand up to scrutiny when contrasted to a realistic view of the interplay of factors at the heart of events.

To begin, we might recall an earlier point I made about fellatio. This was that those—no matter how randy and how many partners they had—who restrict their encounters to protected oral sex do not get AIDS. At that time, I was speaking about this in contrast to anal sex, but S. Byrnes, writing in *Explore*, calls our attention to another avenue of danger that those who restrict themselves to mouth-genitals contact escape. He notes, first, as I've said, that when AIDS first arose, it was noticed that "those with a preference for oral sex, giving or receiving, may have gotten other venereal ailments, but they did not catch the new disease."[10] Nor did exclusive tops (those who were the givers rather than receivers in anal sex), for that matter. Those affected were either passive recipients in anal intercourse or "fistees (those who would have someone's fist anally inserted into them)."[11] Byrnes asked himself why these particular groups were struck and his answers are these: "What was it that bottoms and fistees had in common, besides poppers to relax the smooth muscles of the anus? Lubricant and lots of it if they were promiscuous."[12] He stresses the coincidental (and suspicious) appearance of certain types of lubricants and the beginnings of the AIDS epidemic.

> Were new lubricants introduced to the gay community in 1978? Previously, gay men had used KY jelly, Crisco, or baby oil for anal sex but in 1978 there were new lubricants introduced and heavily marketed to the gay community, viz., Lube and Performance, as advertisements in back issues of gay periodicals show. As a matter of fact, 1978 marked the dawn of special lubricants, both hot and regular, formulated for and used by gay men. They were all oil-based and contained very high amounts of acetone and benzoic acid in them. The oils were, like the bad olive oil in Madrid, denatured. Curiously, as these lubricants became available to gay men in other countries, via mail order, AIDS began to appear in those places. [13]

In Byrnes's view, commercial products employed to ease intercourse are at the root of immune collapse because they contain acids and oils that are inimical to the body.

To move to the effect of intercourse proper, it will be recalled that we earlier dealt with opposing views of whether those in the fast lane were getting sick because of a multiplicity of partners. Those holding that promiscuity did not affect health (in relation to AIDS) highlighted the fact that sperm of positives had been repeatedly found to contain negligible amounts of HIV.

In rebuttal to this point, it was brought up that anal intercourse, for the passive participant, can damage the rectal wall, although at that point we might have gone into more depth on how this damage occurs. In the words of G. T. Steward, "When the thin submucosa [of the anal wall] is eroded and blood vessels damaged, the tissues and blood stream are opened to invasion by all the organisms of the faecal microflora, by the pathogens of all the sexually transmitted diseases, and many others." He goes on that the possibility of rectal injury will naturally be augmented by frequency and "violence of the sexual activity and preference, as with 'fisting' and other accessory, traumatic and contaminating procedures" and with multiplicity of partners.[14]

Steward summarizes studies by Witkin and Sonnabend, as well as Root-Bernstein that point out, further, that once semen enters the bloodstream, it can have allogenic effects, impinging negatively on immune system functions. Moreover, it has been conjectured that N-nitroso compounds from semen may react with volatile nitrites from poppers to "contribute to Kaposi's sarcoma which occurs in this context independently of HIV."[15] According to this author, then, anal trauma during intercourse is especially dangerous because it will allow semen into the blood.

Prof. Root-Bernstein, by the way, remarks that some studies indicate anal lesions are not necessarily typical of male homosexual communities. He says that the first medical report connecting rectal lesions to homosexuality was made by a New York City doctor in 1964. Among a roster of eighteen avowedly homosexual patients, this doctor had seen "4 instances of anal ulceration associated with trauma; 5 instances of granulomatous anorectal lesions; 5 instances of non-specific anorectal lesions; and 5 cases of perianal condyloma acuminata." These cases were reported as quite unusual and noteworthy for that reason.[16]

However, by the mid-1970s such cases were commonplace. In another study done in 1976 in New York City, "134 patients had condyloma acuminata, 43 haemorrhoids, 31 non-specific proctitis, 30 anal fistulas, 18 perirectal abscesses, 18 anal fissures, 17 amoebiasis, 16 pruritis ani and 7 trauma and foreign bodies."[17] So in this new crop of cases, the problems of semen entering the blood were heightened.

To remain with this topic a moment, we might go to another finding brought up in Steward's survey of evidence, which supplements his point about allergic reactions that may follow the placement of foreign semen in a recipient. "Male homosexuals [as well as females who have engaged in anal intercourse] have anti-spermatozoa antibodies in the blood, which cross-react with T-cells and have been linked to the occurrence of... testicular atrophy in homosexual men."[18] The antibodies indicate that the body is, indeed, calling its immune system into action to combat the semen, while the noted cross-reaction with T cells underlines that such "patients, with and without HIV, reject their own

T-cells because they cannot distinguish them from antigens from spermatozoa," and other foreign substances.[19]

Prof. Root-Bernstein picks out the disaster that can develop from this cross-reaction. After semen have gotten into the blood through rectal abrasions, antibodies rush out to eliminate them. However, the antigens (their identifying marks) closely resemble antigens on T cells, which causes a disruption when antibodies are manufactured to eliminate the semen. This "may cause gay men to develop an autoimmune reaction against their own immune systems in which B-cells are pitted against T-cells."[20] He notes, as another liability connected to the intrusive semen and sperm, that they "contain agents that are, in and of themselves, immunosuppressive and act to protect sperm from vaginal and cervical lymphocytes during heterosexual sex."[21]

Still, even though there is considerable evidence concerning the potential health problems associated with anal sex, the reader who attended closely to the earlier discussion of sexual transmission may be ready to chide me for having forgotten to re-mention an important consideration brought up before. A large percentage of heterosexual couples also practice anal sex and yet do not seem to be coming down with AIDS except in very small numbers. How does one reconcile this with any argument for anal sexual transmission?

As I mentioned previously, there is no good comeback to that question at this point, so for a final judgment on transmissibility we'll have to hang fire, only saying that there are a number of health hazards related to this manner of intercourse, though they would be more or less eliminated if during anal intercourse a condom is worn, a safe lubricant is used, and the sex is not violent enough to create lesions.

The same cannot be said, of course, for fisting and other more extreme sexual activities. Fisting is also something that needs to be looked into as a cofactor in disease genesis because, like poppers, it only came into vogue in the 1970s. "Medical experience and surveys by Sohn et al. (1977) strongly suggested that this practice [fisting] was an essentially new and rapidly proliferating activity exclusive to the gay community."[22] Prof. Root-Bernstein actually sees these two components of fast-track behavior as complementary. After noting that the sales of poppers skyrocketed in the same 1970s time period, he comments, "Since one of the major uses of such drugs among gay men is to facilitate anal intercourse and fisting (Newell et al., 1985), it is probable that the two practices evolved together."[23]

From his digging into epidemiological and sociological material, Root-Bernstein has also found evidence that fisting was not the only extreme form of sex that was put on the agenda in this period. Writing in 1990, he noted, "Forms of masochistic and traumatic sex other than fisting, popularized by the Eulenspiegel Society and Hellfire Clubs, also appear to have become more popular during the last two decades."[24] To indicate the relevance of this to AIDS, he cites a 1980 study by Corey and Holmes that indicated fisting and other extreme practices were higher "among homosexual men in cities such as San Francisco and Los Angeles that have high rates of AIDS than in cities in which AIDS is rare, such as Denver and St. Louis."[25]

Cindy Patton, although not endorsing this particular interpretation, has also stressed that understanding the different types of gay culture in different cities is crucial to charting the crisis. She puts this important concept like this:

> AIDS—especially in the gay male population—strikes a set of sub-communities that are bound together by cultural similarities and some migration.... As in many epidemics of the past, community composition and mobility may prove a key to understanding. Each city where AIDS appears has different gay community patterns and histories, sexual mores, and even sexual practices, with differing levels of gay health care and possibly different opportunistic disease pools which affect the secondary disease patterns in AIDS.[26]

Patton is good to bring in at this point, since she has presented a strong case against the promiscuity-as-cause position on ideological grounds. That is to say, she believes it is possible that, looking purely at the science involved, sexual practices play a role in bringing on the disease, but she says we must always be careful to distinguish science from stereotypes. Since there are already so many hostile images circulating among the homophobes about the "sinfulness" and "uncleanness" of homosexual relations, we have to constantly be attuned so that prejudice does not bleed into science. She thinks this may happen in relation to the viral overload theory, whereby such activities as popper use and bathhouse promiscuity encouraged illness. She fills in:

> These theories posited that gay men's bodies had been worn out by too much sex, drugs, previous infections, cheeseburger dinners, quiche and fast-lane life. Their immune systems had simply given out
>
> [She argues that such a theory pandered to prejudices.] The theory touched all the bases: gay men are selfish, upwardly mobile (you have to have a lot of leisure time to get all that sex!), irresponsible, and "dirty." Hell bent on destruction, they don't take the body's signs of minor infection as a hint that perhaps they are overdoing it.... Put another way, types of people get the disease they deserve.[27]

Jan Zita Grover echoes these thoughts by talking of Peter Duesberg's "(homophobic) speculation that AIDS is the result of an immune exhaustion brought on by irresponsible living."[28]

Going further, Patton argues that even the "objective" assertion that it is the passive partners in anal intercourse who are putting themselves most at risk may conceal stereotypes. She says that mainstream culture "considers men who want to be fucked true queers," so the labeling of the passive position in anal intercourse as the threatened one "at once confirms that real queers are imitation women, and that straight men—who might from time to time stick their manhood in queer holes but only dream of getting fucked—won't get AIDS."[29]

Following Schmidt's view on the prevalence and starkness of antigay prejudice in the United States, which has only been accentuated by the rise of AIDS, it is impossible to not address the question of bias when discussing homosexual lifestyle. At the same time, though, let us acknowledge that the researchers who early in the crisis proposed viral load or fast-track theses to account for AIDS in the West were not simply being prejudiced or perverse, since gay males did seem to be a major group under attack, and therefore it would have seemed that something about them made them more liable to the illness.

At this juncture, I merely want to recapitulate three points to give them more substance. First, even if we now say certain behavior of the fast-trackers (Patton calls them "clones") was unhealthy, this shouldn't be taken as a blanket condemnation of gay life. The clones were only a minority within the minority. Patton, for instance, contrasts them to another equally prominent but less re-ported-on fraction of the gay community, the politicos. These were the more puritanical types who spent their time leafletting, getting petitions signed, and agitating for an improvement in gay rights. Also in the gay population were monogamous types, homebodies, and so on. Just as every college has its "party animals," who (on most campuses, at least) are a distinctive minority, so each gay community had their clones. Seeing all males homosexuals as fast-trackers, as the media so often does, is simply adopting the familiar racist tactic of black-ening everyone with the same brush.

Secondly, when nowadays we condemn, say, overuse of poppers in the 1970s and early 1980s, we are not at the same time saying people necessarily should have known better. Compare in your mind the popper users to the people who were smoking in the 1940s, say, and died of lung cancer thirty years later. Do we rail at them for their mindless choices? Poppers are reasonably safe, the ads said back then. The FDA temporarily had taken them off their list of drugs that can only be obtained by prescription. That they were back on the list, it might be surmised, indicated they weren't terribly unsafe. After all, what were considered deadly drugs like cocaine or hashish, even marijuana, were not put on this restricted list but proscribed totally. Back in the World War II era, it was the same for tobacco. Page through a few magazines of the day to look at the advertisements for Pall Mall or Lucky Strike and you will find that smok-ing is not only proclaimed to be safe but even said to promote health! More-over, everybody was lighting up, just as in a certain strata, everybody was inhaling. Poppers and cigarettes were sexy, for god's sake. What is being as-serted, then, is that certain practices that now seem unconscionably risky were once seen as innocent, as innocent as, in days gone by, puffing on a Lucky.

Lastly, let me make a more complex point, one that has already been men-tioned briefly and one I will follow up further as we proceed in our discussion of multifactorial causation theory and AIDS Inc. As we saw, the hedonistic lifestyle of the clones could be considered a collective response to political exigencies. Big victories had been obtained in the Stonewall case and in gaining political clout (especially in San Francisco). Some young men, seeing new space and opportunities for the expression of alternate lifestyles, came to believe a "swinging lifestyle" was the way to embrace this freedom. Yet, although the clone culture was built on the gay subculture, it also incorporated many elements of commercialism. Think about this for a moment. How different is taking a popper before having sex from needing coffee and a donut (that is, sugar) to start one's day in the morning, a butt at break time, or a tranquilizer to unwind at night? In other words, some of the practices that seem most unhealthy in the fast-track lifestyle are little different in kind from those done and considered eminently acceptable in majoritarian life. In short, the most lamentable aspects of the behavior of the disco-hopping, bathhouse-patronizing fast-trackers were not those derived from the gay community but ones closely mirroring vaunted practices of consumer culture.

Nutritional Deficiencies

Another factor that may contribute to immune weakening, and this time one that does not mirror majoritarian practices but simply emulates them, is nutritional deficiency.

One of the keynotes of my own health revitalization and antiaging programs is to change the diet, since the one we tolerate as standard in the United States is disease- rather than life-enhancing. As J. McKenna has it, "A diet high in processed and refined foods, low in whole and 'live' foods has become the staple diet for the vast majority living in the West."[30] The food consumed is laced with pesticides and herbicides, while the meats have their share of antibiotics, hormones, and steroids. Such a diet not only leads to severe vitamin and mineral deficiencies but will eventually eliminate bowel flora and lessen immune strength.

To bring the point home, we might listen to P. Cox, "Many research studies have shown that people with AIDS, and those considered to be living at risk, suffer from wide-ranging nutritional deficiencies."[31] These deficiencies have been shown to involve vitamin A, vitamin E, vitamins B2, B6, and B12, copper, zinc and selenium.[32] "In addition to these, the sulphur containing amino acid related substance glutathione (GSH) has also been found to be typically deficient."[33]

This is an area that calls out for further research. As we will see, when we turn to alternative therapies for AIDS and related illnesses, all the doctors with whom I've spoken see an improvement in diet as vital to treatment.

Sensory Overload

"Sex, drug and rock'n' roll." We've already commented on the first two words in this phrase. A few pages ago we mentioned possible negative reactions of the body to anal sex. Earlier we brought up the down sides of taking serious drugs. Root-Bernstein surmises, "Drug abuse is inherently risky... because shooting dope may create a weakened immune system in which AIDS can take hold."[34] The last phrase (rock 'n' roll) is invoked not because music has toxic qualities but because being exposed to cascades of booming sound can hurt the ears. This might seem a rather frivolous point to bring up, as it was made by I. Young in a piece in *Continuum*, but the underlying point is valid and needs to be made. Sensory overload can lead to immune overload.

Young, who is a musician, has prejudices against the kind of records that were played in fast-track watering holes in the 1970s. He states, "I hate disco. I would almost say that this is a corollary: he who loves music hates disco; he who loves disco hates music."[35] However, when he talks of disco in relation to immune dysfunction, his case has nothing to do with the tin ears of many clones but with the volume at which disco was blared. "At a typical gay disco the sound would be right at pain threshold, all night long. As a consequence, many of the young men who were in the gay disco scene in the 1970's had to wear hearing aids in the 1980's."[36] He goes on to contend that this exposure to loud music played a small role in overstressing immunity.

Admittedly, this would have been a minor role, but it is worth noting that anything that repeatedly overtaxes the senses will contribute to immune system hindrance over the long term.

Blood Transfusions

It may be useful to end this chapter with a discussion of the dangers connected to those who have had blood transfusions and later came down with AIDS. Their cases pose special difficulty for those dissidents who claim HIV is not or is only a minor factor in causing AIDS. The orthodox single out transfusees as proof of their belief concerning HIV's implication as the central motor in AIDS's etiology, since these blood-recipient PWAs often do not have other obvious contributory factors, such as IV drug shooting or popper use, to point to as alternative explanations of why they got sick. Those who have gotten transfusions are an even better example than AIDS-infected infants, insofar as these babies are generally the children of IV drug users and so may have been damaged by drugs received in utero. Thus, the dissident faction has been pressed to explain why blood receivers—especially hemophiliacs, who have a high incidence of both transfusions and AIDS—come down with the immunodeficiency illness. The dissident camp has offered a number of plausible reasons to explain the relatively low level of health of these recipients.

As for hemophiliacs, John Lauritsen at the Alternative AIDS Symposium in Amsterdam in 1992 brought up that "hemophiliacs were born sick," which is to say their condition "goes far beyond the inability of the blood to clot."[37] Improvements in medical care have made it possible for many to live normal lives, but they do so under the burden of a genetic defect. One aid to their existence has been Factor VIII concentrate, which is a batch of blood taken and distilled from many donors. However, there is a downside to this life-saving technology. "Factor VIII concentrate represents a pooling of all the microbes and all the toxins from the blood of many donors. It is not good stuff to take. So, it may very well be that AIDS in hemophiliacs is nothing more than congenital sickliness aggravated by Factor VIII concentrate."[38]

Yet even living under these stressful conditions, with precarious health, the threat of infection via blood concentrates, and the high incidence of AIDS, "the median life of American hemophiliacs has doubled during the last 10 to 15 years [since 1994, even] after 75 percent (15,000) already had been infected by transfusions," Peter Duesberg noted in *Insight* magazine.[39]

This is to say, improvements in health care of hemophiliacs (not, correspondingly, of PWAs) had been extending their life spans even in the face of the new threat of HIV infection. The continual increments in longevity for hemophiliacs, however, did not stay on this upward course; by 1987 life expectancy for this group again dropped. (Duesberg's figures were dealing with the average over the 1979-1994 period, but year by year a downward trend surfaced eight years into the period.) To account for this reversal, David Rasnick stresses, 1987 was "the year that AZT—the drug now openly admitted to be detrimental to AIDS patients—was first widely prescribed to HIV-positive hemophiliacs."[40]

But what about those who are not hemophiliacs and thus do not have a congenital defect, and yet, acquire AIDS after transfusions following, for example, a routine operation? Dr. Gallo, for example, in making an argument for HIV as the primary cause of AIDS, calls our attention to a study that showed nineteen out of nineteen blood recipients who developed AIDS after transfusions were found to have had at least one donor who was HIV positive. (The study was done of those who had received their transfusions before blood was checked for the retrovirus.) And other studies have come to similar conclusions.[41]

Hodgkinson, in reviewing the reports of various researchers, notes that transfusees that later get AIDS are usually those who receive large amounts of blood. "The typical patient who developed AIDS had received an average of twenty-one units of blood or plasma during their hospital stay—four to five times the general average."[42] So, just as the hemophiliac who received a concentrate extracted from numerous donors is more likely to get blood that contains traces of infections that are not screened out by the blood banks, so a blood recipient who receives a large number of blood units is more likely to receive additional infections.

Among those who received blood later thought to contain the HIV retrovirus, such as those who got blood before the blood was checked for the retrovirus, it was found "the disease rate was thirty times higher [for those who got ten or more units of blood]... compared with those who received less than that."[43] Moreover, since, as we've seen, those that have HIV are very likely to have many other infections, the blood they donated "was much more likely to be contaminated with other immunosuppressive viruses transmissible in blood, such as cytomegalovirus, Epstein-Barr virus and hepatitis viruses."[44]

Furthermore, although the infusion of blood may be life-saving, the body is impaired by it in some ways. "It has been established that even properly typed blood causes profound immunosuppression."[45] Paradoxically, this immune depression can be beneficial when it helps the body accept organ transplants; but it also will increase the risk of cancer. As Root-Bernstein writes, "Although the exact mechanism of transfusion induced immunosuppression is unknown, T cells are certainly a primary target, and B-cells and macrophages are also involved."[46]

Moreover, in talking about the debilitating nature of blood transfusions as well as the trauma or disease that leads to a person needing these transplants, it should be realized that those who get blood have a low survival rate. Root-Bernstein in *Rethinking AIDS* cites comparative figures on the deaths of transfusees. A CDC-associated study showed that of 694 people who received HIV tainted blood through transfusions, 331 (48 percent) died within a year of transfusion. Prof. Root-Bernstein notes that a member of the orthodoxy might jump to the conclusion that this shows the destructive power of the retrovirus contained in the blood. However, now match this to a control group. "These researchers also provided data for HIV-negative groups," those who got transfusions of bloods without the virus and who were HIV negative. "Their death rate was just the same. Out of 146 recipients of blood components from a random selection of donors, seventy-three died in the year after transfusion."[47]

The question, brought up a moment ago, of infants who die of AIDS, also needs to be considered in relation to blood exchange. We mentioned that most of these children (80 percent) are children of drug abusers; but, as Hodgkinson notes, "10 percent [including some from the 80 percent] had required blood transfusions, and 6 percent were haemophiliac."[48] He adds that they may need transfusions because they are born with the same immune-suppressing factors in evidence that had long afflicted children of addict mothers. "These include foetal malnutrition, prematurity, low birthweight, small head size, distress at birth, jaundice, underdeveloped thymus, anaemia, and neurological and developmental deficits."[49]

Thus we see that even in the case of those who seemed to have gotten AIDS simply through the infusion of HIV-infected blood, there were many other immune-degrading factors involved, ones that are left out of the cursory thinking that is satisfied with finding HIV in a blood donor's contribution as enough reason for the later appearance of the immunodeficiency breakdown.

So far, we have gone through a diverse tally of factors that are said to contribute to immune system collapse and the advent of AIDS. Others, such as the impingement of environmental toxins, might have been added to the list; but I think we have said enough to begin laying out the dissident theory of AIDS etiology.

The Theory of Multifactorial Causation

I n this chapter, we will examine the ideas of those who have observed the many factors, including the ones just laid out, that seem to be involved in the production of AIDS; these people have gone on to put forward holistic ideas of how a set of cofactors work together to bring down the immune system and bring on the immunodeficiency syndrome.

We can distinguish among these theories by looking at the level of mechanism of the disease's genesis they are primed to focus on. There are (1) those that concentrate on the set of practices, such as those of IV drug users, that form the basis for eventual immune derangement; (2) those that are interested in how immune suppression leads to a liability in the body that lays it open to targeting by HIV, and (3) those who are largely concerned with investigating the biochemical aspect, for example, fixing on oxidative stress as an explanation for immune failure at the cellular level. After sketching in these components of a multifactorial theory of AIDS etiology, it will be time for (4) a few considerations on what this means for treatment. I will include reflections on how the failures of the establishment in controlling the disease may be due to its oversimplified model of how AIDS comes to birth. I will end by the promised look at a bigger picture, both in terms of (5) thoughts about how the multifactorial cause thesis fits in with the emergence of a new paradigm and (6) the way different social factions, not merely doctors and medical researchers, must play a part in the democratic organization of health care. This last section will also give me the chance to correct the one-sided ideas of Schmidt on what role the gay community has taken in alleviating the crisis.

(Let me mention that after conclude this material, I will present a short addendum on the question of whether AIDS is man-made. This does not properly belong in the multifactorial discussion, since whether the retrovirus was first grown in a laboratory is of slim importance if one believes HIV is overrated as the agent that singlehandedly causes AIDS. Nonetheless, a brief examination of the question of whether HIV could have been created in a laboratory will reveal some interesting points about the medical-military complex.)

Risky Lifestyles and Immune Depression

Patton and other gay activist writers on AIDS have argued that we need to distinguish lifestyle from practices. It is the adoption of certain risky practices that may put a person in danger of jeopardizing his or her health. If a writer harps on the "lifestyle" as what is deadly, it is very easy to imagine that everyone who can be in any way tied to that lifestyle is guilty of adopting all the dicey activities in which some in the group indulge. This would be a classic example of stereotyping.

I hesitate to use the following quotation from Michael Baumgarten because he keeps the less discriminating vocabulary of "lifestyle," but if I substitute "practices" for "lifestyle," the citation will serve the purpose of perfectly illustrating how elements of an individual's behavioral repertoire rather than his or her lifestyle as a whole should be isolated as crucial in the disease-stimulating process. Baumgarten writes, after considering factors in a fast-track gay male's activities that may have made him susceptible to immune disease, "If any heterosexual participated for two months in the [same] daily [practices], using the drugs, and keeping the hours, and using alcohol, and other things, I'm sure [he] will succumb to that [set of practices]."[1]

Gay male fast-trackers are often picked to illustrate the thesis that a combined group of risky practices may endanger health, although one might think the drug subculture, whose members are injecting known toxins, would make a better example. The reason this group is selected is probably because, while, for example, heroin users had been around since the forties, presumably with the same folkways, the "clone" group among gay males was of recent vintage, as was the new disease. "Data strongly suggest that the gay liberation movement resulted in a great increase in promiscuity among gay men," Prof. Root-Bernstein writes, "as well as significant changes in sexual practices that made rectal trauma, immunological contact with semen, use of recreational drugs and the transmission of many viral amoebal, fungal and bacterial infections much more common than in the decades prior to 1970."[2] (We will turn to a more nuanced presentation of this point below.)

One thing he seems to miss in a rundown we have seen before is the importance of antibiotics. I think I can explain why Root-Bernstein overlooked this factor, but before I do so, let me recall how Michael Verney Elliot characterized the reliance of the clone subculture on these medicinal drugs. He put it like this when I interviewed him:

> When the gay scene really got into the swing in the seventies it was considered quite okay to get lots and lots of infectious sexually transmitted diseases. It was part of being liberated. People just got blase about gonorrhea or syphilis or whatever—you just went to the quack, got another shot of something, and went straight back to the bathhouse or the backroom bar or whatever. Nobody cared the name of the diseases anymore because of antibiotics. And you can only get so many STDs and use so much antibiotics before your body starts to complain.[3]

Prof. Root-Bernstein may have conveniently missed the role of the doctor's lifestyle in contributing to this downward spiral because it makes his whole model messier. That is, he is trying to pinpoint elements embedded in the fast-track life that make for immune dysfunction but antibiotics are not imminent in this lifestyle. Rather, they only enter because they are part of another subculture, that of medicine. "You just went to the quack," as he puts it. Thus, once we invoke this factor, we have to begin talking about the relations between two groups, not one in isolation.

I merely want to remark on that. At the end of this chapter, I will sort this out on a deeper level, when, drawing on the research of the Birmingham School of cultural studies, I will try to say what a subculture actually is.

The salient point at present is that a host of constituents in the fast-track culture were hazardous healthwise. As Camille Paglia describes the scene, in characteristically colorful terms, "Poppers were coming into the scene at that very moment when the bars suddenly went wild. Men were staying up all night, and drinking." From her unique perspective, though, the riotous living was not so much an attribute of gays as a "male thing." As she puts it, "I think women by virtue of the menstrual cycle and so on, learn very early on how to conserve the body," whereas men in general and gay men in this instance tend to excess. "I've always felt that gay men were pushing the limits of the human body throughout that period, also keeping thin and trim at that period, eating very little and drinking a lot."[4]

Hodgkinson fills out this portrait by noting the emotional underpinnings of the clone culture, which contained both beauty and a terror, "owing to a war waged against gay men by the Moral Majority." They also knew "loneliness, alienation and depression; they experienced shame and self hatred.... And as the epidemic developed, they experienced grief: 'they were in perpetual mourning; their hearts broken by the loss of their closest friends.'"[5]

To leave these anecdotal accounts and return to Root-Bernstein's argument, we can say his position is that anyone in the male gay community who adopted certain sexual practices—for him the most identifiably risky are promiscuity, receptive anal intercourse, and fisting, as per Marmor et al., 1982 and Darrow et al., 1987)—is acting to suppress his immune system, which may lead to the total breakdown of immunity and AIDS.[6]

A corollary of his argument would be that whenever such practices have been adopted, AIDS should have been the end product. Indeed, he writes, "I maintain that AIDS is almost undoubtedly a very old disease among gay men, but that it is only recently" that the practices we have already designated "became widespread enough to affect a significant proportion of gay men." He remarks, insightfully, that not only did society's new tolerance for gay self-expression make bathhouses and other gay establishments viable, but "the willingness of gay men to identify themselves as such was crucial evidence in recognizing AIDS, evidence that was not available prior to gay liberation."[7] In other words, if a homosexual had fallen prey to AIDS in an earlier, more closeted era, he would have concealed his sexual preference from his doctor and so the few cases of the disease that did appear in the past would not have been labeled "gay related."

When I talked with Elliot, he was careful to broaden the discussion of how practices can impact disastrously on health to the other groups that are most often caught in the toils of AIDS. When discussing drug addicts, furthermore, he argues that, although this was not registered on the screens of the CDC and other agencies studying disease prevalence, they had suffered from an AIDS-like illness many years before the named thing made its appearance. "Drug addicts were starting to die of unspecified pneumonias as early as the sixties, and they had all the appearances of people with full-blown AIDS, like weight loss, ghostly thin, very pale, malnourished."[8] The possibility of IV drug users showing evidence of immune breakdown is plausible in that, as Stewart remarks, "Addictive drugs are by themselves highly damaging to the body's defenses and vitality."[9]

In the opinion of Dr. Eleopolos, even in the fast-track group it is drug use rather than sex or other components that tipped the balance of the body toward illness. She notes that gay males who have AIDS "are characterized by a variety of factors that distinguish them from the vast majority of gays who do not develop AIDS, *street drug use being chief among these*."[10] She goes on to speak about a study, that showed "street drug consumption is more prevalent among gay men with AIDS symptoms than is HIV." She also mentions that of the gay men who made up 55 percent of U.S. AIDS cases in 1988, 34 percent were self-identified as IV drug injectors. If one made a recalibration of the AIDS data, whereby heavy drug users, whether gay or straight, would be put in the same batch, those who use drugs "would officially account for 89% of all American AIDS."[11]

When Elliot talked to me, he emphasized that all the high-risk groups, such as hemophiliacs, had obvious threats to their immune conditions.[12] Root-Bernstein goes so far as to bring up a usually unidentified group that he feels should be appended to the list of those at risk, non-intravenous drug users. We need to evaluate the possibility that non-IV users, he says, "who are exposed to

HIV or other immunosuppressive agents by sexual routes will be at as great a risk of AIDS as are intravenous drug abusers." The use of these drugs, rather than heterosexual transmission, "may help to explain why many sexual partners of IV drug abusers—people who are almost all drug users themselves—are developing AIDS."[13]

To circle back to a point raised earlier, where it was noted that many researchers used the fast-track subculture to explain the rise of a new disease, since this subculture itself was relatively new, we might refer to what Root-Bernstein says about the relative newness of all the affected groups. For, as I'm sure the perspicuous reader may have noticed, definitive proof (if it were found) that newly introduced clone subcultural practices, such as inhaling poppers, correlated perfectly with the rise of AIDS in this group, would still leave the rest of the argument hanging. If new practices were responsible there, how explain the ascent of AIDS in other groups, such as IV drug users, whose practices were not new?

A partial answer was provided by Elliot's mention, a few paragraphs earlier, of the fact that he feels a pre-AIDS AIDS was known to drug addicts but unnoticed in this marginal group.

Root-Bernstein picks up on this amendment to dissident theory while expanding the thought to encompass other (although not all) groups. He begins by saying that an important problem to address is "why AIDS has emerged as an apparently new phenomenon during the 1980s when the immunosuppressive factors listed have existed, in some cases, for centuries."[14] He then argues, as did Elliot in a more restricted sense, "that a significant number of cases fitting the pre-HIV definition of AIDS (1987) do exist in the literature for at least 100 years before 1980."[15] He says that we should also consider that "techniques of diagnosis for many rare, opportunistic diseases, such as Pneumocystis pneumonia, in living patients became available only during the 1970s."[16] The idea here is that previously these rare diseases may have been missed by less subtle diagnostic tools. Moreover, he continues, another improvement in medicine has allowed more people to get the rarer opportunistic diseases that are connected to AIDS. This is "the widespread use of antibiotics and vaccine in the past few decades, [which] has allowed immunosuppressed patients to survive previously more prevalent diseases (tuberculosis, smallpox, typhus, typhoid, etc.) that undoubtedly killed such patients in the past."[17]

The conclusion we can draw from studying each and every one of the groups that seems especially susceptible to AIDS is, in Root-Bernstein's summary, "that non-HIV immunosuppressive agents play a significant role in debilitating individuals in these high-risk groups."[18] Therefore, the monocausal theory of AIDS causation has to be faulted. "It is impossible to maintain that any individual in a high-risk AIDS group was infected only by HIV and had no other immunosuppressive factors at work concurrently."[19]

One advantage to this multifactorial approach would be to explain the known divergence between subgroups affected by AIDS in time between infection with HIV and disease. Root-Bernstein lays out the different time lags and a believable explanation for them.

> The average number of years between HIV infection and AIDS is greater than 20 years for mild hemophiliacs, 14 years for young severe hemophiliacs, 10 years for old severe hemophiliacs, 10 years for homosexual men, 6 years for transfusion patients of all ages, 2 years for transplant patients, and 6 months for perinatally infected infants. These differences can only be explained in terms of risk-group associated co-factors.[20]

This is to say that insofar as it is the interplay of a number of factors that are all inimical to the human immune function that can be held accountable for the entrance of AIDS, different subgroups will see their immune systems worn down at different rates dependent on the weight and perniciousness of the stress components involved. Although individual cases may be nonnormative, one can average both the amount of stress and longevity of HIV latency.

If we wanted to shift from different subgroups and try to construct, in order to grasp better the pattern of the disease, an ideal-type person with AIDS, we might listen to Leon Chaitow, who sums up how the multifactorial theorist would see the profile of a person who succumbs to AIDS. His ideal type, one that, admirably, does not give in to stereotypes, also has the virtue of suggesting that many of the unhealthy factors involved in the individual's activities are ones promoted by Western culture.

Chaitow begins by arguing that such a person, as a member of a circumscribed minority, suffers above-average levels of stress—and remember, even hemophiliacs, classified as chronically ill by the mainstream, have to deal with prejudice and attempts at disenfranchisement. Reminding us a little of Schmidt, he says, in his portrait of the ideal type of a person with AIDS, "Imagine someone in whom there is a degree of constant or recurrent emotional stress… involving perhaps relationships, employment, social factors and health worries."[21] There is also the burden of "nutritional imbalance," which will be caused both by exceptional patterns, such as the use of recreational and medicinal drugs as well as a heavy intake of alcohol, and by following social norms, such as relying on tobacco and coffee. He elaborates this second connection in this formulation, "This individual's problems might also involve nutritional deficits which are incidentally widespread in western society—with even organically grown vegetables now containing only half the zinc they did a century ago."[22] Tack on the presence of recurrent minor infections: "herpes, candida, various viral, yeast, parasite and bacterial infections," as well as the STDs, generally ill managed by antibiotics. The result of this panoply of strains is "defense, immune, repair, eliminative and other functions and systems working overtime to maintain a semblance of reasonable function."[23]

Immune Suppression Can Be Caused by Multiple Factors

If we are willing to (tentatively) accept the idea that AIDS is produced by way of the immune system's cracking under the strain of a set of interrelated stresses—the mix of stresses being different for people depending on the pathway that leads them to the disease—then the next subject to settle would be the exact mechanism by which the undermining of function by distress eventually translates into disease breakdown.

One view of this process would give HIV a leading role but sees it acting something like herpes. As we brought out, nongenital herpes is a virus that takes up long-term residence but only sallies forth to cause cold sores and other irritations when the body's defenses have been weakened due to lack of sleep, crisis, other sicknesses, and so on.

In Joseph Sonnabend's view, as reported by Neville Hodgkinson, initial reception of HIV retroviruses causes a mild immune system retaliation, which forces the virus (relatively benign at this point) to retire back into host cells "where the viral DNA... [is] maintained in the cell in a dormant or latent state." Like Sleeping Beauty, it will remain in limbo, perhaps forever, unless awakened by the "kiss" of some form of chemical respirator. "Sonnabend cited several mechanisms that had been shown to trigger into action the cell DNA associated with HIV production."[24] These would include chemicals such as interleukin-1 and tumor necrosis factor, which are created by the body in the course of battling against a number of infections. Herpes viruses themselves have been experimentally shown to be capable of bringing HIV into the open. He also says "antigens displayed on the surfaces of foreign cells such as from semen and other people's blood" may arouse HIV.[25]

Let's say a few more words about semen. Sonnabend argues that that compound, aside from calling HIV out of hibernation, can cause an autoimmune reaction. To Sonnabend, there is a certain validity to the establishment view that part of the destructiveness of AIDS arises from the illness triggering an autoimmune reaction. According to orthodox thinkers like Dr. Gallo, this happens because of the similarity of surface markers (antigens) holding for HIV and a subset of T4 antibody cells. Primed to attack HIV, some antibodies begin killing their own fellows.

In Sonnabend's perspective, while there is a devastating autoimmune reaction to be found in association with the immunodeficiency syndrome, instigators of the reaction other than HIV can be located. For the PWA subgroup that contains those who have been passive partners in anal intercourse, Sonnabend argues, sperm may be the trigger to immune function confusion. He cites evidence that MHC proteins, which act to guide immune system reactivity, "can act as receptors for sperm."[26] Given the danger that these sperm cells may be hooked into the immune system, "Sonnabend postulated that repeated exposure to semen," which, as we saw, may enter the bloodstream through rips in the wall of the rectum, "was causing the immune system to lose sight of which proteins belonged to the self and which to others, eventually leading to the appearance of antibodies and killer cells which destroyed the body's own T-cells."[27]

Of course, this broaching of the immune system would only come into play in a fraction of people with AIDS; however, it leads us back to a more general point that Sonnabend makes. In his theory, it is not enough for HIV to be called back from latency, since, he believes, "a healthy immune system kills cells that start to produce HIV, limiting virus production." However, if the immune system were functioning far below peak, because of a degree of autoimmune reaction caused by semen and/or other reasons, HIV could begin to make inroads. As he puts it, "A damaged immune system might not be able to exercise that restraint" on HIV that would force it back into latency."[28]

He argues that antibodies only appear as a second line of defense against HIV after the killing of cells releasing HIV, the first line, has failed. This is a notion supported by Root-Bernstein, who writes, "Antibody, far from being protective against HIV, appears to be highly diagnostic of loss of immune regulation of HIV."[29]

To make this as clear as possible, we might present the kind of temporal schema we used when discussing Koliadin's position on how antibiotics featured in the illness. As Sonnabend sees it, the history of an individual with AIDS would run as follows. (Note: I will bring in a few direct quotes from a passage where Sonnabend is centering his examination on fast-trackers.)

1. Infection with HIV. ("Because of their lifestyle, most of these men had encountered HIV.")
2. HIV is partly killed off by immune reaction and remaining portion is left in latent stage in cell's DNA.
3. The individual goes about weakening his or her immune system through use of toxic drugs or by other means.
4. Presence of an eliciting chemical brings HIV out of hiding. ("Some of them had also at some point been subject to the repeated infections, viral reactivations and multiple rectal inseminations which had been shown capable of bringing HIV out of latency.")
5. Encountering, this time, a much weaker immune system than it found when making its earlier incursion, the retrovirus begins really wrecking the remaining immune machinery. ("AIDS patients and most promiscuous gay men were immune-deficient, and so unable to block HIV activation directly with a cellular response."[30])
6. At this point, when the body has already lost the first round, antibodies against HIV are produced.

It should be clear, by the way, that one may find Sonnabend's concept about the immune-incapacitating abilities of foreign semen implausible, while still accepting his overall thesis that HIV is only able to do its dirty work on an already trashed immune mechanism. As Lynne McTaggard put it when I spoke with her, "AIDS is immune suppression, in a word… and immune suppression can be caused by one of a number of factors," not including HIV.[31]

Lest it seem I am swinging too far in one direction, let me repeat what may seem obvious, that none of this eliminates the crucial role of HIV. The perception that should be conveyed is simply that HIV is not the Renaissance bug the establishment portrays it to be. You might recall Duesberg's complaint that the composition of HIV made it seem much like all the other known retroviruses, none of which had any but the most mundane capacities. According to this alternative viewpoint, HIV does not have to be able to proliferate and destroy the human immunity that would stop its proliferation. All it has to do is reproduce under the shadow of an immune function that is operating very suboptimally. Eva Sneed explained the new place she sees HIV occupying in the course of examining AIDS progression: For AIDS to exist

> [o]n the one hand, there is an inducer. Some type of virus, of which HIV may or may not be a part…. we know that most of the people afflicted do suffer, with high levels of chronic fatigue, cytomegla viruses, and others. So, number one, a viral inducer. Number two, some type of toxic promoter—environmental, industrial, internal, combination thereof. And then third, above everything, abnormalities and undesirable habits in lifestyle and diet. These appear to be the most important for the development of a full-fledged illness.[32]

The "most important" but, as Dr. Currant told me, often the ones that are last to enter the average person's awareness. "People are in denial about the real cause of AIDS," he explained. "For one, they don't want to give up the idea of drugs and things, drugs and toxins which are so present in our environment."[33] His conclusion may seem rather surprising at first. He ended, "It lets people off the hook—to believe in the virus—it really lets people off the hook."[34] What he meant was that even if one, under the assumption that HIV is solely responsible for AIDS, made tremendous changes in lifestyle—for the clone, giving up bathhouses and promiscuity; for the drug user, not sharing needles—some of the more unhealthy (and attractive) aspects of the life could be held onto: for the fast-tracker, poppers; for the IV user, the drugs themselves. Thus, accepting monocausal HIV theory would mean one would definitely have to "clean up one's act," but perhaps not have to clean the whole house. This is why it may remain the more compelling argument psychologically, despite all the evidence we have so far accumulated that it does not do the trick in satisfactorily demonstrating how the disease is generated.

However, before we declare the multifactorial causation theory a decent replacement for the monocausal one, let's take another step in exploration and see how dissidents explain the impact of different factors at the molecular level.

Oxidative Stress at the Cellular Level

Moving down a notch to look at a theory of AIDS that considers disease production at the level of the interaction of tissues and molecules brings us to the theory of AIDS developed by Eleopulos and the Perth group. They hold that at the cellular level the breakdown of the immune system is signified by oxidative stress.

Oxidation in terms of the human body refers to the process where in the course of chemical reaction when atoms or molecules are combining, an electron or electrons are lost, no longer fitting in the ensemble. This free electron, now needing a mate, will combine with one from another molecule, destablizing that entity, which now may combine with other molecules, forming in some instances a harmful compound. This is a reaction that the body's defenses are prepared for, having at their command antioxidants that may capture the free-spinning electrons or sidetrack the development of harmful compounds.

Moreover, certain substances that a person can take—vitamin C is a well-known member of this category—will boost his or her antioxidative capacities. On the other side, there are substances, such as amyl nitrite, that go in the other direction and either encourage oxidation or weaken antioxidative response. Dangerous oxidizing agents were often found to form a part of the PWA's environment.

Taking members of the fast track by way of illustration, Hodgkinson notes, when summarizing the Australians' ideas, "Many of the agents [such as the antibiotics] used for treating the infectious illnesses that were much more common among homosexuals than heterosexuals... were oxidising."[35] He furthers the point by continuing, "On top of that, viruses themselves caused oxidation of their host's tissues, and an accompanying immune suppression, in the course of their growth process."[36] He singles out the cytomegalovirus and Epstein-Barr viruses as widely found in PWAs and notorious for their oxidizing propensities. As alluded to, the nitrites, such as poppers, were also encouragers of the development of free radicals.

We know, certainly, that a number of practices engaged in by subsets of those who later became afflicted with AIDS, practices such as shooting heroin, are maximally unhealthy; but up to now we haven't scrutinized exactly how such practices lead to physical deterioration. From Hodgkinson's perspective, the Perth group have shown where the damage occurs. "As far back as 1986," he says, "Eleopulos had been offering a multifactorial theory of AIDS causation in which the various contributory factors were unified by their shared ability to put the body's tissues under a chronic oxidative assault."[37]

To illustrate this more specifically, let's go back to Sonnabend's theory that sperm, entering the bloodstream through rectal tears, creates problems. The risk here "could come from the challenge to the immune system presented by the seminal fluid and sperm itself, synergistically increased by repeated exposure to sperm heavily infected with known microbes" contributed by viral diseases."[38] As we saw, Sonnabend feels the risk here is that semen, which carries antigens that might confuse a key immune system protein, will bring about an autoimmune reaction. According to the Perth group, what should be looked for through the intrusion of this semen is instead that microbes from venereal diseases and other infections will be found in the sperm and will create oxidative damage.

In her own words, then, Eleopulos argues that for people with AIDS, "the one major significant variable is the concurrent exposure of the patients to oxidizing agents including sperm, nitrites, opiates and factor VIII."[39] She holds that these agents "induce malignancies, immunosuppression and increased susceptibility to infection."[40]

So far we have laid out what looks like a coherent theory, but we have not given a smidgen of hard data to support it; thus, giving such evidence should be our next step.

An article in *Free Radical Biology and Medicine* has pointed to a number of changes in PWAs that indicate reduced antioxidant immunity and evidence of oxidative stress. The writers of the article note that they find in their patients "perturbations to the antioxidant defense system, including changes in levels of ascorbic acid, tocopherols, carotenoids, selenium, superoxide dismutase, and glutathione" in various organs.[41] There is also some indication that a number of enzymes of the oxidant defense system have altered their activities.[42] At the same time, other physicochemical readings as well as an increased use of oxygen in the body attest to oxidative stress, which, significantly, can be observed "in asymptomatic HIV-infected patients early in the course of the disease."[43] Thus, this stress precedes any outwardly observable illness conditions. The authors conclude that oxidative stress may act in concert with other factors in the illness and contribute to "viral replication, inflammatory response, decreased immune cell proliferation, loss of immune function, apoptosis, chronic weight loss, and increased sensitivity to drug toxicities."[44] Moreover, they say, "Available evidence supports the conclusion that HIV-infected patients are exposed to chronic oxidative stress."[45]

This study gives some credence to the view that oxidative stress is one means by which AIDS is causing deterioration at the cellular level. Adding this to our overall understanding, we see that the multifactorial causation theory is comprehensive. It singles out a group of behaviors that are compromising to health, indicates how these behaviors lead to biochemical events in the body that chisel away at the immune system, and gives an idea of how damage to some parts of immune response may be rooted in molecular processes that are setting free electron radicals while reducing the body's ability to contain their damage.

As we have seen, a doctor working with PWAs who proceeded from this outlook would have to prescribe a totally different treatment regimen than would one who held to the HIV etiological perspective. Next, we want to look at both possible ways of proceeding, highlighting some of the failures accruing to the orthodox method and then seeing what alternative therapies are suggested by this new idea of cause. (We will only point out these new directions here, reserving a fuller exposure of alternative treatment for later in the book.)

Treatment Failures

Root-Bernstein feels strongly that as long as the AIDS orthodoxy relies on what he considers a faulty causal explanation of the origin of AIDS, money spent trying to find a cure may as well have been poured down a rabbit hole. It's an idea we've seen expressed before, though perhaps not with such passionate eloquence. He storms, "We're talking about a disease into which we have poured more money than any other disease in the history of mankind, and we haven't saved anybody." He goes on stridently. "And it turns out we have huge holes in our understanding on the most fundamental levels. Given the effort, that's inexcusable."[46]

If, as the orthodox see it, the cause of AIDS is reducible to one bug, HIV; then the means of eliminating the disease is equally simple: Kill the bug. When a doctor finds his or her patient has syphilis (which is produced by the bacterium *Treponema pallidum*), the recommended treatment is pour on the antibiotics, penicillin, specifically, and kill the bacteria. Orthodox doctors, when they are confronted with PWAs, turn to the same strategy. But since HIV is much hardier than *Treponema pallidum*) (or so it seems), the dosage of antivirals has to be heavier.

As Koliadin summarizes the situation, "With the promulgation of the HIV=AIDS formula in 1984… the number of potential victims of deadly practice of administer[ing] broad-spectrum antibiotics as a permanent prophylaxis skyrocketed."[47] Koliadin, believing that HIV is not the main culprit, sees the results of this chosen treatment as inevitably disastrous. Psychological stress and the proscribed "toxic medication led to gradual decline of CD4-lymphocytes in individuals with a positive test for HIV. This apparent progression of immunodeficiency forced physicians to intensify prophylaxis by broad spectrum antibiotics."[48] But the more antibiotics taken, the more the imposing side effects, such as the killing of indigenous benign microflora, whose disadvantages we presented above. The process is one of iatrogenesis, that is, disease caused by the medical treatment itself. And it is a "self-accelerating process," in which the more medicine, the more physical collapse of the patient, hence the need for a stronger dose.[49] It's as if to cure a patient's alcoholism the physician told him to drink stronger liquor.

All this may recall the heroic medical treatments routinely administered in the nineteenth century. Nowadays such treatments as bleeding—by applying leeches, which were meant to get rid of bad blood and ended up killing or making the patients anemic—or forcing tubercular patients to sit outside in freezing-cold weather to brace them up, putting them in danger of pneumonia while offering little of therapeutic value, are looked back on as quaint and irrational. We think of those doctors who offered them as palliatives with pity for their lack of scientific understanding. Many alternative health practitioners, looking at today's own heroic regimens for AIDS patients, shudder with the same revulsion they feel when reading about medical barbarities of yesteryear.

When I brought up to Nobel Prize winner Kary Mullis, the topic of how valueless these treatments would be if the one-agent causal theory of AIDS origin were disproved, he became incensed as he thought of the damage these treatments had done. "I mean it is criminal behavior to set off a bomb in a federal building. Is it criminal behavior to start passing out poisons to people who have no real diagnosable diseases, to babies, for instance?"[50] I countered that, no matter how depressing the results, these establishment types thought they were helping. His response to Celia Farber on another occasion more or less echoes what he said to me:

> The horror of it is every goddamn thing you look at... seems pretty scary.... The whole medical profession... It's just a bunch of people that have become socially important and very rich by thinking about the fact that they might be able to cure the diseases that actually cause people in our society to die. And they can't do shit about it. It's scary, that's what it is.[51]

The reason medical professionals can't do anything about it, according to the interpretation we are sounding, is because they have started off on the wrong foot. "It is a tremendous mistake to base our policy decisions concerning AIDS on an exclusive HIV basis... the recognition of non-HIV immunosuppressive factors in AIDS suggested that these programs are failing because they are too narrow."[52]

A doctor whose understanding flows from the multifactorial causal theory would equate blocking the disease with taking the measure of a host of influences, beginning with devoting more study to see how these myriad factors are implicated. "AIDS will only be understood," Root-Bernstein writes, "when we begin to explore the ways in which anal sex, infections, drugs, blood products, anesthetics, antibiotics, and malnutrition interact. At present, we know almost nothing about such interactions."[53] Even before this work is done and the method by which AIDS comes to exist is elucidated in detail, we can suggest, "Those who wish to avoid contracting AIDS should avoid all potential causes of immunosuppression"; those who are positive but not ill "may find that if they, too, avoid this lengthy list of immunosuppressive co-factors, they too will stay healthy."[54]

To get more specific as to what can be done—and keep in mind, we will fully present recommended treatments later—we should look at Root-Bernstein's outline of how to counteract the encroachment of AIDS. His treatment would include "nutritional elements, prophylaxis against non-HIV infectious agents, elimination of addictive and recreational drug use, treatment of autoimmune processes... and careful analysis of possible immunosuppressive results of medical procedures."[55] Dr. Hiram Caton adds that the overall plan must be to reduce oxidation by removing toxins and stressors that contribute to damage to systemic functioning.[56] Both Caton and Root-Bernstein agree that following such treatments, and expanding on them as new elements of the multifactorial picture are seen, will lead to a degree of success that has eluded the orthodox camp.

Whatever the exact recommendations, the multifactorial causal model does call for a practical approach that is much different from the one being carried out by the establishment. There are also theoretical advantages to looking at multifactorialism, which I will investigate in the next section.

Alternative Paradigms

REIGNING PARADIGMS

At the end of Part One, I argued that not only had the AIDS controversy exposed venality, close-mindedness, and weak research in the current organization of medicine, but that is has laid the road open for the replacement of the reigning paradigm.

Remember we said that, like all of us, those working in the health field are creatures of habit. It was suggested that certain thought patterns were familiar to biologists, so they tended to cradle new theorems in this already accepted language. For example, the canonical nineteenth-century discovery by Pasteur, Koch, and others that some diseases are caused by microbes may have fueled the acceptance by today's scientists that AIDS is caused by some similar vehicle. Laurence Foss and Kenneth Rothenberg, whose work we mentioned previously, justify this tendency in the following elegant aphorism: "Explanatory frameworks, not individual findings, are minimum units of meaning."[57] In other words, any finding, such as that PWAs have low T cell counts, is only meaningful within a grid of suppositions. In this instance, the drop of T cells could be because (1) HIV is knocking them off or (2) an overabundance of cortisol is shifting the TH mix of T cells in a direction that is wrecking overall T cell economy. And either of these premises is bottomed by basal conceptions, namely, (1) belief in a monocausal Pasteurian etiology, and (2) a multifactorial perception of disease creation.

It is the position of Foss and Rothenberg that the assumptions that make up the second grid, adopted thoroughly, represent the overturning of the currently governing paradigm. The multifactorial approach to disease, as differentiated from the unifactor one, would hold:

It is not the host-microorganism interaction that accounts for the disease, but a particular microorganism interacting with a host under special conditions. The particular mutagen (microbe) taken singly is not a pathogenic agent, as implied in the physical theory of disease causation. Rather pathogenesis is the product of the interplay of mutagen(s) and host-in-a-particular-state, a state induced by way-of-life factors.[58]

This is to say—and this applies even to syphilis, TB, malaria, and other illnesses for which the presence of a germ is a known causative factor—the germ must be given the green light by an impaired bodily system before it can proceed to bring on a disease.

Taking this to a deeper level, we can see that this view signals a more disjunctive relation between self and world than exists for the monocausal understanding. In orthodoxy there is a seamless transition from environment to body—a toxin gets in and affects the human body, whatever condition is presently maintained in that body. This concept, in which the toxin or germ expresses itself no matter what the initial state of the medium, is akin to the thought of many American Fundamentalists, such as those who have fought against AIDS funding on the grounds that sodomy is outlawed in the Bible and therefore shouldn't receive encouragement by having its diseases coddled. Of course, the Fundamentalists' views are textually dubious—their mainstay Bible passage, that Sodom and Gomorrah were destroyed by fire from heaven because God objected to homosexuality, is disproved by any close reading of the story, which shows that Jehovah's real cause for ire was ill treatment of strangers, a violation of an important social convention in the Middle East. This point is ignored by Fundamentalists who act as if the tribes of Jews were, for all intents and purposes, U.S. citizens a few generations removed. However, to get back to the issue, the Bible was written in Greek and Hebrew. The very fact that it is being translated into English irretrievably changes its tenor, since, for example, many of the Hebrew words have no equivalent in our tongue. However, Fundamentalists blithely ignore this discrepancy, just as, the dissidents assert, monocausalists ignore the fact that once a disease agent enters a particular body, its further actions unfold in unique dependence with the state of the body receiving it.

In the disjunctive, multifactorial paradigm, a disease stressor never has a direct effect on the body but always finds its action mediated by the complex, interlaid bodily systems it is attempting to disrupt. Putting this more technically, we could say:

> The pathogenic action is intrinsic to the circuit, linked only in an indirect way to the stimulus (microbe) that set if off. Rather than a substantive mode whereby a pathogen invades a body, pathogenesis is described [by the new paradigm] in a relational or communications systems mode. Instead of the enemy, the microbe is a trigger, a critical "fluctuation" made critical owing not to its intrinsic "virulence" (force) but to the state of the host.[59]

A word about "critical fluctuations" before we end this discussion. The body exists in a state of flux, moving through various levels of health as it fends off various troubles and faces obstacles. No one, not even pop star Michael Jackson who lives in a glass bubble, can avoid regularly meeting threats to optimal physical functioning; and it is the mark of the healthy person not that he or she is never incapacitated by anxiety or disease, but that the person can shrug off the imposition quickly. Thus, even the condition of a person who lives a healthy life fluctuates between upper and lower states; it is when it is toward the lower end that disease has a chance to get a purchase.

But what about the adjective "critical"? Here Foss and Rothenberg are alluding to a principle of the study of organized systems in disequilibrium. Economist Hazel Henderson sets out the laws of this kind of setup. It is a "system internally dynamic and structurally dynamic, governed by positive feedback loops, which can amplify small initial deviations, which sometimes break through thresholds and push the system to a new structural state."[60] Systems of this type, from the human body to a marsh ecosystem, are dynamic and continuously adapt to newly arising conditions. However, where this system is different from a stable one is that under select primers, the whole system can, touched by small-impact factors, mutate and go in a new, unexpected, and unpredictable direction, as from health to disease. The "critical" fluctuation is the one that touches off this novel move by the whole system. Keep in mind, though, that the fluctuation is completely dependent on both a sudden change in the system's environment and the then presiding state of the system.

We can say that, when it wants to assess how an illness was generated, the new paradigm looks at two intertwined ranges of factors. So, even if HIV were guiltier than it seems, from this innovative paradigm's perspective, it would never be enough to study this particular retrovirus's action alone to get a handle on why someone fell ill. The new outlook tells us that AIDS originates by way of the intersection of environmental inducers and an internally overwrought body.

IN DEFENSE OF SIR FRANCIS BACON

It might be salutary to talk, very briefly, about how our guide through these dark woods, Sir Francis Bacon, stands in relation to this new configuration of knowledge.

If the dissenters were to build a zoo of bete noires—where we would find the beasts who most contributed to building up a linear, monocausal world view—right next to the king of the beasts (Descartes), most would probably put Bacon in that zoo. This is because Bacon said many times that nature is made to be manipulated to increase the happiness of mankind—and, so it seems he was looking to find, by sifting through prerogative instances, cases of simple cause and effect that could be turned to our advantage.

Yet, just as people can be betrayed by their times, so a person can be upheld by them. If we assay for a moment the nature of Bacon's language, which is that of all Elizabethans, we see it suggests a much more qualified endorsement of simplistic monocausal action than his direct statements would make one think.

Let's look at how he talks of the attractions and resilience of bodies. Under the head of attractions, he is thinking, for example, of how water is drawn up a pump by suction. He puts it in these terms, "Bodies do not suffer themselves to be separated at any point from contact with another body, as delighting in mutual connection and contact." Discussing resiliency, he waxes even more eloquent: "Let the third motion be what I call motion of liberty, by which bodies strive to escape from preternatural pressure or tension and restore themselves to the dimensions suitable to their nature."[61] An eraser, for example, crushed by a hand, when released, will spring back to its original shape.

What he seeks to establish in such discussions is "the occult virtues and sympathies of things," and we can see from his choice of words that "things" (stones, water, minerals) are given a lively, participatory being, delighting in partnering with some forms of matter, while fleeing constraint, almost as if they had a primitive consciousness.[62]

An establishment honcho, on examining such passages, might say these are lingering traces in Bacon of alchemy and other pseudoscientific fields that would have to be extirpated before real science could proceed. However, from the new paradigm, Bacon's unexpectedly animistic language might be taken as evidence of a more multifactorial view of events.

If we turn to the biology of AIDS and contrast the two perspectives on this front, we might come to this. While an establishment doctor might allow a shade of anthropomorphic language when referring to a virus (which will act like a "killer" when it "invades"), the body itself and all its subsystems will be treated as passive receptacles to receive the intruder. In the multifactorial approach, each element involved as one moves into a disease state, from HIV to other toxins to the antibody apparatus, plays an active role, and none is dictated to by other players but makes an individual response as it relates to interlacing factors that touch it. This second view, then, would be smoothly compatible with the type of language chosen by Bacon, which seems to equip each type of matter with autonomous, albeit schematic, instincts.

DRUG SUBCULTURE: TRAINSPOTTING

Let's back up to Schmidt's jaundiced views on how AIDS became culturally prominent in the gay male society. It will be recalled that he believes the community became structured around the illness in a self-defeating way. Those who had HIV became an elite of glamorous victims; the dying became martyrs and Camilles, and the left behind, tearful mourners. Thus, in Schmidt's view, the group had fashioned a morality play out of the crisis, a development that both was emotionally flawed and misdirected attention away from more constructive approaches.

Although, as I remarked previously, Schmidt has his share of fresh insights, this particular view is dead wrong. Before I attempt to correct it, though, let me say a few words about subcultural communities in general. To illustrate my discussion, rather than discussing the gay community I will talk about a group that is even more despised by the straight majority, heroin addicts, among whom AIDS is also rampant. The novel *Trainspotting* will provide us a large fund of data; while its ideas about HIV transmission and other facts of the disease are as cockeyed and erroneous as the worst put out by the establishment, its portrait of how a group of Edinburgh "junkies" face the devastation wrought by AIDS in their numbers is amazingly perceptive.

(I label the author Welsh's ideas about AIDS cockeyed in light of such incidents as those presented in the chapter "Bad Blood." There a woman becomes infected with HIV after one night, one "shag," they call it, with an infected man who rapes her. Next the narrator becomes infected after one night with her. If, as we discussed earlier, the likelihood of the first infection is one in a million, the likelihood of the second—one-shot, female-to-male transmission—is more like one in 2 million.)

The addict community, as depicted in the novel, pivots around a core contradiction. On the one hand, junk users, fixated on getting their daily fixes, can be outrageously selfish. For instance, on a bus ride down to London, Renton (or Rents), the main character, conceals from his best friend, Spud, that he still has a supply of heroin, even though they are both suffering withdrawal pains. "Renton's slide into the misery of withdrawal continues apace. He knows he has to act. This means holding out on Spud. However, there was no sympathy in business, and much less in this particular one [taking drugs] than in any other."[63] This leads to a hilarious scene in which Rents has to "cook" up his shot and inject himself in the toilet of a careening bus; but, more importantly for us, this is a representative example of how egotistic one becomes when the dope is talking.

On the other hand, however, one can hardly hold out against the world, especially when engaged in constantly breaking the law, let alone procure smack, speed, hash, or the other "good things in life," unless one has a full and relatively reliable network of friends. Buddies not only offer support and companionship but cover for you in the frequent contretemps with cops and barroom brawlers. As one of Rents's group puts it as an unquestionable law, "Ye back up yir mates."[64] This is a rule that transcends the ingrained prejudices of normal society. So, when Spud is in a bar with his half-caste brother, Dode, and that brother, is maddened by taunts of "black bastard" from a group of Orangemen, "real Nazis," Spud has to rally to his side. Spud thinks (in dialect) as he sees the fight erupt, "Ah'm sortay shakin wi fear, raw fear man, and one guy jumps at Dode, n they've goat um doon, so ah huv tae steam in."[65]

The increasing encroachment of immunodeficiency disease among the addicts does not resolve this tension between individuality and community but rather heightens the contradiction on both ends, so those who had been merely selfish become treacherous. But, in the same moment, bonds that had been merely weakly sustaining are tightened and strengthened.

For an example of the latter, one has only to look to the story of Dave, estranged from his family and bent on a bizarre revenge on the man who indirectly transmitted the virus to him—this is the one in 2 million case—whose world changes when he admits he is HIV positive. His family and friends rally to his side, overcoming earlier indifference and alienation. As Davie puts it, "As with Janet [and his other friends], so with my family. We have found an intimacy [now that he has revealed he is HIV positive], which may have otherwise eluded us. I wish I hadn't waited so long to become a human being."[66]

But I want to end by remarking on the simultaneous increase of perfidy in the group; because, paradoxically, this also indicates how the bonds among members of the set have become more powerful.

This novel, in which the characters stand so many conventions on their head, ends by playfully tweaking James Joyce, riffing off *The Portrait of the Artist as a Young Man*, one of the key texts of Modernism. The Irish novel ends with the alienated hero forsaking Ireland because its cramped parochialism and religiosity make it impossible to be an artist, since such a vocation demands that one be able to distance oneself from all the trappings of convention. He flees to Paris, where he imagines conditions are more congenial for creativity and Bohemianism.

Trainspotting closes with Rents making a break for Amsterdam. He is leaving Edinburgh, however, because he is not alienated enough. He gets on the ferry with a huge bundle of cash (15,000 pounds) that he has ripped off from his mates. He too has been seeking an opening whereby he can leave the close atmosphere of his homeland, not as a strike for intellectual freedom but to get off drugs. Rather ironic that he would choose Amsterdam, where heroin and other hard drugs are legal, as a place to kick the habit! We can let that pass, however, since we are concerned with Rents's intentions not his rationality. Besides, there is a deeper irony than this. Rents robs his friends royally, taking from them the largest amount of money they are ever likely to see—it was acquired from a lucky drug deal—and probably the only chance any of them will ever have to realize their dreams. Yet, frankly, he is not doing so because he is greedy. The depth of his betrayal is necessitated because this is the only way he can assure he will be a permanent exile. He has gone away before but always been drawn back to his comrades, and once with them he is back on drugs. (This is what I meant by saying he is "not alienated enough.") However, now that he has beggared them, he is an outsider. He thinks of Begbie, the most violent of the gang.

It was Begbie who was the key. Ripping off your mates was the highest offense in his book, and he would demand the severest penalty [that is, death]. Renton has used Begbie, used him to burn his boats completely and utterly. It was Begbie who ensured he could never return. He had done what he wanted to do. He could now never go back to Leith, to Edinburgh, even to Scotland, ever again.[67]

Tying together the last two points, we can say that the drug subculture responds to AIDS by increasing the closeness and interdependence of its members, who we see bond at funerals and the raucous parties that follow them. This change proves claustrophobic to some, such as Rents, who ultimately sever ties completely—which action itself can be seen as beneficial to overall group coherence, since disloyal elements have been expelled.

CULTURE AND ITS SUBCULTURES

To return to the fast-trackers Schmidt mentioned, and to bring us closer to our point about the need for a multifactorial organization of health care, we need to distinguish more rigorously between culture and subculture.

We have been calling the Edinburgh drug users in Welsh's book a subculture, in that their grouping is a delimited part of the larger society, one that lives by its own rules and has its own habits. The difficulty with this terminology is that "society" is a vague and undefined concept. A more advanced understanding of subcultures can only come into play if the idea of the society of which they are a part is narrowed and refined, so that we see them in relation not to some amorphous society but to distinct regions within the overall culture. We can look, for instance, at the fast-trackers as a subculture within the gay male stratum. This type of focus is how the Birmingham Centre for Cultural Studies (which founded the cultural studies discipline), to whose work I will now turn, approached an understanding of the topic.

In the introduction to *Resistance Through Rituals: Youth Subcultures in Post-War Britain*, they write, that "subcultures are sub-sets—smaller, more localized, and differentiated structures, within one or other of the larger cultural networks."[68] Characteristically, in a hierarchized society, with dominating and dominated groups, these subcultures arise among the oppressed. To understand them we need to examine two viewpoints. "Subcultures then must be related to the 'parent' cultures of which they are a sub-set. But subcultures must also be analyzed in terms of their relation to the dominant culture."[69] In the case we are concerned with, then, this would entail that the "clones" must be analyzed both in terms of wider gay society and in connection to the straight culture that persecutes the minority.

So why do these subcultures arise? If we confine ourselves to those that grow up in subordinate groups, such as those in racial or sexual minorities, we can see they offer a manner of solution to the broader minority's quandaries. According to Stanley Cohen of the Birmingham group:

The latent function of subculture is this—to express and resolve, albeit "magically", the contradictions which remain hidden or unresolved in the parent culture. [Turning to an examination of specific subcultures, he continues that] all represent in their different ways, an attempt to retrieve some of the socially cohesive elements destroyed in the parent culture [by the incursion of the dominant culture... . The original mod style could be interpreted as an attempt to realize, but in an imaginary relation the conditions of existence of the socially mobile white collar worker.[70]

The concepts we have in hand, then, are these: The subculture of a subordinate group arises as a means of confronting one of the obstacles of the larger group, which, as dominated, suffers various forms of stereotyping and exploitation from the more powerful groups. However, as opposed to, say, a political action committee formed within the group, this subgroup treats the minority's grievances obliquely.

Let's look at Cohen's mod example. The mods may be unfamiliar to some readers, but the recent Hollywood films about Austin Powers, who is portrayed as the ur mod, may be of some help. The musical group the Beatles was the most famous product of this formation. The lower middle class youth who entered this subculture had little chance, except for individual exceptions, of moving up in the world (due to lack of educational opportunity and so on), but they could do so in imagination by creating dress, music, and a cultural style that mimicked elements taken from the upper middle class, while leavening it with elements from their own milieu. (The Beatles, who we can use as exemplars, mixed elements of the middle class pop song with rhythms drawn from the Motown popular with the petty bourgeoisie.) Thus, the mods soothed their frustrated aspirations—ones they shared with their whole group—toward moving up in the world, by copying the style of the group they wanted to enter.

Applying this closer to home, we may recall Andrew Ross's discussion (from his book *No Respect*) of 1960s gay camp. He argues that the male homosexuals who developed this fashion, were men who were debarred from many professional and academic posts (or, if they gained them, had to conceal their sexual orientation) and who, in camp, created their own system of values, which overturned the judgment criteria of the establishment (Warhol was on the leading edge here). Nonetheless, insofar as they did have a criteria, a canon of highly valued works and so on, their deployment of a camp aesthetic mirrored the hierarchization and hierophantism of the straight system. Again, the solution was imaginary. Since they could not be enrolled in straight careers as tastemakers, they set up their own aesthetic realm, which compensated them to a degree for the lack of a place from which to speak to the wider society. But this solution left the actual fruits they hungered for untasted.

Schmidt hangs his analysis on the fact that the rise of neoconservatism, with its diatribes against the "homosexualist" threat to our moral fiber, created hysteria and increased stress in the gay male culture. Now, at risk of seeming to endlessly defer certain subjects I keep promising to take up, we will have to wait a little longer for a full consideration of—to put it most broadly—the connection between changes in the United States's political economy and the rise of AIDS. This I will undertake when I engage in table talk at the end of my discussion of AIDS Inc.

For the present, though, I want to offer this first flyer at explaining why the fast-track subculture arose as it did. Rather than agreeing to the idea stated earlier that the fast-trackers were acting out a presumptuously premature belief that gains at Stonewall and in other battles for gay rights meant they were now liberated, I would argue that the rise of the fast-trackers represent the self expression of a faction of a blocked subculture. Although various struggles had given gay males more breathing room, they still existed in a stigmatized and excluded position in the wider society. As Dan Beauchamp comments, "Gay sexual practices [in the '70s] may have been shaped in part by societal pressures and laws forcing gay men to associate secretly in bars and bathhouses out of view of the majority community, while at the same time proscribing gay association by cohabitation and marriage."[71] What was staring the minority in the face was the fact that, although token acceptance of gay rights had been achieved, the rigorously sexist U.S. society cannot allow gays any real semblance of equality, because this would put at risk the rigid masculine/feminine dichotomizing—which gay culture problematizes—that stabilizes the hierarchical organization needed to maintain power differentials.

As did the mods, the fast-trackers took up elements of the straight consumer culture, in almost a parodic form, as a "magical" way to appear to be participating in a culture from which they were excluded. To use terms more familiar from civil rights struggles, we could say that the clones, moving through anonymous cubicles in the bathhouses, practiced fantasies of integration in a segregated society.

AIDS Sparks a Paradigm Shift

The weakness of this imaginary way of coping with relative powerlessness came to a head in the AIDS crisis, which told the gay community—as Randy Shilts so powerfully documented—that the majority cared little or nothing about homosexuals and that the government and medical complex were not interested in their welfare.

This caused a radical "rethinking" of the strategy that had been imagining progress would come through the seeking of minor concessions, ones that piled on top of each other would eventually yield an adequate incorporation into the wider society. Those who "rethought," something that was not done conceptually but in practice, the position of their minority, realized (1) they would have to set up a temporary autonomous zone (to use social theorist Hakim Bey's term) of their own creation to solve their health problems (while still striving to garner support from majoritarian organs) and (2) that these zones would have to be structured in a questioning mode.[72] By this last, I mean that they could not simply reproduce the ranked order of place characteristic of both commercial and scientific structures. They would displace these structures by breaking with such paradigms both *between* and *within* their organizations.

(Important caution: To simplify matters I am not at this point evaluating other communities' equally vital, and equally paradigm-shifting, responses to the crisis, such as that of African-Americans, but we will look into these responses when we return to this topic later.)

If we look at the case of San Francisco in the early days of the crisis, we see that the gay community developed a mesh of dispersed organizations, that did not arrange themselves into an inferior/superior order, but worked as interdependent, equal agencies, each handling different inputs to build a synergistic pattern. Some of the major components of this distribution were the Shanti Project, which acted as a counseling agency; the San Francisco AIDS Foundation, which arranged medical treatment; the group behind the AIDS Candlelight March, which raised public awareness while memorializing those that had passed; Project Inform, which studied and disseminated knowledge about drug treatments; Open Hand, a food service for PWAs; and SANE (Stop AIDS Now or Else), an ACT UP style group that engaged in public protest, such as blocking the Golden Gate Bridge. These and other groups, working on varied components of the situation, created a network in which the members, though often in conflict, had created a new model of the multidimensional integration of differently oriented people working efficiently in concourse.

Moreover, quite a few of these separate agencies carried out a wide number of tasks in a way that integrated in an egalitarian way professional and nonprofessional volunteers with PWAs. In illustration of this, take the case of New York City's Gay Men's Health Crisis, an organization that offered crisis intervention (in which clients received "one-on-one support in order to deal with the social, behavioral, practical, and emotional consequences of their illness"), support groups, recreation, and financial and legal advice.[73] Each sector of the work was carried out by fraternally and sororally interacting professional and non-professional volunteers, who, significantly, did not dictate to but listened to and empowered the PWAs. A discussion by Lewis Katoff and Susan Ince explains how this last attitude, which held that PWAs should be helped to work out their own solutions, came to be in the financial department.

> When financial services started at GHMC, the goal was to remove the client completely from having to deal with bureaucracies and agencie.... This system sometimes created more stress for clients than it reduced. The less clients knew about entitlement agencies, the more frightening the systems became... [Realizing the error in their original approach]

the financial advocacy department... reorganized to meet these goals more efficiently and with greater empowerment of clients. The approach has shifted from application processing to the present approach of client education with step-by-step guidance and individual advocacy.[74]

Such a setup is in marked contrast to that of the average orthodox medical or governmental bureaucracy, where the client is made to feel unvalued and anonymous.

Philosopher of science Richard Whitley of Manchester University in his essay "Changes in the Social and Intellectual Organization of Science" has made the argument that "particular social arrangements are consonant with particular theoretical ideals and lead to the production of particular types of knowledge."[75] In other words—to highlight one point—there is an alignment between the content area of a subject field and the social organization of workers in that field. As an example, Whitley points to physics, the queen of the hard sciences. In that field it is theoretical work (mathematical formula creation), that is held in highest esteem, with the carrying out of experiments being consequently downgraded in prestige value. "Because the ideal is the reduction of all the special fields of research to a coherent formalism [which can only be carried out in theory] work directly focusing on this goal is more central to the discipline as a whole than research in any derivative field."[76] This means that power, money, and honor distribution are slanted toward the more theoretical fields—departments in these fields only exist at the elite universities—and those working in lesser fields are disesteemed and disempowered. The overall picture is that a physics that concentrates on elaborating an understanding of atomic particles that interact in a rigorously determined way—although this majority view is disputed by Capra and others—mirrors its subject in the social arrangements of science, in which hierarchization and privilege reign supreme.

If we accept this position, we can then more easily account for the seemingly unscientific behavior of Dr. Gallo and other establishment figures who engaged in witch-hunting and the blackballing of any scientists who broke with their views. As much as Dr. Gallo's branch of virology has the same monocausal and deterministic view of nature, the social arrangements of this science value deference to authorities and strict views of protocol, little to the benefit of finding the truth. In the same way, a medicine that is interested in multifactorial causality and an interactive relation between host and disease-causing factors will tend to sponsor a more open-ended, balanced, free social organization of treatment and research. Therefore, to make a point that will be completed, finally, at the end of this book, an alternative medical perspective forms a good fit with emancipatory social organizations.

Was AIDS
Man-made?

L et's take up again the variation of the idea of multifactorial cause of
AIDS, which holds that HIV plays a significant but reduced roles in the
course of the disease. According to this view, it will be recalled, once the
immune system has been destabilized by the influence of toxic drugs, STDs or
through other means, HIV begins to raise its head, only able to do its dirty
work of infecting and destroying cells once the body has been weakened by factors unrelated to the retrovirus.

Going along with this view, the question where HIV came from arises.
Surely, throughout history there have been people whose immune systems have
been weakened, but there was little evidence of the AIDS disease before the
1970s. Of course, we've already indicated that some dissenting scientists believe
HIV is not new. They say that in the past, people didn't live as long and didn't
live past the first incursion of syphilis or other infections. So, AIDS would seldom be manifested because people weren't hardy enough and medicine wasn't
advanced enough to allow any but a rare few to live long enough to come down
with the disease.

Although there is merit to that view, in this chapter I would like to deal
with those alternative thinkers who believe that AIDS and HIV are new additions to the disease spectrum. It is necessary for these dissenters both to knock
down claims such as that the HIV virus has been floating around in sub-Sahara
Africa for centuries, as well as indicate where and how HIV came to be. Theories

handling the latter concern range from ones that say HIV is the product of contaminated polio or other vaccines to the thought that it has crossed to humans from other species—or other animals—to the idea that it is the escaped byproduct of chemical warfare experiments.

In examining the credibility of these theories, we will be able to recap a number of themes we have already introduced, such as the unmerited bad press generally given Africa by the Western media in a journalism resonant with centuries of racism, as well as broach topics that have not yet been covered in any depth, such as how commercial interests have penetrated and warped medical research in some areas.

Laboratory Viral Experiments Going Awry

Earlier, I mentioned that scientists have long warned of the downsides of biotechnological research, one being the potential escape of viruses from laboratories. However, given that biotechnology is a growth industry, that is, one from which entrepreneurs expect to garner the super profits usually available in a newly emerging field, such warnings have been unheeded as lucrative research has plowed ahead. Dr. Richard Strecker, who I interviewed in depth on this topic, and others have set themselves the goal of alerting the public to these dangers.

In an article in *Rolling Stone*, T. Curtis provides a pedigree to the fear expressed by scientists on this type of mishap, noting that since the early 1960s this peril has been acknowledged in biological circles. Curtis goes on to report that in the 1970s Jay Clemenson rang the alarm at a cancer research conference in Tokyo, where he said new viral epidemics were on the horizon. "This new plague would be the result of experiments with animal viruses in the laboratory that were being artificially adapted to man," Clemenson felt. "Since they would settle in man before a vaccine antidote was found, the epidemic would be uncontrollable."[1]

Whether Clemenson was being an alarmist or realist at this point, it is true that laboratory experimentation that leads to the creation of new viruses has been around for years. When this research was new, it used the hit-or-miss strategy of infecting a cell with two different retroviruses and hoping the two mate. If they do, "retroviruses [become] like a mother and father getting together and reproducing millions of children, each of whom are both like the parents, and only slightly different."[2]

With advancements in technique, since the later 1970s genetic engineers have been able to attach new genes to viral DNA. This allows the scientists "greater control over the type of virus they are creating."[3]

Another way to obtain novel viral entities is to adapt presently existent viruses so that they can act upon other species, ones that are normally immune to them.

Altering Species Specificity

This third technique of creating novel viruses is obviously relevant to the discussion of HIV, since many orthodox scientists hold that HIV has jumped species from monkeys to humans. Although it is possible for viruses to move between species, such movement seldom outlasts a single interspecies communication, because each species has peculiarities of cellular chemistry and organization to which its viruses are closely adapted and on which they are dependent.

John Seale, writing in *New Scientist*, refers to the single-contact transmission of a virus in this statement. "This happens very frequently in nature with the zoonoses, when animal or insect bites transmit viruses which cause rabies, yellow fever and other haemorrhagic fevers."[4] However, such viruses are passed only from the animal hosts to humans, not between humans. As Seale puts it, "The insurmountable problem for most viruses becoming permanently established in new host species in nature, is failure to infect regularly other individuals in the new host species." A man infected with rabies after being bitten by a mutt will not "shed the virus in large quantities... and infect other people," so his infection will not lead to an "epidemic, human-adapted virus like the AIDS virus." Although it has been argued that such a cross-species transmission and then adaptation is theoretically possible—and some scientists believe smallpox originated in that manner—it will only happen "only once in hundreds, or in thousands of years."[5]

However, scientists who are willing to play god have been pressing forward to overcome viruses' squeamishness. "According to scientists, species specificity can be altered by growing an animal virus in human tissue."[6] Curtis explains how this would be done. "If you have a cow virus and you want it to grow in human beings, it is very simple to manipulate the virus to make it grow in human tissue. After the virus is grown in human tissue, it will be able to grow in human beings."[7]

Bovine Visna Virus

Proceeding inductively from what has been established so far, that is, taking the two general facts about contemporary laboratory procedures—that new viruses are concocted by combining two existent ones and that animal viruses can be adapted to humans through their cultivation in human tissue samples—and we arrive at the following argument, as advanced by Curtis.

He had wondered "whether it would be possible to construct an AIDS-like virus by crossing known animal viruses. The answer we came up with was that it was not only possible but that it had, in fact, occurred."[8] He notes, in making this claim, that HIV shares a similar genetic structure with two animal retroviruses: the visna virus found in sheep and the bovine leukemia virus. "Every paper that has been published on the genetic structure of the AIDS virus compares those genes to one of these viruses."[9]

The next thing Curtis looked into was "whether it was possible to make an AIDS-like virus by crossing the bovine leukemia virus of cattle and the visna virus of sheep."[10] This, too, was a reality in the shape of the bovine visna virus (BVV), "which caused AIDS-like diseases in primates and which had a structure virtually identical to the AIDS virus."[11]

Strecker, writing in the *Journal of the Royal Society of Medicine* endorses Curtis's point and even goes ahead to debunk other theories of cross-species origin, particularly one we will scrutinize later, that HIV may have come from SV40 or another monkey virus. "It is the gospel of the United States NIH that the AIDS virus arose spontaneously in monkeys—animals not commonly known to harbour visna-like viruses," Strecker says. But he puts his money on Curtis's proposition. "Most likely the AIDS virus arose by hetrodimer recombination of bovine leukaemia virus and visna virus in a commonly infected host cell." Furthermore," he extrapolates, "it seems... the virus expanded its host range... by culture growth in malignant bone marrow tissue."[12]

Curtis takes his theory one step beyond where Strecker is willing to go in arguing there had been the "artificial adaptation of the virus to people... made possible by growing it in human tissue."[13]

He offers no evidence for this final speculative claim, but there are good reasons to take his other opinions seriously. Let's look into bovine visna virus, which has been renamed, due to its similarities to AIDS, bovine immunodeficiency virus (BIV).

In 1987, Jeremy Rifkin, noted science writer, petitioned the Department of Agriculture to study the possible links between AIDS and various animal viruses. He notes that BIV "was first isolated in domestic cattle in the early 1970s."[14] A *New York Times* report on this research identifies the isolators as "scientists at the National Animal Disease Center in Ames, Iowa, a facility operated by the Department of Agriculture."[15] Specifically, Martin Van Der Maaten found the virus in 1971 in cattle from Louisiana. However, "the virus remained in frozen storage until 1984, when tests were begun to determine its properties and its relationship to other animal and human viruses." M. A. Gonda, a colleague of Van Der Maaten, said, "the virus looks like the AIDS virus under the microscope and shares about one-third of its genetic material with the AIDS virus."[16] In the course of research on this virus, "scientists successfully infected human cells with BIV (BVV) and concluded that 'BVV may play a role in either malignant or slow viruses in man.'"[17]

What stirred Rifkin to write were two suggestive associations between the human and bovine diseases. For one, AIDS is said to be a lentivirus. We've seen that many dissenters reject this category, because few such viruses have been shown to exist and none definitively. It is also disputable whether AIDS undergoes a long latency period, as do the lentiviruses, since even many in the establishment believe it simply undergoes an infective assault on the immune system for a few years before it becomes full blown, rather than retiring altogether.

Even so, given the controversial nature of lentiviruses, it is certainly enlightening to note, as Rifkin did in 1987, that "there are only three other known lentiviruses and they are all found in domesticated animal herds in the U.S. and other countries."[18] (Others have been subsequently identified, including "the cat AIDS virus, called feline lymphotropic lentivirus or FTLV and a monkey AIDS virus called simian immunodeficiency virus or SIV."[19]

Secondly, and this is the more frightening prospect:

> It has also been reported that bovine viruses are "a fairly common contaminant of fetal calf serum," and have been "demonstrated [by Georgiades in 1978] in a high proportion of batches of commercial fetal bovine serum for cell culture purposes." Fetal bovine serum (FBS) universally used in the creation of cell tissue culture for subsequent use in the production of vaccines mayhave been contaminated with BIV (BVV) through a number of routes.[20]

To repeat and amplify, fetal bovine blood, which is commonly used as the base for laboratory cultures of cells that will be used to create vaccines, such as that used on polio, has been found to be fairly commonly infected with BIV.

The point is echoed by J. Grote in a piece for the *Journal of the Royal Society of Medicine*. "Worryingly, BVV has been shown to be present in high proportion of batches of foetal bovine serum for cell culture."[21] He explains the significance of this: "The seriousness of this becomes apparent when we consider that the manufacture of vaccines requires the growth of virus in cell cultures using foetal calf serum in the growth medium."[22] If these vaccines become contaminated through their medium, then, Grote agrees with Georgiades, who unearthed the presence of the virus in the serum, that "BVV may play a role in either malignant or slow virus disease in man [and]... exposure of genetically predisposed individuals may result in apparent or latent infections."[23] (When we discuss monkey viruses, I will bring out the possible connections between BVV [BIV] and another cattle virus in disease causation. I am reserving that discussion until we can more adequately treat the idea of helper viruses.)

Perhaps it is not unexpected that the Department of Agriculture ignored Rifkin's warning, probably consigning his petition to the circular file, and that other agencies have been equally indifferent. The CDC, for example, has pooh-poohed any possibility of the two diseases being connected. "'There are some zealots who have made the suggestion that animal viruses are related to AIDS,' said Dr. Peter Drotman, an epidemiologist in the AIDS program at the CDC.[24]

Bovine Leukemia

Rifkin noted, without any explanation of how the viruses combined, that BVV was the product of a union of bovine leukemia virus and the sheep visna virus.

For our purposes, it is not necessary to establish whether BVV is a hybrid or singular creature, nor would it seem to lead us any further in the study of AIDS to examine the possible parent viruses.

However, as a sidelight, I would like to say a few words about bovine leukemia virus, simply to bring out the possibility of disease transference from cattle to humans and to note the general negligence of the public's health by the dairy industry and its government backers. Anyone who is familiar with Shilts's magnificent reporting on the foot-dragging of the blood banks over instituting screening tests against HIV-infected blood (even as the evidence of people infected by transfusions piled up) may note that the arrogant disregard of their customers' well-being is a red thread that runs throughout U.S. industry's record; itis not simply confined to the action of corporate bad boys such as those in the tobacco or asbestos business. We will see more of it when we come to the behavior of companies producing recommended treatments for AIDS.

An excellent study of the presence of bovine leukemia in cow's milk has been done by R. M. Kradjian. He notes how common this virus is in milking cows and emphasizes that in the West the United States has distinguished itself by being in the rear guard when it comes to addressing the problem. He states:

> The bovine leukemia virus is found in more than three of five dairy cows in the United States! This involves about 80% of dairy herds. Unfortunately, when the milk is pooled, a very large percentage of all milk produced is contaminated (90 to 95%). Of course, the virus is killed in pasteurization—if the pasteurization was done correctly.... This is a world wide problem.... Several European countries, including Germany and Switzerland, have attempted to "cull" the infected cows from their herds. Certainly, [you would think] the United States must be the leader in the fight against leukemia dairy cows, right! Wrong! We are the worst in the world with the former exception of Venezuela, according to Virgil Hulse, MD, a milk specialist. At present, [Hulse says] Venezuela is doing better than we are.[25]

Curtis notes that the differing approaches to contaminated milk in Europe and the United States stem from different convictions of scientists on the two continents. The Europeans are wary of diseases being passed through milk and, "as a result, European dairy farmers have carried on a campaign of eradication," culling from their herds the sick animals.[26] "In the United States, veterinarians came to an opposite conclusion and stated that these diseases posed no threat to humans due to the species barrier."[27] The predictable reaction of farmers, on hearing this opinion, is that they mix healthy and diseased cattle together, using both to provide milk. "As a result... nearly half of all dairy herds in the United States are contaminated with these viruses."[28]

The possibility of the leukemia virus being passed across species from cattle has been studied by experiments that transmitted the virus to other animal subjects through such means as injection and ingestion of milk from leukemic cows. (Milk contains white blood cells, which might include cancerous cells.) Sheep, goats, rhesus monkeys, and chimpanzees have developed leukemia after being infected in some of these ways.[29] For instance, "In a 1974 experiment, U.S. doctors took newborn chimpanzees away from their mothers at birth and weaned them on milk obtained from virus-infected cows."[30] A number of the monkeys came down with diseases previously unheard of in their species. These diseases were leukemia and (one you'll recognize) PCP.

Of course, similar research has not been conducted on humans, but experiments have shown that "the virus can infect human cells in vitro."[31] Moreover, some samples of human blood have shown evidence of antibodies to bovine leukemia virus. Kradjian comments, "This is disturbing. How did the bovine leukemia virus particles gain access to humans and become antigens? Was it as small, denatured particles [in milk]?"[32]

Less direct evidence of the crossover of leukemia from cattle to humans can be gathered by way of epidemiological studies. An examination of leukemia rates in the United States shows that "Iowa, Nebraska, South Dakota, Minnesota and Wisconsin have a statistically higher incidence of leukemia than the national average," and these are the states that hold the infected herds.[33] Similar studies from Russia and Sweden indicate that there, high rates of leukemia in human populations correspond to areas in which infected cattle are raised. Moreover, "Dairy farmers have significantly elevated leukemia rates."[34]

One can also look at correlational studies in which subjects who have similar lifestyles but who differ in their consumption of milk are compared for their cancer rates. We've already made clear, in the case of the relationship between HIV and AIDS, that correlation by itself does not prove any type of causal connection between the two concurring phenomena, but it does suggest that the overlap should be studied further. To cite just one such study, I can mention that done in Norway and reported in the *British Journal of Cancer*, in which 1,422 people were followed for twelve years. In general, no correspondence was seen between milk and various cancers; "however, in lymphoma, there was a strong positive association. If one drank 2 glasses or more daily (or the equivalent in dairy products), the odds were 3.4 times greater than in persons drinking less than one glass of developing a lymphoma."[35]

All this is germane to the topic under discussion, which is that bovine visna virus may have slipped into vaccines through the fetal calf blood that was used to produce them. It presents another example of how a bovine disease possibly may be transmitted from cattle to other species, also through blood, though in this new instance blood is transported and concealed in milk.

The Role of Smallpox Vaccine

To return to the question of vaccines, we may recall all the hoopla—or should we call it hubris—that surrounded what was said to be the total extirpation of one of mankind's greatest scourges, smallpox. The World Health Organization's aim was to inoculate everyone on the planet that needed to be immunized so that the disease, with no one left to infect (in a way that would lead to disease), would vanish. Here would be one pest that would never trouble us again.

The nobility of the goal is unquestionable, but some scientists have noted the suspicious coincidence that non-Western AIDS seems to have surfaced in those areas that were the major targets of the smallpox vaccination campaigns. The seven central African countries most affected by AIDS—Zaire, Zambia, Tanzania, Uganda, Malawai, Ruanda, and Burundi—are the same ones that were the most thoroughly vaccinated by the health organization. Meanwhile, "Brazil, the only South American country included in the smallpox campaign, has the greatest incidence of AIDS on that continent."[36]

If the vaccinations increased one's susceptibility or even led to AIDS, this would also "provide an explanation of how the infection was spread more evenly between males and females in Africa than in the West and why there is less sign of infection among five-to-11 year olds in Central Africa [who were born too late to be vaccinated for smallpox]."[37]

On top of such geographic correlations, there have been studies that indicate the smallpox vaccine is associated with immune-damaging diseases. De Latte mentions a 1968 article by Fulginiti et al.: "We learn from it that children who were suffering from immune diseases, and were given smallpox vaccines, developed 'acquired immune deficiencies,' AIDS-like diseases, long before this term was coined."[38] Symptoms observed in the sickened children included "lung infections (*pneumocytisis carinii*), yeasts (*candida albicans*), spleen inflammation, pancreas and liver dysfunction, multiple staph and strep infections, drop of the T-cell counts," as well as "gangrene in some of the vaccinated limbs, which subsequently were amputated."[39] All of the children in the study died. HIV had not been heard of at that time, "so tests for it were not performed."[40]

Scientists, looking at the geographic dovetailing and the evidence (scant as it is) of possible links between vaccination and illness, have wondered out loud whether it is possible that while keeping people from getting one disease, smallpox, the vaccinations did not assure them of staving off another, AIDS.

I'm not forgetting that many believe that the existence of large-scale African AIDS is a myth, but—as indicated—for the sake of this discussion we are going with the HIV-as-cofactor concept. According to this idea, if the vaccine disseminated HIV, this itself would not cause AIDS, unless other immune-debilitating factors were concurrent. The opinions of these two subsets of AIDS dissenters—those who believe HIV was seeded in Africa by vaccines and those who believe that there is no HIV in Africa (evidence of it is based on a faulty blood test and on presumptive diagnoses, they say)—are obviously incompatible, yet I believe both deserve to be heard. In starting this book, you may recall, I stated as my purpose the airing of credible dissenting views, perhaps noting my disagreement when (as in the case of Schmidt) some of the ideas were inconsistent or conceptually inadequate, but not censoring them if on the whole they were based to some degree on sound evidence.

Further, when we were discussing what rules should be adopted for declaring HIV properly isolated—since Duesberg and others were saying Koch's traditional postulates should be jettisoned, while others said they still deserved qualified allegiance—I said that the question should not be settled prematurely insofar as a definitive answer did not yet exist. There were plausible arguments on each side, so the best anyone could do was sift the evidence for him- or herself while keeping an open mind that will be receptive to facts that are not yet in. This same mind-set is worth cultivating in relation to the question of the presence or absence of AIDS in Africa. Note well: As Stanley Aronowitz often remarks, in order to honestly enter into either side of a question, one has to be able to play devil's advocate, which is to say, one has to tentatively take up a position—as I am doing now—in order to see what it entails and where it leads. And this goes double when confronting a side of a dispute that you would naturally tend to dismiss. Offering provisional sympathy is the only way to see into a stance's depths.

Returning to our exposition, let me pose this question: What connection could there possibly be between getting a shot to shield one from a disease and coming down with another disease? For smallpox, two theories have been advanced to limn this connection. The first possibility is the one we have already broached. To begin manufacture of a smallpox vaccine the virus is cultivated on living tissue. Then the virus is extracted from the tissue and used to infect an animal. But it is impossible to remove just the virus, when moving to this second step, since tiny disjecta from the living culture are inevitably swept up in the transfer. "Sometimes the DNA, RNA, small bits of protein that are removed are contaminated. These particles are too small to clean out. Therefore, they become part of the vaccine preparation."[41] As Curtis puts it, "When you go to make your vaccine [from the virus you have cultivated], you suck up that conglomeration and extract not only the virus that you want but any other thing that is growing there."[42]

We already stated that fetal calf serum is commonly used in these virus-growing tissue cultures, and when it is, material from calf's blood, including BVV virus particles and cancer cells, will be included with the virus preparation. The existence of this problem is widely recognized, but many scientists say these contaminants to the selected virus will be "sterile" in the vaccine, which means ineffective in humans. However, this is more of a fond hope than a proven truth.

The point of a vaccine is to give the human a dose of a weakened virus or cousin virus to the one being guarded against, either of these being used to invoke antibodies to the feared viral agent. However, there is no logical reason to hold that the weak focal virus in the vaccine couldn't be cut with traces of another, still robust virus.

This is not a mere hypothetical case. Curtis singles out polio vaccine preparations from the 1960s. (We will handle this topic more fully later.) "Frequently the contaminating virus, SV-40, grew in the same tissue culture, as the polio virus, and often outnumbered its population. People were getting a far greater load of the SV-40 virus in their vaccine preparation than the polio virus."[43] SV40 virus may cause kidney or brain cancer, although infection precedes disease by twenty or thirty years. Sixty-five million Americans received this contaminated vaccine.

The second way the smallpox vaccine may become a double disease carrier—it is made up to carry a watered down pox—has to do with what happens to the virus when it is grown on the animal host. According to a description by Curtis:

> Cattle would have their bellies shaved and then sliced open. A virus would be applied. They would place the cow in stanchions, come back a few days later, scrape off the scabs and collect the scabs and the effluent in a stainless steel container. They would dry it out and that would be their next batch of smallpox vaccine. Any virus, including the bovine visna virus or the bovine leukemia virus, could have potentially contaminated any of these vaccine preparations. It is possible that that is how AIDS was introduced accidentally.[44]

In other words, just as some believe leukemia is reaching people through milk, this view sees the vaccine as containing cattle diseases that are piggybacking on top of those that the vaccine is meant to contain. To restate, the vaccine made from the cowpox from the scabs of an infected cow may contain free riders that can potentially cause disease.

It is implicit in this argument that either BVV or another cattle virus, once given through a human subject via inoculation, translates into HIV. However, aside from this view, there is also a second theory on what part these freeloading viruses play. This second view is put forward by the orthodox camp and has the advantage, from their view, of preserving HIV as a separate entity, that is, not a cattle disease that has crossed to humans but an individual virus in its own right. From their perspective HIV was already waiting, dormant in the African population, but not able to cause disease unless awakened by a biological trigger. They say the vaccinations pulled this trigger.

In a book on immunization, Christopher Miller notes, "According to Dr. Gallo... the use of live vaccines such as that used for smallpox can 'activate' a dormant infection such as HIV."[45] With Dr. Gallo, establishment figure head, raising this hue and cry about contaminated vaccines, the WHO had to take notice. Pierce Wright, in a story in the *Times* of London, quotes a WHO adviser as saying, "I thought it [the geographic overlap of vaccination campaign and AIDS outbreak] was just a coincidence until we studied the latest findings about the reactions which can be caused by Vaccinia [the name of the smallpox vaccine]. Now I believe the smallpox vaccine theory is the explanation to the explosion of AIDS."[46] The *Times* piece continued by stating, "Many experts are reluctant to support the theory publicly because they believe it would be interpreted unfairly as criticism of WHO," and because of its potential impact on "other public health campaigns with vaccines such as against diphtheria."[47]

The Role of Hepatitis Vaccine

A reader who accepts that some blundering in the manufacturing of the vaccine may have led to the offsetting of the benefits promised by the vaccine—it is argued that the campaign to plow under one illness ended up preparing the ground for another—still may be led to ask how this touches on the situation in the West where a large smallpox vaccination program was never instituted. If one were looking for parallels, cases in the West where a vaccination campaign appeared in the same populations who were later brought low by AIDS, then one would have to turn to the trials among gay males of a hepatitis B vaccine.

"In the United States, the hepatitis B campaign was an experimental study of young, white, healthy, male homosexuals who were recruited in 1978 in New York City by the city's blood bank. In 1980, the same thing occurred in Los Angeles and San Francisco," Curtis writes.[48] Here is an eye-opening geographic coincidence. AIDS appeared first in the cities where the hepatitis tests had been conducted. Again, only a correlation, and one that would have to be counterbalanced with other considerations, such as that these three cities had the most developed fast-track cultures. Still, it is jarring to learn, for example, "The morbidity mortality weekly reviews noted that the first six out of ten cases [of AIDS] in San Francisco came directly out of the hepatitis B vaccine project."[49]

Let's look at more exact figures. The hepatitis vaccine trial started, as noted, in November 1978, "largely supported by grants from CDC and National Institutes of Health." The first to be given the vaccine were "1,083 white homosexual men (average age 29) in New York."[50] Critic Tony Brown notes, "In January of 1979, two months later, the first case of AIDS appeared in a homosexual in New York."[51] Two years later, in March 1980, the vaccine was administered to 1,402 gay men living in the five scattered cities of Chicago, Los Angeles, San Francisco, Denver, and St. Louis. Brown continues, "Seven months later, gay men were struck with AIDS in Los Angeles and San Francisco."[52] Writer Waves Forest adds to this that by 1981, 25 to 50 percent of those who were inoculated had AIDS, and by 1984 the numbers were up to 64 percent. "The figures for subsequent infection rates of participants in that suspicious study are allegedly 'unavailable,' safe in the hands of the Justice Department."[53]

The fact that so many men in the vaccine trials later came down with AIDS hardly creates a watertight argument that there is any causal relation between the two factors. One might think that if a man was in the vaccine program, it was probably because he felt himself in danger of getting hepatitis and was trying to protect himself. So, one could further extrapolate, he might consider himself in jeopardy because he was a member of the fast crowd and/or (most probably "and") he was an IV drug user. Thus, it might be that his immune system was already under assault due to his lifestyle, so his enrollment in this program was not an important or deciding factor, in his later getting AIDS. (In a moment we will have to revise these suppositions.) Keeping this in mind, but also remembering that I want to play devil's advocate here, let's follow the logic of this hepatitis causation theory a little further.

We saw that one bit of evidence used by those who believed poppers played a role in weakening the immune system preparatory to AIDS was the historical confluence between the rise in trendiness of the nitrite inhalants and the appearance of AIDS approximately in the same time period. Those who want to charge the hepatitis vaccine with the same collusion with the disease see an even closer matchup by date between the trials and the immune deficiency disease. As the article "AIDS: New or Old" brings out, "There is general agreement that old (pre-1978) American blood samples all test negative for HIV antibodies. This is the reason why most medical epidemiologists believe HIV was introduced into the U.S. sometime around 1978."[54] Going back and studying blood samples taken of gay men who were to participate in the vaccine trials, but before the immunizations had begun, showed them all to be HIV negative. But in the 1978-79 blood samples of those who had received the vaccine—the blood had been preserved at the New York City Blood Center—it was seen that 6.6 percent were positive.

The pamphlet from which I extracted these figures raises some pointed questions about this chronology

> How did these gays become HIV-positive in 1978? Was HIV present in New York City before 1978?... It is unlikely that the source of HIV was the pooled gay blood that was used to make the experimental vaccine because the vaccine took 65 weeks to make. This means that vaccine production for Szmuness [who was in charge of the trials] began in 1977, [so if HIV got into the vaccine from donated blood] then some gay blood specimens should have tested positive. But, as stated, 1977 gay blood tests were negative for HIV.[55]

This is all rather perplexing. It's not a (relatively) simple case such as what we saw with the smallpox vaccine, in which it is posited that a virus from cattle was accidentally included with the drug. Here it has been established by retroactive tests that the human blood used for making the vaccine was HIV free. And there is another problem, one that strikes at our earlier hypothesis about hepatitis inoculees belonging to the fast lane. Whatever their lifestyle, those getting the vaccine were healthy.

June Goodfield writes: "In those (gay men) who received all three injections, 96% developed antibodies against the (hepatitis) virus. Overall, the vaccine was shown to be 92.3% effective in protecting high risk individuals against hepatitis B."[56] For the vaccine to be effective, it must find in the body an efficient immune defense system that will produce antibodies on demand. A. Cantwell writes, "Recent studies have shown that hepatitis B vaccination is not very successful in immunodepressed people. In HIV-positive individuals, the success rate of the hepatitis B vaccine is about 50%, only protecting one out of two people infected with the AIDS virus."[57] The conclusion that naturally follows is that "the experiment would have been a failure (never 96% effective) if the immune systems of the men hadn't been working at full capacity."[58]

This originally healthy group was highly liable to become infected by HIV and then get AIDS? As we've already mentioned, in ten years more than half would have the immune illness and many of the very first PWAs came from this group. It should also be pointed out that the rate of infection in this group of inoculated was phenomenal.

> [By 1982] 30% of the men in the Szmuness experiment were HIV- positive! For that year, this extremely high infection rate is the highest rate of HIV infection ever recorded for any "high risk" group in the AIDS medical literature. This 1982 rate of 30% far exceeds the rate for any African population, where Dr. Gallo and others claim the disease has been around "for decades" or "for millennia.[59]

By now it should be clear that writers such as Cantwell, who hold the view I am highlighting, are not in accord with the group of dissenters who point to the fast track as the road to perdition. The latter see the unhealthy components of the clone lifestyle as weakening the immune system, which leads to AIDS. On the other hand, Cantwell, working on the hepatitis thesis, argues that it was men whose health was basically sound who were getting infected with HIV during the hepatitis trial. I imagine it would not be impossible to reconcile these two perspectives. One could hypothesize, for example, that the vaccinated later turned to fast-track, unhealthy lifestyles or, perhaps, were living them already but had not yet suffered their more damaging, long-term effects. However, there is a conflict, and I simply want to acknowledge it. This is not the first time we've noticed that the various strands of dissident theory do not all jibe. This is something we will have to live with until an alternative conceptual synthesis arrives. While this book does not present such a synthesis, it tries to provide the raw materials from which such a synthesis could draw.

Let's get back to the perplexity mentioned a moment ago concerning how the vaccine might have passed HIV to those who were injected. This doesn't seem right. If it were made by using the blood of gay men who were not HIV positive, then where did the retrovirus come from? Are the proponents of this viewpoint painting themselves into a corner?

To be completely forthright, to my knowledge those who support this idea have not worked out their ideas in enough detail, not with the backup of, say, those who hold smallpox vaccinations played a part in the disease. Many believe the government was using the vaccine for a hidden-agenda biological experiment within the hepatitis trial. However, that can more properly be discussed in the next chapter, where I will take up biological warfare.

Other writers argue that the vaccine could have been contaminated by sick monkeys who were being used in the study. When I interviewed Leonard Horowitz, he propounded that idea, laying out the following scenario:

> All it would have taken was one monkey, perhaps sitting next to another monkey in a cage in Litton Bionetics Laboratories in Northwest Uganda, that was infected with an AIDS virus, to be shipped across the ocean, to come to New York City, location of the New York City Blood Center and Krugman and Hilliman's labs which were involved in these experimental vaccine programs.[60]

The Role of Polio Vaccine

In marked contrast to the paucity of evidence and explanation of how the hepatitis vaccine might have been contaminated with retroviral material, the fact that one of the polio vaccines used in Africa was contaminated with a monkey virus has been abundantly acknowledged and verified. As Curtis describes, "There was, in fact, an almost forgotten mass vaccination campaign in which an oral polio vaccine was administered to at least 325,000 people, and perhaps more than half a million people, in equatorial Africa from 1957 to 1960."[61] Later it was found that one of the two vaccines used in the program was laced with monkey virus.

When I say this contamination has been "acknowledged," I mean at the highest possible level. The revered developer of the vaccine, Jonas Salk, has confessed this was a tragic error, though one he feels is excusable. "We took all the precautions that we knew of at the time," Salk says, but "sometimes you find out things after the fact."[62] He goes on to talk about SV40, the monkey virus, which, as we will investigate below, was mixed in with all the early vaccines.

> SV40 virus… was a contaminant in monkey kidney cell cultures. The last thing in the world one would want to do now is to make vaccines out of the tissues of monkeys that come from the jungle. That was a learning experience you might say. At this point, we are much more sophisticated and can use continuously propagating cell lines, such as the Vero cell. The SV40 story alerts us how to avoid these things in the future.[63]

Salk's admirable willingness to eat crow when he knows he was wrong does not translate into a willingness to imagine there may be further unexplored troubles with the vaccines given out in the early period. When Curtis asked him about some researchers' investigations into whether there may have been a connection between polio vaccines distributed in Africa and later AIDS outbreaks, Salk "flatly refuses to discuss the subject." His response was: "I don't think I can be helpful to you, other than to try to dissuade you from pursuing that kind of a hypothesis, because what value is it? What value is it to anyone to try to imply such a cause and effect relationship?"[64]

One objection Salk might have made to those who are pressing this connection is that since the vaccine was given out years before AIDS appeared, the dating does not offer the immediacy of cause leading quickly to effect that we saw in relation to the hepatitis link; rather, it depends on a belief in the slow evolution of the virus from an earlier (possibly simian) form to one that is now hazardous to humans. A calculation of this transformation was made by Gerald Myers, the federal government's chief expert in genetic sequencing. He says, "HIV dates from about 1960 assuming it arose from a single, common ancestor."[65]

Let me say a few words about how these calculations are made. I rely on L. Garrett's book *The Coming Plague* for specifics. Myers's method, working with the GenBank group, which have been studying HIV, does not discover when HIV itself originated. What it tries to determine is the common ancestor of all the presently existent strains. "When the GenBank group looked at viral strains collected in a given geographic location over a period of years they could see that HIV-1 was evolving—or mutating—at an overall rate of 1 percent per year."[66] Assuming this mutation rate has stayed constant—one could first gauge it by studying changes in HIV RNA over a period of time—it was possible to figure out how long it would take the existent variations to have branched out from a common progenitor. If the 1 percent rate is accepted, "that would mean the clades had a common HIV ancestor that existed just thirty years prior, perhaps around 1962."[67]

Just to follow out GenBank's theorization a little further, we can mention that, still analyzing with this technique, they decided "after ten years of mutating along a single course, the HIV family tree had spread out suddenly, yielding the six distinct clade lineages."[68] Myers denominates this sudden emergent mutation from another viral form the Big Bang, saying "sometime in the early 1970s a biological event occurred that resulted in the sudden and explosive divergence of what had been a virtually linear evolutionary path for HIV-1."[69]

For us, though, it is not the bang that is most suggestive but the dating of the progenitor, since it seems to coincide with the period when the polio vaccine was introduced. Thus, the common ancestor of HIV comes on the world stage just when new viral strains (from monkeys) are getting into the human bloodstream. Though this certainly doesn't prove there was any connection, it does give us pause.

However, if HIV had arisen in this way, it would indicate that a simian virus could somehow make its way to a human host. A few pages back, though, it was urged that species-hopping viruses seldom continue beyond a single infection, being able to cross once, as when a mosquito transmits to a human, but not going any further. This is something that also bears on the smallpox vaccine theory, since there also a species jump is presumed.

The first point to keep in mind is that, while rare, such crossings are not impossible or unheard of. However, before saying more about that, let me remark another rare scenario.

When I mentioned species-jumping viruses earlier in our discussion, I dropped what might have seemed like dark hints about how scientists playing with viruses may give them greater adaptability. But this stray thought should not be interpreted as an indication that viral manipulation would only be done for diabolical reasons. "Teasing" viruses is part of creating vaccines. As previously stated, the virus can be a weak form of a commonly more devastating one. Infected with the harmless form, the inoculee will develop antibodies that will allow him or her to stave off any incursions by the stronger versions. To obtain the weaker strain, the virulent virus may be manipulated. It's in the realm of possibility that the newly created, disabled form has new qualities, such as the ability to aid other viruses that are associated with it in crossing species boundaries.

Admittedly, this is a far-fetched idea, but I have been thinking of it for two reasons. For one, viral manipulation as part of vaccine production does seem to have created species vaulting. Curtis notes, "In recent decades, some scientists believe, live-virus, vaccines may have actually helped transfer viruses across species lines."[70] He uses as an illustration canine distemper parvovirus, or CPV. This is a new disease that had never been seen in dogs prior to 1977. Then, suddenly, within months of its first appearance, it was found throughout the world, striking dogs with intestine and heart disease. "CPV is intriguingly similar in its genetic structure to a cat disease called feline panleukopenia virus (FPLV), but it's even more similar to the vaccine for this disease."[71] The obvious possibility, one that a number of scientists have put forward, is that "the cat virus most likely was introduced into dog cells in the laboratory, where the strain adapted itself to the new host."[72]

The second point that has encouraged me to discuss this idea is that SV40, the known contaminant of the polio vaccine, has been shown to aid the growth of other viruses. Although it is not said to make its fellow viruses more compatible with other species, Dr. G. D. Hsiung, who has studied SV40, says that it not only acts on its own but "increases the development of human adenoviruses in cell cultures, and has been found in cells infected with different recombinants or hybrids, such as SV-40/measles virus, SV-40/reoviruses, and SV-40/rabies viruses."[73]

The most frightening part of this finding is that it notes a connection between SV-40 and the adenoviruses. These are the very viruses that one so often finds in AIDS patients. Snead goes so far to say, seeing that AIDS is usually found in a person who has already been beset by multiple viral infections, that "HIV can only reproduce when accompanied by another virus. One of the effective assistants is the adenovirus."[74] Thus, it is known that the simian virus that was being delivered along with the vaccine was capable of promoting the growth of other viruses, particularly ones associated with HIV, and was also found combining with other viruses.

To flesh out the discussion of viruses working in pairs to damage the body, let me bring up a topic alluded to in my discussion of bovine visnu virus, which, as we saw, was renamed bovine immunodeficiency virus, or BIV. We also learned in discussing cattle diseases that some posit that BIV, if it has been passed to man, perhaps through smallpox vaccine, can be linked to HIV. We also remarked that BIV had been singled out for study because it is one of the rare lentiviruses.

At the time I held off discussing another finding about this virus because I wanted to discuss the interplay of sympathetic viruses in a single section. It seems, according to an article in *Virology* by Y. Geng et al., that it has a symbiotic relation with bovine herpes virus (BHV-1). "It is currently not clear what mechanism is employed by BHV-1 in transactivation of BIV," they write, "whether it transactivates BIV directly by inducing a bovine cellular NF-KB-like protein which interacts with the BIV-LTR, as has been proposed with HIV-1, or indirectly via other cellular intermediates."[75] What is clear is that the herpes virus plays some part in activating the immunodeficiency virus. The authors of this article are most concerned with the potential parallels to the activity of HIV. "During HIV infection in humans many infected individuals undergo a period of latent infection.... Infection by herpes virus may then play an important role in activating the HIV promoter expression, and resulting in increased virus production."[76]

This article does not see any tie between BIV and HIV except insofar "as the activation of BIV by BHV-1 may provide us with a useful animal model to study possible interactions between two pathogenic viruses in vivo," and thus its findings are not relevant to our examination of the connection of HIV and BIV-contaminated smallpox vaccine.[77] It does fit into our exposition insofar as it shows that one virus playing a part in another's awakening is a documented happening. Thus, we are not simply spinning cobwebs of thought when we suggest that one possible form of animal viral crossover would be for the animal virus to not so much do damage to the new species as put into action another latent virus, either already in the host or transported in the same vaccine.

But let's go back to the less remote possibility. An undetected virus from an animal, in this case, a monkey, has been included in the vaccine preparation and goes along for the ride. Once injected into the person seeking help, it crosses the species barrier. Looked at in one way, being a "remote possibility" means the event is bound to happen eventually. As Curtis says so powerfully, "There are natural obstacles preventing a virus from crossing the barrier to become established and thrive in a new species. But it happens."[78] He adds that when the unlikely happens, "the virus frequently becomes much deadlier in the new species than it was in the original hosts."[79]

Viruses, then, tend to stay in one species, but if they are going to cross, it will generally be between animal groups that have comparable makeups. With that in mind, we can see that the possibility that a viral transfer was made from another animal group to humans would be more probable through the polio vaccine manufactured by using monkeys than through the smallpox vaccine, which is dependent on cattle. "Because monkeys are genetically similar to human beings, some simian viruses can leap the species barrier with devastating effect."[80]

A second point that increases the plausibility that the polio vaccine is implicated in cross-species transfer—versus a belief that smallpox is involved— is that, as we've recognized, the BVV contamination of smallpox vaccine is a conjecture, while the tainting of polio vaccine by Simian Virus 40 (SV40) is documented and accepted.

However, where the polio thesis falls flat when compared to the smallpox and hepatitis vaccine theories is that with the two latter vaccines there is a correspondence between those who got AIDS and got the vaccine. As we will see, though I want to hold this discussion in abeyance for the moment, the polio vaccine argument is only strong when made concerning African infections. Before pressing forward with an investigation of that element of the exposition, it would be useful to lay out more facts about the case.

We've already examined how vaccines are made. First the viruses that will be used to make them are cultured in live tissue. The cowpox virus culture was stimulated by calf blood. Polio virus, in contrast, is grown in primate cells. "Either monkey or human cells will work, but researchers selected monkeys because their tissue was more available and there were fears that human cell lines might spread cancer."[81] Specifically, in the 1950s, polio viruses were cultured on monkey kidneys.

Some of these kidneys were infected with viruses. "Scientists knew about some of these viruses and developed tests to identify and then eliminate the tissues that contained them."[82] But, as we know now, at least one of these viruses, SV40 (the fortieth monkey virus identified) evaded detection. It was later identified and animal carriers barred from use in vaccine manufacture.

> Between 1954 and 1963, an estimated 10 million to 30 million Americans and scores of millions of people around the world were exposed to a virus that infected the kidneys of the Asian rhesus monkeys imported mainly from India. The virus survived the formaldehyde that Salk used to kill his polio viruses.
>
> SV40 was delivered straight into people's bloodstreams along with their Salk shots and via sugar cubes in field trials of the weakened living virus developed by Sabin.[83]

Since we know those getting inoculated for polio in this time period were also getting a dose of SV40, the obvious next question to ask is whether there is any indication that SV40 can affect women and men. At this point, we hit a brick wall. "Remarkably, considering the large numbers of people who received the SV40-contaminated polio vaccines, no one has conducted a major epidemiological study in the U.S. to discover whether there is any pattern of illnesses caused by the virus."[84]

A carefully designed, large-scale epidemiological survey would be the surest way to test whether SV40 has affected those to whom it was administered, but saving that we can look at individual cases. As far as current research has proceeded, there is no hard evidence that SV40 can cause illness in humans. We also know, though, that when SV40 is given to another species, mice, it can cause sickness. At Johns Hopkins, it was injected in mice who were bred to lack immune systems, and they "developed Kaposi's sarcoma-like tumors, similar to those afflicting many AIDS victims."[85]

As for the simian virus's effect on humans, a number of inconclusive statistical associations have been found. Although they are not sufficient to prove SV40 can cross species, they certainly indicate the need for sustained study of this possibility. Research done in Australia in 1968 found a correlation between those who received polio vaccinations and cancers in children past one year of age. Some years later, "German scientists found evidence of SV40 in 30 out of 110 brain tumors [examined], and later reports indicated a jump in the frequency of brain tumors among those [people] who had received vaccine contaminated with SV40."[86]

Another possible link with human illness, AIDS in particular, can by seen by looking at genetic structure. In 1985 *Science* published research identifying "the location of the human T-Cell receptor alpha-chain gene [which] was located on chromosome 14."[87] This chromosome also contains other genes that control immune processes. "Considering this scenario, it would be safe to assume that impairment to the expression, of chromosome 14 (that is of the performance of its functions), would manifest by symptoms usually considered diagnostic of AIDS," according to Snead, thus indicating HIV may damage this gene.[88] The similarity here is that SV40 also "has a particular affinity for attacking chromosome 14." Snead asks the weighty question about the comparable proclivities of the simian and human viruses. "Would this fact [their similar attack patterns] be the undiscovered link connecting the polio vaccine, immunosuppression, cancers, leukemias and AIDS?"[89]

Admittedly, all this evidence is rather spotty, more suggestive than evidentiary, and still leaves us in the dark about the gaping hole in the theory concerning why AIDS in the United States, if it had some connection to the widely distributed polio vaccine, was so selective in who it struck. To deal with these problems, we need to introduce further hypotheses.

The difficulty with the theory as applied to the U.S. vaccine program has not been ignored by those who work from the idea that the vaccine program had some role in the development of AIDS. "If HIV were one of those numerous anonymous monkey viruses contaminating the early Salk and Sabin vaccines," Curtis states, "presumably there would have been an explosion of AIDS in the U.S. outside the currently defined high-risk groups: male homosexuals, intravenous-drug users, hemophiliacs and the sexual partners of those people." He acknowledges, "Of course, that sort of eruption hasn't happened."[90]

The only place such an "explosion" might be seen is in Africa. But this means there must have been some unique quality of the vaccine, not merely that it was contaminated with SV40, or some other simian virus, that gave it the power to be an AIDS precursor. Curtis points out that "equatorial Africa was the site of the world's first mass trials of an oral polio vaccine—a vaccine cultured in monkey kidneys but different in at least one important respect from the Sabin vaccine ultimately adopted worldwide."[91] Tantalizingly, and irritatingly, Curtis leaves unclear what this difference was; but we can get a little more information about this particular leg of the vaccination campaign from Snead, who writes:

> The world's very first mass oral polio vaccination campaign took place in Ruanda-Urundi and parts of the northeastern Belgian Congo in 1957-58. Ruanda-Urundi was a small U.N. trust territory bordering the much larger Belgian Congo on the east and administered together with the Congo, as a unit, by Belgium. A quarter of a million people were vaccinated, mostly in the Ruzizi Valley of Ruanda-Urundi, 200 miles west of Lake Victoria. Today Ruanda-Urundi has been divided into the independent countries of Rwanda and Burundi, both hard hit by AIDS. Rwanda, in fact, may well be the hardest hit area in the world—its capital city of Kigali enjoys the dubious distinction of having 25% of its citizens infected, the highest percentage yet measured."[92]

This is the type of lead, the coincidence of one place being both hardest hit by AIDS and the first mass vaccination locale, that needs to be followed out epidemiologically to see if there is a real connection here. But there is another possibility Curtis has noted, contained in his phrase "if HIV were one of those numerous anonymous monkey viruses contaminating the early Salk and Sabin vaccines." He is not confining the possible HIV transmitter to SV40, which has gotten a lot of press simply because it has been acknowledged as a contaminant; rather, he is suggesting there may be an as yet undetected second contaminant at the bottom of the problem.

In proof of this thesis, he might have discussed a case he cites (but without putting this interpretation on it) the investigation of virus researchers in Washington, D.C., who in 1988, restudying the data on pregnant women collected for a 1959-1965 medical analysis, found "the incidence of brain tumors in children of mothers who'd been injected with the Salk vaccine was thirteen times greater than that of offspring of mothers who hadn't had those polio shots."[93]

However, tests on the blood of these mothers showed that SV40 was not the cause of the tumors. Thus, there might be another cause, such as a second contaminant, at the root of this problem. Obviously, this is not directly connected to AIDS, but it is being referenced to show that viruses besides number 40 may have found their way into the batches of vaccine.

G. D. Hsiung, whose research into SV40 was tapped earlier, has made a project out of studying the degree to which cell cultures used in laboratories are virally contaminated. His group suspected there would be some contamination. In an older study, his researchers found, "to their great surprise, from Feb. 1966 to Feb. 1967, on 417 different lots, all of them were contaminated."[94] They found 227 rhesus and 190 green monkeys "were producing viruses at the rate of 50% every month." Yet these were cultures that the researchers using them "believed to be safe and normal." Moreover, the contaminations often were not evident at first. When studied three weeks after the cultures had been established, only 3 to 4 percent of cultures exhibited contamination; but if one put in the time and observed them after "29 to 55 days, the contamination rate increased dramatically. The SV-5 showed only after 90 days, while the SV-40 required even more than a year to show."[95]

We saw already, when discussing using cattle tissue and blood as bases for cell cultures, that it is very easy to get contamination, but Hsiung's work shows not only that it is theoretically easy but that it is happening. Add this to Salk's own admission that mistakes were made in producing the early vaccine, and one will find the idea that the vaccines were contaminated well within the realm of possibility.

More directly relevant matter on other possible monkey viral contaminants can be found by following the polio vaccine program in Africa a little further along in time. As noted, eventually SV40's presence in the vaccine was detected, and kidneys of monkeys infected with this virus were discarded. (This doesn't mean, by the way, that vaccine already produced was written off as a loss. "The faulty vaccines were usually not withdrawn from the shelves, even when contamination become known to regulating authorities."[96])

Nonetheless, to cultivate viruses for new batches, "in 1961, vaccine producers shifted to kidneys from African green monkeys, which in the wild were free of SV40.... Unfortunately, green monkeys were infected with something else."[97]

Let's jump ahead a bit. In 1982 and 1983, veterinarians at the California Primate Research Center and at Harvard's New England Primate Center saw that many of their macaques were being killed by diseases that resembled AIDS. This disease had been observed since 1969, but these scientists were the first to observe the similarities between it and the immune deficiency syndrome. They also were able to lay their fingers on the cause. "The monkey's illnesses, the researchers discovered, were triggered by a previously unrecognized retrovirus called simian immunodeficiency virus (SIV)." If that weren't startling enough, "Among the natural hosts for this virus were none other than African green monkeys, but in that species, typically, SIV didn't cause serious disease."[98] Although this virus is not genetically very close to HIV-1, " Gallo says some versions of this monkey virus are virtually indistinguishable from some human variants of HIV-2."[99]

As you might suspect, Gallo is not making these connections because he wants to heap blame on WHO or the vaccine producers. He does not subscribe to the vaccine transfer thesis but is interested in genetic similarities between simian and human viruses, because he feels there may have been some crossing between species. Going beyond SIV, with its resemblance to HIV-2, he notes, "The same is true, Dr. Gallo says, of HTLV-1," the variant found in the West, which also has its monkey cousins.[100] His associate Genoveffa Franchini has found some simian viruses, and in particular simian T-cell leukemia viruses (or STLV-1), "which are, Dr. Gallo says, as close to most of the human HTLV-1 viruses isolated from the Caribbean islands, southern most United States, southern Japan and equatorial Africa as some STLV-1s are to one another."[101]

If it were found that one of these viruses were also a contaminant in polio vaccines—this has not been found to date—it would make the cross species transfer possibility even more alarming in that SIV produces illnesses resembling AIDS in a way that SV40 does not.

But we haven't reached the end of the complexities of this issue. We have been proceeding as if SIV were a single virus, but this is not the case. L. Pascal explains, "There are many different varieties of SIV infecting the various monkey species," and although only three types of monkeys are routinely used in vaccine manufacture, "it is entirely possible for many different SIVs to exist within a single monkey species."[102] Thus the problem of virus varieties is much more complex than we had supposed.

> Within our own species, there are three separate varieties of polio, requiring three separate vaccines, and dozens of other known viruses in the same genus at polio, plus an unknown number of still-undiscovered close relations. Our methods of detection of SIVs are very crude. They would have missed all or nearly all the rare varieties, and would probably have missed even very common varieties if those were antigenically distinct enough not to react to the tests for the known members of the family.[103]

The least we can say when we discuss the very first polio vaccines that were tested, as noted, not in a metropolitan center but in Africa, where disastrous results would have fewer repercussions, is that the developers were playing with fire, in wagering not their own lives, of course, but those of indigent people in undeveloped countries. Further, they were doing so on the hope that contaminations from the cultures in which they were culturing the viruses, contaminations for which they had no means to screen, would either not be included in the vaccine package or, if they were, would not damage those who were vaccinated.

None of this is entirely satisfactory. That is to say, taking all we have covered in this chapter on the hypotheses concerning the implication of vaccines in the original causation of AIDS, we have seen that none are thoroughly convincing, though more from a lack of in-depth research than from fallacies locked into the premises. There is limited evidence that viruses might make the leap from other species to human; proof that some vaccines were contaminated; evidence that production of vaccines is fraught with dangers of adding unwanted viruses to the drugs; overlap between groups given vaccines and groups who later got AIDS. None of these as yet have enough grounding in evidence. And the situation is not helped by the nonfunding or attempts to sidestep studies that have been done. I already cited a *Times* article by Pierce Wright that quoted a scientist "hired by the WHO to investigate the possibility" of the smallpox vaccine being tied in with the later appearance of AIDS in Africa.[104] When quoting this report, I didn't mention the fate of this adviser's study. "He did his work, presented his report, was paid, and the report was buried because," according to Wright, his conclusions were not welcome. Basically, "the author stated in his report that the WHO was responsible for the spread of AIDS" in Africa via the vaccine.[105]

The best case that could be made for the dissenters—one that cannot yet be made—which would be the worst thing that could happen, would be one that followed out one of three possible paths. A viral contaminant (from cattle or a monkey) in one of the vaccines, when it entered a human, either (1) crossed species and became HIV, (2) awakened a latent human virus, or (3) altered existing endogenous human viral material to activate or create HIV.

Although a dissenter still has a way to go in digging out evidence if he or she wants to prove that one of these routes is indeed the highway HIV traveled to reach the human species, it would be macabrely ironic if one of these possibilities were true, precisely reversing the claims of those racists who see AIDS as unleashed on the innocent West due to strange African practices. We saw that the racists claim it was bizarre animistic blood rituals carried out by natives in the African bush that gave birth to AIDS. But if the vaccine as precursor thesis were sound, it would show evidence of another, equally gullible and bizarre animism, that of scientific medicine. I call it animistic in that, as we have seen, the early producers of vaccines fetishsized science insofar as they had faith that their vaccines would not be contaminated or liable to carry unwanted animal viruses in their load, a faith as hollow as that of the most arrant superstition. So the cause of AIDS, from this perspective, would be the idolatrous rituals of virologists, carried out in immaculate laboratories, in the WHO's sterile wilderness from which the retrovirus was spread to innocent Africa.

Is HIV an Old Virus or an Acquired Interspecies Virus?

One prominent countertheory to the idea that HIV was the product of a contaminated vaccine holds that the retrovirus is an ancient African virus, one mired in latency until certain altered modern living conditions brought it out of its lair. We saw hints of this theory previously. At this point, it might be appropriate to briefly review some of our earlier thoughts in light of the present, quite different discussion.

As should be obvious, the dissenters who hold to the vaccine theory, unlike those we discussed earlier, believe AIDS did arise in Africa, though not in the way the establishment imagines. They say HIV was seeded by aberrant vaccines, while the dissenters examined in our discussion of Africa believe AIDS comes from the West and was brought to the continent by tourists.

I will now deal in different ways with both groups who are in opposition to those who hold the vaccine-genesis theorem. I will raise and debunk the orthodox idea that HIV is a virus of longstanding on the African mainland, and I will raise and weigh the objections of the fraction of dissenters who hold that the vaccine, cross-species transfer idea is flawed.

First, what can we say against the position that "HIV-1 is an ancient virus which has infected an isolated group of humans in central Africa for millennia, relatively harmlessly, but spread recently to the rest of mankind because of urbanization and related social changes."[106]

None other than Luc Montagnier has poked holes in this theory. As quoted in an article for *New African* by Cantwell, Dr. Montagnier notes that "the Africa origin of AIDS has been debunked by several epidemiological studies." He goes on to cite a few, for example:

> When a team of scientists led by J. W. Carswell tested the blood of old, sexually inactive people living in geriatric homes in the Ugandan capital Kampala (Uganda is said to be the epicenter of AIDS), the team discovered that none of the elderly people tested positive for HIV antibodies. This 1986 study concluded that the virus had not been around Uganda for a long time. On the contrary, the study suggested that HIV arrived in the country only recently.[107]

Another study he reports on was done in 1989 and consisted of HIV-testing the San bush people who reside in the central Kalahari desert in Botswana. "The San are considered to be the oldest race currently living in Africa, and San-type skeletons date back 15,000 years or more." San adults were given the blood test and "not one person tested positive for HIV."[108]

These two investigations are merely the tip of the iceberg. This is one topic that, fortunately, has been gone over in depth, and the evidence has piled up showing that no trace of HIV can be found going back beyond the last few decades. "Extensive studies in Africa and elsewhere fail to reveal any isolated group of infected humans, and examination of stored serum fails to show HIV-1 infection was endemic anywhere before the 1970s," Seale notes.[109]

Having shown that to date there are no facts confirming the idea that HIV is of ancient lineage, let's look at the objections that one important dissenter, J. Seale, raised, not so much directly against the vaccine thesis but against two of the possible routes by which it is supposed to have traveled into humans: either the leap of a monkey virus across the species barrier or the combination of a monkey and human virus to form a new, powerful retrovirus.

Seale's argument against the first of these provisional propositions is that those who hold to this point of view have been overly fixated on a one-way transmission from animal to man. That is, they have centered their attention on whether the movement from monkey to human is possible, which everyone says can take place in rare cases, and then on calculating its probability. But, according to Seale, the basal "hypothesis is defective because HIV-1 is so species specific for humans that it has been impossible to infect animal species with the virus other than the chimpanzee."[110] Great efforts have been expended on what is called "a back infection." Yet, "even with injections of the virus, in massive quantities, into the blood and brains of a great variety of adult and newborn primates, and many other mammalian species," AIDS and HIV positivity has not been produced. According to Seale, this failure offers nearly a death blow to the species-jumping theorem in that "viruses which have recently crossed the species barrier can, experimentally, almost always be made to reinfect the species from which they originated."[111] The fact that the so-called AIDS retrovirus doesn't work that way makes it too anomalous to be true. "HIV-1 appears to have infected man for less than two decades, yet it is already superbly adapted to infect man alone." For Seale, this is sufficient to disprove this route of transmission.

The second hypothesis, which believes "HIV-1 was recently created by chance, natural, recombination and mutation of a virus which previously existed in man, with a virus from another animal to produce a new virus," also contains substantial difficulties. The problem here is that it is dependent on the preexistence of a virus in humans that could act as a template for the intrusive animal virus to latch onto in the process of forming a new entity. But as Seale explains, "this hypothesis is weak because no trace has been found of a virus antigenically related to HIV-1, and endemic in humans before the 1970s."[112] The obvious objection to Seale here would be to bring up the human T-cell leukemia/lymphotropic viruses (HTLV-1 and HTLV-2), which, it may be remembered, Dr. Gallo originally thought were near matches for the AIDS virus. Wouldn't Dr. Gallo's original confusion indicate a homology of genetic structure between these viruses and those of HIV? Seale says, on the contrary, "The nucleotide sequences of the human T-cell leukaemia/lymphotropic viruses show insufficient homology with HIV-1, or HIV-2, for them to have evolved recently, by chance mutation and recombination of HTLV with a lentivirus from another animal species"[113]

This second objection is obviously much weaker than Seale's first in that the fact that a human precursor to HIV has not yet been identified is hardly proof that it doesn't exist. After all, until the 1960s retroviruses had never even been seen or heard of, either.

The frame-shattering blow against the picture of HIV as a species-vaulting monkey virus is stronger, although we should keep in mind that the arguments on all sides of this controversy, from Seale's to those of the vaccine-thesis upholders to the orthodox makers of the vaccines, are based on statistical probabilities. The polio vaccine designers, for example, argue that the risk of an animal virus accidentally included in a vaccine crossing between species is so low to be left out of the calculations. Those who say this did happen, a virus was brought along by the vaccine and took hold in human tissue, calculate the risk as higher. Seale says we must refocus the risk assessment to study the possibility of a virus jumping species and then not being able to jump back. In other words, where one lays one's chips depends on how one judges the odds.

Blood Plasma as HIV Disseminator

While we are on the topic of human error and how it is possible that scientific mistakes transmitted HIV through contaminated vaccinations, it would be apropos to look at a less widely held opinion about the distribution of HIV, which does not concern how HIV originated but how it was spread. This is the idea, which was cogently put forward by Michael Verney Elliot when I interviewed him, that blood plasma was the source of infections. This is not so much a worked-out theorem as the indication of an area crying out for examination, one not being ventured into because it might step on the toes of too many countries as well as of the medical blood distributors.

Elliot states that little has been exposed by the media about "the international trade in blood and plasma throughout the 1970s."[114] This "sordid" but profitable business had involved the bringing in of blood by pharmaceutical companies from "Chad and Zaire which is the center of AIDS in Africa," as well as from Senegal. Other operations brought blood "out of Brazil labeled as orange juice."[115] We saw that Brazil was the one South American country infected with AIDS. As he recounts it, the Caribbean was also a major supply source. I will let him put it in his own words, both because this will allow the reader to appreciate the process by which intellectual theses grow and are refined as new information is added, and to see how frustrating this can be when such theses touch upon business interests who have the means to conceal many of their doings and make the public reporting of any information found out about them difficult.

I remember that one of the very first units to actually start exporting plasma to the United States was set up in Haiti. So I thought if they send it from Haiti with the cradle of HIV, could it have anything to do with the import of Haitian plasma?

The principal driving force [in the blood business] in the early seventies was the need for agumin and gamma globulin.... I heard that Dr. Matilda Quinn, one of the founders of AMFAR... was quoted as saying... that it was possible that HIV had found its way into the United States by imported batches of gamma globulins.

Haitians were paid three dollars a donation unless they agreed to be vaccinated, in which case they would have valuable gamma globulin in their plasma and they were paid five dollars. So it was all looking... bad, but nobody wanted to talk about it.[116]

Since Elliot's attempt to find a connection between blood imports and HIV distribution has not gone much beyond the first formulation stage, I will end discussion of it here. This might seem a rather offbeat note on which to close, in that Elliot's ideas are still in the most primitive stage, while those who are working on the hypothesis that HIV is the unexpected product of vaccines have been worked out with some degree of sophisticated elaboration—both anticipating objections and drawing on a wide range of literature to buttress their claims.

Yet, we have also seen that formidable objections, ones that they haven't answered, have been made to some of their hypotheses. Moreover, as I outlined earlier, each of the possible ways that have been outlined by which the vaccine could have transferred HIV to humans is weak or incomplete in one or more dimensions.

Perhaps, though, we could see Elliot's and the vaccine theorists' ideas as two stages that need to be gone through in arriving at a plausible interpretation. Bacon's views on how science is to be conducted might be appropriate here, although from what I have said so far about his method—about the prerogative interests and so on—it might seem what he offers is more of antiquarian interest than of any use to the present. (In the next chapter, we will discuss the instance of the fingerpost.) However, what we have offered so far of his thoughts in this compass have been very partial. We need to get a better idea of their full scope.

Charles Whitney, in his *Sir Francis Bacon and Modernity*, lays out Bacon's ideas on scientific method. The first step where the prerogative instances enter in is inductive negation. After arranging every important and distinguishing fact about a phenomenon, such as heat, one begins to frame hypotheses about, say, how heat operates. The idea is not to construct and then test a single theorem, but to come up with all possible explanations. Once these have been arranged, the scientist, constantly checking them against the table of facts, eliminates them one by one, until the last one standing will be found to be unassailable and, thus, true. Although there is theory involved here, in coming up with all possible ideas that would make an aspect of a phenomenon comprehensible, Whitney emphasizes that for Bacon the major theme is that it is the scientist's constant reference to the facts gathered that brings him or her to knowledge of nature.

> After systematically collecting all pertinent facts in the investigation of any problem, the scientist must resist the vain or curious desire to construct his own speculative theory and instead must, by an elaborate system of cross-references, eliminate all possible wrong answers until left with the only possible answer.[117]

Applying Bacon's idea of what science should be to the AIDS situation, we would say that it is absolutely essential to examine every reasonable avenue. Thus, when we come to look at how HIV got to Africa, we would need to look at every means through which it might have arrived, from having already been there in latent genetic material in the population to arising from transmuted monkey virus to its being carried in by vaccines or vacationers. We also would have to consider the more foundational possibility that what is often labeled AIDS in Africa is really something else altogether.

By shrugging aside any but their own, sometimes strong, occasionally flimsy theses on major questions about AIDS, orthodox science has certainly fallen far from the Baconian ideal. Yet, as this book indicates, it is a disservice to knowledge to throw this ideal, along with Koch's postulates (and some would say even the Hippocratic oath) on the scrap heap.

Because I find Bacon's ideas still valuable, across the centuries, I want to interrogate any reasonable thought on the issue. That is why I took up the ideas of Schmidt, finding much wrong with them but also much of value. For this same reason, I want to move to the final chapter that deals with the possibility that HIV is an accidental or even planned product of biotechnology or germ warfare, knowing full well that even examining such an interpretation opens me to the charge of being a loony conspiracy theorist or paranoid. Yet if we rely on the early directives of Bacon, we must carefully scrutinize every possible way to understand a situation, from ones that are outwardly attractive to those that have little to recommend. Only thus will we be worthy to find the truth, which is so often found lingering on the byways, off the beaten track of establishment science.

Biological
Warfare and
Genetic
Engineering

There is always a honeymoon period when a new technology is introduced. One sees it in the new century with the Human Genome Project and the mapping of human DNA. Perhaps we can explain such a period of euphoria by referring to the dreamtime interpretation introduced previously.

According to that thought, future scientific advances are only joyfully anticipated because it is naively imagined they will be connected with accompanying social changes that will make their benefits widespread. In a similar way, when genuine breakthroughs are achieved, such as the unraveling of the DNA sequence, all the pundits act as if their value was not only within the narrow realm of science (whereby geneticists will understand certain internal operations better) but also in offering a boon to the wrecked and marginalized who increasingly make up U.S. society.

The new century in this country has not offered anything but a continuation of the government's unconscionable social policies of the last twenty years. On the one hand, there was the shielding or ignoring of the most rampant corruption in business, as, for instance, bailing out (to the tune of a few billion of the taxpayers' money) the savings and loans, whose mismanagement and bilking of depositors became legendary, and in the overlooking of the most incorrigible polluters. On the other hand, the same government became stingy when it came to needs of the less well off, leading up to President Clinton's eviscerating of the welfare program, and other means of pulling one strand after another from the social safety net.

In the realm of medicine, to which the Genome Project ostensibly relates, more and more people are without health insurance, so they are forced to rely on emergency rooms for what little care they can scrounge. Meanwhile, those who are a little better off join HMOs, which do everything they can to dump them if they ever should fall so sick as to need expensive care.

You might say I'm just generalizing, so let's zero in, briefly, on one area. Perhaps you're familiar with the lamentable record of the medical insurance industry in relation to AIDS. Its response to the epidemic, seemingly dictated by an insatiable greed, has been highlighted by two policies: trying to make sure it will not issue insurance to anyone who is positive or has AIDS, and trying to drop anyone or refuse to cover the expenses of anyone who comes down with AIDS when he or she has insurance already.

Karen Davis and Diane Rowland have carefully documented this situation. As to how to avoid insuring people that might be positive, the most obvious method is to test for positivity. "It was recently estimated that 50 percent of health and life insurers test applicants for the [HIV] antibody."[1] Those who register positive can kiss their chances of getting insurance good-bye. In fact, the discriminatory possibilities of this testing are so great that some states, such as Massachusetts and New York, have prohibited HIV testing for insurance application purposes. The insurers have not found themselves without recourse when faced with such laws. Ever since Washington, D.C., enacted such a ban on testing, to provide just one example, "many insurers have refused to write new policies in the city."[2] Less blatant methods of discrimination have also been practiced, including refusing to write insurance for people who live in certain zip codes (that are perceived as gay neighborhoods), the refusal to make out policies for males who live with same-sex roommates, and even the hiring of private detectives to determine an applicant's sexual orientation. Most outrageously, as is contended in a case brought against the Great Republic Life Insurance Company, the insurer, it was charged, would not grant policies to males who worked in suspicious vocations, "such as floral, fashion, and interior design."[3]

Posing a more difficult problem for the companies is what to do if a PWA already has insurance. "Known techniques [to avoid paying] include canceling the policy or refusing to pay—citing a preexistent condition, fraud, or the experimental nature of treatments."[4] Here the outrageousness award should go to Circle K company, which announced to its eight thousand employees that for any new employees, the company plan would not cover AIDS.

After negative press, this tactic was abandoned by Circle K, but it brings up the most efficient way to avoid helping AIDS patients. Many insurers will not extend a PWA's policy if he or she leaves employment—and most AIDS patients eventually have to leave their jobs because of illness—even if the person losing the job is willing to pay a new, higher rate to maintain coverage.[5]

The point being made here is that whatever the advances being made in medicine in the United States of late, they are coterminous with a greater and greater debarment of people from ever taking advantage of those advances.

As noted in an earlier aside, one spectacular component of the AIDS crisis, for those who are willing to look soberly at the situation, is that it exposes the increasing hollowness of the American medical dream. As Nancy McKenzie puts it in a damning indictment, the excessive interest in how AIDS is transmitted is itself a clue to the mockery that American medicine has become.

> The almost single-minded focus on transmission allows us to continue to internalize medical technology as The Answer and to continue to deny that our institutions are largely unable to offer comfort or care to the HIV-affected.... It allows us to disregard the fact that our health care facilities are, for fully one-third of the population, impoverished and worn out. It allows us to ignore our predominantly two-class system of health care; to ignore the medical triage that substitutes for sustained medical care in most urban hospitals. Wholesale, it allows us to fail to focus on the criminal inadequacies of our health, education, employment, and justice systems.... By focusing only on issues of transmission, we give ourselves license to ignore the relation between homelessness, drug abuse/mental illness, and HIV infection and to fail to understand how the three relate to the worsening economic condition of our cities.[6]

So taking into account that life for at least the lower third is diminishing and getting harder year by year, a more honest way to describe such scientific projects as the Genome Project, in terms of their impact on society is that they will be of great benefit to rich people and should be celebrated as such. The wealthy can all look forward to healthier futures, but to pretend anyone but a narrow elite will be affected by any new scientific breakthrough, except for the worse, is a cruel hoax.

It seems I've taken the long way around to come to the purpose of this chapter, but only because I want to rest what I have to say on a larger point. Along with analyzing human DNA, another of science's recent wonders has been the biotechnological engineering of food. Although the aware have seen the possible disastrous consequences of this type of research, the positive waves of excitement that represent the honeymoon period—based on ideas that these will be more "efficient" crops, more bug resistant, and more productive, so as to help feed the world—have swept aside most compunctions.

In this chapter I want to talk a little about the biotechnological fiddling with plants and animals—after all, one of the things scientists are monkeying with are viruses—but of even more concern is how this science is used as a stalking horse.

That is to say, at present, the rosy glow around the triumphs of biotechnology offer a patina of respectability to experiments with human genetic material, such as work involving the creation of new human viruses (in relation to a cancer study) and even work done on biological weapons. As to the last matter, I don't mean that such weaponry is suddenly considered worthwhile, but rather that it is considered safe, safely contained in laboratory walls by experts who assure us that accidental release of these weapons is impossible. Thus, there is a bridging phenomena whereby the overplayed value of one leg of science grants a certain immunity to questions of integrity and security of another.

And this detour can now make it easier to see who is labeled a conspiracy nut. I said already that if I plunged into the topic I am about to take up, which is that HIV may have been a genetic experiment run wild, or even one being tested on humans, I would probably earn the sobriquet "paranoid conspiracy theorist." I don't say there are not tried and true conspiracy buffs who fit the most disparaging use of the term. This would include those poring over the *Decretals of the Elders of Zion*, a trumped-up document, that purports the Jews have been set on world dominance since the year 1066, or those who think the Masons have the same plans. But, classed amid this bunch are also those with a more visionary perspective, such as those who in the 1950s decried the dangers of civilian nuclear power, warning of meltdowns of reactors and worse. Now that some of their forebodings have proven more accurate than the optimistic expressions of establishment spokesmen, it seems their conspiracy theories were in good part accurate historical prognostications.

But, attend, here is the linkup I am trying to make. Thinking of the second variety of "nut" only, we can say that a conspiracy theorist is a person who tries to raise the alarm against a new technology during the sweetheart period. He or she is being ridiculed for not being as credulous as his or her peers

Further, if I can indulge in a little foray into mass psychology of my own (à la Schmidt), the wrath that the public turns against the one who rejects their faith in the latest technological fix comes, I contend, not because most people care greatly about the new science, but because they are clinging so hard to the dreamtime.

In other words, bundled into espousals of the new technology is a sense that this new technology will somehow bring about an amelioration for the majority, even the bottom third. This has become even more important during the last decade as our country has become denuded of hope. For genuine hope can only be for society as a whole to improve, not for this or that individual to cut a bigger slice of the social pie to others' disadvantage. And with the fall of Communism, most anarchist or Communist dreams of a more equalitarian society have been dropped. The scientific version of progress is one of the few places from which a little hope can be gathered.

(Perhaps you understand that I am talking only of the those who are working within the bureaucratic leviathans of society, from universities to big government to corporations. Women and men within the alternative health movement or other progressive social movements who are actively working for change need hope as a motivation. Those among this group, such as people already mentioned in the volunteer agencies that were created in an equalitarian way to handle the AIDS crisis, have an additional modicum of hope in that their new organizational arrangements help them model and experience the way a more equitable and caring society would feel from the inside.)

Let's move on. Whether you reject or accept the ideas presented in this subsection, and you'll see that I don't fall into accord with everything presented, there are certain facts that you should recognize are accepted by everyone.

1. Experiments on manufacturing new human viruses have been carried out in the United States both for benign purposes (as in medical research) and malign ones (as in making weapons).
2. If these viruses somehow escaped or were let loose from the laboratories where they are being invented, they would cause unimaginable havoc.
3. Occasionally things go wrong in laboratories and people are infected.

If we add these together, we have all the premises we need for saying that no matter how statistically remote the possibility may be, it is not inconceivable that HIV was created in and somehow left a science center. We will spend the rest of this chapter seeing how this might have happened.

(Again, the caveat thrown out in the previous chapter applies. My focus on the various ways in which HIV might have appeared does not mean I am suddenly shifting ground and claiming HIV by itself can cause AIDS or even that it is particularly harmful absent the cofactors already enumerated. What I will be doing, though, is listening to many researchers, most of whom feel HIV is a major factor in AIDS. What my own concern will be is the validity of their arguments about HIV's genesis, which can be assayed without going further to fall in with their opinions on the place of the retrovirus in the disease.)

Hybridization of Viruses

In discussing cattle viruses, I briefly mentioned, without enlarging on the means by which this might have been accomplished, Curtis's idea that the visna sheep virus and the bovine leukemia virus may have been combined. At the time, that might have seemed a rather stray and unfounded thought. However, on the contrary, as we will show now, such hybridization experiments have been toyed with enough to scare some scientists, including Robert Gallo himself.

It is not so much that scientists working with viral hybridizations were pushing nature in new directions when they embarked on experiments in this field, but that they were following reactions that occurred in their laboratory. As J. Cookson and J. Nottingham note in their *Survey of Chemical and Biological Warfare*, "It was observed (Heubner et al.) that in tissue culture cross-infection with two different viruses may occur."[7] This is not, by the way, evidence that such reactions occur naturally, since cultures in laboratories are highly artificial; yet it does indicate that the hybridizations do happen spontaneously. "As a result [of this combination] a virus which resembles one of the original viruses in its protein coat (due to alterations in its DNA) and also possesses characteristics of the other virus may be produced."[8] But such a virus could potentially be devastating to the host. Suppose a relatively benign virus to which the host was accustomed—such as the many endogenous viruses we have discussed—which are tolerated by the human body, were combined with a lethal virus. Suppose further that it was the benign virus that provided the coat, hence the antigen markers through which the human immune system identifies intruders. Flagging this virus as an innocuous character, it would ignore it. The virus "would be suitably disguised to the animal or plant it was infecting," and so could wreak havoc unrecognized.[9]

Although such composite viruses have been seen only in lab conditions, one concern has been that biotechnologically manipulated animals, ones who have genes from other species (called transgenic specimens), will be prone to hybridizations. A group headed by Dr. Gallo warned of this danger, saying "the use of transgenic mice for AIDS research could lead to HIV interacting and combining with the mice's own 'natural' viruses."[10] They feared the creation of a "super AIDS," one that, because of the added boost given by its partner, might make the variant even more widely infective and perhaps give it new means by which to spread, such as by "airborne transmission."[11]

While this specific type of accidental viral combination may seem highly improbable, the more general concern of those scientists who are up in arms about the hazards of this type of research is that promoting viral hybridization (experiments where researchers stimulate viruses to combine) might lead to unwanted, unfortunate consequences. Back in 1974, the dangers of such tampering were already scouted by the Committee on Recombinant DNA Molecules of the U.S. National Academy of Sciences. In a report, the group called for the organization of "scientists throughout the world [to] join with the members of this Committee in voluntarily deferring... experiments (linking) segments of the DNA from oncogenic animal viruses. . . to other viral DNAs."[12] In sketching out why the committee wanted to voluntarily clamp down on such research, it noted "a person infected with a hybrid retrovirus has acquired a new, infectious gene which may be immunosuppressive, encephalopathic, or oncogenic—or all three."[13] Such a manufactured hybrid virus would likely be transmissible "via RNA virions" and, as a new entity for which the immune system had not been able to evolve defenses, could be near genocidal in societies. "If infection causes persistent viraemia, and if death is delayed for several years," their report explains, "virtually everybody in a densely populated country could acquire the lethal gene with a few decades."[14]

Commenting on the report in *New Scientist*, J. Seale adds that even the most scrupulous laboratories sometimes have slips. And if the facility were working on viral hybridization, "a single laboratory error, or a deliberate act, infecting an intact person with a newly created human-cell-adapted hybrid retrovirus, could donate an AIDS-like epidemic to the rest of mankind's other diseases."[15]

When I interviewed Leonard Horowitz and we got onto the topic of carelessness and human error in scientific institutions, the type of thing that worries those who are attempting to stop viral hybridization experimentation, he mentioned that accidents at labs are much more common than people realize.

> For example, one of the various monkey outbreaks [where animals that were supposed to be kept away from the virus being tested were accidentally infected] was [in 1969] in the labs in Davis, California, which imported their monkeys from Litton Bionetics.... Then there was several in the seventies, including an outbreak in Philadelphia. Interestingly enough, almost all of them came from Northwest Uganda from Litton Bionetics Laboratories. The Marburg virus outbreak which killed several dozen health care professionals as well as laboratory workers in Europe in 1967 from the Ebola was traced back to a study [where] John London, the Director of Litton Bionetics in northwest Uganda, injected rabdo into 18 Simian monkeys between 1965 and 1970 . . . The National Cancer Institute's number one researcher in Simian monkey virus virology, Sy Colter, stated clearly that 'the [rabdo] virus is a man-made virus."[16]

The last sentence indicates not only that unwanted mishaps take place, but that they may involve newly minted, deadly viruses.

There is even some evidence that such an unplanned viral release once went beyond infecting caged animals in a laboratory and strayed into the general public. The controversial incident involves the surprising "reappearance on May 4, 1977, in Anshan in northern China of the H1N1 [flu] subtype (which had been extinct in humans for more than 20 years), with a molecular structure identical to the virus that caused an influenza epidemic in 1950."[17] Since this virus was no longer in any populations, where it came from for a second go-round was a mystery. It is known that this virus had been kept frozen in virological laboratories. A group of scientists writing in *Virology* postulated that a virus had been taken out of the deep freeze and "had escaped," so the epidemic "resulted from a man-made event."[18] Although a report in *Nature* recorded denials by Chinese and Soviet scientists that such an event had occurred, Seale finds that many virologists are of the opinion that "artificial freezing of this rapidly evolving RNA virus and its subsequent release from suspended animation remains the only plausible scientific explanation for its reappearance genetically unchanged 27 years later."[19]

Biotechonology and Viral Manipulation

Perhaps due to the influence of the Committee on Recombinant DNA Molecules' warning and that of other scientists, viral hybridization has not become a large project in civilian research (although it is being done, as we will see); but viral research itself is a booming business, particularly in relation to agriculture and animal husbandry, where the rush is on to use viruses to protect foodstuffs in the soil, inoculate animals in the wilds, and even vaccinate humans through what is on their plates.

To take only representative examples—the point simply being to establish that biotechnology is deeply committed to virology for beneficial (and profitable) purposes—we can begin with viruses that are being designed to kill the pests that harry plants. In 1991 "Chemical giant DuPont Co. and a Maryland biotech firm announced a partnership that plans to make viruses that kill insects but leave crops beneficial and humans unharmed."[20] Given that chemical pesticides eventually induce the appearance of resistant bugs as well as leech into the water table where they may, by various routes, find their way back into the human body, such use of viruses may only seem a good bet until the inevitable imbalances they cause surface.

Although the previous news bulletin left vague what viruses will be used against the plant pests, a report from *Scientific American* is more explicit about how scientists are working on a vaccine devised by biotechnological manipulation of genes. It will be used to fight against rabies in animals in the wild. "It is a genetically engineered product that should be safer than earlier vaccines because it includes only a piece of the rabies virus, not the whole organism."[21] Here the underlying strategy, which is to infect the target with a weaker version of the virus in order to promote the development of antibodies that will guard the host, is not new. It is the basic method of vaccines. The innovation is how scientists are creating the weak virus by going in and fooling with its DNA. "Researchers at the Wistar Institute in Philadelphia and at Transgene S.A. in France developed the vaccine in 1984 by inserting a single gene from the rabies virus into a weakened version of vaccinia, which is already a rather innocuous virus."[22] Yes, innocuous, but, as we already saw, liable to carry other, less friendly viruses as unnoticed passengers. But such dangers are not mentioned as we read: "The recombinant is now used extensively for foxes in several countries, including Belgium and France, and seems to be as effective outdoors as the attenuated rabies virus vaccine," with blatant demands for increased funding. "In the next several years, oral vaccination could well halt the raccoon rabies epidemic in America and be extended to other species—that is, if research funding, which is already a low priority in the U.S., can be maintained."[23]

Closer to home than this vaccine for wildlife, however, are tamperings with human viruses and vaccines, which, though done in relation to fruits and vegetables, are aimed at providing a palatable version of an inoculation. An Associated Press report on an article in the *Proceedings of the National Academy of Sciences* discusses an experiment on potatoes that portends that "genetically engineered fruits and vegetables that make the body immune to contagious disease may one day replace some vaccination shots."[24] The experiment carried out at Texas

A&M and the Roswell Park Cancer Institute in Buffalo, New York, took the protein gene that creates immunity to hepatitis B and put it in potatoes, which were then fed to mice, who, it is said, developed antibodies to hepatitis. Hugh S. Mason, from Texas A&M, who is coauthor of the study, said, "The idea is to produce an edible plant that contains genes that cause it to produce antigens."[25] These antigens would alert the body to create the needed antibodies against the disease.

Showing their innocence in the face of modern history, the researchers continue, "Plants that inexpensively protect against disease would have great benefit in developing countries," adding that the WHO "reports that about 2 million children worldwide die annually from disease that can be prevented with vaccines, such as diphtheria, tetanus, whooping cough, polio and measles."[26] The highly negative connotations that must accrue to any further attempts to test unproven drugs in the underdeveloped countries, so that any bad effects will not be meted out on the rich countries, are lost on these researchers.

Although their ignorance may not be shared by others, the direction of their research is. Tarting up vegetables with vaccines is a hot topic in biotechnology. As Christopher Miller states, "In fact, biotechnology firms have been experimenting with adding vaccines to bananas, lettuce, potatoes, tomatoes, and soybeans for several years now."[27]

I'm not saying here, like some modern day Luddite, that any type of playing around with viruses is wrong and should be stopped. Rather, I just want to make evident that viral research is a going concern and thus it should hardly be surprising if we find virologists tampering with the kind of viruses that can be tied to HIV.

Biotechnology and AIDS

Although this will be a brief detour from the subject of viruses, to get a broader view of the manipulation and study of genetic material we might mention that such work has also played a role in the study of AIDS itself—for example, in attempts to identify a gene whose presence would make one particularly susceptible to AIDS, and in splicing HIV, as we saw against Dr. Gallo's better judgment, into transgenic creatures.

Taking up the latter subject first, we need to begin by discussing lentiviruses. Although some biologists dispute the existence of these types of viruses, even if they are accepted as realities, they pose problems for those who believe AIDS has moved from an animal species into man. "Lentiviruses are highly species specific in nature, readily transmissible only to closely related species of mammal."[28] However, we went into species specificity in the previous chapter. A trait more relevant to the genetic manipulation of lentiviruses that has been carried on is that "all known naturally occurring lentiviruses [including HIV] are exogenous, being transmitted by infection and not in the germ line [as are some birth defects."[29] Although this is a plus for the health of mankind, in that these lentiviruses will not be passed down to children of the infected in the genetic material of their parents, scientists engaged in the form of research Dr. Gallo discouraged have been seeing whether genetically transferable HIV can be produced. "Recently... a breeding colony of transgenic mice was created in which the genome of HIV-1 had been integrated artificially into the chromosomes of fertilized mouse ova. HIV-1 proviral DNA was present in all cells of the offspring, and in some mice infectious virus was expressed."[30]

Mice seem far enough away from humans, species wise, that their diseases pose little immediate danger; but note of this experiment indicates that while potentially hazardous attempts to create hardier viruses or ones that have new transmission pathways may have been curbed by the warnings of individual scientists and scientific groups, they are still on some agendas.[31]

What can be taken to be a more positive direction of genetic research into the immune-related disease is the inquiry into whether specific genes make one immune to or liable to AIDS. As Gina Kolata reported in the *New York Times* in September 1996, "Scientists have long suspected that some people might be immune to the AIDS virus. But now they are accumulating powerful evidence of the extent and strength of such immunity."[32] One piece of evidence is a study published in *Science* at the time of Kolata's story, which presented data on whites indicating one in a hundred have immunity to HIV infection, while one in five whites have an inherited resistance to the disease's progress. More depressingly, "almost no blacks have the particular protective mutation investigated in this study"; however, the study continues, "scientists say that other forms of genetic immunity almost certainly exist in both blacks and whites."[33] The existence of these further forms of genetic protection are suggested by findings that of 1,850 subjects of the investigation, all of whom were at risk, 600 remained HIV negative. Since the identified genetic factor would not have provided for such a high number, "presumably the other people had some other form of genetic protection."[34]

If we accept the validity of this report, we would then be forced to wonder about where such a trait could come from. Is it merely fortuitous or the product of a former advent of AIDS? Dr. David Baltimore, Nobel laureate molecular biologist, whose discovery of retroviruses we already discussed, theorizes "that AIDS ravaged Europe centuries ago and that people with immunity today are the descendants of survivors."[35]

Another question worth posing in relation to these claims of an AIDS immunity is what accounts for this immunity. Although answers to this question are not forthcoming, on the genetic level factors have been identified whose absence is necessary but not sufficient to guarantee a robust response to the disease.

The first of these is a subtype of a protein designated GC. A *New York Times* story says this protein is either of GC IF, GC IS, or GC 2 variety. "People with GC IF factor have a greatly increased susceptibility for the disease." The article goes on to note that one study, unrelated to AIDS research, found "58% of Central Africans had at least one of the factors which causes greater susceptibility as against 16% occurrence in Europeans studied."[36] This is another indication that race (or possibly ethnicity) may play a part in disease vulnerability.

The second critical factor whose presence makes one liable to succumb to the disease is the CKR5 mutation of the DNA. Research into the effect of this mutation, according to Baltimore, has shown that this gene is connected to "the key receptor through which people get infected."[37] However, the diagnosis of the causative role of this genetic anomaly in making for a propensity to become infected is merely a preliminary finding. Whether scientists will eventually be able to do genetic screenings to see if individuals have a propensity for the disease is in doubt. Even if they could, the value of such tests would be ambiguous. "Stephen Soba, a spokesman for the Gay Men's Health Crisis... said 'There was something chilling, horrifying even, in the idea of testing for such a gene.'"[38] After all, one who took the test might end up finding out he or she was especially prone to the disease while not learning anything new that would help him or her avoid or palliate the condition.

Creating Viruses

Let's return to an examination of where biotechnology is directly concerned with virology, starting with seeming anticipations of the AIDS disease in proposals for new research.

A number of those who have studied the history of AIDS were more than shocked when they unearthed an article published by the World Health Organization in 1972 with the title, "Virus-Associated Immunopathology: Animal Models and Implications for Human Disease," in which research is recommended into studying and even creating viruses that destabilize the immune system. In plain English, the authors were discussing how to produce something like HIV! This article, along with other WHO discussions we will review below, have proven so provocative that some researchers seem to have lost their heads as they surveyed their implications. But before enlarging on that point, let's look at what the article and these other reports actually said.

The WHO authors began by advancing certain premises about viruses, including "1). The properties of viruses are seemingly ideal for producing immunopathological damage and 2). Virus induced immunodepresssion might conceivably be highly instrumental in prolonging certain virus infections."[39] From here further research was recommended, including a fuller plumbing of what role viruses had on immune functions and an assessment of whether they could exert selective effects on different parts of the body's immunity.

Dr. William Strecker, who in his video, *The Strecker Memorandum*, mined this report for the most chilling aspects—indeed they were there to be mined—points out that many of the features presented as potential viral capabilities in this WHO piece turn out to be the very ones that HIV later displayed (or is said to be display).

Moreover, Strecker urges us to pay close attention to what D. B. Amos of Duke University said at a conference sponsored by the John E. Fogerty International Center and the World Health Organization. Fogerty opines, "In relation to the immune response a number of useful experimental approaches can be visualized."[40] Fogerty seems especially charmed by this program: "One [approach] would be a study of the relationship of HL-A type to the immune response... to well defined bacteria and viral antigens during preventive vaccinations."[41] Although it is sobering to read Fogerty's recommendation that the reaction of the human immune system be studied as it is affected by such entities as viruses that accompany vaccinations; surely Strecker is jumping the gun by concluding, "Therefore, it is clear that the World Health Organization and the National Institutes of Health decided in 1970 to inject known virus and bacteria into children [Fogerty mentioned children as good subjects in an unquoted passage] of the same parents during allegedly preventive vaccinations."[42]

The fact that such topics were discussed among peers at a respectable conference indicates it is something that the group finds worth studying, but does not indicate that the experiment went any further, even to the planning stage. Moreover, as the quote on the rabies trials should have indicated, modern scientific reports as well as proposals for further research are as much promotional as scientific. Such a proposal can be seen as a feeler to see whether any funding agency could be approached to back such new inquiry. An even more frightening version of this was noted by Curtis, who has uncovered in *The Bulletin of the WHO* (1972, volume 47, page 257) a proposal to construct what sounds like HIV. "Let's make a virus that selectively destroys the t-cell system" is how he summarizes the content of this part of the article.[43]

Of course, if such proposals, where viruses are added to vaccines to study cross-reactions, were funded, it's doubtful whether they would have gotten much publicity, even if they were done, say, with the knowledge and consent of the vaccinated. With the advent of actual immune deficiency diseases devastating some sectors of the population, scientists and bureaucrats would have even more reason to bury previous discussions of them. Be that as it may, so far we don't have much evidence one way or another as to whether such experimentation was ever carried out. So, contentions like the one made by Curtis— "We don't believe it is merely coincidental that AIDS in the United States first appeared in the exact same epidemiological groups as the hepatitis B vaccine project," which implies deliberate, not accidental, contamination of the vaccine —are not warranted by the evidence we have.[44] The same could be said about the accusations raised by Dr. William Campbell Douglas in *Health Freedom News*, where he claims "the AIDS virus was deliberately manufactured by the World Health Organization in collaboration with the National Cancer Institute... [combining] two deadly retroviruses—bovine leukemia virus (BLV) and sheep visna virus—to create the AIDS virus."[45] Note, though, Douglas's claim here is not as Machiavellian as Curtis's; he is not saying HIV was consciously disseminated, but consciously devised in labs and then possibly accidentally distributed.

None of this means the possibility that such experiments were done should be ignored—given the track record of such reprehensible scientific programs conducted by the government as the Tuskegee project, in which poor, untutored black sharecroppers were given placebos rather than treated for syphilis so that researchers could study the progress of the disease—but we must be stingy with conclusions when the evidence is scarce. (As an aside it's worth noting that, ominously enough, "Dr. David Sencer, the Director of the CDC who approved the continuation of the Tuskegee study in 1969, became New York City Health Commissioner, and a key player in AIDS policy in the early 1980s."[46]

Hepatitis Vaccine

Let's pursue the supposed hepatitis vaccine connection a little further. The biology of it has been worked out by Strecker, who, we should emphasize, is willing to believe the transmission to gays through the vaccine, if it happened, may have been accidental. He argues that HIV follows a very unusual viral transmission pathway, not via infection with the virus itself but through reception of infected macrophages. This is "a mechanism of exchange known as the Trojan horse phenomenon. Macrophages containing the AIDS virus transfer the virus from one person to another. The only other known virus that is transferred in this manner is visna."[47] As we saw, Strecker believes hybridization experiments with bovine visna virus are behind the creation of AIDS. In his view, this new virus was mixed into the vaccine. "Homosexuals who were initially infected [via vaccine], would sexually transmit the virus to other gay men." The virus got out of the cohort taking the vaccine through sexual-intercourse-induced "micro-transplantation," whereby infected macrophages are transferred."[48]

Thus, taking Strecker's investigation as representative, most dissident work on this topic has gone into the ways that accidental or deliberate creation and distribution of HIV feasibly might have taken place, without hard facts showing that it did. The drift toward belief that the heavy hand of government could have been meddling with viruses and decided to manufacture a plague is, of course, stimulated by memories of government chicanery in the past. Historian Blanche Wiesen Cooke is made suspicious of the government, particularly because it has engaged in previous campaigns to change homosexuals. Cook has noted that "in an attempt to convert gay males into heterosexuals, experiments were performed at the prisons of Vacaville and Atascadero in California in the 1960s. These included the use of mind altering drugs and aversion therapies (including electroshock)."[49]

Equine Infectious Anemia Virus (EIAV)

Since we are on the subject of the possible dissemination of HIV in connection to hepatitis, it might be suitable to examine assertions that link the retrovirus to other diseases that have been studied in laboratories.

As we've seen HIV—still taking, as a heuristic measure, the stance that HIV is a co-factor in causing AIDS—resembles a number of animal viruses, including those of cattle and monkeys, on a number of planes. As yet unmentioned in this respect is the equine infectious anemia virus (EIAV), which is, according to a study by R. Squire done at Johns Hopkins University in 1968, a "a model of immunoproliferative disease," and can be temptingly compared to a number of human diseases.[50] The closeness of the two diseases is partly visual. In *AIDS: Origin, Spread, and Healing*, Geisler notes, "Under the electron microscope, HIV and Virus of Infectious Anaemia of Horses look identical."[51] Moreover, in findings of Montagnier, who first claimed and then disavowed the idea that there was a kinship between the viruses, these two viruses serologically cross-react in the sense that "the respective defense mechanisms of horses and human beings develop antibodies which are tailor-made to be effective likewise against EIAV and HIV."[52]

Some dissenting scientists claim the proximity of the viruses may be due to a cross-species transfer engineered in a laboratory and then accidentally released. Thoughts about this, the dissenters think, could have discouraged Dr. Montagnier from pursuing this research avenue, which may have trod upon too many toes. Although there is no good evidence for this lab-based species jump, what is easily attested is that investigation of the horse anemia-HIV connection has been left idle, although, as Geisler suggests, it would seem to be a fruit-bearing field.

"The stubbornness with which international AIDS-research is keeping its distance from the Virus of Infectious Anaemia of Horses is astonishing," he urges.[53] Where animal models of the virus have been located in everything from mice, sheep, and goats to cats and monkeys, "only EIAV, whose core and envelope are similar to HIV as those of no other virus, is hardly ever mentioned."[54] In the 1970s there was a good beginning made at looking to the equine virus for parallels to HIV, which was followed up by "a successful experiment to infect horses with HIV [which] was run in the mid 80s: In four out of the six animals, antibodies were formed against the injected HIV."[55] Geisler contrasts this to ongoing unsuccessful attempts to infect monkeys, such as one that utilized fifty-seven specimens, none of whom produced antibodies. Yet by the 1990s, this direction of research had petered out.

If EIAV was, in fact, turned into HIV in a laboratory, H. Lorscheid thinks this may have happened at the U.S. government veterinary research station at Pullman, run by the Department of Veterinary Microbiology and Pathology at Washington State University, Pullman. In this facility, extensive inquiry had been done into the equine virus. Notably, the government also had run the Regional Smut Research Laboratory at this site, where the Department of Agriculture conducted research into fungi. Lorscheid wonders about the spread of the rarer opportunistic infections that accompany AIDS and the workings of this research center. PCP, which can be transmitted by air, "made its appearance in the USA for the first time in 1964 among several adults without previous immunodeficiency: at Pullman," striking a married couple and their young daughter.[56]

As with the possibility that hepatitis vaccine was contaminated, this discussion of a horse virus has stayed in the realm of conjecture. Indeed, it has less to offer, since the hepatitis vaccine was at least tied to a program that affected a group later infected with HIV.

Cancer-Specific Viruses

Another hypothetical scenario that ties the release of HIV into the population to another disease-related program is that which sees the creation of HIV as connected to programs to develop viruses—programs that seek a cure for cancer.

It may be remembered that Dr. Gallo had begun his career trying to track down viruses that lay at the root of cancer, even finding one, although ultimately this proved an area of research that was largely barren of results. He was only one among a host of scientists at the National Cancer Institute in the 1960s and early 1970s that saw this as a promising orientation. As Horowitz told me, "In order to research that theory, they were first developing cancers," which involved not only developing tumors but "specific cancer viruses to create the tumors."[57]

What Horowitz found especially worrisome about this research is who was sponsoring it. "The administrator of it was an organization called Litton Bionetics and Bionetics Research Laboratories."[58] This is same company that we saw had a notoriously bad track record in terms of accidents at its laboratories, which had killed both animals and lab technicians. If that weren't bad enough, there was also "a biological weapons component," which was referred to as Project MacNaomi (as we will see, this program was initiated much earlier) and funded some of the research into these cancer-causing viruses. Moreover, we are now going the realm of pure speculation insofar as details obtained by the Freedom of information Act show, "a ten-million-dollar grant in 1968 to develop AIDS-like viruses for national defense."[59]

Studying government records, Horowitz has reconstructed this history of one research trajectory in the NCI, in which an HIV-like virus was played with.

> Researchers including Dr. Robert Gallo and Dr. Robert Tain from Litton Bionetics took the SV-40 and they extracted the DNA out of it.... then they infused this blank viral envelope... with cat leukemia sarcoma

RNA, and of course you know that the AIDS patients suffer from a laundry list of symptoms essentially identical to cat leukemia sarcoma. They also infused it with chicken sarcoma RNA, which caused immunosuppression, wasting, and death.... They still had a virus that couldn't infect humans. So to get it to jump species they cultured it in human white blood cells, so that it would develop a GP120 protein on the external envelope and attach to cell membranes.

Now what's the support for these allegations?... You could find this reference in 1970 in the *Proceedings of the National Academy of Sciences* by Robert Gallo et al. He presented it in Mulle, Belgium in front of NATO military scientists, a symposium entitled "The Uptake of Informative Molecules into Living Cells."

We know how much they were paid, we know what they did, we know everybody that was involved.... The Linton group (including the Dr. Gallo group) were paid approximately $2,153,000 a year, starting in 1962 going into the mid-seventies. And what they did do—is actually what I just shared. They took SV-40 and a whole host of monkey viruses and essentially cross-matched the DNA and RNA from one species of animal viruses to another.[60]

At this point, we have moved onto the terrain where one would suspect an HIV-like entity would most likely be produced, if it were to be produced anywhere: research being conducted for the creation of weapons of biological warfare. Let's be clear that we have not shown that Dr. Gallo and his crew created HIV or that whatever they did create ever left the laboratory bench. However, we have now come to the point where we can say that the alarms raised by Strecker and others about proposals to create immune-devastating viruses by the WHO have a solid footing. We see that some of these proposals were acted upon.

Biological Weapons

We can conclude this chapter by turning to an analysis of this same cancer research from a different angle, seeing it now as it fits in the context of biological warfare. We need to look at this in context, learning both what spurred the United States' entrance into this arm of weapon development and what other projects were part of the spectrum of research.

If one wanted to broadly characterize the dissidents versus the orthodox students of AIDS, one might say the dissidents are historians, while the establishment group are phenomenologists. In other words, the orthodox are most focused on how the virus gets in the body, how it acts when there, and how it possibly may be sidetracked. The dissidents, who could hardly focus on HIV, since usually they don't believe it to be a decisive factor in AIDS anyway, do look at transmission, activity, and cure of the disease, but with a special emphasis on how AIDS arose in a particular historical matrix. When discussing poppers, for example, or the hepatitis vaccine trials, we saw that dissidents were very concerned about timing. They were trying to see what was new in the climate of the era that might have helped sponsor the disease. This divides them decisively from our society's self-appointed moralists (like the fundamentalists, secular or religious, skewered by Simon Watney in *Policing Desire*) who blame AIDS on eternal traits of humankind. After all, promiscuity, which these moralizers see at the root of the problem, has been with us since the biblical patriarchs. (See Solomon and his thousand concubines.)

In line with this historical view, dissidents insist we recognize that the United States' biological warfare program (around since World War II) became a major element of our national defense only in the 1960s. Drawing on the findings of the Senate Church Commission Hearings, which investigated the doings of the CIA, and Walter Isaacson's book on Henry Kissinger, Leonard Horowitz told me that the real push into biological warfare came with the presidency of Richard Nixon. "Kissinger, within weeks of taking over as NSC adviser under the first Nixon administration, ordered Blaire to do an assessment of the records of biological weapons capabilities."[61] Kissinger was a proponent of putting up a strong military front against the Russians, not only wanting the country to be able to out-bomb the Soviets nuclearly, but out-poison them biologically. At the time, "cancer viruses were a hot thing and it was cited in the *Congressional Record* that we would develop these before the Soviets for national security reasons."[62] Just as high-priced security analysts at the Rand Corporation and other hawkish think tanks were wildly overrating the nuclear capabilities of the Russians, so they imagined them in possession of great stores of biological weapons. "We feared that the Russians were going to be ahead of us, which they never were. Our CIA intelligence reports were misleading. [In reality] the Soviets were behind us in biological weapons research."[63]

Like nuclear weapons, though, biological ones were first put on the agenda during World War II. And, in another concurrence, just as Nazi physicists were given hasty pardons and transported to the United States where they could help design this country's space program, it seems Japanese biological warfare experts were written the same type of pardons so that they could prime our research in that theoretical theater of war. In an exposure that caused shock waves in Japan, although it went unnoticed by most of the U.S. mass media, "Japan's public television network unearthed new evidence from Russian and U.S. archives on how key members of the [Japanese] army's secret germ warfare unit escaped prosecution as war criminals after World War II."[64]

As the *Los Angeles Times* reported, records of such experiments as purposely infecting prisoners with anthrax and other diseases, translated into English, were "found at the U.S. Army's Dugway Proving Ground in Utah using the U.S. Freedom of Information Act."[65] Although exactly why the Army needed these documents is not clear, they were not obtained by the U.S. military by happenstance but through a hard bargain. "The evidence showed that the U.S. military obtained data from the tests in exchange for shielding the perpetrators from prosecution at the 1946-1948 Tokyo war crimes tribunal, the program said."[66]

The Japanese "scientists" were not doing sophisticated work, but simply watching people die of diseases with the same simplicity and barbarity U.S. doctors displayed in the Tuskegee experiments. By the time of Dr. Gallo's participation, things had progressed to elegant biotechnological work. However, before we reach that exalted level, let's look more at the interim period, between the war and the creation of the cancer branch of Project MacNaomi.

As Scheflin and Opton outline, "In early 1952, the CIA and the Army entered into a Memorandum of Understanding for mutual cooperation [on biological warfare]. The Army Chemical Corps, located at Fort Detrick, Maryland, agreed to do covert research to aid the CIA's own research efforts."[67] The two organizations were agreed on three large aims: "[1] maintaining a stockpile of chemical and biological weapons... [2] developing and stockpiling [other, newly created biological weapons] for operational use, and [3] testing and perfecting delivery systems for these" weapons.[68] The joint project was titled MKNAOMI (spelled slightly differently above).

As far as the records go, MKNAOMI was active from May 5, 1952, to February 18, 1970. Scheflin and Opton comment, "That does not necessarily mean that the program ended at that time. It means only that the records of its being in existence have not been kept or released since then."[69]

However, when these authors note that records end in 1970, they are failing to mention one crucial bit of information. The records likely end there because President Nixon had formally closed down all research into and stockpiling of biological weapons. This may seem to fly in the face of what I said previously about Kissinger's role in developing new biological weapons programs under Nixon; but, in fact, it was this very shutdown, which was out of line with Kissinger's wishes, that explains another part of the puzzle, namely, how cancer research became so intimately tied into developing military viruses. Up to now, the connection between cancer research and biological warfare has seemed fortuitous, but let's consult the record further.

In 1969, Nixon ordered a halt to offensive biological warfare (BW) research and weapons stockpiling by the United States. The U.S. Army was directed to destroy its toxins, viruses, and bacteria (supposedly) with heat and disinfectants by May 1972; the planned "disposal" of the scientific personnel associated with the program was not so easy. Some of these "biowarriors" went to the CIA. Others quickly found new support from the National Cancer Institute (NCI), particularly in its Virus Cancer Program (VCP).

The NCI funded and supervised some of the same scientists, universities, and contracting corporations (supposedly for cancer research) which had conducted biological warfare research.[70]

Not only did temporarily unemployed biological weapons experts quickly find work in fighting cancer, the very same facilities that had hosted the germ warfare laboratories were repackaged as centers at which to conduct cancer studies. Two former BW facilities would play a large part in the VCP. Fort Detrick in Frederick, Maryland, which, as we learned, was the center of the Army Chemical Corps, became the central location for the MKNAOMI program, to the tune of a hundred-thousand-dollar-a-year stipend from the CIA. The site was recrafted as a cancer center. "In 1971, when Nixon announced that Fort Detrick would be converted into a center for cancer research, the buildings were decontaminated and the facilities were turned over to the NCI, which renamed the facility the Frederick Cancer Research Center."[71] (Unfortunately, this decontamination and sterilizing of the environment did not make the center particularly safe, since it continued to be beset by human errors, "over 25 years [both as a biological weapon and cancer research facility, there were] 423 cases of serious infections and three deaths, which on average is one serious accident every three to four weeks."[72] On the West Coast, "in Oakland, CA, the Naval Biosciences Laboratory was involved in early work with plague and collaborated in massive open-air tests of biological warfare 'simulates' in the San Francisco Bay area in the 1950s."[73] The lab also changed tacks and went into medical studies.

But the line between biological weapon research and battling cancer was a fine one. "One of the first new priorities of the Fort Detrick Facility after the ban [on biological weapons study] was the large-scale production of oncogenic (cancer-causing) viruses."[74] Working with the National Cancer Institute, the VCP began mass-producing these newly discovered cancer-causing agents. In a decade, "the VCP produced 60,000 liters of cancer-causing and immunosuppressive viruses."[75]

Along with demobbed scientists and renamed facilities, industry also had to scramble to adapt. Although they had lost their contracts to make biological viruses for warfare, with no retooling they could begin making them for cancer research. For example, Charles Pfizer and Company, a pharmaceutical firm, snatched up a nice contract with the NCI to supply a large quantity of Mason-Pfizer monkey virus as well as work on adapting other animal viruses to human cell lines.

Meanwhile our friend Litton-Bionetics (formerly Bionetics Research from 1966 to 1968) got on the gravy train with lucrative, multiyear contracts. "One of Bionetics Research Laboratories' most important NCI contacts was a massive virus inoculation program that began in 1962 and ran until at least 1976, and used more than 2,000 monkeys."[76] As the description implies, the research involved injecting the monkeys with human and animal cancers to find one that could be transmitted. Those monkeys in this study who received the aforementioned Mason Pfizer monkey virus died from an immunosuppressive illness, product of the first known immunosuppressive retrovirus. By the way, a project officer on this program was none other than Dr. Robert "Finger In Every Pie" Gallo.

This wasn't the only program carried on by the new cancer-fighting agency that would seem relevant to the AIDS issue. Other key research attempted to overcome what we saw was the extreme difficulty of moving viruses between species. As an article on covert research on biological weapons states, "Under these programs, 'species barriers' were routinely breached in efforts to find or create infectious cancer viruses. This illustrates another aspect of the NCI BW program. Zoonoses, disease that can be transmitted from animals to humans, make up the majority of BW agents."[77]

If this weren't bad enough, the effectivity of some of these new viruses were tried out on human subjects. As Prof. Jacob Segal of West Germany, retired director of the Institute of Biology at Berlin University, noted, "In the mid-1970s experiments were carried out at Ft. Detrick on volunteer long-term prisoners who were promised freedom after the tests."[78] If one wanted to find a path for HIV from the laboratory to the outside world, this could certainly be one, whereby a prisoner is given a virus whose deleterious manifestation appears many years down the road, so he is adjudged healthy and released from jail to go on to infect others.

Equally shocking projects—we will only touch on them briefly, since they are not as directly relevant—were attempts to develop vaccines, study ethnic-targeted diseases, and test supposedly innocuous diseases on unsuspecting populations. A U.S. military manual from 1975 stated, "It is theoretically possible to develop so-called ethnic chemical weapons, which would be designed to exploit naturally occurring differences in vulnerability among specific population groups."[79] It was imagined that these killers would be diseases that could be genetically engineered in order to exploit minor ethnic and racial differences such as in distinctive food allergies.

These ideas are only theoretical, although during the Vietnam War, studies were financed by the U.S. military to try to locate genetic and biological specificities that distinguished among Asian ethnicities. It is hard to see under what justification such studies could be categorized as research to improve health, but the "defensive vaccine" program, which the Army also carried out, could be easily touted as medically beneficial. In this program, the military attempted to get the drop on the enemy by developing vaccines to biological warfare weapons an imagined foe would throw at us. However, "the development of the vaccine will in general involve the isolation, identification, modification, and growth of the potential BW agent," as H. Strauss and J. King point out.[80] In other words, to develop a vaccine, the first step must be to develop the disease the vaccine will guard against. This gives a green light to the type of research into battlefield biologicals that was once forbidden.

Lastly, let me mention that the military has not been adverse to springing these diseases on the public, a discussion that will bring me up to the present: a time in which the public reaction against biological warfare, which spurred Nixon to prohibitions, has become muted and the strictures the government once placed on this type of research have disappeared. An article in *Spectrum* magazine reported on the U.S. Army's large-scale dry-run tests of responses to biological warfare carried out on an unnotified public. In 1951, for example, "the Army turned loose an organism called *Aspergillus fumiatus* [which produces illness through infection] at the Norfolk Naval Supply Center. This location was chosen because most workers were black, and, for some reason, the scientists involved believed that an aggressor would target blacks at military bases."[81] But such small target populations were not satisfactory to the Pentagon, which next tried spraying bacterial and chemical particles in bigger areas, including on Washington National Airport in 1965 and the New York City subway system in 1966.[82]

After the public outcry that came once these experiments were revealed, such tests were restricted to military bases, such as in the bacterial release tests conducted in the 1980s at "the Dugway Proving Ground, 70 miles from Salt Lake City [where] they are still conducting outdoor experiments there to this day."[83]

The discussion thus far on military research should have accomplished two things. First, it pointed out that there has been serious research carried out on how to adapt animal viruses to humans, how to construct immune-suppressant diseases, and other strategies that would seem to lead in the direction of creating a disease like AIDS. Second, it has shown that there were a number of doorways opened between the lab and the outside world, such as the use of prisoners and practice exercises on the public, by which a disease might have moved out of the confines of the facility where it was produced. Neither of these points, which provide hard evidence about the general policies of the Virus Cancer Program and the activities at the rechristened laboratories, are more than speculative in relation to the specific generation of HIV.

If you come to think of it, one of the more depressing revelations that is likely to strike anyone who begins to dig through this material is the supineness of the mass media. In discussing HIV, tons of ink have been spilt trying to link the retroviruses' origins to Africa, but nary a line appears in the mainstream about the activities carried on in ostensible cancer research as they relate to the possible genesis of the disease. Although there is no one-to-one relationship here, the general avoidance of the topic of biological warfare in all the standard organs of information may be partially related to the way this branch of science has been able to quietly build itself back up after suffering disgrace during the Nixon years, at which time it hid under the robe of medical investigation.

The public's acceptance of the concept that for our defense the United States may need some elements of biological warfare (at least defensive factors such as vaccines) may have allowed increased funding of this sector as well as encouraged the Pentagon (as we will see) to undertake irresponsible drug trials on its own soldiers.

As to the upped expenditures, "The Committee for Responsible Genetics says that the overall military spending on biochemical research bearing on germ warfare has increased 400 percent" between 1980 and 1985.[84]

As far as the reckless handling of biological warfare-related drugs, this came to a head during the Gulf War, where, supposedly provoked by fears that Saddam Hussein would unleash biological weapons on U.S. troops, the soldiers were inoculated with untried vaccines. The situation was so scandalous that on March 6, 1995, Clinton created a committee to study the illnesses that were still devastating Gulf War veterans after they were subjected to irresponsible vaccine tests. Hearings by the Senate Committee on Veterans' Affairs chaired by Jay Rockefeller (D-WV) had already exposed some of the worst excesses of the military medical corps.

> Gulf War veterans were given the investigational drug pyridgostigmine bromide, a nerve agent, together with toxic bacterial vaccines and now many suffer from immune system and neurological damage including extreme fatigue, memory loss, joint pain, seizures, personality changes, vision and breathing difficulties and some have had babies who are deformed or have died.
>
> Soldiers were ordered to receive multiple injections of experimental vaccines without being informed what kind of vaccine they were receiving or the potential risks. Medical reports were seldom kept on which vaccines were given, making it impossible to track side effects. Those who attempted to refuse the vaccines were threatened with court martial.[85]

The Rockefeller committee documented that soldiers were given, for one, botulinum toxoid, an unapproved vaccine that was meant to protect them against botulism. Others received an experimental vaccine given to prevent anthrax. Studies of the former vaccine, which has been around longer but whose use had been abandoned, showed that it had numerous health-threatening side effects.[86]

Still, the administration of these vaccines had the excuse that they were being done to protect the troops from illnesses with which it was possible the enemy might try to infect them. Even such flimsy reasoning was not present for other experiments. "For example, DoD [the Department of Defense] now admits they conducted 'Man Break' tests, exposing soldiers to chemical weapons in order to determine how much was necessary to 'break a man'"[87]

Going through some of the testimony given to this committee, I feel a heavy heart about media coverage. I think back to the chauvinistic excitement and flag waving in the press when the short-lived Gulf War took place. Then the rallying cry was "Stand behind our boys overseas." Yet, when these same "boys" are testifying in the Senate that they became human guinea pigs for ill-conceived biological warfare tests, their comments are slotted in the back pages if they are covered at all, given that such testimony is highly embarrassing to the top brass.

To get back to our topic, one finds, as I've said, the same pussyfooting when it comes to possible links between HIV and BW. The only way you are likely to hear about it is when a denial of any such link is issued by a government agency after accusations have been leveled from some out-of-the-way media outlet. Tracing the circuitous route by which such a denial finally landed in the *Philadelphia Daily News*, for example, makes a story in itself.

"On July 4, 1984 the New Delhi, India, newspaper, *The Patriot*, published an article making the first detailed charges of AIDS being a CBW agent."[88] Appearing in South Asia, the story, which quoted from publications by the U.S. Army's chemical warfare unit, did not make it into the purview of the Western media, but then it was picked up by the Soviet journal *Literaturnya Gazeta*, which on October 30, 1985, more than a year later, cribbed from the Indian article and used it to make hay against the United States. While this article was debated internationally, it was still a closed book for the U.S. mass media. "American newspapers, predictably, ignored the story until the US government officially denounced the Soviet article as anti-American disinformation."[89] Then they repeated the government denials. "On February 18, 1987 [for example], *The Philadelphia Daily News*, carried a McClatchy News Service report that Col. David L. Huxsoll, chief of UAAMRIID, told a meeting of scientists that the Soviet charges (of US creation of AIDS) were unfounded 'disinformation.'"[90]

It may seem to the reader, having gone through this chapter on biological weapons, that we have started at one place with one goal and ended at another with a different goal. I began by asking whether there was any basis to the charges brought by the dissidents, the ones who see HIV as central to the disease, that the retrovirus may have originated in biological warfare experiments. Even then, I had two reservations. First, even if HIV was shown as having been created in a laboratory crucible, this wouldn't tell the whole story, since, I believe, HIV never comes into its own unless harnessed to other factors. Nonetheless, such a finding would certainly be damning, in terms of exposing the real (but usually disregarded) dangers of bioengineering, the ability of the government to hide what it is doing from the public whose lives it is endangering, and the venality of scientists engaged in such research. Second, I had little faith that we would find any evidence of such government research, not because it is unlikely that it was done but because the wraps around such investigation are so tight. On this second matter, I was a little surprised that we found so much, that we learned, to repeat, both that heavily bankrolled studies of killer retroviruses that targeted the immune system had been undertaken and that there were various routes—accidents, through prison volunteers, or by planned experimental release—through which these viruses could get outside the labs.

However, we also arrived at a second finding for which I had not planned to search, the discovery of an insidious link between cancer research and germ warfare. In a certain sense, it stands to reason. Even conservative historians realize that war-making has been an economic lifesaver in the United States.

When the country plunged into a depression in 1929, even the most dramatic and helpful programs that Roosevelt's government eventually would create, such as the WPA and the NRA, could not reverse the downturn. Many argue the Depression was based on underconsumption (or, the same thing from a different view, overproduction). The greater public did not have the funds to buy all the products of mass production. The problem was solved by the coming of World War II and the massive need for armaments. Weapons had two advantages. One, they were bought by the government so that a falloff in consumer spending wouldn't affect their sales. Two, they had planned obsolescence built into them with a vengeance; they were quickly used up in battle and needed a lot of replacements.

This historical point, about the way the war pulled up the economy, is widely known. But less thought about is that after the war, although there was a peacetime conversion in many industries, the armament makers' sales to the government stayed substantial and soon enough grew bigger, whether there was a war on or not. A number of factors were involved here, but one that certainly had a significant role was the knowledge that without vast expenditures on war craft the economy would not be able to use up its excess capacity. This relates to the symbiosis of cancer and biological armaments research. It's not that cancer research wouldn't get any money if it weren't tied to the military. On the contrary, we have seen that in this instance a type of military science that was repugnant to the public was using cancer as a veil. However, in a government that has so long devoted a large portion of its budget to defense and in which defense contractors are major players in the economy, it's not surprising that our society has become militarized, so even the struggle to conquer a disease gets mixed with the creation of viruses to kill our enemies. Nor does this militarization of biology clash with any aspect of the professional or medical culture, since, as we've seen, as far as it is ruled by a Cartesian, mechanistic worldview, it corresponds rather closely with the black-and-white way of looking at things found in military command posts.

Instances of the Fingerpost

We might, to revert for a last time to this aspect of Bacon's writing, call this linkup between cancer research and warfare an instance of the fingerpost. This occurs in such cases where two (or more) hypotheses present themselves and at first the intellect refuses to choose between them.

> When in the investigation of any nature the understanding is so balanced as to be uncertain to which of two or more natures the cause of the nature in question should be assigned on account of the frequent and ordinary concurrence of many natures, instances of the fingerpost show the union of one of the nature with the nature in question to be sure and indissoluble... and thus the question is decided.[1]

In plainer terms, such an instance comes to the rescue of a researcher when he or she is met with a welter of information that it is difficult to categorize. It seems hard to decide which way an initial foray into the material is to be directed. Then along comes the fingerpost to start one on the right track.

We might see the points about cancer research and the militarized economy as such a post, "which are set up where roads part to indicate the several directions."[2] What it indicates is not so much that to understand AIDS we must comprehend such things as battle tactics. No, the idea is broader: AIDS is tied up with the general direction of the economy, so we must try to fathom both together. This is the way we should next direct our steps.

PART THREE

AIDS INC.

Having talked about the perspectival and actual fraternity established among the military, business, and medical research, perhaps we might pause to say a little more about the last two partners in this infernal brotherhood. We need to ask what part business and greed have played in steering biological research in certain directions and how certain organizations that treat or help PWAs seem to have either entered into it for the money or, more depressingly, seen earlier, humanitarian goals gradually demoted from prominence as making a buck becomes all consuming.

On the surface, the first of these propositions seems the more debatable. "Are you really saying, Gary," you may ask, "that doctors doing basic science, instead of being driven to find the truths of nature, have their eyes firmly fixed on the main chance?"

As we will see, the situation is a little more complex than that. But we might begin by asking whether it isn't strange that among all the possible causal theories of why people get AIDS, the one that was quickly accepted by the orthodox majority and then sold to the public, namely, that the immune deficiency came from HIV, is the very one that would seem poised to make the most money for pharmaceutical companies?

Let's try a thought experiment. Suppose it was shown from the get-go that AIDS was caused primarily by lifestyle elements, particularly by taking drugs. So to deal with it, the concerned government and private agencies would have to embark on programs to get people off these substances. Of course, drug companies could horn in here, as they have in the spectacularly unsuccessful methadone program; but as drug rehabilitation is now conducted, it has been found that the most effective means for kicking the habit is one-on-one counseling and group therapy. Neither strategy offers any avenue for pharmaceutical companies to profit from someone else's misery. In fact, contrary to the whole thrust of these multinational firms, this manner of breaking addictions is labor intensive, that is, it relies on people not machines. The big companies are in favor of pushing as many employees out the door as possible, not hiring more. Even more chilling to corporate managers of health, the most successful program for breaking drug attachment is Narcotics Anonymous, which is all volunteer, free, and has a long sustained policy of avoiding any commercialization. We can see, then, that if it were found drug use was a primary cause of AIDS, this would be poison to those companies who rule the medical environment.

But let's take this discussion out of the realm of imaginary possibility and look at some hard data: First, the intersection of medicine and free enterprise in general, then its influence on AIDS work, finally getting back to why the HIV = AIDS theorem can be said to offer pleasing prospects to free enterprise.

Science as Business

From the first page of this book, where we saw with what pizzazz Dr. Gallo showcased his presentation of the HIV = AIDS idea, it has been kept in mind that money takes a big role in the games scientists play. As mentioned, a discovery has to be hyped so that funding agencies take notice and other scientists will accept it. (We saw that Latour claimed, slightly differently, that other scientists embraced any discovery that they could use in their own work. We might bend his insight to say that even more significantly, pharmaceutical companies may become interested in a new finding inasmuch as they find it cash exploitable We will have more to say about Latour's apologies for scientists at the end later.)

This is not cynicism but simply a reporting of the way things work. When I chatted with Frank Buianouckas, he mentioned seeing Dr. Gallo give a talk at Lehman College for the Italian American Society. By this time, Dr. Gallo admitted HIV might need cofactors to spur it on. He was asked why, if he had long suspected the necessity of cofactors, he had once been so one-sidedly adamant that HIV alone was causing this disease. "And Dr. Gallo's response back to that—he said, well, you know when we first discovered the virus we had to say it was so powerful—we wanted to get a pharmaceutical company interested in this. This entire thing is to get investment."[1]

Dr. Gallo, who can be outspoken, perhaps is too easy of a target. But many in establishment science, though most won't admit it so freely, are going this route, strenuously beating paths to research areas that are business compatible and leaving untrodden any byway that cannot be connected to profit making. In *Get Healthy Now!* I mention that the investigations of herbal remedies, where one herb is assayed for its healing potential, are almost never undertaken here (though such studies are quite common in Europe and the Far East). I believe such studies are not done in that these natural herbs, not being patentable, cannot become the exclusive product of one company that can squeeze the consumers who want or need to buy it. Thus there is little incentive for these companies to sponsor any examination of them, no matter how beneficial they may be.

Stefan Lanka, who broached the same topic when I talked with him, put it in these words, "Science nowadays is not science anymore. It's now about articulation, power, business, and industry.... The spirit of science has definitely been lost."[2] This is not to say one shouldn't make money for a valuable contribution. But, as Lanka argues, currently the cart gets front row in place of the horse. "First they [scientists] have an idea, then they patent the idea, then they develop the experiments and they make a clinical study, and then they sell th product. This is how it happens today."[3]

Since this insight has been elaborated by Aronowitz, Sandra Harding, and others mentioned in our notes, I will leave it to them to fill in the dots and turn to the subject at hand.

The Economics
of AIDS

Money, Prestige and Jobs

As we've seen, each team of scientists and, speaking more of disciplines, each field of research has its own bandwagon, which it tries to propel into the limelight. We noted how Duesberg and others called retrovirologists an aggressive group, whose endeavors in the 1980s to get more funding for a once trendy but now merely passé type of research, finally hit pay dirt when they proved, at least to their own satisfaction, that a retrovirus was at the bottom of the AIDS crisis.

When the wagon got rolling, seats on board were suddenly valuable plums. Reporter Mark Anderson quotes Duesberg, who characterizes all the hotshot retrovirologists as having first-class tickets on the gravy train. "They're either on the boards of biotechnology firms developing vaccines, developing anti-viral drugs, developing AIDS-test kits, selling them, marketing them, owning the companies. They're millionaires. Dr. Gallo alone, even at the NIH, gets a supplement to his salary of over $100,000 a year."[1] Others of Dr. Gallo's celebrity stature reap equally lucrative rewards. "Baltimore sold a company on AIDS a couple of years ago for $30 million," Duesberg continues; and another key player, "Varmus, had to sell millions of investments before he could become director of the NIH."[2]

Moreover, the rewards for going with the flow are not only monetary but visible in skyrocketing reputations. As David Rasnick notes, "Dr. Gallo now is the leader of the Human Retroviral Institute, a totally pointless institute. It has no function at all. Retroviruses are harmless, and yet you've got this whole institute to research them. So they don't want to see this go away, obviously."[3]

Let's not imagine that these scientists are maidens who waited for businesses to come courting. They were not behindhand in pouncing on the moneymaking aspects of research. Writing in *Science*, J. Cohen informs us:

> Thirteen years ago, when the public was just learning about a new disease called AIDS, the scientists tracking down the virus that causes it were already thinking about how their research could be marketed. French and American groups eventually claimed to have "co-discovered" HIV independently and in different ways. But in one respect their approach was the same: shortly before announcing their discoveries, both rushed to file patents that described how to determine whether a person's blood harbored the virus. And thus they gave birth to the HIV/AIDS industry.[4]

He goes on to document the gold rush at the U.S. Patent and Trademark Office, where by 1996 fifteen hundred patents related to HIV and AIDS had been granted. The FDA, on its side, had approved by that date eight anti-HIV drugs and was readying approval of the next.[5]

Duesberg makes the obvious extrapolation from this evidence of profit-taking by scientists that what was once a bandwagon has now become a steamroller that overwhelms any objection to the HIV thesis. He notes the obvious, "It is very difficult if you have 20 or 30 people working for you with a company that makes a couple of millions on the side to tell them, 'Hey guys, we're working on the wrong track. Let's give up that stuff.'"[6]

Of course, there are multiple ironies involved in this. Dr. Gallo and other researchers at the cancer agency were Johnny come latelies who originally dismissed the disease as not deserving serious attention; while others, such as Dr. Montagnier's group, were hard at work. Moreover, as we can find from Shilts, both the general public and the federal government, as well as most city governments, turned a deaf ear to the cries of PWAs through the first years of the crisis, since the illness only affected politically marginal groups. However, once Rock Hudson died, which made a splash in the headlines, and once the cause of the problem seemed to have been identified in HIV, wheels started turning. When one heartthrob had died, other celebrities got involved (as we will discuss below) and there is nothing like a few celebrities to get fund-raising off the ground. But the second factor proved even more decisive. If a virus could be seen to cause AIDS, then making blood tests to detect that virus and discovering a vaccine for it could be profit generating. This made the pharmaceutical companies sit up and take notice.

To pull in another economic metaphor, for a medical specialty to become recognized and fundgenerating, its pet disease must go through a "takeoff period." In economics, this term was used when considering what would be necessary for an underdeveloped country to turn into an industrialized one. The takeoff period theory held that the country's economy had to be bolstered by outside help (as from the government) before it could stand on its own. But once it was under way, such help would be unnecessary. Everexpanding industry, employment, and investment would come forward naturally. Although the validity of this idea has been widely contested, we can say that something like that happens in medical research. Once a field heats up, garnering its first set of government grants, private donations, and celebrity sponsors, it tends to be self-sustaining, expanding as it plows like a juggernaut through the medicine. Of course, this can only be kept up so long, but while the trend continues, the good times roll.

"Since 1984," according to Lauritsen, "there has been the phenomenal growth of a multi-billion dollar AIDS industry, which comprises pharmaceutical companies, condom manufacturers, the latex industry, AIDS organizations... viatical settlement companies (which purchase the life insurance policies of terminally sick people), biotechnology, retrovirology, hospitals, counselors, social workers, and much more."[7] Duesberg spells out the type of numbers that are involved here. "Twenty million AIDS tests are performed per year at $50 apiece. Twenty million times 50 is a nice number for a bio-technology company to start."[8] Naming company names, Rasnick fills in, "Abbott makes $1 million [per year] on their blood tests. They don't want to see that go away. And the companies that make AZT and the protease inhibitors, they don't want to see that go away."[9]

With such profits being rung up, of course Wall Street had to get into the game. "Around this field of experimentation a vast body of business literature has grown up," Cohen states in the *Science* article previously mentioned. Some of this literature is propaganda trying to get investors to buy into new biotechnology firms that concentrate on AIDS. "Other commissioned reports, which can sell for $5,000 or more, aim to inform companies more objectively about how they should position themselves in the volatile and unpredictable HIV/AIDS marketplace."[10] High prices are paid for these reports by wary investors who want to know where to lay their chips. Cohen takes as an example a review of the AIDS market done by financial analysts Frost and Sullivan, which "says that the industry targeted on HIV's nine genes and 17 proteins rang up $1.3 billion in sales last year," and pinpoints where the biggest money is being made, noting half is on drugs for opportunistic infections associated with AIDS.[11]

And the AIDS bonanza translates not only into nifty profits but employment opportunities. According to Rasnick, "There are more than 100,000 scientists and doctors working on AIDS—more than the annual number of AIDS patients in the U.S."[12] To move abroad, we might cite a study done for the *Journal of Public Policy*, which surveyed thirteen areas in England to find "there were 3,400 staff overall and 1,800 AIDS patients, an average of two to one. Separate research showed an even higher ratio in Scotland, with 550 AIDS workers employed by health boards who between them had only a total of 164 AIDS patients."[13] Of course, such ratios are not found in the United States where health allowances are more meager, but we can say with M. A. Sushinsky that this outlay on AIDS is unprecedented. "The amount of money and public resources spent in the last ten years for the fight against AIDS is unique. Last year [1997] a record $2.5 billion was approved by Congress for AIDS prevention and research. No other disease in history, killing relatively so few people, has generated such a large amount of research monies."[14] (Don't forget that in the first years of AIDS, the U.S. government ignored the problem, only giving piddling monies to those working on the disease.)

As to the last sentence of Sushinsky about vast expenditures on AIDS, you might have heard statements to that effect made by the prejudiced whose point is that this money should not be spent on marginal groups. Obviously, that is not what I am getting at in quoting his remark. I don't begrudge money spent wisely for caring for the sick, of whatever group, but this money is often being thrown away, that is, it is obtained from the government under the false pretense of being for AIDS research, then it is turned to other purposes. As L. Garrett made clear in an article in *Newsday* in March 1996:

> Tens of millions of dollars allocated by Congress for AIDS research in 1994 was spent inappropriately—either on studies that have little relevance to the disease or on administrative expenses that are impossible to pin down, according to a slew of experts who studied the issue during the past year.
>
> Three recent reports on AIDS research funding, including a major effort by 100 scientists and advocates to be released tomorrow [March 14, 1996] say that a large proportion of the $1.3 billion spent by the National Institutes of Health that year went to program that have little or nothing to do with AIDS.
>
> The reports suggest some of the money was used administratively to help keep the individual health institutes in business, and some was siphoned off to cover basic research that wasn't related to the AIDS battle.
>
> . . . [According to an earlier study] of 139 cancer institute laboratories funded for AIDS research in 1994.... 92 had published fewer than 10 papers on AIDS during the past decade. Some have never published an AIDS paper, though they received more than $250,000 in 1994.[15]

Moreover, as D. Mertz and U. Schuklenk bring home, given the unbalanced state of health care in the United States a dramatic increase in the spending on one segment of medical research is equaled by a dramatic decrease in spending on other diseases. "In the U.S., where universal medical care is not guaranteed, AIDS treatment is state funded to a large extent, though the poor who die of other causes largely go without medical treatment."[16]

Here, too, an economic metaphor will be illuminating. Consumer business is often called a "zero sum game." There are limited numbers of dollars that the public will spend on, say, fashion. What money is captured by the sales of one store in a mall is lost to its neighbors. So the success of one store is dependent on others' failures. Here's where the parallel stops. In the market, if one type of good gains in popularity, it's possible more money would be funneled into this sector by consumers in preference to what they might have spent in other sectors. For instance, if a new style of wide-legged jeans became very attractive to teens, they might spend less money on CDs and put more into jeans. The game is still zero sum, but money is being gained by one whole sector.

Government spending on health should have this same flexibility, so that, for example, if a new plague were to break out, federal funds could be moved away from the newest Pentagon gizmo and put into health care. But it doesn't work that way, because any kind of program promoting citizens' welfare takes second bench, both in power and in size of the budget, to the military and to debt servicing, the latter being paid mainly to banks. Thus, the zero sum paradigm only refers to the first model, where money spent on one item, like cancer, has to be removed from another, like heart disease. In fact, what Garrett reported in *Newsday* as a scandal, that some earmarked money "was siphoned off to cover basic research that wasn't related to the AIDS battle," can be viewed as a possibly tragic situation in which researchers have to battle over crumbs to support what may be equally valuable studies.

Let's compare cancer and AIDS spending, first to show how AIDS has recently garnered the lion's share of new funding. "For each AIDS death reported in the USA in 1990, the government spent $53,745 in research and education. That's more than 15 times the $3,241 spent per cancer death," according to Mertz and Schuklenk.[17] They further note that in 1990 the money spent on AIDS was much higher than that spent on cancer, though cancer's death toll was 12 times as great. The writers summarize, now talking about spending for all illnesses, "Funding for other diseases has remained stable or even decreased during the period when AIDS funding has increased."[18]

The second point to make, though, in assessing the connection of AIDS to cancer investigation, is the trendiness factor. Sadly enough, we are seeing something of the Fort Detrick syndrome here. It will be recalled that when Nixon pulled the plug on biological warfare research, the biowarriors involved in this work changed hats and became students of cancer. But then cancer research, after hurrying down many blind alleys (such as Dr. Gallo's idea that cancer was a virus) and with little to show as far as effective treatments, became less of a fund gatherer. According to Farber, "It's well known that many scientists who undertook AIDS research jumped straight from the rotting ship of the cancer industry."[19]

Ellison, instead of seeing a cancer connection, in a kind of behind-the-arras, secret history, suggests the development of the HIV thesis is linked to the machinations of another bureaucracy seeking to expand its turf. This agency is the EIS (Epidemic Intelligence Service), which studies the distributions of epidemics in populations and has nearly two thousand trainees in national and international health care. Its graduates include former U.S. Surgeon General

William H. Stewart; Jonathan Mann and Michael Merson, "past and present heads, respectively, of the World Health Organization's Global AIDS Program"; and science reporters such as the *New York Times'* chief medical correspondent, Lawrence Altman.[20]

The training of its cadres consists of six weeks of epidemiological study, then two years of service "on CDC salary, in state and local health departments around the country."[21] Graduates pledge to continue supplying the EIS with any facts they gather about epidemics.

Ellison pounces on this agency—I've given a few more details about it than may seem necessary, since it is, as he says, largely invisible to the general public—because, like the retrovirologists who found in the later 1970s that their subject was less exciting and less a magnet for research funds than they had thought, the EIS, in the same time period, with no epidemics to trace in the United States, was worried that the funding for the agency would be chopped by a cost-conscious federal legislature. Ellison argues that its representatives were constantly fishing for diseases that could be labeled as epidemics.

He tells the story of EIS member Donald Francis, who had gotten his doctorate from work on feline retroviruses and in 1981 worked in the CDC's hepatitis laboratories division. 'Within just 11 days after the first report of AIDS cases… [Francis called a mentor and] insisted the new syndrome must be caused by a retrovirus."[22] Later, speaking at the CDC, he said, "This is the epidemic of the century, and every qualified person should want to have a piece of the action."[23] At first, he found few willing ears for his thesis, but one was his "close colleague," Dr. Robert Gallo, who took an interest as did Robert Biggar, "another EIS member at the NIH, [who] helped mobilize the huge federal institute behind the retrovirus hunt."[24]

Like a good reporter, Ellison gives the reader enough information to draw his or her own conclusions, even ones that go against what he says. For, after going through his evidence, I would say it's a toss-up whether Francis's lonely crusade to prove AIDS was caused by a retrovirus was more fueled, at least in terms of scientific affiliations, by the EIS or the company he kept with fellow retrovirologists. In looking at Ellison's discussion, we have a case where his general idea is actually much more helpful than his specific elaboration of it. That is, that the EIS has many mysterious tentacles throughout the health field does not seem to be proven as satisfactorily as the broader idea that in the "doc eat doc" world of competitive funding battles, each scientist with clout will try to recruit colleagues to help him or her prove that any medical emergency needs to be studied by the field to which that scientist belongs.

AIDS Organizations and Funding

Turning to another component of the AIDS business, let's revert to Lauritsen's note about "the phenomenal growth of a multi-billion dollar AIDS industry," when he listed, among those whose careers associated with the disease, not only hard scientists and doctors but "counselors, social workers," and so on. People in these fields are employed by either government, private business, or the charitable sector. All in all, "There are presently more than 93,000 AIDS organizations in America, one organization for every four people that have ever had a diagnosis of AIDS."[25]

Within this mix, one group that took innovative approaches and pioneered in fighting for such agendas as public access to drug trials, patient empowerment, and the organization of work along nonhierarchical lines, was the early volunteer charities, which Cindy Patton calls ASOs (AIDS service organizations). As we will discuss later, these groups were moving in sync with the emergent, holistic paradigm of health care.

However, no matter how valuable these groups have been, we shouldn't hide the fact that many have been co-opted by the system. As Patton puts it, "There was a major shift in the fight against AIDS between mid-1985 and mid-1986 in the largest U.S. cities…. This shift was away from gay-liberation inspired assistance… and toward an assimilation of activists into the new AIDS service industry."[26] That is, as the change took place, quite a few bought into the commericalist assumption that the value of an organization is to be measured by the dollar amount of its budget, the prestige and respect it has garnered in the mainstream, and the celebrities on the guest list of its annual fund-raising dinner. We have argued, on the contrary, that these groups' true effectiveness depends on nonquantitative measures, such as the quality and compassion of its services as well as the degree to which its organizational setup embodies egalitarian policies, particularly in how its policies engaged PWAs at all levels of planning and execution, effectively destablizing the border between person providing and person enjoying the services.

Without invoking a conspiracy theory, we might at least suppose that the countercultural nature of many of these organizations made the mainstream eager to co-opt them in order to bring them within the bounds of the acceptable. As I will show when discussing treatments, for example, the underground drug supplier networks—and other informally organized methods to get around the constricted way drug approvals and sales are managed—caused a minor panic among conservatives who angled to change the organization of drug trials and other components of the system, which they did in order to not simply defuse demands but derail practices that had moved outside of the recognized property system run by the legal drug lords. Patton has diagnosed more concretely what factors were pulling the ASOs, in her phrase, "from grassroots to business suits." One was, paradoxically, that the gay community that supported the ASOs had a degree of informal infrastructure and tradition of self-help that gave it an independent start in addressing the crisis; while, due to its being at a "relatively early stage of formal political development," it was wary of and not used to working with government agencies. This spurred early self-reliance but also meant that, unlike the African-American community, for example, it was inexperienced at drawing hard lines between its members and the government whenever the two entered partnership. Also, she argues, "The recentness of the 'demedicalization' of the homosexual seems to have left its traces in a narrowing of the gay community responses to AIDS... focusing on treatment, care and individual education."[27] That is to say, since the establishment had, until quite lately, defined homosexuality as a type of medical maladjustment, gays, while fighting against it, were still prone to view their community as medically defined. This is in contrast, for example, to African-Americans, whose community, although also constantly assaulted by discrimination, did not find blackness primarily defined as a biological deficiency and tended to view AIDS as just one among a number of problems.

However, deficient as it was for seeing reality, this medicalized model, still prevalent in the gay male community, jibed perfectly with what government funding agencies sought in those who would benefit from their largesse.

> Government funding patterns from 1985-1986 also promoted the formation of a largely white, covertly gay-community-based AIDS service industry: [because] money went to groups who corresponded to the Public Health Service (PHS) model of AIDS as an epidemic illness, narrowly defined by the acquisition of a virus through specific behaviors— a definition that largely ignores the social and political factors which shape the emerging demographics of AIDS.... To the extent that ASOs could present themselves as competent fiscal agents, working specifically on "AIDS," they received funds.[28]

As much as government partnership and funding led to an abandonment of the liberationist perspective that had originally motivated the ASOs, it pulled the sting from the volunteer organizations. The loss of this direction has been felt most deeply by those who have been involved for years with these groups. Dr. James Currant notes the sea change in these words: "The AIDS foundations... have become... totally self-serving... there is a lot of pharmaceutical money and government money that's funneled [to them]—you have people making salaries, and making money."[29]

Speaking particularly of his own Californian bailiwick, Passante Conlan told me of his own disillusionment.

> One of the saddest things I've seen develop over the last fifteen years in San Diego is the way our AIDS organizations have transformed from organizations of charity, to organizations of big payrolls, big bucks and basically a bureaucracy of government funding. No longer does it seem that its the client that's of concern, but keeping the money coming in so that the massive amounts of salaries can continue to be paid.[30]

There was not only anger but bitterness in his voice as he continued,

The money we are spending here in San Diego on AIDS is tremendous, but very little of it actually winds up getting to the client. The situation, I feel, is the same throughout the country, not just here in San Diego. Many of the organizations that started off as charity organizations have once again lost their way, and have become organizations [sans charity].[31]

AIDS Fund-Raising and Glitz

The next point I want to make may seem counter to common sense at first as well as liable to misinterpretation. So read on through my whole discussion before you pass judgment. My feeling is that once a program becomes glitzy, it loses its ultimate purpose, although it may now be capable of initiating new programs that are highly beneficial to those it serves.

Here is exactly what I mean by that proposal, using Shilts's example as fodder. The death of Rock Hudson brought new funds and attention to the AIDS crisis. It was arguably one of the main factors that got the government moving to support research and treatment—pushing the disease through its takeoff period—while making the media and the mainstream public take notice, so that it would be impossible to hide from its impact and the necessity of addressing it. I allow this without argument. But the long-term, radical goals of the organizations were suffocated by the embrace of the celebrity culture. That is, those groups who were run on volunteer energies, governed by a democratic ethos, ignored the rules of the system when they proved intractable and worked to get PWAs to take as much responsibility as they could in making choices on treatments and care; once they entered the mainstream, they had to begin to pull back from all these directions, so they became, in essence, defanged.

Here is how I may be misinterpreted. It may seem that I am saying avoid publicity like the plague. No, this would be self-defeating. Moreover, some of the astutest users of the media, such as ACT UP, have eschewed the celebrity culture and continued to hold to holistic goals while being in the thick of things in terms of negotiating with the FDA and other agencies on how to conduct drug trials and deal with other globally significant issues. Moreover, I am not saying the situation pre-Rock was positive as a whole, as if I idealized the all-volunteer, small-scale but morally pure and righteous period, a nostalgia Cindy Patton rightly castigates. I am only saying that it is too much to ask of an AIDS organization that it give up its deepest hopes (of shaping a responsive, democratic medicine, which will be as attuned to health needs of minorities as it now is to those of the dominant majority) in order to participate in government programs and work with other medical institutions.

I say this because all too often AIDS purists express dismay over the circuslike atmosphere that surrounds some AIDS-related events without explaining pragmatically how such freewheeling, actor- and actress-studded activities detract from the cause.

Let's look at some of these laments, giving them credit for recognizing the problematic nature of AIDS's recognition by the media. Hiram Caton states, AIDS "won celebrity as a human rights cause, as a lawyer's bonanza and as a media sensation." Being the media's darling means, he argues, "in its short career, AIDS has become the most talked-about, anxiety laden, fiercely contested, lavishly resourced, and withal the most wept-over illness of modern times."[32]

With this amount of recognition, the disease becomes a motif of the journalists. "Finally, the media understands that the public will read any story about AIDS," Plumely notes.[33] This is a note expanded on by Duesberg, who states the illness "was a windfall for the journalists."[34] It is liked as a subject by science writers, Duesberg claims, because they no longer have to take the trouble to understand a dry subject like nuclear physics and then pitch it to their editor as newsworthy. Instead, "now, all they have to do is call Anthony Fauci or some other expert on AIDS research, touch it up with a little anal intercourse and intravenous drugs in the bathhouse, and have their next story for the Sunday *Times*. Everybody would love it and everybody would read it."[35]

In Simon Watney's book on the media treatment of AIDS in Britain, he adds that the public's presumed lust for sensationalized AIDS articles in the newspapers means the press even "manufacture" stories, such as by posing hypothetical situations. "At this moment in time [1987], everything to do with AIDS is newsworthy," he states. "AIDS is caught up in the larger machine of sales wars and takeover battles within the newspaper industry.... Hence, we see stories being invented from scratch."[36] He cites in evidence one article by a *Sun* reporter, who wrote of "My Misery Posing as AIDS Victim: Why the Country Closed Its Doors on Me" when he pretended to be a PWA. Another good illustration is headlined, "I'd Shoot My Son If He Had AIDS, Says Vicar!" In both cases, instead of researching the actual situation of PWAs, the journalists have sought out or manufactured sensational and hypothetical situations.

In *Continuum*, Celia Farber spotlights some of the more outlandish and garish elements of AIDS print culture.

> *POZ* magazine is a microcosm of the madness that has become AIDS: A slick, glossy, celebrity-studded death culture with its devotees perpetually interviewing each other about what it's like to be so concerned, so angry, so committed to "saving lives" so fabulous. The magazine is floated by pharmaceutical ads for AIDS medications, life insurance policies, and even a service called "Lasting impressions" that sends letters to your loved ones on specific dates after your death."[37]

While in *Time*, Rudnick is depressed by the whole magilla. "Here's what I find most shocking about AIDS: it has become comprehensible."[38] He elaborates:

> Friends have been dead for 15 years. Worthy organizations like the American Foundation for AIDS Research and Gay Men's Health Crisis (G.M.H.C.) own buildings with serenity gardens, and the red ribbon

symbolizing AIDS awareness appears on mugs, Christmas-tree ornaments, and beach towels while remaining a fashion challenge for Oscar-night actresses in strapless gowns. Elizabeth Taylor and Sharon Stone impressively share the fund-raising crown.[39]

His complaint is that AIDS is now business as usual. It seems to me the underlying, inarticulate meaning of what he is getting at is precisely what I outlined earlier. If business as usual meant that due to the AIDS crisis we had begun to refashion our health care system from a two-tiered to one-tiered system and made other changes that dealt with the gross inequities now governing, it would be cause for celebration. But what Rudnick must mean to say, consciously or unconsciously, because the logic of his statement depends on it, is that business as usual indicates the framing of AIDS as something from which someone can profit, but not as something that will change a desperately inadequate health care system. In moving onto this even keel, then, GMHC and the other organization mentioned have won on one front, the broad social recognition of the gravity of the crisis, but they have lost on another: the progressive refashioning of the nation's health care.

Come to think of it, I'm not surprised Rudnick didn't spell this out. This is hardly a thought one pronounces in *Time* magazine.

But—and this is not directed at Rudnick in particular—these reflections can be given a larger context related to the politics of the unexplored. Typically when one gives reasons for supporting a position, there exists an unvoiced stratum of underlying assumptions, which (as Habermas formulates in *Theory of Communicative Action*) should be able to be dredged up for examination if necessary when the discussion happens to turn to foundational exegesis. But, I would counterstate, in cases such as Rudnick's, where a liberal is arguing from a radical's substructural premises, the reasoning can never be followed out in its deeper ramifications because it would force on the speaker a position he or she wouldn't dare take.

Given that a pundit or researcher is speaking from an establishment-subsidized social location, it would (to echo Duesberg) be self-destructive for him or her to point to premises that give notice of the rank insufficiencies and injustices of this same establishment, and yet, paradoxically, if the speaker does not covertly build up from such premises, he or she will have nothing charged to say. One then has the spectacle, in cases like Rudnick's, in which a quite common liberal position rests on premises that never dare speak their name. Discussion must always be maintained on superficial, technical questions, since no one dares to descend to first principles—as if we all rode in a boat that could never venture out of shallow water.

HIV=AIDS Becomes Unchallenged

But we are coming to an even more perverse wrinkle in the crisis, one we can again approach through an economic metaphor. In the early 1980s, radical economist and futurist Hazel Henderson wrote a book, *The Coming of the Solar Age* (a tome that deeply influenced new paradigm theorist Fritjof Capra whose ideas we have already sampled). In it, she argues that there is a mirror side to "economies of scale." This term is used in academic studies to praise corporate bigness. The academic argument goes that a larger company, say a chain of retailers, can easily outpace and outprice your smaller mom-and-pop corner store, because it can buy in bulk at reduced rates, centralize its bookkeeping and other back-office services, and do many other things that give it market advantage. Such arguments have not only been applied to sales organizations but especially to production, where the multinational that can shop its factories around—placing them in the region where it can get the lowest wage workers and have the least danger of being bothered by labor laws and pollution regulations—has a major advantage over smaller companies whose ability to bounce from nation to nation is limited.

Henderson's argument runs in exactly the opposite direction. She argues that at a certain level of giganticism, a company's affairs become so entangled and intricate that even if a computer can sort them out, they cannot be optimized because they are beyond the comprehension or control of the line officers. Moreover, it becomes harder for these firms to distance themselves from "externalities." These are costs that establishment economists say do not have to be put onto the balance sheet because they are outside of the company's purview. But, Henderson insists, such ledger-book shifting of costs amounts to nothing but legerdemain. Externalities commonly include such items as the pollution caused by hazardous wastes dumped in streams, the cancers caused in the processing of uranium, and so on, that is, anything down the road the company thinks it can get out of responsibility for, no matter how much it is to blame. Henderson argues that it is the bigger companies that are most known for laying these burdens on society, and as these costs come home to roost, as they have of late on the tobacco industry and asbestos industry, to name only two, we will find that these behemoths have in fact been practicing diseconomies of scale.[40]

However, what is most germane for the topic at hand is her insight that as an industry gets bigger, it gets increasingly cumbersome. It loses its flexibility and gets stuck with ways of doing things that may have been taken on for ad hoc reason but that now must be held to no matter how burdensome. Some dissidents see HIV as a similarly loyally adhered-to losing proposition.

It could be said that when retrovirologists first fought for the classification of AIDS as a virus-engendered illness, it didn't make much difference to them which virus was ultimately found to be the culprit. The main thrust of their argument was simply that it was a virus, they were the only ones who were qualified to analyze it, and therefore they deserved all the funding. But as HIV in particular has been kept on center stage, it seems everything about it grows sacrosanct and unquestionable.

(This may explain, by the way, why Duesberg's argument, which accepts HIV as a cofactor, is so virulently attacked, even though if it were accepted, it would hardly cost retrovirologist jobs. They could research HIV as a vital co-factor in the disease as well as they research it as the prime mover. However, as the HIV doctrine becomes more dogmatic, each minor clause in it must be defended as hotly as its major premises. This is not to overlook, though, the inverse ratio between prestige and fierceness of polemic. The more respected an opponent is—and Duesberg, as we saw, was considered Nobel timber—the more abusive of an attack he or she rates from orthodoxy.)

Johnson points out how convenient the HIV equation first proved. "The HIV theory was immediately triumphant because it was the kind of solution to the AIDS mystery that all the major players wanted to see."[41] Obviously, it was a boon to virologists; but, in the medical field it also redounded to the glory of epidemiologists at the CDC, who "gained new importance and prestige." As identifying this causal agent was seen as marking great progress in moving toward a cure or vaccine, it made the Reagan administration look good. We already saw how this administration tried to stage-manage the announcement of the discovery for maximum publicity. Further, "Organizations of AIDS patients had cause to hope for a cure, and they were assured also that, since 'everyone is at risk'... an unpredictable new virus and not their own conduct was to blame for their condition."[42] Drug companies stood to reap big bucks. "No one had a motive to doubt, and so no one doubted."[43]

Now that many of the dreams of these groups have come true, it would be very hard to back away from HIV, let alone the general concept of a viral cause. When I talked to Mark Conlan, he put it quite eloquently. "If we suddenly woke up one morning and it was on all the morning news shows that Peter Duesberg vindicated HIV as not the cause of AIDS—then a good part of the social, political and economic structure of the gay community [and the larger society] would collapse overnight."[44]

This is the point Henderson made about unwieldy structures. They become impossible to change, as if they were welded shut. In this case, too many people would lose niches wedded to this particular interpretation of the disease. In my talk with Charles Thomas, he made just this point. We were discoursing on why the HIV = AIDS concept seemed to have a stranglehold on establishment thought, and he speculated:

I think that their [holding to the HIV equation] motive is to maintain their jobs. They want to have a job. And, for example, if I were running a laboratory with twenty people in it and my funding was coming [from] working on a particular problem involving HIV, I certainly would not question whether HIV caused AIDS or not; because if HIV can be demonstrated to have nothing to do with AIDS, then my funding would not be very long-lived and I would have to let people go and I would be unable to do the work that I wanted to do. So getting funded is a prime consideration. As a matter of fact, it's the number-one consideration for the working scientist today. He has to have money to do the job.[45]

It is not only employment, though, but a locked-in market. Again, an economic scenario might help clarify this idea. It is often said that Europe so quickly became a strong economic competitor to the United States after World War II (at the very beginning of the peace the United States had been on top of the world and Europe a smoking ruin) because Europe started from scratch to rebuild a razed infrastructure, thus adopting all the newest industrial techniques. The United States couldn't match the continent in latest methods, because our manufacturers were naturally reluctant to abandon their older plants, in which they had so heavily invested. In our parallel case, as more and more medical infrastructure had been geared up to address HIV specifically, the cost of dismantling this productive capacity becomes increasingly daunting.

Let's look for a moment at the range of investment we are talking about here. The aforementioned *Science* article by J. Cohen discussed some of the profits being rolled up in the industry. The first protease inhibitor, saquinavir, that came to sale in December 1995 under the aegis of Hoffman-La Roche. Cohen states that "according to IMS America—a Dun & Bradstreet company that tracks sales of drugs primarily through U.S. drug stores and hospitals—saquinavir [in four months] had already grossed 35 million in revenues."[46]

Although the AIDS drug market is volatile, since new additions may deflect sales in different directions, in 1996 economic forecasters saw a bull market for this type of drug, which is still being favored in establishment therapy. In 1996, one market analyst predicted "that six protease inhibitors will be on the market [soon], each selling for $4,000 a year, bringing in revenues of $755 million."[47]

Another growth area is assays given to patients to track disease progress. We saw already that CD4 count was the measurement of choice—even though, as dissidents have noted, exactly what a healthy person's count should be has never been determined. In 1995 the general assay market rang up "$92.3 million in revenues... two-thirds of which came from tests that measure levels of white blood cells with a receptor known as CD4."[48]

Another type of assay measures "viral load," based on how much of one or more HIV-specific proteins can be found in the blood. One marketer of these tests is Chiron. "Mike Richey, Chiron's vice president of diagnostics sales, estimates that the potential U.S. and European market for all viral load tests is $80 million to $120 million" per year.[49]

And we shouldn't leave out of the picture the blood tests that are said to tell you if you are positive or negative. "The bulk of this market, which Frost & Sullivan estimates at $83.2 million a year [as of 1996] in the United States, belongs to companies that use... the HIV antibody test."[50] Its proceeds were split among Abbott, with 55 percent of the market; Sanofi Anthrop, with 21 percent; and Ortho and Organon Technica dividing the rest.[51]

Thus, with all these resources committed to the HIV-alone theorem, the medical industrial complex is not likely to take kindly to alternative causal theories. In 1997, Duesberg and Rasnick nicely summarized the array of forces committed against anyone who wants to upend the HIV thesis.

1. The HIV/AIDS orthodoxy's annual budget of $8 billion from the U.S. taxpayer alone.
2. The thousands of AIDS organizations, including countless public health and activist careers and tens of thousands of scientific reputations that are exclusively built on HIV.
3. The numerous medical and social benefits available to HIV-positive activists and patients,
4. The staggering commercial interests in HIV-tests, over 20 million tests per year at $50 or more in the US alone, HIV vaccines and anti-HIV drugs.
5. The prospects of numerous complaints and malpractice suits against the HIV/AIDS orthodoxy for those who were told they are destined to die based on HIV tests or were helped to die with AZT.
6. The prospect of a profound loss of confidence of the American public in its medical and scientific elite.[52]

The wide gulf in views on the significance of HIV between dissidents and the establishment depends, as this solidly brings home, not only on holding different paradigms on health and disease but also on one's ties to AIDS Inc.

DISSIDENCE SCORNED

A moment ago, invoking the name of Duesberg, we mentioned that, given the way careers, funding, and prestige have become tangled up with holding to the HIV thesis, it should come as no surprise that those who rock the boat face hostility. The hostility is not only directed at them but serves as a way to enforce the orthodoxy's community norms. The public pillorying of Duesberg, which, as we saw early on, was disquieting to those who believed scientific disputes should be settled by debates, not threats, served to keep others in line. According to Frank Buanoukas, who raised this topic when I talked with him, "I think it is true that scientists will not challenge the [HIV] hypothesis because all they have to look at is Duesberg, see what has happened to him, all his funds have been cut."[53] Even if scientists with government AIDS grants might be tempted to entertain the hypothesis that HIV does not equal AIDS, "they won't say anything for fear of being scratched off the list."[54]

And one aspect that seems particularly unsettling in the handling of Duesberg's case was that according to documents obtained from the Department of Health and Human Services (HHS), the assault against him was mounted by that agency with no regard for scientific grounds. To put that in other terms by way of an example, we might assume that when Dr. Gallo attacked Duesberg, as we saw, sometimes he resorted to "street fighting," that is, assaulting Duesberg's ability to reason or his credentials, rather than basing his case on scientific merits. But I think we have a right to assume that, while the fiery Dr. Gallo might have occasionally flown off the handle, he was ultimately defending a scientific thesis. He was fighting for an intellectual position.

On the other hand, the materials from the Office of the Secretary of HHS and from other government agencies have no perspective on the medical issues whatsoever and concentrate only on public relations damage control.

The mentioned document from HHS, a memo by Chuck Kline headed MEDIA ALERT and dated April 28, 1987, combines inadvertent absurdity with a patronizing view of science, all with sinister undercurrents. After mentioning that Duesberg has published an article debunking the HIV = AIDS position, we hear, "Inexplicably, the paper was published in the March 1 edition of *Cancer Research*... but nobody within the department or the news media seems to have been aware of it until it was disclosed Monday, 4/27, by a gay publication in New York City [through the efforts of playwright and gay activist Larry Kramer]."[55] The implications of "inexplicably" seem to be that, according to this media flak's understanding of science, a responsible journal will not publish anything that breaks with the dominant paradigm, no matter how senior the writer and how well documented his or her facts. The revealing black humor is, of course, that the health agencies don't even keep abreast of material on their subject in the major scientific journals in the field. The task of being scientifically au courant is left to activists.

The first understanding, namely, that the job of scientific institutions is no longer to search for truth but enforce conformity is, of course, the more repellent, but one Kline repeatedly reinforces. For example, when describing Duesberg he appears abashed to find, "Dr. Duesberg has been an NCI grantee doing research in retroviruses and oncogenes for 17 years and is highly regarded. He is the recipient of an 'outstanding researcher' award from the department."[56] Here the connotation is that anyone receiving funding from the NCI should be goosestepping along with the others parading behind the HIV bandwagon, and the fact that Duesberg isn't makes Kline incredulous. After all, Kline seems to think: "Doesn't one give grants to buy silence?"

The next point he raises is more ambiguous. He says that Duesberg's "article apparently went through the normal pre-publication process and should have been flagged at NIH."[57] It could be that he simply means the NIH should be aware of any up-to-date research or discussion; but the suggestion is again that there is something essentially wrong with "the normal pre-publication process," if articles damaging to HHS's views, the ones on which it bases funds allocation, are let into print.

With such a mind-set, we can well imagine that Mr. Chuck Kline is not going to recommend more investigation be done into whether HIV is really behind AIDS. Instead, as expected, he tells the Secretary, the under secretary, the surgeon general, the White House, and others to whom he CC'ed this memo that "we need to be prepared to respond," adding, "I have already asked NIH public affairs to start digging into this."[58] Similar responses were being floated in other departments, such as in the NCI by public relations officer Florence Karlsberg, who memoed top officials in her organization, pointing out peer pressure wasn't working—"Bob Gallo and others have tried to educate Peter re HTLV-3 and AIDS—but it's hopeless"—and so counseled the organization of a kind of medical SWAT team. Karlsberg recommended forming "a response team comprising Blattner, Dani Bolognesi, Anthony Fauci and Robert Gallo."[59]

The upshot of such tactics, where putting a good face on things takes priority over getting to the bottom of vexing scientific issues, and coercion replaces debate, is a science of fear rather than enlightenment. Hear what Rasnick has to say: "It's fear all the way through. Even scientists, people like me, are subjected to fear and intimidation."[60] It's not that all science has suffered diminishment, he feels, but that one area of science has become a no-go zone: "I can talk to my own colleagues, freely and openly, on virtually every topic in science—even about whether there are deities or not—except alternatives to HIV as the cause of AIDS. It's astounding! It's incredible! When I went to school, I was taught that only happened 1,000 years ago!"[61]

As noted, though, this fear is not a product of mass hysteria based on nebulous qualms and anxieties. There are real material punishments for those who don't toe the line. "We have totalitarian science directed entirely from Washington in hypothetically free universities," Duesberg said. "You can survive with tenure but you certainly cannot run a centrifuge or pay your graduate students or write a paper if you don't have a government grant."[62]

And when it comes to discouraging AIDS dissidents, not only the universities but the FBI has gotten into the act. According to a report in the *New York Daily News*, "The FBI has secretly spied on New York-based AIDS activists—monitoring their demonstrations."[63] The surveillance extends not only to the unusual suspects, such as ACT UP, which engages in nonviolent civil disobedience, but "according to FBI files, it also watches other groups, including the Gay Men's Health Crisis, the Coalition for Lesbian and Gay Rights and SAGE… an organization for gay senior citizens."[64]

We've titled this part of the book AIDS Inc. On glancing through our table of contents, a reader might take this to suggest that once a medical issue becomes commodified, that is, translated into a product or service that can be taken to market, it will tend to take on the coloration of commercialism and lose the "markers" of scientificity. It may be expected, for instance, that research into dealing with a disease will then be predisposed to directions that would yield a salable product rather than realms that would be less obviously capital enhancing.

Although this is true to an extent, it can only be said with two heavy qualifications. The first is one that was remarked on when we were discussing biological weapons production. At that time, it was suggested that the large military budget, dwarfing other federal outlays, excepting for debt, tended to curve all research into directions that dealt with national defense. However, we can put this in more general terms, simply to say that the economy is deeply swayed by government interference. This, though, is not the kind of interference that conservatives get exercised about, where the state is making available such things as unemployment insurance and Medicare, things conservatives think should be left to parsimonious charities. No, I'm talking of such interference as, for example, price supports for agribusiness, in which the government will pay to lift prices on farm products when in the open market they fall below a certain floor, or subsidies and tax breaks to businesses to help them innovate or break into foreign markets, or any of the cornucopia of handouts that the government gives according to its generous "corporate welfare" plans. We have to see AIDS Inc. as an entity that is a Keynesian hybrid of government agencies, pharmaceutical labs, Wall Street investors, and other business buccaneers.

The second qualification is simply that we need to recognize that every "Inc." has a set of actors who work for their company's or agency's advancement, as a way to both add to their own power in the developing formation and to stabilize the wider field in a way advantageous to the actors' reference group. Thus, Dr. Gallo fights to advance his HIV causal thesis not only because it represents a gaudy feather in his cap as the discoverer of the cause of AIDS, but because if this thesis flies, then the arena of biological science will tilt in the direction of virologists over, say, as in our much earlier example, bacteriologists.

Then, to try and draw things together here, we can see that retrovirologists, in their planning of a strategy, were not unlike the fast-trackers, who evolved a compromising strategy to take advantage of the toleration for gay male self-expression within some urban settings, particularly those of nightlife, but also in full awareness of the continuing hostility and prejudice that kept them from open expressions in such venues as jobs, where asserting a gay identity would close most career paths. So, retrovirologists also had to follow a winding road to keep themselves afloat. If we concentrate on those who tried to identify AIDS with cancer, back in the 1960s, we see that though this thesis would have had tremendous returns if proven, it was so far-fetched that it would have been unlikely to raise many backers. However, when biological weaponry was under assault due to public outcry, cancer research into transferable cancers—ones that could be transmitted to the enemy—became an excellent cover for graying biowarriors, who could transfer their expertise to this field. After all, other theories of cancer, ones that are on the surface more plausible, such as that it may stem from years of exposure to toxic chemicals, could not be used as rationales to make arms. Thus, there was a powerful symbiosis going on. The unemployed biological warfare experts needed a respectable field in which their skills at making toxins would still be useful and the virologists needed a way to get funding for a theory of cancer causation that was ill-suited to the facts but well-suited to toxin creation.

If we take that to be the first act—that is, virologists moving from pure cancer research to the development of biological weapons— the second act would be AIDS. Farber has suggested that virologists in cancer research hurried into AIDS because their work in cancer wasn't panning out. However, we don't need to believe that to see that this strategy was feasible for them under any circumstances, in that it would provide them with another win-win alliance with the pharmaceutical companies. Here they could take a leaf from the book of their erstwhile colleagues in the cancer industry (those who didn't hold with the viral cause theory). These colleagues, with their chemotherapy and other drug-based treatments, had an emphasis (as would be so with AIDS) not cure but staving off the inevitable end. For them, profitable alliances had been made with the legal drug lords. Here the virologists could make their own alliances, first by doing the spadework for an HIV-identifying blood test and later by developing prophylactic drugs.

However, in making these points, I want to steer clear of the type of argument we saw from Ellison, when he talked about the influence of the EIS in pushing AIDS study down the viral path. His presentation smacks of those made by conspiracy theorists insofar as the EIS seems near invulnerable as it advances its plans unopposed on a pliant world. To counter such suggestions, I need to stress that the retrovirologists, as they first put forward the HIV thesis, were fighting for their views against formidable foes and had no sure sense of prevailing.

Cindy Patton's discussion of the differences between the virologists and the immunologists can bring out one skirmish in the virologists' struggle. As she depicts them, the immunologists have a perspective that accords with the emergent paradigm we have been describing. "According to both holistic and immunological accounts, it was our bodies and not outside invaders that were the problem.... The chief evil was a modern society that wore our bodies down."[65] In keeping with this worldview, when immunologists examined AIDS, they found it to be based in "immune overload due to drugs, 'fast living,' and other stressors, while treatment under this discipline would be more concerned with strengthening the immune system than fending off aggressive viral interlopers."[66] As Patton sees it, there was a running battle in the early days of AIDS science between these two biological subfields, with virologists the eventual champions. "The greater technological and financial commitments (and potential rewards) of virology linked data and dollar to become the dominant way of thinking about AIDS. AIDS research ended, for now, a struggle between two subdisciplines."[67] Although she is sketchy on the skirmishes and counterskirmishes that took place in this war, the salient point is that virologists, until they passed the takeoff period, faced an uphill drive to gain ascendancy .

It may seem by now that in concluding I have forgotten the subject of AIDS Inc., going on instead about the politics of intergroup competition. Yet if the society is structured so that in the science it sponsors, honor accrues only to those who obtain money, not in the form of high paychecks but in lump-sum research grants, then all science falls under the sway of two gods. While worshiping Athena, who is the lover of wisdom, it must also bend knee to Pluto, holder of all wealth. Scientists angle for grants not only for their own prestige but for the prestige of the discipline. Each act of scientific discovery pays a double homage, one to each jealous god.

We have left out of the picture a second group of social actors: those who pilot pharmaceuticals. But the study of their twists and turns can await the examination of establishment AIDS treatments.

An Imaginary Conversation with Prof. Stanley Aronowitz

B y now, we have winnowed through enough concepts, myths, and worries on our journey to come to a brief halting place.

Through much of this I have contrasted two paradigms lately identified as that of immunology (in an idealized form) and that of virology. Yet in a very significant way, I have not presented these paradigms in a satisfactory light. By inadventure, I have made it seem they are purely intellectual artifacts that make for more or less elegant theories. This is erroneous. Paradigms begin and reside in praxis.

By this I mean that they arise as coherent responses to invidious situations, not as cognitive exercises. For example, those who first set up community drug trials, ignoring the medical establishment and working to enfranchise and listen to people with AIDS in designing their program and selecting drugs to test, did not do so out of allegiance to a particular intellectual program. Rather, these were necessary, ad-hoc measures to counteract the slow pace and closedmindedness of the drug-testing establishment, which refused to study drugs that PWAs were interested in and perhaps had experimented with through underground networks.

Another point that may have deserved greater stress is the one about literary composition that was laid out near the outset of this book. Many books silently present themselves as the product of a woman or man chained to a desk. This is a distortion. In fact, *all writing is workshopped*, although this may not be done formally. In other words, any text comes out of a practical negotiation with one's community.

And this takes places on two levels. On one level, an author constructs a text by emplacing his or her personal interpretations on top of a bulk of material gathered from one or more interest groups. For example, in *Second Opinion*, I have given my spin on a host of materials gathered from AIDS establishment and dissident writers, as well as from those involved in the alternative health movement. But even that personal portion of the writing, the author's inflection, *is not so much an assertion of self as a dialogic fragment*. By this, I mean that in each thing a person writes, he or she is in implicit dialogue with one or more people, living or dead, friend or foe.

The Russian literary critic Mikhail Bakhtin has explained this passionately:

> A speech [he is talking of writing] in any essential and profound dialog is analogous to the hidden polemic. Every word in such a dialog speech is directed towards its object [what is being discussed] but at the same time reacts intensely to the word of the other person, answering it and anticipating it. The element of answer and anticipation penetrates deeply into the intensely dialogical word. Such a word envelops and draws into itself the speeches of other people and intensely reworks them.[1]

So, my thoughts have been shaped by an implicit conversation with my compeers, mentors, and students as well as being clarified and chiseled through rambling, wide-ranging talks with my collaborator, Feast. These conversations with Feast while they sometimes became dispersed and vague, at times had a resonance close to music.

In the opening of my book *Who Are You, Really?* I tried to show how such dialogue can spur one to deeper reflection and soul-searching. I had been complaining to my brother about how my intensive study groups didn't seem to be getting results. The same participants who nodded their heads in agreement over the need for a vegetarian diet and would carefully work with me planning vegan meals could be caught a half hour after our session ended walking out of McDonalds holding a Big Mac. Disturbed as I was by this, it was only after a long talk with my brother—who told me I should not waste time on melancholy plaints and devote my mind to unraveling what it is that makes people such backsliders—that I began the psychological inquiry that led to that book. As I wrote it, I constantly talked to my sibling as a means of spurring me to clarify my ideas and also as a lawful check on any unreality or falsity that will inevitably creep into an abstract presentation.[2]

It comes to me that it might be a good time to pause and look critically at some of the concepts I have ventured to put forth thus far, since neither their shortcomings nor full implications have yet been plumbed. It also appears that I might use this brief section to "show rather than tell" in relation to my thoughts about writing. I want to revive an older form, the imaginary conversation. In the nineteenth century, writers such as Walter Savage Landor would invent discussions that were pictured as taking place between distinguished intellects who had never met in reality.

Although I have never met Stanley Aronowitz, many points in his books on science have been fundamental for my understanding of the subject. Moreover, my collaborator, Feast, who studied under Stanley, knows well his incisive and biting intellect. While, a man of infinite generosity and humility around persons, Aronowitz has a never-sleeping vigilance when it comes to detecting logical inconsistencies or marks of pretension around ideas. In this short digression, I am going to picture what he might have said were he given a draft of the earlier part of this book to critique. (Needless to say, he has no responsibility for the words we are putting in his mouth.)

Bruno Latour

NULL

Give us your take on our treatment of Latour?

ARONOWITZ

I think it would be rewarding to more rigorously contrast Latour's ideas with two other key philosophers of science whom you have invoked: Popper and Kuhn.

You noted that Popper argues no scientific hypothesis can be made incontestable. The best one can hope for is a hypothesis that has weathered many attempts to discredit it.

In Latour's hands, this thesis is given a sociological tinge. He says theses that have resisted overturn are applauded by the scientific community simply because other researchers can base their own research on them, hence helping their own careers.

This is put quite baldly by Latour. I quote:

> Let us suppose that scientists are investors of credibility. The result is the creation of a market. Information now has value because... it allows other investigators to produce information which facilitates the return of invested capital. There is a demand from investors for information which may increase the power of their inscription devices, and there is a supply of information from other investors. The forces of supply and demand create the value of the commodity.[3]

To continue, a theory that can be banked on my others becomes part of the architecture of the field. Eventually, it becomes what he calls a black box, something that is no longer even questioned.

FEAST

But why do you say AIDS compromises his position? After all, he is giving an accurate view of establishment science.

ARONOWITZ

For all Latour's élan and iconoclasm, he never comes near fathoming the depravity of modern science.

NULL

Strong word.

ARONOWITZ

Perhaps duplicity would be better. For Latour every scientific article is read avidly to see if it can provide the reader a return on his or her investment (i.e., money spent on acquiring credentials and so on). But when Latour says this, I wonder if he is willing to face the swaggering mountebankism that Epstein recounts. You'll remember that that author analyzed citations of Dr. Gallo's original paper that connected HIV and AIDS. In that paper, Dr. Gallo says HIV may be the cause of AIDS but that so far the question of causation is undecided.

Gradually, year after year, more and more citations attribute to the article the opinion that HIV equals AIDS. This becomes one of the major building blocks of the establishment worldview.

This is how a Latourian approach ultimately collides with the Popperian perspective from which it seemed to spring. Here a theory is widely embraced that has not only never been put to any tests to verify it, but has never even been asserted in the first place.

NULL

But there have been attempts to ground this equation.

ARONOWITZ

Yes, after the fact. But the founding document, the type of accepted fact that Latour claims others can use in their investments, the founding fathers' constitution of AIDS science, doesn't exist.

But let's move on to Kuhn. As you've remarked, Kuhn sees a period's establishment science as part of one worldview. This viewpoint will only come under fire in cases where another, alternative view, put forward by a large group of dissident scientists, is ready to replace it. Sustained challenge to orthodoxy only takes place when an alternative network is waiting in the wings.

Now, Latour's opinion is not congruent with this, because he sees the progress of science as the result of the efforts of individualistic entrepreneurs: Diesel, Pasteur, Edison. If one scientist, working virtually by himself, is both an inventive thinker and a good organizer who can convince others to fall in with his program, he can overturn previous views.

Yet, it seems to me, if we abandon one of postmodernism's most beloved fetishes, these views can be brought into line in a way that will help us understand the AIDS conflict.

NULL

Explain.

ARONOWITZ

A key paradox of postmodernism was touched on by de Certeau, who said, in effect, Michel Foucault (leading postmodern thinker) proclaimed the death of the human being as an object of interest to science and received a chair at the College de France. But if the individual is so meaningless, why does Michel seek to heap honors upon his own insignificant example of this individuality?

NULL

In my book I refer to this type as a Dynamic Aggressor. "In the end," I write, "most Dynamic Aggressives define success by the amount of power, control, and money they amass."[4]

FEAST

Very well-said. But, what is the effect of all this celebrity on the writing itself of these postmodern egoists?

ARONOWITZ

Take Latour. His overvaluation of this aspect forces him to overload his project with unnecessary encumbrances.

Let me illustrate by talking about *The Pasteurization of France*. Here he brilliantly explains how through logrolling, back-scratching and finagling, Pasteur got his vaccination program accepted in both agriculture and science, making each group see its own goals as obtainable only by way of a detour through Pasteur's distillery. Yet, Latour sabotages his own formulations by constantly throwing in one out-of-keeping element that if abandoned, would make his whole Pasteurization infinitely better..

NULL

What is that unnecessary element?

ARONOWITZ

Pasteur.

NULL

Come again.

ARONOWITZ

Let me put it like this. Pasteur doesn't go about accumulating allies by interesting one farmer, then one biologist, then one public-health authority, and so on. If he (and his organization) had to enroll cross-adherents (those whose purposes are made to seem to cross through Pasteur's lab) one at a time, he would never finish and never succeed.

Rather, there are two parallel tracks. First, Pasteur is addressing societies, chambers of commerce, and such. Now, these organizations already have a sprinkling of Pasteur's disciples among them. After the great man's lecture, they fan out through the group and work to move it, particle by particle, into the cross-adherent camp. Second, Pasteur is only effective as the head of a party, that is, an association of like-minded scientific workers who propagandize for his ideas. Their work is also crucial to clenching deals with other parties in society, who are being pushed to accept his methods and products.

Latour cannot capture any of this. He cannot see into the group actions, which are, in fact, the real means by which a scientific breakthrough is spread. This is why Kuhn, who does depict science as a collective creation, is so much more farseeing than Latour.

NULL

We discussed the importance of viewing activities in terms of group projects when we discussed the fast-track lifestyle.

ARONOWITZ

Let's move onto that topic, since I have a bone to pick with your characterization.

Toby Marotta

NULL

Pick away.

ARONOWITZ

You argue that the more promiscuous gay males who frequented baths, discos, and so on in the seventies were kind of a compromise formation. While differing from homosexuals of the previous period in being more openly gay—at least in these venues, and thus (to a degree) eschewing the earlier Mattachine-type conception, which felt gays would eventually blend into straight society without developing separate institutions—they still did partake of the earlier, integrationist perspective in that they adopted many of the excessive traits of the (straight) consumer society.

I think this idea could be criticized on a couple of fronts.

First, let's remember that the perking up of gay nightlife in the seventies, though it couldn't have happened without Stonewall and the wider acceptance, or, at least, prominence of people living an openly gay lifestyle, was not an aberrant occurrence among a subgroup but part of a general social loosening of sexuality.

In Charles Kaiser's *The Gay Metropolis*, for example, he writes about New York, "The effects of the sexual revolution were hardly confined to the homosexual community in Manhattan. In the seventies, there was an explosion of massage parlors, thinly disguised brothels where scantily clad women satisfied their male customers."[5] That there was a new promiscuity in straight as well as gay culture is hardly arguable.

NULL

One might turn to your own writings to deduce that the depoliticization among young gays in the decade after Stonewall was also part of another trend, a general souring with the voting game. As you put it:

> Just as the candidates in the [1972] election were perceived as fundamentally similar, young people suspect that all politicians are alike.... Beneath these judgments is the sense that to be "apolitical" is the proper response to the charade of national elections. Here the term apolitical should not be understood as apathy .. [but as] contempt for the political forum which seems so completely monopolized by political and cultural figures devoted to the conventional wisdom. Young workers and students... are more conscious of the reality of corporate domination of American life than ever before . . , [but are no longer] convinced that small measures can alter the situation.[6]

ARONOWITZ

What this amounts to is that clone culture was part of a generalized fast-track lifestyle embraced by both straights and gays.

The second point I want to note is that the baths and bars were not simply places for round-the-clock orgies and nonstop libidinal pursuits. All the anecdotal evidence we have, which is considerable, would suggest these venues were forums for many types of behavior. To take a sample account, look at Dennis Altman's description of a bath.

> Men go to the baths to fuck. Or so it seems. In reality most of us go for all sorts of more complicated reasons—because we are lonely or bored, or restless, because we are looking for him, because we want to see and feel other bodies and/or be seen and admired.[7]

FEAST

And such sociological essays could be complemented by fiction.

NULL

Say on.

FEAST

Take the description of a visit to the baths by David B. Friedman in *Eighty-Sixed*. The hero is looking, as Altman characterizes it, "for him," passing up many overtures until he finds a man that fulfills his ideal. After this one tryst, he is satisfied and goes home. And note, this novel has been praised for its authenticity.

I'm trying to say that not everyone went to the baths for a free-for-all in the orgy room. Moreover, as Friedman explains, the baths were as much a place of waiting as dallying.

> The baths were no different from the bars—a lot of waiting.... Everyone was looking for someone more attractive than himself. Everyone thought he was more attractive than he was in fact.[8]

ARONOWITZ

And there's yet a third objection I have to your explanations.

NULL

Shoot.

ARONOWITZ

Drawing from Cindy Patton, you characterize gay male culture as divided between militants (gay activists fighting for an end to prejudice) and hedonists (bar-hopping clones). Now, for one thing, if you really delve into the literature, you'll see that this distinction between politicized and apolitical is not an accepted one.

What one finds in most discussions is that gay male culture in the United States in the seventies was divided between those who wanted to engage in political activity and those who thought the best way to change society was through a longer term strategy of cultural revolution. By this latter method, its proponents argued, a gay subculture was created, as epitomized in New York City, for example, by such institutions as the Gay Activist Alliance's "Firehouse." This was a former fire station in Soho, leased in 1971, which quickly became a community center, film society, and dance hall as well as host to discussion, therapy, and organizing group meetings.

From this vital subculture, it was thought, would arise men who were proud of their heritage and open about their orientations, and this would lead to a more tolerant society. Perhaps, unfortunately, this originally political view of culture tended to make its holders give up on group activities. As Marotta puts it in *The Politics of Homosexuality*,

> By defining self actualization and social validation as touchstones of liberation... the first gay and lesbian liberationists discouraged homosexuals from enduring conventional political and organizational activity that was constraining and encouraged them to indulge in more pleasant pastimes and to move on to more satisfying institutions.[9]

By this light the clone culture was originally as political in origin as was that of the more reform-minded politicos, and so its supposed apolitical nature is being overemphasized.

Sir Francis Bacon

FEAST

Your remarks make me think about what was said about Bacon. Just as you've shown that one must trace a particular development (such as the fast-track culture) out of a social matrix, we should, conversely, see individual thinkers as part of a broad milieu.

NULL

There's a related point about milieus that I want to throw in. Thinkers are only honored if they have had some effect on future developments. Thus, to understand Bacon's significance, we should look not only at what he said but at how his ideas were picked up later.

Historians of science note that Bacon was the first to understand the possibilities of harnessing science for human benefit, both by going about research systematically and by having the state organize study. But, hindered by skepticism and factions, Bacon himself could do nothing to embody his ideas. They only entered the real world after his death when adopted by his Puritan followers. The Puritans came to state power when their army beat the king's troops in the English civil war. Once this faction among the Puritans could get near the levers of power, they could begin to experiment with Bacon's recommendations.

FEAST

So, Bacon's ideas had lived on in his writings?

NULL

Undoubtedly. Historian Charles Webster writes that Bacon's book *Instauratio Magna* came to be regarded by Beale's generation [the Puritans of the early

1600s] as the authentic guide to intellectual regeneration. Accordingly, the fragmentary philosophical system bequeathed by Sir Francis Bacon became for Puritan intellectuals both the basis for their conception of philosophical progress and the framework for their utopian social planning.[10]

> These activists convinced the new government to develop smelting scientifically, carry out a geographical survey of Ireland (under William Petty), and move toward a more efficient way of minting money, among other innovations in which science was productively applied to industry and commerce.
>
> An important thing to realize, though, is that a gigantic change in the social structure—when the Puritans came to power the king was beheaded and a republic was established—allowed the new Paracelsian medical paradigm (based on chemical analysis) to overthrow the moribund but well-entrenched Galenian one (based on the humor theory of medicine).

The sudden surge of Paracelsian medicine and philosophy was intimately related to the social and religious trends of the Puritan Revolution. By 1640 an influential audience existed which was intrinsically suspicious of classical authorities and receptive to new approaches.... It was realized that the established church and traditional medicine had not in their centuries of dominance succeeded in ministering to man's spiritual and physical needs.[11]

ARONOWITZ

And I might remark, to tie this into something Dr. Null has often noted, there seems to be a natural connection between progressive views on health and a positive, nonmaterialistic way of living. So his followers showed the developer of the new paradigm, Paracelsus, to have castigated decadent scholastic medicine, sacrificing fortune for a wandering existence in search of true knowledge; he gathered his information from whatever source seemed promising; he communicated both with humanist scholars and the illiterate peasantry.... Private gain had been scorned, and the sick poor were treated without charge.[12]

But before this conversation turns into an all-nighter, let me halt it. We will return to these thoughts about Sir Francis Bacon in the conclusion of the book. Both these and other interesting ideas that have been churned up in this invented but not unrealistic discussion deserve further comment. Nonetheless, I don't want to summarize them here but, rather, go over why we discussed them.

I said the point of creating this dialogue was to uncover a dimension of collaboration. Of course, a book does not make it onto the page this way, by meandering conversations, but rather through drafts and redrafts exchanged back and forth. But a conversation serves to unlock certain crucial aspects of composition.

A dialogue serves to reveal a human situation in its many-sidedness, while exposing any decided opinion about that situation as necessarily tenuous. Thus, when our earlier ideas about fast-track culture in the 1970s were taken up in our free-wheeling colloquy, it was seen that some sides of the question had been overlooked, such as the generalized—homo and hetero—promiscuity among select social groups in the period. It was also seen that there might have been flaws in our very formulation of the question, where we saw a hard divide between politicos and hedonists. The division might instead more properly be seen as between two ways of doing politics.

As we will take up more fully later, one prime problem with the whole AIDS dilemma is a violent lapse from the fundamentals of conducting an honorable argument, that is, one in which each side gives the other credit for good sense and takes heed of its opponent's position, modifying its own stance in line with rational objections.

I personally believe part of the reason for the incapacity of those in the establishment (and sometimes those in the alternative health movement) to enter into good-faith argumentation is psychological. There is an intense desire to hold onto one's individuality as somehow a center of everything. Thus, to use this as an illustration (but not to assert that what I am going to say is accurate), Dr. Gallo will so tenaciously hold to the HIV thesis, not only because—as we noted—it would interest pharmaceutical firms, but because he has somehow connected this belief to what he takes as the deepest part of his self. To doubt HIV = AIDS is to doubt his ego. But, as our imaginary dialogue begins to indicate, ideas are threshed out in a group context. No single person is responsible for any of them, though one person may take credit for a certain personal formulation of them.

Further examination of this issue will be held for the conclusion; a discussion of what treatments have been put forward by the orthodox and the dissidents as means of overcoming or mitigating the effects of AIDS is long past due.

PART FOUR

CONVENTIONAL TREATMENT PROGRAMS

Failure of Conventional AIDS Treatment

O nce virologists' views had won the day, at least in the large camp of the establishment, their understanding shaped the search for treatments. They felt the key thing was to balk the virus, whether by interfering with its reproduction (with the protease inhibitors) or by devising a vaccine that could be used to inoculate people with a weakened form of HIV. Under this regime, regimens such as those recommended by immunologists, who spoke out for strengthening natural immunity, were rudely shoved out of the foreground.

Perhaps it would not be out of place—though this has been a theme constantly harkened to in these pages—to mention that the general agreement within the mass media, and among granting agencies and establishment science programs, that the virologists' understanding of AIDS was the proper one does not only mean that funding and encouragement will flow to treatment projects appropriate to this reading of the disease, but that other options will be actively discouraged. Before offering a critical look at vaccine programs and, in a more extended way, AZT, we might once again dip into the subject of dogmatism's restricting of options.

Restricted Options

We might surmise that virology beat out immunology in the AIDS sweepstakes, not only because, as Patton urged, it held greater promise of "potential rewards," but because its monocausal premises were deeply in tune with the older paradigm, still dominant in medicine, while immunology was for opening a conversation with a newer viewpoint. In this newer outlook, health and disease were the products of multiple factors, coming from the environment and the body's inner chemistry, as well as from the more traditional bugs. Even when a virus, say, entered the body, it could not have things all its way. Whether it was effective depended on the atmosphere pertaining to the individual, not on its own viral nature.

When I talked to Dean Black, who counts himself a defender of the new paradigm, he argued that a "workable recovery program involves methods that are unacceptable to traditional medicine:"[1] The emergent paradigm, which sees the individual as (within limits) negotiating his or her own health, thus turns to lifestyle as one aspect that must be evaluated in disease formation. Black continued that from this position, any treatment program must include measures that touch upon daily living. "This opens up the door to all the lifestyle measures that medicine has so long discounted… that medicine cannot accept because it is not based upon their model and doesn't use their drugs or rely on their expertise."[2]

Martin Walker went further than Black in pointing a finger at how medical authorities and their business partners expressed their displeasure at the non-conventional. He told me, "I think one could easily say that there's been a campaign by orthodox medicine and by the medical establishment against alternative and complementary medicine in this country."[3]

The latest episode in this campaign came in 1988, soon after the licensing of AZT, when the drug's manufacturer Wellcome helped in "setting up an organization which was called the Campaign Against Health Fraud," whose mission was to discredit "people involved in alternative and complementary medicine and the field of AIDS."[4] Having seen the methods employed by the establishment to bash dissenters, it will come as no shock to readers that the anti-health fraud zealots hardly had time to spend on disproving alternative practitioners' claims, but, according to Walker, concentrated on such items as character assassination and "trying to get them struck off from professional organizations."[5]

I asked Prof. Philpott, who we've met before, how it is that so many doctors still adhere to this older, increasingly questionable paradigm, and he turned my query to a larger area. The problem is not so much that scientists and physicians are locked into one perspective—since, after all, adopting a point of view is a prerequisite to processing information—but that they enter their careers with a smug, closed-minded approach. He attributes this typical mind-set to early training.

At the universities, we are supposedly looking at a problem and trying to examine every possible explanation…. In fact, particularly in biology, what happens is the professor writes things on the chalkboard. What he writes is basically chiseled into the board. It is not to be questioned except for "Well, I don't understand that. Please explain it more clearly, Dr. So-and-So." But it's never, "Excuse me Dr. So-and-So. That doesn't seem right to me at all."

It's not only the professors who enforce this sort of censorship but fellow classmates. They become angry and upset at you if you slow down the class because they want to know what they have to know or rather what they have to memorize on a test.... There is no examination in the laboratory or in the classroom of what is really going on. It's just memorizing of the dogma.[6]

Once a physician, for instance, graduates and completes his or her internship, he is likely to take up practice with a conservative attitude, being partial to establishment views and disdainful of those expressed by dissenters. However, what if he or she is among the lucky few who have managed to keep an open, flexible approach? Is the physician then likely, when faced with an illness that doesn't respond to accepted treatments, to move onto treatments that fall beyond the pale?

In a discussion with Bill McCreary, I found that there are many pitfalls that line the path of a doctor who decides to venture onto the unhallowed ground of alternative treatments. For one thing, such a doctor isn't covered, in case of failure, by the excuse that he has been taking recommended steps. The doctor may hear good things about a drug used in alternative therapy but be afraid to prescribe it. So often "doctors say it's not approved by the FDA so we can't touch it."[7] McCreary recognizes the self-preservation tactics behind such timidity but gets fired up when he thinks of the implications. "Doctors usually do a better job because they're there, they know the patient." However, this cuts no ice with the current setup. "Today, the doctor works from a cookbook that is issued out of federal government. [Doctors] have to abide by that cookbook or they may be found guilty of malpractice."[8]

In a complementary way, a patient will not be likely to direct his or her doctor's attention to our hypothetical new drug, because, as McCreary went on to explain, people generally do not want to take responsibility for their own health and are therefore willing to accept whatever cure is sanctioned by the medical establishment.[9] In his words, "People turn their lives over to someone else.... [They] don't want responsibility.[10]

Thus, the accepted HIV causation theory and the treatments that flow from it are hedged around with multiple barriers to keep people from getting too close and questioning them, just as many fences keep visitors from touching the polar bear. I detest zoos, by the way, since I think polar bears should be allowed to live in the Arctic and not serve as spectacles for sated tourists. In the same way, I think the HIV thesis should be allowed to manfully face its critics, rather than being hidden in the holy of holies behind establishment-woven veils.

As long as this thesis and others are held to be unquestionable, we get, as Lord Baldwin, chairman in Britain's Parliament of the Group for Alternative and Complementary Medicine, outlines, not science but travesty. We hear the hype but don't look at a treatment's actual results. Baldwin explains, "I'm a great believer in looking at people's track record. The track record of the drug approach to AIDS is absolutely abysmal."[11] He compares the outlay here to that on cancer, which also, he argues, is not very good.

"You look at people who've done hard epidemiological work on this [studying cancer treatments' successfulness] and they say that the battle against cancer is a, quote, qualified failure."[12] But billions have been poured into cancer research. He concludes, speaking of both diseases, "There is something wrong with a society that abdicates so much power to scientists who are going round in circles. If you look at the track record of these things it is very very bad. Common sense says it's time to look in other directions."[13]

But we don't have to take Lord Baldwin's word for the "abysmal" state of AIDS treatments, we can examine them ourselves.

AIDS
Vaccine

From the start the search for an AIDS vaccine has been ill-starred, for not only are there special difficulties to making an inoculation against a retrovirus (which has never been done before), but companies supporting the effort quickly got cold feet when they didn't see profits on the horizon, while the drug trials that were conducted seemed plagued with problems. Let's begin by looking at what hurdles the creators would have to leap in the steeplechase to find an AIDS vaccine.

Obstacles to an HIV Vaccine

Assuming for the sake of argument both that there is an HIV retrovirus and that it has the traits ascribed to it by true believers, then the retrovirus's volatility, mode of action, and latency period all put particular obstacles in the way of those trying to create a vaccine. One problem that has already been noticed in these pages is that HIV is said to have a high mutation rate. This is vexing for the vaccine hunters for obvious reasons. As L. Pascal summarizes the situation, "New varieties of AIDS have been entering the race at a rate of one every year or two or three. Each of these varieties will almost surely require a separate vaccine."[1]

Even more problematic is the fact that more than one variety of HIV is found in each individual. "The putative virus is alleged to mutate so rapidly that a single individual is said to produce many different strains of virus simultaneously,"

Michael Verney Elliot notes.[2] This raises the question: "Which one do you vaccinate against? How could any vaccine protect a constantly moving target?"[3]

Another problem is that to date, effective vaccines have been made only against straightforward viruses, not retroviruses. DNA viruses, as will be remembered, simply move in and get the cell machinery working for them, while retroviruses, such as the AIDS virus, actually integrate their own DNA into the cells of humans. This means that if anything goes wrong with the vaccine and a strong form of HIV is given in the inoculation—we've explained that vaccines are made from weakened versions of the virus or from benign cousins of the original—then it would be well-nigh impossible to turn back. This is why pharmaceutical companies are nervous about vaccines' possible misfires. As Cynthia Cournoyer comments, "Potential manufacturers are extremely concerned about liability. California is under pressure from them to formulate policy on compensating victims who could be damaged by a new vaccine."[4] Cournoyer feels that because of the unpredictability inherent in developing a type of vaccine never obtained before, "the likelihood is quite high that the vaccine will become the most dangerous we have ever seen."[5]

Moreover, the long latency period between infection and disease expression adds another exceptional difficulty. Since, as Verney Elliot states, "the average length of asymptomatic infection [is] some ten years," testing of the vaccine will be long extended.[6] Although the need for this vaccine is desperate, given the latency period, "any trial of 'anti-HIV' vaccine must run for at least ten years to [determine] whether the vaccine will prevent disease after infection with the alleged viral cause of AIDS."[7]

Two other difficulties facing vaccine makers also come to mind here. "Unlike other viral diseases for which successful vaccines have been made, recovery from HIV infection has not been documented." This doesn't mean we can't find HIV positives who never succumb to AIDS—we already have—but that we don't know of people who once they get full AIDS, throw it off and return to health, as a small percentage could always throw off smallpox and other infectious diseases. But, as the *NIAID Fact Sheet* makes clear, without such a model, making a vaccine is much harder. "Researchers have no human model of protection to guide them when constructing candidate HIV vaccines. Indeed, whether a natural protective state against HIV can exist remains unknown."[8]

Secondly, and a bit more technically, a vaccine has to go against the invading virus where it hurts, boosting immune system strength in those areas where the virus first finds a home. Accepting that HIV is sexually transmitted—remember, we are following the beliefs of the would-be vaccine inventors—then "an HIV vaccine also may need to stimulate mucosal immunity," so that mucosal immune cells would coat the digestive and reproductive tracts and would be "the first line of defense against infectious organisms" that are sexually transmitted. "Unfortunately," the fact sheet notes, "relatively little is known about how the mucosal immune system protects against viral infection."[9] This means there is yet another missing piece to the vaccine puzzle, which must deal with long latency, volatility, lack of a human model, the mystery of retroviruses, and now this.

Economic Feasibility

With so many difficulties facing those who embark on the endeavor of finding a vaccine, it stands to reason that many would not be able to stay the course, and in the last few years the number of companies working on the prospect have narrowed. However, it is not the arduousness of the task alone that has daunted many, but also their realization that, even if a vaccine was found overnight, it may not give its proud owner the bonanza profits that most firms expect.

In a comprehensive article in the *Village Voice*, M. Schoofs summarized the situation in the mid 1990s:

> Early on, at least four large pharmaceutical companies and a bevy of biotechnology firms threw their hats in the ring, in part because the hepatitis B vaccine proved that the safer and supposedly cheaper recombination technology could work. But as an AIDS vaccine has proven more stubborn, the private sector has fled.
>
> Of the big companies, only Pasteur Merieux—the world's preeminent vaccine producer—has made a substantial effort. Private funding has completely dried up for the effort at Therion Biologics.... Therion's program, now half its former size, subsists solely on U.S. government research funds. Three years ago, Bristol-Meyers Squibb substantially cut back its program.... Last year [1994], Gerentech and Biocine... slashed their programs by more than 80 percent, after a controversial decision by a federal scientific panel nixed large human efficacy trials of their vaccines. The two companies had each spent between $50 million and $100 million shepherding their products, both based on a viral envelope protein called gp120, through Phase I and II safety trials. Last year was also when Merck and its smaller partner suspended work on a product they finally concluded wouldn't work, after spending perhaps $30 million.[10]

Three years later, science seemed no closer. According to an Associated Press dispatch, in a press conference Dr. David Baltimore, head of the United States's AIDS Vaccine Research Committee, "predicted Sunday [February 1, 1998] that a safe vaccine to prevent AIDS could still be more than a decade away from testing."[11] Moreover, at this time the danger that such a vaccine might expose people to was reiterated. A critic noted that "the experiment could cause the volunteers to become infected with AIDS, risking their health and possibly their lives." The group International Association of Physicians in AIDS Care, who are arguing for the immediate testing of vaccines, said they had "lined up three pharmaceutical companies to provide free AIDS-fighting drugs should volunteers [who are vaccinated] contract the disease."[12] Not a very comforting suggestion for anyone who might want to volunteer, given the quite circumscribed effectiveness of these "AIDS-fighting drugs," which at best add a few years to the lifespan.

Thus, as there has been minimal progress in developing a vaccine—below, when discussing vaccine trials on monkeys, we will find that the same pessimistic view was being pronounced (about the nearness of vaccines) in July 2000 at the time of this writing!—companies have continued to be reluctant to sponsor work. As Dennis Panicali, who heads Therion, the company that we saw has had to rely on government support, complains, "The investment community... has become increasingly disillusioned about AIDS vaccines."[13] AIDS vaccine researcher Dani Bolognesi of Duke University backs him up: "The perception out there is this stuff is so difficult and the likelihood of anything working is so slim that it's not worth it."[14]

The just-mentioned article contrasted worldwide pharmaceutical business investment in vaccine research, which added up to under $25 million, with the U.S. government's contribution of $111 million to show that if one had to rely on corporations to do the job, one might as well throw in the towel.

And businesses' cold feet haven't been warmed as it becomes clearer that the vaccine market, though substantial, won't be as large as first predicted. A report in *Science* noted that "the reason pharmaceutical companies aren't pouring dollars and energy into AIDS vaccines the way they would into a hot new mood-elevating drug is simple: The AIDS vaccine market in developed countries is likely to be much smaller than early estimates indicated."[15] Not only don't the people affected in Africa nor their governments have the funds to pay for expensive vaccines, but by summer 2000 they had begun protesting about price-gouging on AIDS drugs, causing a truckload of bad press for the pharmaceutical vampires, who the Africans were saying are eager to use African bodies as vaccine-trial guinea pigs, but who they then don't want to deal with when these same bodies appear as charity cases needing low-priced drugs.

Again, this has been taken as a signal for the legal drug lords to withdraw from costly research into this field, and few companies remain interested, leaving the best research to take place in university laboratories under government sponsorship.

Promising Venues

I don't mean, though, that the search for this particular holy grail has been abandoned. There are still doughty knights on the prowl, ambling down various scientific highways. Let's look at two of the paths chosen by vaccine researchers: one on which trials have recently begun, and another that has lately shown disastrous consequences.

Probably the most hopefully awaited vaccines are those being developed by VaxGen. One drug, Aidsvax, started a three-year trial in June 1998; another of VaxGen's vaccine, based on the same principle, was approved by the FDA for testing in mid-2000. The Aidsvax study would "include 5,000 U.S. volunteers and 2,500 volunteers in Thailand."[16]

VaxGen is "headed by virologist Donald Francis, who successfully battled the Ebola virus in the Sudan and smallpox in India," and the company had been a division of Gerentech Inc. until spun off in 1996.[17] Gerentech cut it loose because of problems with the gp 120—based AIDS vaccines on which the division had focused.

Glycoprotein 120 (gp 120), which is found on HIV's shell, is in the region where it binds to CD4 cells. Studies of HIV antibodies show that they are generally aimed at blocking this area on the virus. For this reason, according to the *NIAID Fact Sheet*, "vaccines based on genetically engineered HIV envelope proteins—gp160 and on its cleavage products, gp120—have been the most well studied to date."[18] However, (as Schoofs noted) Gerentech saw little progress in its work on gp 120 vaccines and "shelved the product after spending $50 million trying to develop it and after FDA funding was denied for further testing amid doubts of its effectiveness."[19] Still, scientists continue in this direction—walking the same circle, Baltimore might say—in their attempts at vaccine development.

What has, until lately, seemed like a more promising avenue has been working with simian immunodeficiency virus, hoping that as in the case of smallpox, a variant of a virus found in animals can be found to be beneficial to humans. Rosy reports of this strategy came in 1995. "Recently, encouraging results from AIDS vaccine studies in animals.... Vaccines made from... (SIV) have succeeded in protecting the monkeys from developing simian AIDS."[20] However, if we roll the story forward to the present (2000), we will see that this early hope was brutally dashed.

Problems with Vaccine Trials

On July 2, 2000, an Associated Press report on work with SIV opened with the lead, "A weakened virus that many consider the best hope for an AIDS vaccine suffered a serious setback Thursday when tests in adult monkeys showed *it may actually cause the disease it was meant to prevent*."[21] This is the type of vaccine that researchers have said offers the best chance for obtaining results quickly. They were understandably crestfallen. Gallo expressed this by saying, "If the live attenuated approach fails, it means *AIDS vaccine development is no further ahead than it was in 1984*, when little was known about the epidemic's cause."[22] Moreover, this wasn't the first setback for this approach. In 1995, Dr. Ruth Ruprecht of Dana Farber Cancer Institute had shown that a weakened SIV would eventually mutate into the virulent version in baby monkeys. However, until these latest results appeared, it was thought that the same reaction, of the SIV mutating into a more powerful version, would not take place in adults.

This is only the latest vaccine trial to spring leaks. Those that are coming at the problem from the gp 160/120 end also have been fruitless so far. As an article in *Continuum* notes, "a five-year test for an HIV vaccine which cost the U.S. Defense Department $20 million has ended in failure. The vaccine, based on the supposed HIV envelope protein gp160, had no effect on the 304 allegedly infected volunteers studied between 1990-1995."[23] Other work on gp 120—as noted a moment ago, gp 120 is a derivative of gp 160—this done by virologists Steven Wolinksy and David Ho, reported equally bleak results in the *Journal of Virology*. Wolinksy concluded about his own study: "The bottom line is that we had neither beneficial nor adverse effects in any of the individuals we studied. The results are very disappointing." Anthony Fauci, director of NIAID (the National Institute of Allergy and Infectious Diseases whose fact sheet we have been studying), after examining these findings, said, "This fortifies the decision I made three years ago not to fund gp120 vaccine trials."[24]

Even the best trials have shown only spotty results. The government's AVEU trials, for example, could not sustain even a moderate level of antibodies in the blood. Keep in mind that , as mentioned, the point of a vaccine is to stimulate antibodies to the virus that is to be counteracted. These antibodies must remain in the blood to steal a march on any of the infectious virus that happens to get into the body.

NAIAD described its own experience with vaccine trials conducted between 1988 and 1994 on more than fifteen hundred men and women, run out of "six medical center AVEU sites located in Baltimore, Nashville, Seattle, St. Louis, Birmingham and Rochester, NY." On the good side, the agency reported, "To date, all vaccine candidates tested have been well-tolerated, generally producing only mild side effects typical of most vaccines." However, overshadowing this was the bad side, "The first candidates tested stimulated production of antibodies, although levels decreased within a relatively short period of time."[25] Thus the vaccine wore off rather quickly. Moreover, "Initial formulations and dosages of these vaccines produced few or low levels of neutralizing antibodies and rarely elicited cytotoxic T cells, which are invoked through cell-mediated immunity to kill HIV-infected cells."[26] Perhaps with the reasonable fear that too strong a vaccine-used virus might turn into HIV, researchers used viruses that proved too mild to greatly stimulate the immune system.

However, with a weak vaccine at least the inoculated stay healthy. With a stronger one, that is, one using a stronger version of HIV to bring out the antibodies, a much more terrifying result might take place, and, in fact, has. In 1994, "researchers were forced to acknowledge that 'at least five volunteers in the government's principal AIDS vaccine study have become infected with the AIDS virus after receiving the vaccines.'"[27] In other words, as happened with the monkeys receiving the simian version of an immune deficiency virus, here it may be that the inoculated, weakened virus turned into the real thing. I say "may" because the volunteers simply may have become infected along a usual route; although this would show the vaccine was impotent. In their own remarks on what happened, the researchers raised the possibility not that the vaccine had mutated but that it had paved the way for HIV. The results "raised researchers' concerns 'not only about how well the vaccine works but whether it may have increased the likelihood of their infection and... even accelerated the progression of disease.'"[28]

Criticisms: Historical, Theoretical, Economic

Having witnessed the repeated failures with which those scientists devising vaccines have been hit, skeptics have been questioning many of the premises on which vaccine production as a viable option have been based.

For one thing, as Verney Elliot had highlighted, some dissenters are beginning to look askance of the records of earlier vaccines. As you will remember, we have already made clear that the supposed original triumphs of the virus hunters against smallpox and tuberculosis were nothing of the sort. That is— and this bears repeating, since the virologists are constantly waving around this false claim—these two diseases, and others that were devastating in the nineteenth century, were already on the wane when vaccines and drugs were found to control them. They had been beaten back by better sanitary methods. The doctors had about as much right to claim they wiped out these diseases as Gallo has exclusive title to finding HIV.

Professor Gordon Stewart, who has been studying the history of diseases, notes that the conquest of polio via vaccine, which we mentioned as the virus hunters' last great victory, also is not as clear-cut a triumph as it seems. Verney Elliot reports his findings:

> In the USA during the late forties, there was a noticeable increase in polio cases. This prompted the authorities to pay a bounty of $25 to GP's reporting any suspected case of polio, treating it as a notifiable disease. The numbers of cases of polio shot up, causing a national panic. Any stiff neck or slight limp was reported. Curiously, at the same time, the official number of cases of asceptic meningitis, which shares some symptoms with polio, and previously reached some 25,000 annually nationwide, disappeared completely. A whole disease just vanished. Subsequently, when the polio epidemic had abated... the numbers of meningitis cases returned to their previous level.[29]

Applying the lesson to the present situation, we can say that vaccines are not all they are stacked up to be, so even if an HIV vaccine were found—and that's going along with the idea, for the sake of the argument, that HIV is key to the disease—it might not be that earth-shattering in terms of stemming AIDS.

Funny, I just alluded to Sabin, one of the developers of the polio vaccine, because he has made himself heard also in this latest controversy on vaccines, but now he is plumping for the dissenters and saying a vaccine won't work! As a *Village Voice* reporter opens a story on this eminence grise of virology, "At the age of 86, microbiologist Albert Sabin could be expected to rest quietly on his laurels, contemplating the glass case in his parlor filled with dozens of medals bestowed on him for developing the oral polio vaccine."[30] Rather than basking, however, he "has emerged from retirement to deliver a harsh critique of AIDS vaccine research."[31] His reproof is not that vaccine research is barking up the wrong tree, but rather that on this particular tree, the vaccine makers have not even reached the first branch. Less metaphorically, he feels they are jumping the gun by already doing trials on humans with an insufficiently developed drug.

Verney Elliot elaborates a similar position, arguing that even if one believed in HIV as the chief causal agent—he doesn't—one would be blocked from coming up with a vaccine by the same difficulties that, as we recounted, vexed those who were trying to identify HIV as a separate entity. He begins, "the theory of vaccinating against a disease requires that one must find its true cause, and if it is an organism like a virus or a bacterium, a vaccine perhaps may be constructed which will raise prophylactic antibodies to neutralise the bug."[32] But how is one to accomplish that if "the so-called AIDS-virus, HIV, has never been properly isolated, let alone shown to cause immunodeficiency?"[33]

As we established earlier, the most that has been found by orthodox scientists are pieces of HIV that have been "cobbled together" to present a theoretical (some would say "wishful") picture of what the retrovirus is imagined to be. But without a surer idea of it, the type that had been obtained for polio, for example, how can a workable vaccine be obtained? This is the hard question both Sabin and Verney Elliot ask.

The more cynical, or those who have taken to heart the lessons of Part Three, AIDS Inc., will reply that the race to begin trials is as much motivated by the desire to cash in on the small amount of government and pharmaceutical funds flowing into such research as by a desire to find anything. John Moore, of New York's Aaron Diamond AIDS Research Center, for example, criticizes the VaxGen United States-Thailand trials we have already alluded to as "a total waste of time and money."[34] In his reading, a trial may be embarked on, even before an adequate vaccine is perfected, because this is a way for its director—even if the vaccine is unsuccessful—to gather monies and prestige.

Boosting Immunity

Recently, we brought up Patton's thought that the triumph of the HIV equation also represented the victory of the virologists over the immunologists, the latter being purveyors of the view that AIDS is caused by the interaction of multiple factors, not a lonely, lowly virus. At the beginning the virologists seemed to have carried the day, but their failures in treatment have meant that for some the the issue reopens along the lines of this cleavage. Nina Ostrum puts it like this: "Some... scientists [disappointed with the lack of progress in vaccine development and other treatments] are even suggesting that HIV not be targeted at all in vaccine efforts; instead, they argue, the body's own immunity should be boosted."[35] You will recall that this was the treatment originally scouted by the immunologists, which was shelved when their position on the disease was beaten back in the face of the virologists' stance.

The return to this insight was launched in 1995 by NIH AIDS researcher Dr. Gene M. Shearer, who quickly found himself at odds with Gallo and others who harp on HIV's primary role in disease causation.[36] Research in this direction would also want to look at reported cases of natural immunity to AIDS. Schoofs reports that "a few people seem able to ward off the virus. Separate research teams have found prostitutes in Kenya and Gambia who have been reportedly exposed to HIV for years but have not become infected."[37] Similar findings have been noted in "exposed but uninfected gay men, drug users, health care workers, and babies born to HIV-positive mothers."[38] It would seem that these people either have extraordinarily capable immune systems or some unusual element in their systems that eliminates HIV or other damaging factors, and that a study of their physiology would yield answers to questions of how to boost immunity.

We will return to considerations of immune system enhancement when we turn to alternative treatments. Now, however, it's time to explore an even greater scandal and boondoggle in the establishment's attempts to treat AIDS. In fact, after reading the following discussion of AZT, current establishment treatment of choice, you may thank your lucky stars that a vaccine—were it to chart a course in any way parallel to that of AZT—was not developed and foisted on the public.

PART FIVE

THE AZT HOAX

T o date, AZT has been medical orthodoxy's primary approach to treating AIDS. Although other drugs exist, they are complementary to AZT both in that their modes of operation are similar and in that they are often prescribed to be taken in rotation with it.

AZT's action is not dissimilar to that of chemotherapy agents used to combat cancer. It goes in and destroys cells that are dividing. From our point of view, this is a doubly flawed tactic. On the one hand, inasmuch as I see AIDS as product of a synergy of interactive factors, with HIV being only a minor contributor to the illness, any strategy focused exclusively on eliminating the retrovirus, even if it were effective and low in side effects, would not be solving the problem. On the other hand, as I see immune system deterioration as a precursor to AIDS, then it would be worse than misguided to prescribe a drug like AZT, since the cells it targets (any that are dividing) include a good portion of those in the immune system. As we saw, immune cells are created at an extraordinary rate. So while killing HIV virus particles, AZT also rids the body of a good portion of the agents that the body has created to fight off such viruses.

Because of these flaws, whose effects have become increasingly evident (as we will document below), a number of authors have shot down AZT's pretensions with full barrels. John Lauritsen, for example, a man who, as we will see later, unearthed some startling facts about how drug trials proceeded, studied AZT research literature and reported the following:

There are supposed to be a hundred fifty studies showing benefits for AZT. I can't say I've read all of them, but I have read the major studies and I can say, without exception, it is not good research. AZT is an incredibly toxic drug, so toxic that it is incompatible with life.[1]

Duesberg, whose position is in line with our beliefs on the minor role of HIV, claims:

[AZT] is AIDS by prescription. Nothing could cause AIDS more directly than a substance that is doing only one thing, killing cells, particularly the cells in the bone marrow, which are the immune system. You are killing off your immune system every six hours with AZT.[2]

Geneticist Richard Strohman is in accord with Duesberg.

There's every reason to think that AZT will kill you faster from an AIDS-related disease than if you were taking AZT because it's a DNA chain terminator. It will kill every cell in your body that's making DNA.[3]

A word is necessary about outdated terminology. Both here and later AZT will occasionally be labeled as a DNA chain terminator. That has been the usage until recently, for it was assumed that, given AZT is a synthetic version of the DNA ingredient thymidine, once in a cell AZT would integrate itself into the DNA and screw up the works. However, Anthony Brink called my attention to recent work by Eleopulos et al., drawing on earlier studies by others, that has established the fallacy of this claim.[4] This work indicates that AZT cannot be functioning in the way supposed. Without giving too many details, we can say that, in Brink's summary, "AZT cannot take the place of thymidine triphospate [the form thymidine assumes in the DNA] in a growing proviral HIV DNA chain, and thereby terminate it, unless it has been triphosohorylated," that is, "converted by an intracellular enzyme into its active form."[5] The new work shows that AZT is not converted in sufficient amount to affect DNA. Therefore, calling AZT a "DNA cell terminator" is no longer accurate, although it will appear in this book in quoted material.

This change in understanding of AZT's effects should not be taken as letting the drug off the hook since its toxicity to cells, as we will see, has been extensively documented.

Let me say, further, that the opinions I am reporting about the downsides of AZT were expressed after the drug was used by PWAs for some years and so are profiting from hindsight. In fact, the early results of drug trials made the drug look much better than it eventually turned out to be. In fact, there was a kind of unholy synchronicity in this. The trials were ended prematurely, since the drug seemed to be startlingly effective against AIDS. However, it turned out these results were AZT's only strength, in that the drug stemmed the disease's progress early on, while its damning weakness was that these results did not last. AZT did not seem to increase overall longevity and even may have decreased it. Compelled by the public's demand for answers and other pressures, those involved in approving AZT were so eager to come up with a useful measure that they were led down a garden path by the first positive results they found.

But we are getting ahead of the story. Before listening to my claims, you need to hear my evidence, the same evidence that motivated the condemnations issued by Strohman, Duesberg, and the others quoted. I will begin by talking of AZT's prehistory as a cancer drug as well as looking at the flawed drug trials that were used to prove its viability. In my first discussion, I also want to say more about the pressures that disposed scientists to sometimes chime in as backers of AZT before enough was known about the substance. Then, we will examine how patients have fared who have gone the AZT route. Third, we will make a critical study of how AZT affects the immune system. Fourth, we will note how the media has continued to ignore problems with the drug. Fifth, we will manage a peek at how the drug company making AZT has attempted to buy off its critics. Sixth, we will criticize the drug cocktail treatment, which is an attempt to compensate for some of the disasters now being attributed to AZT by using other drugs whose value is just as dubious as that of AZT.

Marketing, Toxicity, and Motivation

In this chapter, I will proceed partly chronologically, partly topically. First, we will look into how AZT was discovered as a possible contributor to the war against cancer and then abandoned when it proved too toxic. I will look at the first trials (Phases I and II) of the drug as an AIDS combatant, which brought it speedy approval. Then, before looking at Phase III trials, which took place after AZT was being marketed, I want to provide some background as to why what in hindsight look like rather careless decisions were made to give AZT the stamp of approval. It's easy to fall into one of two mirroring traps when considering why historical events take place. For critics, a conspiracy is often detected. In such a reading of events, Burroughs-Wellcome was so hungry for profits, it rammed AZT through the approval process despite obvious shortcomings. Defenders fall back on haplessness as explanation. Things that should have been caught about the drug were innocently missed due to the fevered atmosphere surrounding the trials. I believe neither of these viewpoints is appropriate. We need to think a little more about what pressures and motivations lay behind the choice of this drug.

Development of AZT

Pharmaceutical companies often send their representatives, hat in hand, to government committees, where they justify the high prices they have slapped on their drugs. Typically, they say the big markup is needed because of the millions the company has spent on research and development. Whether there is any truth to what they say in general, it certainly doesn't apply to AZT, which is now generating megaprofits for the conglomerate Glaxo-Wellcome. (In referring to that company, we will call it by the name it had in most of the period under discussion, which is Burroughs-Wellcome.) This drug was originally created by government-funded research into cancer.

Its life as a leukemia fighter was brief in that it quickly came to be considered too dangerous for human use. It was strongly toxic and it may be (this point is under dispute) that it was found to cause cancer itself.

One of those early involved with the drug, Richard Beltz, is at least reported to have put this last claim as simply as, "AZT was shelved for two reasons: My studies showed that it caused cancer at any dose and it was too toxic even for short-term use."[1] However, he later said that this statement was falsely attributed to him. It's best, then, simply to say that the drug's toxicity put it out of the running as a cancer fighter.

To continue with AZT's background, we might note that work on the drug was initiated in the early 1960s by Jerome Horwitz at the Detroit Institute for Cancer Research, using funding from the National Cancer Institute. It was created as a weapon to turn against leukemia.

> Like other scientists, Horwitz had been screening other drugs to see if they held any value as cancer therapies. He became dissatisfied with that approach and soon turned to a more intellectually rigorous one—designing a compound specifically built to stop malignant cells from going beserk.[2]

AZT's method of operation was to "to destroy dividing cells which were producing tumors"; but it could not be kept selective. "AZT was… indiscriminately cytotoxic. It could kill any dividing cells by interfering with the reproduction of DNA."[3] That is, while it was hoped that the drug would only kill cancerous blood cells, it tended to wipe out anything that moved. To be a little more specific about its operation, AZT eliminated "growing lymphocytes by termination of DNA synthesis," but since it couldn't distinguish between cells that were growing naturally and those that were spurred to grow by cancer, it didn't seem much good.[4] It was taken as far as animal trials, which proved "AZT failed to prolong the lives of leukemic animals."[5] It was "never patented by its creator,… [and] never used for its intended purpose as a cancer chemotherapy."[6]

What about AZT's second shortcoming, its cancer-causing facet? Although more recent studies show this propensity in the DNA terminator, there is some debate over whether this was known early on. It is an important point in that Burroughs-Wellcome said it had no knowledge of this danger of AZT until after the drug had been distributed. If earlier studies had been done, the company could be charged with negligence and unconscionable ignorance.

John Lauritsen attempted to uncover the prehistory of knowledge of AZT's cancer-causing component, beginning his search by talking with Horwitz. "According to Horwitz, the drug failed to prolong the lives of leukemic animals," and this was the only reason it was dropped like a hot potato.[7] When Lauritsen zeroed in on the cancer issue, Horwitz backed away. In the reporter's words, "I then asked whether cancer had been observed in any of the animals. At this point Horwitz became quite defensive, and said he 'categorically denied' that cancer had been found."[8] This doesn't mean the animals didn't get cancer, but simply that they hadn't been tested for this illness. However, Lauritsen contrasts this to other reports, such as that from a colleague, "who has requested anonymity," and informed him "that Horwitz told him in an interview that AZT was abandoned because of its extreme toxicity... [and because] not only did AZT not cure cancer, it caused it!"[9]

I would say this lets Burroughs-Wellcome off the hook on this problem, since the evidence isn't clear either way, although it leaves standing the current finding that those who take AZT have an increased propensity for cancer.

AZT In Phase I and II Trials

Let's move on to AZT's recent history. Martin Walker moves the story forward, "After development of AZT was dropped, it became an orphan drug... and languished, on the shelves of the National Institutes of Health."[10] As an orphan drug, any company had the right to pick it up. In 1985 the agency decided it should be tested in relation to AIDS, but "not due to farsightedness or any sixth sense ... inside NIH research establishments everything which came to hand was being tested for antiviral qualities."[11]

What actually happened is this. The NIH offered to test any drugs sent in by drug companies for antiviral qualities. Sam Broder, NIH director, went around cajoling companies to use his facilities. At a meeting on October 5, 1984, "Broder spent a lot of time... convincing Burroughs-Wellcome that an effective drug [if it were found] had commercial potential, that a reasonable return on investment could be secured." All this simply to get them to submit their drugs to trials that would be paid for and carried out by the NIH! "There is money to be made in AIDS, Broder said to Burroughs-Wellcome. Perhaps big money."[12] BW did use the NIH service, and when AZT, which BW had sent in, looked promising, Broder was back begging the company to develop the drug and start testing trials. What was making Burroughs-Wellcome reluctant was that in 1985 it didn't see a big market among AIDS patients. There weren't enough of them. Its foot-dragging showed a noticeable lack of foresight, in that the drug would turn out to be the company's cash cow.

To highlight the dollar value of the drug, in relation to the trials, before moving on to the intricacies of these patient studies we might mention that in 1987 when the first AZT study (designated 002) was favorably finished, BW's corporate parent, British-based Wellcome Foundation, saw its stock jump "24 percent after reports that the drug would gain FDA approval."[13] Profits began to roll in as PWAs paid up to ten thousand dollars per annum; but, as Kilzer and

Wilmsen wrote in the *Denver Post*, this was a mere fraction of what the drug could net. What 002 showed was merely that the drug would be beneficial to people who already had AIDS, and it was only approved for their use. Another study from August 1989 indicated that AZT might be helpful to patients with mild symptoms. This study was rushed through—"what was supposed to be a two-year study ended at 11 months"—and its positive results meant that the drug would have a much larger customer base. "By November, Wellcome stock was priced at over twice its pre-AZT level."[14]

Let's get back to our chronology. Exactly how did these studies go? BW moved into researching the drug after repeated pleas from Broder. Once convinced that it had a winner, Burroughs-Wellcome had to both carry out human trials and get the license to use the drug from government regulatory agencies. For help with the second set of tasks, BW "put the drug in the hands of Sam Broder."[15] To supplement whatever scientific zeal might have motivated Broder to work hard on shepherding AZT through the regulatory process, "Burroughs-Wellcome gave the NCI $55,000 in 1985 and $25,000 in 1986."[16]

Phase I trials, which are small and preliminary in nature, covered nineteen patients at NCI's facilities in Maryland and at Duke University. The analysis of these patients' progress was scheduled to be done jointly by the NCI and the sponsoring company, as is the usual procedure, but at the last minute BW withdrew. "Days before the trials were scheduled to start, BW threw a wrench into the whole works when its officials announced they were backing out of that assignment."[17] The firm was leery of handling patients' blood samples, afraid its own lab workers would get contaminated. The NCI took responsibility for these trials itself and once these panned out, moved to the larger Phase II trials.

Between February and June 1986, Phase II trials were conducted. People with AIDS or advanced symptoms of HIV infection were enrolled at twelve medical centers around the country."[18] The trial was completed in record time. Whereas such studies can go on for up to twelve years, although this would be the outer limit, the ones on AZT took no more than eighteen months. Those in charge said closure came quickly because of the extremely positive results: In nine months only one PWA died in the group of 145 administered the drug, while those in the placebo-given group of 137 suffered 19 losses. It was this and not the fact that the trials were especially carefully done and organized that garnered kudos for AZT.

The type of good things said about the drug could not be said about the second-stage trial. According to Walker, "The trial was so badly organized that no follow-up information was recorded on any of the trial subjects, making it impossible to see what might have happened in the long term."[19] Thus, once the group was disbanded, it was never learned whether AZT did much more to prolong life in those who had taken it. This is especially significant in that one contention is that AZT makes for short-term improvement but has no long-lasting effects against AIDS.

However, how poorly conducted these trials were would never have been known to the public except that in 1992 John Lauritsen, using the Freedom of Information Act, obtained documents about the trials from the FDA. As Walker reported in *Continuum*:

> These documents, although heavily censored, revealed that the trials had become unblinded, with trial subjects crossing between groups; that serious adverse reactions to the drug had gone unreported, including 19 cases of anemia requiring life-saving transfusions, and that trial records had been altered to reflect better results for the drug. The trial was so chaotic at its Boston centre that in January 1987, the FDA was forced to hold a special meeting to decide whether or not to allow through the data from this and two other centres—which it did.[20]

But let's hear Lauritsen's own words, since it was he who got the documents and released them to the public.'

> It was apparent that midway through the study, things which had previously been attributed to the drug, in other words, side effects of AZT, all of a sudden were then attributed to HIV disease instead of AZT. Recording forms were erased many months later and changed in such a way as to favor AZT. But the absolute black and white issue on which I would base my charge of fraud is indicated in a memorandum which was written by the case officer of the FDA who reviewed the new drug application of Burroughs-Wellcome for AZT. In this memorandum, it is admitted that FDA investigators used data that they knew were false. They gave the excuse that doing so didn't change results very much. And a further excuse was that if they didn't use bad data they would have hardly anybody left in the study, meaning, of course, that they had bombed. Well, this is just something that cannot ever be done. I don't think there is a scientist in the entire world who would say that it is ethical to knowingly use data that is no good.[21]

Keep in mind, his scathing evaluation is based on evidence that is now available to the public, but which the FDA had no intention of making known and certainly kept well hidden when it was approving the drug.

We should also be careful to note, though, that Lauritsen's demand for scientific rigor should not be immediately translated into a condemnation of the drug. The fact that the trials were done in a sloppy and unprofessional manner simply means that they are unreliable. Whether AZT itself has benefits or not is still to be assessed.

Another thing worth examining at this point is what is behind this sloppiness, that is, the question of motivations we touched on previously. However, since we will be giving fuller details on how a trial can go wrong when we discuss Phase III trials, we will put by that discussion for now.

One thing Lauritsen and others have brought out about this trial, however, that does need further immediate comment is the record of "trial subjects crossing between groups" and other examples of anarchy reigning among those being given the drug. While Lauritsen, quite rightly, frowns on this behavior as destructive of the trial's credibility, we should not overlook that there is another side to the matter. Part of the reason for the chaos plaguing the trial was not only the shoddiness of the administration but the fact that "the trial subjects were in revolt."

Note their situation. They all had the disease for which nothing had been found to even prolong life, let alone cure it. They knew that AZT, whatever its drawbacks, could help them eke out a few more months or possibly years of life, yet they were in a study in which they didn't know if they were getting the real thing or a useless placebo. To many, with death only a few steps away, this seemed criminal and they began trying to short-circuit the blinding that kept them from knowing which camp they were in. "Word got out that capsules containing AZT tasted different from the placebo, and a number of patients found out what they were taking before the discrepancy was corrected."[22] Once this way of circumventing the rules was detected and the placebos were made identical to AZT, some enterprising patients sent their pills to a laboratory for analysis to determine their status. Others "played a mix-and-match game, exchanging pills with one another to increase the odds of receiving the active drug."[23]

Although these activities added to the ultimate unreliability of the tests, they also testify to a formidable challenge to "business as usual" within the scientific establishment. These patients were doing two extraordinary things, both of which produced profound changes.

They were questioning the ethics of using placebos in such situations as this, where the drug being tested is the only one known to do anything against the disease. Their behavior shook the medical establishment, which began to rethink the value of placeboed studies and dropping that manner of research in similar situations.

Moreover, these subjects of a drug trial asserted their strength. By refusing to obey orders and thinking for themselves, they seized a small amount of power. Where science is used to having things its own way in such circumstances, treating those it is using in its tests like guinea pigs, these recalcitrant subjects were making known that they could resist what they saw as inhumane aspects of the experiment. They were serving notice that the affected community could demand changes in the way science conducts itself. What was most frightening to medicine was to realize the trial subject had substantial leverage. Those who ran experiments would have to listen to those who would enroll in their trials or (possibly) suffer disruption and even, if things went far enough, the invalidation of their experiment.

Certainly, though, in this specific struggle between the experimenters and their subjects, there are multiple ironies. On the one hand, I am suggesting that the unruly behavior of these people with AIDS was striking a blow for patients' rights. On the other, the drug they were scheming to get, if they learned or thought they were on placebo, was not all that helpful. If fact, data has shown that the less AZT they got and the more placebo, the better off they may have been. And I say this not so much from the point of view of condemning the drug outright as based on the later finding that the company started out recommending too high of a daily dose.

Looking cynically, we might imagine that the high dose given in the trial was due to the fact that Burroughs-Wellcome reasoned: the higher the dose, the higher the drug bill for a patient. But that is just speculation. All we do know is that the amounts of AZT per patient first given have since been seen to be too high. Walker states, "After almost four years of licensed use, it was accepted that AZT had a 1,000 times higher toxicity than had been quoted by Burroughs-Wellcome in the Data Sheet Compendium." He then argues the cynical view, "At an end cost of 10,000 [British] pounds per patient per year, Wellcome attempted to keep the dosage as high as possible," although most doctors, viewing the toxic side effects with their own eyes, "eventually more that halved the recommended daily dosage."[24]

This high toxicity was even noted in the government before Stage II trials started, in a report that was apparently brushed aside. This came out in a report Lauritsen obtained titled "Review and Evaluation of Pharmacology and Toxicology Data for the Drug Retrovir (generic name Zidovudine, also known as AZT or azidothymidine)," written by FDA toxicology analyst Harvey I. Chernov, and dated December 29, 1986. After reviewing several dozen studies, both on animals and human beings, Chernov concluded that "the single most important finding was that AZT was toxic to the bone marrow, causing anemia."[25]

Despite these misgivings, expressed but not publicized, the quick tempo moving this drug toward approval continued after the Phase II studies. The requirement that the company complete Phase III trials before applying for a license was waived, although these trials would eventually have to take place—with another pile of irregularities, as we will reveal in the next section.

Speedy Approval and Early Prescribing of AZT

Before moving to that trial, I want to pause and say a few words about the pressures driving doctors to prescribe AZT. This will allow us to deal further with the issue of motivation, which has been hanging fire for a time, and which we will begin to fill in here.

A little chronology first. Given the green light by the FDA, BW moved quickly both to get the drug to AIDS patients and to the next set of trials that would establish, to its satisfaction, that even HIV-positive patients who had no symptoms should be given the drug. As noted, this second group was where the real market was.

As soon as the government approved AZT for wider dispensation, a vast PR blitz was orchestrated to make positives "get with the program." When I interviewed Lauritsen, he explained how positives were indoctrinated.

> They were told that they should go for... early medical intervention. There were slogans put out, "Put time on your side." The early intervention meant purely and simply AZT. And rather than putting time on the side of these people, what the drug did and is doing is to terminate their lives.[26]

(To introduce a brief aside, let me mention that although no accusations have been made against the company on this next score, it certainly appears compromising that Wellcome, which was urging anyone in the UK who was positive to take its expensive drug at the same time was making the tests that determined whether one was positive or not. "In the early days of AIDS, Wellcome had a monopoly franchise on HIV testing kits [in England] which were produced by its subsidiary company Wellcome Diagnostics."[27]

Specifically, Wellcome developed the antibody testing kits along with Dr. Robin Weiss, the executive director of the Institute of Cancer Research, which is the British equivalent of the U.S. National Cancer Institute. Work was begun in 1984 and by "1985, it was estimated that the British market for diagnostic kits was worth between 3 and 4 million pounds."[28] By 1987, Wellcome launched its second-generation test kit, called the Wellcozyme HIV Monoclonal, from which it garnered more impressive profits.

In such a setup, the danger of a conflict of interest is evident. As Walker put it, "Wellcome knew that ultimately they would be able to sell AZT to almost every individual that they had diagnosed as HIV antibody positive," and this certainly wouldn't encourage them to develop tests that erred on the side of finding too many people negative. To repeat, no improprieties have been charged against the company on this score, but, as Walker comments, "the fact that one pharmaceutical company had control over the diagnosis and treatment of a major illness was ethically wrong."[29])

To return to our thread, as BW pulled out all stops in promoting its drug, a flood of money swelled into physicians' organizations. Walker gives one instance. He tells how in 1987 the British Medical Association (BMA), a professional trade union for doctors, set up the BMA Foundation for AIDS. "In March 1988, Wellcome gave a covenant to the Foundation, a sum of 36,000 pounds annually for four years, totalling 144,000 pounds."[30] As you might imagine, the group became a strong advocate of AZT treatment.

Doctors began and continued to prescribe AZT in face of growing evidence that the drug was associated with perverse side effects. Even Wellcome's own *Physician Desk Reference* entry on AZT eventually had to admit, "it is often difficult to distinguish adverse effects possibly associated with AZT administration from underlying signs of [what is called by contagious-AIDS proponents] HIV disease."[31]

The generality of doctors didn't seem to take this or even broader hints of the drug's toxicity as seriously as was warranted. They kept right on prescribing it, albeit at smaller doses and sometimes in a mix with other drugs. Currant's opinion of this, that is, of continuing with what he considers a bankrupt treatment, is that it is not good. He told me, "American medicine has gone completely out of control."[32] His own distress over sick friends has fueled his anguish, he said, mentioning, "my friend being in the other room so ill, trying to desperately rebuild his body from the destruction of our doctors who poisoned him—in the name of their medicine."[33]

Even the doctors, Neville Hodgkinson told me, can't face the horror of what they have done when they rushed to prescribe this drug. There were "so many tragic consequences of this mistaken idea that I think... there's a sort of collective pride in the scientific community that just can't bear facing this horrendous possibility [that prescribing AZT is murder]. I think that's probably even more important than the financial side [in explaining why doctors are not facing up to the downsides of AZT]."[34]

However, to begin the discussion of motivation, let me say that in taking up the question, I do not mean so much to introduce new material as remind readers that a point raised earlier about science can, with equal suitability, be applied here.

We said that one ideological function of medicine was creating a dream time. Money was put into medical research with the understanding that any benefits discovered would accrue to the health of the general population. (How would it look if a proposal were couched like this: We are trying to devise a vaccine for AIDS for rich people.) Yet, given that—as I think we have already established—in the United States we have a two-tiered system with one type of expensive care for the rich and a cut-rate version for the poorer majority, the dream-time ideology of medicine illicitly creates a future world, in which the inequities of the distribution of services have somehow disappeared.

The new point I want to make here is this. Taking the future worldview from the side of the doctors themselves, we might suppose that they have taken it deeply to heart. In my experience, those involved in medicine always have an element of idealism. Although they may expect as their right a good living, they also need to feel that what they are doing will benefit the world. If they were told, in our previous example, they were to work on drugs that would be exclusively for the benefit of the wealthy, they might balk. Thus, an admirable highmindedness, the desire to think one's work is for the universal good of others, becomes blindness, since it refuses to recognize that current medicine has no such intention of helping any but a small group.

So how does this relate to AZT? After all, government subsidies have made it widely available, at least in the West. It is simply that the propaganda surrounding the rollout of the drug draws upon this public-spiritedness found in the future-world scenario. Burroughs-Wellcome kept hitting the note that AZT was the best hope for the poor victims of the disease. The refrain was that prescribing AZT was humanitarian, not a profit maximizer for the firm. BW was playing upon doctors' idealism, which is also tapped by the dream-time belief. By contrast, very often dissidents' remarks, such as Walker's cynical statement about why Burroughs recommended a high dosage disturb the idealized image of medicine establishment doctors hold—even if such statements are true.

What I'm trying to say is that what might be considered the most valuable attributes of individual humans, such as their desire to live a life that partly is of service, can end up closing them off from reality, insofar as facts expose their profession as being discordant with their ideals.

Most certainly, there is more to it than this. What about cases where health workers do seem to be acting from greed? I have something to say about this also, but we can better handle this after we have looked at the Phase III trials, where pecuniary motives, at least in one case on which we have evidence, seemed to play a fundamental part. We will take this up in the next chapter.

Testing Today

It might be thought that even if corners were cut and errors of judgment made in the very first studies of AZT because the government was so eager to find something or anything that would be effective against AIDS, later workups on the drug—now that some of the pressure was off—would be characterized by a more meticulous approach.

Instead, we will find that even as AZT became accepted by the establishment, shoddy studies were still done, both funded and trumpeted by BW. At the same time, other studies, such as the Concorde trials, cast AZT in a much harsher light. Such debunking studies, however, came under heavy fire not so much from other scientists, although there was this, too, but from the pharmaceutical giant, which was not adverse to underhanded methods in questioning examinations that didn't have results that it liked. Let's look at some later findings on AZT's effectiveness and toxicities, both positive and negative, to see how the story unrolled after AZT was already licensed.

We can begin with the Stage III trials of AZT, which took place from 1987 to 1990. As we saw, these were set in motion after AZT was already licensed. We have a good eyewitness to the irregularities going on at one site, at least, in Lynn Gannett, who was data manager for two and a half years at the Syracuse clinic involved in the study. Her savage characterization of how the medical side of the trials were conducted runs to adjectives such as "unethical," "corrupt," and "immoral."

She arrived at this from observing, over her extended tenure, that the directors of the program were primarily interested in "enrolling as many patients as possible on studies (which would entitle them to more money and perceived prestige)."[1] In consequence, "patients were ROUTINELY enrolled who failed to meet eligibility requirements."[2] She cites as an example that patients would be enrolled with flimsy evidence, such as a doctor's note saying that they were HIV positive. One woman accepted on such meager documentation was given AZT for three weeks before it was found she was negative. She is highlighted because it took three weeks to uncover the error, not because it was the only such case.

The screwups didn't stop there, but reached deeper into the preparation of accurate data on the value of the tested drug. Most devastating in terms of the value of the results, "There were COUNTLESS unreported (meaning unreported on the research forms) diagnoses, opportunistic infections, symptoms, concomitant medications and adverse reactions."[3] The parenthetical statement was included because she indicates that symptoms presented on a patient's medical chart often were left off the forms used for reports on the reaction to the drugs. This type of lackadaisical attitude obviously nullifies the value of the study.

Perhaps the only thing more shocking in her report is the casual cruelty of the trial's conductors. This was shown, for instance, in the case of a black female patient who, six weeks into the program of taking AZT, ended in the emergency room with "severe shortness of breath, fatigue and weakness [which] required her to be hospitalized for a total of five days."[4] Once she was well enough to continue her treatment—in complete violation of the study's protocols, which said that after such an episode a subject should be taken off the drug—she was put back on at a reduced dose.

We might pause to note that here scientists *do* seem more interested in purely financial and careerist payoffs than possessed by the idealist goals I was assigning them a few pages ago.

Let us note, first off, that the ethics of an individual represents the flow-together of a concatenation of professional, religious, and family obligations upon which financial and other selfish considerations can impinge to a greater or lesser degree. Now, when I mention professional obligations, I am thinking not primarily of such concepts as allegiance to the Hippocratic oath and such general ideas but of the specifics of a workplace.

As is well known, in the 1950s sociologist Irving Goffman clarified an idea that was long recognized intuitively. This was that all bureaucracies create an in-house morality, which sets them off from the general public and stablizes a set of ideals that become guiding lights to the occupants of positions in the organization. Anyone who either reveals the internal workings of the bureaucracy to outsiders or goes against the ideals will be branded a traitor. What I am leading to is simply the thought that the urge to enroll as many subjects as possible to increase the income and prestige of the organization can be ascribed to the demands of the organization, not individuals, demands that are rooted in the ascendant aims of the system.

This doesn't exculpate those who went along with the bad practices, but it does indicate that when assessing their malfeasance we may find that people acquiescing in such activity are often doing so as upholders of the group spirit rather than for more tawdry reasons.

This also means that we cannot necessarily generalize from the Syracuse study and say that all Phase III trials were equally faulty, since each site will have its own group aims. That the trials were all conducted under the aegis of one large directional hand, though, could lead one to suspect that similar improprieties may have been rife.

Gannett has kept annotated copies of materials that record the lunacies and scientific misconduct that went on during this Phase III study, and this puts her in a better place than most to contest the value of some of the studies done to prove AZT's credibility. She was working from the inside. Others, however, have raised a hue and cry about their insufficiencies as they look from the outside at flawed methodology or exaggerated interpretations.

In 1990, for example, an article entitled "Survival Experience Among Patients with AIDS Receiving Zidovudine (AZT)" was printed in the *Journal of the American Medical Association (JAMA)*. Lauritsen notes the article's weaknesses:

> Researchers from the government and Burroughs-Wellcome studied 4,805 PWAs treated with AZT. [Somehow] they lost track of 1,120 patients, not knowing if they were even alive or dead. The researchers then used statistical projection methods to guess what results they might have obtained if they had not lost the 1,120 patients, and came up with a 10-month survival estimate of 73 percent. They then wrote their report in such a way that the 73 percent guess appeared to be an actual survival statistic.[5]

A piece by Philip Pizzo et al. and titled "Effect of Continuous Intravenous Infusion of Zidovudine (AZT) in Children with Symptomatic HIV Infection," which was published in the *New England Journal of Medicine*, 6 October 1988, also tries to tout the powers of the drug. This study, however, is more embarrassing than convincing. The authors studied twenty-one children given AZT and said the drug had "boosted their IQ's by 15 points, which would be one standard deviation." The conclusion was that even though "five of the children died, in a very short time," [since their IQ scores went up], "even [HIV-infected] newborn babies should be given AZT."[6] I call this embarrassing because it is making AZT out to be some kind of wonder-working panacea that does miraculous things it wasn't even designed to do. There has never been any drug discovered that increases intelligence, nor does it seem logical that a drug meant to destroy DNA would also increase brain power.

If we can say that the last two mentioned studies promoted AZT in a buffonish way, we should underline that usually AZT boosters are subtler. One way to get unwarrantedly positive results for a drug is to complete its trials quickly.

We've already noted this point in connection to Stage II AZT trials. Investigative journalist Tom Bethel notes the self-serving nature of such alacrity. "They always discontinue the tests after a short run because [drugs] initially show a good effect. If you put a toxic substance into people, the body's immune system responds with a fight back and you get an increase in the t-cell count." That won't last forever, he continues. "If you take [AZT] long enough, it wears you down, and eventually kills you."[7] The patients take the drug for a longer time than those in the studies did, so the patients are exposed to risks that have never been assessed.

Here is a second promotional method. If scientific findings appear that reflect adversely on AZT's ability, publicize an already produced investigation that counters the negative report.

This happened in relation to the Concorde study, although this was only one of the resources Burroughs-Wellcome brought into play in an effort to derail the effects of the investigation.

The Concorde study, conducted in the early 1990s on behalf of the British and French governments (with its first results released in 1993), had as subjects 1,749 asymptomatic patients with CD4 positive T cell counts less than 500/mm3. It was the largest study of AZT ever conducted and, unlike the hastier studies done in the United States, followed the subjects for three years. Thus the researchers were able to better assess the long-term value of AZT administration. *The findings were damning.* "In that study, *no statistically significant differences in progression to advanced disease were observed after three years between individuals taking AZT immediately and those who* deferred AZT therapy or *did not take the drug*," according to the Concorde Coordinating Committee."[8]

The Concorde did not portray AZT use as valueless. The drug slowed the rate of the disease progression for those who had AIDS during the first year of taking the drug. (AZT's viral-fighting abilities seemed to wear off over time.) It also "clearly demonstrated that AZT was not harmful to the patients in the immediate AZT group as compared to the deferred AZT group."[9] This last finding doesn't say much, in that it doesn't compare AZT takers to those who abstained from the drug. However, despite a number of positive side lights, this was the first major study to cast doubts on the basic validity of AZT prescription in that results indicated the drug had little long-term effect in reversing the disease.

Burroughs-Wellcome couldn't be happy with such a finding and, according to an anecdotal report by Celia Farber, had tried to head off the study by getting the researchers to cushion their blows. She writes, "I had heard that the Concorde team had been under tremendous pressure from AZT's manufacturer, Burroughs Wellcome, to soften its results."[10] One member of the research team told her, when she was asked about BW trying to influence how they presented their findings, "There has been pressure, and it has been placed at the very highest level." Being less forthcoming about who was applying the pressure, Dr. Ian Weller, a chief investigator of Concorde, followed the other researcher's remarks with: "We've carried out this study against incredible adversity, but we are not going to cave in to any pressure." He continued, perhaps naively, "We'll win the battle in the end. We show the science, that's all that matters."[11]

One result that Burroughs-Wellcome was able to achieve by its pressure was to have a clause inserted in the Concorde trial protocols that would allow, if the findings were damaging, for a waiting period before results were made known, so that BW could organize a response. Since the results were so bad, the interim clause took effect. "During the waiting time, a number of Wellcome directors cashed in their own shares in the company."[12]

When the results did surface, not only did the company argue the study was flawed, but many doctors who had been using AZT treatment "took the stand," as it were, to testify to the good character of the drug.

Meanwhile, as these results were first being publicized, an article appeared, amid much ballyhooing, whose results ran directly counter to those found in Concorde. "The article at the centre of the row was published… in an American medical journal [in 1993]. It was based on a study carried out for Wellcome in Australia and a number of European countries, mainly in the late 1980s."[13] The researchers for this examination found that AZT use cut in half the rate of AIDS-related diseases experienced by PWAs and, so, in conclusion, they recommended giving the drug to all HIV positives, with or without outward symptoms. It was not lost on readers that the researchers' conclusions contradicted findings "released four months… [before] the Concorde trial [which] found that AZT (or zidovudine) had no effect in preventing AIDS."[14]

Some saw the sudden taking of this Australian/European investigation out of the mothballs as part of a transparent ploy urged forward by the company behind the study as a way to counter Concorde. "Leading doctors expressed outrage at what some regarded as a public relations stunt by Wellcome. The company's study, they said, was old, flawed in its methodology and had been terminated before anything meaningful could be reported."[15]

The fact that Burroughs-Wellcome funded the study is not itself cause for alarm, since companies are expected to pay for such inquiries to determine if their drugs are viable. If they don't do so, many promising drugs may be left in the shadows.

The U.S. government also has conducted independent studies of possible AIDS drugs, particularly through NIAID's clinical trials program, which got off the ground in 1986. Since the government wouldn't be marketing any drugs it examined, it might be felt its reports would be less likely to be swayed by inappropriate commercial considerations. Although this is true, other dangerous biases have been known to creep in. Martin Delaney, among others, has argued that NIAID should recommend areas of study rather than wait on its hands for proposals. Allowing scientists to guide research, it has been said, has kept work fixed on puzzles whose solution would bring the largest share of fame. Specifically, this means that scientists fixate on finding a drug that might stop HIV altogether. Equally important riddles—ones more likely to be worked out in short order, such as how to halt opportunistic infections—are disregarded. Many say this is myopic.

> [A] major conflict between activists and researchers centered on the ACTG's [NIAID'S AIDS Clinical Trial Group] system's penchant for studying only antiviral therapies rather than drugs to treat opportunistic infections.... The scientific rationale for trying to disable the AIDS virus was that a drug that stopped the deterioration of the immune system would also prevent the onslaught of opportunistic infections. But the spectacular professional payoff that would accompany the discovery of an antiviral was also an enticement. The researcher who could defeat the AIDS virus... was sure to bask in glory, most likely walking away with a Nobel Prize.[16]

However, aside from this all too human motivation, there was another problem with giving researchers free reign. Instead of being guided by government initiatives, they tended to take their lead from drug companies. Scientists would propose research into anti-AIDS drugs that belonged to firms that were funding their research. As Mark Harrington from ACT UP put it, "Sure a scientist may think he is being objective, but what happens [to his objectivity] just knowing that a drug company is giving him thousands of dollars?"[17] The improprieties that might be involved in such conflicts of interest were of such concern that in 1989 the NIH drafted "guidelines that would have barred researchers from conducting NIH-sponsored clinical trials if they were consulting for an involved pharmaceutical company at the same time."[18]

Protests from NIH scientists shot down these guidelines, which were never adopted, but the fact that they were discussed shows the concern over this industry/researcher dyad. In *Against the Odds*, Arno and Feiden bring out how costly this dependence on business sponsorship can be. They cite the case of Peptide B, a drug that had shown promise in combating HIV attachment to T cells while also demonstrating zero toxicity. However, it took five years for the ACTG to approve it for trials. This unconscionably long delay probably had several causes, but one mentioned is that "several ACTG investigators had financial interests in therapies that competed directly with Peptide B's mode of activity."[19]

When a new drug has no company sponsorship, it may be left in the lurch, as was Peptide B, but if it is a focus of a corporation's interest, it will be made prominent. "Commercial and industrial interests have helped launch a large number of vested interest journals which print the research work which they have funded," Walker reports in *Continuum*.[20]

Meanwhile, concurrently with Wellcome's continuing advertising of the advantages of AZT, studies have appeared, ones we will touch on only briefly here, indicating that AZT possesses more toxicity than it had earlier been credited with, that it may be less effective than it was supposed, and that, in any case, it is often prescribed wrongfully.

Toxicity

To return to the Concorde study, Graham Ross notes that beyond the fact that the investigation voided AZT's ability to combat AIDS, it also recorded higher deaths among AZT takers than among those on a placebo. "Despite the conservatively worded conclusions, the actual figures show a 30 percent increase in death for those taking AZT prophylactic whilst asymptomatic. (8 percent taking AZT whilst asymptomatic died during the trial whilst only 6 percent died who took the placebo)."[21]

These facts are in tune with the constantly reassessed and "upped" measurement of the drug's toxicity. Back in 1985, a benchmark was established that measured the believed toxicity of the drug, according to the weighing of LD50 (LD = lethal dosage), that is, a measure of how much of the substance would kill 50 percent of cells in a culture. Ten years later, it was known "that this estimate of lethality underestimated the killing power of AZT a 1000-fold."[22]

We may note also, in reference to changing attitudes toward the drug, that early claims that AZT was low in toxicity, such as those presented by Burroughs Wellcome, were based on a partial understanding of what toxicity is. As Lauritsen explained at the Amsterdam Alternative AIDS Symposium in 1992:

> Always they [AZT boosters] look only at what are called "acute toxicities," meaning short term. But any toxicologist will tell you that there are also "chronic toxicities" or long-term ones. And one of the truisms of toxicology is that you can't predict chronic toxicity from acute toxicity. Some of the classic carcinogens have no acute toxicities at all, but in the long term they cause cancer.[23]

As soon as longer-term studies appeared, these beliefs in AZT's inoffensiveness were shot to ribbons.

This is not surprising, since, as we have mentioned, AZT tends to mow down dividing cells at random, showing little discrimination in its onslaught. In an article by Kremer and others, it is argued that AZT also interferes with the oxygen-supplying process. "AZT has, in countless cases, brought about the inevitable and slow asphyxiation of the patient's body cells, which are in particular need of oxygen, and hence the equally inevitable death by poisoning of those persons" taking the drug.[24]

Another place where AZT causes disturbance is in mitochondria function. Mitochondria are small bodies found in cells where oxygen is processed and raw materials, such as proteins and fatty acids, are converted into energy sources, particularly ATP. Use of AZT is known to cause myopathy, a breakdown, of mitochondria as is even mentioned by BW on materials it provides where it is stated, "WARNING:... Prolonged used of Retrovir has been associated with symptomatic myopathy [in mitochondria and other areas] similar to that produced by human immunodeficiency virus."[25]

A number of studies give details of how AZT is assumed to interfere with the activity of these internal cellular mechanisms. A piece in *Lancet* by Aranaudo tells us, "Long-term zidovudine [AZT] therapy in patients with human immunodeficiency virus (HIV) infection can cause a mitochondrial myopathy."[26] Aranaudo's team studied HIV positive people, contrasting those who took AZT with those who did not, and found the nine takers had "severely reduced amounts (up to 78 percent reduction vs. normal adult controls) of mitochondrial DNA." Neither the two HIV positives who did not take the drug nor the controls showed any similar dropoff in mitochondrial action. The scientists say this difference is "probably due to zidovudine-induced inhibition of mtDNA replication by DNA polymerase gamma and is not a secondary effect of HIV infection."[27] Such an effect, the abrupt collapse of DNA replication in these cell areas, is to be, expected since, as we've insisted, AZT stops DNA proliferation in all parts of the cell, not only in those supporting the retrovirus.

Aranaudo's findings are echoed by experiments on rats conducted by Lewis et al. and published in the *Journal of Clinical Investigations*. After supplying the rodents with AZT for thirty-five days, the researchers found pathogenic changes in the animals' skeletal muscles. "Selective changes in rat striated muscle were localized ultrastructurally to mitochondria.... Decreased muscle mitochondrial DNA and mtRNA, and decreased mitochondrial polypeptide synthesis in vitro were found in parallel."[28] The conclusion here was the same as that of the previous study: that there was "AZT-induced inhibition of mtDNA replication," which in this case resulted in the muscle damage.[29]

AZT has been known not only to inhibit mitochondrial function but to disable the immune system. Just as antibiotics kill benevolent microflora in the human intestine and elsewhere while killing virulent bacteria—thus possibly proving in the long run more hurtful than helpful—AZT, which can be said to collaborate with the immune system as much as it inhibits HIV replication, concurrently weakens that same natural immunity. For one thing, the drug is reported to be a "potent inhibitor of lymphocyte proliferation. The result of this inhibition is immunosuppression."[30] It will be recalled that these lymphocytes are essential components of the body's defense system.

Narrowing down the type of lymphocytes affected, More, Costello, and Gill, in separate studies, argue that AZT "attacks and kills CD4 cells—the very thing it is supposed to prevent HIV from doing."[31] We've seen that a PWA's CD4 cells increase after the person begins taking AZT. So it may be thought that even if AZT eliminates some CD4 cells, it doesn't do so to such a degree to check their growth. However, a study by Caufield and Goldberg argues that such an increase is not necessarily a measurement of health given that "research has shown that AZT suppresses the immune response, significantly affecting proliferating CD8 cytoxic lymphocytes."[32] It is the duty of these CD8s to hunt and shut down infected cells. "It is natural for the immune system to destroy infected self-cells such as CD4s. Inhibiting the process can only result in uncontrolled replication of HIV." That is, according to their view, the sudden invigoration of CD4 counts after AZT is not evidence of the superiority of this drug but its inferiority, since the greater number of CD4s is not due to their ability to reproduce better but to the inability of CD8s, now inhibited by AZT, from destroying damaged, infected CD4 cells![33]

These findings on the suppression of lymphocytes' generation have made a number of doctors draw the obvious conclusion that it seems rather ass-backward to shovel into a patient whose immune system is already damaged a drug that will, as it were, administer the coup de grâce to that same immune system. Martin Feldman notes, "The current drugs on the market, primarily AZT, tend to severely weaken the immunity and make the body have to work harder to have immune strength. The body uses up its basic nutrients in the process. Really, the body is fighting against the AZT. It's a detriment."[34] Richard Strohman adds, "I think people are in a desperation mood to use a drug that will produce some sort of a short-term benefit. But the cost is a further encroachment on the individuals' immune system. It makes no sense."[35]

Let's revert for a moment to our findings on cancer. When I was first discussing the early reason that AZT was abandoned as a cancer remedy, I alluded to the disputed question of whether, at that time, it was known to be carcinogenic. I also said that regardless of how we decide that historical question, nowadays AZT's cancer-causing qualities are recognized. "According to the FDA analyst who reviewed the AZT toxicology data—and recommended that AZT not be approved for marketing—AZT 'induces a positive response in the cell transformation assay' and is therefore 'presumed to be a potential carcinogen.'"[36] The mentioned cell transformation assay is an examination in which the tested drug or other substance is given to cells in vitro (that is, ones being cultured in a laboratory) and its effect studied. If it induces cell transformations similar to those undergone in vivo (in living creature) by cells that are becoming cancerous, it is suspected that the administered substance is a cancer-promoting agent. A close correlation has been shown by those chemicals, such as AZT, that score high in transformations on the assay and those that are known to promote cancer.

Remember that the time lag between the continued exposure to a carcinogen and the actual onset of the disease is quite lengthy, reaching up to twenty years and more, and therefore it is impossible to count human cases to indicate the cancerous potential of a newly used drug such as AZT, but its carcinogeneity can be gauged to a degree by using animal models, where a cancer will develop more quickly. In one study, "the offspring of mice given the toxic drug developed at later age tumors in the lungs, liver, and reproductive organs. The study confirms an earlier study which showed AZT caused vaginal cancer" in baby rodents.[37] Furthermore, in an investigation in which pregnant monkeys were given the drug, the offspring "incorporated the drug into their DNA," indicating it may surface as cancer or other cellular abnormalities later in life.[38]

However, it's not so much the danger of cancer, which lurks a ways down the road, but the other, immediate effects we have noted, including the disturbance of oxygen transfer and the havoc wreaked on the immune system, that mean AZT is not only hazardous but intolerable for many. "The toxic effects of AZT, particularly bone marrow suppression and anemia, are so severe that up to 50 percent of all AIDS and ARC patients cannot tolerate it and have to be taken off it."[39]

The effects of the drug on children are also deleterious. For example, preliminary findings from the study labeled ACTG076, which was "giving AZT to pregnant women to assess its ability to interrupt transmission of HIV to a fetus," as reported in an ACT UP publication, "showed that 10 percent of the children had extra digits (fingers, toes)… [and] five babies were born with neutropenia (a blood cell abnormality) but apparently it reversed itself."[40]

Kremer's article, mentioned a few paragraphs ago, notes that newer combination therapies, in which AZT use is supplemented or alternated with that of other drugs, has so far proved little better than use of AZT alone. His group's simple point is that if shooting a toxic substance like AZT into the body with little understanding of how its operations affect body chemistry and function is risky, then adding a few more drugs of toxic capacity (such as protease inhibitors) is likely to increase the risk. "It is to be feared that sooner or later the unphysiological intervention in the complex interplay of body cell growth factors through artificial protease inhibitors," Kremer et al. write, "will disturb equally vital functions of the basic tissue and cells, together with their mitochondrial energy centres, as is already the case when AZT and allied nucleoside-analogues are administered."[41]

Longevity

As we've seen, the brunt of the Concorde study's criticism of AZT was not that it was a hazardous substance, but that it did little to prolong PWAs' lives. Later studies have supported and refined this conclusion, many centering on whether early or late administration of AZT offers differing benefits. This last issue has become prominent because, as we saw, early claims about the benefits of the drug said it is more helpful the earlier it was administered.

Even members of the establishment camp, such as UCSF dermatologist Marcus Conant, "a long-time advocate for the use of nucleoside analogue drugs," although not willing to forgo the use of the drug "lament[s] the dismaying lack of evidence that these drugs prolong survival in AIDS—regardless of where their use is initiated in the progression of the disease."[42] He cited a host of studies that came after Concorde, such as a European-Australian Collaborative Group trial and the United States Veterans Administration Cooperative Study, which reported that AZT and similar drugs will not increase the overall longevity of a PWA, although when they are taken they will retard the development of symptoms for up to eighteen months. In Conant's words, "You're left with deciding when you want your eighteen months of benefit—at the beginning or at the end." He means being that whenever you take AZT, you will produce a lessening of symptoms, but no matter when you take it, your end will be as soon as it would if you hadn't taken it at all.[43]

Perhaps one of the most surprising studies that examined the effectiveness of early administration of the drug was that published in 1995, surprising not for its conclusions: "that giving AZT to symptom-free individuals who test positive on the HIV antibody test does not stop progression to AIDS," conclusions that we see were becoming commonplace, but for who had made the report.[44] As Ostrum reported in the *New York Native*, the study had been done by scientists connected to the AIDS Clinical Trials Group (ACTG), which is in charge of government-sponsored trials of anti-AIDS drugs. The startling note was "the team [who did the study] consists of many of the same individuals involved in earlier studies that claimed to show that AZT delays progression to AIDS," including being led by "the same clinician who published that opposite result six years ago [that AZT slows progression], Dr. Paul Volberding of the San Francisco General Hospital."[45]

(To say a little more about Volberding and his associate Donald Abrams, it must be noted that already in 1993, they had turned a corner on their beliefs. Both wrote editorials for the *Bulletin of Experimental Treatments for AIDS*, where they said "that while the currently used drug therapies AZT, ddI, and ddC may show a transient increase in CD4s, they do not extend survival time, but merely delay the onset of symptoms."[46] Abrams, the more outspoken of these early champions of AZT, said, "We really must face the critical issue and not pooh-pooh survival in favor of modest transient increases in CD4 counts, or even significantly decreased short term disease progression."[47] (The critical point so is that far nothing in the way of drugs has been developed that has much effect on the illness long-term.)

A study analyzed by Drs. Lawrence Corey and Thomas R. Fleming of the University of Washington collaborated this point about AZT's ability to temporarily forestall symptoms, but pointed out the liabilities of taking the drug early on in the course of the disease.

During 2-1/2 years of follow up [of those who took AZT, the study] found that those with early treatment reduced the development of full-blown AIDS by nearly half. But once AIDS developed, they tended to go down hill faster. And their survival at the end of the study was 77 percent, slightly worse that those who began treatment later....

Just why the early treatment grouped lived no longer, even though they delayed the start of AIDS, is unclear.... [Corey and Fleming] speculated that AZT's benefits decreased with time. So AIDS is more quickly fatal when it arrives.[48]

The Veterans Administration study, which was headed by Prof. Hamilton and to which we just saw Conant refer, had an interesting anomaly in the data which was spotted by Elliot. While the investigation backed other studies that indicated taking AZT early or late made no difference in a patient's life span, there was a discrepancy between how PWAs died. "The people who received late medication (when they already had most of the AIDS symptoms) all died" of typical AIDS-related diseases, but "of the 20 who died in the early study, 10 didn't die of AIDS disease."[49] One committed suicide and three others died in non-illness-related ways, but "that still leaves you with 6 people out of 20," which he concludes might have died because of the toxicity of the prophylaxis.

Wrongfully Prescribed?

Even if, as we will see below, there is some logic to short-term use of AZT—which is to say, it may not be helpful but it is not out of keeping with the type of drastic and occasionally successful cures that are used for cancer—AZT's long-term use is unjustifiable, as is its use on those with weak or undeveloped immune systems, such as newborns. Yet, as we also know, the drug has been prescribed wholesale.

One group that certainly hasn't seemed to benefit from the prescription is hemophiliacs. We've already seen that they as a group are prone to ill health due to their debilitating condition. We also had reason to note that statistically the survival rate of those hemophiliacs with HIV has been lower after the introduction of the drug. "The death rate for hemophiliacs was actually going down during the first decade of AIDS. However, that trend reversed itself sharply in 1987 when HIV positive hemophiliacs were treated for the first time with AZT," according to David Rasnick.[50] He noted that the establishment blames this lifted mortality on the march of AIDS rather than on what, in Rasnick's opinion, is the true culprit. "No one has ever looked to see if treating HIV positive hemophiliacs with AZT—the most toxic drug ever given to people—may actually be killing them at ever increasing rates."[51]

Moreover, the questionable practice of giving AZT to presymptomatic positives, which we already saw assailed by the Concorde study that argued the drug did little to extend life, has been questioned even more fundamentally by Eleopulos, who in 1998 looked at more recent studies to argue that "evidence exists showing that AZT causes 'a significant increased risk of death among the patients treated early,' that is, in patients who commence treatment when asymptomatic. In other words, instead of curing patients or prolonging life, AZT shortens it."[52]

Also in line with our earlier discussions, where we saw that Africa has often been the locale of some of Western scientists' most dubious charades, AZT has been provided to pregnant women and newborns in Africa to test its potency. An article in *Continuum* reports that the drug is handed out "in African countries without proper information on the drug's lethal effects."[53] AZT was made available "as part of a programme of 15 trials which have been conducted for the past two years [from 1996 to 1997] on more than 12,000 women in 11 developing countries."[54] Nine of these investigations were financed by the U.S. government, and others were backed by France, Belgium, Denmark, and the UN AIDS program.

If this were not enough, some leaders of AIDS orthodoxy, such as William Haseltine (of the Harvard School of Public Health), "have gone so far as to advocate giving AZT to perfectly healthy, HIV-negative members of 'high risk groups,' such as gay men, to prevent them from becoming 'infected.'"[55]

We have seen, then, that contrary to the encomiums of the drug's manufacturer, AZT doesn't have a lot going for it. It does tend to reduce symptoms but these reappear later with even more virulence. It doesn't seem capable of prolonging the lives of PWAs and may even cut their wicks shorter.

In a moment, I will show what logic propels the use of this drug, beyond the logic of despair, which also plays a part. The latter is the logic which makes some embrace it, saying, "We need something, anything, to fight against this disease."

Michael Callen, a long-term positive, says this manner of thinking can often be found behind the most strident in the movement, whose slogan is "Drugs into Bodies." The movement itself has laudable goals, such as speeding up the FDA and other agencies' bureaucracies in their systematic study of drugs, and pushing for such innovations as parallel tracking, whereby PWAs (or those with other diseases in relation to other therapies) who do not qualify as subjects for trials that test new, promising drugs can get those drugs for compassionate use. However, as Callen describes it, there is a danger that one will fall into a "drugs-into-body frenzy," which limits one's view of the effect of new drugs to two possibilities. He notes that some of "my AIDS activist friends... only seem to talk about two possible outcomes to taking experimental drugs: one is that it works and one that it doesn't." But, as he emphasizes, "There is a third, apparently much more common possibility, which is that you will be worse off than if you did nothing at all."[56]

The pathos of this position and the tragedy of the AZT has not been lost on dissidents. Dr. Joseph Sonnabend, whose practice early brought him into contact with many PWAs and who has spoken out against AZT as a deficient stopgap, nonetheless has to face patients who see no other option. "It's beyond belief," he says, "I don't know what to do. I have to go in and face an office full of patients asking for AZT. I don't know what to do as a responsible physician."[57]

Well-meaning physicians, who want to help their patients but do not question authority, prescribe the drug and then attribute their patients' deterioration to the wasting disease, but others see the treatment as dealing the death blow. James Whitehead, when I interviewed him, put it perhaps more poignantly than most.

> In the late eighties friends of mine [were] suddenly becoming ill, and it turns out, I knew they were perfectly healthy men, they tested positive, and you don't find out they're sick until a few years later, after a few years of taking these toxic drugs, and so it's too late. They were poisoned, they got sick, and they died from AZT.[58]

James Currant echoed this view when I talked to him, saying "AIDS is caused by the drugs used on people.... the drugs are so toxic they fulfill the prophecy of the AIDS."[59]

As Kary Mullis told me, all these tragic outcomes could have been predicted given the nature of AZT, which had been first scouted as a chemotherapy agent to be used against cancer.

> If you start taking any other chemotherapeutic agent for the rest of your life—it would be that agent that probably would kill you. You know when you give chemotherapy to somebody with cancer you give them a round of it for as many as fourteen days or just a few days and, hopefully, you will not kill the patient, but just the cancer. But you don't keep giving it to him until he dies. Because he certainly will! And AZT is just like those things, [except] it's more lethal![60]

Rationale and Long-Term Effect

Two points that Mullis made in the last quote call for fuller discussion. He notes that while AZT was developed and acts in the same manner as certain, anticancer chemotherapeutic drugs, it is not being used in the true and sometimes effective spirit of cancer-battling agents. This insight can properly be elaborated by examining the rationale behind the AZT prescription. Mullis also notes that the basic negative effects of the drug are only visible over time. It's a point made already, but it might be interesting to pursue it in another direction by looking at the long-term survivors of AIDS, who are often those who have shunned the AZT solution.

Rationale

As Duesberg explains, AZT had been designed to work as would any other chemotherapeutic drug. "Chemotherapy," he says, "is a rational but desperate treatment for cancer."[1] The toxic drug given in the process will kill any and all growing cells in the patient. After a short round of chemo, "the hope is the cancer is going to be totally dead, and you are only half dead and recover." Duesberg points out that the dangerous violence of the method, which is slaughtering cells wholesale, not targeting only cancer cells, is evident in the side effects. "You lose your hair, you lose weight, you get pneumonia, you get immune deficiency... because it's severe cellular intoxication. You kill a lot of good cells, too."[2]

Writing in *Continuum*, Huw Christie underlines that the logic of cancer chemotherapy, and the reason it can be effective, is that cancer cells are more proliferative than average cells. If one took AZT as a cancer therapy, its possible benefit would be as follows. "Cancer cells... grow faster than normal tissue cells, the idea therefore being that when incorporating AZT, they die more quickly than normally replicating cells too." If everything goes as planned, "when the treatment is finished and the chemotherapy stopped, the normal tissue cells can set about making up for their own lower rate of loss."[3]

By this point, we can see the insanity of trying such a process against AIDS. If AZT is to be given indefinitely, this presumes HIV particles are not being completely wiped out in the way cancer cells would hopefully be in a limited time. Thus, the differential separating abnormal (here, retroviral) particles and normal cells does not obtain, and the advantage of chemotherapy is lost altogether. In other words, the rationale for AZT use has been imported from cancer therapy, while the differences between cancer cell proliferation and (presumed) HIV proliferation are so great that this rationale has lost all meaning.

According to establishment gospel, "the introduction of this DNA chain terminator [AZT] in the presumed presence of HIV is justified by the propaganda... that preventing either the virus or the cell in which it's incorporated from reproducing is helpful to the individual's immune system, even at the expense of gradually killing all other reproducing cells in the body."[4] For cancer, such tactics might work because the body is given time to recuperate from chemotherapy and rebuild where it has lost cells. If a chemotherapy agent is given over a long time span— "Not just for two or three weeks. Every six hours, your HIV-positive person takes 250 mg of AZT"—then as Duesberg explains, "they lose weight, they become anemic, they lose their white cells, they have nausea, they lose their muscles. Like Rudolph Nureyev, they cannot even stand on their own legs."[5]

One might predict the consequences of AZT simply by noting where the most new cells are being created in the body and thinking of what would happen if cell generation were halted in these precincts. Duesberg remarks of AZT, "It is hell for the bone marrow, which is where the T and B cells are made. It's hell for that."[6] Thus, if we follow the establishment and believe that HIV particularly targets T cells, then by using AZT we are dosing people with a drug that specializes in killing the very same cells. The establishment acknowledges this, justifying it by noting that many of these T cells have been infected with the virus. Nonetheless, they don't deny that the drug also eliminates many healthy cells. One might say, then, that the drug is a double-edged sword that both kills HIV (presumably) residing in T cells and helps the disease progress by aiding it in killing healthy T cells. As Duesberg puts it, AZT's "cytotoxic effects on the fast growing cells of the bone marrow, the source of T-cells, are equivalent to AIDS by prescription."[7]

We might recall that T cells are a type of blood cell, so, on top of the health problems for PWAs created by AZT's attack on these cells, the creation of PWAs with shattered blood-generating centers in the bone marrow will have social consequences. As John Lauritsen outlines, "If AZT continues to be administered to thousands of patients"—in 1990, for instance, there were ten thousand patients on AZT—"this will mean an intolerable drain on the blood supply, with many AZT patients requiring transfusions as often as every other week."[8] He elaborates: "It is one thing when someone becomes seriously ill or has an accident or major operation." Such a patient will have a short-term need for blood to replace what has been lost. "But AZT is now creating entirely another category of patient—those whose bone marrow becomes irreversibly damaged, whose continued existence is forever dependent upon the blood of others."[9] Thus, these people with AIDS—we might call them PWAZTs, since it is their taking of AZT that is more responsible for their bodily collapse than their putative AIDS disease—would be sapping the nation's blood supply, a situation that might be justified on humanitarian grounds if the treatment did any good, but that seems reckless since the need results from a drug that has yet to show any but transitory benefits.

Long-Term Survivors

Taking what was reported in the last section about AZT's inability to prolong life, though it may temporarily curtail symptoms, we can extrapolate that if anyone is to be a long-term survivor of AIDS, he or she will not be someone who grabbed AZT as a lifeline. Both physicians who work with PWAs and long-term survivors have reiterated this point.

James Whitehead put it straightforwardly. "Every long-term survivor I know with so-called HIV has not used AZT or any of these so-called antiviral drugs."[10]

When I brought up the topic to Michael Bomgardener in 1996, he emphasized that he both noticed the danger of AZT use as it affected PWAs and saw that doctors who realized the drug's toxicities often were afraid to go against the orthodox grain in openly admitting that AZT had been found wanting.

> I never liked the drugs because the clients that I worked with who were on AZT got sick—when they got off AZT they got better—if they stayed on AZT most of them died, while those people who did not—who refused to take AZT—stayed well.
>
> ... [Moreover, he saw] doctors in the clinic who secretly refuse to treat patients with AZT [because AZT had the same effect as AIDS].... They did that secretly because they were afraid of losing the job if they would have openly suggested not to use AZT for people who are HIV positive.[11]

When I interviewed James Currant, he made some of the same points.

Every time I have seen somebody and I've seen quite a few now, over the last five or six years, I've been investigating this, every time someone has gotten off the drugs, turned their life around, rejected the HIV hypothesis, they have always improved tremendously, and in most cases, depending on how long they had taken these toxic drugs, turned around and went back to life, and became healthy again. Unfortunately, it's too late for many—they were too poisoned.[12]

Finally, we should listen to Michael Callen, who survived AIDS (for seven years at the time of the following statement) without taking the DNA-inhibiting drug. "I wouldn't take AZT if you paid me" is his forthright statement. His position has not always been an easy one. As he puts it, "I've gotten the shit kicked out of me for saying this, but I think using AZT is like aiming a thermonuclear warhead at a mosquito."[13] What has been most encouraging to him, though, is that he is not alone in his uncooperative attitude toward AIDS orthodoxy. In a speech in Amsterdam, he said,

I have been pretty prominent in my opposition to the use of AZT. And I didn't expect to find similar opposition among the long-term survivors, but I did. It was pleasantly surprising. Of the four dozen [I know], only four had ever used it at all, and [of them] three are dead and one is dying of AZT-induced lymphoma. The overwhelming majority of long-term survivors had somehow managed to resist the enormous pressure to take AZT. This was very much a distinguishing characteristic of long-term survivors, by which I mean a skepticism about experimental medications. Long-term survivors, for what it's worth, take a very skeptical wait-and-see approach.[14]

At this point, you may well be asking, "If AZT is a drug fraught with so many problems, why in hell would anyone take it in the first place? Why doesn't every PWA join the skeptical long-term survivors that Callen mentioned and eschew AZT in favor of lifestyle changes, healthy diets, and other immune-enhancing solutions?"

The short answer would be that the producers of AZT have had to face up to the mounting exposure of its drawbacks, and the new idea of a drug cocktail, which we will discuss in the last chapter Part Five where AZT (with some of its shortcomings acknowledged) is put in rotation with other drugs, can be seen as a limited acknowledgment by the orthodox of AZT's doleful impact.

But this doesn't explain why with all its negative features AZT is kept as an ingredient in the drug cocktail at all. So, the long answer will have to explain what forces have been marshalled to downplay scientific evidence and anecdotal reports of AZT's insufficiency. Always remembering our point about the desperation and grasping at straws that motivates some PWAs and their advocates to back AZT, since it's the only horse running, and our earlier commentary on the influence of institutions and the future-world idealist ideology, we next want to look at other factors that have come into play in making AZT look like more than what it is, a dead horse.

In the next chapter Wellcome's business plan will be reviewed so that we can see in more detail how the firm has tried to keep AZT viable and salable by undermining dissidents' views while taking various tacks to extol the drug. From there, in the following two chapters, we will look at how the media has become the willing patsy of the drug industry (with a side glance at a recent myopic article in *Newsweek*); and then see how patients have been traduced into supporting the drug. We will end this examination, as noted, with some observations on drug cocktails.

Big
Business

In "AIDS Inc.," we discussed the confluence of money, prestige, employment, and perks that have accrued to the those working on the disease, and which has made them reluctant to question any of the pillars of orthodoxy. We don't need to rehearse such data in relation to Wellcome, although a couple of lesser points, such as the one I made about how pharmaceuticals are indifferent to the value of organic compounds, no matter how promising their medicinal properties, will be repeated briefly from slightly different angles. There were also points made earlier, such as the connection between the AIDS and cancer establishments, that will be further fleshed out.

However, for the bulk of this part of our discussion, we will look at how Wellcome met the practical task of building sales in an environment where it had to increasingly fend off reports of AZT's dangerousness and lack of utility; and we will look theoretically at the homology among the monocausal (exploded) paradigm, orthodox science, and the commodity form of medicine.

Before any of that, though, a word on Burroughs-Wellcome as an entity and the profits it has won through the exploitation of its drug.

Burroughs Wellcome: Background and Profitability

In 1935, Sir Henry Wellcome put in his will the stipulation that profits from the Wellcome Foundation were to be invested in the philanthropic institution he established, the Wellcome Trust. This trust—something like the Rockefeller

Foundation—is a charitable institution that distributes monies for medical and other research under the direction of retired academics. (Coincidentally, the book we quoted earlier on the foundations of experimental science, *The Great Instauration*, was written with a grant from this trust.) The trust was not simply the product of Wellcome's fortune but had been jointly founded with the American William Burroughs, so it has deep roots in both countries. "By the mid 1980s this trust was one of Europe's biggest medical research funders," Dr. Bruce Halstead said, "The trust was linked with the other major European medical research trusts, and consequently Wellcome-connected scientists staffed many of the university departments and regulatory bodies across the world."[1] The trust not only dispensed largesse but created less public-spirited divisions, such as Burroughs-Wellcome, which is "their largest subsidiary and major profit earning company," according to Walker.[2]

One of the ways BW earned these major profits was by sales of AZT. Arno and Feiden's *Against the Odds* provides a useful chronology: "Milestones in AIDS Drug Development," which includes, among other things, an accounting of the money made by Wellcome in hawking AZT. Consulting this text, we find that in March 1987, BW received FDA approval to market the drug. Shortly thereafter, it set the price of AZT at ten thousand dollars a year. Five months after the drug had been approved, worldwide sales were at $25 million. A year later, in August 1988, yearly sales stood at $183 million, and by the following year they were $408 million. The next year, 1990, they had climbed to $696 million. In 1991 they hit a billion and in 1992, $1.4 billion.[3]

With such a record of profits constantly moving skyward, we can say that, in commercial terms, Burroughs-Wellcome must have been doing something right. We already know this wasn't improving on or making safer its cash-cow product.

How to Increase Sales

The most important way the company boosted its profits has already been touched on: it moved beyond niche marketing by obtaining FDA and other countries' approval to sell to ever-widening groups. The original FDA approval of AZT in 1987 was only for sale to those who already had AIDS. Subsequently, it was approved for those who did not yet have the disease but were HIV positive with low CD4 counts. This was in 1989. In the following year, it was approved for administration to children. With each approval, the market expanded.

Not only did Wellcome trumpet studies that proclaimed the value of AZT and spread largesse to certain key doctors' groups, as we've already noted, but, as sales increased and more money was sloshing around in corporate coffers, Wellcome began giving out funds wholesale to any group involved with the AIDS question, manuring the field, as it were, in hopes of even greater harvests to come. As Hodgkinson noted in the London *Times*, "As sales of AZT have grown last year [1992] reaching 213 million pounds [in England] the company has extended its own funding to a huge range of AIDS organizations, including a parliamentary group to which it has contributed 650,000 pounds."[4]

The company also started programs to give AZT nearly gratis to PWAs in underdeveloped countries. "As part of the 'bouquet of assistance' the drug producer is offering governments of African countries large quantities of AZT against 'a substantial discount.' This way a large group of Africans, who can't afford the expensive drug, can be prescribed 'cheap' AZT."[5] Not only does this practice give the firm's image a sheen of humanitarian benevolence as well as propagate the idea that AZT is the way to go in dealing with AIDS, particularly in its putative heartland, but it dampens criticism in advance, such as that often made of Western pharmaceutical purveyors who are accused of proving drugs on African subjects, then pricing them out of the range of the same people who were used to test them.

However, even with so many effective ways to promote AZT (both those just alluded to and ones we noted earlier) and with such excellent and expanding sales, Wellcome, as would any firm with foresight, was planning for adverse contingencies. In 1990, it had already hired market analysts BdZ (Barclays de Zoete Wedd) to investigate the prospects for the continued marketing and production of AZT. "According to BdZ, the FDA decision to license AZT for asymptomatic antibody positive subjects, that year, had stabilized the market."[6] Nonetheless, according to BdZ, there were a number of not unlikely scenarios that might rock this stability. The analyst noted that the following events might devastate the AZT marketing picture: "the sudden obsolescence of AZT with the discovery of a cure for AIDS; any publicity about debilitating toxicity; any publicity about questionable efficacy; the possible rapid approval for competitor drugs; [or] a decline in the projected AIDS epidemic."[7]

Although some of these possibilities, namely, the cure for AIDS, now seem as probable as Wellcome's board of directors being carried off by space aliens, revelations of AZT's toxicity and uselessness have become increasingly common. We noted this previously but have yet to make clear the complete scope of Wellcome's response.

"In 1993, the delayed publication of the Concorde trial results showed conclusively that asymptomatic antibody-positive individuals who took AZT, died more quickly and in greater number than those simply affected by AIDS-defining illnesses."[8] An article by Walker entitled "HIV, AZT, Big Science" argues that as the immediate (if transient) result of this Concorde bombshell, "the drug went out of fashion as a mono-therapy."[9]

A two-pronged approach was needed to combat this disillusion. On the one hand, "Wellcome, its scientists, and public relations staff… worked hard to rehabilitate the drug and in large part succeeded in burying the implications of the Concorde Trial results."[10] We've already seen, for example, how Wellcome heralded an outdated Australian study that boosted AZT's benefits as a rejoinder to Concorde. On the other hand, a new tactic for selling the drug was promoted that seemed to compensate for some of the glaring deficiencies associated with taking AZT by itself. This was the drug cocktail strategy. (Perhaps we would do better to call it a "drug sandwich," since the medications used did not come in liquid form but pills. Besides, the word "cocktail" gives the combination an unmerited glamor.)

When I say that this new therapy was part of Wellcome's strategy, as it began espousing combos over AZT solus, I do not mean to imply that Wellcome was behind the original creation of this therapy. Rather, what happened was that it took advantage of new treatment possibilities put forward by others.

To add just a word about the drug sandwiches, let me quote an article by Garrett that appeared in *Newsday* on March 5, 1996. "[Establishment] scientists strongly believe that the most dramatic effects on patients will result from taking three or four protease inhibitors and more traditional anti-HIV drugs (AZT and other nucleoside analogs) at the same time."[11] If you thought the cost of using AZT was high, your mind will boggle at the projected price tag for these sandwiches. Garrett states, they "could well cost more than $40,000 a year for drugs and another $20-40,000 a year for lab tests and doctor visits."[12] Not that this would mean wealthy PWAs would be the only ones to utilize this therapy. It's governments that will be stiffed with the bill. In the United States, according to a *New York Times* article appearing around the same time as Garrett's, "President Clinton said he would ask Congress for $52 million more in money to pay for prescriptions for the new drugs."[13]

Later we will have time to consider whether the drug sandwich is actually of benefit to those who take it. We will also say a few words about the screwy logic that seems to run that one can mitigate the harmfulness of one poison (AZT) by mixing it in with a variety of other poisons. The point here is simply to state that it seems the drug mixture treatment was invested in so heavily because it deflected some criticism of AZT while not removing the drug as a vital component of AIDS therapy.

New Business Directions

While Wellcome was able to withstand the buffeting AZT underwent when its weaknesses were exposed by the Concorde and other studies, it also faced the problem of sales expansion. To keep on track with ascending profits, even after positives and children were made eligible for the treatment, the company would have to keep uncovering untapped potential consumers.

In 1990, John Lauritsen sketched, after recounting some of the problems the company faced, what he saw as Wellcome's ongoing strategy to increase AZT's rollout to customers.

It is estimated that there are from 500,000 to 650,000 potential customers for AZT in the U.S.—people who are HIV-positive and have T-4 cell counts below 500. However, there is a serious, though not insuperable, marketing problem here. Most and perhaps nearly all of these targeted consumers are unaware that they carry antibodies to HIV. On top of that, they don't feel sick (probably because they are in fact perfectly healthy). How is Burroughs Wellcome to persuade them to take an expensive drug, with no scientifically established benefits, that will give them violent headaches, destroy their bone marrow, and cause their muscles to shrivel up? How, indeed?

The answer is to conduct a massive propaganda campaign among members of "high risk groups" (meaning primarily us: gay men) to per-suade them to take the HIV antibody test. Those who test "positive" will then be counselled to have T-cell tests done regularly, under the care of an enabling physician. Those whose T-4 cells drop below 500… will be told that they are suffering from infection with a deadly virus.… How-ever, the "good news" is that AZT will "delay the progression," and that with luck the patient may be able to survive for a number of years.[14]

As we know, Lauritsen's envisioning of Wellcome's tactic was both an accurate depiction of reality ten years ago and a prescient prediction of what the general lines of the firm's strategy for the future would be. The only thing he left out was the way the drug producers and establishment figures would have to com-bine an extolling of AZT's virtues, with a warning to all PWAs and positives to ignore the siren songs of dissenters who were pooh-poohing the value of the wonder drug. In May 1997, *AIDS Treatment News* noted the establishment: Drug companies, medical centers and pro-drug community organizations were warning the public of "an organized, well financed campaign which encourages people with HIV to reject lifesaving medical care [AZT or sandwiches]," a cam-paign led by Duesberg.[15] *Treatment News* calls attention to an open letter en-dorsed by "about 20 drug-promoting AIDS organisations" to the U.S. National Academy of Sciences, "complaining about the continuing public campaign of Peter Duesberg," which they claim is influencing PWAs to "put down" (rather than "down") their drugs.[16] Taking a leaf from the dissidents' book, pharma-ceutical company-sponsored groups went so far as to distribute flyers outside a HEAL meeting, attended by Duesberg to decry his baleful influence.

As the dissenters against AZT gathered strength and lung power, a more extreme plan was concocted to pump AZT into bloodstreams, though not by the pharmaceutical firm itself. This idea was hatched in the *New England Jour-nal of Medicine* in 1998. It said there "needed to be mandatory reporting of HIV infection to U.S. authorities. This would allow health agencies to find the part-ners of infected people, which can then be tested and—when tested positive—be prescribed expensive anti-HIV drugs too."[17] Given the downtrending of AIDS cases in the West we have noted and the problems with all these drugs, not to mention the civil liberties hornet's nest such a policy would stir up, it's unlikely that it will be enacted, although it does show the lengths to which the orthodox will go in attempting to prop up a failing prophylaxis.

Cure or Commodity?

As our earlier discussion of Rene Dubos's ideas has shown, the West has entered a vast medical gray area, where few major diseases appear as immediate reactions to infections or other stressors as does a cold, though, like the cold, they doggedly resist all treatment. Diseases such as heart illness or cancer take a long time to manifest themselves and cannot be attributed to any one cause. So far, they have neither been cured nor successfully treated. What I want to point out is, if we look from the perspective of medicine as a business, the irremediableness of these diseases is not a bad thing.

Before explaining that last statement, let me emphasize that decades of medical research and buckets of money have scarcely affected these health problems. Dr. Seymour Brenner notes: "In 1950, the year I started in my practice, 50.6 percent of all people diagnosed as having cancer died," that is, as a direct effect of the disease. He goes on, "In 1990, forty years later, after approximately forty billion dollars have been spent in research, 49.7 percent of all people diagnosed as having cancer die. Seven tenths of one percent improvement."[18]

Writing in the *Wall Street Journal*, M. Waldholz notes that by 1992, drug companies had come to view AIDS as falling in the same category, being an illness that at best can be staved off temporarily but not eliminated. Waldholz writes, "Today, drug makers concede that it may be a long time—if ever—before there's a cure for AIDS."[19]

As I brought up a moment ago, this may not be considered the end of the world. "'AIDS looks to be a disease like cancer or tuberculosis,' says Ralph Christoffersen, senior vice president-research at SmithKline. 'What we're hoping for are drugs that extend life.'"[20]

We've seen that Wellcome and other companies are concerned with protecting their profits, and one of BW's great fears was that a cure would be found. That would make AZT's value go up in smoke. Think further about this, from the viewpoint of such as Christoffersen. For these execs, a palliative that extends life is preferable to a cure.

Why do I say that? Let's say there were a cure that worked as a cancer-busting chemotherapy is supposed to function. After one month or two of sessions, the patient arises from his couch (where he or she received the treatment) cured. The profits accruing to the drug company are at an end. Say, instead, he or she has to take the treatments indefinitely because the disease can't be cured. Here the profits stack up much better. So, there is a real commercially minded yearning for life-extending but not disease-eradicating treatments. The longer the PWA lives without getting better, the longer he or she can be milked.

I don't mean to suggest that a pharmaceutical business wouldn't be happy to market a better treatment. After all, a disadvantage of AZT is that, basically, it doesn't really extend life. If it did, it would be preferable for a drug company, since life extension is the key to profit enhancement. From the purely economic standpoint, prolonging life is preferable to curing anything. And thus, at bottom, the company's primary motivation, to earn bales of money, and the PWAs', to cure their ill health, are at loggerheads. The patients' and the company's goals can be but are not necessarily congruent.

One place where the two sets of actors' goals diverge is an area we have already mentioned. As noted, the desire for a profit from a patentable product excludes research into safe, natural substances. Robert Cathcart, M.D. explains:

> It has been estimated that it takes anywhere from fifteen to fifty million dollars to get a drug through the FDA. If a drug is not patentable, it does not pay for the drug company to do the double blind studies and the toxicity studies on animals or whatever is required by the FDA to prove it. So, we have this interesting situation where nutritional substances, which are not patentable, are excluded from being looked at by the FDA.[21]

When I talked to Dr. Joan Priestley, she backed Cathcart on the same note, saying that "nutrients in high doses are very safe, cheap, non-toxic and effective. They would be the perfect drug if they were drugs."[22]

But here's where the two motivations I talked about, the PWAs' desire to get better and companies' lust for profits, are not parallel. Nutritional substances "cannot be patented. No one can get any commercial benefit from them"; and, in consequence, pharmaceuticals wear blinders around them.[23] This had led Burkett to question the organization of research, where it seems all funded scientists cruise down the same main street—let's call it AZT Avenue—while side streets are avoided as if they were impassable. As he puts it, "Can a research program dependent on the initiative of pharmaceutical companies—and researchers working closely with them—pursue all possible therapies objectively?"[24]

This disjunction is given flesh, as it were, in the form of the counter commodity. I coin this term for AZT-like drugs, not because they go against the general structure of the commodity but because they push the implicit contradiction in the form to such a pitch that they undermine its existence.

What do I mean by commodity? Using the definition provided by classical economics, we can see how it aligns with our conception of the disparate, overlapping interests of two groups, drug sellers and PWAs. We can call the first group's interest in developing any therapy its exchange value (that is, how many dollars it will generate); the interest of people with AIDS in the treatment is its use value (that is, how effective the drug will be in taking away or lessening their pains).

These two values may but probably won't match. (I've already shown that drug companies would tend to favor life prolongers over cures, since the first would register high exchange value. PWAs would prefer the second.) Now, one of the vast undertakings of the drug companies (or our whole economy, for that matter) is to convince everyone that exchange value equals use value. The commodity is that carnival mask that pretends its price tag is exactly equivalent to its value to the consumer.

In explaining why these values are often incongruent, for our purposes it is not necessary to consider such contentious questions as how either of these values are determined in the first place. It is worth noticing that the equating of these values is made difficult by the fact that exchange value is a quantitative measure while use value is basically qualitative. In arguing for a discrepancy between the two values, we can simply point to items such as prestige goods. These are things like designer jeans, where the quality may not be better than that of no-name jeans, but the exchange price is much higher simply because a famous designer's logo is slapped on the hip pocket. Certainly, it could be said that the name adds prestige and thus does have a use value higher than the no-names. We need not go into these subtleties at present. It's enough to say that most economists have agreed that there can be a gap between the two values, especially in an economy so dependent on advertising, which has among its purposes attributing nonexistent values to shoddy products. Typically, this gap yawns when use value is strongly displaced in favor of exchange value. In such cases, the minimal benefits that the consumer gets from the purchase are far outweighed by its high price.

But, if we can set aside the issue of exchange value for a minute, what happens when the good reverses its putative use value? What if one purchases AZT in expectancy of its use value in increasing health by some parameters, but what is actually being paid for is illness? The most duped patient, who keeps taking AZT even as it makes him or her sicker, is in a paradoxical situation. The more AZT he or she buys, the more that will be needed, since each purchase subtracts from the use value (health) presumably being purchased. This defines the counter-commodity, an object whose exchange value can no longer have any connection to its use value in that its attributed use value is a mirage over its opposite.

Burroughs-Wellcome has been put in a rather delicate position, then, in proposing to sell a counter-commodity, needing to reverse the polarity to make it seem that AZT's death's-head was a face of benevolence. Yet impossible as such a task may seem in the abstract, a number of concurrent developments have made this easier.

Walker in an essay in *Continuum* explains, " The fact that people were ill with greater frequency and died more quickly when they took AZT, did not affect the public perception that users of AZT got better, or lived longer lives of better quality than people who unfortunately did not have access to the drug."[25] He attributes this misperception in part to the simultaneous development of better techniques of treating opportunistic infections, which was making PWAs live longer just as AZT was being marketed. The success of the new techniques redounded to the glory of AZT. He also enumerates some of the ways Well-come used to put its drug in the best light and concludes, "Wellcome's strategy of hegemony [of making AZT the drug of choice for AIDS treatment]… was highly successful. In 1992, five years after AZT was licensed, the 44.7 tons of AZT produced that year returned Wellcome over 250 million pounds profit."[26]

When I interviewed Dr. Ralph Moss, he made the additional point that the idea of extravagantly touting a product of dubious value was pioneered by the cancer industry. "The paradigm that was laid down for how to milk the cancer problem is basically the same paradigm which is being followed in milking the AIDS problem."[27] This, he feels, is hardly surprising, since many of the AIDS drug producers (those who followed BW into the marketplace with the drugs that go into the cocktail) had seen service in the cancer war. Moss states, "Here's how it breaks down. The largest producer of chemotherapy [for cancer therapy] in the world is Bristol-Myers Squibb. They make between forty and fifty percent of all the chemotherapy. They also make the AIDS drug DDI."[28]

We also have to bear in mind the point raised earlier. Espousing AZT is, de facto, espousing the viral origin theory of AIDS and all that goes with it. Walker stresses this connection, "Today [in 1997], ten years after licensing, AZT is still used as a gold standard by scientists and doctors who believe HIV is the sole cause of the AIDS-defining complex of illnesses."[29] It is such a mainstay, he be-lieves, both because other drugs have been made in its image and because it serves as "circular proof that HIV causes AIDS."[30]

He put this same thought more fully in this extract:

AZT specifically, and ongoing work by scientists on attempts at anti-viral therapies generally, confirmed in both the public and scientific mind, that HIV was the sole cause of AIDS. AZT was marketed as the cure for a viral condition and, lay thinking went, scientists would not have invented an anti-viral cure if the illness was not caused by a virus. AZT may well have been the first drug in history which defined the ill-ness it was meant to treat, rather than the other way around.[31]

In a discussion with Dr. Bruce Halstead, I learned another value of the AZT doctrine. He put it like this. "Creating one drug to cure AIDS must be prompted by economic considerations, since AIDS is actually a complex of ap-proximately twenty-five different diseases."[32] His basic idea is that belief in AZT's efficacy will distract the public from asking indelicate questions about one of the difficulties with the orthodox view: that people don't die of AIDS, but other diseases, and so it is difficult to say if it is the AIDS or the killer dis-ease that really is to blame.

But let's get back to Walker, who is really a man after my own heart, in the way he has of distilling home truths from reams of material. His final insight is that the intertwining of AZT with HIV doctrine, along with the cozy relation-ships of drug companies, researchers, and physicians has created a kind of large-scale "altered state." He writes, "The production and marketing of commodities creates certain realities and truths which are often far more persistent than the scientific assumptions upon which the commodities themselves are based."[33] Case in point, the selling of the drug under discussion.

In the process of producing and marketing AZT, the Wellcome Foundation set in chain a powerfully persuasive machine which created information, culture and social relations with one purpose, to sell the drug. This network had a life force which would have continued to drive it forward, even if it had occurred that the drug quickly killed everyone who took it.[34]

Although the last clause may be an exaggeration, the general truth of his statement has been verified by all we have come across in our study of the drug and the ideology that it both props up and is propped up by. As it has been incorporated as an element in the core of establishment AIDS theory, it "has achieved a materiality of considerable proportion, and it has spurred an industry."[35]

Media
Power

While Wellcome was benefiting from the way its drug received a cordial reception from the establishment because it capped the reigning theory of viral causation, it would also be the beneficiary, in the United States at least, of a mainstream media that was naively willing to accept the opinions of the establishment at face value.

We've documented the general supineness of the mass media already, so in this chapter all we have to do is apply what we have learned. We'll start by noting some of the pressures on journalists, which, like those on scientists, keep them in line by offering rewards for compliance with orthodoxy and punishments for deviation. Next, we will note the general air of conservatism that pervades the press rooms and broadcast studios, making most people in mass communication eager to conform. We'll end by looking at a couple of cases of how the media slants stories or leaves unmentioned unpleasant facts that do not fit the party line.

Media Manipulation

One way a reporter is evaluated is by the quality of his or her sources. The journalist will be especially respected if he or she can get "inside dope" about what is going on at "important" institutions, such as the White House or Gallo's laboratory; and to get this data involves courting establishment figures.

By contrast, a sure way to lose this important route to the inside track is by questioning the institution's press releases or spokespersons. Tom Bethel explains that if journalists "do not report the story correctly, which is to say, see things our way, then the phone calls will not be returned.... And if you are on deadline, you are worried that your phone calls are not going to be returned."[1] Nina Ostrum, a reporter for the *New York Native*, supports Bethel's notion, saying that if a reporter questions what establishment scientists are saying, "they have the power to cut off access to you so that you don't get the interviews, you don't get the quotes for the stories, you don't get the papers ahead of the time that they're published."[2] Even a reporter working for a progressive outlet, which allows for questioning of establishment opinions, will suffer if he or she can't get easy access to the orthodox, who, even if they are wrong, are having a tremendous impact by their activities, which make up a big part of the AIDS story. Ostrum acknowledges this, commenting of journalists, "Once they're cut off by government scientists, they have a difficult time doing their own job. It is very difficult for them."[3]

Earlier we suggested that science writers loved AIDS because, since it had to do with topics like drug use and promiscuity, it brought excitement to a beat that often focused on such dry topics as the discovery of a dinosaur egg in Utah. Ostrum, to continue with her for a minute, makes the further point that these journalists are happy with the establishment version of AIDS in that it simplifies the complexity of this perplexing ailment. "If they only have one microbe to warn you about, if there's one virus to produce disease, and there's one thing that you should do to prevent it, and that is to have safe sex or not share needles," then one doesn't have to scratch one's head over puzzling ambiguities. So the orthodoxy catechism is "a very easy message for the press to pick up on."[4]

When I interviewed Martin Walker, he applied this rule to AZT. "As soon as Wellcome began to market AZT in 1987 the question of whether or not there were any cofactors was never discussed. AIDS was caused by a virus. The virus could be treated with AZT."[5] Where previously we discussed this situation in terms of AZT's acceptability as a bolster to orthodox positions, here Walker is underlining that it also kept things simple for journalists and was embraced by them for this reason.

Bethel adds that most journalists, like the general public, still hold scientists in an awe that, given the scandals in the last thirty years, no longer surrounds politicians (or for that matter evangelical preachers). He feels that many reporters, given the revelations of government misdeeds and cover-ups since the days of Nixon's Watergate, will take a skeptical attitude toward government statements in several areas. However, "the problem is [reporters are] only ready to apply it [their skepticism] to particular fields like national security, foreign policy, domestic policies." This doubting does not go across the board. "When it comes to health and science they say, 'Uh-oh, I'm not qualified to question this.' They just go along with the official version."[6]

Remember that the vaunted sophistication of reporters, after all, stems from their knowledge of the machinations and scandals that are concealed behind the closed doors of the establishment. It does not arise from any ability to see cracks in the foundation of the house in which that establishment resides.

Lastly, when I interviewed Mark Conlan, and questioned him about why the gay press, with some outstanding exceptions, also hones to the mainstream line on AZT; he reminded me that Wellcome and the other heavy-hitting drug makers can profit from the same type of chummy relationship that developed between these publications and the popper makers. He put it this way: "I would say that the information being published in most of the gay press with a very few honorable exceptions is totally establishment information. Totally down the line."[7] We agreed on that, and he expanded, "The reason for that is partly due to the immense amount of advertising that AIDS clinicians and pharmacies selling AIDS medications [do]… they are a tremendous source of revenue for the mainstream gay press."[8]

Lauritsen made the same point, speaking of BW's AZT campaign in particular. "Gay publications all over the world, from local bar rags to those with international circulations, are now carrying the [Burroughs-Wellcome] Living with HIV ads."[9] After noting that the *New York Native* and the magazine *Outlook* were the only notable publications resisting BW's ad dollars, he quoted an interview with Burroughs-Wellcome marketing representative Joe Disabato, who said he "found little opposition to the ads from the gay press" and continued, "In a way Burroughs Wellcome is giving money back to the community through the campaign. Economically, this will be great for the gay press."[10]

The press generally takes no umbrage at such self-serving pseudo-humanitarianism. It's easy to believe in the honorableness and intelligence of a company that is filling your coffers.

Self-Censorship in Journalism

Many of the topics raised in the last section suggest that journalists, though they may prefer simple, connect-the-dots stories, do have some backbone left when it comes to questioning the government on certain political issues. While that may be true (though events such as the Gulf War suggest it is not), many informed people I have talked with characterize science journalists as a breed that is particularly conservative, hinting that if one were making a chart of the political leanings of typical journalists, science reporters would be slotted right next to the obligatory right-wing pundits writing essays for the editorial page.

One conversation that led me in this direction was with Michael Elliot. After noting, "I've been working in television for most of my thirty years in journalism," he stated, "It's amazing how many… people who call themselves medical journalists only want to recycle official government handouts, breaking it down into bite-size garbage to the general public."[11] Journalists like these don't need to be threatened with the stick of being cut by sources or enticed with the carrot of getting inside dope. They have swallowed whole the establishment viewpoint. "They're very, very wary of interviewing people who don't have the orthodox view. They say it confuses the public or it misleads the public or it may encourage the public not to carry on having safe sex."[12]

Not only do most establishment journalists suddenly grow squeamish when offered a chance to interview a dissident, they ignore other journalists who bring out a heterodox opinion. Lauritsen lamented his own experience. "Those of us who have been trying hard to get out the truth about the 'health crisis'… have been demoralized and disgusted, time and again, by the dishonesty and cowardice of the media."[13] The media's refusal to face embarrassing facts about AZT is what he uses to make his case. "When an article of mine in the *New York Native*," whose results we sampled earlier, "and later an NBC documentary— revealed that AZT had been approved for marketing on the basis of fraudulent research, why didn't the mainstream media pick up on the story?" he asks.[14]

We can answer for him that judging by what we have just detailed, your average mainstream story resembles more a confirmation class in which students win merit if they correctly parrot the minister's exact words, than a debate where all sides are heard and weighed. As Lauritsen summarizes, "With regard to AIDS coverage, the media are exquisitely cognizant of [and acquiescent in] an 'elite consensus'—the consensus of the AIDS Establishment."[15]

Journalistic Bad Faith

To complete this foray into media criticism, let's look at two instances of reporting, one lax and one slanted, both of which favor the establishment position on the value of AZT.

What we might call a sin of omission only indirectly relates to the use of AZT but does bring out more about the celebrity culture, which we saw earlier has become deeply involved with AIDS. While this involvement has helped move the country's attention to the AIDS scourge, it also tends to put conformist and non-questioning attitudes front and center, as the celebrity world is largely one of elitism and establishment thinking.

One of the more distressing ploys of the media—although in this case there is some sense to establishment journalists' statements that they are only giving the mainstream readers or viewers what they want—is that whenever journalists try to draw sympathetic pictures of PWAs, they inevitably portray some hapless white heterosexual middle-class, middle of the roader who supposedly received the illness via a blood transfusion or other innocent route. This is to say, to arouse viewers the media will choose the most unrepresentative possible case, since journalists feel the average audience member cannot possibly feel anything for a person who doesn't closely resemble him- or herself.

However, to create this type of sob story—since real life is a little too quirky to fit such patterns—often calls for cutting and trimming or, at the very least, the omission of embarrassing facts. You may remember what we said about the

Kimberly Bergalis case. This "virgin" who said she had been infected with AIDS by her dentist during routine dental work showed signs of anal intercourse at her autopsy. This fact was conveniently overlooked by the media, whose "story" consisted of creating hysteria about infection being passed by physicians.

The Ryan White story also had some intrusive and inconvenient sidelights, which Lauritsen unearthed. Ryan White was a charming and spunky white youth who died in 1990 at age eighteen after having contracted AIDS five years before via a blood-clotting agent used for his hemophilia. He suffered discrimination from the beginning of his sickness; he was barred from going to school in his hometown, Kokomo, Indiana, because authorities were afraid he would infect his fellow students. Once the courts mandated that he be allowed back in school, he suffered overt hostility from his peers.

His case was a cause célèbre, highlighted on March 3, 1988, when the boy spoke to the White House AIDS Commission, saying, "I came face to face with death at 13 years old."[16] He was surrounded by the famous, including Michael Jackson, Elton John, Vice President Dan Quayle, Senator Edward Kennedy, President George Bush, and Donald Trump.

For the story to work, of course, he had to be as innocent as Bergalis—he was innocent—and he had to die tragically of AIDS. It's in relation to the last part of the plot that the press had to play fast and loose with the facts. For what really happened, according to Lauritsen, was that Ryan White "did not die of 'AIDS' but rather of hemophilia aggravated by Factor VIII concentrate and AZT poisoning."[17]

Here is Lauritsen's version of what happened. Craig Schoonmaker, founder of Homosexuals Intransigent and a friend, told Lauritsen he'd heard on the radio that White "had been admitted to the hospital with uncontrolled internal bleeding."[18] However, after combing the print media for the following days and after White's death, he could find no mention of this bleeding. Remember that hemophilia is a defect that manifests in the blood's lack of a clotting factor and that can lead to death through unstaunchable bleeding. Lauritsen managed to contact, through an intermediary, the Hemophilia Foundation of Indiana. "The people there knew Ryan White very well, and confirmed that hemophilia itself was his major health problem and the cause of his death."[19] As we can guess, Lauritsen lays some of the blame for White's death at Wellcome's door. Since AZT interferes with the production of new blood cells, it disturbs the already fragile internal ecology of hemophiliacs. However, by the press ignoring evidence that it was hemophilia, not AIDS, that killed White, they help AZT maintain its clean bill of health.

If we grant the truth of Lauritsen's findings, then we have to say the media are handy at disposing of facts that don't fit their fairy tales. But what about complaints about dissidents that can't be ignored? An article in an August 2000 *Newsweek* by David France gives us a sample of how hotly, and underhandedly, AZT will be defended by "objective" journalists.

Ironically, the AIDS dissident Christine Maggiore, who the article profiles, has gained notoriety (and thus can't be ignored) because she has her own set of celebrity supporters, mainly rock musicians. It's Maggiore's contention that AZT and other prescribed drugs are decimating PWAs. Further, she believes AZT and other, newer AIDS drugs have "side effects [that] are suspiciously similar to AIDS symptoms, and the drugs are said to be lethal poisons."[20] Note France's telltale "said to be lethal," as if anybody, even the orthodox, doubted this.

However, such subtle digs are the least of Maggiore's worries in this article, which is short on the merits of her opinions and long on innuendo, character assassination, and pop psychology. We find that Maggiore has "no scientific training or college degree"—presumably France has a Ph.D. in retrovirology; otherwise, according to his own norms, how would he be qualified to discuss the subject? Moreover, Maggiore's "emotional" articles "appeal to people with a beef against the establishment." He doesn't get any more specific about who these embittered readers are; perhaps he means everyone in the United States who has AIDS or has been told he or she is HIV positive, since it is they who have been repeatedly lied to by the government, stigmatized by the media, and milked by the drug makers.

France has a lot of space to fill, but none of it is devoted to assessing opposed claims. Apparently, for France AIDS dogmas are equivalent to the Ten Commandments, unquestionable truth straight from God's drawing board. Instead of spending time on what Maggiore says, France goes on about "'flat earth' types who fervently doubt the conclusions of science."[21] France is willing to jettison all of science's basic principles, which turn around letting the best man (or woman) win as contending theories fight for acceptance. Instead, for France, science is what the well-heeled, famous names tell us to believe. Anything else, even if it consists of counterhypotheses set forth by other eminent scientists, is nonscience.

From there, he goes on to a divagation on the psychology of dissidents, quoting a professor of psychology who says, "The basis of denial [of such things as the opinions of AIDS orthodoxy] is a need to escape something that is terribly uncomfortable."[22] The professor Joseph Tecce, we learn, "has studied Holocaust deniers and AIDS dissenters."[23] It's a pity France didn't quiz him further and ask about the meaning of "projection," the concept that guilty people imagine others embodying the negative traits that most characterize themselves. If you were given a multiple-choice test and told to identify the person who is trying to escape the uncomfortable nature of AIDS through pathological means, would you fill in: (a) a person who thinks a new anti-AIDS wonder drug is a great step forward, even though the drug does not lengthen life and tends to produce the same symptoms as the disease, or (b) the person who thinks the drug has little value and says other means of therapy must be tried?

However, such specifics are far less important than the general tone of the article, in which an exercise in avoiding the issues passes for journalism. A profile of an AIDS dissenter, though mentioning what her opinions are, goes no further than that in weighing evidence but proceeds to (1) assail her credentials, (2) tie her to the group that denies science, (3) tie her to the psychologically weak who can't face the fearsome nature of reality. Such underhanded practice could be pulled on anybody. God forbid France were asked to do a piece on Gallo. I guess he would begin by (1) airing Gallo's dirty laundry (such as the attempt to steal credit form Montagnier's group, (2) then link him to misguided scientists, such as those who battled against Pasteur in holding with the theory of spontaneous generation, and then (3) blacken his name a little further by citing some authority on why people conform. The expert would be a college professor who had studied Nazi concentration-camp guards and the National Cancer Institute!

As I've tried to make clear throughout, neither side deserves such shabby treatment, although it's par for the course for the suborned science journalists who populate the U.S. media.

Buying Patient and Professional Support

I've already spoken of how different groups, from parliamentary bodies to physicians' associations to community organizations that support PWAs, have fallen prey to Wellcome's blandishments. As the company's profits have risen, it has doled out more and more money to those who could help it push AZT down patients' throats. It might be appropriate to mention a few more nuances of this strategy, although the most important point that needs to be made is how this marketing fits into BW's innovative fighting strategy.

At first glance, it may seem that there is nothing new in Wellcome's generosity. "Don't drug companies routinely give donations and possibly discounted or free drugs to doctors as well as contribute heavily to politicians who legislate on medical funding?" you might ask. "What's so innovative about that?"

To a degree, this objection is valid, since Wellcome would have shown similar generosity to doctors and legislators no matter what new drug it was bringing to market. However, if we stop at this observation we miss what was qualitatively new about the AIDS situation. Walker brings this out quite well. "Those who suffered AIDS-associated illnesses or who had been diagnosed HIV antibody positive, mainly gay men, were an unknown factor. Pharmaceutical companies had no real experience of dealing with large, youthful, cultural identity groups."[1] In other words, when it came to marketing drugs to large groups, Wellcome, like other drug manufacturers, only had experience in addressing the elderly or late middle-aged. For selling to older people, it would have ties to organizations for the retired and geriatric hospitals, and other such institutions. So, when faced with a large body of the young who were terminally ill, the pharmaceutical business had to think creatively. Its solution was this. It found "the greatest potential for drug pushing was… in the plethora of self-help organizations which were springing up throughout the country." These became key targets, because "here… not only gay men gathered but specifically those who had tested HIV antibody-positive."[2]

Walker lays out what became the firm's business plan once it had pinpointed this crucial audience.

> Wellcome set out to buy up all the self-help groups which had contact with gay men who tested HIV antibody positive in Britain and America. Where they were unable to fund them directly, they gave grants for journals, papers and magazines or for specific projects. There were no overt strings attached to such money but recipients had to adhere to the medical model of AIDS…. The grant funding of self-help groups in the field of AIDS, by vested interest organisations, is perhaps one of the greatest scandals of AIDS medicine.[3]

By "medical model" he is referring to the HIV = AIDS equation and therapies that follow from it. Thus Wellcome would give grants, for example, to hold conferences on AIDS, not suggesting to the organizers that they recommend AZT in any way but insisting that the panelists espouse respectable positions, that is, that they hold AIDS is due to a retrovirus. The company well knew that once this virological stance is taken, drugs like AZT become the only envisionable avenue of treatment.

Let's say a few more words about how Wellcome, once it had happened upon this groundbreaking approach, promoted its wares. For one thing, the company aggressively moved forward to court mainstream AIDS advocacy groups. As Tom Bethel told me in an interview, "Project Inform was warning about [the dangers of pharmaceutical] drug use. Then Burroughs-Wellcome came through with a hundred and fifty something thousand dollars for a whole new computer system. Ever since then, they've been blowing their [Wellcome's] horn."[4] Celia Farber gives a few other illustrations, mentioning the Chicago-based Test Positive Aware Network, which was given "$350,000 from Burroughs Wellcome, which amounts to half of its budget."[5] Money also flowed into the hands of the Treatment Action Group (TAG), an offshoot of ACT UP, "which received $10,000," while ACT UP Golden Gate nabbed two thousand. As TAG founder Peter Staley put it, "It's hard to find an AIDS organization that hasn't taken money from Burroughs Wellcome."[6]

Some may be surprised to see the name of ACT UP or, for the more knowledgeable, Peter Staley, on BW's gifts list; but the savvy company (along with others in the fold of selling AIDS drugs) has gone out of its way to cultivate groups that once poured scorn upon it. "Part of the drug companies' strategy is to court advocacy groups like the Gay Men's Health Crisis that the companies once battled over the pace of research and the price of medications," according to a *New York Times* article by W. Dunlap and L. M. Fisher.[7]

The story mentions another drug company, Agouron, which followed Wellcome's lead in trying to sell its commodity, a protease inhibitor, to groups that help PWAs, including some that once were thorns in the side of the drug-selling establishment. In June 1996, the reporters note, Agouron played host in La Jolla to representatives of groups including Act Up (Paris); the National Minority AIDS council; Women Organized to Respond to Life-Threatening Diseases; the Latino Commission on AIDS in New York; Gay Men's Health Crisis; and Project Inform of San Francisco.[8] Remember that ACT UP is famous for its principled opposition to drug companies, calling them price-gougers and inhumane—in 1989 in a protest action, some members of the New York branch, led by Peter Staley (yes, the same just mentioned rationalizing TAG's accepting of money from BW) barricaded themselves in a Burroughs-Wellcome office in a protest against the high cost of AZT. Although we need to recognize that each branch of this organization is autonomous, it still seems ironic that a drug company would be schmoozing with members of a group that had invaded its offices to carry out protests.

Although Agouron's action may seem a trifle surprising, perhaps Wellcome's courting of the same militants, mentioned earlier, comes as less of a shock, given what we have seen about the company's aggressive and proactive marketing policies, in which it leaves no stone unturned in scampering after profits.

I might add that, aside from approaching various community AIDS organizations directly, giving them funds with the (usually unvoiced) assumption that they would adopt a viral causation view of the immune deficiency disease, Agouron also went a step above these agencies and had amicable dealings with their funding agencies. As Walker notes, "From an early stage they [Wellcome outreach personnel] managed to gain influence on the committees and boards of the major fund-dispensing bodies which acted as gate-keepers for voluntary sector funding."[9] By making friends with "strategically placed individuals on the boards" of the important foundation and grant-delivering organizations, Wellcome "made sure that funds were channeled only to organisations which believed in the use of antiviral drugs."[10]

Another part of its programs for getting its drug accepted as the drug for AIDS and positives had to do with getting countries around the world (once the United States, Britain, and a few others were already in the bag) to endorse AZT. As Walker tells the story, academic seminars served as beachheads in marketing to governments. He begins by noting that "to give the marketing bandwagon maximum publicity, Wellcome organized the biggest world-wide media campaign that had ever been carried out by a drug company."[11]

The thrust of this media blitz was not to interest individual patients or doctors but the state. "The idea was that if Wellcome could sell governments in bulk, the fine tuning of AZT marketing could be left to in-place networks of doctors and scientists in those countries."[12] Each time a government scheduled a hearing on whether AZT should be approved for sale, Wellcome set up a conference, where some would extol AZT, to precede the hearings. "Licensing hearings in European countries were preceded and followed by symposia, geared to attracting maximum press coverage." Walker mentions, for example, such a symposium in Paris that went on in September 1987 and was attended by 180 doctors and journalists with "all expenses paid meeting at the Hotel Sofitel, at Sevres."[13] Later that year, there were conferences in Naples and Ecuador, also timed to government hearing. As a matter of course, the conferees would adopt the HIV = AIDS model of the disease and find good things to say about AZT. Also, as a matter of course, the seminars "were paid for in every last detail by Wellcome, Burroughs Wellcome or one of the other drug companies which made up the Wellcome... group."[14]

Lastly, in discussing BW's marketing scams, oh, sorry, I meant marketing schemes, in which a drug of dubious use value is turned into a commodity with great exchange value, we might turn to the arena where BW and other drug makers lose any aura of scientificity, the kind that is bestowed by such ventures as academic symposia where doctors read papers on experimental research. Of another, very unscientific patina are the AIDS Expos, three-ring circuses where drug companies hawk their AIDS-fighting pharmaceutical products like hardware stores selling flypaper.

The *Times* piece by Dunlap and Fisher mentioned earlier describes the POZ Life Expo of 1996. We've already been alerted to *POZ* magazine's morbidity chic by Celia Farber, but here the publication has worked, along with its many cosponsors, to translate its ethos into something with cash value. Also behind the event were Bristol Myers Squibb, Pharmacia and Upjohn, Roche Laboratories, Glaxo Wellcome (BW), Janssen Pharmaceuticals, Abbott Laboratories, and Agouron.

In two days of activities it brought in seven thousand visitors. The writers comment, "While it was all about life with HIV, it had the high-sheen sparkle of an auto show."[15] Roche Laboratories' booth, for instance, "had large back-lighted panels promoting" its latest breakthrough as well as a TV tape and sales pamphlets in English and Spanish." The Expo also featured "a lounge, with upholstered chairs and a soft-drink bar... and a doctor on call: Paul F. Weber, medical services manager for Hoffman-La Roche... [while] at the Abbott booth, visitors helped themselves to literature, ballpoint pens and 5,000 nylon tote bags."[16]

In a way, that puts a cap on the story of AZT, which started as the brainchild of a government researcher desperately seeking a cure for cancer and ended as just one of a number of AIDS-related drugs (many of them valuable). But all these drugs have had their image degraded in such a glitzo-rama, where the suffering, pain, and frequent nobility of PWAs is hidden by a tide of commercialization.

However, before closing this chapter and summarizing our findings, we need to move to a short discussion of drug cocktails. I've already noted that sandwiching AZT into a rotation with other antivirals can be seen as an attempt to lessen some of the harmful impact of AZT without abandoning it altogether. However, aside from noting that simply lessening the dose of AZT does not solve the problem of the drug's dreadful side effects, I have not considered the value of this type of treatment. Along with the fact that these other components of the drug cocktail also have some pretty devastating collateral effects—and I admit the jury is not yet in on how bad or acceptable these effects are—the use of the cocktail opens a whole new can of worms on civil liberties issues. For, as we will see, certain dangers of the drug cocktail (ones that AZT doesn't have) have made many in the establishment demand that people on this regimen be monitored to see that they don't skip any of the doses. And many say those who miss their pills must be punished, to the point of imprisoning any PWA who misses too many doses!

AZT
Reborn

B y now I imagine that you may be thinking something that can be put like this: "Gary, you claim to be even-handed, willing to seek positive approaches to health wherever they may be found, even, you have said, in the camp of the most rigid orthodoxy. You say that, while you find the dissidents generally have the more sensible and genuinely health-enhancing views, particularly in that many adhere to the emerging paradigm, you stand ready to take elements from traditional medicine when they are valuable. So you say. But when it comes down to it, you are nothing but a sourpuss naysayer, who seems to condemn every bright idea the establishment comes up with, from vaccines to AZT. Now I suppose you will have something bad to say about drug cocktails."

Let me begin by saying it's true I have found little of value in the establishment's antiviral-drug approach to combating AIDS; but, bear in mind, this doesn't at all mean that I think the situation for PWAs is hopeless. On the contrary, following this discussion, I will outline some of the treatments that I believe can successfully cope with the illness. Note also, though, that my criticism of establishment therapies may have been unrelenting simply because they all start from the flimsy premise that the disease can be reduced to the work of a single retrovirus.

Let's take the possibility, which I believe can be argued, that HIV is a co-factor in the disease. Taking this position, we could say that the elimination of it might, for a moment, resurrect certain failing portions of the body, failing because they are under insult by this virus. Yet, the patient would not be on the road to recovery because the collapse of the immune system, which is attributable to causes other than the virus, has not been repaired, nor have those other causes been confronted. Thus, the virus can never be eliminated.

This is an important point. The cocktail solution is much lauded, as we will see, because at first it seems to get rid of all HIV in the blood. Still, even the pillars of the establishment caution that in these situations HIV has not been totally eliminated and will eventually surge back. But where is the logic in that? If all these viral killers, from AZT to protease inhibitors, are awash in the body, why can't they get rid of the virus altogether? How hardy can HIV be? After all, in parallel cancer therapy, where chemo is given to kill cancer cells, the treatment does work in a percentage of the times. The cancer cells are totally wiped out by the toxin. So, why can't these antivirals, which are even more poisonous than cancer drugs, show the same type of success? However, if we adopt the strand of dissident opinion that holds HIV is a cofactor, we can see the solution. If HIV is itself the "opportunistic" response to an immune system weakened by other means, then as long as the other means continue to be active, there is no chance that HIV will go away. Even if some ultraviolent viral drug were to kill every existing HIV particle in a PWA's body, the retrovirus would probably reappear—according to the dissident concept—since it is as common and prevalent in the environment as nongenital herpes, which, as we know, almost everyone has but from which few suffer ill effects.

To go further, to try to reverse the sense that I am totally pessimistic about antivirals, let me say that if some antiviral drug were invented that had minor side effects, it would have a place to play in AIDS treatment. It could be used in a different type of combo, combined with nondrug therapies such as nutritional enhancement, ozone therapy, and lifestyle modification. These last-mentioned therapies would be doing the brunt of the work, though, since they would be addressing the root of the problem, while an antiviral would be handling one of the disease's subsidiary manifestations.

There's no denying, by the way, that the combination method is more effective in reducing AIDS mortality than is AZT. According to a *Lancet* piece, which reported the results of the randomized, double-blind Delta I study, "mortality was lower in the combination groups up to 40 percent as compared to AZT alone."[1] This study compared those groups taking AZT alone with ones taking AZT and ddI and ones taking AZT and ddC. Moreover, recent examinations that have divided subjects into those taking no therapy, those taking AZT, those taking two combo drugs, and those taking three have found the lowest mortality in the last named group, with this set of PWAs showing a three to four times lower death rate than those in the no therapy category.[2]

Granted, that an antiviral could conceivably have a role to play, and even that—if the only choices of therapy that existed were AZT and the combo scheme, then taking the combo would certainly be the better option—it must be added immediately that the protease inhibitors and other components of the drug cocktail (or sandwich) would not be the best candidates for this role because they have multiple negative side effects and limited efficacy.

In this last section on antivirals, I want to look at the drug cocktails, particularly the protease inhibitors, the only new components. (By this I mean that beside the protease inhibitors, the other drugs given are, like AZT, blockers of reverse transcriptase). We will begin by (1) looking at how protease inhibitors and the drug cocktails work and are rationalized. The second part of this explanation will bring out the new theories of HIV progress developed by Dr. David Ho, which are used to justify the cocktail approach as well as the concept of checking viral load to assess disease progress, a method newly minted to fit in with this new therapy. Following that we will (2) look at the problems with the PCR measuring method used to assay viral load, (3) examine some of the adverse physical reactions caused by this approach, and then (4) move on to a study of some of their presumed genetic effects on HIV, which are very dangerous for the PWA. Perversely enough, these genetic effects have so alarmed many researchers that they are calling for monitoring of PWAs' adherence to the drug regimen. with jail or other punishments for repeat offenders.

Multidrug Regimens: HAART

The drug cocktail treatment, which in its latest incarnation is called HAART (highly active antiretroviral therapy), consists of a combination of two different types of drugs, both of which are said to stop HIV replication. The first group, called either nucleoside analogs or nonnucleoside analogs, are those which interfere with reverse transcriptase. We saw early on that this substance is needed by any retrovirus to transform its RNA into DNA. We also saw that it is much more widely used in the body than its discoverers thought, and so any drug that stops its production will be quite harmful to overall human biology. AZT is the most well known of these RT inhibitors, but their ranks now also include ddI, ddC, D4T, 3TC, and Ziagen. The second group, the protease inhibitors, as the name implies, stop production of protease, a protein that is thought to be essential at a later stage of HIV expression, after it has already created its own DNA with host-cell machinery. These include Norvir, Crixivan, Viracept, and Agenerase.

As we saw, AZT was approved for positives after some reports showed the drug would delay the onset of AIDS. However, the rationale behind early intervention with the drug sandwich went this pragmatic sanction one better, in that it was derived from a new understanding in the realms of orthodoxy of how HIV acted when first entering the body. Further, just as the progress of AZT in combating the retrovirus was measured with the new standard of CD4 count—don't forget, it would be a little hard to study the effects of the drug based on appearance or nonappearance of health-threatening symptoms, since, as noted, AZT itself caused sicknesses that were indistinguishable from those caused by

AIDS—so in the same way was the drug cocktail's ability to fight the virus assayed with a new measurement, viral load.

Some details on this new concept of HIV's activity as well as on the intricacies of the viral load theory are certainly in order. *Time* magazine reporter C. Gorman explains the new view of HIV's life story as the flipside of former ideas. "Once not so long ago, researchers believed nothing much happened after HIV gained entry into the body."[3] The virus would be beaten back by the human immune system but like the herpes virus would find some unprotected site in the body in which to conceal itself and vegetate until "something, no one knew what, spurred the microbial invader to awaken…. In this picture, the AIDS virus spend most of its life hibernating before starting its final deadly assault."[4] However, since 1994, Dr. Ho and his colleagues argued that there was no such dormancy in the HIV virus, which shouldered arms the moment it got into a host. "Dr. Ho showed that the body and the virus are, in fact, locked in a pitched battle from the very beginning."[5]

Dr. David Ho

According to Dr. David Ho, billions of copies of HIV are being made every day, which infect billions of T4-cells. These T cells are destroyed not by HIV, but by the immune system. They are replenished every day, but over the years, the immune system loses ground and HIV finally wins. This process was likened to a kitchen sink with the drain open, the water pouring in from a tap (new T-cells being made) at a slightly lower rate than it drained away (infected T-cells being destroyed).[6]

This new view implies a new therapeutic approach. If HIV doesn't do much damage until years after first infection, there is little urgency to fighting the disease at an early stage (not that this stopped many doctors from advocating giving AZT to positives who seemed years away from full-blown AIDS). However, if HIV is carrying on a heavy assault from day one, then early intervention is more than justified. Gorman, imagining he can see the thoughts in Dr. Ho's head, puts them like this: "Dr. David Ho wondered. What if you didn't wait until the end stages of the disease but started combination therapy during the first few weeks of the infection, before too many billion viral particles had formed?"[7]

(There is also another consideration that entered into Dr. Ho's calculation, which has to do with the guesstimated mutation rate of HIV. According to Johnson, "Essential to Dr. Ho's theory is the idea that HIV mutates so rapidly that within days or weeks it has become resistant to whatever 'antiviral' drug the patient is taking."[8] At this point, we simply have to note that this provided further ammunition for the argument that a set of antivirals, not a single one, was preferable in treatment. "In order to prevent this [that is, the development of resistant strains of HIV], it is recommended that the patient take three-drug 'combos' which theoretically hit HIV from all angles simultaneously, thus reducing the chance that a resistant strain will survive."[9] We will not dwell on this at the moment, but later we will see that it becomes of utmost importance when physicians consider the danger of PWAs not taking all their pills on schedule.

To go back to how Ho and his team tried to answer the question about the ability of early drug combo intervention to squelch HIV, Gorman explains that the researcher and Dr. Martin Markowitz tested drug cocktails on two dozen men in the earliest stages of infection. At the time of writing, "Some of the men have been treated for more than a year. None of them show any trace of HIV in any of their blood."[10] Ignoring the fact noted early in this book that it is always hard to find much HIV in the blood, even when a person is dying of AIDS, Gorman sees this as cause for great celebration. "Dr. David Ho believes that prospects for success are good. Assuming that nothing has been overlooked, combination therapy should burn the virus out of the body in two to three years, according to Perelson's latest mathematical models."[11]

Aside from certain perplexing unaddressed questions that the apparently half-asleep science writer let slip by, there is a certain logic to Ho's discovery. (By unaddressed questions, I mean, for instance, these: If this treatment is able to "nip HIV in the bud," as it were, before it gets a toehold in the body, why does it take two or three years to wipe the bugger out? Moreover, isn't it a little premature to base a positive evaluation on "Perelson's mathematical models" rather than the slightest proof derived from studies of actual positives?)

When I say that Dr. Ho's concepts are logical, I don't mean in relation to biology or virology but in terms of career advancement! Let's be quite clear that I am not taking this scientist to task for creating a theory that will jibe with certain prominent interests. As Latour has shown, this is how all science works. It is always created in line with definite extra-scientific interests. Even dissident scientists, I daresay, tend to think along lines that will fit with the interests they hold dear. What separates them from those in the establishment is that they are working in connection with different interest groups. Where an establishment researcher may be acting in accord with the agendas of drug companies, a doctor like Joseph Sonnabend, an early dissident, acts in the interest of the affected gay community that he saw as a practitioner in New York's West Village. Moreover, what I am noting here does not relate to the truth of his claims. Whether, as Ho claims, HIV, if it exists, does multiply at an extraordinary rate from time of entrance into the body is a factual question that is not affected one way or the another by what interests Ho's theory supports. But whether his theory will get a hearing from such establishment organs as *Time* magazine is deeply touched by this question.

I do not have Gorman's ability to read minds, so what I have to say is merely a hypothesis; but let's look at the options facing an intelligent young scientist who in 1994 wanted to make a splash in AIDS science without rocking the boat. We've already made the point that as orthodoxy has gained more strength, it has become more hidebound, jealous of the slightest tampering with its doctrines. If Dr. Ho, then, were to propose a theory that upset some of the faithful's ideas, it wouldn't seem that he would get much of a hearing, that is, unless he could find a counterbalancing group (also among the orthodox) whose influence would fall on the other side of the scale.

Even stating the situation in this way probably makes the solution clear. Although the new vision of HIV both unsettles dogma (which saw HIV as a rather lazy virus that did little till called into action after years of dormancy) as well as moves even further from common knowledge about retroviruses (which, as we saw, are usually quite unassuming benign entities), it would certainly offer a boon to drug companies. The theory spoke to them by saying that these drug cocktails should be given full force from the outset of the disease, and it probably made them see, as they digested the innovative thought, make them see—to paraphrase a famous poem—"visions of dollar signs dancing in their heads."

So, we can speculate that Dr. Ho's theory received so much respectful attention because if accepted, it would be of great benefit to the drug makers. Further, although it was revisionist, it didn't tamper with any of the fundamental building blocks of the viral cause theory. In fact, it might be said to help strengthen them. If Dr. Ho's theory were accepted, it would seem to show that the traditionalists were open to new ideas, and thus it would give them further credibility, although they had to be new ideas that bloomed on a very narrow patch of ground.

And not only did Dr. Ho's theory well please drug makers, but it also would be a boost to those who made certain types of tests. By this statement, I'm referring to my early point that to evaluate the progress or retardation of the disease in this early stage and so measure the effectiveness of early treatment with drug sandwiches, a new assay was called upon, that of estimation of viral load, which was done with a PCR test. If Dr. Ho's theories became generally accepted, the number of expensive PCRs administered would go through the roof, with consequent tidy improvement in the profits of the test makers.

So what is viral load and how are the tests used to determine it? You'll notice I didn't ask what a PCR test is, because, if you think back, you'll recall we already discussed these tests in our investigation of the AIDS blood tests. Then, we noted that these were highly discriminating tests that were used to identify HIV by finding DNA or RNA associated with it in a patient's blood. We noted at that time that using PCR was controversial because scientists couldn't agree on how many different elements from HIV had to be found in the blood for the presence of the retrovirus to be proven. This point was open for dispute, because, for one thing, the discovered elements could be contaminants, and, for another, the elements were not necessarily exclusive to HIV.

Problems with PCR and Viral Load

Let's move to more critical reflections on the verifying tests, before noting the health problems these combination treatments have caused.

Using PCR to find the amount of RNA in a blood sample and then using "mathematical models" to calculate the number of HIV particles in the blood—the last number being what the viral load measurement amounts to—seems par for the course for HIV science. By that I mean, a test that is already disputable in its original form is taken up and pronounced to have extended validity, being used for a purpose it had never been designed for, as the centerpiece of a whole new treatment strategy. Perhaps this is why "although being used to evaluate the success of protease inhibitor therapy, the viral load tests have not even been approved for use by the FDA."[12]

The real wacky part of this is that even with its flaws, when given as a blood test to establish whether a person is positive or negative, the PCR is at least being used in accordance with its designated purpose: to detect whether proteins are in existence in the blood. On the other hand, when it is used to find the size of viral load—that is, when used as a primitive abacus—it is being pressed into service doing something it was not meant for and, as many contend, is not capable of doing.

Dr. Kary Mullis, who you'd think would know something about PCRs, since, as we've noted, he won the Nobel Prize for inventing them in the first place, lays out this objection forthrightly. "Quantitative PCR is an oxymoron." As Lauritsen reports Mullis's opinion:

> PCR is intended to identify substances qualitatively, but by its very nature is unsuited for estimating numbers. Although there is a common misimpression that the viral load tests actually count the number of viruses in the blood, these tests cannot detect free, infectious viruses at all. The tests can detect genetic sequences that are from viruses, or theorized to be so, but not viruses themselves.
>
> What PCR does is to select a genetic sequence and then amplify it enormously. It can accomplish the equivalent of finding a needle in a haystack; it can amplify that needle into a haystack. Like an amplified antenna, PCR greatly amplifies the signal, but it also greatly amplifies the noise. Since the amplification is exponential, the slightest error in measurement, the slightest contamination, can result in errors of many orders of magnitude.[13]

There are three points to bear in mind. First, the PCR was not developed as a quantitative measure, and so to try to use it as such without altering it to fit its new purpose seems like a dubious procedure, to say the least. This perhaps would not be a decisive objection if Ho or others could prove that the PCR assay, whatever its original end, was a tool that could do double duty, also helping with viral load count (not that any such rigorous proof has even been offered). However, the next two points cast doubt on anyone's ability to make such an argument, anyway.

The second is that since the PCR test does not count HIV but proteins said to be in HIV, those using the test not only make estimates of the virus according to a minuscule sample but have to hope the proteins are from the virus and not from another source. An article in *Continuum* quotes Mullis on this difficulty, saying that he "states the problem is PCR is too efficient—it will amplify whatever DNA is in the sample, regardless of whether that DNA belongs to HIV or a contaminant." Moreover, to add another wrinkle, he adds the question, "And how do you decide which part of the amplified material could be HIV and which part the contaminant(s), if you couldn't detect HIV in sample without using PCR?"[14]

Third, and most crucially, the whole point of PCR was to greatly magnify the findings of a tiny sample; therefore, it is very likely that a procedure that is slanted in the direction of enlarging findings will consistently overcount whatever it sets out to enumerate. In this matter, Lauritsen cites an analysis by molecular biologists Peter Duesberg and Harvey Bialy, which "analyzed the 1995 papers by Ho and Wei (*Nature* 373), [the ones] that launched the whole viral load bandwagon," in order to show that such distortions in using PCR tests as we have just warned against will and do occur. The two scientists "found the estimates of free virus had been overestimated by several orders of magnitude. In the Wei study, 100,000 so-called 'plasma viral RNA' units really amounted to less than two infectious viruses per milliliter of plasma. And in the Ho study, 10,000 'plasma virions' corresponded to less than one infectious virus."[15]

Beyond these difficulties, there are a number of more specialized concerns about the applicability of the tests for measurement purposes. One was noted by Luc Raeymaekers in the journal *Analytical Biochemistry* in 1993. Without going into the details of how the PCR is done, we only need to know that it involves comparing two blood samples (one from the subject and one from a control). The critic points out that by studying the published papers on how these tests are being given, the reader learns that "the fundamental assumption that the relative sizes of the samples remain constant is not met in practice."[16]

You'll see that, here again, for a qualitative study—the attempt simply to find out if a certain protein is in the blood or not—this inconstancy is not important. Yet, if a numerical count becomes the object of the PCR test, such a discrepancy is of extraordinary salience. As Miller puts it, commenting on the Raeymaekers article, puts it, "HIV researchers continue to use PCR to quantify viral load [while] there is simply no way of knowing whether a given estimate is correct or is 100,000 times too high!"[17] Todd Miller, another scientist researching this aspect of the PCR, says even more forcefully, "If the relative amounts of your test DNA and your known control are not equal, there is one thing you can say for sure about the estimate of your starting target (the amount of proposed HIV RNA in the patient's blood sample): It will be wrong."[18]

A second problem for the tests has to do with the question of the infectivity of the HIV that is detected by the viral load test. It has long been accepted (by all camps) that not all virus particles are infective. Whether because they are damaged, have mutated into a noninfectious form, or for other reasons, many of the viruses that enter a body during infection are nonstarters. A percentage of these "decommissioned" particles are always mixed with live ones, the size of the fraction varying with the type of virus and due to other considerations. With HIV, this in-mixing of noninfectious forms is considered to be quite high. Again, the problem is evident. This PCR count, even if it worked perfectly, could still not distinguish between active and inactive retroviruses; and so, if we consider "viral" in this instance to mean "infectious viruses," the test would not be so much assessing "viral load" as "viral plus assorted deactivated viruses load."

To the nonscientist reader, this might seem to be an overly fine point, until he or she learns that in the Ho/Markowitz paper in which they set out the premises and rationale for using PCR testing to gauge viral load, they estimate, as an aside, that the ratio of inactive to active is 99.8 percent to 0.2 percent. No, not the other way around. They calculate that 99 percent of HIV particles found are noninfectious.

During the discussion period of an AIDS conference, Dr. David Rasnick brought up this rather glaring deficiency with the test to measure viral presence as well as its implications for the whole concept of the early onslaught by a potent retrovirus. Dr. Mellors, a doughty champion of the combo cocktail, was on the platform when the following interchange transpired, as retailed by Rasnick. He began by citing the large amount of noninfectious HIV reported in this key paper and then extrapolated:

> "Non-infectious HIV, then, is the source of RNA and proteins, includ-
> ing protease, from which the genetics and other characteristics of HIV
> are derived [in the PCR test]." Mellors agreed. (How could he not?)

Now I had him. "Since non-infectious viruses have no conceivable clinical relevance [that is, they play no part in causing disease] then neither could any data be derived from them. "What's the significance of all the non-infectious HIV?" I asked. I had no idea how he could work himself out of this corner, but even I was stunned by his response:

> "The non-infectious particles (HIV) are pathogenic."

Now here was a first. I don't think that anybody's ever gone on record before proposing that non-infectious virus could cause disease.

I sat there flabbergasted, noticing the murmur that had broken out... [and] which continued even after the lecture had been dismissed. On the way out of the room an Indian scientist grabbed my arm and asked, "Did you hear that?"

Indeed I had. AIDS was caused by a deadly army of viral corpses.[19]

It's an exemplary story, although at this moment in our critique, I'm not so much interested in the overall argument against the early-proliferation-of-the-virus hypothesis—we've already gone over the arguments about the lack of HIV in the blood reported by many research workers—as in stressing that the test has many untrustworthy features.

Basically, it all comes down to the fact that new assays of the viral presence are being continually devised because the old ones don't work with a virus that appears invisible to other measures. As Johnson puts the issue, "One of the main arguments against the HIV/AIDS hypothesis is that, when employing traditional methods of virus detection, HIV has never been inferred in significant amounts in people with AIDS."[20] He elaborates, "Virus culture, for instance, has been adequate to find other viruses, but not HIV. Why not?"[21] He feels all the hullabaloo and excitement over a new test would never have occurred if HIV were actually such a hot-blooded, prolific virus as Ho and his group contend, "Indirect methods would not be necessary if a significant amount of HIV were there to begin with."[22]

Others, though, see a more mercenary intent behind the way these viral load tests are being promoted, despite the evidence that they are woefully inaccurate. In the words of Johnson, "Even though no one has shown them to be accurate, viral load assays are being vigorously promoted as state-of-the-art necessities for PWAs, and it's not hard to figure out why": the price tag.[23] "Measuring viral load will cost about $200 per test," while the price of the drug treatments that these tests validate and are taken in tandem with is also astronomical.[24] According to the *U.S. News and World Report* of February 2, 1996, "the yearly cost of a protease inhibitor [is] at around $6,000, and the cost of triple-drug combinations at up to $12,000 to $18,000."[25] In sum, Johnson feels difficulties are being ignored, because for the makers of the PCR tests (and those who make the drugs in the combo), "it is obvious what a cash cow this is going to be."[26] The corporations' attitude, then, is "Damn the dissenters, full speed ahead."

Shortcomings of Expedited Drug Approval Process

There are those in the alternative health movement who see nothing but evil in the streamlining of the drug approval process that came in the wake of the AIDS crisis. Changes were made after the protests and lobbying of both patient rights advocates and drug companies, strange bedfellows. Those who rightly see that this system has quickly led to manifold abuses also need to recognize that there are some positive features to the changes. For one, extricating drug approval from the morass of red tape that had surrounded it was a plus. Moreover, any change that is made due to input from patients helps establish a more democratic flavor to the health care environment, which is all to the good.

However, in our culture, positive reforms have a way of going astray as they are hijacked by special interests. Think of the system by which the major U.S. political parties choose their presidential candidates. In the 1960s, under the banner of progressive forces, the selection system, which had been dominated by cronyism, with each candidate handpicked by backroom party bosses, was restructured to break this influence. What is the result? Now these bosses have little effect on things; but instead of making things more democratic, the selection process is under the heel of corporations and big donors, whose contributions are all important for fighting a primary campaign. A reform that was meant to get rid of wire pullers who controlled things from behind the scenes has merely put the wires in new hands.

In an analogous way, the speeding up of the drug approval process, which was targeted at cutting bureaucratic interference without diluting the objectivity and rigor of the testing process, has often been perverted by slipshod and indecently hasty trials. We already saw that AZT, as it went through Stage II and III trials, fell short in many ways; but at least AZT, if it screwed up, had the opportunity to go through the whole dynamic. "None of the recently lauded HIV protease inhibitors approved by the FDA has yet completed a Phase III clinical trial."[27]

Compounding this laxity in the examination of drug cocktail treatments is the fact that since there are so many drugs, each with different effects, and the goal of the therapy is to give them together in a synergistic way, the only way to determine the best possible panoply of drugs to use would be to try every possible combination. But this would be impossibly time-consuming, expensive, and onerous, yet it would be the only foolproof way to arrive at the best treatment. Rather than follow this route, the FDA has approved the different protease and RT inhibitors one by one with the unexpressed understanding that if a drug is effective by itself, it should also be effective when combined with others.

That at least is the working hypothesis, whose truth value is not to be found in carefully controlled trials, but by letting individual doctors mix and match the drugs and hope for the best. Waldholz put it like this in the *Wall Street Journal*:

> Protease patients are, in effect, guinea pigs in one of the largest and most expensive medical experiments of our time. The new AIDS drugs won Food and Drug Administration approval so rapidly the researchers still don't have a clear understanding of which drugs work best together, when it is best to initiate therapy, and which patients will benefit best from which drug combinations.
>
> For AIDS researchers, there's an upside: A huge "test" population of tens of thousands of patients can now be tracked closely to see whether any of the new drug combinations can stave off HIV for months, years, or even decades.[28]

The upbeat ending cannot hide the alarming nature of these "tests," which the author puts in quotes because they are not tests in the normal sense of the term, which are run with strict monitoring and record keeping, nor are there even controls to indicate whether a patient's progress or lack of it is connected to the drug administered or to other circumstances. These "tests" are simply what casual records a doctor may keep on his patient as he tries different drugs, depending on "what feels right." Dr. Andrew Carr of the Centre for Immunology at St. Vincent's Hospital, Sydney, deplores such a lax idea of testing. "It is therapeutic chaos. Doctors are prescribing what patients ask for, or they're guessing, adding different drugs when they feel like it. I've never seen anything in medicine quite like it."[29]

Even those who stand behind the drug sandwich as a valuable tool in fighting AIDS have expressed dismay about the lack of consensus on how this cocktail is to be mixed. In December 1997, for example, an information packet was mailed out by the Minority AIDS Council, Project Inform, AIDS Treatment Data Network, and Gay Men's Health Crisis that gave facts about the new therapy. After naming the various drugs and saying that the new method was our best hope, the text ended rather lamely, that "we still do not know how to use these drugs most effectively. We do not know when is the best time to start therapy, which combinations are best to use first, how long the effect of these drugs can last, and what are the long-term side effects."[30] This adds a rather glum note to this attempt to be upbeat.

The packet imagines that what will be learned are the best ways and combinations in which to give the pills, but what is also being found, points we can only survey briefly, is that the drugs cause adverse physical reactions, may not dampen the viral load long term, do not seem to affect death rates of PWAs for the better, cause damage at the cellular level, have unknown, long-term toxicities, are expensive, need to be started impossibly early in the disease, and are logistical nightmares. They also cause the danger of mutations, a subject we will discuss below.

Adverse Physical Reactions

Like AZT, all these drugs are toxins, which will inevitably cause some degree of physical damage as they go about their work of poisoning HIV. The drug companies freely admit that the protease inhibitors and other pills prescribed in a typical cocktail can provoke strong negative reactions, although they temporize by saying that some of the more drastic reactions, though they have been observed, are rare.

As Ostrum brings out, the very first approved protease inhibitor was Hoffman-La Roche's Invirase. When unveiling the drug, the company mentioned diarrhea, abdominal discomfort, and nausea as the most common side effects of the drug. However, La Roche also noted the drug might "rarely" hit a user with effects such as "confusion, weakness; acute myeloblactic leukemia; hemolytic anemia; attempted suicide; seizures; clotting problems; exacerbation of chronic liver disease, jaundice, and right and left upper quadrant abdominal pain."[31]

Invirase and the other drugs used in the cocktail also have peculiar effects on weight. "Paunches, buffalo humps in the neck, puffy cheeks and other unusual accumulations of fat are changing the body shapes of any number of people taking combination therapy.... Patients call the effect protease paunch."[32] These manifestations of redistributed weight seem particularly apt to strike women. Toronto AIDS activist and PWA Maggie Atkinson, speaking to women, warned, "Unexpected effects from protease inhibitors can include bizarre fat redistribution in which patients' arms and legs become thin while their torso gains weight."[33] Atkinson had taken the drugs and found, "My arms and legs started to waste but my body weight stayed the same," and she added, "I'm not alone, other women are experiencing these changes."[34] She also noted that women taking protease inhibitors suffered from severe acne, menstrual disruptions, breast enlargement, and diabetes.

In a moment, we will provide a more extensive list of side effects that have been experienced with two of the most common pills given as part of the drug cocktail, but first let's look at more precise studies of selected downsides to these pharmaceuticals, concentrating on the protease inhibitor, since those that block HIV through retarding reverse transcriptase, the nucleoside analogs, tend to resemble AZT in their effects on the body.

A study from the University of California at San Francisco reported in May 1998 found that people with AIDS who took protease inhibitors were developing cytomegalovirus retinitis (CMV), a disease that leads to diminishing sight and then blindness. "Five patients with CD4 cell counts lower than 100 cells/cm3 were administered PIs."[35] Their CD4 counts did improve to above 200 cells/cm3, but "this did not protect them from CMV, and in fact PIs may even trigger it, say the researchers. It was previously thought that only patients with CD4 cell counts below 50 were susceptible to the disease."[36]

Another disease that those who take protease inhibitor must be wary of is diabetes. This information comes out of a report from the FDA. In May 1998 on the Rethinking AIDS Web site, it was reported, "The FDA recently discovered 83 patients who contracted diabetes or hyperglycemia, high blood sugar, or had those diseases suddenly worsen after they began taking protease inhibitors."[37] Five of this group had ketoacidosis, "a dangerous diabetes complication that often results in coma," while twenty-one had to be hospitalized. The FDA went on to remark, "On average, diabetes symptoms struck about 76 days after patients began taking protease inhibitors, although some patients had the first symptoms a mere four days into treatment."[38]

Now, as promised, let me insert an abbreviated list of a number of the pathological states that have occurred in relation to two of the more used of the drugs in this deadly cocktail shaker. I do this not so much to be exhaustive as to suggest the range and severity of the reactions that have occurred.

ROCHE'S INVIRASE
[leaving out the adverse physical reactions already mentioned]

BODY AS A WHOLE: allergic reaction, chest pain, edema, fever, intoxication, external retrosternal pain, shivering, wasting syndrome, weight decrease.

CARDIOVASCULAR: cyanosis, heart murmur, heart valve disorder, hypertension, hypotension, syncope, vein distended.

ENDOCRINE/METABOLIC: dehydration, dry eye syndrome, hyperglycemia, weight increase, xerophthalmia.

GASTROINTESTINAL: cheilitis, constipation, dysphasia, eructation, gastritis, gastrointestinal inflammation, gingivitis, glossitis, hemorhhage rectum, hemorrhoids, hepatomegaly, melena pancreatitis, salivary glands disorder, stomatitis, vomiting.

HEMATALOGIC: anemia, microhemorrhages, pancytopenia.

MUSCULOSKELETAL: arthritis, back pain, muscle cramps, musculoskeletal disorders, stiffness.

NEUROLOGICAL: convulsions, heart rate disorder, hyperesthesia, hyperreflexia, hyporeflexia, dry mouth, facial numbness, facial pain, paresis, poliomyelitis, progressive multifocal leukoencephalopathy, spasms, tremor.

PSYCHOLOGICAL: agitation, amnesia, anxiety, depression, hallucination, insomnia, reduced intellectual ability, irritability, lethargy, speech disorder.

REPRODUCTIVE SYSTEM: prostate enlarged, vaginal discharge.

RESISTANCE MECHANISM: abscess, angina tonsillaris, candidiasis, hepatitis, herpes simplex, herpes zoster, staphylococcal infection, influenza, lymphadenopathy.

RESPIRATORY: bronchitis, cough, laryngitis, pneumonia, rhinitis, sinusitis, upper respiratory tract infection.

SKIN AND APPENDAGES: acne, dermatitis, eczema, hair changes, hot flushes, pigment changes, skin rash, skin ulceration.

SPECIAL SENSES: earache, decreased hearing, taste alteration.

URINARY SYSTEM: micturition disorder, urinary tract infection.

MERCK'S CRIXIVAN

BODY AS A WHOLE: Abdominal distention, chest pain, chills, fever, flulike illness, fungal infection.

CARDIOVASCULAR SYSTEM: cardiovascular disorder, palpitation.

DIGESTIVE SYSTEM: acid regurgitation, anorexia, constipation, dry mouth, dyspepsia, eructation, flatulence, gastritis, gingivitis, infectious gastroenteritis, jaundice, liver cirrhosis.

HEMIC AND LYMPHATIC SYSTEM: anemia, lymphadeopathy, spleen disorder.

METABOLIC/NUTRITIONAL/IMMUNE: food allergy.

MUSCULOSKELETAL SYSTEM: back pain, leg pain, myalgia, muscle cramps, muscle weakness, stiffness.

NERVOUS SYSTEM: agitation, anxiety, decreased mental acuity, depression, dizziness, neuralgia, somnolence, tremor, vertigo.

RESPIRATORY SYSTEM: cough, halitosis, pharyngeal hyperemia, pneumonia, respiratory failure, sinus disorder, sinusitis, upper respiratory infection.

SKIN AND SKIN APPENDAGE: body odor, contact dermatitis, dermatitis, dry skin, flushing, herpes simplex, herpes zoster, night sweats, skin disorder, skin infection.

SPECIAL SENSES: blurred vision, eye pain, eye swelling, orbital edema, taste disorder.

UROGENITAL SYSTEM: premenstrual syndrome, renal colic, urinary frequency, urinary tract infection, urine abnormality, urine sediment abnormality, urolithiasis.[39]

This is a rather chilling list, and similar ones can be provided for each drug being put into this new regimen; but the drug companies can certainly, and with some justification, stress that not everyone will suffer these reactions. The firms argue that the worse reactions only occur in exceptional cases.

Moreover, bear in mind that such lists of adverse reactions as are provided in package inserts accompanying drugs (from which some of these reactions were culled) tend to err on the side of caution. As Koliadin states quite reasonably,

> Read [the] annotation even to innocent drugs (which are known to cause almost no side effects in real medical practice). You might find a frightening list of side effects.... It is because manufacturers are prone to EXAGGERATE drastically, not underestimate, [the] side effects of their drugs... to avoid problems with legal suits in the future.[40]

Not that we are talking of benign drugs here, nor that many of the adversities noted in relation to the drug cocktail ingredients are not quite prevalent. However, these lists do tend to scramble prevalent and rare effects. In fact, I might ask the reader to treat these caveats retroactively, applying them to my occasional previous uses of drug insert messages to prove points.

The dissident might rebut this justification by posing the following question: "If it is admitted that each of these drugs alone can have side effects that rival the notorious ones documented for AZT, isn't the chance of being stricken down by an adverse physical reaction multiplied when one takes a whole slew of these drugs rather than one?"

The pharmaceutical companies that supply the drugs for the sandwich have an answer to this objection—a rather nutty one—but before mentioning it, let's record that a number of researchers have adduced that mixing these toxic chemicals does provoke a greater tendency to illness. In *Positive Health News*, Mark Konlee writes as follows about the effect of combing AZT and DDI: "The combinations have, in several anecdotal reports, led to a gastrointestinal distress, lymphomas, brain tumors, lung cancer, melanomas, MAC, candidiasis, retinitis, PML, wasting syndrome, beer bellies, buffalo hump, and many other AIDS related opportunistic infections and several deaths."[41] He argues that the abrasive effects of either AZT or DDI alone are not enough to account for these health problems. "There must be some toxic drug inter-reaction occurring between AZT or DDI and the protease inhibitors."[42]

It goes without saying—to digress for a moment—that if mixing these prescribed drugs may cause unwanted damage, taking any other drugs with them, whether other prescription items or illegal substances, can lead to disaster. "When illegal drugs are combined with protease inhibitors, the combined effects become more dangerous, as Phillip Kay found out. He died after taking ecstasy (MDMA) while he was on protease inhibitors."[43]

One reason for these sorts of reactions is that the protease inhibitors weaken the body's ability to cushion the effect of powerful substances. "These highly toxic AIDS drugs inhibit certain enzymes the liver uses to metabolize other drugs, often causing much higher levels of those drugs to build up in the blood."[44]

To return to the general problem of the hazards that may arise from the concatenation of the drugs in the sandwich, let me mention the excuse with which the pill pushers turn away fears of negative occurrences due to the piling on of their products in one cocktail. "The drugs' manufacturers point out that there is no overlap in toxicity between the protease inhibitors and the AZT-like drugs, so it is reasonable to take the drugs in combination."[45] Ostrum explains what this innocuous explanation really boils down to: "In other words, while AZT is destroying the bone marrow, a protease inhibitor is causing kidney failure—there's no overlap so, according to the drugs' manufacturers, no problem."[46]

Death Rates and Viral Load

Perhaps the drug companies' facetiousness is rooted in a belief that might be expressed like this: "Of course there may be side effects, even, rarely, extremely harsh ones when these cocktails are quaffed, but with a terrifying disease like

AIDS, such strong drugs are the only ones that can do the job. The fact that these combos quickly decrease viral load and make people healthier is reason enough to take them and praise them. Don't forget, nothing else works."

But are these cocktails as good as they are cracked up to be? We've already seen the unfortunate fact that the long-term effects of AZT fell far short of predictions. AZT regenerated the body and decreased symptoms when it was first taken, but the effects eventually wore off and the illnesses came back. The reappearing opportunistic infections and other debilities were now worse than they generally were for those who hadn't taken the drug, and life expectancy wasn't extended. Moreover, it was said that a lot of these grimmer facts about AZT hadn't been realized because the drug's long-term effects hadn't been sufficiently studied. Certainly, these findings about the cocktail's predecessor should make one wary of embracing this new treatment, especially as its long-term effects have been given even less attention than AZT's!

For God's sake, Robert Gallo himself has reprimanded fellow scientists for recommending cocktails without being aware of the possible consequences they will entail down the road. In an interview in *Newsday*, he said,

> The caveat in all this is that these drugs are toxic.... And the lesson of cancer chemotherapy is that the longer you take the drugs, the greater the toxicity. It's cumulative.... That may not be a problem, if it turns out you only have to take these drugs for two to three years to get full [HIV] suppression. But if the time scale is three to forty years, you will reach toxicity.[47]

Lauritsen gives us a little more detail on just what was left out of the testing of the components of the cocktail insofar as they were approved in a streamlined fashion. He begins that when the FDA gave them the stamp of approval, "their toxicological profiles were far from complete. To my knowledge no reports have been published on animal studies or on such tests as the Cell Transformation Assay, so the carcinogenic potential of the drugs is unknown."[48]

Since the protease inhibitor and other elements of the combos were rushed through the approval process without sufficient time to investigate long-term effects, it is only now, a few years down the road, that we are learning whether they are useful over the long haul. As with AZT, many of the rosier predictions about the cocktail's benefits are turning out to be hollow.

For one thing, the vaunted viral load is not being kept down. (I'm not forgetting that we've argued viral load is a porous and unreliable measure. Here I only mean to suggest that the combination therapies seem to fail on their own terms.) H. Christie interviewed a number of doctors about the continuing efficacy of the sandwiches. "'Viral load is creeping back up even in patients on triple therapy,' said Clive Loveday, a virologist at the Royal Free Hospital, London."[49] Christie also talked to Prof. Ian Weller, who had been one of the key movers in the Concorde trials. Weller's reaction: "My gut feeling is that we've gone over the top again."[50] That is to say, early viral load depression by the cocktail has been erroneously projected as something that will continue until the retrovirus is wiped out. Dr. Anthony Fauci corrects this optimistic impression. With the cocktails, he says, "people do quite well for six months, eight months or a year, and after a while, in a significant proportion, the virus starts to come back."[51]

As viral load increases again, accompanying symptoms also move back to peak levels. An editorial in *Continuum* discusses the case of a PWA named Roemer, who had great success with a cocktail... at first, but then saw all his symptoms reappear. The writer's conclusion, based on this and other cases, is that

> a year and a half after protease inhibitors came into widespread use in the USA, they already seem to be failing 25 to 30 percent of the 150,000 people using them. Increasingly doctors are seeing people like Roemer... diligent determined patients who tolerate the drugs and take them religiously, but for whom the apparent benefits do not last.[52]

And here we are discussing cases where the drugs do not prove so obnoxious to the body that the individual patient can't swallow them. Those whose bodies reject the cocktails are a substantial percentage. "Estimates of how many patients show no improvement or can't tolerate the side effects vary from 15 percent to about 33 percent."[53] In one investigation, German researchers have come up with even higher levels, counting both those who can't tolerate the drug and those who may tolerate it but see no improvement once they are taking it. The Germans saw "a 44 percent failure rate from 198 HIV+ patients treated with PIs [protease inhibitors]."[54] Their conclusion is that this "previously unreported high failure rate indicates that: "the favourable results from controlled studies with antiretroviral drugs containing protease inhibitors cannot simply be translated into everyday clinical practice."[55] Citing the PIs by name, they list a treatment failure rate of 64 percent for Saquinavir, 38 percent for Ritonavir, and 30 percent for Indinavir. Similar reports were forthcoming from a 1997 conference of the American Society of Microbiology in Toronto: "Combination therapy is failing in over 50 percent of patients.... Data from San Francisco General Hospital and elsewhere is 'far different from drug company trials,'" members of the conference claimed.[56]

We do not yet have any scientific studies that compare survival rates of those who drink the cocktails and those who do not, but there is at least one study that has indicated the protease inhibitors do not add much to a PWA's life length as compared to AZT-type drugs. Christie reports that on February 25, 1997, the results were released of a study done by the National Institute of Allergy and Infectious Disease, which compared 1,156 people with AIDS. One group of 579 participants took two nucleoside analogues (AZT, 3tc, or D4T); while the second group, made up of 577 PWAs, took three drugs, adding a protease inhibitor to the mix. "After nine months to a year of treatment, in the two drug group, 63 people either had died or had new AIDS illnesses. In the three drug group, the number was 33. In the former there were eighteen deaths, in the latter, eight."[57] At first glance, we might want to offer qualified praise to the three-drug combo. However, "The study officially concluded that it was not possible to say the difference in death rates was statistically significant" after all factors were taken into account.[58] (For instance, all PWAs chosen were already showing symptoms, and it would be impossible to balance each group so that each had people with AIDS at exactly the same stage of the disease, given that each person's health history is highly individual.)

Any type of definitive opinion about whether the sandwiches will improve the mortality rates of those with AIDS cannot be rested on one study. However, this experiment is certainly suggestive in the following sense. We've already seen that a number of studies, such as the Concorde, indicate that AZT does not prolong life. So, we can extrapolate that if this NIAID study shows little difference between the mortality (and new symptom formation) of those taking AZT and a drug with similar properties and those taking the AZT-type plus protease inhibitor, then the new combo doesn't do much better in keeping people alive.

As we mentioned, this is mere speculation, although we can say, concerning the decline in AIDS deaths that occurred around the time of the introduction of PIs, that the involvement of the inhibitors in this positive outcome is questionable. For instance, some have attributed the 30 percent drop of New York City AIDS-related deaths in 1996 (and New York had 16 percent of U.S. AIDS diagnoses at that time) to the greater use of cocktails. But city health officials said the PIs came on the scene too late. Assistant commissioner of the city's Department of Health, Mary Ann Chiasson, stated that "'the AIDS death rate began to fall before the main drugs were introduced.'… [According to a report of her opinion] she suggested the decline may be linked more closely to better general health practices and more effective treatment of opportunistic infections."[59]

Effects at Cellular Level

We've already gone into details on how AZT, viewed in terms of its effects at the cellular level, disrupts the working of the mitochondria and disturbs the body's natural immune response. How the drug cocktails with protease inhibitors act cellularly is still largely unknown, though judging by how they debilitate the body, they can hardly be benign. While waiting for the jury to come in on this question, we can say that indications so far are that combining AZT with other nucleoside analogs in a cocktail does more or less the same thing as AZT alone: quells immune function and plays havoc with the mitochrondria. No surprise there.

As to the former action, if we may be permitted to add some fresh material to a point already made, which is necessary to a contrast I would like to suggest, we might note a study by B. Tindall et al. After bringing out that HIV infection calls forth a strong immune counteraction with the production of CD8 positive lymphocytosis (cytoxic lymphocytes), which combat the retrovirus, the group presents their findings. "Here we report seven subjects with primary HIV-1 infection who were treated with zidovudine (AZT) and in whom this acute CD8+ response was significantly depressed compared with an untreated control group."[60]

But what is of more concern to us at this juncture is whether similar effects occur with the cocktail of RT inhibitors. It seems that the exact impact of such combos is different, although the broader result, a profound weakening of immune response, is the same. "In one study, AZT was used with Norvir and 3TC. In the first two weeks, there were significant increases in both CD4 and CD8 counts but a 15 percent decrease in Natural Killer cells."[61] That is to say, though a nucleoside analog sandwich does not reduce CD8 counts in the manner AZT does when given alone, it does damp down activity of another vital component of immune fightback, the proliferation of natural killer cells, which would normally be called forth to repel viral penetration. Not only that, as Mark Konlee highlights, "depressing Natural Killer cell activity leaves you with little or no immunity against cancer and lymphomas."[62]

To take up the second cellular level manifestation of the nucleoside analog combo's workings, that is, the effect on mitochondria, we find that here there is little difference from AZT, which suppressed and weakened the action of these cell "factories." E. Benbrik et al. state: "AZT, DDI and DDC all exert cytotoxic effects… and induce alterations of mitochondria…. All 3 compounds can inhibit mitochondrial mtDNA polymerase."[63]

This impairment, joined with the action against killer cells, means that taking the anti-RT cocktail doesn't bode well for the functioning of natural immunity. The most the cocktail prescribers can hope for is that the drugs take the place of as much immune function as they eliminate.

Challenges of Multidrug Regimens

Aside from the health risks of the new cocktails, there are also nonphysical difficulties. For one, although the cocktails are said to work better the earlier they are taken, nipping the virus in the bud as early as recommended is impossible. Moreover, the logistics of taking upward of twenty pills a day is mind-boggling; and the expense of these combo is staggering.

As to when to first start the regimen, Dr. Ho and others insinuate that the cocktails will not be 100 percent effective unless they are taken at the very onset of the illness. Gorman comments, "Even if the treatment works, it isn't practical. HIV positive patients would have to start taking the drugs immediately after infection, before they realize they're sick."[64]

It sounds as if, in order to be safe, the first time one has unprotected sex or shares a needle one should start downing the pills. Crazy as that sounds for more than one reason, some authorities have recommended it. In 1996 San

Francisco's city health director Sandra Hernandez said that "protease inhibitors should be used as 'morning after' pills to prevent new infection."[65] She states blandly, "You can potentially eliminate new infections if you get people started on protease inhibitors plus one or two other drugs." Putting aside the basic fact about the disease—that even if we accept needle sharing and sex as primary transmission routes, few (outside of journalists) think you are likely to get the disease after one such episode—but remembering that the drug combo is one that delivers terrifying side effects, so therefore no one should take it unless he or she is sure of infection, this means tests. Not one but multiple blood tests, since these tests are so fallible. However, Hernandez ignores the elementary need for tests to discover status, since one could hardly have test results on, for example, the morning after a night of unprotected sex. As Beswick states, putting the case mildly, "The problem is how to identify non-health care workers who have been exposed to the virus through sexual or needle sharing situations."[66] Beswick subtracts health care workers from this uncertainty, assuming if one is poked with an HIV-laden needle, he or she will be more aware of the timing of the transmission of the retrovirus. Such heatlh-care-worker cases, though, are an insignificant portion of the infected.

Another point is that taking all these drugs is daunting. "The new drugs can be enormously difficult to take, logistically, even for those who can tolerate them (and afford them)," says Farber. "Patients take handfuls of pills that can make them feel awful day and night, rearrange their lives so their stomachs are empty or full when it's time to take this or that drug."[67]

The use of beepers to remind a PWA that it's pill-popping time dates back to the rigid schedule that had to be maintained for AZT. These strident "alarm clocks" continue to be used for the cocktails, though they have to be set to go off more frequently. Ian Young has written eloquently of the intrusive and nagging power of these little boxes. His editorial was written as remarks on aspects of AZT use, but it can be applied with equal justice to the combos.

> People, both sick and healthy, who have been persuaded by their doctors to take AZT, carry with them 24 hours a day a smooth, slick smartly designed plastic box in a tasteful shade of off-white... equipped with a beeper which goes off every four hours, night and day, ensuring that the carrier never gets a good night's sleep.
>
> [He quotes from a *New York Times* article by a man who carries a device.] "The beeper has a loud and insistent tone, like the shrill pips you hear when a truck is backing up on the street. Ask anyone who carries one—these devices insidiously change your life. You're always on the alert, anticipating that chirp, scheming to turn off in time before it can detonate. It's relentless."[68]

Just as today's harried executive equipped with beeper and/or cell phone can never escape the reach of his or her office, so the harried PWA can never escape the ever-reoccurring beep, which can sometimes sound like death pounding his sickle against the door.

Lastly, to take up a matter already alluded to, we need to emphasize again that since the price of these drugs has gone through the roof, paying for them will put an intolerable strain on any user's budget, that is, for those who have a budget left. As a *Time* writer puts it, "The cost of the cocktails (up to $20,000 a year) puts them beyond the reach of all but the best-insured patients—and out of the question for the 90 percent who live in the developing world."[69]

The last remark is another bitter irony of the AIDS story. It seems that what the establishment considers its best available remedy, these cocktail concoctions, are least available to those it considers in most need of them, third-world AIDS sufferers. "For the majority of those with HIV outside the U.S. and Europe, the cost of the new 'cocktail' treatments seems a cruel joke. The average Kenyan would exhaust his annual income in less than a week on the regimen," Purvis notes. He gives these specifics, "In India, where the government imposes 100 percent to 150 percent customs duty on pharmaceuticals brought from overseas, even a two drug treatment can run to $3,500 a month."[70] This is about seventy-five times the monthly earnings of a poor worker.

Although health organizations, such as the United Nations, have developed programs to make drugs available for lower costs, the reduced prices themselves are still unreachable by the world's poor. In 1997 the *Financial Times* reported on a drug-relief program involving Glaxo-Wellcome, Roche, and Virco, with each company deciding on what lower prices it would charge. "Some products are expected to be 50 percent or more [of the normal sales price]. A world bank report recently warned that it was [still] too expensive to use new anti-HIV drugs in poor countries," even with more affordable price tags.[71]

For dissenters, though, there is one more irony in the situation. Since these drugs are so prohibitively expensive that it appears the pharmaceuticals don't even want to give them away on test subjects, for the first time it is United States and other Western world PWAs who are the guinea pigs, in roles once played by the less financially fortunate.

The Danger of Mutations

We've already noted, without comment, that another worry about protease inhibitors is that they may cause HIV mutations. This, too, was a worry accompanying the adoption of AZT as a treatment. The careful reader will recall that doctors were arguing about when in the disease's course a patient should be started on the Wellcome drug, since they foresaw the pill losing its potency as AZT-resistant strains of AIDS inevitably arose. But the mutation dangers with protease inhibitors are greater, while the measures that have been proposed to safeguard against such mutations border on those usually associated with a police state.

Ostrum explains, "Protease inhibitors act by cutting an enzymatic link in the reproduction of HIV; that is, they change the virus's composition as new generations are produced."[72] As we've seen, these drugs interfere with a particular HIV enzyme's functioning. Ostrum is noting that while this in some cases kills the virus, in others the virus will live on to reproduce mutated forms. If you recall Ho's idea that it will take years to kill off all the HIV, you'll realize that this must entail that a few direct hits with the protease inhibitors are not capable of wiping the viruses out completely, so many must live on and reproduce during the early stages of treatment with the cocktails. "Because of this mode of action, researchers who believe HIV alone causes AIDS fear that discontinuation of the drugs could result in the evolution of new types of HIV that could, potentially, be more easily transmitted."[73]

The logic is getting a little spotty at this point; but there are two points worth bearing in mind. Before mentioning them, let's register some of this argument's deficiencies. Given that mutations are random, it would be just as likely that a less transmittable form evolved under the assault of the drugs than the more transmittable one feared. Further, given that any mutated form will flourish to the degree that it is superior (in some way) to its progenitors, the most likely form of HIV to evolve is one resistant to protease inhibitors! (And of course, this one could develop while the drug is being taken.) Putting the last two points together, we can then say that the evolution of a resistant strain would seem an inevitability, but this new form is not necessarily more transmissible, since its being more resistant to drugs has no known connection to its ability to pass to new hosts.

However, let's imagine for this next part of the argument that the new form is more transmissible. If the hapless protease-inhibitor taker is going to infect someone with a more transmissable HIV—let's say by sneezing on that bystander—after the infected person stops taking the cocktail, why couldn't the infected do the same while slurping the cocktails, since, it seems, these new "Super HIVs" (actual term used by orthodoxy) are just as likely to grow up while the body is being dosed with PIs?

However, laying aside such qualms, which are not pursued in orthodoxy, we can note two other crucial items. First, we see that in one central way protease inhibitors are worse even than AZT, since the latter didn't encourage HIV to mutate (by interfering with protein manufacture), even though it may evolve as the virus desperately tries to get out of the path of the AZT steamroller.

Second, swallowing, as usual, the more alarmist predictions of the drug makers and allied scientists who have been ringing the tocsin about the danger of stopping treatment once started—and regardless of the fact that their arguments are as watertight as Swiss cheese—many public-health officials have begun mulling over (or shouting from the rooftops) that we are going to need strong and sometimes punitive controls to keep these possibly infective PI-taking PWAs in line.

Quarantines

Imagining along with the orthodox scientists (for the sake of our discussion) that new easily transmitted HIV viruses will appear if a patient stops drinking the drug cocktail, then it would seem the PWA better not stop taking the drugs if he or she doesn't want to become a new-fangled Typhoid Mary, spreading AIDS far and wide. Moreover, still entertaining this view, it would seem to be the responsibility of physicians to make sure the PWA doesn't suddenly break off treatment.

Ostrum has summarized the dire course of action seemingly dictated by this understanding of the mutation possibilities. As she has written so trenchantly on this topic, I will quote her at length before moving on to show that a number of health officials and agencies are getting behind coercive programs to ensure that individuals stay on their pill regimens. Ostrum explains:

> In other words, patients who begin taking the drug cocktails and then stop become threats to the public health.
>
> Following this logic, some experts are suggesting that AIDS patients who begin taking the drugs might have to take them forever, to avoid becoming public health threats.
>
> It will, therefore, be necessary to mandate that treatment continue without interruption once it has begun. To that end, some clinicians are now openly suggesting using the tuberculosis "direct observation" treatment method as a model for treating AIDS patients with protease inhibitor cocktails.
>
> In the TB "direct observation" treatment model, patients are required to attend a designated clinic where a public health worker watches him or her swallow the required number of pills... on an established schedule.
>
> If the TB patient doesn't comply with "direct observation"—doesn't show up at the clinic, or refuse to take the drugs—the patient is put in jail and forced to take the treatment until he or she is declared cured.
>
> This is legal; TB patients who refuse to comply with treatment are considered to be public health threats.[74]

We've already noted the half-baked but headline-grabbing ideas of San Francisco's city health director Sandra Hernandez, who said the cocktails should be taken the morning after one contracts HIV. Knowing her penchant for dramatic (and simplistic) responses, it's hardly unexpected that she would be beating the drum for "direct observation." As reported in the *Bay Times*, "Hernandez has a plan to prevent the problem of 'mutant' HIV strains from emerging: the instigation of 'Directly Observed Therapy' (DOT), following the tuberculosis models."[75] Although she didn't mention that this model includes jailing recalcitrant patients, she calls for the adoption of DOT, since it "has been shown to be the most effective means of treating tuberculosis as well as thwarting the emergence of multidrug-resistant strains of the bacterial infection."[76] According to her plan, as reported by ACT UP/San Francisco, "if the patient missed a scheduled dosing or refuses to take the pills, he or she would be jailed and forced to comply."[77]

Joining the rambunctious Dr. Hernandez in her call for the adoption of programs modeled on direct-observation TB drug administration is Dr. El-Sadr at Harlem Hospital's AIDS clinic, who "believes her medical center's success in treating tuberculosis [via DOT] may serve as a model for what's needed to make proper use of the new [cocktail] drugs."[78] The doctor told a reporter for the *Wall Street Journal* that to make sure they show up for their drugs, "as an incentive patients receive subway tokens and free certificates to McDonald's."[79] Again, no mention is made of how to handle recalcitrant patients.

While these experts have been airing their views about the need for coercive pill pushing, others, though these are people on the fringes, have begun taking the law into their own hands, as it were, in designing or putting into practice either underhanded or punitive ways to make sure PWAs are getting their full daily dosage.

Jeanne Bergman, a senior policy analyst for Housing Works in New York City, said to a *New York Times* reporter, "We've actually thought of a kind of adult day care, where you can offer a whole plate of services while overseeing someone's drug regimen."[80] Being a bit more scrupulous than the other health professionals we have been hearing from, she expressed some doubts about the practicality (not the morality, though) of this program. She explained, "But the tricky thing would be getting people to volunteer to something that approximates self-inflicted house arrest."[81]

A far more ethically slippery method is proposed by another medical worker. This method of getting patients to take their medicine lacks the tincture of humanitarianism displayed by that of the Housing Works advocate. A March 2, 1997, *New York Times* article by Deborah Sontag and Lynda Richardson portrays the practices of one doctor who is a rabid defender of the cocktail. "Because one patient 'misses doctor appointment and fails to take medications,' a Dr. Jeanne Carey refuses to prescribe drugs to her."[82]

And let's keep in mind that even the analogy between TB and AIDS is a perniciously false one, in that the pills given to combat TB have been proven to work and eliminate the disease, while the cocktails are hardly more than experimental drugs with known devastating side effects but unknown long-term benefits. As Johnson puts it scathingly but not unjustly:

> Every article on the subject [of the drug cocktail regimen] so far has a different expert guess about how this whole program is supposed to work: no one knows if you can get cured or merely hold the line; no one

knows the long-term prognosis for those who take this triple-toxic triple combo.... Anyone foolish enough to sign up will become a test animal for people who don't know what they're doing.[83]

One last point and then we will mark the subject closed until more information is forthcoming. Civil libertarians have been quite exercised about the dusting off of the direct observation model as a way of pressing cocktails upon balking patients; because it seems to lead to incarceration. However, even if prison were not part of the equation, the idea of forced consumption of medicine itself should be taken as repellent.

To turn once more to an economic metaphor, let's think of the justifications that have been offered for the expense of advertising. After all, with some products, such as Hollywood films, the money devoted to selling the product is greater than that spent on making it and getting it to market. In Baran and Sweezy's *Monopoly Capital*, it is argued that advertising is a core irrationality of our society. To simplify their statements, they hold that if advertising were abolished and people just bought things on their unpromoted merits, a great waste of money would be stopped, and those funds that would otherwise be spent could be devoted to useful purposes. Moreover, these salesmanship techniques, the authors say, can load the gullible down with white elephants that, after the euphoria felt at purchasing the monstrosity has passed, are finally seen to be irredeemably worthless. Thinking along these lines, though, one might wonder how a company would sell a white elephant (like protease inhibitors) whose demerits were becoming increasingly evident? The Orwellian answer establishment health professionals are now proposing to solve the problem of "underconsumption" is one that says in which if you don't "go shopping" (that is, take your expensive pills) every day, you will be put in irons.

We've now completed the (admittedly long) characterization of the medical and ethical collapse of establishment AIDS science. In the conclusion of the book, we will take up and renew some of the grand themes broached so far, such as the rise of the new paradigm and the necessity for a more democratic system of health care, one that puts patients first and pill pushers last.

Although perhaps not succeeding as well as the dispassionate reader might like, I have tried not to paint everything in black and white, as if everyone in the establishment were blind and everyone who dissents from them were seers. Often enough, we've catalogued how key orthodox experts, from Montagnier to Gallo, have spoken out against certain hallowed ideas; and, at the same time, I've expressed strong reservations about many thoughts of Duesberg and other dissidents. Again and again, I've argued that key questions about AIDS—such as whether HIV plays a role or whether it exists at all—cannot be settled with the evidence we have. This means—although I may not have always have put it like this—that the establishment may be right on every disputed point. (Of course, on other topics I have shown that the dissenters hold all the trump cards.) Thus, I have tried to suggest that though I stand with the dissenters, there is still some truth on the other side, and some exaggeration, falsity, and sharp dealing among those who advocate an alternative perspective. I might echo Coleridge's words in "Fear in Solitude" where he speaks his equally biting advice to his country on its political situation.

O Britons! O my brethern! I have told
Most bitter truth, but without bitterness.
Nor deem my zeal or factious or mistimed:
For never can true courage dwell with them
Who, playing tricks with conscience, dare not
Look at their own vices.[84]

PART SIX

ALTERNATIVE TREATMENTS

The New Atlantis

B acon was dead set on moving the king or other influential members of society to sponsor scientific research as a means to develop new inventions that would redound to the health of the commonwealth. Instead of the open hearts he hoped to find when he offered his plans, however, all that his proposals met were deaf ears. Yet he was not downhearted; if he was, it did not deter him from constantly trying different avenues of persuasion.

We've seen that in his *Advancement of Learning*, he dared to assail all the learning of his time, pointing to huge gaps in knowledge, noting that current methods were not producing anything of salient value (books were begetting more books, but nothing to the people's benefit) and assailing scholars' obsequious adherence to dead authorities, even when these past masters were patently deficient. However, as we've also seen, this demolition work was not the only means he used to put forward his program. In *Novum Organon*, he laid out the principles of experimentation, insisting that this method was the best way to learn more about nature on the road to increasing mankind's control over it. Moreover, in *The New Atlantis*, a book we have not yet considered, he attempted to rouse interest in his ideas through the world's first work of science fiction!

The New Atlantis is a utopian romance, but unlike what happens in other books in this favorite Elizabethan genre, the reader does not find peculiar customs or bizarre social arrangements; instead he or she takes a voyage to a society whose center is neither the royal court nor the countinghouse, but the laboratory. (Thus, I dub it a founding work of science fiction.) A ship runs off course and ends up on a mysterious but well-run island. The book is a direct rejoinder to More's *Utopia*, which it refers to sneeringly. In More, the described country is vastly superior to others in the world because of new social arrangements. There is no private property and the rulers are elected democratically from among the learned. Bacon, on the other hand, leaves a replica of the English social structure intact, with a king and an aristocratic class, but founds Bensalem's superiority (in wealth, fealty of the people to their monarch, health and good disposition of the populace, and so on) on its creation of the House of Solomon (a kind of National Academy of Science)—which devotes its time to scientific experiments, many of which eventually issue in new inventions that make life easier, happier, and more healthful.

This house also bears another name, the hero is told.

> It is sometimes called "the College of the Six Days' Work"... [for, according to the Bible] God has created the world and all that is therein is, in six days, and therefore he [the founder of Bensalem], instituting that house for the finding out of the true nature of all things, whereby God might have more glory in the workmanship of them, and men the more fruit in their use of them, did give it [Solomon's House] also that second name.[1]

I might add that where More's *Utopia* is still enjoyed today, Bacon's foray into science fiction gathers dust in libraries or is grudgingly taken up by students as assigned reading. Bacon lacked the common touch, writing here with an uncharacteristic wooden style and seeming (if he wanted to catch the popular imagination) to place the wrong emphases. For instance, he quickly passes over the scientific wonders of the island (the type of things readers want to hear about, such as submarines and airplanes) with a brief word or two, then devotes pages to minute descriptions of pompous ceremonies.

I bring all this up to prepare a large analogy. If the first part of our book can be called a combination of *The Advancement of Learning* and *Novum Organon*, since it charted both the deficiencies of mainstream science and suggested how science should be conducted and judged, then these final pages combine *The Advancement of Learning* with *The New Atlantis*. It will be recalled that *Advancement*, while largely disapproving of contemporary science, did highlight a few bright spots where genuine knowledge existed and was being created. So, immediately following this chapter, I will begin examining treatments that avoid the morass that so much AIDS therapy has fallen, treatments following the lead of the new paradigm, concentrating on building up the body's natural powers rather than focusing on fighting a germ. At the very end, I will trace out my own utopia, though not by hatching an ideal arrangement of medical care from a nest in my brain. I will simply note how both the three types of alternative therapy outlined and some of the political struggles around AIDS I've already mentioned suggest how we might create a radically different and saner way of conducting health care in our society.

Examining a New Approach

How can the fear and suffering of those who become infected by HIV be lessened? Results of conventional treatments tell us that there is a need for a shift of emphasis. Uncommon physicians report success with treatments that are getting less attention than they deserve. It is our belief that additional resources should be devoted to certain aspects of this public health problem, to increase the likelihood of ameliorating it.

This part of the book will look at examples of alternative ways to understand the health disturbances that now fall under the rubrics of "asymptomatic HIV infection," ARC, and AIDS. We then will look at examples of alternative approaches to treating these disorders. The objective is to show some current directions for therapy that seem deserving of more intensive and extensive testing.

We've already said that the new paradigm directs us to put aside narrow ideas of helping the sick that focus simply on exterminating some disease-producing agent. Instead the way forward in this new way of thinking is to look at the patient's whole environment in making the diagnosis, not only analyzing infections and obvious illness-creating components but also looking at toxins in the environment, unhealthy lifestyle practices, and stresses, even up to the mental attitude with which a person views her or his own body. A totally holistic therapy would work in all areas at once, aiming to direct a person toward positive health practices while helping the body make the best possible use of its own defenses and healing capacities. The three doctors we will meet shortly all practice holistically, though for purposes of exposition I will highlight a particular key element of their treatment that has proven particularly successful. I begin by isolating each of these special components, but this does not mean that any of these natural healing methods alone will do all the work of repairing a weakened body without complementary lifestyle and attitude changes (to the degree these are needed).

The first method that will be examined is aimed primarily toward supporting the organism's own mechanisms for healing and defense, by bringing deficient nutrient levels back toward optimal levels (which for the diseased organism may be higher than those considered "normal," if there is a need to compensate for lowered abilities to absorb and/or utilize those nutrients). This supporting approach may also use herbal products that have been found to strengthen immune responses or other body functions, even where the exact mechanism of these effects remain unknown.

The second direction involves a sophisticated but highly nonconventional approach to normalizing the body's metabolic balance, mostly through use of special fatty acid and trace metal preparations.

The third direction is a therapy that uses an elevated yet apparently entirely safe blood level of the triatomic form of the most common element in our body, oxygen, employing it for its effects on (a) the immune system, (b) the metabolism of other cells, and (c) viral activity and replication.

Then, examples of the application of these therapies to HIV infection will be given. Nutritional therapy will be exemplified by Dr. Christopher Calapai, metabolic balance adjustment by Dr. Emmanuel Revici (who originated and is the main exponent of this approach), and ozone therapy by Dr. "Z" and Dr. John Pittman. In each case, we will look at patient records to help us assess the benefits of each therapy.

Nutritional and
Lifestyle Changes

When HIV-infected patients who had been asymptomatic or had early ARC are evaluated, they usually exhibit decreased body mass, body fat, and total protein, which may result in part from the frequently reported gastrointestinal dysfunction, diarrhea, and malabsorption.[1] This malabsorption may be an early and continuing consequence of HIV infection, for "small intestinal malabsorption is a major component in the severe wasting seen in some HIV-infected patients with chronic diarrhea."[2] If this is so, and the virus creates problems in nutrient uptake at the same time, these problems will stimulate the virus. As an article in the *Annals of the New York Academy of Sciences* puts it, "Once occurring, malnutrition leads to immunosuppression, infection, and mucosal damage, with failure of normal intestinal mucosal turnover and healing, resulting in further malnutrition."[3]

(You'll note that this therapy assumes HIV plays a cofactorial role in AIDS.)

Malabsorption in HIV-infected patients, at an early or a later stage, might also result in part from defective gastric acidity (possibly from an autoimmune reaction against gastric mucosa). For example, in forty-eight AIDS patients the mean fasting pH of gastric juice was 5.9, compared to 2.9 in controls.[4] The resulting multiple depletions are insidious and progressive.

Malnutrition in HIV-infected patients can result not only from malabsorption of nutrients but also from anorexia (high in HIV positive patients), and from high resting energy expenditure (REE), which also occurs with HIV infection.[5] Whereas in malabsorption, the body is not drawing sufficient nutrition from ingested food, in REE, though the nutrition supply may be adequate, more energy is being expended than is warranted by the processes taking place. The latter may be quite important in AIDS, as Melchior et al. put it in the *American Journal of Clinical Nutrition*: "In stable and malnourished HIV patients, the progressive wasting may be partly related to an increase in REE," even in the absence of opportunistic infections.[6]

This is very interesting, for an increase in resting energy expenditure may characterize other kinds of tissue damage that have no direct relation to AIDS, yet share an impairment of immune responses. These metabolic imbalances may also be aided by nutritional therapy.[7]

Malabsorption associated with HIV disease has other effects. It may be that the cognitive deficits ("AIDS dementia") that often characterize later stages of the disease result at least in part from B12 malabsorption.[8] Vitamin B12 malabsorption is common in AIDS and may be a very early manifestation of HIV infection.[9] Correspondingly, cognitive deficits may occur early in HIV infection.[10] Psychological functions are quite sensitive to vitamin deficiencies; and in early AIDS patients, various measures of cognitive function were found to correlate with B12 levels.[11, 12] Further, neural damage from demyelination may be a consequence of B12 or folate deficiencies in HIV-infected patients.[13]

Thus, boosting nutrient levels would seem to be a called-for treatment. This therapy can begin early. While most of this data has been gathered from patients with advanced disease and opportunistic infections, Dr. Calapai thinks it is critical to look at the nutritional state in a person as soon as he or she is diagnosed. He looks at this state as soon as a patient comes to his office, "rather than waiting for some medical problem to occur, and then looking at how to play 'catch-up' medicine to help the person back to health." As he sees it, the whole concept of preventive medicine is to make periodic assessments of the healthy person; and when you find things that are abnormal or problematic, you must correct them before they end up causing major problems.

Nutritional therapy was pioneered in working with patients who had multiple organ failure following trauma or surgical infection. Cerra et al. noted that after major surgery, shock, or acute infection, what often occurred was "an abnormal metabolic regulation that appears to result in persistent inflammatory response with associated immune suppression."[14] In their work in combating such metabolic weaknesses, Cerra used "substances such as arginine, n-3 polyunsaturated fatty acids, and RNA... for their ability to modulate inflammation and improve immune function." His group found a good response to this program. The results "suggest that the administration of the added nutrients is associated with the return to more normal responses of the immune system."[15]

This type of treatment has also been used successfully with the elderly. When older people are institutionalized, they are at risk of both malnutrition and infection. "To determine whether supplementation with vitamins A, C and E would improve cell-mediated immune function in elderly long-stay hospital patients, researchers from St. James University Hospital, Leeds, conducted a controlled trial."[16] A group of thirty was divided into those who received the supplementation and those on placebo. "Several indices of immune function... showed significant improvement in the treated group but not in the placebo group."[17]

Nutritional support of immune function with antioxidants also appeared in a geriatric study, which concluded: "Improving the antioxidant status of the aged might be very beneficial in slowing the decline of immune response, which could then decrease the incidence of severe age-related diseases."[18]

If we look into studies that have measured the effect of nutritional supplementation on people with AIDS, we will find good work indicating that such treatment can not only be useful in helping the body gain the capacity to fight the incursion but aid health in other ways, such as by directly countering the retrovirus and helping the body absorb the harsh antivirals.

It has been established that vitamins E and B6 are necessary for immune response, "and supplementation at higher than RDA levels may be necessary for optimal immune function."[19] Beta-carotene may also prove useful for raising immune competence.[20] The usefulness of this last vitamin was shown in a "trial with eleven HIV-infected patients [which] showed rises in numbers of natural killer cells and activated lymphocytes after three months treatment, with no evidence of clinical toxicity."[21] A second experiment has shown that beta-carotene will generally augment immune function for healthy people; while a third study, this one again with HIV positives—twenty-one were in the study—"showed a 17 percent higher CD4 (helper t-cell) count during a four-week period of beta-carotene than during the placebo period."[22]

While, as we have been discussing, several sorts of nutrient augmentation improve immune function, and thereby aid the body in coping with infection (including that by HIV), some nutrients—if in adequate amount—may have more direct effects upon infecting organisms. For example, some trials indicate HIV is suppressed by ascorbate.[23] One of the most interesting investigations of the subject summarizes comparisons of the anti-HIV activities of ascorbate (vitamin C) and thiol-containing reducing agents like glutathione and n-acytl cystiene in chronically HIV-infected cells in vitro.[24] They report several studies showing that vitamin C both inactivates HIV and inhibits its growth. After showing the last using infected T lymphocytes, the researchers conclude: "These results further support the potent antiviral activity of ascorbate and suggest its therapeutic value in controlling HIV infection in combination with thiols."[25] This is collaborated by another examination of ascorbic acid alone, which was shown to inhibit the activity and growth of the AIDS virus.[26]

Besides increasing immune system potency and striking against HIV, nutrients may aid the use of more conventional antiviral medications. It has been suggested, for example, that vitamin E has clinical potential in ameliorating the fetal and bone marrow toxicity of AZT.[27]

Granted that certain vitamins or other nutrients can help the body tolerate antivirals and hinder the proliferation of HIV, still the main focus of the treatment we are going to examine is how various nutrients enhance immunity and improve absorption. Nutritional treatment must start from the knowledge that individuals that have full-blown AIDS have overwhelming nutritional problems resulting from malabsorption—whether from secretion defects or gut infection. One way Dr. Calapai addresses this problem—we are only going to take a representative example of his treatment now, since we will profile it more fully below—is by putting patients on total parenteral nutrition (TPN). He does this when the PWA is not eating and not absorbing terribly well. TPN is a way to try to bypass the gut and get nutrients into the bloodstream.

Others have dealt with the problem of poor absorption by using diet supplements given directly into the intestines (enteral diet). Some nutrients, whether given enterally or orally, have been shown to enhance immunity by directly affecting the lymphoid system and immune cell function. These would include certain amino acids as well as omega-3, omega-6, and short-chain fatty acids. Arginine has been shown to enhance the function of T cells and reduce infection. "Fish oils containing EPA were shown to have beneficial effects in inflammatory skin disorders.... They may also be useful in treating autoimmune diseases."[28]

Nutritional Depletion and Augmentation

Vitamins and Other Nutrients

The main object of this chapter will be the examination of Dr. Calapai's practice; however, in discussing his extraordinary work, we will be looking at the select vitamins and other nutrients he has employed in reempowering the immune system and aiding absorption. We will thus be leaving in the shade many other supplements that have shown some effectiveness in fighting AIDS but are not featured in this practice. The use of supplementation is an area in which more work needs to be done. Still, many studies that prove the benefits of different nutrients in combating the immunodeficiency disease have already appeared, but they have received little of the recognition they merit.

Having immersed myself for many years in the scientific literature that examines the health value of different vitamins, minerals, and other supplementation, I thought it might be of use if, before we examined Calapai's practices, I provided a compendium of some recent findings on different substances in relation to their alleviation of AIDS or AIDS-related diseases. I will also include my notes on substances such as bifidus, which have not been tested against AIDS but have been proven to support the immune system or aid absorption, properties that likely would be helpful in combating immunodeficiency diseases.

So before outlining Calapai's work, I will make a brief mention of studies reported in refereed journals concerning these nutrients. Notes refer the reader to the exact citation so that investigations can be consulted further at the source.

Bear in mind that one study should not be given too much weight; but as the positive findings pile up about a nutrient, as they do in relation to many listed here, one's level of trust in the value of the supplement ay increase.

ACETYL-L-CARNITINE

Results of one study show that six grams (g) per day of L-carnitine for two weeks improved lipid metabolism as well as immune response in AIDS patients receiving AZT.[1] This indicates that L-carnitine will dampen some of the toxic effects of AZT; although our stressing of such a type of investigation, here or later, should not be taken to be a backhanded endorsement of AZT and other antivirals, whose demerits have already been amply illustrated. Rather, they indicate that PWAs who feel they must take these pills can supplement them with other nutrients that will act to counteract some of the known side effects.

Acetyl-L-carnitine has also been shown to help in such AIDS-related diseases as dementia and senility.[2] As to the latter, one double-blind, placebo-controlled study examined the effects of 1,000 mg per day of L-acetylcarnitine on the senile human brain. Results showed treatment led to significant improvements relative to controls.[3]

ACIDOPHILUS

Acidophilus is a beneficial bacteria that is normally found in the intestine, where it transforms various substances into lactic acid, consequently creating an acidic environment that keeps down the existence of various illness-causing bacteria. A report in *Gut* testify to this antibacterial power, showing that human Lactobacillus acidophilus strains had an inhibitory effect against *Escherichia coli* (aka *E. coli*), *Salmonella typhimurium*, and *Yersinia pseudotuberculosis*. [4, 5]

The bacteria's ability to stop Candida in its tracks is also well known, as is its general aid with other intestinal problems. For instance, an article in *Digestive Disorders* notes that preparations containing viable lactic acid bacteria of human origin may be of value in restoring normal microbial function and reducing symptoms in patients suffering from gastrointestinal infection and related conditions.[6] A similar study was reported in *Clinical Teratology* where a mix of *Lactobacillus acidophilus* and *Bifidobacterium bifidum* was shown to be effective against gastritis and duodenitis.[7]

Since the elderly are prone to digestion difficulties, particular attention has been paid to their response to acidophilus therapy. Results of one study showed that a multibacterial combination consisting of *L. acidophilus* and *B. bifidum* administered to elderly patients with bowel disorders proved effective with respect to restoration of duodenal bacterial flora and subsidence of clinical symptoms.[8] A second study looked at sixty elderly patients with symptoms of diarrhea, abdominal pain, and meteorism. Following a seven-day washout period, all patients were given six capsules per day consisting of live lyophilized *L. acidophilus* and results showed significant relief of symptoms in all patients.[9]

Moreover, acidophilus can help restore the microflora in the intestine that have been weakened or wiped out by antibiotics. This is directly relevant to AIDS; as we remarked earlier, one dissident theory has it that the killing off of beneficent bacteria in the gut due to constant dosing with antibiotics or by other means creates the internal environment in which harmful bacteria can become prominent inhabitants. An investigation of the effects of *L. acidophilus* on gastro-intestinal side effects of oral broad-spectrum antibiotic therapy in outpatients with ear, sinus, or throat infections indicated that *L. acidophilus* in combination with amoxicillin/clavulanate was correlated with a significant decrease in patient complaints of gastrointestinal side effects and yeast superinfection, with 89 percent reporting the elimination of infection. Similar findings were reported in a separate inquiry in *Nahrung*.[11] Beyond that, studies have noted acidophilus's effectivity in countering side effects of specific antibiotics.[12]

Knowing that a breakdown of digestive functioning and absorption in the intestines is a symptom of AIDS, and then seeing the value of acidophilus in combating a number of intestinal problems, such as the increase in malign bacteria in the region, it's a shame more studies haven't been commissioned on how this bacteria might affect PWAs. We do know studies have appeared showing that acidophilus enhances immunity. An article in *Immunology* concluded that *L. casei* and *L. acidophilus* enhanced the activation levels of immune systems, while an animal study showed that mice given 50 micrograms per day of viable cultures of *L. acidophilus* and *Streptococcus thermophilus* experienced significant enhancement in their immunity.[13, 14]

Further we possess one study that was reported in the *Journal of Experimental Medicine* that reached the tentative conclusion that the survival of HIV in the female genital tract and possible transmission may be inhibited by *L. acidophilus*-peroxidase-halide system activity in the vagina.[15] Nonetheless, there is a crying need for more work on this bacteria to find out what bearing it may have in alleviating AIDS-related problems, such as the adjunct proliferation of harmful bacteria in the GI tract.

ALPHA-LIPOIC ACID

This is an acid manufactured by the body in small amounts to help with carbohydrate metabolism. It has antioxidant properties; that is, it will act to deactivate free radicals. It is widely used, for example, in Germany, in treating diabetes, because it decreases the body's resistance to insulin and decreases the symptoms of the disease. A handful of studies have been done suggesting that supplementation with this acid increases resistance to HIV. The two most widely cited are those done by Baur and by Suzuki.[16]

BIFIDUS

Bifidus is a supplement that contains *B. bifidum*, which, like acidophilus, is a gut bacteria that plays a helpful role by eliminating toxins. It also manufactures a number of B vitamins and lowers the pH level. Like acidophilus, it comes to the aid of the normal microflora in the intestine when they have been depleted by antibiotics or disease. You may have noticed that some of the articles cited in connection to acidophilus, such as the one in *Clinical Teratology*, discussed studies where patients were given a mixture of acidophilus and bifidus to effect improvements. Thus a number of the articles brought up in that discussion are equally applicable here.[17]

In other studies that we have not mentioned, bifidus with acidophilus or other substances has constantly proven itself a friend to beneficent microflora and an enemy to various invasive agents such as the rotavirus. For instance, a double-blind, placebo-controlled study published in *Lancet* in 1994 found that infant formula supplemented with *B. bifidum* and *S. thermophilus* decreased the incidence of acute diarrhea and rotavirus shedding (production) in hospitalized infants.[18] A further study looked at bifidus's effect on mice who were infected with rotavirus, which commonly causes diarrhea. Onset of diarrhea was significantly delayed in pups treated with Bifidobacterium relative to placebo litters suffering from the infection.[19]

Other investigations have recorded that bifidus is equally resistant to salmonellosis infection.[20]

Bifidus has been shown to resemble acidophilus in its effects by aiding the elderly in a positive modulation of the immunological and inflammatory response and by counteracting the sometimes baleful effects of antibiotics on digestive functioning.[21, 22]

Unfortunately, nothing has been done to study this substance's effect on AIDS. Here is another nutrient whose immune-mobilizing, antiviral, and intestinal microflora-enhancing properties should make it a prime candidate for research into its possible usefulness in combating AIDS.

BORON

Boron is a trace mineral that is found in many fruits and vegetables. It appears to decrease the excretion of calcium and thus a deficiency of it is implicated in diseases such as arthritis and osteoporosis, where calcium is deficient. More work is being done now into how these bone diseases can be treated by using boron as a nutritional supplement.

A study published in the proceedings of the 1993 International AIDS conference showed that amino acids modified by boron inhibited HIV protease.[23]

BROMELAIN

Bromelain refers to a group of enzymes that have a place in the body's metabolism in digesting proteins. Unlike most other enzymes, these have been shown to aid digestion. They also possess anti-inflammatory capabilities and so have proved important in treating muscle injuries: pain, swelling and bruising. Thus, they are prominently used in sports medicine. Not much work has been done in assessing what value they may have in treating AIDS, but an article in the *Journal of Immunology* reports an investigation showing bromelain supplementation boosting T cell proliferation.[24]

COENZYME Q10

Coenzyme Q10 is another substance found in the human body. Its role is to help in the creation of ATP, which supplies energy for cellular functions. Studies have shown its value in improving cardiac function, reducing tooth decay, and bolstering immunity.[25]

The study of this enzyme's relation to AIDS has become imperative, because it has been established that AIDS patients have a significant blood deficiency of CoQ10 relative to controls and relative to ARC patients. In an examination by Folkers et al., it was established that ARC patients show a significant deficiency in the substance relative to controls, as did patients infected with HIV. CoQ10 deficiency increased with the increased severity of the disease.[26]

Since AIDS leads to depressed levels of the coenzyme, administration of it would seem to be a helpful countermeasure. A later study by Folkers looked at just this possibility. The study profiled two ARC patients who have survived four to five years with no symptoms of adenopathy or infection while taking CoQ10 continuously. The authors also report results on fourteen newly found normal subjects that experienced increased T4/T8 ratios in response to CoQ10 administration.[27] Two years later the Folkers group went beyond observation and did an investigation in which CoQ10 and vitamin B6 were administered together and independently to three groups of subjects. Results showed that blood levels of IgG increased significantly when CoQ10 and B6 were administered together as well as when CoQ10 was administered alone. T4-lymphocyte blood levels increased significantly when CoQ10 and B6 were administered together and independently as did the T4/T8 lymphocytes ratio.[28]

Hopefully, in the future more scientists beyond this pioneering team will take an interest in the coenzyme's qualities.

DHEA

DHEA is a substance produced by the adrenal glands, which is later converted to either testosterone or estrogen. It is reputed to increase energy and build up the immune system; but whatever the merits of these claims, it has been shown to reduce the viral load of AIDS patients and stem HIV replication. It is also noteworthy that one study showed that high-risk homosexual males who were HIV negative had significantly higher DHEA levels than age-matched HIV-positive males.[29]

(I should pause to insert another caveat here. I have already shown that the concept of viral load is not the most useful measure of disease level, but again I am citing such studies not because I have suddenly become an advocate of using this test but to show the effectiveness of these natural supplements according to the establishment's own criteria. Further, since most scientists have accepted these criteria, some of the best studies of supplementation's effect on the disease are couched in these terms.) Studies in such journals as *AIDS Treatment News* and *AIDS Res Human Retroviruses* as well as ones delivered at various international AIDS conferences have brought out DHEA's effectiveness in such areas as killing HIV in vitro and in lowering viral load.[30, 31]

One investigation indicated that there was an inverse association between DHEA levels and increased HIV progression in patients with CD4 counts below 300. Supplemented DHEA to levels greater than 400 mg/dl increased survival in such patients.[32] A second study showed that the administration of an average oral dose of 75 mg qd of DHEA coupled with standard antiviral treatment to patients infected with HIV led to significant increases in CD4 and CD8 counts.[33]

As was the case with CoQ10, the fact that AIDS correlates with deficiency of DHEA suggests the need for supplementation as one means of fighting the illness.

GLUTATHIONE

Glutathione is a substance that is found mainly in the liver where it detoxifies many harmful compounds. It is the primary antioxidant in the body, protecting against free radicals, which cause cellular damage. It also stimulates lymphocytes and helps to defend the organism against viral illnesses, including HIV.

Its effectiveness against viruses has been pointed to by the conclusions of a study by Palamara et al., which found that glutathione acted against human herpes 1, inhibiting the replication of the virus by as much as 99 percent.[34]

As with a number of other substances whose presence in PWAS we have canvassed, glutathione appears in reduced levels in AIDS patients.[35] Since studies show glutathione is important in the regulation of T cells, disruption of its production will represent a breach in this important line of defense of the immune system.[36]

Taken as a supplement, glutathione may reduce HIV activity. Research by Kalebic et al. suggests that HIV activity is profoundly decreased by the presence of antioxidants, such as glutathione and ascorbic acid.[37] Their study posits, "Glutathione may be essential in stopping the virus from activating."[38]

Specific studies of glutathione's effect on in-vitro HIV have exposed its extinguishing capabilities. An examination by David Ho and S. Douglas looked at such effects and found that HIV-1 infection was blocked or substantially reduced by glutathione or NAC in cultivated cells. Reductions of 90 percent and higher in the amount of virus released were seen when cells were treated for four hours with a minimum of 10 mm of glutathione or NAC.[39] Similar findings were made known by Palamara's team at the 1993 International AIDS Conference.[40]

NAC (N-acetyl cysteine), by the way, is a substance that converts into glutathione in the body; this is why in the aforementioned experiment we made reference to "glutathione or NAC." However, other, less directly related substances also act to remedy glutathione deficiency, including selenium and vitamin C. In addition, silymarin, a flavonoid complex from milk thistle, has the ability to increase glutathione in the GI tract, liver, and erythrocytes.[41]

As you have seen, most of the lab work on glutathione has looked at its effects in vitro, but a number of complementary physicians are using intravenous glutathione as a part of their HIV and AIDS protocol. Dr. Joan Priestley, for example, says that:

Glutathione is a wonderful substance. It deserves all of the attention that our government and the NIH are giving to it right now. I have all my clients take NAC on a daily basis. NAC [as we know] becomes glutathione in your body. Glutathione specifically attacks the AIDS virus in about four different steps.[42]

Dr. Calapai had this to say:

When we add glutathione intravenously with vitamin C, we see a significant turnaround in the [AIDS] patient's comfort, attitude, and well being. Their T cells seem to stabilize or increase. NAC taken intravenously can inhibit reverse transcriptase activity better than ninety percent. There is no drug available for any treatment or disease that can do better than ninety percent with minimal or no side effects.[43]

Recall that glutathione is used by the body as an eliminator of toxins. It makes sense, then, that it would not have the hazardous side effects we find with artificially created substances, such as AZT, which have not been refined over many generations to do the detoxifying job in harmony with the body's liveliness.

IODINE

Iodine is an element that is needed to make the thyroid hormones. It is found in many vegetables, such as kelp. These hormones are necessary for normal cell metabolism to be carried on.

The shame about iodine is that it has proven again and again to be an effective fighter against viral infections, yet nothing has been done to test whether it will have any devitalizing effects on HIV.

Let me note some of the many benefits that have been found from iodine treatment.

Iodine has been used in the treatment of goiter, hypothyroidism, fibrocystic breast disease, and other diseases. For example, a trial of the value of povidone-iodine vaginal pessaries given twice a day for two weeks to women with candidal, trichomonal, and nonspecific vaginitis discovered the treatment to be effective and free of side effects.[44] Other studies have gauged it and found it successful in fighting conjunctiva of newborns and adults, reducing the occurrence of intestinal bacterial infections after colorectal surgery and bacteremia after tooth extraction, as well as in eliminating nasal methicillin-resistant *S. aureus* from the nasal cavity and other bacteria from the mouth.[45-48] Further, a study reported in the *Journal of Internal Medical Research* indicated that a povidone-iodine paint administered to patients with a fungal infection resulted in the patients' complete or significant improvement; a study in *Obstetrical Gynecology* recorded that patients with vulvovaginal and cervical herpes virus infection, who had received external and intravaginal povidone-iodine preparations, showed a reduction in duration of symptoms in nine of the ten cases.[49]

In vitro studies note that the povidone-iodine solution decreases herpes virus, type II, by upward of 99.99 percent and has like devastating results in vitro against *S. auerus*.[50-52] Animal studies show an iodine preparation having antibacterial/cariostatic effects on rats' teeth as well as guarding rabbits against staphylococcal infection.[53]

With all these studies showing iodine preparations as effective countermeasures against varied infective agents, then, the lack of research into how it would impinge on HIV replication is sad.

ISOFLAVONES/GENISTEIN

More excusable is the underutilization of the isoflavones in AIDS studies, because these are a variety of phytochemicals, bioactive plant chemicals, which have only lately been noticed and studied. The isoflavones are phytoestrogens, exhibiting weak estrogenic activity, and are found especially in chick peas and legumes. One study shows that one of them, genistein, inhibited the replication of herpes simplex virus type 1 in vitro.[54] Little else has been done with them to date.

L-DOPA

L-dopa is a drug that, it is surmised, converts into dopamine in the brain. Dopamine is a neurotransmitter chemical, whose absence results in Parkinson's disease. Understandably, supplementation with L-dopa is used in working to alleviate Parkinson's. It is not available over the counter.

The one study done that relates administration of L-dopa to AIDS indicates that it has alleviating symptoms of nerve dysfunction arising in relation to HIV infection in children between the ages of four and thirteen years.[55]

This is another chemical whose potential AIDS-fighting qualities should be delved into further.

MELATONIN

Melatonin is a substance produced by the pineal gland when the eyes send signals that it is getting dark. Thus, it helps the body regulate sleeping and waking and can be used as a sleeping pill.

While it has not been studied for its relevance in AIDS treatment, its marked assistance to the immune system in animal studies make it a candidate for such study. It ability to restore depressed immune function following soft-tissue trauma and hemorrhagic shock in mice was noted in an article in the *Journal of Surgical Research*.[56] A couple of other studies of mice who had weak immunity due to other conditions also came to positive conclusions about melatonin's restorative abilities.[57] Moreover, one study showed it counteracted the effects of acute anxiety stress on immune response in sheep red blood cells.[58]

These good results indicate that this substance deserves a fuller treatment by AIDS scientists.

NAC

We saw earlier that NAC is substance converted into glutathione in the body, so what we said previously about glutathione's HIV-diminishing capacity, as well as our point that those with AIDS have glutathione levels below the norm, should be read in conjunction with NAC.

As we mentioned earlier, doses of NAC can help restore depleted glutathione. An inquiry reported in the journal *AIDS* found that this restoration was evident in mononuclear cells of HIV patients.[59] A Stanford University double-blind, placebo-controlled study of two hundred volunteers who were HIV positive made the same point: the administration of NAC increased glutathione levels, which may have enhanced survival rates.[60] A piece in the *FASEB Journal* argued further that NAC's ability to increase glutathione levels accounted for the enhanced number of CD4 positive T cells in patients infected with HIV, as found in its study.[61]

Where we saw with glutathione a number of studies showing the substance's anti-HIV effects, so with NAC there are a sheaf of reports on how it restricts HIV proliferation. One by David Ho et al. indicated that NAC significantly reduced HIV infection in in-vitro cell cultures, and this is backed up by a study from Roederer in *AIDS Research* and one by Raju et al. that states NAC was found to be a more powerful antiviral agent than oxothiazolide with respect to HIV.[62-64]

Another favorable comparison with the establishment-recommended antivirals can be drawn from a study reported at the 1992 International AIDS Conference, which assayed the safety of large doses of NAC on patients infected with HIV. "Patients with under 500 CD4 cells/mm3 received either 6 weeks of NAC doses of 3.7, 11, 33 and 100 mg/kg IV or 6 weeks of oral doses of 600, 1200, 2400 and 4800 mg qd. Results showed no major side effects and found NAC to be safe at the doses noted above."[65] Unlike the antivirals, while depressing HIV, NAC also builds up T cell function and growth.[66]

PHOSPHATIDYL SERINE

Phosphatidyl serine is an organic compound that is found particularly in the brain. It supports the membrane proteins that facilitate nerve function. Although its ability to aid a PWA in his or her struggle with the immunodeficiency disease is unknown, we do know that this compound inhibits other viruses, such as the vesicular stomatitis virus, and that it works to offset declines in the humoral immune response.[67, 68] In consequence, it would be an excellent candidate for further study in relation to effectivity against AIDS.

SELENIUM

Selenium is a nonmetallic element that activates an enzyme in the body. Its presence is necessary for healthy immune functioning. "Food sources high in selenium are brewer's yeast, brown rice, cod, lobster, oysters, scallops, shrimp, wheat germ, bran and brazil nuts."[69] As we made mention, selenium increases intracellular glutathione levels. Mark Konlee notes that a further aspect of selenium is that it lowers beta 2 microglobulin levels. "It has been long established that HIV+ persons with low beta 2 microglobulin levels either do not progress to AIDS or progress very slowly where persons with high levels of beta 2 microglobulin levels are rapid progressers."[70]

Furthermore, as with glutathione and a number of other supplements in our survey, a low level of selenium correlates with an advanced stage of AIDS. Konlee refers to a study of 125 HIV positive persons at the University of Miami's School of Medicine, which "showed that persons with below normal levels of selenium were 20 times more likely to die of AIDS-related opportunistic infections than those with normal levels."[71]

Unfortunately, where recognition that low levels of glutathione in PWAs led to some work on trying to discover the effect of glutathione supplementation on those with the disease, this has not happened with selenium. But we do have a few investigations of the element's ability to improve immune reaction.

One study showed that selenium produced a significant increase in the activity of natural killer cells.[72] Another studied its effect on hemodialysis patients and concluded that the supplementation brought about an improvement in T cell response.[73] Lastly, an inquiry whose results appeared in *Nutrition*, noted that supplementation with 200 micrograms of selenium per day for two to four months led to enhanced immune response in patients suffering from short-bowel syndrome.[74]

VITAMIN A/BETA-CAROTENE

In brief, beta-carotene is a safe form of vitamin A known to stop damage from pollution, to enhance the immune system, to inhibit viruses, and to prevent premature aging. In addition, it is a powerful anti-oxidant that may protect the body against cancer, AIDS-related complex and other diseases.[75]

Let me qualify what I might have implied in the last section: The fact that AIDS correlates with the body's loss of a given nutrient does not prove that re-supplying the body with that supplement will do anything to weaken the hold of the disease. I did not directly say it would, but some of my remarks may have been construed in that direction. Yet there is an implication in a deficiency, which is that AIDS or one of its related conditions is depleting or halting production of the nutrient in some way and that understanding why this is so would help us comprehend the path of the illness better.

In this part of the book, we are not concerned so much with this understanding. However, we can highlight that there are a few nutrients, such as vitamin A, both which are known to be lacking in those with AIDS and whose administration has helped sufferers. These nutrients must be centrally involved in the immunodeficiency syndrome's invasion.

Let's look at this deficiency in some depth. Dr. Gregg Coodley, speaking at the Third International Conference on Nutrition and AIDS in Philadelphia in October 1994, emphasized, "Vitamin A deficiency increases the rate of HIV infection."[76] He stated that one investigation done in Africa of three hundred pregnant mothers who were deficient in vitamin A found that the rate of transmission of HIV to their babies was 32 percent. By contrast, those who had been taking vitamin A supplements had a 7 percent rate. Of note, this supplementation "more favorably reduced HIV transmission than [taking] AZT."[77]

This particular investigation was done by researchers from the Johns Hopkins School of Medicine in Baltimore and their colleagues in Malawi in a team led by Dr. Richard D. Semba of Hopkins, who reported the findings at a meeting sponsored by the American Society for Microbiology.

He criticized those who "in the high-tech rush to develop vaccines and other therapies… have been ignoring some of basics," such as looking at the role of poor nutrition in causing people to acquire the disease.[78]

As a science writer for the *New York Times* noted:

A link between vitamin A deficiency and AIDS has a plausible biological explanation. Studies in several laboratories have shown that when deprived of vitamin A, T cells and B cells... fail to function properly. T and B cells are critically important in helping the body fight HIV. A lack of vitamin A also leads to impaired production of cytokines, substances produced by cells that are important in immune reactions.

"Thus, vitamin A deficiency can be a double whammy on the immune system," Dr. Semba said, because pregnancy itself increases demand for the vitamin for both the woman and fetus and HIV infection can interfere with intake and absorption of vitamin A and can deplete existing stores.[79]

In general, low levels of vitamin A are associated with disease in both children and adults. For children, we might mention a study of vitamin A deficiency on T cell subsets of Indonesians between the ages of three to six years. Results found that children deficient in vitamin A had underlying immune abnormalities in T cell subsets.[80] A review of twelve studies showed that the adequate levels of vitamin A is a key factor in the prevention of mortality and morbidity of children in developing countries.[81]

As for adults, the links are there, not only between a number of diseases and abnormally low levels of A but between such low levels and AIDS in particular. An article in the *Journal of Infectious Diseases* reports that fifty adult injection drug users who died from AIDS were matched to 235 controls who survived. Results found that vitamin A deficiency and wasting were associated with mortality and common during HIV infections.[82] The *Archives of Internal Medicine*, meanwhile, featured a study examining the effects of vitamin A plasma levels on immunologic status and clinical outcome in patients infected with HIV-1. It was seen that a deficiency in vitamin A was associated with lower CD4 levels as well as increased mortality in seropositive individuals. The authors conclude that vitamin A deficiency is a serious risk factor for the progression of disease in those infected with HIV-1.[83]

We know from this that PWAs are depleted in vitamin A. What happens if a doctor tries to reverse this state and supplement with A? Will this have any effect on the course of the illness or the immune system's staying power?

As to the vitamin's aptitude for increasing immunity, this is well documented both in relation to people with AIDS and those suffering from other diseases. For one example, we can look to a study reported at the 1989 International AIDS Conference. Thirty mg per day of beta-carotene were given for four months to eleven AIDS patients. Results showed significant increases in the number of immune system lymphocytes with NK markers and markers of activation after three months.[84] An investigation by Coodley studied the effect of a daily dosage of 180 mg of beta-carotene for four weeks on the CD4 counts in HIV-infected patients. Results showed significantly increased CD4 counts and improvement in CD4/CD8 ratios relative to controls.[85] Garewal et al. gave eleven HIV-infected patients 60 mg of beta-carotene per day for four months. Increases were found in the percent of natural killer cells, among other improvements.[86]

We might note that A supplementation is known to ratchet up immunity also in relation to other diseases and immune-weakening stresses. Cigarette smokers, for instance, improved immunity with 20 mg of beta-carotene per day for two weeks, as did those exposed to overdoses of ultraviolet light.[87, 88] It has also helped those with already developed serious conditions, such as those with chronic pneumonia, following extensive surgical treatment, and for those with leukemia.[89]

Animal studies have pointed in the same direction, showing immune-boosting capacities of A dosage in relation to various depressing conditions. As one example, Watson et al. studied mice with LP-BMR murine leukemia virus (which produces an AIDS-like condition) and found immune system improvement after vitamin A supplementation.[90] Others can be consulted in the notes.[91]

Let's get to our second question on whether the administration of vitamin A will, besides improving PWAs' immunity, be of any benefit in alleviating symptoms or onset of the full-blown version of the disease. Here we also have some hopeful results. In a single blind pilot study by Bianchi-Santamaria et al. ARC patients supplemented with beta-carotene experienced a decrease in the progress toward AIDS, as well as recoveries from asthenia, fever, nocturnal sweating, diarrhea, and weight loss.[92] A pilot study by Pontiggia worked with ten patients infected with HIV who had just discontinued use of either AZT or ddI. Along with other treatments, they received 120 mg per day of beta-carotene. Nine patients (one died shortly after treatment began) experienced an HIV burden diminution, clinical improvement, and amelioration of negative laboratory data, and reported subjective improvements in overall quality of life.[93]

Since this last study did not use A by itself, it is not a definitive look at the value of the vitamin, and both this and the previous study were only pilot projects; hopefully more full-scale investigations of this vitamin, whose promise is evident, will be undertaken.

Work has also been done on children born to women infected with AIDS. One study found that children receiving supplements of A had a lower overall morbidity than controls and that diarrhea in HIV-infected children was significantly reduced.[94]

Further, taking A seemed to help maintain intact immunity for adult PWAs on AZT.[95]

VITAMIN B1

Along with A, the B vitamins have been the object of a few studies, where their utility in treating AIDS or in revving up the immune system were gauged, although only B12 has garnered relatively extensive interest.

An article on vitamin B1 was published in 1994 by Shoji et al. and showed that the substance inhibited the production of progeny HIV-1 in chronic and acute HIV-1-infected cells in vitro.[96]

VITAMIN B2

Vitamin B2 has been featured in one important study as well, although this merely indicated the vitamin's ability to stimulate the immune system, here in response to *E. coli* infection.[97]

VITAMIN B6

Jumping to B6, we have one study of how the normalization of vitamin B6 levels in HIV-infected patients suffering from a deficiency led to significant improvements in CD4 cell number as well as other functional parameters of immunity.[98]

VITAMIN B12

More has been done with B12 than with the other B vitamins. As was the case with A, it has been noted that many PWAs are low in B12. "Decreased Vitamin B12 levels occur in up to 20 percent of AIDS patients and may adversely contribute to hematologic and neurologic dysfunction."[99] This deficiency has been tied to decreased secretion of acid, pepsin and gastric juice, which contributes to vitamin B12 malabsorption.[100]

Along with studies that show B12 augments immune function in vitro, work that centers on PWAs has shown some remarkable results with supplementation.[101] An exciting investigation recorded in the *American Journal of Gastroenterology* followed an AIDS patient with advanced dementia complex associated with low serum vitamin B12. After treatment with vitamin B12, the symptoms resolved over a two-month period. The researchers conclude that the AIDS dementia complex is a reversible adverse synergistic interaction between the human immunodeficiency virus and vitamin B12 deficiency.[102]

The researchers' final generalization may be too broad, since it is based on a single case, but another study that dealt with a large population discovered that normalizing the status of vitamin B12 by supplementation of 250 mg/ml in HIV-infected patients produced significant improvements in cognitive function.[103]

We might add that vitamin B12 dosage has also been seen to counteract some effects of senile dementia.[104]

VITAMIN C

Vitamin C has a grand (some might say "overblown") reputation as possessing potent immune-stimulating capacities. You may have heard, for example, of Linus

Pauling's assertion that massive doses of vitamin C can overcome cancer. Equally strong claims have been made for its AIDS-battling characteristics. And these claims are not only supported by pilot studies and in animal experiments, but by the findings of doctors who are using C in their treatments of AIDS patients.

Although it is not yet the time to begin a complete, in-depth examination of therapy strategies, I thought that since the use of vitamin C is a subject I have brought up repeatedly on my radio program and in interviews I included in *AIDS: The Untold Story*, I might note the viewpoints of a few doctors who are immersed in using C in working with PWAs. We will move on to a summary of relevant scientific literature on C's capabilities. Note that vitamin C is also known as ascorbate.

"Vitamin C can double the life expectancy of AIDS patients," according to Dr. Robert Cathcart. His study entitled "Vitamin C in the Treatment of AIDS" concludes that massive doses of buffered ascorbate (50 to 200 grams per twenty-four hours) in combination with other treatments for secondary infections often produces a clinical remission that shows every evidence of being prolonged if treatment is continued.[105]

When I questioned him about the basis of his treatment, he outlined the properties of vitamin C that made it a particularly hardy foe of free radicals, whose presence is increased by immunodeficiency disease. In his words, "Massive amounts of ascorbate—which I like to call ascorbate to distinguish it from vitamin C in tiny doses—neutralize free radicals."[106] We've seen that these radicals are offshoot electrons that careen around the body, causing tissue damage. Cathcart continued, "We have within the body this system called the free radical scavenging system which consists of electron donators…. When the free radical scavengers are overwhelmed, then we have inflammation and other diseases."[107] Vitamin C acts to support the free radical scavenging system, which, we are assuming, is put under stress by HIV or other internal environmental changes associated with AIDS.

Free radicals, however, are not solely destructive agents but are harnessed by the immune system. "When a white cell engulfs a virus by the process of phagocytosis, it puts the virus in a vacuole and then secretes these free radicals called superoxide, peroxide, [and] hyperchlorous acid, into the vacuole killing the virus."[108] A second valuable attribute of C is that it aids the white cell in this killing process.

How does it do this? We've said certain white cells have free radicals in store to use in eliminating viruses. However, these free radicals may damage the host cell. If "a white cell makes too many free radicals it actually poisons itself…. The ascorbate provides the reducing equivalents necessary to protect it against the toxic products that it makes."[109] So C acts to both help the body eliminate these free radicals and harness them.

Let's run through the comments of a few other doctors who have extolled the use of this supplement. Dr. Calapai, whom we met earlier, sees the use of C as part of an all-around natural therapy: "We need to be aggressive in a non-toxic fashion to be able to inhibit the replication of viruses. Research has shown that large doses of ascorbate intravenously and orally can inhibit intra- and extracellular viruses," he states.[110]

Dr. Joan Priestley has also seen the research into ascorbate's antiviral utility and has proven its practical value on PWAs she has treated. "I use very high doses of certain nutrients which have direct anti-viral and immune-system-booting or immunomodulatory effects," she told me. When I asked her which nutrients she used in particular, she expanded, "Specifically [I use] vitamin C in high doses á la Linus Pauling, industrial strength doses. I feel it is a cornerstone of an AIDS management, AIDS prevention program."[111]

Nutritionist Dolores Perri also sees the administration of megadoses as proper therapy. "We can't possibly get enough vitamin C from what we eat as it is a very unstable vitamin. Anyone with a compromised immune system needs a vitamin C drip." In the clinic where she works with AIDS patients, patients are getting from 25,000 to 100,000 units in a vitamin C drip that also includes B complex, glutathione, a multiple vitamin, and minerals. Taking the drip, she explains, is "bypassing the digestive system, going right to the blood stream and getting incorporated into the cells."[112]

With such benefits (as we have seen and will see below), one would think more practitioners would look into C as a complement to other treatments, but too many doctors are uninformed or misinformed. Joy DeVincenzo, an HIV-positive AIDS activist, says, "A lot of doctors including mine do not believe in giving vitamin C just because it's not approved." It was her own study of health that convinced her of the value of ascorbic acid. "Without it I know I wouldn't have stayed stable for this long."[113]

Other doctors may have heard that if people take large amounts of vitamin C, they will develop kidney stones. Dr. Cathcart disputes this claim. He notes that judged purely by biochemical theory, ascorbate should partially break down to oxalate, and thus could contribute to kidney stones. In reality, the stones don't appear, even with maximum intake of C. "So theoretically, people should develop oxalate kidney stones—[but] they don't," Cathcart states. "And I think in science the facts should determine the theory, not the theory the facts. So while the theory advanced that people who take ascorbate should have kidney stones despite the fact that they didn't produce kidney stones, that idea still keeps perpetuating."[114]

Going further, he argues that theory does not jibe with facts in this case because ascorbate is a mild diuretic. With more water leaving the body, "that makes the stones less likely to deposit."[115] Moreover, the ascorbic acid flowing out in the urine is "binding calcium, which makes it unavailable to the oxalate acid" that is used in the creation of the stones.[116]

Let's move on to a quick survey of scientific papers that have documented ascorbate's ability to improve immune function, help the body respond to various debilitating conditions and arm itself against AIDS. In contrast to most of the other supplements examined in this chapter, there is a substantial body of work testing the mettle of this nutriment in direct relation to AIDS. Before handling those materials, though, I want to briefly note how C has proven itself against other conditions.

Cathcart, whose reliance on C in treating AIDS patients we just learned about, has also seen good effects from this vitamin when it is given to combat other illnesses. To his knowledge, a topical C paste is highly effective in the treatment of herpes simplex and Kaposi's lesions. Furthermore, massive doses of ascorbate in combination with lysine will reverse viral illnesses such as hepatitis A, B and C, mononucleousis, shingles, and herpes, he told me when I queried him about C's properties. Moreover, he said, "When drug therapies are indicated, such as the use of Septra or Bactrum to treat pneumocystis, large amounts of ascorbate will help to prevent allergic reactions from the drugs and will usually stop the pneumocystis from re-occurring."[117]

To go to particular studies, we can begin with genital herpes. Two examinations have indicated the ability of ascorbate to combat this viral condition. One by Fitzherbert showed the administration of 100 mg of zinc sulfate and 250 mg of vitamin C twice daily for six weeks to patients with recurrent herpes simplex type I resulted in either total suppression of the eruption or limited eruptions.[118] A study led by Terezhalmy came to similar positive conclusions.[119] Large dosages have also been effective in lowering the incidence of symptoms of upper-respiratory-tract (URT) infections.[120]

Jariwalla and Harakeh in the book *Nutrition and AIDS* state, "A striking property of ascorbic acid is its ability to inactivate viruses and inhibit viral growth in their host cells. Ascorbic acid has been known to be an antiviral agent since 1935."[121] They go on to refer to a host of studies that have uncovered ascorbate's power against various viral diseases, noting such important studies as that of Jungeblut in connection to the inactivation of polio virus by crystalline vitamin C; Holden and Molloy on herpes viruses; Kligler and Bernkopf on vaccinia virus; Langenbush and Enderling on the virus that causes foot-and-mouth disease; Amato on rabies; and Loijkin on tobacco mosaic virus.[122]

Moreover, C not only takes a hand in directly attacking viruses but also facilitates the work of other medicines. For example, one double-blind study reported in the *Annals of Allergy* found that two grams of vitamin C administered to patients with allergic rhinititis produced a significant positive effect one hour after treatment on bronchial responsiveness to inhaled histamine relative to controls.[123] That is, the C made the body more sensitive to the histamines used in treatment. Similar findings about increasing responsiveness were found in relation to the weakened bronchial cavities of heavy smokers.[124] A work in the *Journal of Antimicrobial Chemotherapy* attested to the fact that ascorbic acid enhanced the lethalness against the infection but not the permeabilizing effects of the medicine amphotericin B on *Candida albicans* and *Cryptococcus neoformans* cells.[125]

Finally, before we look at AIDS and C, it should be noted that there are plenty of studies of ascorbate's general immune-boosting potency. An extensive literature review by Banic found, among other advantages, that ingestion of C increased resistance against infections and stimulated phagocytic activity of granulocytes and macrophages. You may recall that these last-mentioned entities are key components of the antibody system.[126] Delafuente, meanwhile, conducted a placebo-controlled study of vitamin C supplementation's effect on immune enhancement among the elderly, discovering that vitamin C may be an important factor in correcting faulty immunologic functions.[127]

When a drug or nutrient is being tested for its effectiveness against a virus, as can be surmised from what we have been recounting, it is brought through a number of stages. First, it has to prove itself against cultured cell lines in the laboratory. These are called in vitro studies. Next are trials in which living subjects are enrolled, in vivo studies, whether of animals or humans. In the case of vitamin C, this supplement has repeatedly shown its HIV-suppressing abilities in the lab, though not as much work has been done in vivo. Along with the anecdotal testimonies of practitioners, we also have a number of studies of patients to rely on.

One in vitro examination of vitamin C and HIV was presented in the *Proceedings of the National Academy of Sciences*. It concluded that nontoxic levels of ascorbate suppressed virus replication in HIV-infected T lymphocytic cell lines. The article put it like this: "In chronically infected cells expressing HIV at peak levels, ascorbate reduced the levels of extracellular reverse transcriptase (RT) activity (by > 99 percent) and of p24 antigen (by 90 percent) in the culture supernatant."[128] Further, unlike AZT and other strong antivirals whose noxious effects we have assayed, the ascorbate showed "no detectable inhibitory effect on cell viability, host metabolic activity, and protein synthesis."[129] A second study by the same authors found HIV-inhibiting power in both ascorbic acid itself and in another form of C, its calcium salt (Ca-ascorbate). Further, they tried C in combination with two thiol-based reducing agents (glutathione, GSH, and N-acetyl-L-cysteine, NAC), both of which we mentioned earlier, and found that these pairings amplified the suppressing agency of all the nutrients. The authors feel these positive results "further support the potent antiviral activity of ascorbate and suggest its therapeutic value in controlling HIV infection in combination with thiols."[130]

By the way, such lab culture studies should not be seen as necessary but nontherapeutic first steps that will have no immediate effect on practice. As we saw much earlier, HIV contamination of blood products was an early problem in the AIDS crisis. It can be seen that although HIV-infected blood would not be consciously given in transfusions, it would certainly be useful to have a substance, such as C is proving to be, that could decisively suppress the action of the retrovirus in blood products. A study in *Biologicals* looked into this question. It examined cell-free human immunodeficiency (CFHIV) inactivation by the treatment of blood products with ascorbic acid. Results demonstrated that 500 mcg/ml of ascorbic acid in a culture medium, whole blood or leukocyte-depleted blood, inactivated CFHIV in vitro.[131]

While these studies make it clear that vitamin C stops the reproduction of HIV, others have asked how it does so. Let us keep in mind as we look over these investigations, by way of contrast, how AZT and the protease inhibitors work. It is believed that AZT stops the action of RT as HIV tries, upon entering a cell, to transform its RNA into DNA. The protease inhibitors are effective later in the virus life cycle and inhibit HIV from replicating.

Findings on C make a number of points about its probable action. One study Penn found that in HIV-infected cells, there was no significant differences in the synthesis or processing of individual viral RNA and polypeptides, which indicates that ascorbate's inhibitory effect is not focused on steps of viral transcription or translation. Rather, the authors argue, C interferes with enzymatic activity after HIV has settled in the cells.[132] A further examination of this topic by Jariwalla and Harakeh, agreeing with the postulation that C interferes with enzyme production by the virus, suggests that "the biological activity of an HIV LTR-directed reporter protein made in ascorbate-treated cells was reduced to -11 percent relative to that of untreated controls."[133] To restate that, once the HIV came in contact with C, the HIV's ability to manufacture protein was cut down. This dampening effect means the HIV would have diminished infectivity, that is, lessened ability to invade host cells.[134]

More general ideas about the antiviral abilities of the vitamin—we have already heard of its skill in hindering free radicals—come from the already cited chapter in *Nutrition and AIDS* where it is affirmed that "important metabolic functions of ascorbate include stimulation of collagen and carnitine synthesis, reduction of molecular oxygen during respiratory burst in phagocytic cells [acting as an aid in antibody activity, that is] and neutralization of toxic free radicals."[135] These functions, the authors claim, "may aid in the prevention/ alleviation of opportunistic infections and wasting syndrome associated with advanced stages of AIDS."[136] They point out further that beyond what Cathcart said about C's effectivity against free radicals, the C in the body addresses the fact that when a person gets sick, the radical scavengers remain unreduced. In other words, the repair mechanism cells, once they are finished with their work, are normally broken down by the body and their parts reintegrated into other operations; but under stress, they are left to clog the infrastructure. To go on, this can lead to deficiencies, and AIDS patients who experience this "often exhibit symptoms of acute-induced scurvy characterized by life threatening weight loss, brittle bones and swollen glands."[137] However, vitamin C can get the scavenging system back on track as well as help the body eliminate these radicals in other ways.

It might be mentioned at this point that, in contrast to the drugs favored by the establishment as ones that interfere with HIV, vitamin C is non-toxic even at very high dosages. As the piece in *Nutrition and AIDS* says illuminatingly,

"The advantage ascorbate has over other anti-HIV agents is that it can be taken at large doses without producing adverse side effects."[138] And this is serendipitous in that there are qualitative differences in effect at certain high doses. The authors of the chapter summarize:

> The unique ascorbate functions that become manifest at high doses, permitting it to act as a powerful scavenger of free radicals/oxidants... empowers it with the dual ability to keep oxidant-induced HIV activation in check as well as protect tissues from cellular damage. In addition, its ability to suppress HIV replication and inhibit microbial growth (directly or via stimulation of phagocytic cells) affords an additional line of defense against proliferation of the AIDS virus and opportunistic pathogens associated with HIV infection.[139]

At least one comparative study has shown, by the way, that in vitro vitamin C has a stronger HIV-suppressing power than AZT. In this study, reported in the *Journal of Nutritional Medicine*, "the effects of ascorbate (vitamin C) and Azidothymidine (AZT) were examined on HIV expression in permanently infected and reporter cell lines."[140] On all the different parameters examined, C appeared the better anti-HIV agent. With HIV-infected HXB cells, for instance, "ascorbate suppressed HIV production and reduced the yield of infectious virus released into the culture supernatant. AZT, which has been reported to block de novo infection of freshly infected cells, did not inhibit constitutive virus production in HXB cells."[141]

Turning, at last, to the more limited work that has been done on surveying the effects of vitamin C supplementation on people with AIDS, we can start with a report Cathcart published in *Medical Hypotheses* based on a careful study of his own work. He reports, "Preliminary clinical evidence is that massive doses of ascorbate (50 to 200 grams per 24 hours) can suppress the symptoms of the disease (AIDS) and markedly reduce the tendency for secondary infections."[142] He goes on: "In combination with usual treatments for the secondary infections, large doses of ascorbate will often produce a clinical remission which shows every evidence of being prolonged if treatment is continued."[143]

An investigation of PWAs by Jariwalla was focused on the correlation of AIDS and "abnormal [i.e., low] blood levels of key micro-nutrients and sulphated amino acids (thiols) that are prevalent early in infection."[144] He argues, then "micro-nutrient imbalance may contribute to increased oxidative stress... and abnormalities in immunologic/neurophysiologic functions underlying HIV/AIDS." Although this is taking us beyond the effects of ascorbate alone, evidence from studies of asymptomatic HIV-positive patients who have corrected these micronutrient deficiencies, by taking ascorbate as a main supplement along with other vitamins and compounds, showed "micronutrient consumption or multi-vitamins used at baseline was associated with significant reduction in rate of progression to AIDS."[145]

To this study should be added one done by Tang et al. in which 108 seropositive AIDS homosexual or bisexual males were studied for daily micronutrient intake relative to the progression rate to AIDS over a period of 6.8 years. It was learned, "Progression rates for those with highest total intake of vitamins C and B1 and niacin were significantly slower than those ingesting lower doses."[146]

While Tang's study did not intervene in the PWAs' treatment (if any) but simply observed what the PWAs were doing, an investigation by Kodama et al. looked at vitamin C's effect when given to four patients with autoimmune disease. Results indicated that vitamin C induced an increase of plasma glucocorticoid activity with a two-hour delay. We learned earlier that the glucocorticoids were steriods that play a role in reducing inflammations. So, this study can be taken to indicate that vitamin C supplementation will boost immune system activity in PWAs.[147]

You may find that this last section is not quite satisfactory, in that although a start has been made in using vitamin C in trials with people with AIDS, on the whole little has been done, especially considering how much evidence there is that C suppresses HIV in vitro. Sadly enough, reams of information are generated about AZT and other establishment, legal drug lord-sponsored drugs, while substances such as C from which little profit can be hoped (since they are not patentable) get far less attention.

VITAMIN D

The other vitamins and minerals have not been given the attention C has received, although a lot of good work has been done here also, especially on vitamin E and zinc. For vitamin D, one significant study has been published in proceedings of an international AIDS conference, which showed that the nutrient given to PWAs would correct certain immune system cells macrophages) that suffered deformation under the disease's scourge.[148]

VITAMIN E

With vitamin E, which has already been found to be an essential antioxidant that can prevent many diseases caused by environmental stressors, including cancer, cardiovascular disease, atherosclerosis, and cataracts, considerably more work has been done.

This vitamin is available in many foodstuffs, such as fish and vegetable oil, nuts, and whole grains; however, research suggests that even the average person is not getting enough vitamin E in the diet. In one survey, 70 percent of women nineteen to fifty were getting less than 100 percent of the RDA of vitamin E, and 41 percent were getting less than 70 percent.[149] Other studies suggest that higher vitamin E intake is correlated to a lower incidence of cancer.[150]

In noticing what research has been done on this vitamin's ability to help the body cope with AIDS, we find that, in contrast to work on C, which has been mainly in vitro studies, work on E has concentrated on animal models.

From our perspective, the problem with this is that no adequate animal models have ever been found for AIDS, a point we've come back to a number of times. However, by looking at some of these experiments, we can at least learn how E acts in a body affected by an immune-weakening condition.

For mice, the virus that is considered roughly similar to AIDS is murine AIDS, caused by the LP-BM5 retrovirus. A number of studies have gone forward that test the effect of heavy doses of E on mice laboring under this illness.

One published in *Thymus* looked into the vitamin's ability to improve thymus function, since this organ's manufacture of immune system components is impaired by the disease (as it is in humans), specifically by interfering with thymocyte cytokine production, which impacts on T cell differentiation. Results indicated that dietary vitamin E supplementation at extremely high levels can modulate in a positive way cytokine production by thymocytes.[151] A study by Wang et al. saw similar results in terms of recording marked improvement in immune function in murine-sickened rats who were put under a heavy regimen of E.[152] A second study of infected mice, also headed by Wang, noted that after giving the mice large amounts of E, the group witnessed, among other effects, restored concentrations of vitamins A, E, zinc, and copper in the liver, serum, and thymus, as well as partially restored production of IL-2 and IFN-gamma by splenocytes. Both of these last had been undermined by the murine AIDS. They also found the vitamin prevented a suppression of splenocyte proliferation and natural killer cell activity, which suppression would normally have taken place as a result of the virus.[153]

These findings are in line with non-AIDS-related examinations of the enhancing effects of E on debilitated rodent immune systems. Such aggrandizements of weakened immune systems have been found, for example, in mice exposed to chronic alcohol consumption, whose compromised immune responses experienced a complete restoration of immune status following three days of supplementation with 5 IU of vitamin E.[154] Supplementation was effective also with mice whose immunity was decreased by aging.[155] Hypertensive rats, too, were found to improve immunity when fed vitamin E.[156]

Of more direct bearing than these studies, though, is one done by Gogu et al. in which E was given along with AZT to mice and was found to protect their bone marrow by 75 percent and 86 percent, respectively, relative to controls.[157]

Studies with other animals have brought in similarly encouraging findings, at least in terms of E's value as an immune system support. This has been noted with calves, guinea pigs, rabbits, and sheep.[158]

I stated earlier while E in vitro studies seem to have been bypassed in general in comparison to those focusing on animals, a few have appeared. One published in 1989 indicated that growing cells in a vitamin E-laced medium increased these cells' antiviral stamina. The authors concluded, "Such findings support the possible use of vitamin E derivatives as a treatment for AIDS."[159]

Let's turn to the effect of E supplementation on humans, both in connection to AIDS and other situations of immune weakness. Bear in mind that general studies of the vitamin show that, like C, it is a powerful antioxidant. "AIDS research shows that antioxidants, such as vitamin E, may help normalize retrovirus-induced immune dysfunctions, undernutrition and other pathological symptoms, thereby retarding the progress of the disease to AIDS."[160]

As with animals, it has been found that humans who are given large doses of E find that their immune capabilities, which have been under assault from the immune disease, are enhanced. In a literature review of E's capacities vis-à-vis the immune dysfunction, the authors note that the immune abnormalities surrounding AIDS are not unlike those stimulated or restored by high intake of vitamin E. They call the reader's attention to other studies in which vitamin E supplementation also has been shown to result in a decrease in the progression of disease to AIDS.[161]

The positive effects of E supplementation have also been experienced by patients who have liver transplants, are on dialysis, or undergo heart surgery.[162] Taking vitamin E has also been recommended for older adults. A double-blind, placebo-controlled study found that healthy older adults taking 800 mg of vitamin E for thirty days experienced an improvement in some in vivo and in vitro parameters of the immune function relative to controls.[163]

Again, we could wish that the effectivity of this substance against AIDS had been delved into further by scientific researchers. Yet from what we have observed, especially from suggestive animal studies, it can be seen that vitamin E could play a yeoman role in helping the body fight off the immune damage caused by AIDS.

VITAMIN K

Little has been done in studying what effects vitamin K might have on immune deficiency disease. One study has appeared, nonetheless, by Qualtiere, which indicated that vitamin K compounds inhibited HIV replication in vitro.[164]

ZINC

In strong contrast with the characterizations we have offered of the scientific studies of the other nutrients in this chapter, most of the work I have seen on zinc concerns studies of people (not animals or cell cultures). Work in this line has been on the effect of zinc supplementation both on AIDS and other diseases.

As to other conditions, those most commonly studied to show zinc's benefits are childhood diarrhea and herpes.

A double-blind, placebo-controlled study done in India examined the effects of 20 mg per day of elemental zinc on diarrhea in children between the ages of six and thirty-five months. "Results showed significant reductions in both the severity and duration of diarrhea relative to controls."[165]

'Extensive work has also looked at zinc's effects on herpes. A piece published in the *British Journal of Dermatology* found that zinc sulphate solution had preventive effects on recurrent herpes simplex of the skin and oral mucous membrane.[166] Other work has shown its ability to alleviate the symptoms of herpes labialis recidivans.[167]

Supplementation with this mineral has also been helpful for those suffering from age-associated immune dysfunction.[168]

Our main point here is to see what zinc has accomplished in combating AIDS. Fortunately, we have a few excellent investigations on this topic.

The paper "Effect of Zinc Normalization on Immunological Function in Early HIV-1 Infection," which was given at the 1991 International Conference on AIDS by Beach, showed that normalization of deficient zinc status was associated with significant increases in CD4 count and improved functional immune parameters among homosexual male patients with HIV.[169] This was followed by a paper at the 1993 conference that backed up Beach's findings by indicating that the administration of 200 mg per day of oral zinc sulphate had immunostimulant T cell effects in asymptomatic patients with HIV infection.[170]

Zhang et al., perhaps noting Beach's happy results, did an in vitro study that indicated the HIV-inhibiting effects of zinc. Results of this study demonstrated that zinc can inhibit renin and protease from HIV-1. Such inhibition, the authors suggest, may explain some of the benefits seen in AIDS patients following zinc therapy.[171]

Although there aren't all that many studies on zinc, at least concentrating on PWAs, they cut right to the heart of the matter. You might well ask, in connection to some of the other substances we have examined, why so much time and expense has been spent on in vitro and animal studies—even though these are ultimately necessary—on substances that have few or no side effects at huge dosages. I'm thinking of such supplements as C, particularly, for not all the nutrients mentioned in this chapter are hazard free. Where in vitro and animal model experiments are de rigueur for chemicals such as AZT, whose side effects are so devastating, it turns out that it is drugs like this that have often bypassed the safer, nonhuman studies to rush into human trials. Meanwhile, it seems that benign substances like C are dallied with when they should be quickly put into clinical trials.

Herbal Treatments

To complement such programs of nutritional supplementation, herbal therapies also deserve exploration. Here there has already been much done in working with AIDS, particularly in the devising of herbal combinations. Such combinations (called formulas) have long been a high point of Chinese medicine, based on a balancing of the various effects of different plants in relation to a bodily imbalance, which is traditionally seen as due to an excess of ying or yang energy. Translated into the terms of alternative health, a translation that is easy to do, we would say that a disease like AIDS arises through the kicking in of an otherwise innocuous retrovirus—because internal conditions have been deranged by long-term disruptions, such as would come about through addiction to powerful substances, or the destruction of intestinal microflora through overreliance on antibiotics, or in one of the other ways we have outlined. The herbal formula would work to offset this imbalance—given that the destructive behavior is no longer continued—by bolstering and restoring the activity of deteriorated sectors.

As we did when we looked at vitamin C, I think we should begin by hearing the stories of some practitioners who have had success treating people with AIDS, this time by the herbal route. Here especially I will concentrate on some herbal formulas that seem to be paying off in forestalling the progress of the disease and alleviating symptoms. Once we have dwelt on these topics, we will provide another listing, this one of herbs that have shown themselves implacable foes of the immune deficiency disease.

One place where a diligent search after herbs that may benefit people with AIDS has been carried out is at the Institute for Traditional Medicine (ITM) in Portland, Oregon. For the last several years the Institute has been sponsoring studies at AIDS clinics around the country, studying the effect of Chinese herbs on AIDS.[172]

Subhuti Dharmananda (ITM's founder) has developed a formula for HIV-infected patients combining strong tonic herbs and other herbs used to directly inhibit infection and inflammation. In addition to the basic formula, ITM herbalists have developed a variety of adjunct formulas tailored to an individual's unique symptoms and constitution.

Dharmananda commissioned a study with 150 PWAs who took the herbal formula for twelve weeks in San Francisco and Chicago centers. Questioned about their energy levels, 76 percent of people who had described their pre-study energy levels as "poor" or "fair" reported an increase in energy. Sixty-eight percent of the patients increased their activities. For patients suffering from diarrhea, 62.5 percent reported that this condition ended and 12.5 percent said it was improved.[173] There was a general improvement in health across the group, with only one of the 150 progressing to a more severe stage of AIDS.[174]

Another doctor who has been trying to devise herbal formulas to deal with AIDS is Dr. Chang at Sun Yat Sen Medical Center. In one of his published studies, eleven herbs used as traditional Chinese anti-infective medications were seen to suppress HIV. "Extracts of the eleven herbs had inhibitory activity against HIV in the H9 cell line… and five of the extracts produced nearly complete inhibition of the synthesis of HIV antigens (97-100 percent inhibition)."[175]

(See Appendix I for HIV-inhibitory herbs and an in-depth list of immune-building Chinese herbs.)

A third practitioner who has devoted his life to the study of herbal medicine is Dr. Quingcai Zhang. When I interviewed him, he made a compelling point about the rationale for using herbal treatments against AIDS in place of the orthodox antivirals. What he said had to do with the nature of the body's own response to infection. "We find that our body does not fight a viral infection by interfering with one step of the virus's life cycle." That, of course, is precisely what AZT and the protease inhibitors do. "Instead," he went on, "our body uses cell-mediated immunity, which consists of natural killer cells, phytotoxic cells, and antibody-dependent cytotoxic cells."[176] In other words, when facing a virus, the body draws on a whole panoply of antibody defenses; and, equally important, these defenses do not consist of merely throwing a spanner into the virus's life process while leaving unattacked the cells the viruses have commandeered. Rather, the body's immune cells "can recognize the viral-infected cells and then destroy them."[177] Whereas such treatments as are now preferred by the establishment, such as AZT, simply stop the virus from carrying out a part of its life cycle, thus leaving the stymied virus and the infected cell intact, but at the same time also stopping the life cycles of all kinds of innocent cellular mechanisms in other cells, the human defense system wipes both the virus and virus-laden cells out completely. As Zhang told me, "Our body has established this mechanism over millions of years. We should follow the body's example of fighting viral infections. In this way we can do it much more naturally and not harm noninfected cells."[178] As he outlined, the proper herbal formulas will mirror the body's immune actions, eliminating completely the viral intruders while leaving untouched cells alone.

Many of the plant remedies used to counter AIDS, ARC, or HIV are a special category of herbs known as adaptogens. These herbs have a wide repertoire of capabilities and work to help the body create biological homeostasis. They are known to help normalize the system regardless of the pathology and are especially good at stimulating the body's own natural immunity. For example, adaptogens have been shown to increase CD4 counts, interferon production, macrophage activity, and natural killer action.[179] Adaptogens are often combined with each other for a more potent synergistic effect. They have long been used in Russia and the Orient with cancer patients.[180]

Certain adaptogens that researchers find beneficial in aiding the immune system are described in more detail below. As with my considerations of vitamins, we will find that as the study of herbs is off-limits for most establishment researchers, there will be many places where no work has been done on a particular herb's possible effectivity against AIDS. However, if the herb has been shown to be useful in treating related diseases or acting to stimulate immunity, we will take that as good enough reason to include it in this list, since it may be a contributor to an anti-AIDS program.

ALOE VERA

Readers will be familiar with this herb as being an ingredient in many natural beauty preparations; indeed, its use for this purpose dates back to Cleopatra. However, it also has less cosmetic effects and has been shown in some studies to have antitumor effects. In *Natural Health, Natural Medicine*, Dr. Andrew Weil reports one study where an extract of aloe was found to promote anti-cancer activity.[181] Fine and Brown note the leaf's value for treating radiation injuries.[182] Although this tells us little about how it would affect an immune deficiency disease, it is a plant whose full medicinal properties have not yet been explored.

AMERICAN CREOSOTE BUSH

This shrub, which grows in the Mojave Desert, is, surprisingly enough, believed to be the oldest living thing on the face of the earth. The shrub grows by stem tops branching off into different segments. Although the original stem eventually dies, the younger parts, which lived for many years connected to the earlier parts, can be said to be a still living part of the first growth. And these Mojave shrubs are estimated to have come from a first growth, twelve thousand years ago!

To have lasted such a long time, such a plant must be hardy. A scientific study reported in the *Journal of Ethnopharmacology* shows that *Larrea tridentata* [American creosote bush] exhibited strong antimicrobial activity against numerous organisms, including *L. monocytogenes*, *C. perfiringens*, *S. dysenteriae*, and *Y. enterocoliticia*.[183] It remains to be tested against HIV, but its suppressant effects make it a plant that merits more attention.

ANGELICA

This plant has been considered to have medicinal properties since ancient times. According to one story, it received its name because an angel revealed in a dream that this plant would cure a plague that was then raging. Like the creosote bush, it has been shown to have antimicrobial features. An investigation by Inamori found that compounds isolated from its root "showed antibacterial activities against gram-positive pathogenic bacteria."[184] So, also like the creosote bush, this plant deserves study.

ASTRAGALUS

Astragalus membranaceous root has been used medicinally in China for centuries. They put it into many formulas because it was felt that rather than heightening one element of immunity, it is an all-around worker that stimulates all aspects of the immune system. Today scientists are learning why the Chinese so revered this plant. Modern research shows astragalus to possess powerful immune-strengthening properties. Of particular note is its ability to foster normal immune response in cancer and AIDS patients (most work has been done with cancer), to correct T cell deficiency and to promote antiviral action.

The plant has generally been researched by fractionating it into compounds of different molecular weights. These fractions are then tested for their antiviral properties. Some of the research conducted on astragalus in this manner is cited below.

Dr. G. Mavligit performed an experiment at the University of Texas in Houston and found that one fraction of a semipurified extract of astragalus stimulates T cells in healthy animals and restores immune function in cancer patients with impaired immune function.[185] His findings suggest that something within astragalus—possibly a polysaccharide—may be able to completely restore immune function of cells in cancer patients. Mavligit further demonstrated astragalus fraction 3 to be a potentializer of interleuken-2 (which, as noted, is a blood cell in the immune system).[186]

In another study, Da-Tong Chu et al. confirmed that some derivatives of astragalus were capable of fully correcting in vitro T cell function deficiency found among cancer patients.[187] Their data indicates that the crude extract of *Astragalus membranaceous* can completely restore a failing immune system and bring T cell function in vitro to the level of normal, healthy people. Since this data is coming from an in vitro study, quite rightly the researchers called for clinical trials of astragalus with AIDS sufferers and recipients of chemotherapy or radiation therapy.[188]

Another compound extracted from astragalus that may benefit the immune system was examined by M. J. Humphries and colleagues and reported in *Cancer Research*. This was the alkaloid swainsonine, which was found to inhibit cancer cells from colonizing the lungs in mice.[189] "The authors concluded that the activity of swainsonine [in suppressing cancer cell development] is apparently due to its ability to enhance Natural Killer cell function."[190]

We see that astragalus has shown utility in backing the immune system when it is fighting cancer. Another example of this is to be located in an experiment noted in *Cancer* . It involved nineteen cancer patients and fifteen healthy subjects given an extract of astragalus root. Its conclusions were that "people whose immune systems were devastated by cancer experienced full restoration of immune function."[191]

Some research also indicates that astragalus is of benefit to the liver. In a 1986 experiment conducted by J. W. Jiang and Q. S. Xao, astragalus was shown to protect animals against the liver toxin carbon tetrachloride.[192] Z. L. Zhang et al., meanwhile, found in animal studies that astragalus protects the liver against stilbenemide, a chemotherapy drug.[193]

By itself, then, astragalus can help fight cancer or the dolorous effect of toxins on the liver by aiding the immune system. This doesn't mean a person's immunity has to be under assault for astragalus to be of benefit. Studies of the effects of astragalus on normal immune systems have also been carried out. One conducted on rats found that the herb "exhibited a marked immune potentiating activity."[194] Work on humans has produced similar findings. In one study, subjects with weak immune systems (as evidenced by depressed natural killer cell activity) were examined. "Ten patients were treated with intramuscular injections of astragalus extract at a dosage of 8 grams per day for three to four months.... At the study's conclusion, NK activity was tripled in the astragalus group from 15.6 percent to 44.9 percent."[195]

Another approach to studying astragalus is to look at the effects of combining the whole plant with other herbs rather than to look at its isolated components. According to Donald Brown, N.D., astragalus combined with ligustrum lucidum produces immune-stimulating effects five times as great as using the astragalus alone.[196] Benjamin discovered that astragalus and ligustrum combined were able to inhibit the growth of transitional cell carcinoma in mice.[197]

At the Institute for Traditional Medicine, whose work was mentioned earlier, twenty immunodeficient ARC patients were given astragalus as part of their treatment protocol along with other immune-enhancing Chinese herbs. After eight months, nearly all the symptoms of ARC cleared up and patients were able to dramatically reduce their use of antibiotics.[198]

BASIL

Not many plants have been studied to the extent that astragalus has, though quite a few have established themselves as immune benefactors in the limited work that has been done. Basil, which you will know from the kitchen, is considered a sacred plant in India, and that is where its health-related properties have been studied. Researchers at Sardar Patel Medical College in Rajasthan, India, examined immunoregulatory effects of basil, where the herb was shown to increase antibodies in mice.[199]

BEE PROPOLIS

Bee propolis is a substance collected by bees from trees and smaller plants and placed as an antiseptic coating on the walls of their honeycombs. A number of in vitro studies have brought out its antibacterial characteristics. One study showed the propolis acted forcefully against the tuberculosis bacillus.[200] It also acts vigorously against *Staphylococcus aureus*; *Trichomonas vaginalis*; and herpes simplex I.[201]

Another study I've seen that tested the bee's compound on humans used propolis films "applied behind the lower eyelids before bed over a period of 10-15 days [which] cut recovery time in half in patients with postherapeutic trophic keratitis."[202] Keratitis is an inflammation of the cornea.

BUPLEURUM FALCATUM

Bupleurum is a Chinese herb containing saikosaponins, which stimulate phagocytosis (the consumption of foreign cells) by macrophages in vivo.[203]

It has also shown itself helpful, like astragalus, in protecting the liver by increasing protein synthesis there.

BURDOCK

Burdock is a genus of coarse biennial weeds that feature bristly burrs. Scientists have isolated chemical constituents of some varieties of the plant and these have proven to be antibacterial, antifungal, tumor protective, and desmutagenic (inhibiting of cancer-causing agents).[204]

CARNIVORA (DRONAEA MUSCIPULA)

In 1923, the West German oncologist Helmut Keller discovered that the juice from the carnivora plant (better known as the Venus flytrap) was an effective, nontoxic cancer treatment. It has become something of this doctor's hobbyhorse, although this doesn't mean his findings are not crucial.

In his original study, Dr. Keller implanted human tumor tissue into the cheekfolds of Syrian hamsters and then treated the animals with intraperitoneal injections of carnivora juice. In fourteen days the tumors became necrotic, that is, the cancers were dying. Controls themselves died of the effects of the spreading localized tumor.

Dr. Morton Walker, who has studied Keller's work, saw "82 to 87 percent remission rate for most types of carcinoma when the patient's immune system has not been compromised by conventional, allopathic chemotherapy or radiation therapy."[205]

After interviewing fourteen AIDS and cancer patients in Germany who were taking carnivora infusions, Dr. Keller concluded that carnivora comes "closer to curing cancer and AIDS than anything else ever uncovered by members of the medical community." All these patients, he reported, were thriving. "Their cancers or other pathologies were either in a state of remission, regression or were gone altogether."[206]

I can't resist relating a story here of a patient who was at his last gasp and found Dr. Keller to be his final hope. I read the tale in a report by Walker in the *Townsend Letter for Doctors* (May 1992).

> There came a time... in April 1991 that Collin Rasmussen acknowledged to himself that his body was steadily worsening.... The continuum being traveled, he recognized, was taking him to the point of dying sooner than he had hoped—perhaps a matter of months.... He decided to take one last stab at life by selling all his possessions in order to accumulate

enough money for a trip to Paris and consultations at the Louis Pasteur Institute....

[As Rasmussen narrated] "It was the eve of my departure.... By attending a meeting of my underground AIDS care organization I was able to say goodbye to many of my caring friends. I knew that it might be the last time ever for me to see any of them.... just as I was going out the door, one... handed me an article [about carnivora]. The article's information caused me great excitement. When I landed in Paris, instead of going to the Louis Pasteur Institute, I telephoned Dr. Helmut Keller in Germany to make arrangements to visit his chronic disease clinic.

"I arrived at the center feeling a great amount of distress and fear. By the end of the second week of IV carnivora, however, my T-helper cell count quadrupled to 272. Then I began detoxifying—breaking out in pimples and urinating more—to rid myself of the HIV blood debris. The carnivora kills the HIV and leaves behind their dead bodies floating in one's blood. At the end of my fourth week of treatment, my T-helper cells elevated to beyond 500 while the suppressor cells decreased constantly. Whatever carnivora does in the body, it seems to work fine for me."[207]

Carnivora can be acquired from the product's manufacturer, Edgar Fischer, manager; Carnivora-Forschungs-GmBH; Postfach 8; Lobensteiner Strasse 3; D-8646 Nordhalben, Germany.

Acknowledging that Rasmussen's singular success may be a fluke, until further study is done of Keller's therapies we can't say if he is really a seer, pointing the way to improved AIDS treatments, but his work does cry out for more attention and evaluation.

CASTANOSPERMINE

Castanospermine is a plant alkaloid from an Australian chestnut. Studies indicate that it inhibits HIV growth, possibly by interfering with the retrovirus's enzyme production. This, however, has not been given sufficient attention.

CATERPILLAR FUNGUS

Caterpillar fungus is a Chinese adaptogenic herb used since ancient times. Its name comes from the fact that the fungus begins to grow on the body of dead caterpillars, once leading people to think it was a form of worm. Chinese doctors have been prescribing it for bone marrow fortification, among other uses. The fortification builds immunity, since the bone marrow is an essential part of the immune system. It also inhibits bacteria and fungal infections. The Chinese use caterpillar fungus in combination with other herbs to treat malignant tumors, anemia, and, most recently, AIDS. This course has not yet been pursued or studied in the West.

CHLORELLA

Chlorella is a green algae, which, in a study by Fukada, was observed to have immune-stimulating qualities.[208] It's another herb whose healing properties have not been of much interest to Western medicine.

CRUCIFEROUS EXTRACT

Cruciferous vegetables (cabbage, cauliflower, and broccoli, among others) contain a substance with remarkable protective properties. The active agent in these vegetables is a dithiolthione compound. When I questioned Dr. Bruce Halstead about the dithiolthione isolate, he noted that it had shown some evidence of protecting normal cells of cancer patients from the radiation used in therapy and also appeared to protect the liver.[209] Here is another substance that should be tested to reveal if its positive qualities may be of benefit to AIDS patients.

CITRUS SEED EXTRACTS

More directly relevant to AIDS, or rather to dealing with the common fungal infections that spring up in its wake, are citrus seed extracts. These are derived from oranges, grapefruits, lemons, and limes. Immunologist and Nobel laureate Jacob Harrish first discovered their tremendous antimicrobial activity. He found citrus seed extracts to be broad-spectrum, nontoxic antibiotics that are antifungal, antibacterial, antiprotozoal, and antiviral.

According to Dr. Alan Sacks, a former medical researcher at Downstate Medical Center now in private practice, who was speaking at an AIDS symposium, citrus seed extracts eliminate over thirty types of yeast infection. Since in AIDS patients *Candida albicans* and other fungi, such as oral thrush, invade the body, causing grievous damage, there is good reason to look to citrus seed extracts as one means to handle the trouble. Sacks explained that the extracts are "probably the most powerful antifungal[s] on the market. I've seen many people with chronic *Candida albicans*, cases that were resistant to very strong medicinal antifungals, improve dramatically within a couple of weeks of using the citrus seed extract." Thus, their usefulness is clear. "My research and... feedback from many physicians in the field are very encouraging about the use of these products as an immune support to the secondary infections that are so threatening to the lives of AIDS patients."[210]

COMPOUND Q (TRICHOSANTHINES KIRILOWII) AND BITTER MELON

Compound Q is the purified extract of Chinese cucumber and has been featured in Chinese medicine for centuries. At one time, it was the new "wonder drug," not for the establishment but in underground circles.[211] Eventually, grandiose claims for its curative powers had to be scaled back. The story is told well in *Against the Odds*, a book to which we have already referred.[212] Of note is that it was the testing of this substance, after unconscionable foot-dragging over studying the herbal extract on the part of the FDA, that gave birth to the first community-based drug testing program, run by Project Inform. A couple of deaths of PWAs in the program caused some reaction against the treatment, although published findings so far have seen some positive effects from administering the compound.

In vitro studies have signaled that use of the purified form of Q is a mixed blessing. It does "block HIV replication in infected t-cells and kill HIV-infected macrophages in cell cultures" without affecting healthy cells. "Unfortunately, there are some undesired side effects from using the purified protein," which are said to be absent when the crude herb is used.[213]

Dr. Zhang, taking note of these in vitro studies and also aware of the history of the herb's use in Chinese tradition, employs compound Q in his anti-HIV therapy. When I discussed this with him, he gave me some of the specifics. He uses two related herbs to attack HIV infection. One is compound Q itself and the other is bitter melon, which is in the same cucumber family as Q. "Both these herbs," he told me, "have proteins that can selectively kill HIV-infected cells and not harm noninfected cells."[214] Moreover, this combination does not only eliminate virus particles, as would the good orthodox therapies, but "can help the immune system to recover."[215]

He then went on to talk about the success he has had in administering this mixture to PWAs. He expounded, "I have one person who started with a hundred thousand copies of virus. After three months of taking bitter melon, that number dropped to around sixteen thousand." Moreover, what he is shooting for, and achieving to a degree, are results that last. As he put it, "In [the treatment of] conventional medicine, after half a year, the virus load would be back to the baseline or over the baseline. If people can take this [compound Q] treatment persistently, it can keep the virus load at a lower level."[216]

Again, to repeat one of the threads of this discussion, one should not extrapolate grand claims about the value of a particular treatment from one doctor's anecdotes; but one can acknowledge that they indicate a promising direction for fuller study.

DASM (DEHYDROZNDROGRAPHOLIDE SUCCINIC ACID MONOESTER)

Another herbal extract, DASM, a compound found in the Chinese medicinal herb *Andrographis paniculata*, which has shown up in Chinese medicine for years, has been tested for its effect in vitro on HIV, where it inhibits the virus's replication.[217] So far, an investigation of its effects on AIDS has not been pursued further.

ECHINACEA

You will have grasped by now that much of this compendium does not point the way to proven AIDS fighters. Later in this chapter we will turn to specific treatments that have worked, but for now I am trying to both shed light on a few substances, such as C, that have shown they have the capability to combat the immunodeficiency illness, and give notice of those substances that have relevant qualities, such as being immune system boosters, but that have not been considered or tested in relation to AIDS. These latter substances are ones that would seem to offer good promise as anti-AIDS agents, although how they would actually perform remains to be seen.

Such a substance, one that has shown an immune-enhancing character but has not been looked at for its effect on HIV, is echinacea purpurea, the purple coneflower. We know that it has immunological-stimulating properties and enhances macrophage activity, cytotoxicity, and phagocytosis. In plainer language, it boosts the action of immune system cells and aids the system by helping in the killing of infected body and blood cells. In vivo testing showed polysaccharides of the plant dramatically increase the number of leukocytes (white blood cells) in circulation, and the number of developing cells in the bone marrow and spleen. From in vitro tests, we know it helps cells resist infections from *Listeria monocytogenes* and *Candida albicans*.[218] Many other in vitro and animal studies of the herb have gone on to indicate that the plant increases nonspecific immunity.[219]

ECLIPTA ALBA

In *For Women Only*, I mention that eclipta alba grows wild in the Southwest of the United States. Herbologist Letha Hadady calls it a strengthener, which will "build up the blood" (i.e., increase the production of the white blood cell component of immunity) without causing inflammation.[220]

Whereas Hadady comes to the herb via Chinese medicine, it is also well known in ancient Indian medicine. More recently, Indian doctors have tried to assess whether its value can also be proven in the laboratory. T. Chandra et al. studied its effects on liver injury from carbon tetrachloride. They found that liver weight decreased (increased liver weight results from fatty cirrhosis due to injury) and levels of enzymes caused by liver injury were reduced to near normal levels.[221]

FEVERFEW

Also in *For Women Only*, I speak of the work of Dr. Mary Gordon, who has spent some energy working on the alleviation of migraines. She has found that feverfew is a useful antidote to these headaches because of its sedative qualities.[222] Scientific studies tells us that it is also an antiviral, having displayed vigor in killing candida in vitro.[223]

GARLIC

As with vitamin C, garlic's reputation as an antiviral precedes it. Perhaps because it has already been tried and found valuable in a number of scientific studies, it has also been given limited experimental exposure in AIDS treatment.

Garlic has been proven effective against opportunistic microbes, like Herpes virus hominus type I, crytococcal mycobacterial, and candida organisms.[224] A study by Guo et al. also brought up garlic's ability to suppress human cytomegalovirus in vitro.[225] In fact, garlic is so effective against bacteria that it has been found to work when antibiotics don't. Experiments by Singh and Shukla found garlic outshining penicillin, ampicillin, doxycycline, streptomycin and cephalexin in its ability to fight against eight out of nine strains of antibiotic-resistant Staphylococcus, Escherichia, Proteus, and Pseudomonas.[226]

Another way garlic resembles vitamin C is that it is a free radical scavenger, which helps to eliminate highly reactive oxidation products implicated in aging, arthritis, and cancer.[227] Moreover, when used against cancer, the herb has given evidence of being toxic to some tumor cells, able to prevent implantation of others, inhibit tumor enzymes, and alter tumor cell antigens.[228]

Because garlic possesses these laudable attributes, some doctors have chosen to test it against AIDS and use it in personalized AIDS treatments.

Dr. Zhang, whose encomiums of herbal medicine we read earlier, recommends garlic to PWAs because it can be used to treat many opportunistic infections, even those with no conventional treatment. "Garlic is very easy to absorb through the gastrointestinal system and it can even be used as an inhalant to treat respiratory infections. It has a low toxicity, can be used to treat a wide spectrum of infections, and is relatively cheap," he told me.[229]

The rationality of his recommendation is backed up by a study published in the German medical journal *Deutsche Zeitshrift* in 1989 by T. H. Abdullad and others. It reports on the results of seven AIDS patients taking five grams of garlic daily. Six of the seven patients had normal natural killer cell activity (NK cells are inevitably reduced by the illness) after six weeks and that all had normal NK cell activity after twelve weeks. They also reported a lessening of diarrhea in one patient with Cryptosporidia, and fewer outbreaks of herpes, thrush, candidiasis, and sinus infections, among other positive findings.[230]

A study by Abdullah, which was publicized at the 1989 AIDS conference, showed comparable results. Here seven out of ten HIV positive patients received five grams, then ten grams a day of aged garlic extract over three months. They "experienced significant improvements in immune parameters and opportunistic infections."[231]

It's a shame more studies of this type are not being done.

GINKGO BILOBA

In *Get Healthy Now*, I point out that ginkgo, long regarded as a memory tonic, also has been seen to increase circulation, and so has been useful in dealing with impotence in men and other problems that have to do with blood flow. Studies in other areas reveal that it is capable of coping well with some infections, such as that of *P. carinii*.[232] This is one herb whose healing qualities have not yet been tapped.

GINSENG

Another standby of herbal medicine is ginseng. Although to date I know of only one study that has challenged the herb to display its abilities in fighting AIDS, that is, by conducting experiments with it, it has repeatedly shown itself a strong booster and husbander of immune function.

Siberian ginseng is said to increase overall vitality, thus helping the body to ward off autoimmunity disorders, fatigue, stress, and degenerative disease. It also helps to normalize arterial pressure, decrease cholesterol, and improve elasticity of the arterial walls.

If you have time and would like to really plumb the depths of research into ginseng's health-giving qualities, you might consult Norman Farnsworth's *Economic and Medicinal Plant Research*. In the book, the American researcher collected and translated many original studies performed in the Soviet Union in the sixties and seventies. These studies found evidence that use of ginseng offers protection against environmental pollution and radiation, and helps normalize body temperature and regulate blood sugar levels. Other experiments saw the herb offering liver protection and enhancement of its ability to eliminate drugs from the body, as well as charting improved adrenal function and increased cellular ability to dispose of lactic acid and other byproducts of metabolism.[233]

Of more direct relevance to our topic are recent studies that indicate ginseng abets immune functioning. For instance, a double-blind, placebo-controlled investigation by F. Scaglione et al. on the effects of Panax ginseng extracts on cell-mediated immune functions in healthy volunteers came to the conclusion that consumption of one capsule every twelve hours for eight weeks led to significant enhancements in chemotaxis, phagocytosis index, phagocytosis fraction, and intracellular killing.[234] Another study showed that ginseng extract might have potentiating effects on T cell immunity when combined with hydrocortisone.[235]

Experiments in which mice were given the herb indicate that it has beneficial effects on natural killer cell activity and other components of the immune system.[236]

The one direct testing of the plant's effect on AIDS that I have seen was presented at the 1994 International Conference on AIDS. Results showed that Korean red ginseng exhibited positive effects on immune markers in subjects infected with HIV.[237]

GOLDENSEAL

Goldenseal is another much-extolled herb, dubbed by Jethro Kloss as "one of the most wonderful remedies in the entire herb kingdom." Its remarkable healing qualities cause proponents to liken it to a cure-all.[238] In fact, in *Get Healthy Now*, I warn against believing any overly extravagant claims about it, using these words, "The root is quite powerful, and should be used sparingly and for limited periods of time; the consumer should also be wary as goldenseal is expensive and may be sold in adulterated preparations."[239]

That said, we can acknowledge that the herb seems to have restorative effects on the immune system and is capable of performing broad-spectrum antimicrobial, antiprotozoal, antifungal actions. According to Joseph E. Pizzorno Jr., N.D., it is effective against a variety of organisms, including *E. coli*, Shigella, Salmonella, Klebsiella, Giardia, Candida and *Cryptococcus neoformans*.[240] Further, work reported in the *Anti Microbial Agents and Chemotherapy* journal indicated that berberine sulfate (one of goldenseal's active factors) blocked the replication of the streptococci virus.[241]

Along with berberine sulfate, the plant has two other especially valuable components, all of them derived from alkaloids in the roots and rhizomes of the plant. These are hydrastine—a colorless alkaloid with astringent properties—and canadine. Hydrastine inhibits certain microbes, while canadine has been shown in vitro to destroy human carcinoma cells. Berberine, whose antimicrobial abilities we just related, is a mild immune system stimulator.

As an attested antiviral, goldenseal is another herb that should be studied for its ability to curtail HIV.

LICORICE ROOT

A little more thought has been given to the role licorice root might play against HIV, probably because it has known antiviral capacities as well as having a history of being used extensively in both Oriental and Occidental herbal medicine to enhance the immune system.

According to Pizzorno's book on herbal medicine, licorice root "helps the immune system by increasing macrophage activity and the endogenous production of interferon-gamma." He continues, "It has also been found to inhibit Herpes viruses and to stop the growth and cytopathology of several unrelated DNA and RNA viruses while not damaging the organism."[242] An article in *Nature* gave greater meaning to these findings by isolating the active component

in the root, which is said to be glycyrrhizic acid. It goes on to collaborate Pizzorno's opinion about the herb's antiviral capacity.[243]

Influenced by the root's known virus-suppressing skills, scientists did an in
vitro study that showed the herb kept "HIV virus from replicating. No toxicity
to normal cells are noted."[244]

Here is another area in which promising beginnings need to be followed
further.

LIGUSTRUM LUCIDUM

I would also recommend that some attention be paid to the Chinese herb ligustrum lucidum, which has been used traditionally to increase immune system
function and vitality. Clinically, it is used to treat leukopenia (low white blood
cell level) caused by chemotherapy or radiotherapy. Preparations containing
ligustrum lucidum are used to treat such conditions as hepatitis, early cataracts
and Parkinson's disease.

Perhaps it might have the same blood-stimulating effects on those taking
AZT or other antivirals as it has on those affected by chemotherapy.

MAITAKE MUSHROOMS

Mushrooms have long been enrolled in the pharmacology of the Japanese, and
recently some work has been done on the anti-HIV qualities of maitake mushrooms. This variety has already been tested on cancer in animal studies. For
evidence of this, we can refer to an inquiry by Aduchi et al., which showed that
cancerous tumor growth was suppressed 40-50 percent in mice fed this food.[245]
Another study attributed maitake's antitumor effects to two factors. "It was found
to directly activate various effector cells that attack foreign cells (macrophages,
natural killer cells,... T cells, etc.), and to potentiate the activities of various
mediators, including lymphokines and interleukin-1."[246]

In a review article, Dr. Hiroaki Nanba states, "Although present analysis of
its effectiveness is still at the animal experimental level, there are already many
cases of people with cancer whose lives have been prolonged by taking the powder." He then extended its potential effectiveness to AIDS, noting, "Just recently, anti-HIV effect has also been recognized in in-vitro tests, both in the
U.S. and Japan."[247]

Hopefully more details will be forthcoming about this mushroom's capacity to hinder HIV expression.

MAYAPPLE (PODOPHYLLOTOXIN)

The next few herbs I would like to briefly characterize are known for their toxic
effects on various viruses. They all lack the recognition they deserve from the

medical community as possible supplements that might combat the retrovirus while leaving healthy cells unaffected.

Podophyllotoxin is a compound derived from a species of mayapple. In vitro tests by Dr. Zhang et al. at the University of South Florida on cultures infected with leukemia have shown that this compound suppresses human lymphocyte proliferation and stimulates the production of interleukins as well as increasing macrophage activity. "This cytotoxic (cell-killing) formula is more toxic to leukemia cells than to normal ones."[248]

MISTLETOE

Of two notable studies of mistletoe, the plant hung up at Christmas, one indicated that iscador, an extract from European mistletoe, when combined with lactobacillus, doubles the ability of natural killer cells to destroy malignant cells; while a second examination, reported at a 1992 AIDS conference, indicated one extract from the herb "had anti-HIV, immunomodulating and anti-cancer activities in 12 symptomatic HIV disease patients followed for 6 years."[249]

OSHA ROOT AND LEPTOTANIA

Observations by Pizzorno indicate that osha root and leptotania, which are indigenous to the American West and important healing plants in Native American medicine, should be studied as elements of a protocol for viral diseases, such as AIDS. So far, scarce attention has been given to these plants.[250]

PEPPERMINT

This well-known plant has been looked at in a few scientific studies, one indicating that it kills the herpes simplex virus, among many other microorganisms; another showed it "to be effective in reducing abdominal symptoms associated with irritable bowel syndrome."[251]

REISHI MUSHROOMS

Where maitake mushrooms have been revered in Japan for their healing qualities, reishi mushrooms have been valued in China since ancient times. Herbalists say the vegetable helps with blood sugar regulation and blood oxygenation, while protecting the body from free radicals and radiation.[252]

RHUBARB

The tasty vegetable rhubarb is another common plant that has been established as containing factors that inhibit viral activity. In a study of 178 Chinese herbs for their antibacterial activity against *Bacteroides fragilis*, a major microorganism in the intestinal flora of humans, "results showed that rhubarb root was the only one... to have significant activity" in reducing the bacteria's activity.[253] It has also been suggested in other experiments that "rhubarb may enhance immune response" due to its action on spleen cells.[254]

ROYAL BEE JELLY

Earlier we saw the beneficial properties of the bees' propolis. These valuable insects also provide us with royal bee jelly, which contains a full spectrum of vitamins and minerals. It is the only naturally occurring source of pure acetylcholine, a substance vital to the transmission of nerve impulses in the body and brain. According to work by Wang and Watson, royal jelly has strong immune-enhancing properties and also helps kidney, liver, and pancreatic function. It has been of some help to those suffering from asthma, liver disease, and digestive problems.[255]

SCHIZANDRA

This tonic herb plays a vital role in traditional Chinese medicine. The small red fruit of the plant helps to balance body systems. When combined with Siberian ginseng, schizandra helps to detoxify the liver.[256]

SPV-30

Let's move on to some herbals that have been taken seriously as possible candidates in the battle against AIDS. The first is SPV-30, which is a boxwood evergreen extract that is in clinical trials in France. A less formal study of the extract was begun in the United States in 1995. A phase I trial, completed in France in 1992, had inarguably positive results. As the trial was described by David Stokes, who has been taking the extract, "After 30 weeks, a group of 22 HIV positive individuals had an average increase of approximately 100 CD4/mm3... while the placebo group had an average drop of approximately 50 CD4/mm3 during the same period." Stokes says that the same or better results were seen in the informal U.S. trial. "Most participants... report increases in CD4s after two months, and even greater increases in CD8s."[257]

ST. JOHN'S WORT

Another herb that can be put in the category of having already been studied, with good preliminary results for its effect on HIV, is St. John's Wort.

Studies performed at New York University's Medical Center by Dr. D. Meruelo have looked into the herb's ability to inhibit HIV both in vitro and in vivo. Meruelo has found that St. John's Wort contains two potent antiviral chemicals: hypericin and pseudohypericin.[258]

Both substances have been found effective in preventing the spread of retroviruses, such as that causing leukemia, in vitro and in vivo. Of particular note, Meruelo found that pseudohypericin can reduce the spread of HIV. Moreover, it is a substance that can pass the blood-brain barrier, which prevents most medicines from entering the brain and attacking any virus that may be lurking there. He sums up his overall findings:

> Hypericin and pseudohypericin display an extremely effective antiviral activity when administered to mice after retroviral infection. Their marked activity can completely prevent the rapid splenomegaly induced by aggressive viruses....
>
> Availability of the St. John's Wort plant throughout the world and the relatively convenient and inexpensive procedure for extraction and purification of hypericin and pseudohypericin further enhance the potential of these compounds.[259]

TANNINS

Less well known than St. Johns Wort, the tannins have been noted in several studies to effectively inhibit virus-cell interactions. These vegetable compounds are a group of glycosides, such as the one that gives the reddish quality to oak bark. Ghee T. Tan et al. studied their effects on HIV and concluded that tannins are "potent reverse transcriptase inhibitors."[260] In another experiment R. E. Kilkuskie et al. described tannins as "a new class of compound that could inhibit HIV reverse transcriptase and HIV replication in vitro."[261]

TURMERIC

Turmeric is the bright yellow spice that gives the coloration to curry powder. As I note about this herb in *Get Healthy Now*, "tumeric has been getting a lot of attention lately from evidence of its ability to stabilize mast cells, which line our trachea and intestinal tract, and serve as a first line of defense for the immune system."[262] These cells are tightly bound together to prevent the intrusion of unwanted materials. Turmeric helps to strengthen these cells. Beyond helping the immune system in this way, the spice has been revealed (in an experiment conducted by a team led by Mazumder) to inhibit HIV replication.[263] Similar findings were arrived at by Sui.[264]

VIOLET

Lastly, in another comparative experiment, such as the one noted above under the rhubarb heading, R. Shihman et al. experimented with twenty-seven extracts of Chinese herbs and found eleven effective against HIV.[265]

The top contender here was violet, breaking up "HIV growth in repeated experiments.

PHYTOCHEMICALS

You may have noticed that in my discussion of herbs, I often refer to research that aimed at isolating active factors in the nutrient, since researchers often found that these chemical components are the force behind immune-stimulating or antiviral qualities of the herb. When we examined St. John's Wort, for instance, two anti-viral chemicals were named as probably accounting for the substance's good qualities. Again, when we focused on goldenseal, we pinpointed three chemicals found in the root and rhizomes of the plant, which were said to be responsible for the herb's antimicrobial action.

This brings up what must be called one of the most exciting, emerging health fields in the world today: that of phytochemicals. These are substances found in edible plants that exhibit potential benefits in the prevention and treatment of disease. It is only now that we have the technical skills of modern chemistry that we can isolate these particular substances; and it is only with the rise of alternative medicine that we have the mind-set that is willing to look into nature's cornucopia in hopes of finding ready-made health-enhancing substances.

Identifying plant chemicals and discovering what (if any) health benefits a particular substance may have is a challenging yet rewarding task for contemporary medicine. The *Journal of the American Dietetic Association*, for one, states that research into these phytochemicals should be put on the front burner. The journal editorialized,

> It is the position of the American Dietetic Association (ADA) that specific substances in foods (e.g. phytochemicals as naturally occurring components and functional food components) may have a beneficial role in health as part of a varied diet. The Association supports research regarding the health benefits and risks of these substances.[266]

"Functional foods" is a designation the association gives to any food or food ingredient providing health benefits beyond the traditional nutrients it contains.

So far there has not been a great deal of research into these phytochemicals. Of course, the great number of them would make it impossible to fully examine their range without years of study. However, on pages 000-000 I will provide a list compiled by the Department of Agriculture indicating which phytochemicals seem to be helpful in various health areas impacted by AIDS. Don't forget that since many of these chemicals are found in commonly eaten foods, they do not have the same danger of possible side effects that are often an unwanted adjunct to man-made medicines and even to some vitamins or nutrients.

Before providing the list, though, let me say a little more as to what has been found out so far about phytochemicals and functional food components.

In the same issue of the ADA journal from which we just quoted, the areas in which the use of phytochemicals have so far proved valuable are laid out. "Phytochemicals," it is said, "have been associated with the prevention and/or treatment of at least four of the leading causes of death in the country—cancer, diabetes, cardiovascular disease, and hypertension."[267]

One researcher who has seen the positive effects of phytochemicals on the first-named of those diseases is Dr. Potter, professor of epidemiology and director of the University of Minnesota Cancer Prevention Research unit. He has been studying the relationship between diet and cancer for more than fifteen years. He has found that people whose diets are heavy in fruits and vegetables have lower rates of most cancers. Most scientists would acknowledge that. However, Potter believes one reason for this correlation is that some of the phytochemicals in the fruits and vegetables are actively protecting the body from cancerous eruptions. Limonene in citrus fruits, for example, is known to increase the production of enzymes that help the body dispose of potentially carcinogenic substances.

Working along the same lines, Michael Gould, professor of human oncology at the University of Wisconsin Medical School, has found that d-limonene, the major component of orange peel oil, protects rats against breast cancer.

Important to note here also is that clinical trials indicate the beneficial effects associated with high fruit and vegetable diets cannot be duplicated by nutritional supplementation alone. In other words, there are benefits to be had in eating healthy foods beyond what would be obtained from the common nutrients associated with them, such as vitamins C, E, A, and beta-carotene.

The following is a list of foods and herbs whose phytochemicals are felt to be either palliatives for the noted conditions or boosters of different parts of the internal architecture. Although determinations are still being made in this field about the exact properties of each foodstuff, an emphasis on eating foods that have shown promise in connection to health improvement is obviously to be recommended. So, for instance, if one has been afflicted with diarrhea, the eating of pomegranates and sunflower seeds would be a sound practice to adopt.

The material on phytochemicals that follows has been obtained from the extensive electronic database assembled by Stephen M. Beckstrom and James A. Duke at the National Germplasm Resources Laboratory, Agriculture Research Service, U. S. Department of Agriculture. This database is accessible on the World Wide Web at: http://www.ars-grin.gov/duke/.

ANTIBACTERIAL: Chinese goldthread; generic goldthread; mango; huang-lia; goldenseal; witch hazel; emblic; pomegranate; aleppo oak; barberry; huang Po; prickly poppy; strawberry; rhubarb; soybean.

ANTIDIARRHEIC: red mangrove; rowan berry; arbutus; pomegranate; marshmallow; canaigre; babul; gum ghatti; poison hemlock; sunflower; guava; coconut;

nance; emblic; pomegranate; fenugreek; winter's bark; smooth sumac; simaruba bark; chinaberry.

ANTIFATIGUE: lettuce; asparagus; endive; black gram; cowpea; lambsquarter; radish; chayote; Chinese cabbage; purslane; oat; garland chrysanthemum; dill; dandelion; pigweed; cucumber; kudzu; spinach; borage.

ANTIGLAUCOMIC: camu-camu; pansy; Japanese pagoda tree; pumpkin; acerola; Chinese foxglove; buckwheat; sang-pai-pi; celery; peegee; bitter melon; American elder; pokeweed; parsley; common smartweed; warburghia.

ANTIHIV: rowan berry; licorice; poison hemlock; sunflower; coconut; fenugreek; winter's bark; rice; white willow; corn; giant reed; black Locust; crab's eye; cork oak; evening-primrose; oats; hull husk; lantana; red mangrove; coffee; mango.

ANTIVIRAL: khasi pine; lemon; chilgoza pine; celery; red mangrove; Camu-camu; rowan berry; arbutus; pomegranate; emblic; canaigre; babul; gum ghatti; poison hemlock; guava; sunflower; chir pine; celery; coconut.

BACTERICIDE: lemon; chir pine; khasi pine; celery; chilgoza pine; grape; red Mangrove; Mangosteen; Camu-camu; Copaiba; Hindi Chaulmoogra; Rowan Berry; Arbutus; Pomengranate; Marshmallow; Canaigre; Babul.

CANDIDICIDE: clove; grape; cowberry; coconut; allspice; garlic; cucumber; bayrum tree; carrot; cherimoya; common thyme; basil; carrot; clove; cantaloupe; allspice; pignut hickory; purslane; spinach; oats.

FUNGICIDE: parsnip; grape; lemon; clove; coconut; potato; grape; cardamom; oats; coffee; rice; common thyme; celery; java-olive; bayrum tree; winter savory; celery; turmeric; allspice.

HEPATOPROTECTIVE: cantaloupe; English walnut; red mangrove; avocado; cucumber; safflower; arbutus; Brazil nut; apricot; sunflower; great scarlet poppy; butternut; canaigre; babul; opium poppy; calabash gourd; pomegranate; gum ghatti; guava; sesame.

IMMUNOSTIMULANT: chicory; gobo; elecampane; dandelion; licorice; coneflower; costus; leopard's-bane; crab's eye; coffee; mugwort; Chinese goldthread; generic goldthread; java-olive; huang-lia; goldenseal; beet; malangra; lambsquarter.

VIRICIDE: lemon; celery; red mangrove; camu-camu; rowan berry; arbutus; pomegranate; emblic; canaigre; babul; gum ghatti; poison hemlock; sunflower; guava; coconut; nance; fenugreek; winter's bark; grape.

Other Natural Remedies

Studies performed on the healing properties of natural substances falling outside of the adaptogen category have also elicited some very exciting findings. Here we have categorized remedies according to proven benefit.

AL-721 (EGG LECITHIN EXTRACT)

AL-721 is a formula made from two active phospholipids found in egg yolk, phosphytitle choline and phosphytitle diethylalamine. It is used as an adjunctive therapy for the containment of AIDS and other viral diseases.

The substance was first developed at the Weissman Institute of Science in Israel by a team led by Meier Shinitsky. Researchers found this formula effective in treating memory loss and impaired immune function, and in easing withdrawal effects from alcohol and drug addiction. The formula was found to rehabilitate physiological functioning in drug addicts, especially in those with HIV-positive blood.

AL-721 was also found to prevent human T cells from becoming infected by HIV. It is thought to work by removing cholesterol from the envelope surrounding the virus and interfering with important receptor configurations, thereby rendering the virus ineffective.

There was great excitement at the time AL-721 was first discovered. In fact, Robert Gallo reported in the *New England Journal of Medicine* in November 1985 that egg lipids inhibited cell invasion by the AIDS virus in laboratory experiments. Within months hundreds of patients with AIDS and ARC were using egg lipids. But shortly thereafter egg lipids became unavailable, and AZT became the drug approved by the FDA.

An examination of why the searchlight of establishment medical interest suddenly shifted away from this compound would make for interesting study, although where the real energy needs to be put in relation to this substance is into further evaluations of its anti-HIV power.

DNCB

A number of other natural substances have proven startlingly good at eliminating HIV in vitro. Before alerting you to them, let me mention one chemical that is being tried as a stimulator of immunity. Anyone familiar with the rationale for acupuncture will know that the purpose of inserting the thin needles is to stimulate blood flow in the targeted area. This blood flow, it is posited, will increase immune and other activity in this locale.

DNCB (dinitrochlorobenzene) has a similar mode of action.
According to an article by James, DNCB is swabbed on the skin with a cotton ball. It is believed that the substance penetrates the skin and elicits an immune

response, which then cascades into a full-scale rejuvenation of the immune system. More specifically, the chemical, after getting in the body, migrates to the lymph nodes. Once inside, "the DNCB antigen is presented to the CD4 T-helper cells which then initiate a cell-mediated immune response. The CD4 cells then proliferate and start to circulate, activating other immune cells."[268] The aroused system would presumably also eliminate HIV.

DIOXYCHLOR

To get to a number of natural remedies that have triumphed over HIV and other viruses in vitro, and thus whose study should be moved along into animal trials—or if they have already shown themselves to be active antivirals in animals, as has GE-3-S, they should be moved into small-scale human trials—we can begin with dioxychlor. It is an antiviral, antibacterial, antifungal, and antiparasitic agent that has been shown in vitro to harm HIV, CMV, herpes I and II, polio, Epstein-Barr, and many other viruses. It attaches to any free-floating pathogenic virus in a serum as the virus moves from one point to another through the bloodstream. Dioxycholor kills bacteria by supplying those that thrive in an atmosphere of little or no oxygen with an abundance of it. It has proven so antagonistic to the growth of bacteria, in fact, that if twenty drops of dioxychlor are added to a quart of milk, it is said the milk will be preserved for three to four weeks.

DYSIDIA AND SEAWEEDS

Dysidia, an ocean sponge, produces a compound that attacks the HIV virus. Ocean seaweeds also contain potent yet nontoxic antiviral agents. In one red seaweed the dilution factor at which it is no longer virotoxic has not yet been found. In other words, even at very low dilutions, it still attacks viruses. Obviously, this is an area ripe for further investigation.

GE-3-S

Hirabayashi et al. report that the lichen compound GE-3-S prevents HIV particles from attaching to the surface of T4 cells in mice.[269] The compound is derived from the lichen umbilicaria esculenta. In the experiment under discussion, when cells were examined by immunofluorescence for the presence of viral antigens, those given higher concentrations of the lichen compound showed virtually no expression of the viral antigen. Anti-HIV activity was also demonstrated by the inhibition of visible damage to infected cells in vitro. GE-3-S showed no acute toxicity to the mice, even at very high doses. Obviously, further study of this substance is demanded.

GLANDULAR POLYPEPTIDE EXTRACTS

Highly purified polypeptide extracts from animal thymus and spleen also have shown anti-HIV properties in experiments. Dr. Donald J. Brown from Seattle pointed out to me that the extracts being used in these trials are not the type of glandulars one might purchase in health food stores. "These are very specific, highly purified, polypeptide extracts that are extracted from thymus and spleen." For proof of the value of these extracts, he noted an investigation in which thymus gland extract was given to 130 subjects, twenty-three of whom were HIV-positive: "What was most interesting was that there was a rise in the CD4 count, and a drop in the CD8 count in HIV-infected individuals who were taking this. In addition, there was a correction of a number of other abnormalities, including weight loss and diarrhea, and an overall sense of well being."[270]

Oral splenic polypeptide has been combined with Siberian ginseng to create PCM-4, a formula that was developed by Dr. Nikolaus Weger in Germany. The formula was given to HIV-positive individuals; and, according to a report in *Nutrition Review*, "Within 4 to 8 weeks, subjects had a 95 percent increase in CD4 counts, a cessation of diarrhea, weight gain, and increased energy."[271] This treatment is approved for use with HIV positives in Tanzania and Uganda.

Let me pause a moment to answer a question that may have been at the back of your mind as you have been thumbing through this list. The reader might put it like this: "Why do so many natural compounds prove effective for human use against viruses?" Although I have no definitive answer, I do ask you to remember one thing. Viruses are ubiquitous in the animal and plant world. Some of them have evolved into peaceful coexistence with their host, so they thrive in a modest way without damaging the overall well-being of the animal or plant on which they are parasiting. Be that as it may, there are also harmful viruses, and because they are so widespread, protection against them, which either eliminates or thins out the intruders in the organism, is also widespread. I believe this is an important factor explaining why we often find helpful antiviral attributes in plant or animal extracts, because all plants and animals have safeguards against the imposition of these parasites. Thus, I might add, there is a certain rough and ready truth in the belief of the Puritans. They held that in the Garden of Eden, humans had a complete knowledge of medicine along with marvelous recuperative powers. Once expelled for eating the forbidden apple, they lost these skills. However, the Puritans held, "The lower animals were left with an intuitive ability to cure their own ailments, as a constant reminder to man of the powers sacrificed at his transgression."[272]

Moreover, getting back to our main strand, it stands to reason that substances derived from animal or vegetable protective systems will not have the toxicity to healthy cells that man-made concoctions like AZT might have, because they have been developed within living bodies in the first place and have been tailored to their ongoing optimum functioning.

HYPERTHERMIA

Let's interrupt our honor roll of substances that have proven potent against HIV to note another type of treatment, hyperthermia, or the administration of an artificial "fever" to a patient to literally burn the virus out of the blood. This may seem like other "heroic cures" we mentioned earlier. Recall that "heroic cures" refer to nineteenth-century treatments, like bleeding, which seemed to put the sick person under so much new stress it was as likely to kill as cure. We cited AZT as the modern-day equivalent. In fact, inducing a high fever in a patient was used in the old days against syphilis.

Though this therapy may seem to readers a relic of the past, recent success with its uses has made even the stodgy FDA sit up and take notice. The federal agency has now allowed testing of the practice after pressure from Senator Lautenberg of New Jersey. One of the reasons for the agency's willingness is the success story of Chuck DeMarco. He was in an advanced state of AIDS, with Kaposi's sarcoma, a persistent cough, weight loss, and (according to his doctor) less than a year to live. I talked to him three years later! He was now in good health and told me about his hyperthermia treatment in Italy.

> I went through the procedure where the blood was removed from my body in dialysis fashion, heated up to about 116 degrees and put back in hot so that a fever could be raised in me above 108 degrees, which is the temperature that viruses begin to die off at.
>
> The very next day, the cough that I had was totally gone. And to this day, 32 months later, it is still gone. The KS lesions I had disappeared. The thrush in my mouth was gone within about 48 hours. Incredible amounts of energy. Energy to the point where instead of sleeping 18 to 20 hours a day, I was sleeping now two to three hours a day.
>
> My CD-4 count, the t-cell count rose. Twenty-one days after the procedure, when I was back here in the U.S., it had gone here from 220 to almost 890 and has remained that way ever since.[273]

Again, as I've said repeatedly, since each case of AIDS is unique, a single case of dramatic improvement like this does not mean a cure is in sight; but what it does mean (something that even the FDA acknowledges) is that researchers should declare full speed ahead on studying how effective hyperthermia would be given to PWAs in various stages of the disease.

LIGNOSULFONATE (LS)

To return to my discussion of supplements that have been shown to be effective against HIV and so demand further study, let me briefly mention lignosulfonate (LS), a water-solubilized lignin obtained from the waste liquor in pulping wood, that is, the wood crushing in the process of making paper and other such products. One study shows LS to have antiviral activity against HIV in vitro. An article by Harumi Suzuki et al. reported that "LS completely inhibits the HIV-induced cytopathic effect, the HIV-specific antigen expression, and syncytia formation to the concentrations of >50 ug/ml." The authors further underline, "LS inhibits the reverse transcriptase activity in a cell-free system."[274] To put that in slightly simpler language, LS stops HIV from killing cells (cytopathy) or making them clump together (syncytia), which is eminently harmful. It also stops the retrovirus from replicating (inhibiting RT) and from locking onto healthy cells (antigen expression). Obviously, this is a formidable foe to HIV at least in cell cultures.

PENTAMIDINE

The story of pentamidine, well told by Arno and Feiden, is another one of the lackluster, bungling performances of the pharmaceutical industry and government regulators when it comes to AIDS drugs. As cases of AIDS began to be seen, it was quickly evident that a rare form of pneumonia, PCP, was one of the opportunistic infections that commonly accompanied it. Twenty years of research had already shown that pentamidine was effective against it, but no company was producing it in 1983 and the world's only stockpiled supply was discovered to be water-damaged and unusable! Moreover, ignoring the requests of the CDC, no U.S. firm was interested in making it; companies said that with the AIDS crisis at that time so small, they couldn't make a big enough profit on selling the commodity. Eventually, the government paid a reluctant company, Lymphomed, to make it. This company's owner would eventually become a millionaire many times over due to the money he made from pentamidine.[275]

The problem with injected pentamidine was that most of it went to the liver and other internal organs, not the lungs, where it was needed. Animal studies showed that putting the drug into a spray was a feasible way to attack pneumonia most successfully. However, the red tape was piled so high that, even though companies, seeing Lymphomed's skyrocketing profits, were now eager to make the drug, it took the FDA two years just to approve the plans for drug trials—two years, that is, before the studies could even begin!

As expected, it was shown that inhaled pentamidine can reduce the rate of relapse by 50 percent in patients with (PCP).[276] If and when PCP returns, it is mild and much less fatal. This drug is available by prescription only.

RICIN

Another plant derivative that has anti-HIV qualities is the chemical ricin, a constituent of castor oil. Its effectiveness in extirpating leukemia in laboratory animals made it seem a substance that would have an impact on HIV. Later, its ability to kill cells infected with HIV was demonstrated.[277] Scientists have called it "a sort of biological missile" as it seeks out virus-infected cells and kills them. It works by destroying the ability to manufacture proteins, which are the basic building blocks of cells," presumably shutting down the cells' operations, so its parasites (like HIV) are put out of action.[278]

ROVITAL-V (R-1103) AND CARCIVIREN-V (C-1983)

R-1103 and C-1983 are composed of a group of plant extracts that appear to possess a broad spectrum of antiviral effects and are completely nontoxic. Dr. J. Roka investigated the inhibition of viral replication for several DNA and RNA viruses with different cell cultures, and found a significant degree of inhibition produced by the R-1103 and C-1983 formulas in cell cultures.[279] Most relevant to our purposes, these substances slow down or even stop the progression of HIV from Stages I, II, III, and IV.

SHARK CARTILAGE

In an interview, Dr. Morton Walker suggested that shark cartilage may have a role to play in the treatment of AIDS, particularly Kaposi's sarcoma (KS). When I asked him why this substance would be of such value in hindering KS, he told me it was because the cartilage contains a great amount of angiogenesis-inhibiting factor. What is angiogenesis inhibiting factor and how could it interfere with KS? Angiogenesis, he first explained, is the formation of new blood vessels; certain cancer cells are known to secrete cytokinins that will form a local capillary network to bring blood to a developing tumor. For KS, this will be located on the surface, since KS is a vascular cancer. He continued, "Angiogenesis inhibiting factor causes Kaposi's sarcoma to stop spreading over the surface of the body because it prevents the capillary network of blood supply from entering into the tumor sites.... The tumor stops growing and stops spreading.[280]

KS is a frequent accompaniment of AIDS. Shark cartilage may provide a necessary tool in helping the body cope with this threat.

Lifestyle Changes

At the heart of the new medical paradigm—a description of its properties has appeared as a submotif of this book—is the idea that one's state of health or illness is the product of multiple, interconnected factors, both in the person's surrounding environment and internally. Moreover, to identify the cause of a viral disease, for example, one has to sort through not only what "bug" has followed what route of transmission to get into the body, but, equally importantly, learn what state of susceptibility has made the person ready to receive the bug with open arms, as it were. This is to say, a disease does not get the opportunity to infiltrate and begin lording itself over a human body unless that body already has a defense system that is functioning suboptimally.

This is not to say a person should be blamed for his or her disease (in that the patient allowed his or her immune system to deteriorate). Rather, I am saying a person is quite often "partially" to blame in that he or she has adopted an unhealthy lifestyle, taking drugs, for example, which has ended up weakening immunity. It goes without saying that in some instances external factors may far outweigh lifestyle as causes behind the eruption of a disease. Moreover, in some cases, one's lifestyle may not be under one's own control. (Think of our prisoners in our scandalous jail system, who are often given a diet of unrelentingly starchy, unwholesome foods. Or think also, to repeat an earlier thought, of smokers in the United States in the 1940s, when even the vast majority of doctors told them cigarettes would not affect health.) Taking all this under advisement, I still would have to say that in my experience, for most people who get sick, and especially those who are struck down in their prime, their adoption (perhaps unwittingly) of unhealthy habits played a major role in leaving them open to disease. So as another component of the alternative way to deal with AIDS and HIV infection, we must look at what lifestyle changes can be put into practice to help move a person toward health and self-sufficiency.

Let me say first off that unlike the types of therapy discussed in the preceding few pages, supplements that are meant to be taken only by those who are suffering under disease conditions, the practices to be outlined in this chapter are not even properly classified as therapies, since such things as exercise and the avoidance of sugar should be adopted by everyone, even the healthiest specimen of humanity, insofar as the person wants to remain healthy and live life to the fullest.

These practices are ones I have also spent much time enumerating in such books as *Get Healthy Now*, where I outline the components of a healthy lifestyle. In this book, then, I will only skim the surface in describing needed lifestyle changes, since more details on each point can be gathered from disparate works by me and other alternative health writers. My stress here will be on how common prescriptions one gives for health have been shown to have ameliorative effects on PWAs.

Dietary Modifications

Nutrition is significant in minimizing the intensity of HIV infection and AIDS, and is infrequently recognized and cited as such. "AIDS patients need to be made aware of the importance of eating only highly nutritious organic foods," Patrick Donnelly, program coordinator of the Whole Foods Project, told me when we spoke of this topic.

I queried him on what foods he saw as the best for a PWA's diet. His reply, as I expected, pointed not only to foods that those ill with AIDS should be focusing on, but ones that should be at the center of everyone's diet. What he said was both succinct and enlightening:

> We're working with what the Physician's Committee for Responsible Medicine calls the new four food groups which are whole grains, legumes, fruits and vegetables. These are foods which are very high in antioxidants. They have nutrients like beta carotene, vitamin C, vitamin E, zinc, selenium, which help to boost the immune system.
>
> For the most part, these nutrients are not found in the high fat, high protein standard American diet. So, we're offering a different point of view. We're trying to get people to look at a new way of eating that is not about providing calories so much as it is about supporting the immune system.[1]

From there, I interviewed Richard Pierce, the project's director. He put a more philosophical spin on Donnelly's prescription: "You have only to look at this food to see that it's full of life. These greens are full of the life that came directly from the earth that grew them. The vitality of that food is what nurtures us."[2]

Remember that attention to the problems of malnutrition in patients with AIDS is of paramount importance because the timing of death in these patients may be more closely related to degree of body cell mass depletion than to any specific underlying infection.[3] Research supports the long-term efficacy of diet counseling and enteral nutrition supplementation in AIDS patients to increase or maintain weight, restore lean tissue, and lessen the ravages of HIV/AIDS.[4] These factors could help to decrease mortality. In addition, enteral nutrition supplementation (that is, putting the supplements directly into the intestines) improves bowel function.[5]

While it is especially essential for people with immune-compromised conditions to eat pure whole foods and take supplements, it is just as necessary to know which foods to stay away from.

FOODS TO AVOID: SUGAR AND YEAST

Several doctors I have interviewed spoke to me of the immunosuppressive qualities of sugar, which presents obvious dangers to people with AIDS, as they knew from working with their own patients.

Dr. Marjorie Siebert told me, "One of the key things in a person having a good immune system is that they're not consuming sugar. Sugar is a definite immune suppressant. A hundred grams of sucrose will cut antibody production by 50 percent for 24 hours."[6] Joan Priestley added, "The average sugar consumption [in the U.S.] is in the hundreds of pounds per year and a high sugar diet is very clearly immunosuppressive."[7]

When I broached the subject with Robert Cathcart, he more directly related the way sugar depressed immune function to the reactions of people with AIDS, particularly in relaying to me how sugar can cancel out any positive effects drugs may provide: "Patients are sent to me who have thrush and who have been put on Nystatin or Diflucan but not told to stop eating sugar," he explained. "This is terrible because it [sugar consumption] breeds resistant strains.... I've already seen patients with Candida that Diflucan does not kill off anymore because of the fact that they were eating sugar."[8]

As we know, those who have AIDS are often beset with yeast infections in the mouth, gut, and other places. Thus the consumption of yeast is to be avoided for positives or people with AIDS. Priestley, whose opinions on sugar we just heard, believes yeast is an unappreciated cofactor in AIDS. While warning her patients of the trouble they might be causing themselves by drinking alcohol and eating white flour and other processed foods, which many people are allergic to, she put particular stress on the PWAs getting "yeasted foods out of their diet in an effort to keep their yeast under control." She further noted, "There is some evidence that the PCP organism is a yeast and not a parasite, so an anti-yeast diet is actually a self-empowering way to control PCP"[9]

Thus, ideally, the PWA should be following the same whole foods, vegetarian diet that any intelligent, health-conscious person would embrace.

Exercise

Another prime component of a well-tempered lifestyle that is also relevant to PWAs is exercise. While exercise's facility in helping those with heart disease and other disabilities is well known, less publicity has been given to the fact that studies indicate exercise is therapeutic for those with AIDS. Schlenzig et al. studied the long-term benefit of physical exercise on the biological condition of HIV and AIDS patients as well as on the course of illness. They found that physical exercise improves biological condition at all stages of HIV infection, including AIDS. "Furthermore, a delay of AIDS-related complications seems possible through exercise. Regular exercise also seems to be correlated with a more stable T4 cell count."[10]

Emotional Resolution

Let's recur for a moment to the thoughts of Schmidt on the immune deficiency disease. It will be remembered that, while finding weaknesses in his overall theory of hysteria and AIDS, we had cause to concur with his feeling that gay males suffering from AIDS, as from any other disease, found their weaknesses compounded by the stress they were under living in a (in many ways) homophobic society. Schmidt mentioned that based on his experience treating patients, some gay males had internalized the negative stereotypes of homosexuality prevailing in U.S. society. The psychotherapist then argued reasonably that a PWA who was burdened with such emotional baggage would profit by not only treating his disease but trying to overcome internal blockages. (A similar case could be made for drug addicts, who also may accept negative stereotypes of themselves that cause health-diminishing stress.) Other medical practitioners have heartily agreed on this last point concerning the need to do emotional clearing as part of a comprehensive AIDS treatment.

I talked with Joan Priestley about this issue. "Another area people need to look at," she said, "is what I call nutrition of the mind." Dr. Priestley believes an essential element in her patients' getting well is their clearing emotions. "I ask my clients who are gay to handle their emotional issues or their emotional issues will handle them." This would mean, in part, people with AIDS coming to grips with and accepting themselves, confronting others who attempt to unjustly stereotype and discourage them, and dislodging themselves from dysfunctional relationships. It would also mean stepping outside themselves. She puts it like this: "I encourage them to develop a spiritual kinship with some kind of benevolent power beyond themselves so that they don't feel like it's just them against the world with this infection."[11]

Dr. Paul Epstein concurs and looks to a combination of emotional and physical healing as preparing the way to regenerate the immune system on the road to besting the disease. "By scientifically strengthening their immune systems and resolving deep issues and conflicts and stuck places inside, people may be improving their immune response. When we put the whole piece together, it really happens," he says.[12]

Stress Reduction

The last cited doctors, when they are laying emphasis on emotional clearing, do not mean that the person with AIDS should sit in a corner and try to reconfigure his or her attitudes through strength of will. Various changes in attitude, such as the ability to deal with stress or escape disempowering relationships, will grow as the person talks over situations with supportive friends and therapists as well as participates in group activities, whether protests or dances, that integrate the individual into his or her community. The ability to individually reflect on one's choice of lifestyle and make changes will be nurtured by these discussions and actions. Moreover, the stress reduction that is a necessary prelude to facing one's own hangups—for it is impossible to think lucidly when one is beset by constant anxieties—can be cultivated by becoming a practitioner of one of the techniques that have evolved for damping stress and channeling and focusing one's energy.

Let's pause to look at some of these recommended practices, beginning with those that arose over long periods in what was once called the celestial kingdom.

For centuries, the Chinese have understood that one's inner energy or chi makes life possible. Through practices such as tai chi and meditation, and treatments such as acupuncture and reflexology, this life force can be freed, strengthened, and enhanced. We will also note the importance of yoga, an Indian technique.

TAI CHI

Tai chi involves the learning of a set of sequential movements that are neither strenuous nor quick, but which are graceful, flowing, and precise. On the surface, such routines would seem to be ideal for increasing flexibility, poise, concentration, and ease. However, as Eric Schneider, director of the Northeast Tai Chi Association, explained to me when I quizzed him on other health benefits of this practice, perhaps tai chi's greatest overall result is the boosting of energy. "The way one can increase their storehouse of energy is by practicing something like tai chi because the environment has energy and that energy is brought into the system."[13]

What is he getting at? It is a commonplace that classical Chinese painting is of landscapes absent of any people or with people depicted on the margins. However, given traditional Chinese beliefs, this is not because the human is considered of little importance. In fact, these landscapes can be seen as profoundly humanist if viewed in the right way. The Chinese perspective has been that the human and natural world are in profound congruence. When the human world is proceeding smoothly—the Emperor is ruling benevolently, for example, farms will be productive and the forests, hill, seas, and mountains will be pacific and thriving. Thus a painting of a tranquil landscape is evidence not only of that slice of nature but of a tranquil state in the human community. In other words, for the traditional Chinese, the picture by a great artist of a bush on a mountain says as much about the human soul as about a species of vegetation. We should add, moreover, that while this may not be the case with men and women, the normal state of nature is one of peace and harmony; so if a human is out of sync, nature will be pushed out of the tranquility that is its common resting point.

Taking this to the exercise we have been considering, we can say that it is believed tai chi is a way of helping the body gravitate into harmony with the surroundings. That is why, by the way, tai chi purists, such as Teacher Linn, who has a class in a park in New York's Chinatown, says that true tai chi must be done out of doors.

Schneider explained further that this gravitation has to do with achieving proper vibrations in the body, by drawing energy from the environs "vis a vis the simple breathing apparatus. The body goes through an alchemical process converting that to useful chi." He noted that according to ancient thought the body has five systems, each with a separate vibration pattern. In the same way, the immune system consists of a number of overlapping, interlocking parts. "When you start practicing tai chi," Schneider mentioned, these systems "start to resonate with the core of the vibration in the body. Then they harmonize themselves."[14]

As Nhi Chung, my coauthor's wife, explains, "The concept of chi relates to air." Chung has practiced tai chi for thirty years, since her youth in Saigon, where hundreds used to start their day by going through different routines in public parks. "The word chi," she says, "in Chinese is literally air. The ideogram for chi contains the sign for rice. Above the rice, you can see two horizontal lines, which represents the steam rising from it."

"When the rice is half cooked, the steam comes up. Steam means chi. In tai chi, you try to move your chi to circulate your body and to circulate through your organs."

A more philosophical explanation is offered by longtime practitioner and renowned sculptress Yuko Otomo.

Tai chi is to honor your body, which is a tool to connect to the universe, which is you. Tai chi helps me to get in touch with who I am which is part of the cosmos.

It helps me to slow down, to be rooted and, at the same time, uplifted. It also helps me to appreciate what we have.

Acupuncture

While Schneider has seen PWAs profiting from tai chi, this practice has not really caught on among those with the disease. Acupuncture, on the other hand, is one of the more popular treatments that HIV positives patients have turned to since 1982.

Acupuncture, as is well known, involves inserting very thin needles into select points in the skin. As with tai chi, the philosophy behind it focuses on energy. Where tai chi is set on getting the various energetic vibrations to move in harmony, acupuncture is centered on dealing with energy that is not flowing fluently because of internal obstacles. Nurse Abigail Rist-Podrecca told me: "Wherever disease processes are occurring you'll find a blockage. Acupuncture needles inserted in these different points... open these pathways and let that energy flow smoothly through those meridians."[15] Explaining this in terms of physiology, she said that sticking in a needle "dilates the blood vessels so that... you get more circulation, more of the nutrients, more oxygen flowing through."

Rist-Podrecca has proved the value of this treatment in practice. I asked her to outline her general approach to PWAs. "With people that have HIV, we're working on some of the immune system points and also just symptomatically working to... energize... the various areas that are not working smoothly."[16]

Dr. Michael Smith, who had worked at Lincoln Hospital in the Bronx, which had pioneered in developing acupuncture in a hospital setting for both those coping with drug addiction and AIDS, notes that after a number of treatments, there is a decrease in diarrhea, night sweats, and other symptoms.[17]

Other treatment centers that have reported symptomatic relief through use of this time-honored therapy are the Somerville Acupuncture Center in Boston, the AIDS Alternative Health Project in Chicago, and the Quan Yin Herbal support program in San Francisco.[18]

Aside from these testimonies, there have been scientific studies on the effectivity of acupuncture. In fact, one investigation was carried out at the same Lincoln Hospital. The conclusion here was that for most of the two hundred

AIDS patients treated from 1982 to 1987, there was "a reduction in fatigue, abnormal sweats, diarrhea and acute skin reactions after 4 or 5 treatments. Some experienced a 15 to 25 pound weight gain and others a decrease in side effects to AIDS-related medications."[19]

Another study, published in the papers of an international AIDS conference, found that people with HIV who use the acupuncture have extended survival rates. "In addition, they regularly report a substantial reduction in symptoms and side effects to HIV related drugs. CBC blood counts frequently normalize." The study concludes by suggesting that acupuncture be made a part of standard treatment for HIV infection.[20]

Yoga and Meditation

While tai chi and acupuncture act to restore a natural balance in the body's flows of energy, they are more oriented to general well-being than stress reduction. Yoga and meditation, though also broadly health producing, have been seen in the West as particularly helpful in reducing the load of stress we all are apt to shoulder under conditions of "civilization." While yoga is ultimately rooted in East Indian practices, meditation has been a standard religious exercise in most faiths.

To say just a word about these techniques, which are covered widely in other sources, let's note that yoga and meditation do more than help the body on a physical level; they also quiet the mind and allow rebalancing and healing on the spiritual level.

When I asked Dr. Majid Ali to define the basis of his work with these ancient practices, he used these words: "The way that you look at the world around you determines the state of the biology under your skin.... You can be in a stress that wears you down or in an even steady state energy mode which is restorative."[21]

As he describes it, you can glimpse the parallel with Chinese thought. Where tai chi seeks through restrained movement to tap into and harmonize with natural energies, yoga and meditation act to bring the mind into an awareness of the inner energy pulsations. This is not done directively, that is, so that the adept can influence this activity, but as a way to create a healing temperament, as the meditator or yoga practitioner "allows this energy to guide... [the student] into a healing mode." As Ali summarizes, "We want a transition from an ordinary thinking stressful mode which causes disease into a non-thinking meditative deep restorative healing mode. That's our goal."[22]

Reflexology

Let me conclude by mentioning a couple of Western alternative therapies that are being applied to AIDS patients with a modicum of success.

The first is reflexology. As I explain in *For Women Only*, "Reflexology is based on the principle that people have areas in their feet that correspond to every part of their body.... Massaging these specific areas on the feet thus helps to improve the functioning of particular organs and glands."[23]

This therapy has been used as a relaxation treatment for AIDS patients in Uganda. According to a report given at the 1994 International Conference on AIDS, "As of November 1993, four months after this program began, 85 percent of AIDS patients reported some pain relief, relaxation and better sleep after the exercise."[24]

Reflexology has also been employed as part of a comprehensive program for HIV positives and people with AIDS at the John Bastyr College of Naturopathic Medicine. (I will provide more information about this innovative program in appendix III where I present the school's supplementation and botanical protocol).

Aromatherapy

Lastly, we come to aromatherapy, the use of plant essences in healing. One might see this as the precursor to the development of phytochemical therapy, which isolates different chemicals found in plants and acts to mobilize them for purposes of building health. In aromatherapy, the distilled oils of different plants are used.

Because a number of these oils have already been seen to have antiviral properties, it is not unreasonable that doctors would try them against HIV. The antivirals include essences of palmarosa, cinnamon, and cloves, which have been used to counter tuberculosis bacillus, and lavender essence that has been effective against gonorrhea and syphilitic sores and chancres.[25, 26]

Taking up the task of exploring whether similar antiviral effects could be useful against HIV, Teopista did a study that was published as "The Use of Aromatherapy in the Management of People with AIDS." The investigation worked with eighty randomly selected AIDS patients. Half received psychological counseling with aromatherapy; the other half received counseling but no aromatherapy. "Results showed that the aromatherapy group reported less aches and pains, faster healing of wounds, restored physical strength and a better ability to cope."[27]

This brings to an end our survey of various changes in lifestyle, such as eliminating certain unproductive foods from the diet and stress reduction techniques that are of proven benefit to PWAs. It is fitting that we ended with a therapy whose benefits are both physical, such as helping healing, and psychological, in that those who used the plant essences seemed to having less problems dealing with stress. All of the recommendations that have appeared in the foregoing section are of the same nature, since they are found to be active both physically and psychology.

Sir Francis Bacon
and Ancient Wisdom

I mentioned that Bacon was a consummate persuader, though one who, due to recalcitrant historical circumstances, found most of his advice falling on fallow ground. At the head of our discussion of therapies, I noted that in *The New Atlantis*, he envisioned a refurbished commonwealth in which a state-supported scientific institution, the House of Solomon, busied itself with experimental investigations that provided labor-saving and health-promoting devices and discoveries.

It was appropriate to mention it there because, as in Bacon's fiction, much of what was looked at were AIDS therapies whose full value, if they truly possessed any, will only come out in the future. For one, we talked about supplements and alternative therapies that often had well-attested histories of fighting other viruses, and so would seem to be well suited for combating HIV. In these instances, though, the substances or methodologies had not yet been used in any experiments to see whether, indeed, there would be any AIDS-fighting utility to these techniques or supplements. Use of future tense, then, was quite appropriate there. Secondly, we talked about therapies and substances, which had been shown to militate against HIV replication or act in some other way to block the disease or build up the power of the immune system. Though this second group of treatments already has shown promise, it either had not been taken far enough in studies or, as in the case of vitamin C, though it had repeatedly shown its use value, had been generally ignored by practitioners and writers on the AIDS issue. Again future tense was deemed suitable in that we have to await for a later time before these substances are fully proved out or more widely adopted.

Now, Bacon, wily arguer that he was, thought not only of the future as supplying ammunition for his cause—and the Atlantis correctly foresaw state sponsorship of scientific endeavor—but of the past. He called attention to the Greek pre-Socratic philosophers, who he noted indulged in a more experientially based scientific theorizing than Plato and later thinkers of the classical age. This was a time-tested means of recruiting allies, poring over history until writers whose ideas seem to jibe with the arguer's are found. However, Bacon also had a more ingenious trick up his sleeve. In his book *The Wisdom of the Ancients*, he claimed that Greek myths, such as the love affairs of Jove or the tale of Orpheus, were disguised allegories of (along with other things) the methods of true Baconian science, as far as its principles were understood in olden days. "Upon deliberate consideration," Bacon writes, "my judgment is that a concealed instruction and allegory was originally intended in many of the ancient fables."[1] The reason discourses on scientific method and its findings had to be dressed in this way, he goes on,

> is because it is sometimes necessary [to use myths to convey information] in the sciences, as it opens an easy and familiar passage to the human understanding, in all new discoveries that are abstruse and out of the road of vulgar opinions. Hence, in the first age, when such inventions and conclusions of the human reason as are now trite and common were new and little known, all things abounded with fables, parables, similes, comparisons, and allusions, which were not intended to conceal, but to inform and teach, whilst the minds of men continued rude and unpracticed in matters of subtlety and speculation.[2]

In *Wisdom of the Ancients*, Bacon lays bare what he sees as the inner sense of certain fables. For example, he shows that the story of Proteus, who readers of *The Odyssey* will recall was the shape-shifter that Ulysses had to beat in wrestling in order to get him to disclose a secret, is actually an allegory of what the ancients knew about matter. Bacon notes, at one point, that Proteus is said to be a servant of the sea god Neptune, because "the various operations and modifications of matter are principally wrought in a fluid state."[3] Again, the way Ulysses made him yield by binding him, holding tight as he changed from octopus to tiger to eagle and so on till he eventually tired and gave in has the following meaning in Bacon. "If any skillful minister of nature [i.e., experimental scientist] shall apply force to matter and by design vex it… [then nature] changes and transforms itself into a strange variety of shapes and appearances." Gold, for instance, is heated till it liquefies, then vaporizes, then combines with another element, and so on. "At length running through the whole circle of transformations and completing its period, it in some degree restores itself if the force be continued."[4] It is the same with those trying to identify a new chemical in the brain (this is precisely what the scientists Latour observed in *Laboratory Life* are doing) run various elaborate tests on it—seeing what it combines with and so on—to determine if it is indeed a new substance and, if so, what its properties are.

(We might note that according to the research of Rossi Bacon adopted this procedure of claiming to find allegories of scientific method in old chestnuts when he realized his ideas were not likely to get a very kindly reception from the intellectuals of the day. As a preliminary flyer he sent an unvarnished presentation of his views to two friends. The one whose reply survives reproved Bacon for his audacity. Rossi, commenting on this friend's negative letter, says, "These lines undoubtedly impressed Bacon, who had already begun to feel the hostility of scientific circles to his programme and to suffer from intellectual isolation."[5] Rossi argues that these early rebuffs determined Bacon to find a less direct way of presenting his viewpoint.)

By now, you're probably wondering how this relates to the work at hand. Well, we are going to look at the work of a quartet of doctors who have already been working with people with AIDS for a number of years using some of the alternative methods we have been discussing. We will see their hope-inspiring results and hear their individual philosophies. (While a doctor doling out the AZT and going the establishment route does not have to evolve his or her own way of viewing AIDS but merely parrot the party line, a doctor who is striking out with alternative methods has to carefully consider and elaborate his or her understanding of what AIDS is.) Although these doctors' opinions and treatment styles cannot be conceived of as allegories in the usual sense, that is, they don't stand for anything other than what they are, they can be seen as metaphors of the emergent medical paradigm. To say this another way, each of these doctors has a keynote element in his practice that can be appreciated not only as beneficial to the patients under his care but as exemplifications of one of the principles of the new paradigm. After detailing each doctor's philosophy, treatment, and results, then, I will pause and note how his work underlines one component of emergent science. In our finale, we will sum up what I take to be the parameters of this new way of seeing health and disease.

Dr. Christopher Calapai

Nutritional Support: Dr. Calapai's Philosophy

The first doctor to receive an in-depth profile here is Christopher Calapai, whom I interviewed at length. He was kind enough to walk me through all his protocols and share his healing philosophy, which in a nutshell is this: Immune function is strongly influenced by nutritional status. In treating immunological disorders—AIDS currently being the most prominent—it seems eminently reasonable to pay close attention to nutrition.

Regrettably, Calapai told me, this rather elementary fact is usually rated as of low or no importance by conventional medicine. "We see many patients who, as soon as they were determined to be HIV positive, were put on antiretroviral medication, with no attention to their diet, or other aspects of their lifestyles that might affect their immune function." This in spite of the fact that "there is considerable published research that supports our view that nutrition is highly relevant and should not be neglected."[1]

Approach, Method, Results

Dr. Calapai asks new patients to bring in a week's worth of their diet history, in which they jot down any food they consumed over the period as well as noting other intakes, such as cigarettes. He goes over this history meticulously. He then tells them how important it is to institute certain lifestyle changes that are beneficial for them.

The first strong message he wants to convey is to stop habits that are contributing to ill health. One of these contributors is smoking. He is convinced that smoking is very damaging in many ways—it helps to wipe out vitamin C (interfering with its actions) and generates free radicals. He makes sure his patients don't drink any alcohol—again, because that interferes with the uptake of nutrients as well as generating free radicals. He informs them about other dietary factors that also have strong effects on immune function. He points out, for example, that high-fat diets depress immunocompetence.[2]

Some people, I told him, might find it surprising that a person with a life-threatening disease such as AIDS is still smoking and drinking, doing things that most would recognize as not conducive to good health. He then spoke of the motivational root underlying such habits. They take these things "to try to change their symptoms or their mood, to feel better." And this is a symptom of a wider problem. "We're a little bit of an overindulgent society, and we like to have a very quick response to our problems; and we're not really concerned with what's causing it."

Again and again, he has seen the tragic consequences of this tendency toward self-indulgence. He mentioned a patient who almost made the changes that seemed essential for her health and survival, but she couldn't stay with it and resumed smoking. She had a recurrence of bronchitis and expired in the hospital. He said that only reinforced his low opinion of cigarettes.[3]

Procedures and Protocol

This doesn't mean that Dr. Calapai is a drill sergeant, who comes down hard on any patient who may want to take a substance that is not healthful, strictly speaking. He is flexible, acknowledging that he has to accept the place where the patient has arrived and try to influence, not control, him or her. For example, if a patient wants to use the antiretrovirals—and many of the patients that do come in either have taken them or want to take them—that's fine. While the crux of what he is doing is nutritional assessment and adjustment, if PWAs do want to take the prophylactic Bactrim or antibiotics, that's fine, too. He has recommended that for patients and has written prescriptions for it. He has no objection if a patient wants to see other doctors.

Besides, his program is not composed simply of don'ts. Equally or more important are the do's, getting the patient on a diet and supplementation program where the intake is augmenting immunity, not tearing it down.

To prepare a proper program, Calapai has to precisely judge the patient's present status. This involves both aggressive testing for common infections that can contribute to the progression of the disease—for example, CMV, EBV, herpes, mycoplasma, TB, toxoplasma—as well as detailed blood evaluations of cellular, mineral, vitamin, hormone, and viral and antigen levels, plus immunologic indices.

(Dr. Calapai's general diagnostic protocol is outlined in greater detail in appendix II.) It also includes a physical examination.

The results plus the history are used to devise dietary recommendations, recommendations for lifestyle changes, and programs of oral, intravenous, and intramuscular supplementation. (Also see appendix II for more specifics.)

He outlined for me how he proceeds. "I sit down with my patients, give them a five-page handout as to what they should eat and what they shouldn't eat. I like to try to get them away from red meat, get them more toward a vegetarian diet." In fact, in his book, even eating chicken or shellfish is not safe insofar as both can present that body with arsenic, which is damaging to immunity. "I like to have people as close to a vegetarian diet as possible," he finished.

To recur to any earlier point, he ended by saying, "I don't expect the people to be a hundred percent or thousand percent on the diet or the protocol. But when people are doing well, that's reassuring and reinforcing."

Clinical Experience

Now for the ten million dollar question. What results is he seeing?

> Many of the patients that I put on the protocol after two weeks or so say they're feeling very good. They have a lot of energy, they're sleeping better, many say that some of the skin problems that they have are improving. I have not had people say this is causing me a problem or I feel worse. The overwhelming majority of people say that they feel better in a number of different ways.

Although Dr. Calapai is probably aware of many of the studies that have shown problems with AZT and other antivirals in terms of the health of those who take them and in terms of their theoretical justification, he doesn't talk about them but rather the evidence he has seen personally. This can be broken into two parts: first, cases of patients who shun the antivirals and with a suitable natural treatment program do well; and, second, those who have taken the antivirals and seen a precipitous decline in health.

Although he doesn't prohibit his patients from taking antivirals, most of those who come to him "don't want to take prescription medication, they don't want to take antivirals." It seems that without these pills, "they're doing extremely well just on this protocol."

You may remember that some pages ago, we mentioned that the exact relevance of T cell count to health has never been established. That is, the benchmark of what a normal, healthy person's T cell count should be is undetermined. Thus, basing a drug's ability to combat AIDS on this measure seems rather dubious. From the experiential side, Calapai has noticed a similar anomaly. He noted, "Some of them [his patients] do have very low T cells, and I don't know whether or not that's a great indicator of how well a person is doing." He doesn't know in that he has seen patients with direly low (by orthodox standards) counts who seem to be striving. "I have patients," he said, whose "T cells may be five or twelve total helper cells—but they've been for a year free of coughs, colds, or any other opportunistic infections." Moreover,

> Some of them are exercising, gaining weight—some are body builders. It's hard to get a good handle on everything that's going on with them, but these people are coming back and saying that they're doing wonderfully! So, I think that's certainly an important picture.

As to his observations on the second point, the poor health of those on antivirals, he related, "Patients say to me that all of their friends and people that they've known that have gone on some of the strong antivirals—especially AZT—are individuals that either have expired, or have gotten sicker and sicker continually." Admitting this is anecdotal evidence, he added, "I don't know whether that's a wonderfully accurate statement but this observation seems to be relatively common."

Although you might say his positive results speak for themselves, a scientist might argue that he is just lucky in his patients. Our imaginary scientist will not be satisfied, as indeed he shouldn't be, until scientific studies have been published on the value of such a program. Although Dr. Calapai would not gainsay the need for more research, he does emphasize this:

> AIDS is a syndrome where people are dying relatively rapidly, so I don't know how appropriate it is to wait until there's firm and hard data before recommending some of these safe, nontoxic treatments. A lot of people are not doing well with the conventional approach. I think we have to try to use the safe things—the herbs and nutrients—and see how much help we can offer the patient, whether they do or don't want to take the antivirals.

Case Summaries

Dr. Calapai provided me with four brief case summaries so I could see his actual results.

PATIENT #045

This is a 32-year-old Caucasian female presenting, that is, coming to the doctor, January '92 with a HIV+ history since testing Sept '91. She has had headaches, sinus congestion and fatigue.

Initial testing revealed low cholesterol level. CD4 T cells prior to presentation here (recorded October '91) were 616.

The patient was put on protocol.

CD4 cells went up to 1,000 by September '92; P24, B2 and neopterine were normal as of February '93.

The patient feels very good and hasn't had any opportunistic or other infections.

C PATIENT #046

A 33-year-old male Caucasian presented January '93. He had tested HIV+ in August '91. History of IV substance abuse 11 years ago. Patient experiencing fatigue. Patient had been put on Bactrim three times weekly.

Initial blood testing revealed elevated liver enzymes, low WBC (white blood cell count), RBC (red blood cell count) and hemoglobin, high monocytes and low vitamin C levels. CD4 was 2 percent.

Patient was put on IV protocol and after three weeks felt significant improvement in energy and strength. Patient exercises regularly, including weight lifting, and has gained weight since on protocol.

P24 test was negative February '93, while B2 micro, measured in the same month, was normal. Neopterin 12.5H. Repeat chemistry showed a decrease in LFTs. CD4 count was 3 on April 4 '93.(All tests that were abnormal are in the process of being repeated.)

The patient clinically feels "great," hasn't had any symptoms, or infection history, and continues to exercise regularly without problem.

PATIENT #541

This is a 32-year-old Caucasian male who presented February 10 '93 with a three year history of HIV+ test. He had hepatitis B in the past.

Initial testing revealed high protein, high LFTs, herpes and CMV. There were magnesium, vitamin A, B6 and vitamin D deficiencies. CD4 were 62 at 3 percent.

The patient was put on protocol. LFTs decreased, vitamin and mineral deficiencies were corrected, P24, B2 and Neopterin were normal. CMV turned negative.

Clinically, the patient is doing very well, weight is stable. Patient exercises frequently and is very energetic.

PATIENT #559

This is a 33-year-old male Caucasian presenting February '93 with a four-year history of being HIV+. The patient has had no apparent opportunistic infections; however, has some fatigue.

Initial blood testing revealed hypocalcemia (lack of calcium in the blood), hyperproteinemia (subnormal level of protein in blood), low cholesterol, low

WBC, high monocytes, positive for herpes and cytomegalovirus. B2 micro was 3.0, Neopt. was normal, P24 was negative, and beta-carotene was deficient.

The patient was put on IV protocol. Repeat testing revealed B2 of 2.8, CMV negative (IgM), and a slight increase in WBC.

The patient feels great, has a lot of energy, and weight is stable.

Clinical Results

All four of the above patients showed a substantial improvement in health status in a very short time.

Dr. Emmanuel Revici

I n contrast with Dr. Calapai's therapeutic approach, which (as we have tried
to show) rests upon a broad framework of research by investigators in nutritional and herbal methods, Dr. Emmanuel Revici's therapeutic approach
is highly individual, and his theory of health and disease is largely his own creation, carefully worked out both in the laboratory and in the clinic over more
than fifty years.

Thus, it will be in order, before we move onto an assessment of his protocol
and results in treating people with AIDS, that we get some idea of his framework.

Metabolic Imbalance and AIDS:
Dr. Revici's Philosophy

Dr. Revici's theory of health and disease concerns the balance and imbalance of
opposing processes. The body undergoes two basic processes: the anabolic
(building up) process, and the catabolic (breaking down) process. The metabolism
cycles through both processes. "The two antagonistic intervening factors... act
alternately, each being predominant for a period of time."[1]

As in the Chinese theories we glanced at a few pages ago, the ebb and flow
of these processes need to go on in an even rhythm for health to be maintained.
If they fall out of sync, disease will come about. "Abnormal changes can take
place in either of two opposite directions ... resulting from the exaggerated pre-
dominance of one of the coupled factors over its antagonist," Revici writes.[2]

So, how does he see AIDS in relation to this theory of imbalances? He relates AIDS to the loss of specific phospholipids (which he calls "refractoriness lipids") that normally play an important role in disease resistance. Phospholipids, as you may know from high school biology, are fats that are important in the architecture of human cell membrane.

Revici sees the route of AIDS as the type of step process we earlier identified in theories such as the one that holds AIDS arises from immune collapse brought on by overusing antibiotics, which itself comes about usually because the patient has been repeatedly infected with venereal diseases, which itself—last point—can be attributed to promiscuity. Or so that theory has it.

Revici's idea is that AIDS comes about via four steps: (1) a primary viral infection inducing (2) a deficiency of the body's natural lipidic defense, followed by (3) secondary opportunistic infections or specific neoplastic conditions, consequent to the lack of the refractoriness lipids, resulting in (4) an exaggerated manifest imbalance, usually catabolic.[3]

The most original point here, since the first three are found in a number of hypotheses we have already examined, is that the immune system failure leads to an imbalance in the metabolism's functioning.

Revici's approach then encompasses interventions in all four steps. He uses antivirals to clear up the infections. "For the refractoriness deficiency, the refractoriness lipids are administered by injection."[4] Opportunistic infections are dealt with using traditional antibiotics or other antimicrobial/antifungal agents. "For the imbalances... the appropriate anticatabolic or antianabolic agents are used."[5]

Case Summaries

Again, as with Dr. Calapai, the proof of the pudding, an evaluation of the adequacy of his four-pronged attack on the disease state, must come from a dispassionate evaluation of patient records. For Dr. Revici, we examined ten treatment records for HIV positive patients. Their ages at the time of HIV infection ranged from twenty-four to fifty (averaging thirty-three years), and their periods of treatment by Dr. Revici ranged from one to ten years (averaging a little less than five). There were seven men and three women.

PATIENT #1

A 31-year-old man was diagnosed HIV positive in March 1986. In August, when therapy was started, there were swollen inguinal lymph glands, and a rectal discharge, with a slight elevation of temperature (99.5), but he was otherwise asymptomatic. Before therapy his T4 count was 668, and since it has fluctuated between about 700 and 850 (November '88 through July '91), then dropping slightly to a range between 500 and 650 (January through October 1992). T8 has also decreased correspondingly, so that his T4/T8 radio remains between 0.4 and 0.6 with no downward or upward trend evident.

He continues to "feel pretty good," and is continuing therapy. He remains active and holds down a full-time job.

PATIENT #2

A 50-year-old man was diagnosed HIV positive in July 1988. He had submaxillary and axillary adenopathy. (Adenopathy refers to any glandular disease. The submaxillary gland is more commonly known as the salivary gland, while the axillary glands are in the armpit.) He was treated from May 1989 to spring 1993 (when the chart was read). In August 1989, shortly after the initial diagnosis, his CD4 count was 813; a year later it had dipped to 517. However, since then, it has remained in the range from 750 to 850, and his T4/T8 ratio has varied between 0.6 and 0.9, with no downward or upward trend evident.

At the end of 1991, he said he was "feeling very, very well"—is working and continuing treatment at this time.

PATIENT #3

A 27-year-old man was diagnosed HIV positive in 1986. He had axillary adenopathy and complained of tiredness, headaches and night sweats. He was treated from April 1987 through March 1993 and is continuing treatment. Around the time of his initial HIV-positive diagnosis his absolute CD4 was 625; subsequently, it stayed near 650 until February 1992, when it was reported to have doubled to 1,310. Correspondingly, his T4/T8 ratio has risen from about 0.8 in 1986 and 1991 to 1.6 in 1992 and 1.7 in 1993.

In 1992, he said he was "feeling well" in March 1993, he reported "feeling very well," with "no complaints except mild weakness." In May 1993, he is working and feeling well.

PATIENT #4

A 30-year-old man was diagnosed HIV positive in 1990. He came for treatment in 1992 with memory difficulties and cervical, axillary and inguinal adenopathy. (Inguinal adenopathy refers to diseases in glands in the groin area.) He was treated from March 1992 through April 1993, and is continuing treatment. Around the time of his initial HIV-positive diagnosis, his absolute CD4 was 243. After three months of treatment, it had risen to 636, but by April 1993, though still above the pretreatment level, it had dropped somewhat to 420. Over this time, his T4/T8 ratio changed very little. In April 1993, the adenopathy remained and there was a mild oral thrush.

PATIENT #5

A 35-year-old man came for treatment in November 1988 with a previous diagnosis of being HIV positive. His only reported symptoms were "pain in the left leg" and a herniated disk. He was treated from November 1988 through November 1992 and is continuing treatment. Over this period his T4 count showed a gradual decline, especially in 1992, starting above 600 (July 1988 and February 1989) and ending at 153 (July 1992), while his T4/T8 ratio changed very little. Nevertheless, throughout, he continued to report feeling "very well" (as noted in July and August 1989, and in February and July 1991) and "feeling fine" (mid-June and November 1992). During the period of therapy, no symptoms were reported.

PATIENT #6

A 31-year-old woman was diagnosed HIV positive in April 1988 after her husband had died of AIDS. She complained of fatigue, insomnia, pain in abdomen, weak knees, itchy skin, hair loss, ganglions (cysts in tendons), and hemorrhoids, as well as alternating constipation and diarrhea. She was treated from summer 1988 through November 1992 and is continuing treatment. She reported feeling "very well" soon after the start of therapy, and by 1989 she was gaining weight. Her T4 counts varied over a range from 158 to 953, but showed no consistent trend. Her count at the start was 510, and in November 1992 was 425. (Her T4/T8 ratios also varied widely, but ended just slightly below where they started.) Throughout, she continued to report "feeling well," though in September 1989 there may have been a fever episode, and in August 1991 she complained of "some pain in upper abdomen." In the following February, there was mention of herpes, and nine months later it was reported that she had axillary adenopathy and had had diarrhea for two months. Nevertheless, she said she "feels fine."

PATIENT #7

A woman diagnosed HIV positive in 1985, at the age of 24, came for treatment in 1989. She was treated from November 1989 to April 1993. When she came she had carcinoma in situ of the cervix, bilateral adenopathy, tiredness, nausea, and complained first of constipation, then (after starting treatment) some diarrhea. In October 1991, she reported severe pain in her lower abdomen during her menstrual period. Endometriosis (cysts or adhesions in the uterine lining) was also noted. Over the course of her treatment, her T4 counts and T4/T8 ratios nearly doubled from 600 to 1,016 and from 0.42 to 0.83, respectively. Axillary adenopathy was again noted in March of '92. She reported feeling well in May and June 1992, though there was "tiredness" in September.

In March '93, she had no complaints. As of the following May, she was well, working, and continuing treatment.

PATIENT #8

A 40-year-old man who was HIV positive came for treatment in 1987, with no symptoms except bronchitis, attributed to smoking (which he stopped). He was treated from November of that year until October 1992. His T4 levels fluctuated considerably, but ended (at 560) about where they had started (at 617); however, his T4/T8 ratios more than doubled over the same period, going from 0/74 to 1.67.

He continued to report "feeling well" or "very well" most of the time, and at the end of treatment there was still no adenopathy or other symptoms except a "harsh pulmonary murmur with crackles" and he reported feeling "tired."

PATIENT #9

A 36-year-old man came for treatment in 1983, having been found in 1980 to be HIV positive. He had left axillary and submaxillary adenopathy, a slight elevation of temperature and was "tired." He was treated from March 1983 to February 1993. His T4 levels started at 1305 in August 1983, by January 1985 had dropped to 544, then in October 1985 to 275; but thereafter fluctuated between 500 and 700, ending at 591 in June 1992. Over the same period, his T4/T8 ratios dropped gradually to about half the starting value—from 0.8 to 0.47. In March 1985, he had a "cold" with fever and chills, and then in August of that year herpes zoster was noted. In March 1986, a fever of 102-104 degrees was noted, but was gone a few days later. In September 1987 "some pain in the right upper chest front" appeared. At the beginning of 1989, a "very small lesion on the left arm" was noted, and in April he was "not well." However, by June he was well and in August 1989, he reported "feeling well." In November 1990, herpes is mentioned and in the next month there is a notation of "severe herpes—otherwise well."

Moving forward to July 1991, we find he had a rectal biopsy revealing Kaposi's sarcoma lesions. But the next year (by October 1992) his proctologist said that "The KS was totally gone." In July 1992, he was feeling "well," and finally, it is recorded, "patient says he's doing pretty good" (as of November).

PATIENT #10

A 24-year-old woman whose ex-boyfriend had AIDS was diagnosed HIV positive and treated from May 1987 to February 1993. Her T4 count was 717 when she came, and 479 at the end. The T4/T8 ratio increased slightly, from 0.50 to 0.69. At the start, she had some adenopathy, fever, night sweats, and pains in the legs, neck, and hands. In July 1987, diarrhea was reported. By October, she was "not well," but did not have diarrhea any longer or fever. In February 1988, she was "feeling well" but had a "bad taste."

At the end of therapy, in February 1993, there was no adenopathy, but a low-grade fever, a cough, and ralls at base of thorax were noted; and night sweats were again mentioned.

Clinical Results

Of these ten patients, only three (two men and one woman; patients #5, #9 and #10) seem not to have improved during therapy. Yet, of these three, patient #9, though worsening in immunologic markers (T4 count and T4/T8 ratio declining to about half) nevertheless felt "well" at the end of ten years of treatment, and his KS had disappeared, according to his proctologist. Further, patient #5, though his T4 count also declined considerably (from 625 to 153) over his four years of treatment, was still "feeling fine" at the end of that period.

Of the other seven who showed more clear improvement, two (patients #3 and #7) had considerable increases in T4 counts and T4/T8 ratios—they roughly doubled. Both also showed complete resolution of pretreatment symptoms (which for patient #3 had been fatigue, adenopathy, headaches, fevers, and night sweats; for patient #7, adenopathy and diarrhea).

Two others (patients #2 and #6), though not showing such improvement in immunologic markers, also had diminution of symptoms.

Thus, of the ten patients, eight currently (or recently) report feeling well (or "very well"—patient #3; and "very, very well"—patient #2) after an average period of 5.2 years of therapy. Five of the eight also report no remaining symptomatic complaints; and two of those have had a marked improvement in immune markers. All are currently continuing in therapy.

It appears, then, that some HIV positive patients were certainly helped by Dr. Revici's therapy. However, since we do not yet have full statistics on all the HIV-positive patients he has treated, nor on the relation of this subset to that larger population, we cannot draw inferences about the overall effectiveness of his treatment. We can say, though, judging from what we have seen, that there is reason to believe there is some value in Revici's way of doing things.

Allegorical Significance

Where Dr. Calapai's therapy can be seen as exemplifying the idea developed in the new paradigm that human disease is the result of multiple factors, including some rooted in the stricken's lifestyle, Revici's thoughts tie in with the more complex theory of the process of causation. An HIV = AIDS theory, even in a sophisticated form such as that provided by Dr. David Ho, will tend to have a rather crude vision of how the disease takes command, simply because there is only one factor, the retrovirus, to play with. HIV may lie low in a gland or, as Dr. Ho sees it, wage an unstinting war against the immune system for years, but all one has to look at, in the orthodox view, to understand the progress of the disease is the retrovirus's machinations. However, once we adopt the more complex etiological explanations that are to be conceived under the new paradigm, the picture of the disease's progress becomes much more multifaceted. We saw with Dr. Revici that an infection led to a breakdown of one component of the immune system, which brought about the rise of secondary infections, which ended by cascading into an overall metabolic derangement. Moreover, in these pages, we've seen other examples of such etiologies, where the breakdown of one internal system exposes another to more stress and heightens its own susceptibility.

Thus, Dr. Revici's understanding of AIDS suggests in wider terms the ability of the new paradigm to put forward an idea of the disease process that sees the initial factors leading to a host of new debilitations, each of which must be coped with via new bodily disposition and each of which can strip another layer from health.

Oxygen, Ozone, and AIDS

T he last two physicians whose treatments we will probe both use a form of ozone therapy. Although a few passing remarks have been made in this text about oxygen and free radicals, we have not spoken of the importance of oxygen in health. Before moving to these final presenters, then, let us comment briefly on this subject.

Some Roles of Oxygen and Ozone

When the reader thinks of nutrients that are essential to health, what first comes to mind is probably vitamin C or protein in general or vegetables, anything related to food or supplements. What is probably forgotten is the nutrient more important than any of these, oxygen. After all, one can go a day without eating anything, but one can barely go a few minutes without breathing. This is because the body needs a constant supply of oxygen. Although to stay alive at all, the oxygen intake system must be working, some doctors have wondered if the levels necessary for life are identical with those that support optimal health. To put this in other words, they ask whether it is possible that a given person may be receiving enough oxygen to stay alive but not getting enough to keep in tiptop shape. The answer is that it is possible, since chronic hypoxia (an inadequate oxygenation of the blood) is consistent with continued life but has many deleterious health effects.

Although hypoxia as a general state of bodily discomfiture has been studied, alternative practitioners have called attention to "hidden" hypoxias, which would be highly localized hypoxias, far less obvious than the gross ones we're wary of. Such hypoxias would occur, for example, if a particular organ or region of the body was having a problem receiving oxygen from the blood-ferrying system or, more importantly, if a subset of blood cells were not holding or transferring oxygen properly. The existence of such localized breakdowns of oxygen utilization might explain the sometimes surprising therapeutic effectiveness of oxygenation therapies, in which the patient receives ozone-containing blood.

This type of therapy has shown good results in combating pulmonary epithelium (a disease of the tissues of the lungs and breathing apparatus), and it has also been found to be "highly effective against viruses and has an unusually high degree of tolerance when administered parenterally."[1]

A study by Kief that was reported in a German journal indicated that this manner of ozone therapy shows "efficacy in cases of chronic aggressive hepatitis."[2]

Ozone and HIV Disorders

LABORATORY AND CLINICAL EXPERIENCE

Of more immediate concern to us is whether this therapy will have any beneficial result if used against AIDS. Current research provides no clear-cut answer. Although for use of ozone in vitro against HIV we have positive findings, the data on its utility in vivo is mixed.

In studies, ozone has been shown to inactivate human immunodeficiency virus (HIV) in serum at noncytotoxic concentrations and in whole blood.[3] Carpendale and Freeberg conclude their study, which was published in *Antiviral Research*, by arguing that ozone's suppressive effects means that it would be a good choice for decontaminating blood used for transfusions. They write, "We have demonstrated here an agent which, in vitro, acts to both inactivate cell-free HIV and suppress intracellular replication without apparent cytotoxicity. These results indicate a potential HIV inactivation treatment for blood and blood products."[4]

Turning to what has been learned when this treatment is used on PWAs, we can note a study by Carpendale alone and the one mentioned by Kief that noted "improved clinical and immunologic status found in patients with AIDS or ARC who have undergone ozone therapy."[5]

An inquiry by Steinhart et al., which focused more narrowly on people with AIDS-related chronic fatigue, said the ozone treatment "increases functional work capacity and appears to improve quality of life and decrease sleep requirements," adding that the therapy "also may be helpful in the treatment of HIV-associated painful PN (peripheral neuropathy)."[6]

Kief's investigation discriminated between patients at different stages of the disease, finding that ozone-laced blood transfusion would have maximum effect earlier in its course. He states that "ozone therapy can lead to obvious remittances of the reduced lymphocytic population and to an astonishing improvement in the clinical status provided that the disease has not severely progressed."[7] If, however, AIDS is in a later stage, "partial remissions can still be achieved in 30 percent of the cases," indicating that against a far progressed disease much of the treatment's potency is lost.[8]

I mentioned mixed results earlier. This is due to one in vitro study that indicated that providing cells with ozone to eliminate viruses was counterproductive in that at the level given it also killed healthy cells. This study appears to be at odds with others that hold ozone therapy can be carried out without damage to normal cells.[9] However, the investigation was utilizing extraordinarily high levels of ozone, greater than would be needed to eliminate most viruses. It may better be said, therefore, that the level of cytotoxicity is low relative to its antiviral effectiveness but may not be zero.

More devastating to the case for ozone therapy are reports that in use with patients, while no toxicity to healthy cells was found, there was no improvement in patients' health, either. A report published in *AIDS* by Garber et al. in a Phase II controlled and randomized double-blind study with an eight-week treatment period discovered "ozone therapy does not enhance parameters of immune activation nor does it diminish measurable p24 antigen in HIV-infected individuals."[10]

This means, since we have studies that both find and don't find usefulness in this treatment, that the ozone methodology is still under adjudication in the court of science.

SUGGESTED MECHANISM OF ACTION
Inactivating Viruses Outside Cells

While the ability of ozone administration to defeat AIDS is still a matter of dispute, work has gone forward, done by those who believe the treatment is effective, to consider by what process ozone hinders the activity of the virus or bucks up the immune system.

An in vitro study by Wells looked at the effect of ozone on HIV located outside of the cells. While concurring with other scientists who had found the ozone having a devastating effect on the retrovirus, the study went on to identify what it took to be this form of oxygen's method of viricide. "The data indicate that the antiviral effects of ozone include viral particle disruption, reverse transcriptase inactivation, and/or a perturbation of the ability of the virus to bind to its receptor on target cells."[11]

Inactivating Viruses Within Cells

The findings thus far convinced the group led by Staal, which reported its own investigation in the prestigious *Lancet*, that "ozone treatment must somehow reach and inactivate intracellular virus because otherwise the inactivation could not be as extensive as has been observed."[12] They posit that the ozone's ability in this department arises not from any ability it can display in "inactivating the intracellular virus—but rather by destroying infected cells. Studies indicate that infected cells are in fact more vulnerable."[13]

Effects on immune system

Just as research showing ozone therapy is effective against a number of viruses led some scientists to invest time in assaying the treatment's ability to combat AIDS, so some who have observed ozone's ability to support immune function have theorized that the improvement of PWAs under the treatment comes from this same immune enhancement.

Several studies have shown that ozone treatments can strengthen immune responses. Bukowski and Welsh point out, since "oxidizing agents can induce IFN (interferon) and probably other cytokines [immune cells,]… it appeared to us reasonable to hypothesize that ozone may act mainly via the stimulation of PBMC (peripheral blood mononuclear cells) and release of lymphokines" in combating HIV.[14] These authors argue that ozone is particularly damaging to HIV not because it directly interferes with replication, its attachment to hosts, or in other such direct ways, but because it stimulates the immune system to clear out these retroviruses. (It shouldn't be necessary to repeat that we are here assuming HIV plays a cofactorial but not definitive role in producing the disease.)

HOLISTIC PHYSICIANS USING OZONE

Given the grimness of the prognosis for those who have AIDS and the fact that accepted strategies to counter the disease cannot be used by many patients—the side effects they experience with AZT or another antiviral, for example, are so severe they cannot be treated with the therapy—a number of physicians on the front line, that is, who are seeing people with AIDS daily, have introduced unproven treatments that they have either heard are helpful or they feel should be considered because they have worked against other viral diseases. Making a decision to go with such a therapy is certainly defensible to a degree, particularly if the chosen treatment is nontoxic and lacking in drastic side effects.

Ozone therapy can have unwanted side effects, but only at quite high concentrations; so insofar as there is evidence of its ability to suppress other viruses and even some research indicating it is effective against AIDS, it is to be expected that a number of doctors have taken it up in their work. Let's look at two of them.

DR. Z'S TREATMENT PROTOCOL AND CASE SUMMARY

I will begin with a brief look at the procedures and results of Dr. Z (who for personal reasons not detailed to me wants to remain anonymous), before giving a fuller presentation of how this therapy has functioned in the hands of Dr. John Pittman.

Dr. Z has treated many HIV positives and AIDS patients, but the outcome of one particular treatment stands out. The method he used with the patient is this. About a pint of venous blood was taken from the patient into an IV bottle, to which was introduced an O3/O2 mixture. The blood is shaken up with the ozone mixture and then dripped back into the patient.

Patient Report

A man was diagnosed as being HIV positive in 1988 and then began to experience fevers, night sweats, malaise, muscle aches and swollen glands. His T4 cell count was 153. He immediately started on a seventy-day program of daily ozone treatments, plus adjunctive therapy.

He became asymptomatic within about a month, and his T4 cell count began to rise, reaching 728 a year later. Some time thereafter, he applied for a hospital job and was retested for AIDS. The result showed him to be HIV negative. This was confirmed by repeated testing, including a PCR test indicating no virus in his system.

This case is anomalous in that it is the only such episode Dr. Z has ever seen. It may be that the first positive ID'ing of the patient's positivity was in error. It may also be that the adjunctive therapy was the basis of the seroconversion. However, in reporting this case, I just want to emphasize that the seroconversion results that Dr. Ho, the latest defender of orthodoxy, has hoped vainly to reach by pouring on the antivirals early in a positive's infection seem to actually occur more often when patients are trying alternative rather than establishment methods.

DR. JOHN PITTMAN'S TREATMENT PROTOCOL

We can get a more balanced view of ozone treatment by looking in depth at the work of Dr. John Pittman, whose Carolina Center for Bio-oxidative Medicine has long used this therapy in ministering to patients with AIDS, HIV, and other immune-compromising conditions.

We begin by describing Dr. Pittman's general outlook. He believes that ozone can arrest the progression of AIDS diseases and turn the syndrome into a manageable condition. He values the treatment's ability to help alleviate chronic problems that plague patients, including dermatological conditions and low-level infections. He also sees a connection between the administration of ozone and improved T cell and CD4 counts.

As we saw in the study by Kief, the therapy is of limited effectiveness for people with end-stage AIDS, numerous opportunistic infections, and those who have taken AZT or lots of antibiotics and other prescription medications. The best reactions usually occur in patients who have T cell counts above 100, although there are exceptions to this rule. Pittman states, "Last October, we treated a patient who came to us with CD 4 count of forty-two. His response to therapy was remarkable and his CD 4 count went from forty-two to two hundred and eighty-five in two weeks."[15]

Pittman uses ozone as part of a wider treatment protocol, working with patients for a minimum of three weeks. He explains that he begins with a modified juice fast for the first ten days of the program. "Everyone drinks fresh pressed vegetable juices and no solid food. They go on a very intensive intestinal and metabolic detoxification program. Certain nutritional supplements are prescribed and the patient begins intravenous therapies."

As with Dr. Z, the intravenous ozone is added by withdrawing blood, infusing it with ozone, and then putting it back in the body. "After the first five days of autohemotherapy [ozone therapy], if the patient has tolerated that procedure with no problem," Pittman explains, "we then begin with the direct intravenous infusion of ozone gas and proceed with that on a daily basis, gradually increasing concentrations and volume until we observe a healing crisis in the patient." This crisis manifests as fevers, chills, and sometimes flu symptoms. After this crisis has abated, "we begin to taper off on the dosage and concentration and go into lower doses which have a more immune-stimulating effect." When pressed for details, Pittman gave exact dosages, stating that he began with about 3,000 micrograms and went up to 10,000 micrograms. Post-crisis, the level is brought back down to 3,000 micrograms.

I then recurred to when he had used the phrase "intravenous therapies." Did he also use other intravenous supplementation beside the ozone? "Concurrent with ozone therapy," he replied, "are other intravenous therapies, which include intravenous vitamin C and mineral infusions as well as EDTA chelation therapy." The treatment package also involves colon hydrotherapy, acupuncture, and other dietary therapies.

In other words, all bases are covered. "It's a full day for most patients. I tell everybody before they come, this is not meant to be a health spa or country club. They've got a lot of hard work to do, just as much as we do. We really see this as more of a jump start for their program."

My follow-up question was what if anything was done for the patients once they had gotten this jump start and left the program. Pittman's response was encouraging.

> Once they leave here, we put them on a very stringent home treatment program that they are instructed to follow for at least three months. During that time we have them taking certain prescription drugs and nutritional supplements that are focusing on the problems that we identify during their initial evaluation.
>
> The clinic monitors the patient's progress with telephone consultations and by further laboratory tests.[16]

What we were not able to get from Pittman was a peek at his patients' files, so we could see for ourselves what the results have been. This is unfortunate, for although I believe his claim that he has witnessed marked improvements in the health of his patients, it would make a better case if we had all the facts at our fingertips. What can be said is that one hallmark of his treatment is its eclecticism. In his own words,

> I think the approach for AIDS has got to be one from a multi-modality standpoint. There is no one single approach. It is only through the combination of appropriate antiviral therapy, immune-stimulating therapy, diet, and detoxification programs that a patient is really going to be maintained and have any hope of improvement.[17]

Allegorical Significance

If we confine ourselves to Pittman's approach, we can say that the broader implications of his program are not hard to find. So far we have suggested that the new paradigm, as embodied in these therapy strategies, sees illness as the product of a number of factors that work together to bring down the immune system and overall health. The course of a disease moves through various bodily systems, creating a cascading effect, whereby one dysfunction, if not headed off, brings about other, novel ones as different areas are beginning to show strain. The last point, the one most visible in Pittman's method, is that such a multifactorial, multirouted disease state can only be thrown off by a similarly many-leveled way of going about a cure.

I want to close this part of the discussion by noting some nuances in these physicians' perspectives, for though they all adopt the newer paradigm, there are strong variations on how they interpret this orientation, which should not be abstracted away. After that, I will briefly reaffirm that while they all have their own ways of doing things and understanding AIDS, these perspectives do form part of a greater whole, which is that of the multifactorial worldview built up from a new understanding of the chronic diseases. We will conclude the book with some high points. First, we will listen to stories of long-term survivors, people who have overcome the odds by following an alternative path. Then we will glance at the extraordinary AIDS conference held in South Africa in summer 2000, where we find that, though Africa is probably not the place of origin of AIDS (as was argued earlier), it may well be the place from which is launched a way of dealing with the crisis that leads to a more humane, intelligent, capable medicine and health care. At this point, we will be able to draw some final conclusions on how AIDS and the response to it have made us rethink not only the manner in which health is understood but the social arrangements that play such a significant role in promoting or discouraging it. We end with a coda in which I will offer my final thoughts on Bacon and the possibilities of science in a society that at the moment is largely controlled by the dead hand of orthodoxy.

PART SEVEN

A NEW MEDICAL PARADIGM

Alternative
Health-Disease-
Therapy Concepts

The physicians whose work with HIV infection were examined approach health and illness from different perspectives. We may here sketch four alternative concepts of disease and therapy:

1. Disease as loss of adaptive flexibility and therapy as "retraining."
2. Disease as excess deviation from "normal" balance points and therapy as restoration of balance.
3. Disease as impoverishment of essential resources and therapy as replacement (or enrichment).
4. Disease as invader and therapy as a kind of "war" that removes and/or destroys the invader.

It will be noted right off that the first three concepts can be taken as suitable for the new paradigm, while exclusive adoption of the last should be construed as adherence to the establishment's nineteenth-century framework.

In our estimation, the first physician, Dr. Christopher Calapai, makes use mainly of the third approach (restoring depleted resources); the second physician, Dr. Emanuel Revici, is deeply committed to the second viewpoint (that of restoring balance); while the third and fourth physicians, Dr. Z, and Dr. John Pittman, also use the third (restoration of resources) approach, here employed in combination with the fourth approach (inactivating the viral invader). Their taking up of the last approach, though, in that they do not focus on it alone but combine it with a second mode, indicates that they are also loyal to the newer foundation, which will allow for the viral invader point of view if it is used in combination with other views, just as it will allow HIV to play a part but not *the* part in AIDS etiology.

Moreover, if sufficient thought is devoted to this categorization, it will be seen that, though useful in highlighting important aspects of each treatment, it is disingenuous insofar as it implies that each doctor's methods apply strictly to one or another (or two in the case of the ozone users) of these therapies. It's not so simple: Nutrients are used for more than countering deficiencies; and ozone may do more than kill HIV. Moreover, while the nutritional approach to therapy, with therapy as replacement, does quite precisely determine the means of therapy, the use of megadose nutrients goes beyond that rationale, as it can partake also of the second (rebalancing) and fourth (repulsing) approaches. Similarly, there is no simple correspondence of herbal therapies to any one health-disease paradigm: some herbal therapies can be understood in terms of the second (balancing) perspective; some in terms of the third (replacement) perspective; and others in terms of the fourth (expulsion) perspective.

Similarly, the ozone story is not a simple one, for studies have indicated that various effects of ozone might make all of the four reference frames appropriate. Ozone may (1) exercise the immune system, and thereby increase its adaptive flexibility; it also (2) brings some immune variables back toward a more normal balance; and (3) it has antiviral effects. Only the third reference frame is not obviously fitted; though, since ozone raises oxygenation levels, it might be viewed in the enrichment frame.

Another way to look at these four therapies is to note what picture of the body seems to be behind each one. The four concepts go from depicting an active, self-organizing and self-dependent organism to one that is passive and externally dependent. In the first picture, the origin of disorder is internal to the disordered living system (loss of adaptive flexibility); in the last, it is external to it (invasion scenario); and in the middle two, there is some combination of internal and external source (an impoverished or unbalanced state from deprivation of essential supplies). Thus, the approaches of the physicians, though differing considerably, share a greater emphasis upon supporting the body's own healing powers. We see, moreover, that their views are in accordance with the emergent paradigm as we have described it, in that this view construes health or illness as a joint production of environmental and internal factors, and thus, unlike establishment medicine, in facing illness gives equal emphasis to changes in lifestyle and dealing with parasites or other external causal factors.

How an individual contemplates his or her body depends on the underlying paradigm endorsed. How do I think of my own health and illness? My understanding of health and disease taps very deep attitudes. Am I (a) a congeries of mechanisms, with subsystems that can be adequately understood separately, with little regard to the status of all the myriad other subsystems? Or (b) is the level of interconnectedness far higher than such ideas acknowledge, so in a healthy organism all the subsystems are supporting each other in myriad ways (most of which are not yet observable or understood)?

In the first (allopathic) worldview, most disease originates through the breakdown of some one mechanism, or a few discrete mechanisms; disease then is to be treated by fixing (or bypassing) the defective mechanism(s). In the second (holistic) worldview, however, disease has to do not only with individual mechanisms but with weakenings of many kinds of couplings between mechanisms, and with overstressing of supporting systems through excessive demands for support. In this view, the total system can be strengthened by adding inputs that may rouse relatively quiescent supporting subsystems to become more active, as well as by giving additional resources to many different supporting systems—therapeutic strategies that are emphasized in the work of the three doctors we have used as examples.

Thus, depending upon one's ruling conception, of either (a) mechanistic simplicity or (b) self-organizing complexity of the health state, one's concepts of what is required for maintaining or returning toward health will range from completely passive (comply with doctor's orders, take your prescribed medicine) to highly active (take responsibility for your own health, learn all you can, make your own choices). As we have argued throughout, today's high-tech medicine shows a decided preference in how it organizes its decisions and actions, for the simplifying, mechanistic models that put the "patient" in a passive role. The aim, usually, is to find a single "pathogen" that is the cause of each definable disease. Then the therapy can be aimed at removing or "inactivating" the pathogen, after which the "patient" can be presumed to have been "cured"—or, if not, at least to be "in remission."

In our current health care system, the drive to structure things in this way is very powerful. And according to one prominent virus researcher, it has taken our concept of AIDS and its etiology up a blind alley.[1] The system we envision, as has been suggested and as will be outlined further below, rejecting the one microbe/one disease model, sees the need for a multidimensional understanding of both cause and healing. And this means the patient takes greater responsibility not only for his or her own health, but also for the direction of the health care system.

The Holistic Approach

Theories and methods that seem very different can have broadly similar results. All three of these alternative treatment approaches have, at the very least, reversed some of the symptoms of HIV infection in some patients, and have improved the quality of life, according to the patients themselves, in many. Some of the observed improvements—but certainly not all—have been accompanied by normalization of immune system indices, parameters such as T4 counts and

T4/T8 ratios. But even when these indices did not improve appreciably, immunological status, and general health, were usually better, as indicated by a marked decline in opportunistic infections.

The three approaches—the nutritional therapy of Dr. Calapai, metabolic therapy of Dr. Revici, and ozone therapy of Dr. Z and Dr. Pittman—are not mutually exclusive. Is there any reason why they should not be used together, complementing each other? For example, it was noteworthy that some of the reported actions of the vitamin C infusions used by the first physician were very similar to those of the ozone infusions used by the third two physicians: both were highly effective in inactivating many viruses, HIV in particular, and both seemed to augment the immune system.[2-4] Perhaps if used together, they would potentiate each other, with results better than either alone.

Here it must be noted again that many of these techniques that are widely thought of as "non-conventional" (and in many instances "unproven") are, in fact, supported not only by much clinical experience, but also by a considerable quantity of publications in peer-review journals. In our experience, many, if not most, of those who denigrate these other methods are unaware of this literature. What has been cited in this book is only a very small sample of what is available—much of it in our own professional and scientific journals, and still more from other parts of the world. Good science and good medicine are practiced in many other countries as well as ours, and if this is recognized, we will be more open than we have been in recent years to effective methods that are in use elsewhere, and give our physicians and their patients the freedom to choose whatever, in their most carefully considered judgment, best meets their professional and personal needs.

One of the very important advantages of many of these "non-conventional" methods is that whether or not they are effective (and there is evidence that some are), they are generally safe. Most nutritional and herbal methods cause far fewer side effects than the pharmaceutical and surgical methods now in favor. This is a major argument for permitting them to be used at the discretion of the individual, as an exercise of each one's right to self-determination.

The costs of these alternative techniques are almost always far lower than the treatments approved by organized medicine for the same disorders. Encouraging the use of these other approaches can save the public vast amounts of money, both as taxpayers and as health-care consumers. Since these approaches offer important help not only for the treatment but also the prevention of disease, if their prophylactic use (where appropriate) were to be encouraged, additional vast savings would accrue.

Those are some of the reasons why we feel it is important for physicians to acquaint themselves more fully with the uses of these methods, so we have here presented some preliminary evidence of their potential value for the treatment of AIDS, and of other effects of HIV infection. It is the authors' view that a physician can be most effective by becoming knowledgeable about several schools and traditions of healing, in addition to that in which he or she has been trained, so that conventional allopathic methods can be supplemented by other approaches that have in practice been proven helpful, even when they have not yet been fitted fully within the explanatory framework of today's "scientific medicine."

Survivor Portraits

Before moving to a broader framework in order to make concluding remarks, there is one last thing to say about alternative approaches. We have looked at three physicians who are working outside of orthodoxy, which they seem to be outdoing in terms of bettering their patients' lives. But we also need to hear from people with AIDS who have been longtime survivors to learn what their strategies have been to hold the disease at bay. As it turns out, more often than not their triumph is based in rejecting mainstream advice, which would have them flood their bodies with antivirals, and embracing an alternative therapy that involves healthy living as well as natural prophylactics.

Literature on long-term survivors with AIDS is replete with anecdotal evidence linking survival to such things as holding a positive attitude toward the illness, taking control of one's health, participating in health-promoting behaviors, engaging in spiritual activities, and taking part in AIDS-related events.[5]

Survivors are very pragmatic problem solvers who educate themselves about alternative treatments, nutrition, and exercise. Take the case of Fred Bingham, director of D.A.A.I.R. He recounts, "I was given a diagnosis of full blown AIDS in October 1989." He went into the hospital, where he was dosed with AZT. He continues the narrative, "By the time I left the hospital, I could hardly walk. I was forty pounds lighter than I am now. My thrush was out of control. I was a wreck and I was scared to death."[6]

As far as he was concerned, AZT was a bust. "I needed something that didn't cause the cumulative toxicity of the nucleoside analogs." He decided he had to take his health into his own hands, and he began researching journals and talking to people: "I was trying to find out how to deal with HIV in a very scientific way and how to manage my unstable neurological condition."[7]

This involved a multifaceted protocol that involved supplementation with "a broad array of nutrients, including herbs, vitamins, minerals, trace elements, amino acids and live cell glandulars." Having designed and stuck to his own program, he has gone from the brink of dissolution to a state of much better repair. "Today," he says, "I have a completely normal immune system."[8]

Bingham's protocol is not for everyone, but what in his narrative seems valid across the board for survivors is that they, as he did, tend to eschew AZT and the antivirals, do their own homework on the disease, and pursue an alternative route to improve health. And, as Bingham alludes to when he says he gained information by "talking to people," survivors network to gain insight—and share their insights—with others.

Let's go through these points in a little more detail. As to whether the antivirals offer a practical solution for people with AIDS, let's begin with another thought from Bingham. "The first thing I ask [when faced with a possible treatment] is: Can I take this [drug] for ten, fifteen or twenty years? Certainly with AZT the answer is no.... Its toxicity is obviously unacceptable to me."[9] Another survivor who has avoided the drugs prescribed by orthodoxy, Cliff Goodman, told me, "Most of my friends who went on AZT are dead and gone."[10] Dr. Michael Lange, an infectious disease specialist in New York City, agrees: "I personally do not prescribe AZT unless a patient insists. I have continued to find that patients survive longer without it."[11]

Rather than take toxic antivirals, survivors turn to unconventional therapies. In so doing, many of them expect to reverse or delay the onset of AIDS.[12] There is a large body of evidence to support these beliefs. In *The Pain, Profit and Politics of AIDS*, I interviewed a number of long-term survivors to get their opinions on alternative versus conventional AIDS therapies. Nick Siano told me, "There's nothing like the immune system doing its job against this virus [as boosted by natural supplements]. No amount of AZT can do what your body can do against this virus."[13] Keith Cylar said, "If it wasn't for the vitamins and the nutritional supplements that I'm currently taking, I probably wouldn't be alive today." Three months before speaking to me, and before he had embarked on a rigorous supplementation program, his immune system was bottoming out. He had "T-cell counts from a low of below 500." However, as he told me, "Vitamins and nutritional supplements have normalized my immune system."[14]

While AIDS survivors generally do not follow mainstream beliefs and treatments, they are not loners. They maintain good relationships with their friends, families, and doctors and often belong to support groups. Bill Thompson, an AIDS survivor and activist, says, "It was through the help from some recovering addicts that I was introduced to juicing." As an ex-addict, he had hepatitis C and a nearly incapacitated liver. "Through the help of my friends, I learned about carrot and beet juice and I went out and got a Champion juicing machine. That's how I've been maintaining good health."[15]

Meeting and empathizing with friends and members of support groups helps the PWA to regain lost confidence and trust in others. The person with AIDS learns that others are in the same boat but may have found paddles that he or she hadn't thought of. Peter Hendrickson, author of *Alive and Well: A Plan for Living*, says that "It's easy to be discouraged when people see a lot of their loved ones dying. In community, in sharing pain, in recharging one's spirit and in living life fully, hope follows. I see that happen all the time and I feel it for myself."[16]

On the basis of this knowledge, Dr. Jifunza Wright offers this advice to health professionals working with AIDS patients:

If there is any real message I would really like to give my colleagues, it would be: it's very, very important for us to just honor and respect life, and people who are HIV-positive, and people who we diagnose with AIDS. They have a will. They can have a will to live and our responsibility should be to facilitate those patients and live in the highest quality of life that they can, no matter what their t-cells are, no matter what major AIDS diagnosis they've had.[17]

Until AIDS patients are offered hope and nontoxic therapies, they must continue to follow their own intuition, do their own homework, and seek out help from like-minded individuals.

Goodman, who has stayed healthy by following an alternative route, summarized for me when I discussed staying healthy with him: "I can only say that I'm doing my best. I'm going to continue on this road unless there's a non-toxic cure that comes along. I'm sure not taking the government's standard [treatment]."[18] He then took the conversation to another level, giving advice to all those with AIDS:

Do your homework before you put anything in your mouth concerning AIDS and know that you have a support network that's already set up. Nationally, a lot of the research has been done. We know certain natural non-toxic means that are good for your immune system. We don't claim cure or miracles but I think that you will do better if you follow non-toxic means.[19]

AIDS Conference
in Durban,
South Africa,
in July 2000

The Opening Moments

On July 9, 2000, in Durban, South Africa, Thabo Mbeki, president of South Africa, gave the opening address at the Thirteenth Annual AIDS Conference.

His speech was given in a football stadium two blocks from the center where the conference was to take place. His appearance was preceded by shouts of "ACT UP" as choreographed by a television producer, who led and encouraged the crowd. Also, before the president came forward, there was a stage show of dancers, drummers and singers in traditional costume.[1] His speech emphasized his hope that participants would confront and work through differences. "I believe," he said, "we should talk to each other honestly and frankly, with sufficient tolerance to respect everybody's point of view, with sufficient tolerance to allow all voices to be heard."[2]

This advice, at least in relation to his own speech, was studiously ignored by the Western media, which has consistently misrepresented what he said to the point that what he repeatedly emphasized and spent the vast majority of his time talking about is usually not even mentioned (or done so with a one-sentence dismissal). Before taking up the gist of his remarks, though, let's reflect on the setting in which they were made, just as, many pages back, we pondered the French restaurant where Gallo chose to reveal his findings on HIV and AIDS.

You may immediately say that I should make the same objections to Mbeki that I made in discussing Gallo's context. You might ask, "Aren't both presenters surrounding what should be a solemn scientific occasion with glizty hoopla?" Although there is a modicum of validity in that parallel, one has to distinguish between the type of proceedings under way. Gallo was giving what amounts to a scientific paper. To accompany such an event, a large dollop of stage theatrics is out of place and demeaning. Mbeki was speaking on a more ritual occasion, seemingly there to welcome guests of the nation, although, as we'll see, he also had some scientific assertions to make. Still, even these were not in the nature of as yet unpublished experiments, but simply his calling attention to some long known but unappreciated facts.

Setting aside this surface similarity, I want to call your attention to three striking differences between the events.

First, location. Gallo's press conference took place in Bethesda, Maryland, home to the National Cancer Institute and to well-heeled doctors, who thus lived near the Washington, D.C., power center of the world's richest country. Mbeki's speech was given in Durban, in a country that six years previously had thrown off white minority rule through an uprising of the whole people. The country was on a continent that had been most disfigured by not only AIDS but by many other diseases such as sleeping sickness, malaria, and TB, for which there are cures but no money to afford them (by sufferers) or provide them (by strapped governments). However, to keep to our subject, all health workers agreed (both those in the West and the Third World) that Africa was the country that had been hardest hit by AIDS and where, while the swath cut by the illness is narrowing in the West, it continues to widen in the south and middle parts of this continent.

And yet, notwithstanding the fact that from very early on in the AIDS crisis Africa was the place in which the epidemic had struck with the most force (at least according to establishment estimates) and was where public health systems were most critically overtaxed, this was the first time the annual AIDS conference—the most important one on the subject—was put on the continent. As Mbeki put it, "You are in Africa for the first time in the history of international AIDS conferences."[3] Such a situating of the conference would seem to indicate that attention was finally being given where it was most due.

Second, what about the culture invoked? In Gallo's press conference, we saw it was that of a chichi French restaurant, in the United States. a sure sign of "high life." To us, it seemed to symbolize that those who settled for this location (whether Gallo or the Reagan administration) wanted to convey that in the United States science was still an affair for "gentlemen," that is, white men of the upper social register and upper bank accounts (albeit Gallo himself had risen from the ranks).

The symbolism of Durban, ACT UP and African drums, is quite different. ACT UP, as we have been illustrating (and we will return to this topic in a moment), has been one of the most forward-looking and fearless players in the AIDS situation of the West. The group does follow the establishment views on treatment, which we have found to be of dubious validity. Still, giving credit where credit is due, this group has with creative protests (seemingly modeled on the zaps of the Gay Activist Alliance in the 1970s) put its finger on and forced the public to pay attention to some of the most scandalous sides of the situation, particularly, concerned government agencies' unconscionable foot-dragging and penny-pinching, as well as the greed of many of the pharmaceutical companies.

Married to this were traditional dances and music. Of course, it is possible to view such cultural exhibitions in a cynical vein, seeing them as no more than the typical example of "folklore" that every country interested in pleasing tourists or visitors will roll out. However, another way to look at this is to see it as symbolic of a desire to make knowledge about AIDS, how to prevent it, and how to treat it and live with it, something that is both known by and (for treatments) in reach of the common person. Moreover, as the combination of dances with ACT UP signs suggests, it is to be hoped that the empowered individual will not only have knowledge and resources but a share in charting the direction of research and terms of care. On this we will say a little more below.

What justifies us in taking this less cynical view of the presence of folk music at the opening the conference are statements such as that of Nelson Mandela in his closing address. Although some news reports imply he was perturbed with Mbeki's flirtation with AIDS dissenters, the thrust of his argument was that all sides must be heard. His justification for this stance was: "The ordinary people of the continent and the world—and particularly the poor who on our continent will again carry a disproportionate burden of this scourge—would if anybody cared to ask their opinion, wish that this dispute" be laid to rest in favor of discussing "the needs of the suffering and dying."[4]

The implication of his message is that the little people, especially the ill, should be guiding the decisions and directions taken by scientists and physicians, rather than vice versa. This position is affirmed less obliquely by Mandela when he says that the best way to address the crisis is through "collective leadership, consultative decision making, and joint action toward a collective good."[5]

To reitcrate, the employment of folk music and dance at the conference's beginning can be taken as symbolic of a desire to take many aspects of the AIDS situation out of the ivory laboratories of the elite and, while still utilizing those labs, trust more in steering the direction of research and care to the imagination and spirit of ordinary people. This, by the way, is a principle that has been central to the way African leadership handles crises. On this score, we might notice, for example, a speech by Patrice Lumumba, shortly after his country, the Congo, had achieved independence. As head of the country, he stated, "We do not want a bourgeoisie government that lives at a distance. We are going to go down among the people, speak with the people every time."[6]

Third, we need to invoke a point already brought up, to use it in contrast to the spirit of the Gallo conference. Where Gallo, as you'll remember, structured his meeting as an agon, in which he was challenged by and defeated an antagonist from the CDC who had dared him to correctly identify unlabeled blood samples, Mbeki argues that the solution to the many dilemmas posed by AIDS will be found by seeking common ground. Even if one side is right on an issue, he suggests, that side's strategy should be to carefully convince its opponents of the truth of its claims, seeking areas of agreement and verbal bridges, as opposed to—as has so often happened—belligerently ridiculing or flaying those with the incorrect opinion. Moreover, as Mandela stated, in most areas of dispute it is likely the truth will be composed of an amalgam of ideas from different positions, rather than belong exclusively to one party.

We see that the setting and mind-set of the Durban conference was worlds away from that of Gallo's HIV = AIDS press meeting, which opposes cooperation, a focus on the underdeveloped nations, and a willingness to listen to the people, preferring competition, the rich nations, and the rich country's elite.

President Thabo Mbeki's Position

As noted, this last point made by the conference's sponsors concerning tolerance and willingness to make the most positive possible interpretation of an opponent's views is not one taken to heart by the Western media, who have, for one, presented a serious distortion of Mbeki's view. What the media rail on about is that, they claim, he has recklessly abandoned the universally endorsed dogma that HIV = AIDS and in its place put various ideas cooked up by flaky AIDS dissenters.

We've seen already that this is patently untrue, insofar as he tried to make the conference an open forum, with both establishment and dissident views represented—and with the establishment attenders in the majority!

(Unfortunately, the conference ended up being much less democratic than had been originally planned. As Anita Allen, who attended, writes, "A decision was taken by the organizers… not to allow any dissident papers to be presented, and not to allow any dissident organzation to exhibit. [7])

Moreover, you will have recognized by now that there is no one dissident view. For example, some in the opposition are trying to figure out if the HIV virus came from a vaccine, while others think there is no such virus. However, none of this is to the essence. The really significant issue here is what Mbeki said. Let's listen to him, for I think after reading a selection of his remarks you will see (1) that what he has to say about causation does not relate to biological conditions (such as drugtaking, viral transmission, immune dysfunction, and so on) at all, but to the sociological context of disease creation, and (2) that it is unlikely that such views would be given wide airing in the West because they (implicitly) indict the bad faith, greed, and barefaced villainy of the West's treatment of the continent. Here are a few of his opening words.

This is the story: [he quotes the WHO] "The world's biggest killer and the greatest cause of ill health and suffering across the globe is... extreme poverty." Poverty is the main reason babies are not vaccinated, why clean water and sanitation are not provided, why curative drugs are unavailable and why mothers die in childbirth.... Every year in the developing world, 12.2 million children under 5 years die, most of them from causes which could be prevented for just a few U.S. cents per child. They die largely because of world indifference, but mostly because they are poor....

[As another illustration, he mentions the following.] A person in one of the least developed countries of the world has a life expectancy of 43 years according to 1993 calculations. A person in one of the most developed countries of the world has a life expectancy of 78—a difference of more than a third of a century. This means a rich, healthy man can live twice as long as a poor, sick one....

[But let's get to the bone of contention.] As I listened [to various AIDS experts] and heard the whole story told about our own country, it seemed to me that we could not blame everything on a single virus. It seemed to me also that every living African, whether in good or ill health, is prey to many enemies of health that would interact one upon the other in many ways in the human body.[8]

Have we followed Mbeki far enough to understand why many in the medical establishment and the Western media would be outraged? Earlier in our pages Charles Geshekter and other writers made this same point about the impact of African poverty on health, but now it is being put in circulation from an international platform by the president of one of the most freedom-loving, respected commonwealths in the world. Mbeki's point is that it is not only ludicrous but shameless to spend ink and money studying how HIV is responsible for African AIDS, while ignoring the fact that the continent has no public health, health care, affordable drugs, or proper sanitation to speak of. It is these conditions that, as Mbeki remarks, have brought about a situation in which there is "the collapse of immune systems among millions of our people, such that their bodies have no natural defense against many viruses and bacteria."[9]

I don't mean to hint here that Mbeki says such concerns as identification of a viral cause of AIDS, for instance, are unimportant. And, in fact, his government has put in place safe-sex programs and other practices that proceed as if viral sexual transmission were key to the generation of AIDS. But what I am suggesting is that Mbeki's most outrageous statement, from the viewpoint of Western orthodoxy, is that: WHEN IT COMES TO FACING DOWN AIDS IN AFRICA, LAB SCIENCE IS NEAR VALUELESS. Or, to put that more temperately— and note that we are not talking about all science but only virology—the depressed political and economic conditions in black Africa are such that if they are not ameliorated first through such means as vigorous public health measures, nothing else, even the most sophisticated science, will be of use.

To put this another way, Mbeki is simply saying that Africa must follow the same path as the West to the eradication of disease. As we saw earlier, nineteenth-century Western public health measures, such as improving sewage and garbage disposal, obtaining clean water for citizens, and so on, were what counted in the ending of the reign of many epidemic diseases, whose extirpation preceded the invention of vaccines for them. (His peroration, of course, touches not only on the need for such measures but the need for funds to pay for these improvements.) Even though, arguably, AIDS is not affected by public health measures, in that it appears in industrialized countries where these measures have long been instituted, it is still possible to say, as Mbeki does, that the immune deficiency syndrome's widespread and greater virulence on his continent can be attributed to the lack of such amenities.

Let me add that a November 2001 article in *Rolling Stone* by South African Rian Malan—something of a conversion story in which a reporter who has been a firm believer in orthodoxy comes to doubts its infallibility when he notes the discrepancy between projected and actual African AIDS death—makes a similar point. Malan quotes with approval the statement of Dr. Joseph Sonnabend, early pioneer of AIDS research.

> "There's a place for AIDS drugs and prevention campaigns," he says, "but it's not the only answer [for African countries]. We need to roll out clean water and proper sanitation. Do something about nutrition. Put in some basic health infrastructure. Develop effective drugs for malaria and TB, and get them to everyone who needs them."[10]

And to come back to the way this appears as a smack in the face to the belief structure of mainstream Western media, Mbeki lays a lot of the blame for this situation at the door of the West. "They die largely because of world indifference, but mostly because they are poor."[11]

We need not repeat points made earlier by Castells about the West's involvement, through the World Bank and other agencies, in the destruction of the African economy, but we should spend at least a thought on this. In the United States, one of the main functions of the media in the last decade has been toadying up to the rich, singing the praises of Steve Jobs, Bill Gates, and other Silicon Valley billionaires, and esteeming money making as the highest ideal. At the same time, this same media conveniently ignores the way the gap between the rich and the poor grows, while a largely suborned legislation allows union busting and other reprehensible corporate tactics to go unpunished, tactics that hope to ensure that the rich stay on top and the little gal or guy remain voiceless. Is this a media likely to give much prominence or respect to a statesman who says, "Beneath the heartening facts about decreased mortality and increasing life expectancy [in the West]... lie unacceptable disparities in wealth?"[12]

Although this viewpoint has been a cause for alarm to many in the establishment, it has brought a ray of hope to many in the trenches. This was registered in interviews with participants at the conference that appeared in the *Bay Area Reporter*. David Barr, a New York AIDS activist, said, "I've been coming to this meeting since 1988 and this is by far the best International AIDS Conference, that I've ever been to." Why? "It is the first time that a whole new set of issues have been raised concerning how health care is provided."[13]

A similar assessment was given by AIDS journalist Laurie Garrett, who stated that the conference was "a huge turning point," as it brought into focus "access to care, inequity in North-South relations," and how to get drugs "to everyone."[14]

The Community's Role in the Fight Against AIDS

T wo thoughts have constantly appeared in this inquiry: (1) our sense of a new paradigm and (2) our feeling that individual decisions are shadowed by participation in groups.

As an example of the first, we noted that the differences between the virologists' and immunologists' view of AIDS correspond respectively to an older and newer paradigm of science (as described by Capra and others). In effect, the immunological view holds that the HIV = AIDS hypothesis cannot be strictly true, in that no matter what role the retrovirus plays, no one factor causes a disease. The virologists, by contrast, opt for the monocausal view.

As an example of the second theme, we found, for example, that Dr. Gallo was part of a mesh of cancer scientists whose viral-origin-of-cancer model was wearing thin, with almost nothing to its credit. The time had come to search for greener pastures. Many cancer researchers began looking for new grazing grounds, though it was Dr. Gallo who turned out to be the "ram" who first got to the new grass. To put this non-metaphorically, Dr. Gallo's actions can be better appreciated if visualized as part of a dispersed group effort.

We have even sporadically tried to connect these two themes. We quoted one authority on nineteenth-century physics, for example, who contended that the heavily regulated hierarchy then believed to characterize nuclear particles was mirrored in status-conscious organization of physicists studying the phenomena.

But finally the time has come to bring it all back home. I want to advance two theses about the connection of group effort and the new paradigm, which I will come to in a minute.

Bear in mind, I have quoted Capra's best-selling book *The Turning Point* from the 1980s, where the author heralds the growing importance of the new social-scientific paradigm, which he links to quantum theory. If we wanted to do a genealogy of his ideas, we might reach back to the 1920s when Alfred North Whitehead was advancing quite similar ideas from an even more intimate perspective, in that Whitehead knew the originators of quantum theory. And yet, while Whitehead foresaw a rewriting of all science, including medicine, in line with physics' daring and dazzling new outlines; sixty years later Capra was predicting the same changes. Twenty years on from Capra, the old paradigm is still in force, while writers still talk about its imminent fall from power.

It seems to me that one explanation for the failure of these expectations can be gathered from our earlier thoughts about Bacon. In our imaginary dialogue, it was suggested that no one will argue Bacon's writings on the use of the experimental method and the organization of scientific workers did not presciently lay out the exact direction science would have to take for it to progress. Yet, he was not able to interest anyone in his work during his lifetime and for decades after his death, his works remained read, but of no practical use. It was only after a civil war brought a new and previously disenfranchised group to power that the stuffy, musty scholastic doctrine that had choked off any advance could be swept aside and a real scientific program begun.

Putting these two points together, we might say that the blockade holding the old medical paradigm in place will not be broken by startling laboratory findings but by a power displacement that lessens the hold of the establishment over education, research laboratories, government agencies, and other entities that now form an institutional prison house.

Yet, contrary to what this last idea might suggest, I am not offering a counsel of despair for those who see the value of the new formulation. This brings us to my first thesis, which is that *the eruptive turmoil brought about by the AIDS crisis has challenged the social organization of science.* A redisposition of power in medicine can emerge if such means as protests against high-priced drugs, the conducting of drug trials in the community, and the refusal of subjects to comply with coercive or life-threatening trial guidelines will permit a new paradigm, and thus a scientific revolution.

I do not say the fight to change the medical system and empower patients will inevitably be successful. Ultimately, however, the alternative health movement can give satisfaction to those who are struggling for a more responsive organization of medicine if the way of doing science can be changed. The movement's success in reorganizing this structure can lead to the general embrace in science of the new paradigm.

In fact, to go a little further, although there have been attempts to imagine what organization of society would correspond to the new scientific paradigm—we mentioned Hazel Henderson on this score—I think we can say that a glance at the new organizational forms that have been developed in the community's fight against AIDS could give us a clearer idea of the new paradigm's influence in the social field. This leads to my second thesis.

I further propose that *forms of political protest and underground organizations represent markers from which can be traced the lineaments of a newly materializing social structure.*

Again with the caution that these markers indicate the new structure if (not when) it comes into being, let's briefly examine three activities—those alluded to in thesis 1—that acted to move control of medicine from the hands of the medical system more into the hands of people with AIDS and their community. We can view these as indicators of what a progressive arrangement of medicine, one that would correspond to the new biological paradigm, would look like.

High-priced Drugs

One of ACT UP's consistent, sterling achievements has been to call attention to the price gouging of drug makers. In January 1989, for example, activists barricaded themselves into an office at Burroughs-Wellcome to protest the high price of AZT. "The ACT UP invaders were ready with two demands: a 25-percent drop in the price of AZT and a genuine corporate subsidy for those without resources or insurance to pay for the drug."[1] In the fall of that year another group snuck onto the floor of the Stock Exchange, using fake badges, and repeated the protest against Wellcome.[2]

It seems to me that if we try to pick out the wider implications of this stance, in which the ability of corporations to batten off tragedy and to decide who will stay sick and who will recover is based on who has a larger share of liquid assets, it is that the determination of health care issues on a strictly for-profit basis is unacceptable.

According to the new view, a health care system based on private property, although it may benefit an elite, is massively unjust and antihumanist, completely severing any connection medicine may have once had to healing. To recover the nobility that the medical profession (and related interests such as drug companies) once had, the new social paradigm demands health care (from treatment to drug selling to research) be administered in light of a number of a fuller number of considerations, including the desires of the patients, the needs of researchers, and concerns of the polity (such as the necessity for a greater democratization of all services), with profitability one of the lesser concerns. These considerations must be weighed in dialogue with all involved: physicians, health care companies, patients, and advocates. We argue, for example, that the price of a given drug will be arrived at through consensus, not by the fiat of economic royalists.

Community Drug Trials

While the last proposal may sound impractical, we can say that it is connected to the central demand of AIDS activists, namely, the demand that patients and their supporters have input into the complete medical process. In our discussion of community-based research initiatives, we showed how these trials were designed by doctors working in affected communities, in collaboration with people with AIDS. As Bernard Bihari of New York's Community Research Initiative (a pioneer in this type of investigation) put it, "We take our direction from the community's needs.... They tell us what problems they want to solve. We won't do a study unless it's going to be attractive to people with AIDS."[3]

The goal of this type of community-rooted science is not to replace "pure" research (that is, investigation uninflected by the community). After all, most projects that have been undertaken by community research agencies have tested substances that have been shown to have promise against AIDS or its attendant diseases in pure research. But what this and related developments have pushed for is democratization of medicine at all costs.

Subjects Protesting Against AZT Trials

It may be objected further that such democratization—which I imagine extending not only from the direction and funding of research to the siting of hospitals, clinics, and medical facilities; the education of health professionals; the pricing of services; and the way doctors and patients interact—is not even remotely feasible because whatever its merits, the powers that be would not allow it.

But consider the protests of subjects over their AZT drug trials, which were mentioned earlier, whereby they ignored their protocols, found ways to determine if they were in the placebo group, and disturbed the scientificity of the study in other fashions. The end result was that the rules governing drug trials were rethought with new ways of going about them that would redress some of the long-neglected inequities of the system. For instance, women and minorities had been routinely left out of trials. This was not done for scientific reasons. To take only women, Arno and Feiden comment that "most often, the decision to exclude women from a trial was done simply to save drug companies money."[4] Under pressure from activists, there have been conscious efforts to include the excluded in drug trials.

Moreover, the disruptive actions of the subjects in the AZT trials have brought about a rethinking of the place of placebos in trials for deadly illnesses and changes that try other methods, such as giving all subjects doses of the drug but in different quantities, to avoid what some consider the immorality of giving placebos in such cases.

What these protests and their successful aftermath indicate is that people with AIDS and their advocates do have leverage. A health care system, like a prison, cannot operate without tacit support from its subjects. If these subjects stand up boldly for their rights, especially the right to be heard and have their opinions weighed in all decisions, the establishment must listen on pain of losing all legitimacy. Not only listen, but allow increasing democracy in its structure.

Over the past fifteen years, the AIDS crisis has shown that the whole medical system can be shifted in progressive, humanitarian, dialogistic directions by those who are dying in its hands. Certainly, no one else had the guts or vision to do it.

(These remarks, by the way, shouldn't be taken to imply that the avant garde practices of those who are attempting to change the power distribution within our medical system give a complete profile of how institutional arrangements would have to change to correspond to the new paradigm. Other long-sustained ways of coping with the AIDS situation, such as those we outlined earlier when discussing how the African-American community integrated the handling of this health emergency into its already elaborated helping agencies, would also have to take part in the new system.)

The New Paradigm and the South African AIDS Conference

I t may seem that in devoting so much time to a discussion of the new scientific paradigm, I am going far afield from this chapter section, which was the AIDS Conference at Durban. "Mbeki's remarks," you might comment, "do call for a greater tolerance for new ideas, even for new paradigms, in handling the immune deficiency disease; but beyond that, there are no real links between the African sensibility and your talk of a new perspective founded originally in the theoretical physics of the early twentieth century."

Such a comment has surface plausibility, but before acknowledging its rightness, let's take a moment to look back at some of the writings of those who we associate with the first days of African independence.

Leopold Senghor was an African poet who fought for his country's freedom and became the first president of an independent Senegal. In his essay "Negritude: A Humanism of the Twentieth Century," he begins by underlining how in the early 1900s, "new discoveries of science—quanta, relativity, wave mechanics, the uncertainty principle, electron spin—had upset the nineteenth-century notion of determinism... along with the concepts of matter and energy."[1] This is not news, but his next insight is more startling. He claims, "Negrititude [which is how he labels the essential component of the African geist] by its ontology (that is, its philosophy of being), its moral law and its aesthetic, is a response to the modern humanism [which he says arises from modern physics] that European philosophers and scientists have been preparing since the end of the nineteenth century."[2]

This is not the place to launch into an exposition of his complex ideas, but basically Senghor's premise is that the new Western framework for viewing reality, which arose from relativity theory and other doctrines that presaged the new paradigm, corresponds rather nicely to a perspective long prevalent in Africa, "which conceives the world, beyond the diversity of its forms, as a fundamentally mobile, if unique reality that seeks synthesis."[3]

(This, to take only one of his analogies, is why when modern artists like Picasso and Braque began searching for ways to portray this new reality, they turned to African sculpture. "Of course," Senghor writes, "without the discovery of African art, the revolution [of Cubism] would still have taken place, but probably without such vigor and assurance and such a deepening of the knowledge of Man."[4])

Talk about "ontology" and "moral law" may seem alien to many readers, so let's try and put this idea of an African ontology in more down-to-earth terms, doing so in a way that brings us back to our consideration of how the new mental paradigm calls for a renovated organization of society.

Kwame Nkrumah, a leader of the Gold Coast African freedom movement who was later elected the first president of an independent Ghana, writing in the *African Forum*, laid out his concept of the essence of African politics.

> We know that the "traditional African society" was founded on principles of egalitarianism. In its actual workings, however, it had various shortcomings. Its humanist impulse, nevertheless, is something that continues.... We postulate each man to be an end in himself, not merely a means; and we accept the necessity of guaranteeing each man equal opportunities for his development.[5]

To sum up, which we can do succinctly, since many of these points have echoed and re-echoed throughout the book, the new paradigm, which sees events as multicausal congeries of interacting factors, implies a multifactorial approach to disease treatment. A corollary of such an understanding, one that is not logically necessary but that has been made by Henderson, Capra and other thinkers, is that a social structure complementary to this perspective would be markedly egalitarian and democratic, allowing for the noncoercive interaction of all actors involved in a given social situation to determine the outcome. Thus, as we've suggested, in this setup the price of a medicine or the direction of medical research would be arrived at through a democratic procedure, with all concerned having input. The new wrinkle added to this is that, according to the African thinkers just quoted, both the new intellectual paradigm and (to an imperfect degree) the democratic organization of society it implies have been an intrinsic part of the African world.

Thus, it is more than fitting that the new paradigm, as represented by the thoughts of many of the AIDS dissenters, has been granted a greater exposure on the international stage due to the first AIDS Conference in Africa.

And there's a second point linking the new paradigm and African thought. Just as pre-European African village societies (ideally) allowed for a degree of democracy that seems congruent with what we take to be the social structure most appropriate to the new paradigm, so a number of new social arrangements proposed at the Durban Conference (suggested both by Westerners and native Africans) are aimed at instating a more democratic ethos and structure into the practices of AIDS science.

The fact that the Western mass media has consistently, but contrary to the facts, portrayed the conference as a place where dissidents acted as usurpers who were trying to replace the orthodox view, rather than, as they were in reality, civil rights advocates who wanted to extend the franchise (as it were) so that more voices would be heard in the debate, demonstrates a common failing of the establishment media. It can only visualize scientific conflict as played out in a winner-take-all, competitive mode. This media has difficulty even conceiving another side of science, that of give-and-take, which can result in the gradual unfolding of truth through negotiation, which may occur even in cases where one side's view is largely more correct.

We will add a little more to our discussion of the one-sidedness of the media in our coda; but let's finish this discussion of Durban by looking at some of the proposals and events at the conference that aim to or already have increased the inclusiveness of medicine.

The first purpose of this international AIDS conference, as we've noticed, was to make it possible for the orthodox and unorthodox to break bread.

As we've already noted, relations between the camps were hardly equal, with the dissidents precluded from presenting papers or—with the one exception of the American Foundation for AIDS Alternatives—having exhibitions. However, as lopsided as the setup was, with the establishment clearly in the saddle, there was still an opening wedge of dialogue. Moreover, as Anita Allen has shown me, the not-yet published South African Presidential AIDS Advisory Report, which has drawn heavily on dissident opinion, is shaping up to be one of the most devastating critiques of the orthodox stance yet published.[6]

Dr. Sam Mhlongo, head of the department of family medicine at a South African hospital, points out, "I think for the first time... the two views have come together. All because President Mbeki created this chance which has never been created anywhere in the world."[7] This meant: "The other side—the orthodox view—could not name call the so-called 'dissidents.'... They have called us names but it was not possible to do that in these four walls here. They had to listen to us and we listened to them."[8] The ability to air views, respond to criticisms, and hear opposition to one's own point of view, all in an unrestrained but non-acrimonious exchange is the prerequisite of genuine science.

Moreover, as Mhlongo explained further, in his interview in *The New African*, the upshot of these free-wheeling discussions has been the establishment of scientific studies that acknowledge the dissident view. "A series of studies are going to be set up that will include... scientists like Peter Duesberg," one of which will "look at previous data and data collection... and see whether they make sense." Once such examinations are completed, "it is possible that when they find that some of the things are not adequately explained, they may design new experiments or collect new data to have some access to the questions raised."[9]

To obtain an idea of what such new studies integrating dissident and orthodox views might look like, we can go over a proposal by Dr. Roberto Geraldo, which was posted during the Internet Discussion of the South African Presidential AIDS Advisory Panel. Geraldo wrote:

> I propose... to find out the real role of immunological stressor-cofactors in the causation of AIDS: Have three groups of people: a) symptomatic AIDS patients; b) HIV-positive asymptomatic individuals; and c) very healthy HIV-negative individuals.... The normal individuals to be controls should match as closely as possible the individuals in the other two groups. [After each group is thoroughly evaluated for presence of immunological stressors in life style and with a complete workup of health status,] prospective trial follow the three groups for several years with periodic clinical and laboratory evaluation.[10]

This research would help us see what part immunological, nonviral stress plays in bringing on or exacerbating the disease.

The broader point, though, is that the conducting of such experiments would allow the two paradigms to be juxtaposed, giving each a fair chance to prove its truth value.

The New Paradigm
and Good and Evil

There is an Idol of the Laboratory as well as of the Market Place.

—Joseph Wood Krutch
The Measure of Man (1954)

Whoever is acquainted with the history of philosophy, during the last two or three centuries, cannot but admit, that there appears to have existed a sort of secret and tacit compact among the learned, not to pass beyond a certain limit in speculative science. The privilege of free thought, so highly extolled, has at no time been held valid in actual practice, except within the limit; and not a single stride beyond it has ever been ventured.

—Samuel Taylor Coleridge
Biographia Literaria (1817)

For they who break into extravagant praises... of the arts in vogue, and who lay themselves out in admiring the things they already possess and will needs have the sciences [that are] cultivated among them to be thought absolutely perfect and complete, in the first place show little regard to the divine nature, whilst they extol their own inventions almost as high as his perfection. In the next place, men of this temper are unservicable and prejudicial in life, whilst they imagine themselves already got to the top of things, and there rest, without further inquiry. On the contrary, those who arraign [the current state of science]... and are always filled with complaints against [it] not only preserve a more just and modest sense of mind, but also are perpetually stirred up to fresh industry and new discoveries.

—Francis Bacon
The Wisdom of the Ancients (1609)

T he time has come to write the epitaph for American medicine in the twentieth century.

In terms of its effect on science, the AIDS controversy, if it has done nothing else, has demolished the structure of human-sector biology.

Large claims. So let me qualify them. I don't mean to suggest that important, profound work is not being done in the field. I mean simply that once a dogma is so firmly established as is the HIV = AIDS hypothesis, normal assertions by researchers of claim and counterclaim about the hypothesis's validity are silenced or drastically curtailed. Then progress is finished, for it is on a freedom of interchange that the existence of science hinges.

Further, to reverse a thought experiment that was carried out by a scientist we interviewed who imagined what would happen if Duesberg's ideas had overnight been proven correct, my indictment of AIDS medicine would stand, even if tomorrow photos of HIV appeared on the cover of the *New York Times*, along with irrefutable proof that the retrovirus all by itself was responsible for AIDS! To reiterate, the stilling of debate (or rather the establishment's pretense that debate cannot be conducted in that it has made its opposition out to be a bunch of bogeymen rather than equally principled, evenly met opponents) represents a sad collapse of the scientific imagination, regardless of which side is ultimately right about various issues connected to AIDS.

Three final points remain to be made: one about the place of argumentation in the foundation of science, one about the wider background to this lack of debate, and a last about the way forward toward a healing of science via a new instauration of a democratic ethos.

Many pages ago, we talked about how embarrassing Bacon's actual concept of scientific practice—as opposed to his general ideas—proved to his followers. While in his scheme he did leave a place for the carrying out of experiments, the type of experiments that are now counted as the heart of science, he made this type of activity the handmaiden to a more central component, the drawing up of long lists of the different properties of substances and interactions, which he put under the heads of different perogative instances. As Rossi explains, Bacon "ignored the use of hypothesis for scientific purposes," that is, such hypotheses as would be the basis for constructing an experimental program of verification or invalidation. "The refutation of hypothesis, together with his opposition to deduction, has rightly been considered one of the greatest weaknesses of his method."[1]

It might justly be asked where Bacon came by such a peculiar version of scientific practice as appears in his emphasis on prerogative instances. Since almost no science of the type we would now understand to be rightly designated so was being done in Bacon's day, he could not have borrowed his principles from observations of actual work. Rather, as consummate historian Paolo Rossi has shown, Bacon drew his leading concepts from rhetoric! "To sum up, Bacon's interpretation of nature adapts fifteenth-century rhetorical... mnemonics. In this way certain typically rhetorical concepts and doctrines were transplanted by Bacon into the scientific field of natural research."[2]

Now, to take this just a little further, the rhetorical doctrines Bacon adopted were not stodgy ideas of how to win in a debate—these are what he characterized as simply decorative, a coloring of good and evil to otherwise indifferent material—but the newly developed, innovative, generative rhetoric of Peter Ramus. As Bacon himself described the new Ramian method:

> Ramus attacks Aristotle and the Schoolmen, on the ground that they treat topics according to presupposed, traditionally received rules... whereas he [Ramus] asserts that... a discussion of rules may take place only after the topics have been explained.... Ramus holds that we ought to amass our material before we deliberate upon its proper function.[3]

From this stems Bacon's idea of assembling vast tables of facts that he felt must precede any other scientific endeavors. It may also be noted that, broadly taken, this attitude, that one must study the behavior and reactions of one's material before drawing conclusions, is consistent with present-day science. However, you may now want to ask, "What's the point? If, as you say, Gary, Bacon's theory of prerogative instances is more or less a museum piece, why bring it up in this discussion of medicine and AIDS?"

There is one relevant thing, though. It seems to me that just as Bacon, one of the chief establishers of the scientific worldview, nourished his revolutionary (for the time) opinions by drawing upon an enlightened theory of argumentation, so it can be hypothesized that the fate of science is intimately connected to a society's degree and creativity of public argumentation.

This takes us directly to our second topic. It seems to me that the stiff-necked obstinacy that one finds so prevalent in AIDS orthodoxy, according to which even logically consistent, cogent oppositional positions put forward by reputable scientists such as Duesberg are, to a large degree, refused a hearing, cannot be laid completely at the door of one particular camp in one particular area of Western medicine. Rather, such a failure registers a generalized drying up of serious debate in the councils of the polity. This drying up is evidenced not only by a limitation of debate to restricted subjects (the point raised by Coleridge in our epigraph), but in a certain insufficiency in the prosecution of any line of argument.

We already touched on this last note when discussing an opinion piece by a liberal mass media journalist. It was stated then that the whole force of this journalist's examination of the commercialization of AIDS was derived from radical premises (i.e., the incompatibility of humanely directed health care and a for-profit orientation), premises that the author dare not avow and that, certain comments indicate, he would reject outright.

Taking such a position entails two fundamental disturbances to rational discussion. For one, the author is sailing under false colors: he bases the pathos and reasonableness of his argument on principles that he feels (not having thought through his stance) are irrelevant and untrue. Two, this makes his entrance into a searching debate impossible, since, at some point, such a debate would be bound to delve into foundations, a delving he is precluded from by a willed ignorance of such principles.

Conservatives, who we have not yet looked at, although they do not follow liberals in putting such a distance between premises and line of argument, tend to adopt a curb on reasoning that amounts to the same thing. They believe in their own bottom-line foundational views, but these views would be so abhorrent to the general public that they must be kept permanently in the shadows.

To be more specific, Western conservatives take as their role defending and legitimating the existing system, which combines a capitalist economy and a democratic government. A chief difficulty in constructing a defense, however, is that the two components of the system are incompatible.

Eric Hobsbawm, speaking of an earlier period but in a description that is equally applicable to present circumstances, has laid out these difficulties in the course of explaining why racism began to play such a predominant role in Western establishment thought in the late nineteenth century.

> Apart from its [racism's] convenience as a legitimation of the rule of white over coloured, rich over poor, it is perhaps best explained as a mechanism by means of which a fundamentally inegalitarian society based upon a fundamentally egalitarian ideology rationalized its inequalities, and attempted to justify and defend those privileges which the democracy implicit in its institutions must inevitably challenge. Liberalism had no logical defense against equality and democracy, so the illogical barrier of race was erected.[4]

The thought here not only explains the prevalence of racism in societies that need to justify the unwarranted privileges of the rich, but, more broadly taken, underscores a reason for the fractured consciousness of many conservatives who find themselves called upon to justify both the democratic ideal (the equality of all before the law, for example) and the inegalitarian reality (where, for instance, a criminal with a big bankroll has a much better chance of getting a light sentence than one with empty pockets).

Where the liberal cannot enter foursquare into a debate insofar as he or she must fight shy of exposing the foundations of his or her argument, the conservative must avoid any open discussion, in that it will likely make clear the incompatibility of the elements in the framework he or she is laboring to justify. So, if the besetting sin of the liberal is shallowness, that of the conservative is righteously defending a mishmash.

However, the incapacity of mainstream liberals and conservatives to engage in searching dialogue because they cannot too deeply examine the premises or consistency of their positions is not the whole story.

If we try to fathom the reason behind a lack of serious argumentation, at least in the United States, we also need to factor in the growing influence of mass media monopolies. The continued merging of large firms and the swallowing up of big companies by bigger ones has brought us to a pass where a few behemoths control the lion's share of public communication, while independent media voices grow fewer and weaker. As Robert McChesney so devastatingly illustrated in his *Corporate Media and the Threat to Democracy*, the paradoxical result of the simultaneous monopolization trend with the appearance of new media forms, such as the Internet, is that the explosion of places from which to get information is accompanied by the implosion of number or firms controlling these outlets.

The stifling of debate, in mass media news broadcasts, for example, follows naturally from the fact of elite control. "By defining the news as being based on specific events or on the activities of official sources, the news media neglect coverage of long-term social issues that dominate society."[5] And by keeping such issues, as much as possible, outside of the public eye, they discourage the raising and discussing of the question of how to address them. McChesney adds that this way of viewing the world, which would sweep debate into a corner, is bred into journalists, who see it as the hallmark of professionalism. He notes, in discussing the results that followed when such standards were first promulgated:

> On the one hand, the commercial requirements for media content to satisfy media owners and advertisers were built implicitly into the professional ideology.... On the other hand, corporate activities and the affairs of the wealthy were not subject to the same degree of scrutiny as government practices; the professional codes deemed the affairs of powerful economic actors vastly less newsworthy than the activities of politicians.[6]

We might mention as an aside, taking a hint from the dean of American theater critics, Eric Bentley, that corporations' seizing of our society's sources of information results not only in the curtailing of discussion and scrutiny of major institutions, but in a freezing up or elimination of serious thought on the arts. Bentley argues that this lowering is not due to the fact that journalism is done for cash, but because of the publishers' demand of writers that they expunge profound considerations from their copy.

> There is nothing wrong with money: it all depends on what is demanded of you in return. To earn his living Shakespeare had, for example, to acquire a highly complex literary language, far above the usage of his native Stratford, the modern [Saturday Evening] Post writer [the Post was one of the top U.S. "intellectual" journals in the 1950s] has to unlearn anything he might have learned from acknowledged classics, or from the depths of personal experience, and acquire the crude, vacuum-concealing lingo which titillates the miseducated sensibility.[7]

To summarize, the steps taken thus far in our argument lead to this conclusion. While the genesis of the scientific outlook, in Bacon, was provoked to a degree by the arrival of a new system of rhetoric (or way of debating), so (as far as it can be detected in the AIDS controversy) the degeneration of science has been most clearly evident in an implicit blockade against discussion. Whatever is debatable within the establishment (such as Ho's contention about the early potency of HIV) is constrained by being forced to accept highly dubious premises (such as HIV being the only culprit in producing AIDS), thus only allowed to proceed along a narrow gauge.

We've seen that the obvious way to obviate this deficiency and move medical research back onto the right track is to reinstitute generously wide discussion over the causes and possible treatment of AIDS, a discussion that issues not only in public debate but in the conduct of experiments specifically designed to adjudicate between contending viewpoints. This is the tack already taken by some of those involved in the Durban conference.

I will conclude with three remarks on the need for and form of a more inclusive debate.

To return to Bacon one last time, it's worth noting that another seed of his revolutionary views, along with the inspiration he gathered from the new Ramian rhetoric, came from his observations of the contrast between craftsmen and the learned, which proved much to the disadvantage of the latter. He found that an essential demand of the age "was to rehabilitate the mechanical arts." By this last phrase he meant all craftwork, from metallurgy to leatherwork, navigation, and other trades. According to Rossi, Bacon

> constantly contrasted the fruitlessness of traditional culture with the progressive nature of the arts. These, he said, unlike theoretical sciences, will not be set up as idols of perfection, for they are continually thriving, growing, advancing, and alive to the needs of humanity.... [Bacon noted evidence of this progress] in the case of printing, artillery, and navigation. These achievements were made possible [Bacon argued] because many minds collaborated to one end; in the mechanical arts there can be no dictators but only "senates" of free and equal workers.[8]

What I think we can learn from this is, first, that a rough and ready freedom of debate and a pragmatic approach to problem-solving is the quickest route to advance in science. Moreover, second, in what is implied more than stated, Bacon did not propose that educated protoscientists throw up their background and become craftspeople tout court, but that they work in alliance with mechanics. Bacon foresaw that science would only be viable if it included elements of different professional and interest groups in its organization. By analogy, an AIDS science staffed solely by physicians and scientists will be hobbled. The most productive organization would be one including those who work in the health-care field along with PWAs, HIV positives, and their advocates. A debate across these strata is one that will most surely, pace Bacon, lead to a genuine grasping of essentials.

Next let me draw a distinction from Jurgen Habermas, whose thoughts on arguments were very briefly alluded to previously. He has given a lucid description of the limits of debate in Western polities in these terms:

> We know how to bring the relevant conditions of life under control, that is, we know how to adapt the environment to our needs culturally rather than adapting ourselves to external nature. In contrast, changes to the institutional framework... have not taken the same form of active adaptation.... They are not the result of purposive-rational action geared to its own consequences, but the product of fortuitous, undirected development.[9]

In bringing out the way Western societies have put great thought into shaping the physical environment but done little to affect the reigning institutions, in the sense that the economy or governing structures are not tailored to the desires and benefit of the majority, Habermas marks out another contour of the type of public discussion that we are saying should be instituted in AIDS medicine as well as other fields. Such a discussion would concern not only issues of treatment and genesis of the disease, but what social arrangements are to exist between the affected and health professionals, within the professions, and between medicine and the citizenry.

I want to end with the thought that fruitful, unrestrained discussion can and must be put at the center of everyday life.

To speak personally and to recall my words from our opening pages, this book has gained weight in its the collaboration of two writers with different backgrounds, my own greatest expertise being in health and my collaborator's is in literature. The sparks of intellectual excitement that our contrasting perspectives have generated in our many, thoroughgoing conversations are what, hopefully, have given ardor and vividness to our exposition. Moreover, I must repeat that my understanding of AIDS derived from my own research and from hundreds of hours of interviews with experts who warmly bestowed the fruits of their knowledge and experience on me.

Further, to continue to restate points made at the opening of this book, in the study groups I sponsor the key note is working together to learn about health, each person's experience thrown into the pot for the benefit of all.

Similarly, Feast, my coauthor, belongs to a literary club, the Unbearables, who concentrate on developing performances in which each writer's input displays not only his or her individual talent but abets the group creation. The ultimate form of the event, far from appearing as a haphazard patchwork, is unified because in the working process, each author shows his or her text to the group, who discusses it and aids in a reformulation so that it will resonate with the greater plan.

All areas of life, then, can profit from a deepening of intercourse. In each exigency, it is worthwhile moving beyond one's own or a single expert's ideas, so as to gather a second opinion to round off and enrich one's appreciation of the situation.

In democratic discussion, barriers go down and we find ourselves sharing our humanity, a frail state, after all, beset not only by sickness but by alarms and confusions over how to materially treat and spiritually confront ill health. It is a state captured best perhaps in a short poem by one of the founders of that same Unbearable group I just mentioned, a poem about our basic human condition, with which I close.

#7

all living
      things...

... breaking
      easily
      into flakes.

—Rollo Whitehead
Poems and Fragments [10]

APPENDICES

Chinese Herbs for the Immune System

List 1 prepared by Dr. R. S. Chang

Alternanthera philoxeroides (kong xin xian)
Andrographis paniculata (chuan xin liang)
Arctium lappa (nin ban zi)
Coptis chinensis (huang liang)
Epimedium grandiflorum (yin yang huo)
Lithosperum erythrorhizon (zi cao)
Lonicera japonica (jin yin hua)
Prunella vulgaris (xia ku cao)
Sophora flavescens (ku shen)
Viola yedoensis (di ding)
Woddwardia unigemmata (gou ji guan zhong)

According to Dr. Chang, a possible explanation for Chinese folk remedies being HIV inhibitors relates to their interferon-inducing properties. "These inhibitors are a single chemical that happen to be present in many species of herbs, and Chinese medicinal herbs happen to be a rich source of compounds inhibitory to HIV."[1]

HERBS AFFECTING MACROPHAGES

Increases the Number of Macrophages
Coriolus versicolor
Lentinus edodes
Glycyrrhiza uralensis

Enhances Phagocytosis of Macrophages
Astragalus membranaceous
Panax ginseng
Codonopsis pilosula

Atractylodes macrocephala
Ganoderma lucidum
Polyporus umbellatus
Lentinus edodes
Angelic sinensis
Epimedium grandiflorum
Psoralea corylifolia
Acanthopanax senticosus
Eucommia ulmoides

(Note, the above lists were originally compiled by Q. C. Pan, Professor and Head of Anti-Cancer Drug Research at the Cancer Institute, San-Yat-sen Medical University, Guangzhou, China.)

HERBS AFFECTING T-LYMPHOCYTES

Increases the Number of T Lymphocytes
Panax ginseng
Ganoderma lucidum
Coriolus versicolor
Lentinus edodes
Phaseolus vulgaris
Atractylodes macrocephala
Coix lachryma-jobi
polygonatum sibiricum
Asparagus cochinchinensis
Ligustrum lucidum
Epimedium grandiflorum

Enhances Lymphoblastogenesis
Panax ginseng
Codonopsis pilosula
Atractylodes macrocephala
Ganoderma lucidum
Coriolus versicolor
Polyporus umbellatus
Coix lachryma-jobi
Phaseolus vulgaris
Polygonum multiflorum
Angelica sinensis
Polygonatum sibiricum
Gelatin from the skin of *Equus asinus*

Rehmannia glutinosa
Ligustrum lucidum
Schisandra chinesis
Epimedium grandiflorum

HERBS AFFECTING LEUKOCYTES

Increases the Number of Leukocytes
Panax ginseng
Codonopsis pilosula
Astragalus membranaceous
Ganoderma lucidum
Phaseolus vulgaris
Gelatin from the skin of *Equus asinus*
Human placenta
Millettia dielsiana
Ligustrum lucidum
Cornus officinalis
Acanthopanax senticosus
Cinnamomum cassis

Enhances Phagocytosis of Leukocytes
Panax ginseng
Astragalus membranaceous
Atractylodes macrocephala
Dioscorea opposita

HERBS STIMULATING/INHIBITING IMMUNOGLOBULINS

Stimulates Production
Astragalus membranaceous
Panax ginseng
Coriolus versicolor
Lentinus edodes
Polygonum multiflorum
Rehmannia glutinosa

Inhibits Production
Psoralea corylifolia
Angelica sinensis
Glycyrrhiza uralensis
Fructus ziziphi jujubae

List 2 Prepared by J. DUKE

INCREASES LEUKOCYTES

Astragalus Ganoderma Milletia
Cinnamomum Ligustrum Panax
Codonopsis

PROMOTES LYMPHOCYTE TRANSFORMATION

Angelica Ganoderma Polygonatum
Astragalus Ligusticum Polygonum
AtractylodesLigustrum Rehmannia
Bupleurum Lonicera Rhus
Codonopsis Loranthus Schisandra
Cuscuta Lycium Scutellaria
Dioscorea Milletia Taraxacum
Eclipta Morus Tremella
Eleuthero Paeonia Ziziphus
EpimediumPanax

PROMOTES PHAGOCYTOSIS

Aconitum CynanchumMatricaria
Adina Datura Mentha
Allium Dioscorea Morus
Althaea Dyera Myroxylum
Angelica Echinacea Oldenlandia
Aralia Eleuthero Panax
Aristolochia Epimedium Persea
Astragalus Eucommia Petiveria
Atractylodes Eupatorium Phytolacca
Aucoumea Forsythia Pistacia
Boswellia Ganoderma Platycodon
Byrsonima Glycine Polyporus
Canarium GlycyrrhizaPsoralea
Cephaelis Hedyotis Rehmannia
Chrysanthemum Houttuynia Scutellaria
Clastanthus Indigofera Stephania
Codonopsis Lentinus Styrax
Commiphora Ligusticum Tussilago
Coptis Lonicera Uncaria
Curculigo Manilkara Vinca

STIMULATES T CELLS

Angelica Eupatorium Matricaria
Astragalus Ganoderma Paeonia
Atractylodes Heracleum Panax
Echinacea Lentinus Phragmites
Epimedium Ligustrum Polygonatum

INDUCES IMMUNOGLOBULINS

Angelica Cuscuta Ophiopogum
Asparagus Cynomorium Panax
Astragalus Glehnia Polygonatum
Cornus Ligustrum Scrophularia
Curculigo

IgA: Chinese foxglove (serum), Astragalus, Epimedium, Ganoderma, Polygonum (secretion)
IgE: Astragalus
IgG: Astragalus, Atractylodes, Codonopsis, Lentinus
IgM: Astragalus, Heracleum

INDUCES INTERFERON

Acanthopanax Camptotheca Lens
Artemisia Datura Stepania
Astragalus Gossypium

PROMOTES ANTIBODY PRODUCTION

Arnica Ganoderma Polygonum
Astragalus Heracleum Polyporus
Carthamus Lentinus Psoralea
Catharanthus Ligustrum Rehmannia
Cinnamomum OphiopogonScrophularia
Coix Panax Stephania
Cornus Phragmites Zexmenia
Epimedium Polygonatum

PROLONGS LIFE OF ANTIBODIES

Adenophora Dendrobium Rehmannia
Angelica Ophiopogon Scrophularia
Asparagus Paeonia Tremella

PROMOTES ANTIGENS

Astragalus Cornus Ligusticum

WORKS AS PANACEA IN AMERINDIAN MEDICAL TRADITION*

Acorus Echinacea Plantago
Betula Gaultheria Rumex
Chimphila Morus Sanguinaria
Cypripedium Panax

*Some speculate that Amerindian panaceas and oriental tonics may have earned their respect through immunological activity.

Dr. Christopher Calapai: AIDS Protocol

1. COMPLETE BLOOD EVALUATION:

CBC
SMA 20
Thyroid function panel
Viral titers—CMV, EBV, herpes, adenovirus
Vitamin and mineral levels
Gastrin, total protein, albumin, globulin
IgG, A, M
T cell subsets
IgE, and IgE for food allergens
P24
TB test

2. PHYSICAL EXAMINATION:

EKG
Spirometer

3. DISCUSSION—LIFESTYLE CHANGES:

No smoking, drug use, caffeine, or alcohol
No unprotected intimate activities

4. INTRAVENOUS THERAPY
(infused two to three hours, one to two times weekly)

Vitamin C 15-30 g
B complex
Zinc
Glutathione

Magnesium
B12
Dexpanthenol

5. INTRAMUSCULAR THERAPY (ONCE A WEEK):

Gamma globulin
Iron/B12 injection
Magnesium sulfate

6. ORAL NUTRIENTS:

Vitamins
Multivitamin (B complex, A, beta-carotene) four times daily
Buffered vitamin. C three times daily (10-15 g)
Digestive enzymes one to three times daily
NAC (glutathione) 1500 mg daily
Garlic capsules—three times daily

Herbs
St. John's Wort (Hypericum) 5 drops twice daily
Echinacea
Ginseng
Astragalus

7. TPN

Intravenous therapy consisting of proteins, carbohydrates, and fats if patient has lost 15 lbs or more or has severe malabsorption.

8. INDIVIDUAL BLOOD TESTS ARE REPEATED AS NEEDED.

Patients with bacterial or fungal infections are treated with appropriate antibacterial or antifungal therapy, as well as the foregoing protocol.

NOTE: Almost all of the patients on this protocol have chosen on their own not to use AZT; if they had already started AZT before beginning this program, they discontinued it once on this protocol. Further, the great majority of patients who have been on this protocol feel "great" and are doing very well.

Prof. John Bastyr: AIDS Protocol

Basic Supplements

MODERATE POTENCY

Thiamine HCL—100 mg
Riboflavin 6-phosphate—10 mg
Niacin—30 mg
Niacinamide—130 mg
Calcium pantothenate—450 mg
Pyridoxal 5'phosphate—15 mg
Cyanocobalamin—400 mcg
Biotin—400 mcg
Folic acid—800 mcg
Choline citrate—300 mg
D-alpha tocopherol—400 IU
Vitamin D3—50 IU
Calcium citrate—300 mg
Chromium citrate—200 mcg
Copper citrate—1.5 mg
Iodine (potassium iodide)—225 mcg
Iron citrate—100 mcg
Magnesium citrate—285 mg
Manganese picolinate—15 mcg
Potassium citrate—90 mg
Selenium citrate—200 mcg
Vanadium citrate—100 mcg
Zinc picolinate—15 mg

HIGH POTENCY SUPPLEMENTS

Beta-carotene—300,000 IU in divided doses

Ascorbate acid—1 g every two hours with substantial fluid. Adjust levels to below limits of bowel tolerance

Flavinoids—500 mg three times daily with meals

Zinc picolinate—15 mg

BOTANICALS

Ecinacea purpurea/augustifolia
Freeze-dried root powder—500 mg three times daily between meals

Lomatia species
Freeze-dried root powder—100 mg three times daily between meals

Ligusticum porteri
Freeze-dried root powder—500 mg three times daily between meals

Allium sativum
One crushed fresh clove four times daily with or between meals

George Wedemeyer: AIDS Protocol
Qigong teacher in San Francisco

Whole foods diet

1/2 cup bitter melon juice (Chinese cucumber)

1/2 cup Daikon juice mixed with 1/2 cup carrot juice

1-2 cups green tea daily

Qigong daily

Acupuncture once every two weeks

Dr. Wuan Xi Wang's herbal formula daily

Dr. Wuan Xi Wang: AIDS Protocol

CAPSULES

Astragalus, altractylodes (white), biota tops, ching-hao, codonopsis, isatis root lithosperm, licorice, *Ligustrum*, moutan, rehmannia raw, viola

Aloe vera juice—3 glasses per day (curbs virus spread)
Shark cartilage—740 mg daily

ANTIVIRAL HERBS

Echinacea/pau de arco extract—three times daily
St. John's wort—effective against retroviruses

IMMUNOMODULATORS

Siberian ginseng extract
Astragalus extract or caps
Reishi mushroom caps

REFLEXOLOGY

The liver point located on the outside of the sole of the foot about halfway between your toes and heel

COLONIC

Weekly colonic once recovery is well under way.

Ann Marie Wishard: AIDS Protocol

Author of Herb Talk

Bee pollen—1/2 teaspoon daily, increasing to at least 3 tablespoons daily (reduces stress, provides energy, and builds the immune system)

Propolis—**1** g or more daily (natural antibiotic)

Coenzyme Q10—90 mg daily (builds the immune system and gets oxygen into the blood)

Selenium—200 mcg daily (antioxidant, free radical scavenger)

Cat's claw—1-2 capsules, three times a day (activates the cells of the immune system, aids digestion by cleansing the intestinal tract)

Zinc—50-100 mg daily (promotes healing)

Pycnogenol—30 mg or more daily (potent free-radical fighter)

Vitamin A—30,00-50,000 IUs daily (antioxidant)

Acidophilus—three times daily (aids digestion and provides friendly bacteria)

Seventy-five percent of diet should be fresh vegetables and fruits as well as seeds, nuts, and grains.

Linda Rector-Page: AIDS Protocol

Author of Healthy Healing

PURIFYING SUPPLEMENTS

Egg yolk lecithin
Vitamin C with bioflavonoids—10-30 g daily

REBUILDING SUPPLEMENTS

Raw thymus extract
Carnitine—500 mg twice daily
Internasal B12 daily

ANTIOXIDANTS

Germanium—150 mg sublingual six times daily
Astragalus capsules—four times daily
CoQ10—30 mg three times daily
Quercetin—500 mg three times daily
Pycnogenol—50 mg twice daily
Vitamin E—1000 IUs daily with selenium
Glutathione—50 mg

Castor oil packs—apply topically and apply a heat pad. Take 5 or 6 castor oil capsules or 2 tablespoons with orange juice before bedtime once a day.

Two ozone treatments per day for five days, one treatment per day for five days for fourteen weeks. When ozone is not available or if not affordable, take oral hydrogen peroxide 5 drops of 35 percent food grade in a glass of water three times a day. Increase the dosage by 1 drop three times per day up to 20 drops three times a day. When 20 drops three times, continue for two weeks, then reduce dosage by drop three times a day until down to 5 drops three times a day. Discontinue for two weeks and have your blood tested for T4 counts and NK counts.

DHEA—50 mg daily
Lomatium dissectum (LDM-100)—1/2 teaspoon in water three times daily
Green tea—daily
Colloidal silver—2 teaspoons daily
Grapefruit seed extract—5 to 10 drops in a glass of water every two to four hours
Pycnogenol—120 mg daily

TO RESTORE URINE pH

1 betaine hydrochloride tablet with meal
1 pancreatin tablet with meal
500 mg magnesium opycnogenol—120 mg daily

Dr. Ken Korins:
AIDS Protocol

New York City Homeopath

HOMEOPATHIC REMEDIES

Ignatia—for worry, silence, sobbing, moodiness, brooding, great despair

Gelsemium—for listlessness, apathy, diarrhea

Arsenicum album—for fear of dying, panic, restlessness, periodic fevers

Thuja—for gonorrhea, high sex drive, discharges, warts, herpes, sores that don't heal, weakness, fixed ideas, rapid exhaustion

Mercurius—for syphilis, fevers, night sweats, aggravation with weather changes, swollen glands, sores in the mouth and thrush, paranoia

Arsenicum idodoum—swollen glands, periodic fevers, great appetite even though losing weight, restlessness

Phosphoric acid—for apathy, scattered thinking, extreme exhaustion

Medorrhinum—for discharges, prostitus, restlessness, bladder infections

Syphilinum—for pain and suffering that is worse at night, fears and despair of recovery, extreme fear of germs

Pyrogenium—got delirium, sored and bruises, restlessness with fevers, offensive body odors, abscesses, and sores that don't heal

PRESCRIPTION DRUGS

Glycyrrhizin (licorice root extract)—200-800 mg daily

Glycyrrhizin—(Japanese antiviral) 7cc along with 20cc of glutathione three to five times a week for up to a month or until stabilization (protection for the liver, spleen, kidneys, pancreas, eyes, lung lining, and stomach lining).

Naltrexone—3 mgs daily taken before bed (lowers high alpha-interferon levels, improves beta-endorphin stimulation of immune function, and reduces the incidence and severity of infection.

AMINO ACIDS

N-Acetylcysteine—600 mg three times a day for six to eight weeks, then continue indefinitely at a lower maintenance dose
L-Methionine
L-Lysine

OTHER NUTRIENTS

Coenzyme Q10
Adrenal gland extract
Acidophilus
Digestive enzymes
Adenosine triphosphate
Milk thistle—in the form of legalon 210 mg daily (liver health)

Anti-Chronic Fatigue Protocol

Comprehensive Program

Dr. Daniel Mowrey recommends the following program for overcoming chronic fatigue.

HERBS

Cayenne
Siberian ginseng
Gotu kola
Kelp
Ginger root

These herbs may help the body to overcome fatigue, physical weakness, and lack of strength. At the same time, they increase stamina.

VITAMINS

Vitamin B1, B2, B6
Vitamin C
Vitamin D
Vitamin E
Niacinamide
Folic acid
Pantothenic acid

MINERALS

Calcium/magnesium
Phosphorus
Zinc

Dr. James Balch Avery has a protocol that includes:

Herbal teas brewed from burdock root, dandelion, echinacea, goldenseal and pau d'arco; combine or alternate.

NUTRIENTS (VERY IMPORTANT)

Ascorbic acid
Coenzyme Q10
Egg lecithin
Maxidophilus or megadophilus or bifido factor
Proteolytic enzymes
Vitamin A and E emulsion

(IMPORTANT)

DMG
Garlic capsules
Germanium
Protein supplement
    (free-form amino acids from the company Ecological   Formulas)
Vitamin B complex (stress formula or injectable form)
Liver
Vitamin B12

(HELPFUL)

Black currant oil capsules
Multivitamin and mineral complex (high potency and hypoallergenic with
    beta-carotene, calcium, magnesium, potassium, selenium, and zinc)
Raw thymus and spleen plus raw glandular complex

In *Natural Health, Natural Medicine*, Dr. Andrew Weil suggests these important lifestyle changes:

Exercise regularly: 20-30 minutes of aerobic activity at least five days a week, even if it's the last thing you feel like doing. Keep the intensity of your activity below the level that leads to exhaustion.

Eat a diet low in protein and fat and high in carbohydrates.

Eat two cloves of raw garlic daily.

Take astragalus root for its antiviral and immunity-enhancing properties. A good product is Astra-8, a mixture of astragalus with seven other Chinese herbs. The dose is three tablets twice a day. You can stay on this indefinitely.

Ask your doctor to prescribe oxygen for home use and experiment with inhaling it for 15 to 20 minutes once or twice a day. If it helps, continue to use it until your energy level improves enough to do without.

Avoid support groups for people with chronic fatigue syndrome. They often give you ideas for new symptoms and convey the impression that the disease will be with you for the rest of your life.

Do not despair! Chronic fatigue syndrome is not a lifelong malady. Many patients recover well after one to five years of being sick. Expect to have ups and downs, with downs becoming less severe and less frequent.[1]

ANTIVIRAL PLANTS

Hundreds of plants have been screened for their antiviral activity. Results of studies overwhelmingly support the hypothesis that large numbers of plants have antiviral properties. Vandenbergh et al. (1986) reported that 200 plant extracts from ninety families (30 percent) exhibit some level of in vitro or in vivo antiviral activity.[2] Abou-karem and Shier (1990) observed antiviral activity in 20 percent of sixty-one plants.[3] Van Hoof et al. (1989) found that the extract from the leaves of *Populus cultivar Beaupre* produced antiviral activity.[4] They attributed its salicin and salireposide to activity against poliomyelitis and Semliki forest virus. Tsuchiya et al. (1984) discovered that chrysosplenium plants and axillarin had potent antiviral activity against rhinovirus.[5]

Mark Konlee:
AIDS Protocol

from How to Reverse Immune Dysfunction
and Positive Health News newsletter

Naltrexone (prescription drug with no side effects)—3 mg once a day after 8 P.M.
DNCB—once a week topical application
Beta-carotene—50 mg two or three times a day
B6—100 mg twice a day
Garlic—5 g daily or 4-6 raw cloves with food daily
Oralmat—3 drops three times a day under the tongue
Essiac tea—1/4 cup twice a day

WHOLE LEMON DRINK

Combine in a blender: 1 quartered organic lemon including seeds and rind, 1.5 cups orange and other fruit juice or water, 1 tablespoon cold pressed extra virgin olive oil. Blend at high speeds for 2 minutes. Pour mixture through a strainer to separate the juice from the pulp. Discard the pulp. The remaining liquid may be consumed all at once or divided into two or three daily portions.

IF NEUROPATHY IS PRESENT

Two triple-strength lecithin capsules twice daily. One tablespoon whole leaf aloe vera juice daily added to whole lemon drink.

FOR LOW BODY TEMPERATURE

Two or three cayenne capsules added to whole lemon drink

Dr. Paul M. Holt

Helpful Digestive Remedies

HOLT STATES:

I use the following to help normalize my intestinal functioning and improve digestion and absorption of foods and nutrients. I strongly recommend them because they meet my criteria for useful tools in treating HIV or other serious immune impairment: relatively inexpensive, easy to administer, and effective for the benefits received.

ACIDOPHILUS: THE FRIENDLY BACTERIA

Acidophilus bacteria are considered friendly because they are normal constituents of the digestive tract and aid in digestion. Imbalances occur in immune suppression, intestinal diseases, and antibiotic therapy. Many harmful bacteria may grow without adequate supplies of the friendly bacteria that help to suppress them. I found several studies showing the usefulness of supplementing with extra lactobacillus acidophilus in the treatment of candida, both in the urogenital tract and intestines; and its uselfulness in suppressing harmful bacteria such as salmonella, clostridium, shigella, and camphylobacter, all potentially harmful, especially in immune suppression. Numerous studies also show that lactobacilli can increase the quality, availability, digestibility, and the assimilation of nutrients ingested. Some studies even indicate that this bacteria may have a role in inhibiting tumor growth as well.

Acidophilus is easy to supplement. It is found in fermented foods, such as yogurt or buttermilk with active cultures, or it can be supplemented in pill form, available at health food stores. There are various strains of the friendly bacteria; certain strains may be more effective than others in the treatment of certain diseases.

GARLIC

Garlic is one of the cheapest and most useful natural dietary tools we can use to improve our health. It's easy to use in cooking and is beneficial in enhancing immune functions, inhibiting cancers, and reducing the growth of harmful organisms, such as bacteria and candida, parasites and viruses. It can be helpful with intestinal problems for these reasons.

Research continues to show the many properties, versatility, and usefulness of garlic. It assists the body in detoxification of harmful substances, reducing the harmful effects of radiation, and lowering cholesterol and fat levels, which contribute to cardiovascular disease.

DIGESTIVE ENZYMES

Digestive enzymes are frequently recommended by nutritionists because they assist in digestion and assimilation of foods and they have been shown to reduce inflammation in the body. The enzymes, which are harmless, basically help to break down foods, thereby aiding in the absorption of nutrients. In addition they may help to break down foreign harmful substances and allergy-causing substances, reducing inflammation.[1]

Direct AIDS Alternative Information Resources (DAAIR): AIDS Protocol

B-6—20-25 mg daily along with a B-complex formula (for increased antibody response and enhanced cellular activity

Beta-carotene—180 mg daily for 300,000 IUs (raised T-cell counts without increasing T8 cells)

B12 injections—hastens recovery from viral and bacterial infections and restores appetite and vigor

Vitamin C—12 g daily, divided into four doses (anti-HIV activity, tissue growth, adrenal gland function, healthy gums, and wound healing)

Garlic—5 g daily for six weeks, then 10 g daily for six weeks (increases T4 cells counts, improves T4/T8 ratios, normalizes low natural killer cell activity)

HERBS

Chlorella
Astragalus
Siberian ginseng
Ginkgo biloba

DAAIR-suggested Steps for Evaluating and Monitoring AIDS Therapies

HOW TO SELECT THERAPIES

A cure for AIDS does not yet exist. As a result, many individuals and companies are leaping into the breach with a wide array of products and approaches. Some of these may have some value; many others, unfortunately, only have value to the people selling them. So before you put anything into your body—even products sold by DAAIR—do some study. These guidelines can help you make a decision about whether to invest your time, money, or life in them. The trick is keeping an open mind while not becoming a sucker. Remember, the following discussion can be applied to everything—from herbs to crystal to pharmaceutical drugs to nutrients to surgery. The bottom line is that making these decisions is not easy. But it can be done.

First and foremost, if you have a health care practitioner with whom you have a good relationship, talk to him or her about your concerns. While many mainstream doctors almost automatically pooh-pooh alternative and complementary therapies, increasing numbers are opening up to the possibilities. Consult with your health care provider and, where possible, bring to him or her technical questions you may have as you undertake your research. And when necessary, get a second opinion.

With all the hype and lies out there (in both alternative and mainstream medicine), you have to separate the wheat from the chaff. Here are some key characteristics about products or therapies that should send up warning flags:

1. It is THE cure (or gives "miraculous" results).
2. Your're told not to use any other treatment.
3. The physician (or researcher) is the ONLY one with the answer.
4. It costs a lot.
5. Testimonials are the main source of information about the therapy.
6. The treatment or the practitioner is being persecuted by the FDA or AMA.
7. The ingredients are a secret.
8. It cures every disease known to humanity.

Any combination of these elements should alert you to the possibility of fraud. Points 1, 2, and 3 are truly signs of serious concern. Points 4, 5, and 6 may be legitimate.

HOW DO I MONITOR A NEW THERAPY?

Once you've decided to try a new therapy, you may wish to monitor its effects for a period of time. Since many people with HIV take a wide range of drugs, nutrients and so forth, it may be difficult to figure out what's going on if you change your regimen frequently over a short period of time. So, if at all possible, in order to assess the impact of this new element in your protocol, stabilize your treatment regimen for about a month. Then add the new therapy. Stick with the new therapy for a period of time while keeping everything else the same. In order for this to work well, discuss your plans with your health care provider.

WHAT IS THE NEW TREATMENT SUPPOSED TO DO?

Obviously, you might want to add a new element (or combination) to your regimen for a variety of reasons. Is the therapy supposed to increase T cells? Reduce viral load? Alleviate symptoms such as itching or diarrhea? Is it supposed to modify immune function in other ways? For example, does it increase CD8 counts or natural killer cells or perhaps reverse anergy (where your body doesn't respond to things such as the TB tine test)? Or is it designed to improve other blood work like red blood cell or platelet counts or lower triglyceride levels? Write down all the claims made about the product so that once you start, you can see if you are receiving the claimed benefits or experiencing undersirable side effects.

MONITOR YOUR BLOOD WORK

Try to have basic blood work done before starting a new therapy. Ideally, this includes getting a viral load test, especially if the therapy is supposed to have antiviral properties. Then, after some time on the new therapy, have more blood work done. Pay special attention to any changes in liver or kidney function, triglyceride levels, and mineral levels to assure that the therapy is not causing problems.

KEEP A DIARY

You might want to keep a personal diary, listing your regimen, the doses you took, any symptoms you may have had, your emotional state, and how you generally feel. You can make a list of your concerns and goals each day and see how you are doing. This may also help you in assuring that you take any medicine you may be using correctly, in the right dose, and the right number of times each day.

WHAT ABOUT SIDE EFFECTS? ARE THERE CROSS REACTIONS WITH OTHER THINGS I MIGHT BE USING?

Any side effects you may experience will often occur within minutes of consumption to the first few hours, days or possibly up to two weeks. Immediate response may be the result of a severe allergic reaction. In the case of longer term effects, it may be due to a long "half-life" for some compounds (that means it sticks around in your body long after taking it in). Some therapies can result in buildup of metabolites or other products. (Metabolites are the various molecules formed during the process of absorbing a drug). Check beforehand to see if the treatment you are on may result in allergic reactions or if there are restrictions on how long it should be taken.

With negative effects, keep track of everything. While many therapies have fairly well-described potential toxicities and others are virtually nontoxic, it is prudent to keep track of ANY changes in signs, symptoms, and blood work. Thus, it is important to monitor for symptoms like headache, dizziness; fever, sweats, diarrhea or constipation, rashes; sexual dysfunction (in arousal or desire); appetite loss; weight loss; vision problems; hearing or taste disturbances; bloating, flatulence (farting); tingling; pain; or muscle spasms. Does it increase or decrease urine flow? Are there changes in stool or urine color, volume or frequency? While any of these symptoms singly or in combination may suggest an undiagnosed infection that should be PROMPTLY investigated, they may also be side effects of the new therapy, especially within the first week or two of use.

This is true not only because of the problems associated with immune deterioration but also because of the possibility of adverse, novel cross-reactions with other elements of your protocol. This is deeply underscored by the FDA approval of Abbott's ritonavir (Norvir), which has a lengthy list of drugs with which it must NOT be used. These include common, over-the-counter drugs like antihistamines and other chemicals like antidepressants. Taking these with ritonavir could cause serious problems and may have resulted in the death of longtime survivor and Philadelphia AIDS-activist Jonathan Lax.

If any of these symptoms arise and persist for more than a day or two, you might want to stop taking the new therapy to see if this improves symptoms. If they resolve (and they are not severe), you might want to try the full dose again to see if they reappear (this is known as "rechallenging"). Make sure you get the results from a diagnostic test, of course, to rule out an underlying infection causing the symptoms. If the symptoms crop up again, try cutting back the dose by half and see if that helps. Again, however, if the symptoms are serious, do this only under a physician's care. Maintain a proactive relationship with your health care provider and let him or her know what you are doing.

For some therapies, you might want to keep track of your blood pressure. If it is already elevated, make sure you know whether the therapy in question can increase it further (as with glycyrrhizin). Other therapies, like coenzyme Q10 and garlic, may alter blood pressure.

IS IT WORKING? HOW LONG SHOULD IT TAKE?

From your blood work, you can begin to assess whether the treatment is having the effect claimed. Is it lowering viral load? If so, how much? How about CD4 positive counts? Is it having any adverse effect on other cell counts (such as CD8), white or red blood cell counts)? If it is an immune modulating therapy, is it causing an increase in viral load? Are symptoms improving? Is it doing anything at all??

Some therapies may take some time before positive results are seen. In the case of SPV-30, the boxwood extract, improvements in viral load and T cell count do not generally occur before about four to six months. Thus, you must tailor your assessment based on what is known about the therapy. If it is possible for you to stay on one therapy for this long without changing other parts of your regimen, you may find out more about whether a treatment works for you or not.

IF YOUR T CELL COUNT IS LOW...

If your T cells are low, it obviously makes it more difficult to do this kind of assessment, especially given the dynamic way one must deal with HIV infection. You may have to make changes rapidly if infections develop or other problems arise. It is important to make your health your first priority.

If you are taking antiretroviral therapy such as nucleoside analogs (3TC+AZT or ddI+d4T, for example) and/or the new protease inhibitors, you might want to consider three things.

1. These therapies can have an impact for up to a year, possibly longer. This means that once you've established your regimen, you can add the new therapy (possibly one that improves some other parameters) and follow it for a longer time without necessarily having to change your antiretroviral therapy for six months or more. Thus, you can establish your regimen with a better margin of safety.

2. The downside is that if the therapy you have chosen is one designed to improve CD4 count or viral load, these powerful drugs may well mask any effect of a new therapy. In other words, it may be difficult to tease out any particular difference. Still, you may find improvement beyond what you saw with just the pharmaceutical therapy, hat is, there may be a "synergy" of positive effects.

Remember that many natural products inhibit different parts of the virus than do pharmaceutical drugs. Nucleoside analogs work against reverse transcriptase; protease inhibitors work against the protease enzyme. But many antioxidant and bioflavonoids operate by correcting damage caused by the virus and an overly vigorous and damaging immune response. They may also attach various HIV enzymes like RT, protease, integrate, the long terminal repeat, and so forth. Such therapies that directly fight HIV include glycyrrhizin, alpha lipoic acid, vitamin E, NAC, whey, and several others. Also, improved bodily function may afford you additional viral suppression, which means a further slowing in the development of viral mutations. And as a result, this may also lengthen the period during which pharmaceutical drugs are effective.

3. You may want to use nutritional or other alternative therapies that have the potential to offset the known drug toxicities associated with many pharmaceutical treatments, especially combinations of standard antiretrovirals, as well as PCP and other prophylaxes. Some Chinese herbs reportedly improve red blood cell counts (e.g., condopsis). For another example, astragalus has proven benefits for anyone on chemotherapeutic drugs. Other nutrients can significantly offset toxicities. Good examples of this class are coenzyme Q10 and carnitine, which can lessen AZT's documented damage to the most important area of your cells, the mitochondria. Other nutrients, like vitamin B6 for AZT and molybdenum with ddI, help replenish stores that are lost as your body processes these drugs.

It has been shown by several studies that taking a proactive stance in your health care is associated with long-term survival. Asking questions, staying involved, and keeping an open mind without being a sucker will enable you to maintain a dynamic approach to your treatment. NIAID director Anthony Fauci has called AIDS a multifactorial, multiphasic, overlapping disease. As such, HIV infection requires a treatment approach that is multifactorial, multiphasic, and overlapping. Your active involvement in that process can help improve your odds of being here for the cure.

AIDS Protocol of the Life Extension Foundation

LIFE EXTENSION FOUNDATION can be contacted at 1-800-544-4440.

(A fuller version of ths protocol can be found at the LEF Web site. This fuller version contains more complete documentation of scientific studies. See: http://www.lef.org/protocols/prtcl-059d.shtml.)

GLUTATHIONE

It is now widely understood that AIDS is associated with a deficiency of glutathione that leads to the generation of enormous levels of oxidative stress that damage and kill otherwise healthy cells throughout the body.

SUBSTANCES THAT BOOST GLUTATHIONE LEVELS

HIV-induced free radical oxidation occurs in the presence of low levels of glutathione. Free radicals impair and destroy immune cells, and scientific studies consistently show glutathione deficiency to be a critical factor in the pathogenesis of immune suppression (AIDS). This suggests that supplementation with nutrients to boost cellular glutathione is crucial to protect against a primary mechanism by which HIV destroys immune function.

The following glutathione-enhancing nutrients are recommended in their order of importance:

NUTRIENT DOSAGE

N-acetyl-cysteine (NAC)600 mg, three times a day
Vitamin-C2000 mg, three times a day (take with NAC)
Selenium200 mcg, three times a day
Alpha-lipoic acid250 mg, two times a day
Whey protein isolate30 to 60 g of powder once a day

S-Adenosyl-Methionine (SAMe)2400-800 mg per day
Glutathione 3500 mg, twice a day

N-acetyl-cysteine is well-tolerated by most people; some may be sensitive to it, however, as it is a sulfur-containing compound. It is recommended to consume vitamin C in a dose of three times the amount of cysteine to mitigate cysteine's oxidation to cystine. Vitamin C is included in the appropriate dose in the list suggested above. Please remember that if your liver is in a weakened state, it may not be able to convert NAC to glutathione. Taking NAC under these conditions may further suppress lymphocyte functions. A solution may be to consume glutathione supplements directly even though they are poorly absorbed, and also to detoxify the liver under the guidance of a qualified health care practitioner. NAC can be resumed after liver toxicity has been resolved.

Anyone taking SAMe should also take on a daily basis 800 mcg of folic acid, 500 mcg of vitamin B12, and at least 100 mg of vitamin B6 to protect against excess homocysteine accumulation in the blood.

Controversy surrounds oral glutathione supplementation, as some research has shown that this form of ingestion has not been efficacious, while other research contradicts this. Glutathione is strongly recommended for HIV patients who can afford this relatively expensive supplement.

DHEA TEST AND TREATMENT

Several significant studies have shown that HIV infection progresses when serum levels of the hormone DHEA begin to decline. Many people with HIV have incorporated DHEA into their treatment program because maintaining healthy blood levels of DHEA might prevent immune degradation from progressing to general immune system collapse.

The Foundation recommends that all people with HIV and most people over forty have their DHEA blood level tested to determine their baseline level. DHEA supplementation should then be initiated and monitored to bring their serum level up to that of a healthy twenty-one-year-old. An appropriate dosage for men is 25 mg, two to three times a day.

DHEA can possibly increase liver damage in people who already have hepatitis and cirrhosis. Before starting DHEA replacement therapy, please consult with your physician and refer to the DHEA-Pregnenolone Precautions in the DHEA Replacement Therapy protocol.

After prudent medical counsel, if DHEA replacement therapy is still indicated in those with liver dysfunction, it can be taken in the following way: the DHEA powder capsule should be opened and the powder poured under the tongue, remaining there for ten to twenty minutes before being swallowed. In this fashion, much of the hormone enters directly into the bloodstream before reaching the liver.

TSH TEST AND TREATMENT

Proper levels of thyroid hormones are crucial for optimal immune function.... The TSH (thyroid stimulating hormone) test is extremely sensitive to both hypo-and hyperthyroid conditions, often showing even subclinical disorders. Popular prescription thyroid replacement drugs are Synthroid (synthetic thyroid hormone); Armour, Forest Pharmaceuticals (natural thyroid hormone); and Cytomel (T3 thyroid fraction). One must be careful not to overdose on thyroid hormone, as it has a fairly long half-life in the body. Thyroid hormone replacement therapy requires the supervision of a knowledgeable endocrinologist.

CORTISOL TEST AND TREATMENT

There are many studies that show the correlation of elevated cortisol levels present in people with HIV. Excessive cortisol production from the adrenal glands is immunosuppressive and usually closely correlates with autoimmune dysfunction.... Therefore, it is important to suppress excessive cortisol production.... The European procaine drug KH3 is believed to be the best way to block the effects of elevated cortisol levels in people with HIV and also in cancer patients. One or two tablets of KH3 can be taken first thing in the morning on an empty stomach, repeating the same dose an hour before dinner. Also, melatonin replacement therapy, vitamin C, and aspirin can lower elevated cortisol.

ANTIOXIDANT THERAPY

Thousands of research studies have proven the positive effect antioxidants have in managing immune suppression. The Life Extension Foundation (in 1985) was the first organization to propose that decline in the immune function of people with HIV might be prevented or slowed down by taking high-potency antioxidant supplements.

Most people with HIV use a high-potency multinutrient supplement such as Life Extension Mix that contains the nutrients most often deficient in those who are immune compromised. Upon a diagnosis of immune suppression, the Foundation recommends this aggressive comprehensive antioxidant "cocktail" that comprises high levels of nutrients to prevent the progression of immune suppression due to oxidative cell damage. People with HIV should consume three tablets of Life Extension Mix three times a day with meals.

S-ADENOSYL-METHIONINE

Apart from raising glutathione levels, a particular form of methionine known as S-adenosylmethionine (SAMe) serves as a methyl donor for a host of methylation reactions in the body. In this capacity, it may prevent and reverse liver disease, as studies have shown it may be the most effective substance known to regenerate damaged liver tissue.

COQ10

A popular supplement used by HIV-positive patients is Coenzyme Q10. Studies indicate that Coenzyme Q10 boosts immune function. It is suggested that HIV patients take at least 200 mg a day of Coenzyme Q10.

VITAMIN C

Vitamin C (ascorbic acid) is required for normal host defense and functions. The Life Extension Mix contains water-soluble ascorbic acid, an ascorbate complex, fat-soluble ascorbyl palmitate, and a natural acerola juice powder complete with synergistic bioflavonoids.

BETA-CAROTENE

Besides being very important to vitamin A production, the main function of the lipophilic antioxidant beta-carotene is to prevent damage to membrane and plasma polyunsaturated lipids....

The Life Extension Foundation recommends against high-dose beta-carotene in PWA (people with AIDS) who also have hepatitis. For some people with hepatitis, long-term use of beta-carotene may cause liver enzyme elevation, indicating potential liver damage. Many PWA have also been infected with hepatitis B or C. PWA who do not have hepatitis or other liver damage should consider consuming 25,000 to 100,000 IU of beta-carotene daily.

VITAMIN B12

Vitamin B12 is essential to everyone for good health. Vitamin B12 is crucial to many enzymatic processes in the body and functions synergistically with folic acid and TMG to enhance DNA methylation and reduce toxic homocysteine in the blood. Vitamin B12, together with folic acid, is very effective at increasing the oxygen-carrying capacity of red blood cells. This is of crucial importance to those taking most antiretroviral medications, as these drugs are notorious for destroying the rapidly dividing red blood cells, causing severe anemia. The Foundation's latest, most advanced form of this essential nutrient is methylcobalamin, which research has shown may function better than conventional forms of vitamin B12 and is readily available by calling the Foundation. Three 500 mcg sublingual vitamin B12 tablets taken daily are recommended. If a blood test reveals a continuing vitamin B12 deficiency, weekly vitamin B12 injections may be indicated.

Pharmaceuticals

NALTREXONE

In 1984, the FDA approved naltrexone (Revia) as a "narcotic antagonist," a treatment to keep heroin and other opiate addicts off addictive substances.... Naltrexone, however, has another very interesting effect: it increases the amount of endorphins in the brain. Endorphins are hormonal neurotransmitters and are immune modulators, which function as a major link in the communication between the brain and the immune system.

Dr. Bernard Bihari in New York City uses naltrexone in his practice for HIV treatment. Naltrexone for HIV infection is taken in a very low dose (about 3 mg an evening), enough to sustain the up-regulation of the endorphin system, and it appears to have no negative side effects when taken at these low doses. In higher dosages, such as the dosages used in the treatment of alcoholism and the obesity trials, there were side effects. Some patients experienced nausea, difficulty sleeping, anxiety, nervousness, abdominal pain and cramps, vomiting, low energy, joint and muscle pain, and headaches. Incredibly, no negative side effects have been documented or reported for naltrexone when used as a treatment for HIV/ AIDS. Dr. Bihari strongly suggests naltrexone at the top of the list of any protocol being considered by persons living with HIV. Naltrexone is an FDA-approved medication that must be dispensed from a pharmacy with a doctor's prescription. Call the Foundation for information about doctors who offer this therapy.

ISOPRINOSINE AND DIETHYL DITHIOCARBOLIATE

Isoprinosine is a drug shown to be capable of slowing down the progression of AIDS, and since 1986, the Foundation has recommended that PWHIV add this antiviral to their treatment program. Isoprinosine is approved by every regulatory agency in the world except the Food and Drug Administration in the United States.

In 1985, the Foundation recommended PWHIV take isoprinosine to slow down the progression of immune suppression, which can lead to full immune-system collapse. Isoprinosine and some other immune-boosting drugs work best when taken on an alternating dosing schedule, two months on and two months off.

BIOSTIM

A French drug, Biostim, was studied with regard to its effect on modulating varying immune responses. In response to staph infection, Biostim therapy significantly increased the critical phagocytic component of immune attack. Biostim also modulated synthesis of human polymorphonuclear granulocytes.

People with HIV should consider using Biostim in three-month dosing schedules as follows: 2 tablets daily for eight days, then discontinue for three weeks; 1 tablet daily for eight days, then discontinue for three weeks; 1 tablet daily for eight days, then discontinue for nine months.

Hormones

MELATONIN

Melatonin is a hormone secreted by the pineal gland during sleep. It is considered a master hormone, as it is a neuroendocrine modulator and can exert a regulatory effect over many different areas in the body.

Evidence suggests HIV immune suppression may be slowed by nightly intake of melatonin. However, melatonin appears to benefit PWA in many other ways, including providing protection against AZT toxicity and wasting syndrome. The Foundation suggests that PWHIV read *Melatonin*, by Dr. Russell Reiter and Jo Robinson. It is a timely and valuable work of great benefit to all and especially to PWHIV. The Life Extension Foundation suggests using 3 to 30 mg nightly.

GROWTH HORMONE

When HIV patients lose a significant amount of body weight, usually more than 10 percent, they may have AIDS wasting. Muscle mass in these individuals may constantly be broken down as an effect of dysfunctional metabolism. [Growth hormone treatment may be of benefit to those with this problem.]

Many insurance companies can underwrite the cost of this treatment. A stipulation of eligibility for this treatment is the concurrent use of antiretroviral medication.

Amino Acids

L-GLUTAMINE

Many have discussed the medical causes of immune suppression that have been proven to create a malabsorption syndrome in the intestinal tract. The amino acid glutamine plays a major role in the health and well-being of the gastrointestinal tract and its supportive organs (stomach, small and large intestine, liver, pancreas, and gall bladder) [so it may be a useful supplement].

WARNING: Glutamine supplementation in individuals with severe cirrhosis of the liver, Reye's syndrome, or any other metabolic disorder that can lead to an accumulation of ammonia in the blood is clearly not indicated and can present an increased risk for encephalopathy or coma. Under these conditions, the body is unable to metabolize excess nitrogen, which converts to ammonia and can cause brain swelling and brain death. When the liver is severely damaged or when hepatic coma is imminent, glutamine is not effective and can cause further brain damage.

Wasting is a life-threatening condition that should not be self-treated. If you involuntarily have lost more than 10 percent of your body weight, you should suspect wasting is taking place and seek the care of a knowledgeable physician. Bio Impedance Analysis can determine lean body mass and establish a baseline for you, against which future treatments' efficacy can be judged. In addition to glutamine, please review the information about Life Extension's Whey Isolate, which is extremely bioavailable and tissue-sparing. In addition to these treatments, your physician may wish to employ anabolic steroids such as testosterone and nandrolone and/or growth hormone. These treatments are very well grounded and medically prudent in cases of wasting and should be highly considered when necessary.

L-CARNITINE

Another nutrient often overlooked by people with HIV is L-carnitine, which has been shown to boost immune function, via several different mechanisms, to protect the heart against AZT toxicity and to enhance essential fatty-acid and glucose uptake. The recommended dose is 3 to 4 grams daily in two divided doses on an empty stomach.

Additional Suggestions

OLIVE LEAF EXTRACT

Olive leaf extract is a nonprescription, over-the-counter food supplement that has been used as a natural means to treat viral, bacterial, fungal, and parasitic infections.

Olive leaf extract contains a phenolic glucoside known as oleuropein…. [it] has been taken by PWA in doses ranging from 1 capsule or tablet four times a day, to 1 cup of extract twice a day. If you wish to take more or less, please consult a Life Extension adviser or your health care professional.

Olive leaf extract does not have any side effects per se, though some people may experience a "die-off" effect, also called the Herxheimer Reaction. A die-off effect is caused by a rapid increase in volume of waste material and pathogens being brought into the lymph system. Reactions to the die-off effect include extreme fatigue, diarrhea, headaches, muscle and joint achiness, and flu-like symptoms. These reactions are temporary and will pass once the body has expelled the circulating toxins. If these detoxifying symptoms are too uncomfortable, reduce the amount of olive leaf extract taken or discontinue use temporarily.

SPV-30

SPV-30 is an herbal extract derived from the European boxwood tree (Buxus sempervirens), a species of evergreen.

AIDS Buyers Clubs around the country have anecdotally reported on a specific SPV-30 treatment regime that has produced the most beneficial results and is included here:

Upon rising, on an empty stomach, take 5 to 500 mg bitter melon capsules with 2 large glasses of water. Wait 30 minutes before eating. Consume the following three times daily, with or without meals, and with 1 large glass of water:

GLYCYRRHIZINATE FORTE (licorice)-300 mg, Jarrow Formulas
(Note: This product can elevate blood pressure.)

1-SPV-30 (boxwood extract)-330 mg, Arkopharma

Consume the following at the end of the evening, 2 hours after eating:
BITTER MELON CAPSULES-5 to 500 mg, with 1 large glass of water

THYMIC IMMUNE FACTORS

Thymic Immune Factors is a synergistic formula that contains herbal activators and a full complement of homeopathic nutrients in addition to fresh, healthy thymus, lymph, and spleen tissues, which produce white blood cells to fight invading organisms and cancer cells. The immunological tissue extracts in this product are raw, concentrated, toxinfree, and freeze-dried to preserve their biological activity. People with HIV may consider using a unique product such as Thymic Immune Factors to potentiate white-blood-cell production and activity.

SILYMARIN

Milk thistle extract, or silymarin, is a unique type of bioflavonoid that exerts a protective effect on the liver. Please refer to the glutathione information in the antioxidants section of this protocol for more information concerning products that promote healthy liver function.

ENZYMES

Proper digestion of nutrients is necessary for maintaining good health in all of us. PWHIV and PWA have an elevated necessity for life-sustaining nutrients. In many cases, even the consumption of all the correct nutrients for these individuals simply does not produce nourishment because of malabsorption. A complete digestive enzyme formula should be used to offer the greatest possible digestive activity.... The Foundation carries several excellent digestive enzyme products that can maximize nutrient absorption.

Notes

Introduction

1. Claudia Wallis, "Knowing the Face of the Enemy," *Time*, 30 April, 1984, 66.
2. Wallis, 66. All further quotes from Wallis, pp. 66-67. I emphasize that this is the *Time* account since it doesn't jibe with what I take to be Shilts's version. See Randy Shilts, *And the Band Played On: Politics, People, and the AIDS Epidemic* (New York: St. Martins Press, 1987), 450-51.
3. Shilts, *And the Band Played On*, 534-35.
4. Shilts, *And the Band Played On*, :448.
5. Wallis, "Knowing the Face of the Enemy," 66. The last part of the citation is quoting Dr. Allen Goldstein.
6. Barbara and John Ehrenreich, "Medicine and Social Control," in *The Cultural Crisis of Modern Medicine*, ed. John Ehrenreich (New York: Monthly Review Press, 1978), 54-55.
7. Irving Kenneth Zola, "Medicine as an Institution of Social Control," in *The Cultural Crisis of Modern Medicine*, ed. John Ehrenreich (New York: Monthly Review Press, 1978), 82.
8. Francis Bacon, *The Advancement of Learning*, ed. G. W. Kitchin (London: J. M. Dent & Sons Ltd., 1973; original, 1605), 32.
9. Ibid.

PART ONE: EVIDENCE AGAINST THE CURRENT AIDS HYPOTHESIS
The Media and Medical Establishment Support Their Beliefs

1. Philip Johnson, "HIV and AIDS: The Present State of the Controversy," *Reappraising AIDS* 2 (2) 1 (December 1994).
2. *What if Everything You Thought You Knew About AIDS Was Wrong*, HEAL pamphlet, 2. (No further data available.)
3. K. B Mullis, P. E. Johnson, and C. A. Thomas, Jr. "Dissenting on AIDS: The Case Against the HIV-Causes-AIDS Hypothesis," *San Diego Union-Tribune*, 15 May 1994.
4. John Lauritsen, interviewed by Null, in *The Pain, Profit and Politics of AIDS* (videotape produced by Gary Null Associates, no date).
5. Steven Epstein, *Impure Science :AIDS, Activism, and the Politics of Knowledge* (Berkeley: University of California Press, 1996), 81.
6. Mark Garish Conlan, interviewed by the author, May 1996.
7. Charles Thomas, interviewed by the author, May 1996.
8. David Rasnick, "AIDS and the HIV Boondoggle."
9. Neville Hodgkinson, "False Positive," *Continuum* 4 (1): 339 (May-June 1996).
10. Hodgkinson, "False Positive," 63. Also the following quote.
11. Hodgkinson, "False Positive," 10.
12. Thomas Kuhn, *The Structure of Scientific Revolutions* (Chicago: University of Chicago Press, 1970).
13. Francis Bacon, *The New Organon and Related Writings* , ed. Fulton H. Anderson (Indianapolis: Bobbs-Merrill, 1960; original, 1620), 58.
14. Ibid., 59-60.
15. P. H. Duesberg, "AIDS Acquired by Drug Consumption and Other Noncontagious Risk Factors," *Pharmac & Ther* 55 (1992): 201-77; R. C. Gallo

and L. Montagnier, "The Chronology of AIDS Research," *Nature* 326 (1987),. 435-36; Duesberg, "Human Immunodeficiency Virus and Acquired Immunodeficiency Syndrome: Correlation but Not Causation," *Proceedings of the National Academy of Sciences* 86 (1989), 755-64; J. Cairns, *Cancer Science and Society* (San Francisco: W. H. Freeman and Company, 1978). See also Duesberg, "Talk by Peter Duesberg," Cal Alumni Day, 6 March, 1993, *Rethinking AIDS home page*, http://www.virusmyth.com.aids/data/pdlecture.htm.

16. Hiram Caton, *The AIDS Mirage* (Sydney: University of New South Wales Press, 1994), 8-9.

17. Ibid.

18. Peter Duesberg, interview by the author, 24 January 1995.

19. Johnson, "HIV and AIDS." 1.

20. Bryan J. Ellison, interview by the author, 15 February 1995.

21. Lamar Graham, "The Heretic," *GQ*, November 1993.

22. Bacon, *New Organon*, 65.

23. Karl Popper, *The Logic of Scientific Discovery* (London: Hutchinson, 1968; original, 1934).

24. Charles A. Thomas Jr., Kary B. Mullis, and Phillip E. Johnson. "What Causes AIDS?" *Reason*, June 1994, 19.

25. A. Liversidge, "The Limits of Science,"*Cultural Studies Times* (Fall 1995).

26. R. Schoch, "Professor Duesberg Is Postive HIV Is Negative," *California Monthly*, June 1996, 19.

27. A. Liversidge, "Interview with Robert Gallo," *Spin*, March 1989.

28. Frank Buianouckas, interview by the author in*AIDS, The Untold Story*. (videotape produced by Gary Null Associates, no date). Also the previous quotation.

29. Neville Hodgkinson, interview by the author, May 1996. Also the previous quote.

30. Charles Geshekter, interview by the author, 13 February 1995. Professor Geshekter said that questioning the internal politics of the AAAS resulted in his being stripped of an office he had held for several years. "My heart sank when I discovered that someone else was now the section chair for History of Science. I received no phone calls, no fax, no letter telling me any of that. I just sort of read it in the newsletter."

31. Dr. Nathaniel S. Lehrman, interview by the author, in *The Pain, Profit and Politics of AIDS*.

32. C. A. Thomas Jr., K. B Mullis, and P. E. Johnson, "What Causes AIDS?" *Reason*, (June 1994).

The Establishment View

1. Susan Sontag, *Illness as Metaphor and AIDS and Its Metaphors* (New York: Doubleday, 1989), 108-09.

2. John Lauritsen, interview by the author, 2 February 1995.

3. Lauritsen, "Psychological & Toxicological Causes of 'AIDS,'" in eds. J. Lauritsen & I. Young, *The AIDS Cult* (Provincetown: Asklepios, 1997), 76-77. Also for the other quotations in this paragraph.

4. Lynne McTaggard, interview by the author, May 1996l

5. R. S. Root-Bernstein, "Agenda for U.S. AIDS Research Is Due for a Complete Overhaul," *The Scientist*, 4 April 1994.

6. John Lauritsen, interview by the author, February 2, 1995.

7. Neville Hodgkinson, "Conspiracy of Silence," *Sunday Times*, 3 April 1994, 4.

8. *New York Times*, 29 September 1992, editorial page.

9. Earl Caldwell, "AIDS 'Cure' Deadlier Than Disease," *New York Daily News*, 20 September 1993.

10. Hodgkinson, "Conspiracy of Silence," 3 April 1994, 4.

11. P. Duesberg, "AIDS Acquired by Drug Consumption and Other Non-contagious Risk Factors," *Pharm. Ther.* 55 (1992), 201-77.
12. R. S. Root-Bernstein, "Do We Know the Cause(s) of AIDS?" *Perspectives in Biology and Medicine*, 33 (Summer 1990), 480-500.
13. Ibid.
14. Tom Bethell, "Second UC Berkeley Biology Prof Joins Dissent," *Reappraising AIDS* 4 (5): 1 (May 1996). Also the following quotations.
15. Ibid.
16. Schoch, "Professor Duesberg Is Positive HIV Is Negative," 19.
17. Andrew Scott, *Pirates of the Cell: The Story of Viruses from Molecules to Microbe* (London: Basil Blackwell, 1985), 98.
18. Scott, *Pirates of the Cell*, 54.
19. Christine Johnson, interview by the author, May 1996. Also the following quotation.
20. Celia Farber, "Fatal Distraction," *Spin*, June 1992.
21. P. H. Duesberg and B. J. Ellison, "Is the AIDS Virus a Science Fiction? Immunosuppressive Behavior, Not HIV, May be the Cause of AIDS," *Policy Review* (Summer 1990).
22. H. Baily, "Why HIV Will Always Be One Step Ahead," *Rethinking AIDS* 1 (1993): 8.
23. Thomas, Mullis and Johnson, "What Causes AIDS?" 19.
24. Scott, *Pirates of the Cell*, 125.
25. Peter Duesberg, interview by the author, 24 January 1995.
26. Duesberg and Elison, "Is the AIDS Virus A Science Fiction?"
27. The HEAL Trust, "Deconstructing AIDS" (pamphlet, London: HEAL); P. Simmonds et al., "HIV-Infected Individuals Contain Provirus in Small Numbers of Peripheral Mononuclear Cells and at Low Copy Numbers," *Virology* 64 (1990): 864-72; A. Guyton et al., *Textbook of Medical Physiology* (New York, W. B. Saunders, 1990).
28. Paul Philpott, interview by the author February 13, 1995.
29. P. H. Duesberg, "A Challenge to the AIDS Establishment," *Bio/Technology* (November 1987): 1244.
30. Michael Verney Elliot, interview by the author, May 1996.
31. Peter Duesberg, "Talk by Peter Duesberg," Cal Alumni Day, 6 March 1993. This is also the source of the previous quotation.
32. Johnson, "HIV and AIDS," 1.
33. Ibid.
34. S. Lanka, "HIV Debate," *Continuum* (June/July 1995): 4.
35. M. Craddock, "Doesn't Anybody Read Anymore?" *Reappraising AIDS*, 4(11): 3 November 1996.
36. The HEAL Trust, "Deconstructing AIDS," London; A. Fauci et al., "HIV Infection Is Active and Progressive in Lymphoid Tissue During the Clinically Latent Stage of Disease," *Nature*, 362, (1993): 355-58; H. Sheppard and M. Ascher, "Viral Burden and HIV Disease,"*Nature*, 364 (1993):. 291; B. Elison, "Does HIV Lurk in Your Lymph Nodes?" *Rethinking AIDS* 1 (1993): 4; and Penta 1 Trial. Medical Research Council's Response to Representations from Standing Committees on AZT Malpractice-Scam (available from HEAL).
37. *What if Everything You Thought You Knew About AIDS Was Wrong*, HEAL Pamphlet, 3.
38. M. G. Conlan, "Interview with David Rasnick: A Real Scientist," *Zenger's*.
39. Duesberg, "AIDS Acquired by Drug Consumption," 201-77.
40. Charles Thomas, interview by the author, May 1996.

41. Robert Root-Bernstein, Transcript of videotape: "The AIDS Catch," 1990, Meditel Productions Ltd, Channel.
42. Johnson, "HIV and AIDS," 1.
43. Farber, "Fatal Distraction."
44. Ibid.
45. Alfred Hassig, "New Thoughts on AIDS," *Schweiz.Zeitschrift fur GanzheitsMedizin, Heft*, April 1992, 171-77.
46. Duesberg and Ellison, "Is the AIDS Virus A Science Fiction?"
47. P. H. Duesberg, "How Much Longer Can We Afford the AIDS Virus Monopoly?" in *AIDS: Virus or Drug Induced?* ed. Peter Duesberg. (Netherlands: Kluwer Academic Publishers, 1996), 241-70.

Critique of Established AIDS Hypotheses

1. Lawrence E. Altman, "U.S. and France End Rift on AIDS," *New York Times*, 1 April 1987, 1, A18.
2. Bacon, *New Organon*, 48.
3. Shilts, *And the Band Played On*, 270.
4. Bacon, *New Organon*, 54.
5. Shilts, *And the Band Played On*, 271.
6. Shilts, *And the Band Played On*, 367.
7. Bacon, *New Organon*, 56.
8. Shilts, *And the Band Played On* 367. See also, 388.
9. Seth Roberts, "Lab Rat: What AIDS Researcher Dr. Robert Gallo Did in Pursuit of the Nobel Prize," *Spy*, July 1990.
10. Epstein, *Impure Science*, 70.
11. Roberts, "Lab Rat."
12. P. J. Hilts, "Federal Inquiry Finds Misconduct by a Discoverer of the AIDS Virus," *New York Times*, 31 December 1992.
13. U.S. House Committee on on Energy and Commerce, Subcommittee on Oversight and Investigations,*Investigation of the Institutional Response to the HIV Blood Test Patent Dispute and Related Matters*. Staff Report, 23 January 1995, my emphasis.
14. Ibid, my emphasis.
15. Ibid.
16. John Crewdson, "In Gallo Case, Truth Termed a Casualty," *Chicago Tribune*, 1 January 1995.
17. Ibid.
18. Ibid.
19. Shilts, *And the Band Played On*, 430.
20. John Crewdson, "U.S. Inquiry Discredits Gallo on AIDS Patent," *Chicago Tribune*, 19 June 1994
21. House Committee,*Investigation of the Institutional Response*.
22. Ibid.
23. Shilts,*And the Band Played On,*, 366-67, 387.
24. Shilts, *And the Band Played On,*, 444-45.
25. Bruno Latour and Steven Woolgar, *Laboratory Life: The Construction of Scientific Facts*. (Princeton: Princeton University Press, 1986), 213.
26. Latour and Woolgar, *Laboratory Life*, 207.
27. Serge Lang, "HIV AIDS: Have We Been Misled?" *Yale Scientific* (Fall 1994): 11.
28. Ibid.
29. Ibid.
30. C. A. Thomas Jr., K. B. Mullis, and P. E. Johnson, "What Causes AIDS?" *Reason* (June 1994).

31. Mullis, Johnson, and Thomas, "Dissenting on AIDS."
32. Ibid.
33. R. S. Root-Bernstein, "The Evolving Definition of AIDS," *Rethinking AIDS* .
34. Institute of Medicine, *Confronting AIDS* (Washington, DC.: National Academy Press, 1986); CDC HIV/AIDS Surveillance, U.S. Dept. of Health and Human Services, Atlanta, Ga, 1986-1991.
35. Peter Duesberg and David Rasnick, "The Drug-AIDS Hypothesis,"*Continuum* 4 (5) (suppl.) (February-March 1997): 3.
36. Vladimir Koliadin, personal communication with author, 16 March 2001. For his statistics, Koliadin cites *Nature* 377 (1995): 79-82.
37. Ibid.
38. Thomas, Mulllis, and Johnson, "What Causes AIDS?" 19.
39. Ibid.
40. Ibid.
41. E. Burkett, "Is HIV Guilty?" *Miami Herald,* 23 December 1990.
42. Paul Philpott, "First HIV-Positive Lab Chimp Finally Develops AIDS After Eleven Years," *Reappraising AIDS* 4 (4) (April 1996): 2.
43. Kary Mullis, interview by the author, May 1996.
44. Peter Duesberg, "The HIV Gay in National AIDS Statistics," *Bio/Technology* 11 (1993): 955-56.
45. Ibid.
46. Michael Bomgardener, interview by the author.
47. Institute of Medicine, *Confronting AIDS-Update 1988* (No place of pub.: National Academy Press, 1988); R. Selik, et al., "Impact of the 1987 Revision of the Case Definition of Acquired Immunodeficiency Syndrome," *Journal of AIDS* 3 (1990): 73-82; and Centers for Disease Control, "Revision of the CDC Surveillance Case Definition for Acquired Immunodeficiency Syndrome," *JAMA* 258 (1987): 1143-54.
48. Similar cases have been reported from other institutions including the CDC.
49. C. Johnson, "HIV Antibody Tests," *The Heal Bulletin*, Special Edition, Spring 1995,1; Peter Duesberg, "The HIV Gap in National AIDS Statistics," *Bio/Technology* 11 (1993): 955-56; C. Gilks, "What Use Is a Clinical Case Definition for AIDS in Africa?" *British Medical Journal* 303 (1991): 1189-90; and Peter Duesberg, "AIDS Acquired by Drug Consumption and Other Non-contagious Risk Factors," *Pharmac Therapy* 55 (1992): 201-77.
50. Charles Geshekter, interview by the author, 13 February 1995.
51. Hodgkinson, "False Positive," 308
52. Charles Thomas, interview by the author.
53. R. Root-Bernstein, "AIDS Without HIV," *Rethinking AID S* 1 (1992): 2.
54. Barry D. Schoub, *AIDS and HIV in Perspective: A Guide to Understanding the Virus and its Consequences* (Cambridge: Cambridge University Press, 1994), 125.
55. Stefan Lanka, "HIV: Reality or Artifact?" *Continuum* 3 (1): 5 (April/May 1995); Elena Papadopulos-Eleopulos, et al., "Is a Positive Western Blot Proof of HIV Infection?" *Bio/Technology* 11 (1993): 696-707.
56. Lanka, "HIV: Reality or Artifact?" 5.
57. Peter Duesberg, interview by the author, 24 January 1995.
58. Ibid.
59. Schoub, *AIDS and HIV in Perspective*, 88.
60. Lanka, "HIV: Reality or Artifact?" 5, my emphasis.
61. Scott, *Pirates of the Cell*, 10.
62. Schoub, *AIDS and HIV in Perspective,,* 141.
63. H. Christie, "Do Antibody Tests Prove HIV Infection? A Blood-curdling Interview with Dr. Valendar F. Turner," *Continuum* 5 (2): 14 (Winter 1997/1998).

64. P. Philpott, "The Isolation Question: Does HIV Exist? Do HIV Tests Indicate HIV Infections?" *Reappraising AIDS* 5(6) (June-August 1997).

65. Ibid.

66. Lanka, "HIV: Reality or Artifact," 6.

67. Ibid.

68. C. Farber, "AIDS: Words from the Front," *Spin*, November 1993.

69. Lanka, "HIV: Reality or Artifact," 7; Papadopulos-Eleopulos et al., "Is a Positive Western Blot Proof of HIV Infection?" 696-707.

70. Christine Johnson, "Is Anyone Really Positive?," *Continuum* 3 (1): 12 (April/May 1995).

71. V. Turner, "Do HIV Antibody Tests Prove HIV Infection?" *Continuum* 3(5): 8 (January-February 1996); D. S. Burke et al., "Measurement of the False Positive Rate in a Screening Program for Human Immunodeficiency Virus Infections," *New England Journal of Medicine* 319 (1988): 961-64; and H. V. Strandstrom et al., "Studies with Canine Sera that Contain Antibodies which Recognize Human Immunodeficiency Virus Structural Proteins" *Cancer Research* 59 (1990): 5628s-630s.

72. See studies cited in previous note.

73. Johnson, "Is Anyone Really Positive?" 12.

74. Eleni Papadopulos-Eleopulos et al., "Has Gallo Proven the Role of HIV in AIDS?" *Emergency Medicine* 5 (7): 121 (1993); F. Chiodi et al., "Isolation Frequency of Human Immunodeficency Virus from Cerebrospinal Fluid and Blood of Patients with Varying Severity of HIV Infection,"*AIDS Research Human Retrovirology* 4 (1988): 351-58; and J. Learmont et al., "Long-Term Symptomless HIV-1 Infection in Recipients of Blood Products from a Single Donor," *Lancet* 340 (1992): 863-67. See also E. Papadopulos-Eleopulos, "The Isolation of HIV: Has It Really Been Achieved? The Case Against," *Continuum*, 4(3) (Suppl):5 (September/October 1996).

75. See references in previous note.

76. Johnson, "Is Anyone Really Positive?" 12.

77. Papadopulos-Eleopulos et al., "Factor VIII, HIV and AIDS in Haemophiliacs,An Analysis of Their Relationship," *Genetica* 95 (1995): 30; Papadopulos-Eleopulos et al., "Is a Positive Western Blot Proof of HIV Infection?" *Bio/Technology* 11 (1993): 696-707.

78. Johnson, "Is Anyone Really Positive?" 14.

79. Griner, *Annals of Internal Medicine*, April 1981.

80. Johnson, "Is Anyone Really Positive?" 14; and Langedijk, *AIDS* 6 (1994): 1547.

81. Ibid.; Griner.

82. Philpott, "The Isolation Question: Does HIV Exist?"

83. Ibid.

84. Anita Allen, personal communication, 5 March 2000.

85. Charles Geshekter, interview by the author, 13 February 1995.

86. Johnson, "Is Anyone Really Positive?" 13, 14. See also *Los Angeles Times*, 1 July 1994; Cynthia Cournoyer, *What About Immunizations?* 6th ed. (Nelson's Books, 1995), 123-4; "HIV False Positivity After Hepatitis B Vaccination," *Lancet* 339 (25 April 1992): 1060.

87. "AIDS: New or Old," *New York Times* 19 December 1991, 65.

88. L. Simonsen et al., "Multiple False Reactions in Viral Antibody Screening Assays after Influenza Vaccination," *American Journal of Epidemiology* 141 (11): 1089 (1995).

89. Ibid.; Johnson, "Is Anyone Really Positive?" 13, 14; *Los Angeles Times*, 1 July 1994;

90. Hodgkinson, "False Positive," 240, 258-59.

91. "Prevalence and Characteristics," *Continuum* 4(1): 8 (May-June 1996).

92. Hodgkinson, "False Positive," 9-10.

93. Joan Shenton, radio interview by the author, 15 May 1996.

94. For mass reactivity, see also Christine Johnson, "HIV Antibody Tests," *Heal Bulletin*, Special ed. (spring 1995): 1; G. Hunsmann et al., "HTLV Positivity in Africans," *Lancet* 26 October, 1985; "AIDS Vaccine Efficacy Trial Sites Selected by WHO," *The Blue Sheet* 34(43): 1-3 (1991); R. Weiss et al., "Lack of HTLV-1 Antibodies in Africans," *Nature* 319 (1986)., 794 95; R. Biggar et al., ELISA HTLV Retrovirus Antibody Reactivity Associated with Malaria and Immune Complexes in Healthy Africans," *Lancet* ii (1985): 520-523; O. Kashala et al., "Infection with Human Immunodeficiency Virus Type 1 (HIV-1) and Human T-Cell Lymphotropic Viruses Among Leprosy Patients and Contacts: Correlation Between HIV-1 Cross-reactivity and Antibodies to Lipoarabinomanna," *Journal of Infectious Disease* 169 (1994): 296-304; J. Langedijk et al., "Identification of Cross Reactive Epitopes Recognized by HIV-1 False Positive Sera," *AIDS* 6, (1992): 1547-48; R Weiss and S. Their, "HIV Testing is the Answer—What's the Question?" *New England Journal of Medicine* 319 (1988): 1010-12; K. Meyer and S. Parker, "Screening for HIV: Can We Afford the False Positive Rate?" *New England Journal of Medicine* 317 (1987): 238-241; T. Germanson, "Screening for HIV: Can We Afford the Confusion of the False Positive Rate?" *Journal of Clinical Epidemiology* 42 (1989): 1235.

95. Hodgkinson, "False Positive," 9-10.

96. Johnson, "HIV Antibody Tests," 6-7; Hunsmann et al., "HTLV Positivity," 1985; L. Mutanda, *The Blue Sheet*, , 23 October 1991.

97. Neville Hodgkinson, radio interview by author, 15 May 1996.

98. Christine Johnson, interview by the author, May 1996.

99. Neville Hodgkinson, interview by the author, May 1996.

100. Hodgkinson, "False Positive," 240.

101. M. P. Wright, "Thousands of HIV False Positives in US," Sumeria Web site, http://www.livelinks.com/sumeria, 21 May 1998.

102. Christie, "Do Antibody Tests Prove HIV Infection?" 16.

103. Ibid.

104. Celia Farber, "AIDS: Words from the Front," *Spin*, November 1993.

105. James Whitehead, interview by the author, May 1996.

106. CDC National AIDS Clearinghouse Custom Search Service and AIDS Clinical Trials Information Service, 18 January 1996 (studies cited not available).

107. Kary Mullis, interview by the author, May 1996.

108. Ibid.

109. Ibid.

110. Paul Philpott, "The Isolation Question," *Reappraising AIDS* 5(6): 1 (June-August 1997).

111. Ibid; "The Isolation of HIV—Has It Been Achieved? part 2," *Reappraising AIDS* 5(6): 2 (June-August 1997).

112. Paul Philpott, "Isolation 101: The Basics," *Reappraising AIDS* 5(6): 1 (June-August 1997), my emphasis.

113. Ibid.

114. Epstein, *Impure Science*, 70.

115. Eleni Papadopulos-Eleopulos, "Between the Lines: A Critical Analysis of Luc Montagnier's Interview with Djamel Tahi," *Continuum* (Winter 1997/1998): 44. See also M. A. Rey et al., "Characterization of the RNA Dependent DNA Polymerase of a New Human T-Lymphotropic Retrovirus (Lymphadeopathy Associated Virus)," *Biochem. Biophys. Res. Commun* 121, (1984): 126-33; I. Toplin, "Tumor Virus Purification Using Zonal Rooters," *Spectra* (1973): 225-35; H. M. Temin and D. Baltimore, "RNA-Directed DNA Synthesis and RNA Tumor

Viruses," *Adv Virol Res* 17 (1972): 129-86; F. Sinoussi et al., "Purification and Partial Differentiation of the Particles of Murine Sarcoma Virus (M. MSV) According to their Sedimentation Rates in Sucrose Density Gradients," *Spectra* 4 (1973): 237-43; K. Toyoshima and P. K. Vogt, "Enhancement and Inhibition of Avian Sarcoma Viruses by Polycations and Polyanions," *Virology* 38 (1969): 414-26; S. A. Aaronson et al., "Induction of Murine C-type Viruses from Clonal Lines of Virus-Free BALB/3T3 Cells," *Science* 174 (1971): 157-9; M. S. Hirsch et al., "Activation of Leukemia Viruses by Graft-Versus-Host and Mixed Lymphocyte Reactions in Vitro," *Proceedings of the National Academy of Sciences* 69 (1972): 1069-72; F. Barre-Sinoussi et al., "Isolation of a T-Lymphotrophic Retrovirus from a Patient at Risk for Acquired Immune Deficiency Syndrome (AIDS)," *Science* 220 (1983): 868-71.

116. See references in previous note.

117. Papadopulos-Eleopulos, "Between the Lines: Luc Montagnier's Interview," 35; Robert C. Gallo et al., "Some Evidence for Infectious Type-C Virus in Human," in *Animal Virology*, eds. D. Baltimore et al. (New York: Academic Press, 1976), 385-405; H. M. Temlin and D. Baltimore, "RNA-Directed DNA Synthesis and RNA Tumor Viruses," *Adv Virol Res* 17 (1972): 129-86; H. Varmus, "Reverse Transcription," *Scientific American* 257 (1987): 48-54; H. Varmus, "Retroviruses," *Science* 240 (1988): 1427-35; Robert C. Gallo et al., "On the Nature of the Nucleic Acids and RNA Dependent DNA Polymerase from RNA Tumor Viruses and Human Cells," in *Possible Episomes in Eukaryotes*, ed. L. G. Silvestri (Amsterdam: North Holland Publishing Company, 1973), 13-34; F. M. Tomley et al., "Reverse Transcriptase Activity and Particles of Retroviral Density in Cultured Canine Lymphosarcoma Supernatants," *British Journal of Cancer* 47 (1983): 277-84. See also M. V. Nermut and A. C. Steven, "Retroviridae" in *Animal Virus and Structure*, eds. M. V. Nermut and A. C. Steven (Oxford: Elsevier, 1987).

118. See citations in previous note.

119. Papadopulos-Eleopulos, "Between the Lines: Luc Montagnier's Interview," 35; see also, "Retroviridae."

120. A. G. Laurent-Crawford et al., "The Cytopathic Effect of HIV is Associated with Apoptosis," *Virology* 185 (1991): 829-39; Papadopulos-Eleopulos, "Between the Lines: Luc Montagnier's Interview,"35.

121. Papadopulos-Eleopulos, op. cit.; E. Papadopulos-Eleopulos et al., "A Critical Analysis of the HIV-T4-cell-AIDS-Hypothesis," *Genetica* 1995, 5-24; Luc Montagnier, "Lymphadenopathy-Associated Virus: From Molecular Biology to Pathogenicity," *Annals of Internal Medicine* 103 (1985): 689-93; Peter D. Duesberg, *Inventing the AIDS Virus* (Washington: Regenery Publishing, Inc., 1996).

122. A. G. Laurent-Crawford et al., "The Cytopathic Effect of HIV Is Associated with Apoptosis," *Virology* 185 (1991): 829-839; and Papadopulos Eleopulos, op. cit.

123. See previous reference.

124. Neville Hodgkinson, *AIDS: The Failure of Contemporary Science* (London, Fourth Estate, 1996), 360.

125. Ibid., 369.

126. Ibid.

127. Ibid.

128. Epstein, *Impure Science*, 171.

129. E. Papadopulos-Eleopulos, "Is HIV the Cause of AIDS? Interview with Christine Johnson," *Continuum* 5 (1):15-16 (Autumn 1997).

130. Ibid.

131. Papadopulos-Eleopulos, "Is HIV the Cause of AIDS?" 15-16.

132. Ibid.
133. Christie, "Do Antibody Tests Prove HIV Infection?" 16.
134. Papadopulos-Eleopulos, "Is HIV the Cause of AIDS?" 40.
135. Hodgkinson, *AIDS: The Failure of Contemporary Science*, 376.
136. David Rasnick, personal communication, 21 May 2001.
137. Papadopulos-Eleopulos, "Is HIV the Cause of AIDS?" 19.
138. Philpott, "The Isolation Question."
139. Papadopulos-Eleopulos, "Is HIV the Cause of AIDS?" 37-38; Robert C. Gallo and A. C. Fauci, "The Human Retroviruses," in *Harrison's Principles of Internal Medicine*, 13th ed, eds. K. J. Kisselbacher et al. (New York: McGraw-Hill Inc., 1994), 808; R. Lower et al., "The Viruses in All of Us; Characteristics and Biological Significance of Human Endogenous Retrovirus Sequences," *Proceedings of the National Academy of Sciences* 93 (1996): 5177-84.
140. Papadopulos-Eleopulos, "Is HIV the Cause of AIDS?" 19.
141. "Pictures Explode Isolation Claims of HIV," *Continuum* 5(1): 2 (Autumn 1997).
142. Ibid.
143. Ibid.
144. Papadopulos-Eleopulos, "Is HIV the Cause of AIDS?" 37.
145. Ibid.
146. Ibid., 11-12.
147. Bacon, *New Organon*, 48.
148. Bacon, *New Organon*, 52.
149. Bacon, *New Organon*, 52-53.
150. Bacon, *New Organon*, 53.
151. Ibid.
152. Michael Verney Elliot, interview by the author, May 1996.
153. Ibid.
154. Ibid.
155. Zygmunt Bauman, *Modernity and Ambivalence* (Ithaca, NY: Cornell University Press, 1991), 215.
156. Ibid.
157. Ibid.
158. Neville. Hodgkinson, "The Origin of the Specious," *Continuum* 4(3): 17-18 (September/October 1996).
159. S. Lanka, "Collective Fallacy; Rethinking HIV," *Continuum* 4(3): 19-20 (September/October 1996).
160. Papadopulos-Eleopulos, "Is HIV the Cause of AIDS?" 40; E. Papadopulos-Eleopulos et al., "The Isolation of HIV: Has it Really Been Achieved?" *Continuum* 4 (1996):. 1s-24s; J. P. Vartian et al., "High Resolution Structure of an HIV-1 Quasispecies: Identification of Novel Coding Sequences," *AIDS* 6, (1992): 1095-98; S. Wain-Hobson, "Virological Mayhem," *Nature* 373 (1995): 102; M. J. Kozal et al., "Extensive Polymorphisms Observed in HIV-1 Clade B Protease Gene Using High-Density Oligonucleotde Arrays," *Nat. Medicine* 2 (1996): 753-59; S. Wain-Hobson et al., "Nucleotide Sequence of the AIDS Virus, LAV," *Cell* 40 (1985): 9-17; P. A. Lazo and P. N. Tsichlis, "Biology and Pathogenesis of Retroviruses," *Semin Oncol* 17 (1990): 269:
A. L. Cunningham, "Structure and Function of HIV," *Medical Journal Australia* 164 (1996): 161-73; F. Barre-Sinoussi, "HIV as the Cause of AIDS," *Lancet* 348 (1996): 31-35.
161. Hodgkinson, *AIDS: The Failure of Contemporary Science*, 369.
162. Papadopulos-Eleopulos, "Is HIV the Cause of AIDS?" 42.
163. Papadopulos-Eleopulos, "Is HIV the Cause of AIDS?" 43; and M. J. Kozal et al., "Extensive Polymorphisms," 753-59.

164. Papadopulos-Eleopulos, "Is HIV the Cause of AIDS?" 36; J. L. Marx, "Human T-Cell Virus Linked to AIDS," *Science* 220 (1983):. 806-09.; M. L. Gougeon et al., "Direct and Indirect Mechanisms Mediating Apoptosis During HIV Infection: Contribution to In Vivo CD4 T Cell Depletion," *Immunology* 5 (1993): 187-94; J. Cohen, "Exploiting the HIV-Chemokine Nexus," *Science* 276(5315): 1261-74 (1997); and S. P. Layne et al., "Factors Underlying Spontaneous Inactivation and Susceptibility to Neutralization of Human Immunodeficiency Virus," *Virology* 189 (1992): 695-714.

165. Papadopulos-Eleopulos, "Is HIV the Cause of AIDS?" 11-12.

166. Peter Duesberg, "Duesberg Defends Challenges to Existence of HIV: Article 1 of 2 for Continuum," http://www.duesberg.com/docs_continuum/continu1.html., 2.

167. "Peter Duesberg Responds," *Continuum* 4(2): 9 (July/August 1996), my emphasis.

168. Anita Allen, personal communication, 5 March 2001.

169. Joan Shenton, *Postively False—Exposing the Myths around HIV and AIDS* (New York: St. Martins, 1998), 96.

170. Peter Duesberg, "Duesberg Defends Challenges," 1.

171. Papadopulos-Eleopulos, "Is HIV the Cause of AIDS?" 41. See also the following: E. Papadopulos-Eleopulos et al., "The Isolation of HIV: Has it Really Been Achieved?" *Continuum* 4 (1996): 1s-24s; R. Lower, et al., "The Viruses in All of Us: Characteristics and Biological Significance of Human Endogenous Retrovirus Sequences," *Proceedings of the National Academy of Sciences* 93 (1996) 5177-84; J. L. Marx, "Human T-Cell Virus Linked to AIDS," *Science* 220 (1983): 806-09; M. L. Gougeon et al., "Direct and Indirect Mechanisms Mediating Apoptosis During HIV Infection: Contribution to in Vivo CD4 T Cell Depletion," *Immunology* 5 (1993): 187-194; S. P. Layne et al., "Factors Underlying Spontaneous Inactivation and Susceptibility to Neutralization of Human Immunodeficiency Virus," *Virology* 189 (1992): 695-714; D. Klatzman, et al., "Selective Tropism of Lymphadenopathy Associated Virus (LAV) for Helper-Inducer T Lymphocytes," *Science* 225, (1984): 59-63; D. Klatzman and L. Montagnier, "Approaches to AIDS Therapy," *Nature* 319 (1986): 10-11; R. B. Acres, "Rapid Phosphorylation and Modulation of the T4 Antigen on Cloned Helper T Cells Induced by Phorbol Myristate Acetate or Antigen," *Journal of Biochemistry* 261 (1986): 16210-214. D. Zagury, et al., "Long-term Cultures of HTLV-III-Infected T Cells: A Model of Cytopathology of T-Cell Depletion in AIDS," *Science* 231 (1986): 850-53; O. Scharff, et al., "Effect of Thapsigargin on Cytoplasmic Ca2+ and Proliferation of Human Lymphocytes in Relation to AIDS," *Biochim Biophysica Acta* 972 (1988): 257-64; R. E. Birch et al., "Pharmacological Modification of Immunoregulatory T Lymphocytes. II. Modulation of T Lymphocyte Cell Surface Characteristics," *Clin Exp Immunol* 48 (1982): 231-38; A. G. Laurent-Crawford, et al., "The Cytopathic Effect of HIV is Associated with Apoptosis," *Virology* 185 (1991): 829-39; R. R. Dourmashkin et al., "The Presence of Budding Virus-Like Particles in Human Lymphoid Cells Used for HIV Cultivation," (paper presented at the VIIth International Conference on AIDS, Florence, Italy, 1991), 122; C. J. Ohara et al., "The Ultrastructural and Immunohistochemical Demonstration of Viral Particles in Lymph Nodes from Human Immunodeficiency Virus-Related Lymphadenopathy Syndromes," *Human Pathology* 19 (1988): 545-49; Robert C. Gallo et al., "First Isolation of HTLV-III," Nature 321 (1986): 119; Peter H. Duesberg, "Peter Duesberg Responds," 8-9; J. L. Lauritsen, "NIDA Meeting Calls for Research into the Poppers-Kaposi's Sarcoma Connection," in *AIDS: Virus or Drug Induced*, ed. P. H. Duesberg (London: Kluwer Academic Publishers, 1995), 325-30; D. D. Ho et al., "Rapid Turnover of Plasma Virions and CD4 Lymphocytes in HIV-1 Infection," *Nature*, 373 (1995): 123-26; M. Holodny et al.,

"Diurnal and Short-term Stability of HIV Virus Load as Measured by Gene Amplification," *Journal of Acquired Immun Defic Syndrome* 7 (1994): 363-68; W. O'Brien et al., "Changes in Plasma HIV-1 RNA and CD4+ Lymphocyte Counts and the Risk of Progression to AIDS. Veterans Affairs Cooperative Study Group on AIDS," *New England Journal of Medicine* 334 (1996): 426-31; J. Schapiro et al., "The Effect of High-Dose Saquinavir on Viral Load and CD4+ T-Cell Counts in HIV-Infected Patients," *Annals of Internal Medicine* 124 (1996): 1039-50.

172. See references in previous note.
173. Papadopulos-Eleopulos, "The Isolation of HIV: Has it Really Been Achieved?" 1.
174. Papadopulos-Eleopulos, "Is HIV the Cause of AIDS?" 41.
175. Epstein, *Impure Science*, 167.
176. Ibid., 139.
177. Ibid., 120-21.
178. Lanka, "Collective Fallacy; Rethinking HIV,"19-20.
179. K. Krafeld, "Inventing the AIDS Virus?" *Continuum* 5(1): 22 (Autumn 1997).
180. S. Lanka, "Lanka Replies to Duesberg," *Continuum* (February/March 1997); J. D. Booke, "A Little Help for My Ends," *Nature* 328: 579-81 (October 17, 1996).
181. Krafeld, "Inventing the AIDS Virus?" 22.
182. Ibid.
183. Ibid.
184. Ibid. See also A. Hassig, H. Kremer, S. Lanka et al., "15 Years of AIDS," *Rethinking AIDS* Web site, http://www.virusmyth.com, May 1998.
185. Lanka, "Collective Fallacy; Rethinking HIV."
186. Ibid.
187. Lanka, "Collective Fallacy; Rethinking HIV," September/October 1996.
188. Ibid.
189. Ibid.
190. Ibid.
191. Papadopulos-Eleopulos, "Is HIV the Cause of AIDS?"13.
192. Ibid.
193. Shilts, *And the Band Played On*, 355.
194. Ibid.
195. Epstein, *Impure Science*, 202.
196. Ibid.
197. Ibid., 205, quoting Emke.
198. Ibid., 216.
199. Ibid., 217
200. Ibid., 218.
201. Bruno Latour, *The Pasteurization of France*, trans. Alan Sheridan and John Law (Cambridge: Harvard University Press).
202. Lanka, "Lanka Replies"; and H. Kremer, "Acquired Iatrogenic Death Syndrome (AIDS)," *Continuum* 4(3): 8-13 Nov/Dec 1996.
203. Lanka, op. cit.
204. Kuhn, *Structure of Scientific Revolutions*.
205. Everett Mendelsohn, "The Poltical Anatomy of Controversy in the Sciences," in *Scientific Controversies: Case Studies in the Resolution and Closure of Disputes in Science and Technology*, eds. H. Tristam Engelhardt Jr. and Arthur L. Caplan (Cambridge: Cambridge University Press, 1987), 97.
206. Ibid.
207. Ronald N. Giere, "Controversies involving science and technology: a theoretical perspective,' in *Scientific Controversies*, 146.

What is AIDS?

1. Bacon, *New Organon*, 56.
2. Ibid., 57
3. Lang, "HIV AIDS: Have We Been Misled?" 9.
4. Ibid.
5. John Lauritsen, "Non-Hodgkin's Lymphoma," *Continuum* 5(1): 3 (Autumn 1997).
6. Lang, "HIV AIDS: Have we Been Misled?" 9.
7. Lauritsen, "The Risk-AIDS Hypothesis," www.virusmyth.net/aids/index/jlauritsen.htm.
8. Robert S. Root-Bernstein, "The Evolving Definition of AIDS," *Rethinking AIDS*. (Free Press/Macmillan, 1993).
9. Ibid.
10. Thomas, Mullis, and Johson, "What Causes AIDS?" 19.
11. P. Philpott, "What HIV/AIDS Epidemic?" *Reappraising AIDS* 4(7), July 1997.
12. Root-Bernstein, "The Evolving Definition of AIDS."
13. Celia Farber, *Continuum* 2(3): 6 (June/July 1994).
14. Ibid.
15. V. L. Koliadin, "Some Facts Behind the Expansion of the Definition of AIDS in 1993," *Rethinking AIDS* Web site, http://www.virusmyth.com, May 1998.
16. Ibid.
17. *What if Everything You Thought You Knew About AIDS Was Wrong*, HEAL pamphlet.
18. Molly Ratcliffe, "Continuum Woman," *Continuum*, 3(1): 15 (April/May 1995). See also *Positively Women Annual Report*, 1994.
19. See previous note
20. Christine Johnson, interview with the author, May 1996.
21. Ibid.
22. J. Lauritsen, "Psychological & Toxicological Causes of AIDS," in T*he AIDS Cult* , eds. J. Lauritsen and L. Young (Provincetown: Asklepisou, 1997), 96-97.
23. Ibid.
24. Hodgkinson, *AIDS: The Failure of Contemporary Science*, 126.
25. Ibid.
26. Ibid.
27. David Mertz, "Sex Wars: The New Left's AIDS-Related Scientism," *Rethinking AIDS* Web site, http://www.virusmyth.come, May 1998.
28. "Commentary," *HIV/AIDS Surveillance Report* 11 (1): 3-4.
29. Schoch, "Professor Duesberg is Postive HIV Is Negative," 19.
30. Philpott, "What HIV/AIDS Epidemic?"
31. Ibid.
32. Ibid.
33. Ibid.
34. Kary B Mullis, P. E. Johnson, and C. A. Thomas Jr., "Dissenting on AIDS: The Case Against the HIV-Causes-AIDS Hypothesis," *San Diego Union-Tribune*, 15 May 1994.
35. *What if Everything You Thought You Knew About AIDS Was Wrong*, HEAL pamphlet.
36. Epstein, *Impure Science*, 132.
37. Michael Verney Eliot, interview with the author, May 1996.
38. Root-Bernstein, "The Evolving Definition of AIDS."
39. Ibid.
40. Mark Garish Conlan, interview with the author, May 1996.
41. Ibid.

42. Peter H. Duesberg, "AIDS Acquired by Drug Consumption and Other Noncontagious Risk Factors," *Pharmac & Ther* 55 (1992): 201-77.

43. David Rasnick, "AIDS and the HIV Boondoggle."

44. "Commentary," *HIV/AIDS Surveillance Report* 11 (1): 3-4.

45. Peter Duesberg, interview with the author, 24 January 1995.

46. Root-Bernstein, "Do We Know the Cause(s) of AIDS?" 480-500.

47. Duesberg and Ellison, "Is the AIDS Virus a Science Fiction?"

48. M. Bell et al., "Occupational Risk of H.I.V.. Infection in Healthcare Workers: An Overview," *American Journal of Medicine* 102(5B): 9-15 (May 19, 1997). Cited in Koliadin, personal communication with the author.

49. Root-Bernstein, "Do We Know the Cause(s) of AIDS?"480-500.

50. *Journal of International Health Research*, 19 July 1994.

51. Ibid.

52. Mullis, Johnson and Thomas, "Dissenting on AIDS."

53. Philpott, "What HIV/AIDS Epidemic?"

54. Ibid. See also Mullis, Johnson, and Thomaset al., "Dissenting on AIDS."

55. T. Gallagher, "A Hollow Day for Prophets of AIDS Doom," *Daily Mail*, 2 December 1993.

56. Philpott, "What HIV/AIDS Epidemic?"

57. Ibid.

58. "Heterosexual AIDS Threat Shown to Be Just a Myth," *Continuum* 4(2): 2 (July/August 1996).

59. Hodgkinson, *AIDS The Failure of Contemporary Science*, 9.

60. W. Geisler, *AIDS: Origin, Spread, and Healing* (Koln, Germany: Bipawo Verlag, 1994), 8. See also R. M. Welsh et al., "Human Serum Lyses RNA Tumor Viruses," *Nature* 257 (1975): 612-14; R. L. Shoeman et al., "Antibodies to HIV in Saliva," *New England Journal of Medicine* 320 (1989): 1145-46; P. C. Fox et al., "Saliva Inhibits HIV-1 Infectivity," *Journal of the American Dental Association* 116 (1988): 635-37; A. Levy, "The Transmission of AIDS: The Case of the Infected Cell," *JAMA* 259 (1988): 3037-38; D. S. Newburg and R. H. Yolken, "Characterization of a Human Milk Factor that Inhibits Binding of HIV GP 120 to its CD4 Receptor," *Advances in Experimental Biology* 310 (1991): 281-91; A .J. Mueller et al., "Infrequent Detection of HIV-1 Components in Tears Compared to Blood of HIV-1-Infected Persons," *Infection* 20 (1992): 249-52; Luc Montagnier, "Lymphadenopathy-Associated Virus: From Molecular Biology to Pathogenicity," *Annals of Internal Medicine* 103 (1985): 689-93. 29; Committee for the Oversight of AIDS Activities, *Confronting AIDS: Update 1988*, 2nd ed., ed. National Academy of Science (Washington, DC.: National Academy Press, 1988), 3; and WHO, "Acquired Immunodeficiency Syndrome (AIDS)—Data as at 1 July 1992," ,*Weekly Epidemic Records* 67 (1992): 201-204.

61. Geisler, *AIDS: Origin, Spread*, 29. See also CDC, "HIV/AIDS Surveillance Report: US AIDS Cases Reported Through June 1992," July 1992, 1-8; R. Bakeman et al., "The Incidence of AIDS Among Blacks and Hispanics," *Journal of the National Medical Association* 79 (1987): 921-28; R. M. Selik et al., "Racial/Ethnic Differences in the Risk of AIDS in the United States," *American Journal of Public Health* 78 (1988): 1539-45.

62. Geisler, *AIDS: Origin, Spread*, 42-43.

63. Ibid, 41. See also L. Mascola et al., "HIV Seroprevalence in Intravenous Drug Users: Los Angeles, CA, 1986," *American Journal of Public Health* 79 (1989): 81-82; and L. S. Brown Jr. et al., "Needle Sharing and AIDS in Minorities," *JAMA* 258 (1987): 1474-75.

64. Geisler, *AIDS: Origin, Spread*, 29. See also W. Winkelstein Jr. et al., "The San Francisco Men's Health Study: Continued Decline in HIVSeroconversion Rates

Among Homosexual/Bisexual Men," *American Journal of Public Health* 78 (1988): 1472-74; W. A. Blatner, "HIV Epidemiology: Past, Present, and Future," *FASEB J* 5 (1991): 2340-48; P. S. Rosenberg, "Declining Age at HIV Infection in the United States," *New England Journal of Medicine*, 330 (1994): 789.

65. See previous note.
66. Philpott, "What HIV/AIDS Epidemic?"
67. Ibid.
68. Ibid.
69. Phillip E. Johnson, interview with the author, 14 February 1995.
70. Duesberg, "AIDS Acquired by Drug Consumption," 201-77.
71. Thomas, Mullis, and Johnson, "What Causes AIDS?" 19.
72. Quoted in Thomas, Mullis, and Johnson, "What Causes AIDS?"
73. C. Geshekter, "There Is No AIDS Epidemic," *Chico Enterprise Record*, 26 November 1995.
74. Ibid.
75. Ibid.
76. Celia Farber, "Fatal Distraction."
77. Ibid.
78. Geshekter, "There is no AIDS Epidemic."
79. Barry Craven, interview with the author, May 1996.
80. Ibid.
81. J. Lauritsen, "Latex Lunacy," *New York Native*, 4 July 1988.
82. Ibid.
83. Ibid.
84. Ibid.
85. Peter Duesberg, interview with the author, 24 January 1995.
86. Thomas, Mullis, and Johnson, "What Causes AIDS?" 19.
87. P. Plumley, "AIDS—The Great Numbers Game," *Rethinking AIDS*, October/November 1993.
88. Ibid.
89. Ibid.
90." Heterosexual AIDS Threat Shown to Be Just a Myth," 2.
91. Farber, "Fatal Distraction."
92. Mertz, "Sex Wars."
93. Ibid.
94. Maggie Gallagher, "Uncle Sam's Great AIDS Scam," *New York Post*, 4 May 1996.
95. Ibid.
96. A. Bennett and A. Sharpe, "AIDS Fight Is Skewed by Federal Campaign Exaggerating Risks," *Wall Street Journal*, 1 May 1996.
97. Ibid.
98. Peter Duesberg, interview with the author, 24 January 1995.
99. Ibid. See also Gary Null, "HIV Equals AIDS and Other Myths of the AIDS War," *Penthouse*, 12 December 1995.
100. Hodgkinson, *AIDS: The Failure of Contemporary Science*, 360.
101. Robert S. Root-Bernstein, "Rethinking AIDS," *Wall Street Journal*, 17 March 1993.
102. Robert S. Root-Bernstein, "The Prostitute Paradox," *Rethinking AIDS*, March 1993.
103. Hodgkinson, *AIDS: The Failure of Contemporary Science*, 206.
104. Koliadin, personal communciation to the author.
105. "Sex, Lies, and HIV Transmission," *Reappraising AIDS* 5(9), December 1997.
106. Ibid.
107. See H. J. Eysenck and S. B. G. Eysenck, *Manual for the Eysenck Personality Questionnaire* (San Diego: EdITS, 1975); N. H. Corby et al., "AIDS Knowledge,

Perception of Risk, and Behaviors Among Female Sex Partners of Injecting Drug Users," *AIDS Education and Prevention* 3 (1991): 353-66; W. Winkelstein et al., "Sexual Practices and Risks of Infection by the Human Immunodeficiency Virus," *JAMA* 257 (1987): 321-25; S. Brody, "Lack of Evidence for Transmission of Human Immunodeficiency Virus Through Vaginal Intercourse," 390.

108. See references in previous note. See also Corby et al., 1991.
109. Hodgkinson, *AIDS: The Failure of Contemporary Science*, 206.
110. Geisler, *AIDS: Origin, Spread*, 34; R. L. Stoneburner, et al., "The Epidemic of AIDS and HIV-1 Infection Among Heterosexuals in New York City," *AIDS* 4 (1990): 99-106.
111. R.S. Root-Bernstein, "Non-HIV Immunosuppressive Factors in AIDS: A Multifactorial, Synergistic Theory of AIDS Aetiology," *Res Immunol* 141 (1990): 815-38.
112. Ibid.
113. Root-Bernstein, "Rethinking AIDS."
114. Ibid.
115. Ibid.
116. Null, "HIV Equals AIDS and Other Myths."
117. Geisler, *AIDS: Origin, Spread*, 31. See also Michael Verney Elliot, interview by the author, May 1996.
118. Ibid.
119. Farber, "Fatal Distraction."
120. "Bad Transmission," *Continuum* 5(1): 2 (Autumn 1997).
121. S. Brody, "Continued Lack of Evidence for Transmission of Human Immunodeficiency Virus Through Vaginal Intercourse: A Reply to Carey and Kalichman," *Archives of Sexual Behavior* 25(3): 329-37 (1996); D. Symon, "How Risky is Risky Sex?" *Journal of Sex Research* 30 (1993): 344-46.
122. Ibid.
123. Anderson et al., "Understanding the AIDS Pandemic," *Scientific American* 266 (1992) 20-6; Farber, "Fatal Distraction."
124. Hodgkinson, *AIDS: The Failure of Contemporary Science*, 94.
125. Farber, "Fatal Distraction."
126. M. A. Sushinsky et al., "Women and AIDS: The Ethics of Exaggerated Harm," *Rethinking AIDS* Web site, http://www.virusmyth.com, May 1998.
127. Neville Hodgkinson, interview by the author, May 1996.
128. Ibid.
129. Papadopulos-Eleopulos, "Between the Lines: Luc Montagnier's Interview,"38.
130. "Lifestyle," *Spectrum* (September/October 1995): 9-10; *Archives of Sexual Behavior* 24(4) (1995); S. Brody, "Lack of Evidence for Transmission of Human Immunodeficiency Virus Through Vaginal Intercourse," 384.
131. "Sex, Lies, and HIV Transmission."
132. Ibid.
133. Ibid.
134. Ibid.
135. Ibid.
136. Ibid.
137. Root-Bernstein, "Rethinking AIDS."
138. "Lifestyle," 9-10; Brody, "Lack of Evidence for Transmission of Human Immunodeficiency Virus," 384.
139 Hodgkinson, *AIDS: The Failure of Contemporary Science*, 125; *British Medical Journal* 304 (1992): 811.
140. F. Cline, "Anal Sex & AIDS: A Dissenting View," *Continuum* 4(4): 18 (November/December 1996).

141. Lifestyle," 9-10; Brody, "Lack of Evidence for Transmission of Human Immunodeficiency Virus," 384.

142. Hodgkinson, *AIDS: The Failure of Contemporary Science,,* 123-24; *Lancet* (14 February 1987): 345-49.

143. See previous note.

144. See note 142.

145. Jody Wells, "Modes of Transmission," *Continuum* 2(4): 17 (August/September 1994). See also Van Voorhis et al., "Detection of Human Immunodeficiency Virus Type 1 in Semen from Seropositive Men Using Culture and Polymerase Chain Reaction Deoxyribonucleic Acid Amplification Techniques," *Fertil Steril* 55 (1991): 588-94; Anderson et al., "Understanding the AIDS Pandemic," 20-26; Farber, "Fatal Distraction."

146. Ibid.

147. Farber, "Fatal Distraction."

148. Geisler, *AIDS: Origin, Spread,* 12. See also B. J. VanVoorhis et al., "Detection of Human Immunodeficiency Virus Type 1," *Fertil Steril* 55 (1991): 588-94; L. Gutler, "AIDS—Was Kann der Hausarzt tun?" *Therapie der Gegenwart* 128 (1989): 21-4; D. M. Barnes, "AIDS Virus Creates Lab Risk," *Science* 239 (1988): 348-49; S. H. Weiss et al., "Risk of Human Immunodeficiency Virus (HIV-1) Infection Among Laboratory Workers," *Science* 239 (1988): 68-71; D. Hamblen and G. Newton, "HIV and Surgeons," *British Medical Journal* 301 (1990): 1216-1217.

149. Cline, "Anal Sex and AIDS, " 18.

150. V. L. Koliadin, "Suppression of Non-Specific Immunity, Not T-Immunodeficiency is the Main Cause of Opportunistic Infections in AIDS," *Rethinking AIDS* Web site, http://www.virusmyth.com, April 1998.

151. Sean Currin, interview with the author, May 1996.

152. Lauritsen, "Psychological & Toxicological Causes of `AIDS,'"96-97.

153. Ibid.

154. Mark Garish Conlan, interview with the author, May 1996.

155. Wells, "Modes of Transmission."

156. Mark Garish Conlan, interview with the author, May 1996.

157. Hodgkinson, *AIDS: The Failure of Contemporary Science,* 125; and *British Medical Journal* (1992): 811.

158. Hodgkinson, *AIDS: The Failure of Contemporary Science,* 20.

159. Ibid.

160. Ibid., 15.

161. Ibid.

162. Paulo Freire, *Education for Critical Consciousness* (New York: Seabury Press, 1973), 138.

163. Hodgkinson, *AIDS: The Failure of Contemporary Science,* 15.

164. Ibid.

165. Ibid.

166. Ibid., 19.

167. Ibid.

168. Koliadin, "Suppression of Non-Specific Immunity."

169. Hodgkinson, *AIDS: The Failure of Contemporary Science,,* 125; *British Medical Journal* (1992): 811.

170. Lifestyle,"9-10; Brody, "Lack of Evidence for Transmission of Human Immunodeficiency Virus," 384. See also A. E. Albert et al., "Condom Use and Breakage Among Women in a Municipal Hospital Family Planning Clinic," *Contraception* 43 (1991): 167-76; P. Russel-Brown et al., "Comparison of Condom Breakage During Human Use With Performance in Laboratory Testing," *Contraception* 45 (1992): 429-437; R. F. Carey et al., "Effectiveness of Latex

Condoms as a Barrier to Human Immunodeficiency Virus-Sized Particles Under Conditions of Simulated Use," *Sexually Transmitted Disease* 19 (1992): 230-234.

171. Edward Luttwak, *Turbo-Captitalism: Winners and Losers in the Global Economy* (New York: HarperCollins, 1999), 61.

172. Ibid., 60.

173. Ibid., 47, 48, 50.

174. Ibid., 71.

175. Farber, "Fatal Distraction."

176. Ibid.

177. Ibid.

178. Hodgkinson, *AIDS: The Failure of Contemporary Science*, 217-218.

179. Ibid.

180. Brody, "Lack of Evidence for Transmission of Human Immunodeficiency Virus," 329-337; Symon, "How Risky Is Risky Sex?" 344-346.

181. Walter Abish, *Alphabetical Africa* (New York: New Directions, 1974), 53.

182. Christopher L. Miller, *Blank Darkness: Africanist Discourse in French* (Chicago: University of Chicago Press, 1985), 41.

183. Ibid., 42.

184. Ibid., 47.

185. R. Harrison-Chirimuuta, "Is AIDS African?" *Dissonance*, 22 May 1997.

186. R. Harrison-Chirimuuta, "AIDS from Africa: Western Science or Racist Mythology," in *Western Medicine as Contested Knowledge* , eds. A. Cunningham and B. Andrews (Manchester, England: Manchester University Press, 1997).

187. Ibid.

188. Richard Chirimuuta and Rosalind Chirimuuta, *AIDS, Africa and Racism* (London, Free Association Books, 1989), 133.

189. Ibid.

190. Ibid., 128.

191. Ibid.

192. Ibid.

193. Ibid.

194. C. L. Geshekter, "Myths of AIDS and Sex," *New African*, October 1994.

195. Ibid.

196. C. L. Geshekter, "Reappraising AIDS in Africa: Under Development & Racial Stereotypes," *Reappraising AIDS* (September/October 1997).

197. Harrison-Chirimuuta, "Is AIDS African?"

198. Ibid.

199. Ibid.

200. Chirimuuta and Chirimuuta, *AIDS, Africa and Racism*, 143-44.

201. Ibid., 143-144. See also I. C. Bygbjerg, "AIDS in a Danish Surgeon (Zaire, 1976)," *Lancet* (23 April 1983): 925; Shilts, *And the Band Played On*; letter from Dr. Bybjerg to Dr. Grote, 18 April 1988, my emphasis.

202. Shilts, *And the Band Played On*, 4-5.

203. Ibid., 4.

204. Ibid.

205. Ibid.

206. Ibid.

207. Ibid., 5.

208. Ibid., 7.

209. Ibid.

210. N. Ostrum, "The Tuskegee AZT Experiment," *New York Native*, 19 November 1996.

211. Ibid.

212. E. Papadopulus-Eleopulos et al., "AIDS in Africa: Distinguishing Fact and Fiction," *World Journal of Microbiology & Biotechnology* 11 (1995): 135.
213. C. Farber, "Out of Africa: Part One," *Spin*, March 1993.
214. Ibid.
215. Ibid.
216. Ibid.
217. Chirimuuta & Chirimuuta, *AIDS, Africa and Racism*, 55; and R. J. Biggar et al., "Regional Variation in Prevalence of Antibody Against Human T-Lymphotrophic Virus Types I and II in Kenya, East Africa," *International Journal of Cancer* 35 (1985): 763-67.
218. Ibid.
219. Steve Jonas in video *AIDS The Untold Story* (New York: Gary Null & Associates, 1994); and Biggar, "Regional Variation in Prevalence of Antibody."
220. Geshekter, "Reappraising AIDS in Africa: Under Development & Racial Stereotypes."
221. Chirimuuta and Chirimuuta, *AIDS, Africa and Racism*, 106; *Guardian*, 12 March 1987; "Acquired Immunodeficiency Syndrome (AIDS)—Update (United States of America)," *WHO Weekly Epidemiological Record* 6 , 6 February 1987: 29-36; "Situation in WHO European Region as of 31 December 1986,"*WHO Weekly Epidemiological Record* 17, 24 April 1987: 117-24.
222. See previous note. Also see Chirimuuta and Chirimuuta, *AIDS, Africa and Racism*, 59-60; I. Wendler et al., "Seroepidemiology of Human Immunodeficiency Virus in Africa," *British Medical Journal* 293 (27 September 1986): 782-85.
223. See previous note.
224. See note 222.
225. See note 222.
226. Farber, "Out of Africa," March 1993.
227. Geshekter, "Reappraising AIDS in Africa: Under Development & Racial Stereotypes."
228. Chirimuuta and Chirimuuta, *AIDS, Africa and Racism* , 124-25; P. Ochieng, "Africa Not to Blame for AIDS," *New African*, January 1987, 25.
229. Ibid.
230. Ibid.
231. Geshekter, "Reappraising AIDS in Africa: Under Development & Racial Stereotypes"; and "HIV Origin a Continuing Mystery: Green Monkey Theory Disputed," *Skin and Allergy News*, 28 January 1988.
232. Chirimuuta and Chirimuuta, *AIDS, Africa and Racism*, 133.
233. Ibid., 42.
234. Ibid., 58-59; S. F. Lyons et al., "Lack of Evidence of HTLV-III Endemnicity in Southern Africa," *New England Journal of Medicine* 312(19): 1257-58 (9 May 1985); and S. F. Lyons et al., "Sero-epidemiology of HTLV-III Antibody in Southern Africa," *South African Medical Journal* 67 (15 June 1985): 961-962.
235. Harrison-Chirimuuta, "Is AIDS African?"
236. Ibid.
237. Ibid.
238. Ibid.
239. Geisler, *AIDS: Origin, Spread*, 185; and SAPA-AP, "Aids Scare, Kampala," *Windhook Advertizer*, 2 February 1982.
240. Geshekter, "Myths of AIDS and Sex."
241. Ibid.
242. Neville Hodgkinson, inteview by the author, May 1996
243. Hodgkinson, *AIDS: The Failure of Contemporary Science*, 285.

244. Ibid., 291; and O. Kashala et al., "Infection with Human Immunodeficiency Virus Type I," *Journal of Infectious Diseases* 169 (February 1994): 296-304.
245. Ibid.
246. Ibid.
247. Ibid.
248. E. Papadopulos-Eleopul0s and V.F Turner, "Deconstructing AIDS," *Independent Monthly* October 1992, 50.
249. See Z. Bentwich, A. Kalinkovich, and Z. Weisman, "Immune Activation is a Dominant Factor in the Pathogenesis of African AIDS," *Immunology Today* 16(4): 187-91 (1995); F. Plummer et al., in *AIDS in Africa*, eds M. Essex et al. (no place of publication: Raven Press, 1994), 195-210; M. Laga et al., *AIDS* 5(suppl 1): S55-S63 (1991); J. Nahmias et al., *Israelis Journal of Medical Sciences* 29 (1993): 338-43; R. M. Hodes and H. Kloos et al., *New England Journal of Medicine* 319 (1988): 918-24; A. Annan et al., *Parasitology* 92 (1986): 209-17; P. Hengster et al., *International Journal of STD AIDS* 2 (1991): 180-84; T. C. Quinn et al., *JAMA* 257 (1987): 2617-21; G. Lamoureux et al., *Ann Inst Pasteur/Immunol* 138 (1987): 521-29; J. H. Perriens et al., *AIDS* 5(suppl 1): S127-S133 (1991); and U. Brinkmann and A. Brinkmann, *Trop Med Parasitol* 42 (1991): 204-13.
250. Ibid.
251. Neville Hodgkinson, interviewed by the author, 15 May 1996
252. Ibid.
253. Ibid.
254. Ibid.
255. Ibiid.
256. Ibid.
257. Ibid.
258. Papadopulos-Eleopulos et al., "AIDS in Africa: Distinguishing Fact and Fiction," 135-43.
259. Ostrum, "The Tuskegee AZT Experiment."
260. Kary Mullis, interview by the author, May 1996.
261. Papadopulos-Eleopulos and Turner, "Deconstructing AIDS," 50.
262. Ibid.
263. Farber, "Out of Africa."
264. Ibid.
265. Ibid.
266. Ibid.
267. *What If Everything You Thought You Knew About AIDS Was Wrong*, , 8.
268. Chirimuuta and Chirimuuta, *AIDS, Africa and Racism*, 151-52; and F. I. D Konotey-Ahulu, "AIDS in Africa: Misinformation and Disinformation," *Lancet* 25 July 1987, 206-07.
269. Joan Shenton, interview by the author, May 1996.
270. Ibid.
271. Ibid.
272. Philippe Krynen, "Looking from Inside," *Continuum* 5(2): 21 (Winter 1997/1998).
273. Ibid.
274. Ibid.
275. Ibid.
276. Ibid.
277. Manuel Castells, *The Rise of the Network Society* (London: Blackwell, 1996), 134.
278. Ibid., 135.
279. Geshekter, "Myths of AIDS and Sex."
280. Ibid. See also Hodgkinson, *AIDS: The Failure of Contemporary Science*, 314.

281. Peter Duesberg, interview by the author, Janurary 25, 1995.

282. Bacon, *Advancement of Learning*, 215.

283. Originally in the Sunday *Times* of London, October 1993. Reported by Thomas, Mullis, and Johnson, 22, my emphasis.

284. Ibid.

285. Farber, "Out of Africa." See also Papadopulos-Eleopulos et al., "AIDS in Africa: Distinguishing Fact and Fiction," 135-43.

286. Krynen, "Looking from Inside," 21.

287. Farber, "Out of Africa."

288. Ibid.

289. Ibid.

290. Ibid.

291. Hodgkinson, *AIDS: The Failure of Contemporary Science*, 301.

292. Farber, "Out of Africa."

293. Joan Shenton, interview by the author, May 1996.

294. Geshekter, "Reappraising AIDS", ww.virusmyth.net/aids/data/cgreappraising.htm; J. Deane, *SIDAfrique* 8/9 (1996): 29.

295. Shilts, *And the Band Played On*, 268.

296. Geshekter, "Reappraising AIDS"; "Commentary,"*The Economist* 7 September 1996, 38.

297. Quoted in Chirimuuta and Chirimuuta, *AIDS, Africa and Racism*, 162-63; A. J. Venter, "AIDS: Its Strategic Consequences in Black Africa," *International Defense Review* 4 (1988).

298. See previous note.

299. Shilts, *And the Band Played On*, 222.

300. Ibid, 223.

301. Ibid., 345.

302. Ibid.

303. Ibid., 346.

304. *7 Days*, October 18, 1986.

305. Chirimuuta and Chirimuuta, *AIDS, Africa and Racism*, 5-6.

306. Ibid.; D. Altman, *AIDS and the New Puritanism* (London and Sydney: Pluto Press, 1986), 38; J. W. Curran et al., "The Epidemiology of AIDS: Current Status and Future Prospects," *Science* 229 (27 September 1985):. 1352-57; *WHO Weekly Epidemiological Report* 47 (21 November 1986); *WHO Weekly Epidemiological Report* 3 (6 January 1987).

307. Geshekter, "Reappraising AIDS"; C. Chintu, *Lancet* 349 (1 March 1997): 649.

308. Chirimuuta and Chirimuuta, *AIDS, Africa and Racism*, 164-65; Sylvia. Federici, "White Doctors in Africa (on the trail of AIDS, or in the ;service of the empire?)," *Left Field* 14 September 1988.

309. Ibid.

310. Ibid.

311. Chirimuuta and Chirimuuta, *AIDS, Africa, and Racism*, 93; D. Zagury et al., "Immunization Against AIDS in Humans," *Nature* 326 (19 March 1987): 249-50; *African Times*, 10 April 1987.

312. Ostrum, "The Tuskegee AZT Experiment."

313. Hodgkinson, *AIDS: The Failure of Contemporary Science*, 301.

314. Ibid.

315. Ibid.

Medicine and Dream Time

1. Bacon, *New Organon*, 90.

2. Ibid., 91.

3. John Ehrenreich, "Introduction," in *The Cultural Crisis of Modern Medicine*, ed. John Ehrenreich (New York: Monthly Review Press, 1978), 12.
4. Fritjof Capra, *The Turning Point: Science, Society and the Rising Culture* (New York: Bantam Books, 1982), 265.
5. Ibid., 323.

PART TWO: EXPLORING ALTERNATIVE CAUSES OF AIDS

1. Celia Farber, "AIDS: Words from the Front," *Spin*, December 1994.
2. Dean Black, interview with the author in *AIDS, The Untold Story*.
3. Hodgkinson, "Conspiracy of Silence," Sunday *Times*, 3 April 1994.
4. Francis Bacon, *The New Organon*, 190.

Is the AIDS Syndrome New?

1. Rene Dubos, *Man Adapting* (New Haven: Yale University Press, 1965), 147-62.
2. Robert. S. Root-Bernstein, "AIDS and Kaposi's Sarcoma Pre-1979," *Lancet* I (1990): 969.
3. Farber, "Words From the Front."
4. Root-Bernstein, "AIDS and Kaposi's Sarcoma Pre-1979," quoting from H. P Katner and G. A. Pankey, "Evidence for a Euro-American Origino of Human Immunodeficiency Virus (HIV)," *J Nat Med Assn* 79 (1987): 1068-1072.
5. Centers for Disease Control, "Revision of the CDC Surveillance Case Definition for Acquired Immunodeficiency Syndrome," *JAMA* 256 (1987): 1143-54; K. E. Astrom, E. L. Mancall and E. P. Richardson Jr., "Progressive Multifocal Leuko-encephalopathy," *Brain* 81 (1958): 93; J. Fermaglich, J. M. Hardman and K. M. Earle, " Spontaneous Progressive Multifocal Leukoencephalopathy," *Neurology* 20 (1970):, 479-84.
6. T. Matthews, H. Wisotskey and J. Moosy, "Multiple Central Nervous System Infections in Progressive Multifocal Leukoencephalopathy," *Neurology* 26 (1976): 9-14; C. F. Bolton and B. Rozdilsky, "Primary Progressive Multifocal Leukoencephalopathy: A Case Report," *Neurology* 21 (1971): p. 72-77; L. P. Weiner, R.M. Herndon, O. Narayan, R.T. Johnson, K. Shah et al., "Isolation of Virus Related to SV40 from Patients with Progressive Multifocal Leukoencephalopathy," *New England Journal of Medicine* 286 (1972): 385-90; A. A. Faris and A. J. Martinez, "Primary Progressive Multifocal Leukoencephalopathy: A Central Nervous System Disease Caused by a Slow Virus." *Arch Neurol* 27 (1972) 357-60; D. Rockwell, F. L. Ruben, A. Winkelstein and H. Mendelow, "Absence of Immune Deficiencies in a Case of Progressive Multifocal Leukoencephalopathy," *American Journal of Medicine* 61 (1976): 433-36.
7. R. Root-Bernstein, "Non-HIV Immunosuppressive Factors in AIDS," *Res Immunol* 141 (1990): 818.
8. Bacon, *New Organon*, 164.
9. M. Verse, "Uber Einen Fall von Generale Sierter Blastomykose Beim Menschen," *Verch Deutsch Ges Path* 17 (1914): 275.
10. H. Rubin and M. L. Furclow, "Promising Results in Cryptococcal Meningitis," *Neurology* 8 (1958) 590.
11. M. J. Fitzpatrick and C. M. Poser, "The Management of Cryptococcal Meningitis," *Archives of Internal Medicine* 106 (1960): 151-274.
12. R. D. Diamond and J. E. Bennett, "Disseminated Cryptococcus in Man: Decreased Lymphocyte Transformation in Response to Cryptococcus Neoformans," *Journal of Infectious Disease* 127 (1973): 694-97; J.R. Graybill and R. H. Alford, "Cell-mediated Immunity in Cryptococcosis," *Cell Immunology* 14 (1973): 12-21; S. C. Schimpf and J. E. Bennett, "Abnormalities in Cell-Mediated Immunity in Patients with Cryptococcus Neoformans Infection," *J Allergy Clin*

Immunol 55 (1975): 430-41; W. T. Butler, D. W. Alling, A. Spikard et al., "Diagnostic and Prognostic Value of Clinical and Laboratory Findings in Cryptococcal Meningitis: A Follow-up Study of 40 Patients," *New England Journal of Medicine* 270 (1964): 59-67; A. Spickard, W. T. Butler, V. Andriole and J. P Utz, "The Improved Prognosis of Cryptococcal Meningitis with Amphotericin B Therapy," *Annals of Internal Medicine* 58 (1963): 66-83; R. D. Diamond and J. E. Bennett, "Prognostic Factors in Cryptococcal Meningitis," *Annals of Internal Medicine* 80 (1974): 176-81.

13. R. Duperval, P. E. Hermans, N. S. Brewer and G. D. Roberts, "Cryptococcus with Emphasis on the Signficance of Isolation of Cryptococcus Neoformans from the Respiratory Tract," *Chest* 72 (1977): 13-19.

14. M. L. Littman and J. E. Walter, "Cryptococcosis: Current Status," *American Journal of Medicine* 45 (1968): 922-32. (See also note 13, above.)

15. Centers for Disease Control, "Revisision of the CDC Surveillance Case Definition," *JAMA* 256 (1987): 1143-54.

16. T. W. Wong and N. E. Warner, "Cytomegalic Inclusion Disease in Adults,"*Arch Path* 74 (1962): 403-22.

17. Root-Bernstein, "Non-HIV Immunosuppressive," 821.

18. Ibid., see also 815-838; D. Huminer et al., "AIDS in the pre-AIDS Era," *Rev Infect Dis* 9 (1987): 1102-08; D. Huminer and S. D. Pitlik, "Further Evidence for the Existence of AIDS in the pre-AIDS Era," *Rev Infect Dis* 10 (1988): 1061; H. P. Katner and G. A. Pankey, "Evidence for a Euro-American Origin of Human Immunodeficiency Virus (HIV)," *Journal Nat Med Ass* 79 (1987): 1068-72; R. S. Root-Bernstein, "AIDS and Kaposi's," 969.

19. Root-Bernstein, "Non-HIV Immunosuppressive Factors," 815-838; M. Elvin-Lewis et al., "Systemic Chalmydial Infection Associated with Generalized Lymphedema and Lymphangiosarcoma," *Lymphology* 6 (1973): 113-21; M. H. Witte et al., "AIDS in 1968," *JAMA* 251 (1984): 2657; R. F. Garry et al., "Documentation of an AIDS Virus Infection in the United States in 1968," *JAMA* 260 (1988): 2085-87; S. S. Froland et al., "HIV-1 Infection in Norwegian Family Before 1970," *Lancet* I (1988) 1344-45; A. J. Nahmias et al., "Evidence for Human Infection with an HTLV VIII-LAV-Like Virus in Central Africa, 1959," *Lancet* I (1986): 1279-80: G. Dufoort et al., "No Clinical Signs 14 Years after HIV-2 Transmi:ssion via Blood Transfusion," *Lancet* II (1988): 510; G. Corbitt et al., "HIV Infection in Manchester, 1959," *Lancet* II (1990): 51.

20. See previous note.

21. Dubos, *Man Adapting*, 163.

22. Dubos, *Man Adapting*, 164.

How to Accurately Assess Immunodeficiency

1. "Diagnostic and Therapeutic Technology Assessment," *JAMA* 267 (21): 2952 (3 June 1992).

2. Michael A. DiSpezio, *The Science, Spread, and Therapy of HIV Disease* (New York: ATL Press, 1998), 72.

3. Farber, "AIDS: Words From the Front," 20.

4. Ibid.

5. Ibid.

6. Bacon, *New Organon*, 183.

7. Michael Ellner and T. Diferdino, "Why Are We Counting CD4 Lymphocytes Anyway? Is AIDS a Syndrome of Immuno-Deficiency or the Syndrome of a Preexsiting Psychosomatic Metabolic Imbalance and Toxification?" (position paper, HEAL, New York, 1995). One-time limited used granted to Gary Null.

8. Ibid.

9. Farber, "AIDS: Words from the Front."
10. Ibid.
11. HEAL spokesperson, interview by the author.
12. "Searching for Markers on the AIDS Trail," *Science* 258 (16 October 1992): 388.
13. T. R. Fleming and D. L. DeMets, "Surrogate End Points in Clinical Trials: Are We Being Misled?"*Annals of Internal Medicine* 125 (1996): 605-13; T. R. Fleming, "Surrogate Markers in AIDS and Cancer Trials," *Stat Med*, 13 (1994): 1423-35; M. A. Sande et al., "Antiretroviral Therapy for Adult HIV-infected Patients. Recommendations from a State-of-the-art Conference. National Institute of Allergy and Infectious Diseases State-of-the-art Panel on Anti—Retroviral Therapy for Adult HIV-infected Patients," *JAMA* 270 (1993): 2583-89; J. P. Aboulker and A. M. Swart, "Preliminary Analysis of the Concorde Trial. Concorde Coordinating Committee (Letter),"*Lancet* 341 (1993): 889-90.
14. C. R. Caulfield and B. Goldberg, *The Anarchist AIDS Medical Formulary: A Guide to Guerilla Immunology* (Berkeley, Calif.: North Atlantic Books, 1993), 95; A. Gianneti et al., "Direct Detection of HIV-1 RNA in Epidermal Langerhans Cells of HIV-Infected Patients,"*Journal of AIDS* 6 (1993): 329-33; E. Langhoff and W. A. Haseltine, "Infection of Accessory Dendritic Cells by Human Immunodeficiency Virus Type 1," Journal *of Investigative Dermatology* 99 (1992): 89S-94S; S. E. Mactonia et al., "Suppression of Immune Responses by Dendritic Cells Infected with HIV," *Immunology* 67 (1989): 285-89; C. A. Janeway, "The Case of the Missing CD4s," *Current Opinions in Biology* 2 (1992): 359-61.
15. Caulfield and Goldberg, *The Anarchist AIDS Medical Formulary:*, 42; T. Kingston, "The Concorde AZT Trial: Does It Fly?" *San Francisco Bay Times* 22 April 22 1993, 6.
16. "Searching for Markers," 390.
17. Ibid.
18. "Italian Study Undermines AZT," *Continuum* 4(4): 3 (November/December 1996).

AIDS as Stress Response

1. The National Institute of Allergy and Infectious Diseases, *The Edge of Discovery* (Washington, D.C.: U.S. Department of Health and Human Services, Public Health Service National Institutes of Health, June 1988), 13.
2. Dubos, *Man Adapting*, 169.
3. Ibid.
4. Howard Waitzkin, *The Second Sickness: Contradictions of Capitalist Health Care* (New York: Free Press, 1983), 45.
5. Ibid., my emphasis.
6. A. Hassig et al., "Suggestions for the Prevention of AIDS in HIV Carriers," *Swiss Journal of Holistic Medicine* (April 1993): 188-92.
7. J. Lahdevirta et al., "Elevated Levels of Circulating Cachetin/Tumor Necrosis Factor in Patients with Acquired Immunodeficiency Syndrome," *American Journal of Medicine* 85 (1988): 289; S. Kobayashi et al., "Serum Level of TNF-Alpha in HIV Infected Individuals," *AIDS* 4 (1990): 169; D. Scott-Alagra et al., "Serum Levels of IL-2, IL-1 alpha, TNF-alpha, and Soluble Receptor of IL-2 in HIV-1 Infected Patients," *AIDS Res and Human Retrov* 7 (1991): 375.
8. Scott-Alagara et al., "Serum Levels of IL-2."
9. Ibid.
10. T. Matsuyama et al., "Cytokines and HIV Infection. Is AIDS a Tumor Necrosis Factor Disease?" *AIDS* 5 (1991): 1405; H. Besedovsky and E. Sorkin, "Network of Immunoneuroendocrine Interactions." *Clinical Exp. Immunol* 27 (1977): 1; C. Craddock, "Corticosteroid Induced Lymphopenia: Immunodepression and Body Defense," *Annals of Internal Medicine* 88 (1978): 564.

11. Hassig et al., "Suggestions for the Prevention of AIDS," 188-192.
12. Peter Duesberg, interview by the author, 24 January 1995; G. Vithoulkas, "The Hypothesis about AIDS," in *A New Model of Health and Disease* (Berkeley, Calif.: North Atlantic Books, 1991), 154-88.
13. J. Josephs et al., "Perceived Risk of AIDS: Assessing the Behavior and Psychological Consequences in a Cohort of Gay Men," *Journal of Applied Social Psychology* 17 (1987): 231-250.
14. R. Root-Bernstein, "HIV and Immunosuppressive Cofactors in AIDS," *EOS Revista Immunologica ed Immunofarmacologia* 12 (1992): 4.
15. R. Beach et al., "Specific Nutrient Abnormalities in Asymptomatic HIV-1 Infection," *AIDS* 6, 1992, 701-08.

The Role of Illicit Drugs

1. Shilts, *And the Band Played On.*
2. CDC National AIDS Clearinghouse Custom Search Service and AIDS Clinical Trials Information Service, 18 January 1996.
3. J. Lauritsen, "Psychological and Toxicological Causes of `AIDS,'" in *The AIDS Cult* , eds. J. Lauritsen and I. Young (Provincetown: Asklepios, 1997), 86.
4. Ibid.
5. Ibid.
6. The HEAL Trust, "Deconstructing AIDS," London; Duesberg, "AIDS Acquired by Drug Consumption and Other Noncontagious Risk Factors," *Pharmacol Ther* 55 (3): 201-07 (1992); J. Lauritsen and H. Wilson, *Death Rush: Poppers & AIDS* (New York: Pagan Press, 1986), 12; W. Lange et al., "Nitrite Inhalants: Patterns of Abuse in Baltimore and Washington D.C.," *American Journal of Drug and Alcohol Abuse* 14 (1988): 29-40; H. Jaffe, *Annals of Internal Medicine* 99 (1983,): 145-51.
7. G. R. Newell et al., "Volatile Nitrites. Use and Adverse Effects Related to the Current Epidemic of the Acquired Immune Deficiency Syndrome," *American Journal of Medicine* 78 (1985): 811-16.
8. R. S. Root-Bernstein, "Do We Know the Cause(s) of AIDS?" *Perspectives in Biology and Medicine* 33: 480-500 (Summer 1990).
9. *Continuum* 4(2): 17 (July/August 1996).
10. Ibid.
11. Ibid.
12. The HEAL Trust, "Deconstructing AIDS"; I. Quinto, "The Mutagenicity of Alkylnitrites in the Salmonella Test (trans. from Italian)," *Bollentino Societa Italiana Biologia Sperimentale* 56 (1980):, 8816-20; J. Osterloh et al., "Butyl Nitrite Transformation in Vitro Chemical Nitrosation Reactions and Mutagenesis," *Journal of Analytical Toxicology* (July/August 1984): 164-69; G. Newell et al., "Toxicity Immunosuppressive Effects and Carcinogenic Potential of Volatile Nitrites: Possible Relationship to Kaposi's Sarcoma," *Pharmacotherapy* (September 1984): 235-36; G. Newell et al., "Volatile Nitrites Use and Adverse," 811-16; K. Jorgensen et al., "Amyl Nitrite and Kaposi's Sarcoma in Homosexual Men," *New England Journal of Medicine*, 307 (1982): 893-94; G.Brambilla, "Genotoxic Effects of Drug-Nitrite Interaction Products: Evidence for the Need of Risk Assessment," *Pharmcol Res Common* 17(A) (1985): 307-21; K. Haverkos et al., "Disease Manifestation among Homosexual Men with Acquired Immunodeficiency Syndrome: A Possible Role of Nitrites in Kaposi's Sarcoma," *Sexually Transmitted Diseases* 12 (1985): 203-08; E. Dax et al., "Effects of Nitrites on the Immune System of Humans in Health Hazards of Nitrite Inhalants," *NIDA Monograph Series* 83, 1988.
13. See the previous note.

14. R.S. Root-Bernstein, "Do We Know the Cause(s) of AIDS?" 480-500.
15. The HEAL Trust, "Do We Know the Causes of AIDS?"; E. Watson et al., "Use of Amyl Nitrite May be Linked in Current Epidemic of Immunodeficiency Syndrome," Unpublished Letter to *JAMA*, in J. Lauritsen and H. Wilson, *Death Rush*, 48; J. Neefe et al., "Daily Amyl Nitrite Inhalation Decreases Mouse Splenocyte Response to Concanavalin A," *Federation Proceedings* 42 (4): 949 (1983); E. Lotzova et al., "Depression of Murine Natural Killer Cell Cytotoxicity by Isobutyl Nitrite," *Cancer Immunology Imunnotherapy* 17 (1984): 130-34; P. Gangadharam et al., "Immunosuppressive Action of Isobutyl Nitrite.," International Congress on Immunopharmacology, Florence, Italy, May 1985; J. Ortiz et al., "The Effect of Amyl Nitrite on T-cell Function in Mice," American Public Health Association Convention, November 1985; I. Soderber and J. Barnett, "Exposure to Inhaled Isobutyl Nitrite Reduces T-cell Blastogenesis and Antibody Responsiveness,"*Fundamental and Applied Toxicology* 17 (1991): 821-24; L. Soderberg et al., "Inhaled Isobutyl Nitrite Impairs T-cell Reactivity," *Advances in Experimental Biology* 288 (1992): 265-69; D. Lynch et al., "Subchronic Inhalation Toxicity of Isobutyl Nitrite in BALB/C Mice," *Journal of Toxicology and Environmental Health* (1985): 823-33; E. Papadopulos-Eleopulos, "Reappraisal of AIDS: Is the Oxidation Caused by the Risk Factors the Primary Cause?" *Medical Hypotheses* 25 (3): 151-62 (!988); E. Singelmann, et al., "Erythrocyte Membrane Alterations as the Basis of Chlorate Toxicity," *Toxicology* 30 (1984): 135; B. Meloche et al., "S-Nitrosyl Glutathione-Mediated Hepatocyte Cytotoxicity," *Xenobiotica* 23 (8):, 863-871 (1993).
16. See previous note.
17. See note 15.
18. T. H. Haley, "Review of the Physiological Effects of Amyl, Butyl, and Isobutyl Nitrites," *Clinical Toxicology* (1980): 317-19.
19. Ibid.
20. J. Lauritsen, "NIDA Meeting Calls for Research into the Poppers-Kaposi's Sarcoma Connection," *New York Native* 13 June 1994; Haley, "Review of Physiological Effects of Amly," 317-319; I. Quinto, "The Mutagenicity of Alkylnitrites in the Salmonella Test (Italian)," *Bolletino Societa Italiana Biologia Sperimentale* 56 (1980): 816-820.
21. Lauritsen, "Psychological and Toxicological Causes," 87-88.
22. Ibid.
23. E. Papadopulos-Eleopulos et al., "Kaposi's Sarcoma and HIV," *Medical Hypotheses* 39 (1992): 22-29.
24. Newell, "Volatile Nitrites," 811-16. [See also Hodgkinson, *AIDS: The Failure of Contemporary Science* , 84-5; Root-Bernstein, "Non-HIV Immunosuppressive Factors," 815-38; ; J. J. Goedert et al., "A Prospective Study of Human Immunodeficiency Virus Type I Infection and the Development of AIDS in Subjects with Hemophilia," *New England Journal of Medicine* 321 (1989): 1141-1148; H. W. Jaffe et al., "National Case-Control Study of Kaposi's Sarcoma and Pneumocystis Carinii Pneumonia in Homosexual Men.—Epidemiological Results," *Annals of Internal Medicine* 99 (1983): 145-51; M. Marmor et al., "Risk Factors for Kaposi's Sarcoma in Homosexual Men," *Lancet* I (1962): 1083-87; H. W. Haverkos et al., "Disease Manifestation Among Homosexual Men with Acquired Immunodeficiency Syndrome: A Possible Role of Nitrites in Kaposi's Sarcoma," *Sexually Transmitted Disease* 12 (1985): 203-08.
25. See Lauritsen, "NIDA Meeting," my emphasis.
26. Ibid.
27. Ibid.
28. Ibid.

29. Tom Bethel, interview by the author, 14 February 1995.
30. J. Lauritsen, *Poison By Prescription: The AZT Story* (New York: Asklepios, 1990), 40; J. Lauritsen and H. Wilson, *Death Rush*:
31. The HEAL Trust, "Deconstructing AIDS"; Duesberg, "AIDS Acquired by Drug Consumption," 201-07; Institute of Medicine, *Confronting AIDS* (Washington, D.C.: National Academy Press, 1988); Centers for Disease Control, *HIV/AIDS Surveillance*, Year-end ed. (Atlanta, Ga: U.S. Department of Health and Human Services, 1992b); R. M. Selik et al., "Impact of the 1980 Revision of the Case Definition of Acquired Immune Deficiency Syndrome in the United States," *Journal of AIDS* 3 (1990): 73-82.
32. Lauritsen, *Poison By Prescription*, 179.
33. Duesberg, "AIDS: Non-Infectious Deficiencies," 5-11.
34. Hodgkinson, *AIDS: The Failure of Contemporary Science*, 161; *British Medical Journal* 301, 1362-65.
35. Bacon, *New Organon*, 179.
36. Ibid.
37. Bryan Ellison, interview by the author, in *The Pain, Profit and Politics of AIDS*.
38. Ibid.
39. J. Valnet, *The Practice of Aromatherapy* (Rochester, Vt.: Healing Arts Press, 1980), 53.
40. *The Physician's Desk Reference*, Medical Economics Data section 47 (1993): 829-35.
41. A. Cantwell Jr., "AIDS Is Not African—Say Scientists," *New African*, October 1994, 12.
42. Richard Strohman, interview by the author, in videotape *The Pain, Profit and Politics of AIDS*.
43. Ibid.
44. Paul Philpott, interview by the author, 13 February 1995, my emphasis.
45. Ibid.
46. W. Geisler, *AIDS: Origin, Spread, and Healing* (Koln: Bipawo Verlag, 1994), 8.
47. Ibid, 10; K. Welch et al., "Autopsy Findings in the Acquired Immune Deficiency Syndrome," *JAMA* 252 (1984): 1152-59; A. E. Pitchenik et al., "Tuberculosis, Atypical Mycobacteriosis, and the Acquired Immunodeficiency Syndrome Among Haitian and Non-Haitian Patients in South Florida," *Annals of Internal Medicine* 101 (1984): 641-45; R. G. Hendrickse et al., "Aflatoxins and Heroin," *British Medical Journal* 299 (1989): 492-93; J. J. Goedert et al., "Amyl Nitrite May Alter T Lymphocytes in Homosexual Men," *Lancet* 1 (1982): 412-16.
48. Peter Duesberg, *Inventing the AIDS Virus* (Washington, D.C.: Regenery Publishing, 1996), 217.
49. Hodgkinson, *AIDS: The Failure of Contemporary Science*, 62.
50. Andrew Shoft, inteview with the author, May 1996.
51. John Lauritsen, "Gallo Admits . . . We Have Never Found HIV DNA in Kaposi Sarcoma," *Continuum* 2 (3): 4 (June/July 1994).
52. Ibid.
53. The HEAL Trust, "Deconstructing AIDS"; R. M. Selik et al., "Impact of the 1980 Revision,"73-82; Stoneburner et al., *Science* 242 (1988): 916; Duesberg, "AIDS Acquired by Drug Consumption," 201-07.
54. Gina Kolata, "New Picture of Who Will Get AIDS Is Dominated by Addicts," *New York Times*, 28 February 1995, C3.
55. Hodgkinson, *AIDS: The Failure of Contemporary Science*, 155; R. Pillai et al., *Archives of Toxicology* 65 (1991): 609-17.
56. Tim Haring, interview by the author, 15 February 1995.
57. Duesberg, *Inventing the AIDS Virus*, 260-263; Duesberg, "AIDS Aquired by Drug Consumption and Other Noncontagious Risk Factors" *Pharmacology and*

Therapeutics, 55 (1992): 201-277; S. V. Meddis, "Heroin Use Said to be Near Crisis Level," *USA Today*, 25 May 1994; Gettman, "Heroin Returning to Center Stage," *High Times*, December 1994, 23; L. P. Finnegan et al., "Perinatal Exposure to Cocaine: Human Studies," in *Cocaine: Pharmacology, Physiology, and Clinical Studies*, eds. J. M. Lakowski et al. (Boca Raton, Fl: CRC Press, 1992), 391-409; B. R. Edlin et al., "Intersecting Epidemics—Crack Cocaine Use and HIV Infection Among Inner-City Young Adults," *New England Journal of Medicine*, 331 (1994): 1422-27; G. R. Newell et al., "Nitrite Inhalants: Historical Perspective," in H. W Haverkos and J.A. Dougherty, *Health Hazards of Nitrite Inhalants*, NIDA Research Monograph 83 (Rockville, Md.: National Institute on Drug Abuse, 1988); J. Lauritsen, *The AIDS War: Propaganda, Profiteering, and Genocide from the Medical-Industrial Complex* (New York: Asklepios, 1993), 104; Lauritsen and Wilson, *Death Rush*, 6, 15; R. S. Root-Bernstein, "Do We Know the Cause of AIDS?" 484; H. Coulter, *AIDS and Syphilis: The Hidden Link* (Berkeley, Calif.: North Atlantic Books, 1987), 47.
58. Besides the references in the previous note, see also Geisler, *AIDS: Origin, Spread*, 42; C. Sterk, "Cocaine and HIV Seropositivity," *Lancet* 1 (1988): 1052-53; Ro, in "New York Ist Jedes Hundertste Schwangere Madchen HIV-infiziert," *Arzte Zeitung*, 1992 24(suppl): 3 (July 1992), referring to W. Wilkinson (Kings County Hospital, Brooklyn, N.Y.), VIII International Conference on AIDS, Amsterdam, 1992.
59. Bob Lederer, interview by the author, in *The Pain, Profit and Politics of AIDS*.
60. Duesberg, *Inventing the AIDS Virus*, 260-263
61. See note 57 above.
62. Ibid.
63. C. Thomas, "But What About?" *Rethinking AIDS* 1 (5), 1993 .
64. Michael Verney Elliot, interview by the author, no date.
65. Ibid.
66. Ibid.
67. Duesberg, *Inventing the AIDS Virus*, 260-63.
68. Ibid.
69. J. Lauritsen, "Talk on AZT," Alternative AIDS Symposium, Amsterdam, May 1992.
70. Ibid.
71. Duesberg, *Inventing the AIDS Virus*, 260-63.
72. Tim Haring, interview by the author, 15 February 1995.
73. Duesberg, *Inventing the AIDS Virus*, 260-63.
74. E. L. Snead, *Some Call It AIDS . . . I Call It Murder!* (San Antonio: AUM Publications, 1992), 87.
75. Hodgkinson, *AIDS: The Failure of Contemporary Science*, 70-71.
76. Ibid.
77. Ibid.
78. Ibid.
79. Ibid.
80. Hodgkinson, *AIDS: The Failure of Contemporary Science*, 94.
81. Ibid.
82. Ibid.
83. Geisler, AIDS: Origin, Spread, 42.
84. Ibid, 42; G. H. Friedland et al., "Intravenous Drug Abusers and the Acquired Immunodeficiency Syndrome (AIDS): Demographic, Drug Use, and Needle Sharing Patterns," *Archives of Internal Medicine* 145 (1985): 1413-17; M. Marmor et al., "Risk Factors for Infection with Human Immunodeficiency Virus Among Intravenous Drug Abusers in New York City," *AIDS* 1 (1987): 39-44.
85. See Hodgkinson, *AIDS: The Failure of Contemporary Science*,.

86. "Clean Needles Don't Prevent HIV," *Rethinking AIDS* Web site, http://www.virusmyth.com, May 1998; *American Journal of Epidemiology* 15 December 1997: 3.

87. Ibid.

88. Hodgkinson, *AIDS: The Failure of Contemporary Science,*, 81.

89. P. Duesberg and D. Rasnick, "The Drug-AIDS Hypothesis," *Continuum* 4(5) (suppl.): 20 (February-March 1997).

90. Ibid.

91. Harlon L. Dalton, "AIDS in Blackface," in *The AIDS Reader: Social, Political, and Ethical Issues*, ed. Nancy F. MacKenzie (New York: Meridian, 1991), 126

92. Ibid.

93. Ibid, 124.

94. Ibid.

95. Hodgkinson, *AIDS: The Failure of Contemporary Science,*, 205.

96. Ibid.

97. Duesberg and Rasnick, "The Drug-AIDS Hypothesis," 20.

98. Hodgkinson, *AIDS: The Failure of Contemporary Science*, 205.

99. Ibid.

100. Duesberg and Rasnick, "The Drug-AIDS Hypothesis," 20.

101. ibid.

Psychological Assessment: Prof. Casper G. Schmidt

1. Laurence Foss and Kenneth Rothenberg, T*he Second Medical Revolution: From Biomedicine to Infomedicine* (Boston: Shambhala, 1987), 118-19, 121.

2. Ibid.

3. Kathyrn Anastos and Carola Marte, "Women—The Missing Persons in the AIDS Epidemic," in *The AIDS Reader*, 193.

4. Ibid, 195.

5. Ibid, 196.

6. I. Young, "The Psychohistorical Origins of AIDS: An Interview with Casper Schmidt," in *The AIDS Cult*, eds. J. Lauristen and I. Young, (Provincetown: Asklepios, 1997), 133-34.

7. Ibid.

8. Ibid, 140-41.

9. C.G. Schmidt, "The Group-Fantasy Origins of AIDS," in *The AIDS Cult*, 27-29.

10. James Feast, "Mass Hysteria," in *Magill's Survey of Social Science: Sociology* (Pasadena: Salem Press, 1994).

11. Schmidt, "The Group-Fantasy Origin of AIDS," 64.

12. Feast.

13. Feast.

14. Michael Baumgartner, "To AIDS-Activism—Thanks for Nothing!" *Continuum* 4(2):24-26 (July/August 1996).

15. Ibid.

16. Schmidt, "The Group-Fantasy Origin of AIDS,"27-29.

17. Ibid.

18. Ibid.

19. Ibid.

20. Ibid.

21. Baumgartner, "To AIDS-Activism,"24-26.

22. Ibid.

23. Schmidt, , "The Group-Fantasy Origin of AIDS,"27-29.

24. Baumgartner, "To AIDS-Activism," 24-26.

25. I Young, "The Psychohistorical Origins of AIDS," 140-141.

26. Ibid.
27. Ibid.
28. Ibid.
29. Baumgartner, "To AIDS-Activism,"24-26.
30. Ibid.
31. Ibid.
32. Michael Ellner, "Protective Stupidity: Epidemic Hysteria, Mass Hypnosis and Escaping from the AIDS Zone," *Continuum* 5(2): 47 (Winter 1997/1998).
33. Ibid.
34. Ibid.
35. G. N. Hazelhurst, "Letters from Hiroshima," in *The AIDS Cult*, 117-18.
36. Ellner, "Protective Stupidity," 47.
37. Hazel Henderson, *The Politics of the Solar Age* (Garden City, N.Y."The Psychohistorical Origins of AIDS: : Anchor Press, 1981), 42.
38. Ellner, "Protective Stupidity," 48-49.
39. David Mertz, "Sex Wars: The New Left's AIDS-Related Scientism," *Rethinking AIDS* We bsite, http://www.virusmyth.come, May 1998.
40. Ibid.
41. I. Young, "The AIDS Cult and its Seroconverts," *Continuum* 4 (5).
42. Ibid.
43. Baumgartner, "To AIDS-Activism," 24-26.
44. Young, "The AIDS Cult and Its Seroconverts," 25.
45. Schmidt, "The Group-Fantasy Origin of AIDS," 32.
46. Ibid.
47. Bacon, *New Organon*, 180.
48. Young, "The AIDS Cult and Its Seroconverts," 24.
49. Ibid.
50. Ibid.
51. Ibid.
52. Ibid.
53. Ibid., and previous paragraph.
54. Ibid.
55. Ellner, "Protective Stupitidy,"48.
56. Ibid.
57. Schmidt, "The Group-Fantasy Origin of AIDS," 64.
58. Quoted in John Strausbaugh, "The AIDS Heretics," *New York Press*, 8-14 March 2000, 12.
59. Fred Cline, "The Fear of Losing HIV," *Sumeria* Web site, http://www.livelinks.com/sumeria, 21 May 1998.
60. Ibid.
61. Ibid. I have shifted the order of one sentence.
62. Feast.
63. Simon Watney, Introduction," in *Taking Liberties*, eds. Erica Carter and Simon Watney, (London: Serpent's Tail, 1989), 20.
64. Ibid.
65. Baumgartner, "To AIDS-Activsim," 24-26.
66. Ibid.
67. Hodgkinson, *AIDS: The Failure of Contemporary Science*, 387.
68. Ibid.
69. Ibid.
70. Irvine Welsh, *Trainspotting* (New York: W. W. Norton, 1996), 78.
71. Ellner, "Protective Stupidity," 48; Redotz, "Considering the Psycho-social Aspects of AIDS," *Mi Hosp Jnl*, 22(8) (August, 1986).

72. Cass Mann, "Deadly Counsels: The Necrophiliacs of 'AIDS'," in *The AIDS Cult*, 157; *Positively Healthy News* 2 (March 1989).
73. Ibid.
74. Mann, "Deadly Counsels," 158.
75. Ibid.
76. L. Cooper and M. Walker, "Whose Hysteria?" *Continuum* 5(1): 4-5 (Autumn 1997).
77. Ibid.
78. Baumgartner, "To AIDS-Activism," 24-26.
79. Ibid.
80. Ibid.
81. J. M. Villette et al., "Circadian Variations in Plasma Levels of Hypophyseal Adreno-cortical and Testicular Hormones in Men Infected with Human Immunodeficiency Virus," *Journal of Clinical Endocrinology and Metabolism* 70(3): 572 (1990).
82. Ibid.
83. Ibid.
84. T.L. Wisniewski et al., "The Relationship of Serum DHEA-S and Cortisol Levels to Measures of Immune Function in Human Immunodeficiency Virus-Related Illness," *American Journal of Medical Sciences* 305(2): 79 (February 1993).
85. Ibid.
86. A. Hassig, L, Wen-Xi and K. Stampfli, "Reflection on the Pathogenesis and Prevention of AIDS," *Swiss Journal of Holistic Medicine* (October 1994).
87. Ibid.
88. Ibid.
89. Ibid; M. Clerici and G. M. Shearer, "A TH1 to TH2 Switch Is a Critical Step in the Etiology of HIV Infection," *Immunology Today* 14: 107-11; R. A. Daynes et al., "Regulation of Murine Lymphokine Production in Vivo III: The Lymphoid Tissue Microenvironment Exerts Regulatory Influences Over T Helper Cell Function," *Journal of Exp.Medicine* 171 (1990): 979-96; B. F. Haynes and A. S. Fauci, "The Different Effect of In Vivo Hydrocortisone on the Kinetics of Subpopulations of Human Peripheral Blood Thymus-Derived Lymphocytes," *Journal of Clinical Investigations* 6 (1): 703-07 (1978).
90. F. de Fries, "AIDS Prophylaxis for HIV-Antibody-Test-Positive Persons (Recommendations from the Work of Alfred Hassig, K. Stampfli, Liang Wen-Xi, and Peter Schleicher, Munich)," 1996 (not yet published).
91. Ibid.
92. Ibid.
93. Young, "Psychohistorical Origins," 135-137.
94. Ibid.
95. "Prejudice Drives HIV Test Results," *Continuum* 4(3):2 (September/October 1996).
96. Ibid.
97. Ibid.
98. "Suicide Pact: San Francisco Anti-AIDS Activists Fred Cline Comments on Reports of Gay Euthanasia," *Continuum* 4(3): 7 (September/October 1996).
99. Ibid.
100. Nick Siano, interview by the author, in *AIDS, The Untold Story*.
101. Ibid; Alfred Hassig, interview by the author, in *AIDS, The Untold Story*; G. N. Hazlehurst, "Lessons from Hiroshima," open letter circulated by HEAL, January 1994, 3-4.
102. J. Lauritsen, "The Epidemiology of Fear," *New York Native*, 1 August 1988.
103. Hazelhurst, "Letters from Hiroshima," 103-04.
104. Baumgartner, "To AIDS-Activism," 24-26.
105. Ibid.

106. Ibid.
107. Ibid.
108. Ibid.
109. Ibid.
110. Ibid.
111. Cindy Patton, "The AIDS Industry: Construction of `Victims,' `Volunteers,' and `Experts," in *Taking Liberties*, 118.
112. Patton, 119.

Microbial Activity and Sexually Transmitted Disease (STD)

1. The HEAL Trust, "Deconstructing AIDS."
2. Root-Bernstein, "Multiple, Concurrent Infections and AIDS," in *Rethinking AIDS* (New York: The Free Press, 1993).
3. The HEAL Trust, "Deconstructing AIDS"; The Los Angeles Health Department Epidemiological Studies, Montreal AIDS Conference (personal communication with J. McKenna, T.B.M. Associates, Berkeley, Calif.); J. Potteat et al., "Serological Markers as Indicators of Sexual Orientation in AIDS Virus Infected Men," *JAMA* 256 (6): 712 (1986); C. Joyce, "Syphilis Increase Fuels AIDS Fears in U.S.," *New Scientist* , 29 September 1990; J. McKenna et al., "Unmasking AIDS: Chemical Immunosuppression and Seronegative Syphilis," *Medical Hypotheses* 21 (1986): 421-30; and material reviewed in Virella, Fudenberg, *Human Immunologic Disorders* (1992), 91-124.
4. Nina Ostrum, interview by the author, 14 February 1995.
5. Ibid.
6. Marjorie Siebert, interview by the author, in *The Pain, Profit and Politics of AIDS*.
7. N. Ostrum, "Real Thing: Dr. Konstance Knox Explains Why HHV-6 May Be the Key to Dealing with AIDS," *New York Native*, 15 April 1996
8. Ibid.
9. Lauritsen, "The Epidemiology of Fear."
10. Paul Philpott, interview by the author, 13 February 1995.
11. Root-Bernstein, "Multiple, Concurrent Infections," 149; M. Rogers et al., "National Case-Control Study of Kaposi's Sarcoma and PCP in Homosexual Men: Part 2, Laboratory Results," *Annals of Internal Medicine* 99 (1983): 151-58.
12. Dr. Christopher Calapai, interview by the author in *The Pain, Profit and Politics of AIDS*. (See also Paul Philpott, interview by the author, 13 February 1995.)
13. Snead, *Some Call It AIDS*, 143.
14. R. R. Watson, "Resistance to Intestinal Parasites During Murine AIDS: Role of Alcohol and Nutrition in Immune Dysfunction," *Parasitology* 107 (suppl): S69-74 (1993).
15. Dr. Joan Priestley, interview by the author, in *The Pain, Profit, and Politics of AIDS*.
16. Bacon, *New Organon*, 186.
17. Ibid.
18. Ibid.

Multifactorial Etiology: A Closer Look

1. R. S. Root-Bernstein, "Do We Know the Cause(s)," 480-500.
2. Ibid.
3. V. L. Koliadin, "Destruction of Normal Resident Microflora as the Main Cause of AIDS," *Rethinking AIDS* Web site, http://www.virusmyth.com, August 1996.
4. Ibid.
5. V. Koliadin, "Suppression of Non-Specific Immunity, Not T-Immunodeficiency is the Main Cause of Opportunistic Infections in AIDS," *Rethinking AIDS* Web site, http://www.virusmyth.com, April 1998.

6. Ibid.
7. Ibid.
8. G. T. Steward, "The Epidemiology and Transmission of AIDS: A Hypothesis Linking Behavioral and Biological Determinants of Time, Person and Place," *Genetica* 95 (1995) 173-83.
9. Lauritsen, "Talk on AZT."
10. S. C. Byrnes, "Benzene, Lubricants and AIDS," *Explore* 8(1) (1997).
11. Ibid.
12. Ibid.
13. Ibid.
14. Steward, "The Epidemiology and Transmission," 173-183; P. Ma and D. Armstrong, eds., *AIDS and Infections of Homosexual Men*, 2nd ed. (Boston: Butterworths, 1989); G.T. Stewart, *Times*, 24 January 1990; S. S. Witkin and J. Sonnabend, "Immune Response to Spermatozoa in Homosexual Men," *Fertil Sterilisation* 39 (1983):, 337-42; G. M. Mavlight et al., "Chronic Immunological Stimulation by Sperm Alloantingens," *JAMA* 251 (1984): 237-41; A. R. Moss et al., "Risk Factors for AIDS and Seropositivity in Homosexual Men," *American Journal of Epidemiology* 125 (1987): 1035-47; W.W. Winkelstein et al., "The San Francisco Men's Study: Continued Decline in HIV Seroconversion Rates Among Homosexual-Bisexual Men," *American Journal of Public Health* 78 (1988): 14724; J. Sonnabend, "Facts and Speculation About the Cause of AIDS," *AIDS Forum* 2 (1989): 3-12; Root-Bernstein, *Rethinking AIDS*; R. Schoental, "AIDS and Neoplasias Associated with AIDS," *International Journal of Environmental Studies* 31 (1988): 269-78; G. R. Newell et al., "Volatile Nitrites: Use and Adverse Effects," 811-816; S. S. Mirvish and H. W. Haverkos, "Butyl Nitrite in the Induction of Kaposi's Sarcoma in AIDS," *New England Journal of Medicine*, 317 (1987): 1103; J. R. Vandenroncke and V. R. A. M. Pardoel, "Autopsy of Epidemiological Methods: The Cause of Poppers in the Early Epidemiology of AIDS," *American Journal of Epidemiology*, 129 (1989): 455-57; V. Beral et al., "Kaposi's Sarcoma and Sexual Practice Associated with Faecal Contact in Homosexual or Bisexual Men with AIDS," *Lancet* 339 (1990,): 632-36.
15. Steward, "The Epidemiology and Transmission."
16. R. S. Root-Bernstein, "Non-HIV Immunosuppressive Factors in AIDS: A Multifactorial, Synergistic Theory of AIDS Aetiology," *Res Immunol* 141 (1990): 815-38.
17. Root-Bernstein, "Non-HIV Immunosuppressive Factors in AIDS,": H. L. Kazal et al., "The Gay Bowel Syndrome: Clinico-Pathological Correlation in 260 Cases," *Ann Clin Lab Science* 6 (1976): 184-92.
18. Steward, "The Epidemiology and Transmission of AIDS,"173-83; L. E. Adams et al., "Sperm and Seminal Plasma Antibodies in AIDS," *Clin Immunolo Immunopathol* 46 (1988) 442; H. W. Sheppard and M. S. Ascher, "AIDS and Programmed Cell Death," *Immunology Today* 12, 1990, 423; and Ma and Armstrong, eds. for *AIDS and Infection in Homosexual Men*, 1989.
19. See previous note.
20. Root-Bernstein, "Do We Know the Cause(s)," 480-500.
21. Ibid
22. Root-Bernstein, "Non-HIV Immunosuppressive Factors."
23. Ibid.; N. Sohn et al., "Social Injuries of the Rectum," *American Journal of Surgery* 134 (1977): 611-12; J. J. Goedert et al., "A Prospective Study of Human Immunodeficiency Virus Type I Infection and the Development of AIDS in Subjects with Hemophilia," *New England Journal of Medicine* 321 (1989): 1141-48; and G. R. Newell et al., "Volatile Nitrites: Use and Adverse Effects," 811-816.

24. Root-Bernstein, "Non-HIV Immunosuppressive Factors," 815-38; T. Marotta, *The Politics of Homosexuality* (Boston: Houghton Mifflin, 1981); L. Corey and K. K. Holmes, "Sexual Transmission of Hepatitis A in Homosexual Men: Incidence and Mechanism," *New England Journal of Medicine* 302 (1980): 436-438.
25. See previous note.
26. Cindy Patton, *Sex and Germs* (Boston: South End Press, 1985), 26, my emphasis.
27. Ibid., 6.
28. Jan Zita Grover, "Constitutional Symptoms," in *Taking Liberties*, 151.
29. Patton, Sex and Germs, 10.
30. J. McKenna et al., "Unmasking AIDS: Chemical Immunosupression and Seronegative Syphilis," *Medical Hypotheses* 21 (1986): 421-30.
31. P. Cox, "Why You Don't Need Meat," *Thorsons* (1986): 18-21, 105, 108, 112; R. Beach et al., "The Role of Reactive Oxygen Species. Antioxidant and Phytopharmaceuticals in Human Immunodeficiency Virus Activity," *Medical Hypotheses* 40 (1993): 85-92; F. Stall et al., "Glutathione Deficiency and Human Immunodefiency Virus Infection," *Lancet* 339 (1992): 902-12; C. Sappy et al., "Relative Decrease in Antioxidant Status During Evolution of HIV Infection: Effect on Lipid Peroxidation," International Conference on AIDS, 1992.
32. H. Greenspan and O. Arouma, "Could Oxidative Stress Initiate Apoptosis in HIV Infection? A Role for Plant Metabolites Having Synergistic Antioxidant Activity," 1992, available from HEAL.
33. M. Baum, "Interim Dietary Recommendations to Maintain Adequate Blood Nutrient Levels in Early HIV-1 Infection," International Conference on AIDS, 1992, PoB 3675; and F. Stall et al., "Intracellular Glutathione Concentration in AIDS: Implications for Therapy," International Conference on AIDS, 1992, PoA2400.
34. Lamar Graham, "The Heretic," *GQ*, November 1993.
35. Lauritsen, "Talk on AIDS."
36. Ibid.
37. Ibid.
38. Ibid.
39. P. H. Duesberg, "Results Fall Short for HIV Theory," *Insight*, 14 February 1994.
40. D. Rasnick, "Blinded by Science," *Spin*, June 1997.
41. See Robert Gallo, *Virus Hunting* (New York: Basic Books, 1991), 283-284. Other relevant studies are found in Feorino et al., *New England Journal of Medicine* 12 (1985): 1293-96; Curran et al., *New England Journal of Medicine* 310 (1984): 69-75. I want to thank Vladimir Koliadin for supplying me with this information.
42. Hodgkinson, *AIDS: The Failure of Contemporary Science*, 203-204.
43. Ibid.
44. Ibid.
45. Root-Bernstein, "Do We Know the Cause(s)."
46. Ibid.
47. Hodgkinson, *AIDS: The Failure of Contemporary Science*, 203-04.
48. Ibid.
49. Ibid; Root-Bernstein, *Rethinking AIDS*; J. Ward et al., *New England Journal of Medicine*, 17 March 1994, 738-743.

The Theory of Multifactorial Causation

1. Frank Buanoukas, interview by the author, May 1996.
2. Root-Bernstein, "Non-HIV Immunosuppressive." See also J. Sonnabend, "Facts and Speculation About the Cause of AIDS," *AIDS Forum* 2 (1989): 3-12; J. Sonnabend et al., "A Multifactorial Model for the Development of AIDS in Homosexual Men," *Annals of the New York Academy of Science* 837 (1984): 177-83.

3. Michael Verney Elliot, interview by the author, May 1996.
4. "Counterculture (Interview with Camille Paglia)," *Continuum* 4(2): 18-21 (July/August 1996).
5. Hodgkinson, *AIDS: The Failure of Contemporary Science*, 91-92.
6. Root-Bernstein, "Non-HIV Immunosuppressive." (See also Eleni Papadopulos-Eleopulos, "Reappraisal of Aids—Is the Oxidation Induced by the Risk Factors the Primary Cause?" *Medical Hypotheses* 25 (1988): 152; R. M. Selik et al., "Acquired Immune Deficiency Syndrome (AIDS) Trends in the United States," *American Journal of Medicine* 76 (1984): 493.
7. Root-Bernstein, "Non-HIV Immunosuppressive"; Sonnabend, "AIDS: An Explanation for its Occurrence Among Homosexual Men," in P. Ma and D. Armstrong, *AIDS and Opportunistic Infections of Homosexual Men* (Stoneham, MA: Butterworth, 1989), 449-70; Sonnabend et al., "A Multifactorial Model for the Development of AIDS, 177-83; Marmor et al., "Risk Factors for Kaposi's," 1083-1087; W.W. Darrow et al., "Risk Factors for Human Immunodeficiency Virus (HIV) Infections in Homosexual Men," *American Journal of Public Health* 77 (1987): 479-83; and R. S. Root-Bernstein, "Multiple-Antigen-Mediated Autoimmunity (MAIsMA) in AIDS: A Possible Model for Post-Infectious Autoimmune Complications," *Res Immunol* 4-5 (1990): 321-40.
8. Michael Verney Elliot, interview by the author, May 1996.
9. G. Stewart, "Conspiracy of Humbug Hides the Truth on AIDS," Sunday *Times*, 7 June 1992.
10. Papadopulos-Eleopulos, "Reappraisal of AIDS," 151-162, my stress.
11. Ibid.
12. Root-Bernstein, "Multiple-Antigen-Mediated Autoimmunity," 321-40.
13. Hodgkinson, *AIDS: The Failure of Contemporary Science*, 93-94; and Root-Bernstein, *Rethinking AIDS*.
14. Root-Bernstein, "Do We Know the Cause(s)."
15. Ibid.
16. Ibid.
17. Ibid.
18. Root-Bernstein, "Non-HIV Immunosuppressive," 827.
19. Ibid., 828.
20. R. S. Root-Bernstein, "Five Myths about AIDS that Have Misdirected Research and Treatment," *Genetica* 95 (1995): 111.
21. Leon Chaitow: Interview," *Continuum* (February-March 1997): 15.
22. Ibid.
23. Ibid.
24. Hodgkinson, *AIDS: The Failure of Contemporary Science*, 55-56.
25. Ibid.
26. Ibid., 52.
27. Ibid.
28. Ibid., 55-56.
29. Root-Bernstein, "Five Myths," 111.
30. Ibid., and Hodgkinson, *AIDS: The Failure of Contemporary Science*, 55-56.
31. Lynne McTaggard, interview by the author, May 1996. (See also Sean Currin, interview by the author, May 1996.)
32. Eva Sneed, interview by the author, May 1996.
33. Mr. James Currant, interview by the author, May 1996.
34. Ibid.
35. Hodgkinson, *AIDS: The Failure of Contemporary Science*, 359.
36. Ibid.
37. Ibid.

38. Ibid., 377.
39. Papadopulos-Eleopulos, "Reappraisal of AIDS," 159.
40. Ibid., 152. (See also M. Lemaitre et al., "Protective Activity of Tetracycline Analogs Against the Cytopathic Effect of the Human Immunodeficiency Viruses in CEM Cells," *Res Virol* 141 (1990): 5-16; V. Beral et al., "Kaposi's Sarcoma Among Persons with AIDS: A Sexually Transmitted Infection?" *Lancet* 1 (1990): 123-28.)
41. G. W. Pace and C. D. Leaf, "The Role of Oxidative Stress in HIV Disease," *Free Radical Biology & Medicine* 19(4): 523-28 (1995).
42. Ibid.
43. Ibid.
44. Ibid.
45. Ibid. See also Papadopulos-Eleopulos, "Reappraisal of AIDS," 151-62.
46. Graham, "The Heretic."
47. Koliadin, "Destruction of Normal Resident Microflora."
48. Ibid.
49. Ibid.
50. Kary Mullis, interview by the author, May 1996.
51. Farber, "AIDS: Words from the Front."
52. R. Root-Bernstein, "HIV One of Many Immunosuppressors (letter)," *Policy Review* (December 1990).
53. Ibid.
54. Ibid.
55. R .S. Root-Bernstein, "HIV and Immunosuppressive Cofactors in AIDS," J *Immunol Immunopharmacol* XII(4), 1992.
56. Hiram Caton, *The AIDS Mirage* (Sydney: University of New South Wales Press, 1994), 16-17.
57. Foss and Rothenberg, *The Second Medical Revolution*, 71.
58. Ibid., 97.
59. Ibid., 98.
60. Henderson, *Politics of the Solar Age*, 325.
61. Bacon, *New Organon*, 231.
62. Ibid., 230 f.
63. Welsh, *Trainspotting*, 332.
64. Ibid., 298.
65. Ibid., 128.
66. Ibid., 261-62.
67. Ibid., 344.
68. John Clarke, Stuart Hall, Tony Jefferson, and Brian Roberts, " Subcultures, Cultures and Class: A Theoretical Overview," in *Resistance Through Rituals: Youth subcultures in post-war Britain*, eds. Stuart Hall and Tony Jefferson (New York: HarperCollins, 1976), 13.
69. Ibid.
70. Ibid, 32.
71. Dan E. Beauchamp, "Morality and the Health of the Body Politic, in *The AIDS Reader*, 412.
72. Hakim Bey, *T.A.Z.: The Temporary Autonomous Zone, Ontological Anarchy, Poetic Terrorism* (New York: Autonomedia, 1985).
73. Lewis Katoff and Susan Ince, "Supporting People with AIDS: The GMHC Model," in *The AIDS Reader*, 545,
74. Ibid, 568-69.
75. Richard Whitley, "Changes in the Social and Intellectual Organization of the Sciences: Professionalisation and the Arithmetic Ideal," in*The Social Production of*

Scientific Knowledge, eds. Everett Mendelsohn, Peter Weingart, and Richard Whitley (Dodrecht, Holland: D. Reidel, 1977), 143.
76. Ibid, 154.

Was AIDS Man-made?

1. T. Curtis, "The Origin of AIDS," *Rolling Stone*, 19 March 1992, 54.
2. Ibid.
3. Ibid.
4. J. Seale, "The Origin of AIDS: Alternative Theories," *The New Scientist*.
5. Ibid., for this and previous quotations.
6. Curtis, "The Origin of AIDS."
7. Ibid.
8. Ibid.
9. Ibid.
10. Ibid.
11. Ibid.
12. R. B. Strecker, "AIDS Virus Infection," *Journal of the Royal Society of Medicine* 79 (September 1986).
13. Curtis, "The Origin of AIDS."
14. J. Rifkin, letter to the Department of Agriculture, Center for Disease Control, and National Institutes of Health, 1987.
15. K. Schneider, "Cattle Virus Is Studied," *New York Times*, 27 October 1987.
16. P. Raeburn, "AIDS-like Virus Found in Cows," *News-Journal Papers*, 26 November 1987, A14.
17. Rifkin, letter to the Department of Agriculture.
18. Ibid.
19. Raeburn, "AIDS-like Virus." See also J. Seale, "AIDS as a Cross Species Transfer," *21st Century* (July-August 1989): 18.
20. Rifkin, letter to the Department of Agriculture; M. A. Gonda et al., "Human T-Cell Lymphotropic Virus Type III Shares Sequence Homology with a Family of Pathogenic Lentiviruses," *Proceedings of the National Academy of Sciences* 83 (1986): 4007-11; M. A. Gonda et al., "Characterization of a Pathogenic Lentivirus from Cattle Which is Structurally, Immunologically, and Genetically Related to the Human Immunodeficiency, and Genetically Related to the Human Immunodeficiency Virus (HIV)," International Conference on AIDS, 4 June 1987; M. A. Gonda et al., "The Visna Virus Genome Nucleotide Sequence of Visna Virus" (Program Resources) (not yet published); R. Gilden et al., "Characterization of Molecular Cloning of Bovine Lenti Virus Related to Human Immunodeficiency Virus" (Program Resources) (not yet published); J. A. Geogiades, "Infection of Human Cell Cultures with Bovine Visna Virus," *Journal of General Virology* 38 (1978): 375-381; C. A. Mims, "Vertical Transmission of Viruses," *Microbiology Review* 45 (2): 267-68 (1981).
21. J. Grote, "Bovine Visna Virus and the Origin of HIV," *Journal of the Royal Society of Medicine* 81 (October 1988).
22. Ibid.
23. Grote, "Bovine Visna"; S. Kingman, "AIDS monitor, Evidence for Origin Is Weak," *New Scientist* 116 (1987): 27; S. Connor, "Laboratory Mix-Up Solves AIDS Mystery," *New Scientist* 117 (1988): 32; M. A. Gonda et al., "Sequence Homology and Morphologic Similarities of HTLV III and Visna Virus, a Pathogenic Lentivirus," *Science* 227 (1985): 173; M. J. Van der Maaten et al., "Isolation of a Virus from Cattle with Persistent Lymphocytosis," *Journal of the National Cancer Institute* 49 (1972): 1649; J. A. Georgiades et al., "Infection of Human Cell Cultures with Bovine Visna Virus," *Journal of General Virology* 38

(1978) 375; and P. A. Nuttal et al., "Viral Contamination of Bovine Foetal Serum and Cell Culture,"*Nature*, 266 (1977): 835.

24. Schneider, "Cattle Virus is Studied."
25. R. M. Kradjian, "Milk: The Natural Thing?"
26. Curtis, "The Origin of AIDS."
27. Ibid.
28. Ibid.
29. Kradjian, "Milk."
30. A. Cantwell Jr., "AIDS is Not African—Say Scientists," *New African*, (October 1994): 13; and H. M. McClure et al., "Erythroleukemia in Two Infant Chimpanzees Fed Milk from Naturally Infected with the Bovine C-type Virus," *Cancer Research* 34 (October 1974): 2745-57.
31. Kradjian, "Milk."
32. Ibid.
33. Ibid.
34. Ibid.
35. Kradjian, "Milk"; *Science* 213 (1981): 1014; and *British Medical Journal* 61 (March 1990): 456-59. (See also N. Sagata and Y. Ikawa, "BLV and HTLV-I: Their Unique Genomic Structures and Evolutionary Relationship," in *Retroviruses in Human Lymphoma/Leukemia*, eds. M. Miwa et al. (Tokyo: Japanese Science Society Press, 1985), 229-40.
36. Miller, *Immunization Theory vs. Reality* (Santa Fe, N.M.: The Atlantean Press, 1996), 62; P. Wright, "Smallpox Vaccine Triggered AIDS Virus," London *Times*, 11 May 11, 1987.
37. Wright, "Smallpox Vaccine."
38. Yves De Latte, *Vaccinations: The Untold Truth* (San Antonio: AUM Publications 1990), 50; and V. Fulginiti et al., "Progressive Vaccinia in Immunology Deficient Individuals," *Birth Defects Original Article Series* 4 (1): 1968.
39. See previous note.
40. See note 38.
41. Curtis, "The Origin of AIDS."
42. Ibid.
43. Ibid.
44. Ibid.
45. Miller, *Immunization Theory*; Wright, "Smallpox Vaccine," 1. (See also "Whispers of Fear Surround Last Vials of Smallpox," *Santa Fe, New Mexican*, 20 June 1993, p. A1.)
46. Wright, "Smallpox Vaccine."
47. Ibid.
48. Curtis, "The Origin of AIDS."
49. Ibid.
50. "Tony Brown's Comments," *Criterion*, December 29-January 4, 1989.
51. Ibid.
52. Ibid.
53. G. J. Krupey, "AIDS: Act of God or the Pentagon," 250; Waves Forest, "Designer Diseases: AIDS as Biological and Psychological Warfare," *Now What* 1 (Fall 1987).
54. "AIDS: New or Old," 61-63.
55. Ibid. See also A. Cantwell Jr., *Queer Blood* (Los Angeles: Aries Rising Press, 1993), 25.
56. Quoted in Cantwell, *Queer Blood*, 129.
57. Ibid.
58. Ibid.
59. Ibid.

60. "AIDS: New or Old."
61. Curtis, "The Origin of AIDS," 56.
62. Ibid.
63. "Safety, Efficacy Heart of Vaccine Use; Experts Discuss Pros, Cons," *DVM*, December 1986, 18.
64. Curtis, "The Origin of AIDS," 56.
65. Ibid.
66. L. Garrett, *The Coming Plague* (New York, HarperCollins, 1994).
67. Ibid.
68. Ibid.
69. Ibid.
70. Curtis, "The Origin of AIDS," 57.
71. Ibid.
72. Ibid.
73. De Latte, 39; and Monograph 29, National Cancer Institute and WHO, December 1968.
74. E. L. Snead, "AIDS—Immunization Related Syndrome," *Health Freedom News*, July 1987, 121.
75. Y. Geng et al., "Activation of Bovine Immunodeficiency-likeVirus Expression by Bovine Herpesvirus Type 1," *Virology* 187 (1992): 835.
76. Ibid.
77. Ibid.
78. Curtis, "The Origin of AIDS," 56.
79. Ibid.
80. Ibid., 57.
81. Ibid.
82. Ibid.
83. Ibid.
84. Ibid.
85. Ibid.
 6. Ibid.
87. Snead, "AIDS—Immunization Related Syndrome," 151.
88. Ibid.
89. Ibid. On the SV connection, see also Miller, *Immunization Theory*, 37; B. L. Horvath et al., "Excretion of SV-40 Virus After Oral Administration of Contaminated Polio Vaccine," *Acta Microbiologica Hungary* 11, 271-75; J. Snider, "Near Disaster with Salk Vaccine," *Science Digest* 1963; "Division of Biologics Standards: The Boat that Never Rocked,"*Science*, 17 March 1972; Snead, "AIDS—Immunization Related Syndrome," 1, 36; W. C. Douglas, "WHO Murdered Africa," *Health Freedom News* , September 1987, 42; W. S. Kyle, "Simian Retroviruses, Poliovaccine, and Origin of AIDS," *Lancet* 7 March 1992, 600-01; and Curtis, "The Origin of AIDS," 56.
90. Curtis, 57-59.
91. Ibid.
92. Snead, 215.
93. Curtis, 57-59.
94. De Latte, 39; and Monograph 29, National Cancer Institute.
95. Ibid.
96. De Latte, 17; and K. Shah and N. Nathanson, "Human Exposure to SV40 Review and Comments," *American Journal of Epidemiology* 103: 1 (1976).
97. Curtis, 57-59.
98. Ibid.
99. Ibid.

100. Curtis, 108.
101. Ibid.
102. L. Pascal, *What Happens When Science Goes Bad: The Corruption of Science and the Origin of AIDS: A Study in Spontaneous Generation* (Wollongong, Australia: University of Wollongong Press, 1991).
103. Ibid.
104. Leonard Horowitz, interview by the author.
105. Ibid.
106. J. Seale, "Origins of the AIDS Viruses, HIV-1 and HIV-2: Fact of Fiction?" *Journal of the Royal Society of Medicine*, 81: September 1988.
107. Cantwell, "AIDS is Not African," 11.
108. Ibid.
109. Seale, "Origins of the AIDS Viruses."
110. Ibid.
111. Ibid.
112. Ibid. For this and the last quote in the previous paragraph.
113. Ibid; M. Guyader et al., "Genome Organisation and Transactivation of the Human Immunodeficiency Virus Type 2," *Nature* 326 (1987): 662-69; M. A. Gonda, "The Natural History of AIDS," *Natural History*, 95 (1986): 78-81; R. A. Weiss, "AIDS Public Lecture," The Royal Society, London, 24 November 1987; R. C. Gallo, "The AIDS Virus," *Scientific American* 256 (1987): 47-56.
114. Michael Verney Elliot, interview by the author, May 1996.
115. Ibid.
116. Ibid.
117. Charles Whitney, *Francis Bacon and Modernity* (New Haven: Yale University Press, 1986), 111.

Biological Warfare and Genetic Engineering

1. Staff of the National Health Law Program, "Health Benefits: How the System is Responding to AIDS," in *The AIDS Reader*, 259.
2. Ibid.
3. Ibid.
4. Ibid.
5. Ibid., 260.
6. Nancy F. McKenzie, "Introduction: The Demands of the HIV Epidemic," in *The AIDS Reader*, 3.
7. J. Cookson and J. Nottingham, "A Survey of Chemical and Biological Warfare."
8. Ibid.
9. Ibid.
10. A. Kimbrell, "Biotech Could Create a New 'Super AIDS,'" *Third World Resurgence* 38, 16.
11. Ibid.
12. J. Seale, "The Origin of AIDS: Alternative Theories" *Lancet*, 15 March 1987
13. Ibid.
14. Ibid.
15. J. Seale, "The Origin of AIDS"; and *Science* 26 July 1974, 303.
16. Leonard Horowitz, interview by the author, May 1996.
17. Seale, "AIDS as a Cross Species Transfer"; D. O. White and F. J. Fenner, *Medical Virology* (London: Academic Press, 1986), 114-18;. G. R. Kendal et al., "Antigenic Similarity of Influenza A (H1N1)Viruses for Epidemics in 1977-1978 to 'Scandinavian' Strains Isolated in Epidemics of 1950-1951," *Virology* 89 (1978): 613; F. A. Ennis, "Influenza A Viruses: Shaking Out Our Shibboleths,"*Nature*, 274 (1978): 309.

18. Ibid.

19. Ibid.

20. "DuPont Maryland Firm Develop Viruses as Pesticides," *Quad-City Times*, 28 December 1991.

21. W. G. Winkler and K. Bogel, "Control of Rabies in Wildlife," *Scientific American*, June 1992, 92.

22. Ibid.

23. Ibid.

24. P. Recer, "Plants that Vaccinate Against Disease," Associated Press, 11 April 1995.

25. Ibid.

26. Ibid.

27. Miller, *Immunization Theory*, 94; C. Tant, *Awesome Green* (Angleton, Texas: Biotech Publishing, 1994), 108-15; and "Health Report," *Time*, 24 April 1995, 17.

28. Seale, "AIDS as a Cross Species Transfer," 521.

29. Ibid.

30. J. Seale, "Crossing the Species Barrier-Virus and the Origins of AIDS in Perspective," *Journal of the Royal Society of Medicine* 82 (September 1989): 519; M. Fukasawa et al., "Sequence of Simian Immunodefiency Virus from African Green Monkey, a New Member of the HIV/SIV Group," *Nature* 333 (1988): 457-61; R. C. Gallo and L. Montagnier, "AIDS in 1988," *Scientific American* 4 (1988): 25-32; M. A. Gonda, "The Natural History of AIDS," 78-81; and Seale, "Origins of the AIDS Viruses," 537-39.

31. See also H. Thormar and B. Sugurdardottir, "Growth of Visna Virus in Primary Tissue Cultures from Various Animal Species," *Acta Pathol Microbiol Scand* 55 (1962): 180-86.

32. G. Kolata, "New Studies Offer Powerful and Puzzling Evidence on Immunity to AIDS," *New York Times*, September 27, 1996, A 27.

33. Ibid.

34. Ibid.

35. Ibid.

36. "Genetic Factor of Africans Increase Susceptibility to AIDS," *The PROBE*, December 1988; *New York Times*, May 10, 1987.

37. Kolata, "New Studies," A 27.

38. Ibid.

39. Miller, *Immunization Theory*, 61-62; Allison et al., "Virus-Associated Immunopathology: Animal Models and Implications for Human Disease," *Bulletin of the World Health Organization* 47 (1992): 259.

40. Curtis Cost, *Vaccines Are Dangerous* (Brooklyn: A&B Books, 1991), 11-13.

41. Ibid.

42. Ibid; Robert Strecker, *The Strecker Memorandum* (Eagle Rock, Calif.: The Strecker Group, 1988); *Bulletin of the World Health Organization* 47 (1972): 259; *Fogarty International Center Proceedings* 15 (1972).

43. Curtis, "The Origin of AIDS," 54.

44. Ibid.

45. Miller *Immunization Theory*, 61-62; W. C. Douglas, "WHO Murdered Africa," *Health Freedom News*, September 1987, 42; Curtis, "The Origin of AIDS," 106.

46. Snead, *Some Call It AIDS*, 381; R. Lederer, *Covert Action Information Bulletin* 35.

47. Curtis, "The Origin of AIDS," 54.

48. Cantwell, "AIDS Is Not African," 13.

49. Snead, *Some Call It AIDS*, 381.

50. Geisler, *AIDS: Origin, Spread*, 71; R. A. Squire, "Equine Infectious Anemia: A Model of Immunoproliferative Disease," *Blood* 32 (1968): 157-69.

51. Geisler, *AIDS: Origin, Spread*, 60.

52. Ibid; R. C. Gallo and L. Montagnier, "AIDS in 1988," *Scientific American* 259 (1988): 25-32; R. C. Desrosiers and N. L. Letvin, "Animal Models for Acquired Immunodeficiency Syndrome,"*Review of Infectious Diseases* 9 (1987): 438-46; L. Montagnier et al., "A New Human T-Lymphotropic Retrovirus: Characterization and Possible Role in Lymphadenopathy and Acquired Immune Deficiency Syndromes," in *Human T-cell Leukemia/Lymphoma Virus, The Family of Human T-Lymphotropic Retroviruses: Their Role in Malignancies and Association with AIDS*, R. C. Gallo et al., eds. (New York: Cold Spring Harbor Laboratory, 1984), 363-79; L. Montagnier et al., "A New Type of Retrovirus Isolated from Patients Presenting with Lymphadenopathy and Acquired Immune Deficiency Syndrome: Structural and Antigenic Relatedness with Equine Infectious Anaemia Virus," *Annals of Virology* 135E (1984): 119-31.

53. Geisler, *AIDS: Origin, Spread*, 60.

54. Ibid.

55. Ibid.

56. H. Lorscheid et al., *Der Olskandal*, Weiner, March 1987, 70-73; Geisler, *AIDS: Origin, Spread*, 60; J. M. Watanabe et al, "Pneumocystis Carinii Pneumonia in a Family," *JAMA* 193 (1965): 685-86; G. W. Fischer and C. S. Holton, *Biology and Control of the Smut Fungi* (New York: Ronald Press Company, 1957), 393-421.

57. Leonard Horowitz, interview by the author, May 1996. (See also Snead, *Some Call It AIDS*, 317; B.V. Perbal et al., *Molecular Biology of Polyoma Viruses and Herpesviruses* (New York: John Wiley and Sons, no year referenced), 19.

58. Leonard Horowitz, interview by the author, May 1996

59. Ibid.

60. Ibid.

61. Ibid.

62. Ibid.

63. Ibid; Richard Chirimuuta and Rosalind Chirimuuta, *AIDS, Africa and Racism* (London: Free Association Books, 1989), 7-8; D. Altman, *AIDS and the New Puritanism* (London and Sydney: Pluto Press, 1986), 44.

64. "Tokyo TV Tells How WWII Germ War Unit Escaped Prosecution," *Los Angeles Times*, 6 April 1992.

65. Ibid.

66. Ibid.

67. A. W. Scheflin and E. M. Opton Jr. *The Mind Manipulators* (New York: Paddington Press, LTD, 1978).

68. Ibid.

69. Ibid.

70. "Covert Research in Biological Weapons for Population Management," 171-74.

71. Ibid.

72. Snead, *Some Call It AIDS*, 233; J. Creamer, *Biohazard* (London: The National Antivivisection Society, 1987).

73. "Covert Research in Biological Weapons," 171-74.

74. Ibid.

75. Ibid.

76. Ibid.

77. Ibid.

78. "Green Monkey Theory Discounted," *Sunday Express*, 26 October 1986.

79. R. Harris and J. Paxman, *A Higher Form of Killing* (New York: Hill & Wang, 1982), 240-41; U.S. Army Mobility Equipment Research and Development Center, *Decontamination of Water Containing Chemical Warfare Agent* (Fort Belovir, Va.: January 1975); U.S. Army spokesman, testimony Before a Subcommittee of the House Committee on Appropriations, Department of Defense Appropriations

for 1963, Washington, D.C., March 1962; and testimony before a Subcommittee of the House Committee on Appropriations, Department of Defense Appropriations for 1970, Washington, D.C., 1969.

80. H. Strauss and J. King, "Chapter 5. The Fallacy of Defensive Biological Weapon Programmes," 68.
81. "Germ Warfare—The Enemy is Us," *Spectrum* 36 (May/June 1994).
82. Geisler, *AIDS: Origin, Spread*, 60; *MMWR* 33 (1984): 356, 722; *MMWR* 39 (1990) 430, 936.
83. "Germ Warfare—The Enemy is Us"; *New York Times*, 25 January 1994.
84. "Germ Warfare—The Enemy is Us."
85. B. L. Fisher, "Experimental Vaccines Haunt Ill Gulf War Veterans," *The Vaccine Reaction* 1 (1): 4 (March 1995).
86. Ibid.
87. Miller, *Immunization Theory*, 66; Senator John D. Rockefeller IV, "Is Military Research Hazardous to Veterans' Health? Lessons from the Cold War, the Persian Gulf, and Today: Opening Statement," U.S. Senate, Committee on Veterans' Affairs, 6 May 1994.
88. Seale, "The Origin of AIDS."
89. Ibid.
90. Krupey, "AIDS: Act of God," 248; Lederer, "Precedents for AIDS? Chemical-Biological Warfare. Medical Experiments, and Population Control," *Covert Action Information Bulletin* 28 (Summer 1987): 36-37.

Instances of the Fingerpost

1. Bacon, *New Organon*, 191.
2. Ibid.

PART THREE: AIDS INC.
Science as Business

1. Frank Buanoukas, interview by the author, May 1996.
2. Stefan Lanka, interview by the author, May 1996.
3. Ibid.

The Economics of AIDS

1. Mark Anderson, "What If He's Right?" *The Valley Advocate*, 30 May-5 June 1996.
2. Ibid.
3. Mark Conlan, "Interview with David Rasnick: A Real Scientist," *Zenger's*.
4. J. Cohen, "The Marketplace of HIV/AID$," *Science* 272: 1880 (28 June 1996).
5. Ibid.
6. Anderson, "What If He's RIght?".
7. Lauritsen, "Psychological and Toxicological Causes of AIDS," 97.
8. Peter Duesberg, "Talk by Peter Duesberg, Cal Alumni Day, March 6, 1993," *Rethinking AIDS* homepage, http://www.virusmyth.com.aids/data/pdlecture.htm.
9. Conlan, "Interview with Rasnick."
10. Cohen, "The Marketplace."
11. Ibid.
12. David Rasnick, "Blinded by Science," *Spin*, June 1997.
13. Hodgkinson, *AIDS: The Failure of Contemporary Science*, 129; *Journal of Public Policy* 13 (4): 305-25.
14. M. A. Sushinsky et al., "Women and AIDS: The Ethics of Exaggerated Harm," *Rethinking AIDS* Web site, http://www.virusmyth.com, May 1998. See also Celia Farber, "AIDS, Inc.: Observations of an AIDS Dissident," *Continuum* 2 (4): 7 (August/September 1994); Eva Sneed, interview with the author, May 1996; R.

Lacayo, "Hope with an Asterisk," *Time*, 30 December 1996, 84; E. Burkett, "The Queen of AZT," *Miami Herald*, 23 September 1990. On the money spent, see D. Mertz and U. Schuklenk, "Deathly Doctrine: Christian Churches and AIDS," *Rethinking AIDS* Web site, http://www.virusmyth.com, May 1998; Fred Cline, "The Fear of Losing HIV," *Sumeria* Web site, http://www.livelinks. com/sumeria, 21 May 1998.

15. L. Garrett, "Mi$$pent/Study: AIDS Funds Often Go Elsewhere," *Newsday*, 13 March 1996, A5.
16. Mertz and Schuklenk, "Deathly Doctrine: Christian Churches and AIDS."
17. Ibid.
18. Ibid.
19. Farber, "Fatal Distraction."
20. Ellison, "AIDS: Words from the Front."
21. Ibid.
22. Ibid.
23. Ibid.
24. Ibid.
25. *What If Everything You Thought You Knew About AIDS Was Wrong*, HEAL Pamphlet, 11; *Continuum* 2(4) (August-September 1994).
26. Cindy Patton, *Inventing AIDS* (London and New York: Routledge, 1990), 14.
27. Ibid., 17.
28. Ibid. 18.
29. Mr. James Currant, interview by the author, May 1996.
30. Passante Conlan, interview by the author, May 1996.
31. Ibid.
32. Hiram Caton, *The AIDS Mirage* (Sydney: University of South Wales Press, 1994), 10.
33. P. Plumley, "AIDS—The Great Numbers Game," *Rethinking AIDS*, October/November 1993.
34. Peter Duesberg, "Talk by Peter Duesberg."
35. Ibid.
36. Simon Watney, *Policing Desire: Pornography, AIDS, and the Media* (Minneapolis: University of Minnesota Press, 1987), 94.
37. Farber, "AIDS, Inc.: Observations of an AIDS Dissident," 9.
38. P. Rudnick, "Now It's AIDS INC.," *Time*, 30 December 1996, 86.
39. Ibid.
40. Henderson, *Politics of the Solar Age*.
41. P. E. Johnson, "The Thinking Problem in HIV-Science," in *AIDS: Virus or Drug Induced?* ed. Peter Duesberg (Netherlands: Kluwer, 1996).
42. Ibid.
43. Ibid.
44. Mark Garish Conlan, interview by the author, May 1996.
45. Charles Thomas, interview by the author, May 1996.
46. Cohen, "The Marketplace," 1883.
47. Ibid.
48. Ibid.
49. Ibid.
50. Ibid.
51. Ibid.
52. Duesberg and Rasnick, "The Drugs-AIDS Hypothesis," 20.
53. Frank Buanoukas, interview by the author, May 1996.
54. Ibid.
55. *Rethinking AIDS* Web site, http://www.virusmyth.com, May 1998.
56. Ibid.

57. Ibid.
58. Ibid.
59. Hodgkinson, *AIDS: The Failure of Contemporary Science*, 143-44
60. Conlan, "Interview with Rasnick."
61. Ibid.
62. Duesberg, "Talk by Peter Duesberg, Cal Alumni Day."
63. G. B. Smith, "FBI Eyeing AIDS Groups," *New York Daily News*, 15 May 1995, 2.
64. Ibid.
65. Patton, *Inventing AIDS*, 60.
66. Ibid., 61.
67. Ibid., 62.

An Imaginary Conversation with Prof. Stanley Aronowitz

1. Mikhail Bahktin, *Problems of Dostoevsky's Poetics*, trans. R. W. Rostel (New York: Ardis, 1973), 163.
2. Gary Null, *Who Are You, Really?* (New York: Carroll & Graf, 1996).
3. Bruno Latour and Steven Woolgar, *Laboratory Life: The Construction of Scientific Facts* (Princeton: Princeton University Press, 1986).
4. Null, *Who Are You*, 40.
5. Charles Kaiser, *The Gay Metropolis 1940-1996* (Boston: Houghton Mifflin, 1997), 247.
6. Stanley Aronowitz, *Food, Shelter and the American Dream* (New York: Seabury Press, 1974), 141.
7. Dennis Altman, *Coming Out in the Seventies* (Boston: Alyson Publications, 1981), 86.
8. David B. Feinberg, *Eighty-Sixed* (New York: Penguin, 1989), 21.
9. Toby Marotta, *The Politics of Homosexuality* (Boston: Houghton Mifflin, 1981), 321.
10. Charles Webster, *The Great Instauration: Science, Medicine and Reform 1626-1660* (New York: Holmes & Meier Publishers, 1975), 12.
11. Ibid., 274.
12. Ibid. 283.

PART FOUR: CONVENTIONAL TREATMENT PROGRAMS
Failure of Conventional AIDS Treatment

1. Dean Black, interview by the author, in *AIDS, The Untold Story*.
2. Ibid.
3. Martin Walker, interview by the author, May 1996.
4. Ibid.
5. Ibid.
6. Paul Phillpot, interview by the author, 13 February 1995.
7. Bill McCreary, interview by the author, in *AIDS, The Untold Story*.
8. Ibid.
9. Ibid.
10. Ibid.
11. "Interview with Lord Baldwin (Joint Chairman of Britain's Parliamentary Group for Alternative and Complementary Medicine)," *Continuum* 4(2): 16 (July/August 1996).
12. Ibid.
13. Ibid.

AIDS

1. L. Pascal, *What Happens When Science Goes Bad. The Corruption of Science and the Origin of AIDS: A Study in Spontaneous Generation* (Wollongong, Australia: University of Wollongong Press, 1991).

2. Michael Verney-Elliot, "AIDS Vaccines—The Cruel Delusion," *Continuum* 5(2): 7 (Winter 1997/1998).

3. Ibid.

4. Cournoyer, *What About Immunizations?* 141.

5. Ibid. See also M. Cimons, "New Human AIDS Vaccine Test Rejected," *Los Angeles Times*, 18 June 1994, A24; G. Antonio, *The AIDS Cover-up?* (San Francisco: Ignatius Press, 1986); J. Hyman, "Children at Risk," *The Democrat & Chronicle*, April 1987; "Are Researchers Racing Toward Success, Or Crawling?" *Science* 265 (2 September 1994): 1373.

6. Verney-Elliot, "AIDS Vaccines."

7. Ibid.

8. For everything in this paragraph, see "Challenges in Designing HIV Vaccines," *NIAID Fact Sheet*, 1995.

9. Ibid.

10. M. Schoofs, "An AIDS Vaccine: It's Possible," *Village Voice*, 12 September 1995, 21.

11. "Expert Says AIDS Vaccine More Than a Decade Away," CNN Interactive Web site, http://www.cnn. com/HEALTH/9802/01/aids.vaccine, 25 July 2000.

12. Ibid.

13. Ellison, "AIDS: Words From the Front."

14. "Are Researchers Racing Toward Success," 1374.

15. Ibid.

16. "FDA Allows Large-Scale AIDS Vaccine Trial," Associated Press, 3 June 1998.

17. Ibid.

18. "Clinical Research on HIV/AIDS Vaccines," *NIAID Fact Sheet*, 1995.

19. "FDA Allows Trial."

20. "Evolution of Vaccine Development," *NIAID Fact Sheet*, 1995.

21. Daniel Q. Haney, "Vaccine intended to prevent AIDS may cause the disease," *Augusta Chronicle* online, http://augustachronicle.com/stories/070398/tec, 25 July 2000. My emphasis.

22. Ibid., my emphasis.

23. *Continuum* 4(1): 3 (May-June 1996).

24. "Fauci Slams Vaccine," *Continuum* 5(2): 4 (Winter 1997/1998); *Science* 279, 30 January 1998.

25. "Clinical Research on HIV/AIDS Vaccine."

26. Ibid.

27. N. Z. Miller, *Immunization Theory Vs. Reality* (Santa Fe, N.M.: The Atlantean Press, 1996), 100.

28. Ibid. See also "AIDS Vaccine Study in Peril," *Chicago Tribune*, 29 May 1994, 1.

29. Verney-Elliot, "AIDS Vaccines," 6.

30. S. K. Miller, "HIV Strategy Will Fail, Says Vaccine Veteran," *New Scientist*, 19 September 1992.

31. Ibid.

32. Verney-Elliot, "AIDS Vaccines."

33. Ibid.

34. Ibid.; *Nature* 15(1) (1998).

35. N. Ostrum, "Is Cellular Immunity a Key to Controlling AIDS?" *New York Native*, 21 August 1995.

36. Ibid.

37. Schoofs, "An AIDS Vaccine," 22.

38. Ibid.

PART FIVE: THE AZT HOAX

1. John Lauritsen, interview by the author, in *The Pain, Profit and Politics of AIDS*.

2. Peter Duesberg, interview by the author, in *The Pain, Profit, and Politics of AIDS*.
3. Richard Strohman, interview by the author, in *AIDS, The Untold Story*.
4. Personal communication with Anthony Brink, 27 February 2001. The article referred to in his comments is: Eleni Papadopulos-Eleopulos et al. "A Critical Analysis of the Pharmacology of AZT and Its Uses in AIDS," *Current Medical Research and Opinion*, 1999. Also of note are articles that have established the concentration of triphosohorylated thymidine that would be necessary for effective DNA chain termination to take place. These include pieces by B. L. Robins et al., *Antimicrob Agents Chemotherapy* (1988): 2656-60; E. Font et al., *Antimicrob Agents Chemotherapy* (1999): 2964-68.
5. Ibid.

Marketing, Toxicity, and Motivation

1. *What if Everything You Thought*, 21.
2. Peter S. Arno and Karyn L. Feiden, *Against the Odds: The Story of AIDS Drugs Development, Poltics and Profits* (New York: HarperCollins, 1992), 40-41.
3. M. Walker, "HIV, AZT, Big Science," *Continuum* 4(6) (June-July 1997).
4. Neville Hodgkinson, radio interview by the author, 15 May 1996.
5. P. Duesberg, in J. Lauritsen, *Poison By Prescription: The AZT Story* (New York: Asklepios, 1990), 7.
6. *What if Everything You Thought*, 9-10.
7. Lauritsen, *Poison*, 108-10.
8. Ibid.
9. Ibid.; B. Deer, "Revealed: Fatal Flaws of Drug That Gave Hope," Sunday *Times*, 16 April 1989; C. Farber, "Sins of Omission: The AZT Scandal," *Spin*, November 1989.
10. Walker, "HIV, AZT."
11. Ibid.
12. Arno and Feiden, *Against the Odds*, 39. This and previous quote.
13. L. Kilzer and S. Wilmsen, "AIDS Fear Skews Tests: Rush on AZT Detoured Studies," *Denver Post*, 10 December 1995.
14. Ibid.
15. Walker, "HIV, AZT."
16. Ibid.
17. Arno and Feiden, *Against the Odds*, 42.
18. Ibid., 43.
19. Walker, "HIV, AZT."
20. Ibid.
21. John Lauritsen, interview by the author, 13 February 1995.
22. Arno and Feiden, *Against the Odds*, 51-52.
23. Ibid., 52.
24. James Currant, interview by the author, May 1996
25. Duesberg in Lauritsen, *Poison by Prescription*, 107 and H. I. Chernov, "Review and Evaluation of Pharmacology and Toxicology Data," *NDA* 19-655, 29 December 1986.
26. John Lauritsen, interview by the author, in *AIDS, The Untold Story*.
27. Walker, "HIV, AZT."
28. Ibid.
29. Ibid.
30. M. Walker, "AZT: A Seller's Market," *Continuum* 5(1): 36 (Autumn 1997).
31. Paul Philpott, "Doctor's Claim 'Home Run' with 'Cocktail' Mixture of AZT and New 'Protease Inhibitors,'" *Reappraising AIDS*, March 1996.
32. James Currant, interview by the author, May 1996.

33. Ibid.

34. Neville Hodgkinson, interview by the author, 15 May 1996.

Testing Today

1. Lynn Gannett, "An Eyewitness Account of Gross Irregularities and Medical Incompetence in the Early Clinical Trials of AZT," *Rethinking AIDS* Web site, http://www.rethinking.org/aids/AZT.html., 31 July 2000.

2. Ibid.

3. Ibid., 3.

4. Ibid.

5. Lauritsen, *Poison*, 22: J. D. Hamilton et al., "A Controlled Trial of Early Versus Late Treatment with Zidovudine in Symptomatic Human Immunodeficiency Virus Infection," *New England Journal of Medicine* 326, (1992): 437-43.

6. John Lauritsen, "Talk on AZT," Alternative AIDS Symposium, Amsterdam, May 1992.

7. Frank Buianouckas, interview by the author, in *AIDS, The Untold Story*.

8. CDC National AIDS Clearinghouse Custom Search Service and AIDS Clinical Trials Information Service, 18 January 1996 (studies cited not available), my emphasis.

9. Ibid.

10. Celia Farber, "AIDS: Words from the Front."

11. Ibid.

12. Walker, "AZT: A Seller's," 35.

13. B. Deer, "AIDS Doctors Attack Drug Claims," Sunday *Times*, 1 August 1993.

14. Ibid.

15. Ibid.

16. Arno and Feiden, *Against the Odds*, 193-194.

17. Ibid., 193.

18. Ibid., 194.

19. Ibid.

20. Walker, "HIV/AIDS."

21. Graham Ross, "When Prevention May Be Worse Than Cure," *Continuum* 2 (2): 16 (April/May 1994).

22. Huw Christie, "AZT and Concorde," *Continuum* 2 (2): 12 (April/May 1994). See also Frank Buinouckas, interview by the author, in *AIDS, The Untold Story*.

23. Lauritsen, "Talk on AZT."

24. H. Kremer, S. Lanka, and A. Hassig, "AIDS: Death by Prescription," *Continuum* 4(2): 4-5 (July/August 1996).

25. Michael Bomgardener, interview by the author, May 1996.

26. E. Aranaudo et al., "Depletion of Muscle Mitochondrial DNA in AIDS Patients with Zidovudine-Induced Myopathy," *Lancet* 337 (2 March 1991): 508-10.

27. Ibid.

28. W. Lewis et al., "Zidovudine Induces Molecular, Biochemical, and Ultrastructural Changes in Rat Skeletal Muscle Mitochondria," *Journal of Clinical Investigations* 89 (April 1992): 1354-60.

29. Ibid.

30. C. R. Caulfield and B. Goldberg, *The Anarchist AIDS Medical Formulary: A Guide to Guerilla Immunology* (Berkeley: North Atlantic Books, 1993), 48; W. Heagy et al., "Inhibition of Immune Functions by Antiviral Drugs," *Journal of Clinical Investigation* 87 (1991): 1916-24.

31. C. Costello, "Zidovudine and Bone Marrow," *Lancet* (19 November 1988); P. Gill et al., "Axidothymadine Associated with Bone Marrow Failure in AIDS," *Annals of Internal Medicine* 107 (1987): 502-05.

32. Caulfield and Goldberg,*The Anarchist AIDS Medical Formulary:* 18.

33. Ibid.; Heagy, "Inhibition of Immune Function"; and S. Rowland-Jones and A. McMichael, "Cytotoxic T-Lymphocytes in HIV Infection," *Seminars in Virology* 4 (1993): 83-94.

34. Martin Feldman, interview by the author, in *AIDS, The Untold Story.*

35. Richard Strohman, interview by the author, in *AIDS, The Untold Story.*

36. Lauritsen, *Poison,* 12.

37. "AZT Causes Cancer in Offspring," *Rethinking AIDS* Web site, http://www.virusmyth.com, May 1998; *Journal of the National Cancer Institute* 89 (1997): 1602-08.

38. "AZT Causes Cancer in Offspring." See also Lauritsen, *Poison,* 118-19.

39. C. Farber, "Sins of Omission: The AZT Scandal," *Spin,* November 1989.

40. N. Ostrum, "Nightmare on AZT Street," *New York Native,* 4 September 1995.

41. Kremer, Lanha, and Hassig, "Death by Prescription," 4-5.

42. Caulfield and Goldberg, *The Anarchist AIDS Medical Formulary,* 57-58.

43. Ibid.; D. A. Cooper et al., "Zidovudine in Person with Asymptomatic HIV Infection and CD4+ Cell Counts Greater than 400 per Cubic Millimeter," *New England Journal of Medicine* 329 (1993): 297-303; J. M. Pluda, "Parameters Affecting the Development of Non-Hodgkins Lymphoma in Patients with Severe Human Immunodeficiency Virus Infection Receiving Antiretroviral Therapy," *Journal of Clinical Oncology* 11 (1993): 1099-1107.

44. N. Ostrum, "Early Intervention: An Idea Whose Time Has Gone?" *New York Native,* 21 August 1995.

45. Ibid.

46. Caulfield and Goldberg, *The Anarchist AIDS Medical Formulary*, 12: P. A. Volberding, "Is it Possible to Prove a Survival Benefit from Early Treatment?" *Bulletin of Experimental Treatment for AIDS* November 1992, 12-15.

47. Ibid.

48. L. Fleming and T. R. Fleming, "Study Fails to Say When to Start AIDS Treatment."

49. Michael Verney Elliot, interview with the author.

50. David Rasnick, "AIDS and the HIV Boondoggle."

51. Ibid.

52. E. Papadopulos et al., "A Critical Analysis of the Pharmacology of AZT and its Use in AIDS," *Rethinking AIDS* Web site, http://www.virusmyth.com, May 1998.

53. "Fresh UN Action," *Continuum* 5 (1): 2 (Autumn 1997).

54. Ibid.

55. Lauritsen, *Poison,* 11.

56. Michael Callen, "Long-Term Survival," AIDS: A Different View Symposium, Amsterdam 1992.

57. Farber, "Sins of Omission."

58. James Currant, interview by the author, May 1996.

59. Ibid.

60. Kary Mullis, interview by the author, May 1996; Neville Hodgkinson, interview by the author, May 1996; J. Lauritsen, "Psychological and Toxicological Causes of `AIDS,'" in *The AIDS Cult* , eds. J. Lauritsen and I. Young (Provincetown: Asklepios, 1997), 97-98.

Rationale and Long-Term Effects

1. Duesberg quoted in B. Guccione Jr., "AIDS: Words from the Front,"*Spin*, September 1993.

2. Ibid. It should be said, though, that there is some dispute over whether Duesberg is right in attributing an increase in pneumonia to AZT treatment. Koliadin has asserted that the case study Duesberg uses as the basis of his claim is one in which

the increase in pneumonia of those undergoing an AZT regimen might well be attributable to other causes.

3. Christie, "AZT and Concorde," 11.
4. Ibid.
5. Guccione, "AIDS: Words From the Front."
6. Lauritsen, *Poison*, 140-41.
7. P. H. Duesberg, "Can Epidemiology Determine Whether Drugs or HIV Cause AIDS?" *AIDS-Forschung* 12 (December 1993): 627-35.
8. Lauritsen, *Poison*, 47.
9. Ibid.
10. James Whitehead, interview by the author, May 1996.
11. Michael Bomgardener, interview by the author, May 1996.
12. James Currant, interview by the author, May 1996.
13. Farber, "Omission."
14. Callen, "Long-Term Survival."

Big Business

1. Walker, "HIV, AZT,"
2. Ibid.
3. Arno and Feiden, *Against the Odds*, 251-60.
4. N. Hodgkinson, "How Giant Drug Firm Funds the AIDS Lobby," Sunday *Times*, 30 May 1993.
5. "Glaxo Dumping AZT in Africa," *Rethinking AIDS* Web site, http://www.virus-myth.com, May 1998. Also *SA Mail & Guardian*, 22 August 1997.
6. Walker, "HIV, AZT."
7. Ibid.
8. Ibid.
9. Ibid.
10. Ibid.
11. L. Garrett, *Newsday*, 5 March, 1996. See also Paul Philpott, "Doctor's Claim 'Home Run' with 'Cocktail' Mixture of AZT and New 'Protease Inhibitors,'" *Reappraising AIDS*, March 1996.
12. Garrett.
13. *New York Times* (further information not available).
14. Lauritsen, *Poison*, 120-21.
15. "Drug Pushers Worried," *Rethinking AIDS* Web site, http://www.virusmyth.com, May 1998; *AIDS Treatment News*, 15 August 1997.
16. Ibid.
17. "Drug Industry Seeks Further U.S. Market Growth," *Rethinking AIDS* Web site, http://www.virusmyth.com, May 1998; *New England Journal of Medicine* 11 September 1997.
18. Seymour Brenner, interview by the author, in *The Pain, Profit and Politics of AIDS*.
19. M. Waldholz, "New Discoveries Dim Drug Makers' Hopes For Quick AIDS Cure," *Wall Street Journal*, 26 May 1992.
20. Ibid.
21. Robert Cathcart, interview by the author, in *The Pain, Profit and Politics of AIDS*.
22. Joan Priestley, interview by the author, in *The Pain, Profit and Politics of AIDS*.
23. Ibid.
24. Burkett, "Queen of AZT."
25. Walker, "Seller's Market."
26. Ibid.
27. Ralph Moss, interview by the author, in *The Pain, Profit and Politics of AIDS*.
28. Ibid.

29. Walker, "AZT, HIV."
30. Ibid.
31. Ibid.
32. Dr. Bruce Halstead, interview by the author.
33. Walker, "AZT, HIV."
34. Ibid.
35. Walker, "AZT: A Seller's Market," 38.

Media Power

1. Tom Bethel, interview by the author, 14 February 1995.
2. Neeyah Ostrum, interview by the author, 14 February 1995.
3. Ibid.
4. Neeyah Ostrum, interview by the author, in *AIDS, The Untold Story*.
5. Martin Walker, interview by the author, May 1996l.
6. Tom Bethel, interview by the author, 14 February 1995
7. Mark Garish Conlan, interview by the author, May 1996.
8. Ibid.
9. J. Lauritsen, "HIV Voodoo from Burroughs-Wellcome," *New York Native*, 7 January 1991.
10. Ibid.
11. Michael Verney Elliot, interview by the author, May 1996.
12. Ibid.
13. J. Lauritsen, "Science by Press Release," *New York Native*, 21 August 1989.
14. Ibid.
15. Ibid.
16. J. Lauritsen, "The AIDS War: Lies and Censorship in Coverage of the Epidemic," *New York Native*, 12 August 1991.
17. Ibid.
18. Ibid.
19. Lauritsen, "AIDS War"; R. Pearson, "AIDS Patient Ryan White Dies: Indiana Youth's Plight Touched the Nation," *Washington Post*, 9 April 1990; D. Johnson, "Ryan White Dies of AIDS at 18: His Struggle Helped Pierce Myths," *New York Times*, 9 April 1990.
20. David France, "The HIV Disbeliever, *Newsweek*, 28 August 2000, 48.
21. Ibid., 47.
22. Ibid., 48.
23. Ibid.

Buying Patient and Professional Support

1. Walker, "AZT: A Seller's."
2. Ibid.
3. Ibid.
4. Tom Bethel, interview by the author, 14 February 1995
5. Celia Farber, "Words from the Front."
6. Ibid.
7. W. Dunlap and L.M. Fisher, "Drug Companies Turn Agressive in Promoting New Drugs for AIDS," *New York Times*, 5 July 5 1996.
8. Ibid.
9. Walker, "AZT: A Seller's Market," 37.
10. Ibid.
11. Walker, "HIV, AZT."
12. Ibid.
13. Ibid.

14. Ibid.
15. Dunlap, "Drug Companies Turn Aggressive."
16. Ibid.

AZT Reborn

1. Koliadin, personal communication with the author, citing *Lancet* 348 (1996): 283-91.
2. Ibid., citing *NEJM* 338 (1998): 853-60 (a study carried out in the UnitedStates utilizing 1,255 patients); and *BMJ* 315 (1997): 1994-99, (a Swissstudy based on 5,176 patients).
3. C. Gorman, "The Disease Detective," *Time* 30 December 1996, 59.
4. Ibid.
5. Ibid.
6. C. Johnson, "Viral Load and the PCR: Why They Can't Be Used to Prove HIV Infection," *Continuum* 4(4): 32 (November/December 1996).
7. Gorman, "Disease Detective," 61-62.
8. Johnson, "Viral Load," 32.
9. Ibid.
10. Gorman, "Disease Detective," 61-62.
11. Ibid.
12. P. Elmer-Dewitt, "Turning the Tide," *Time*, 30 December 1996, 54-55.
13. J. Lauritsen, "Protease Inhibitors in Provincetown," *Continuum* 4(5): 8-10 (February-March 1997).
14. Johnson, "Viral Load."
15. Lauritsen, "Protease Inhibitors."
16. Johnson, "Viral Load," 34.
17. Ibid.
18. Ibid.
19. D. Rasnick, "Non-Infectious HIV is Pathogenic," *Continuum* 4(6) (June-July 1997). Also www.virusmyth.net/aids/data/drconf.htm. Note, in a private communication, Rasnick stated that his question was only about viral "particles" not "proteins or protease." The basic import of the question and the answers are the same.
20. Johnson, "Viral Load."
21. Ibid.
22. Ibid.
23. Ibid., 36.
24. Ibid.
25. Ibid.
26. Ibid.
27. David Rasnick, "Blinded by Science," *Spin*, June 1997.
28. M. Waldholz, "Some AIDS Cases Defy New Drug `Cocktails'" *Wall Street Journal*, 10 October 1996.
29. H. Christie, "From Hype to Hesitation," *Continuum* 4(5): 11 (February-March 1997).
30. Mark Konlee, "An Evaluation of Drug Cocktail Combinations for their Immunological Value in Preventing/Remitting Opportunistic Infections," *Positive Health News* 16 (Spring 1998): 2.
31. N. Ostrum, "Apocalypse Now: The New Miracle Drugs for 'AIDS' May Have to Be Administered at Gunpoint," *New York Native*, 691 (15 July 1996): 22-23.
32. "Combo-Buffalo-Humps," *Continuum* 5(2): 3 (Winter 1997/1998); *New York Times*, 5 February 1998.
33. "Female Matters," *Continuum* 5(2): 4 (Winter 1997/1998).

34. Ibid.; *Globe* 17 January 1998.
35. "'New' Protease Inhibitor Effects," *Rethinking AIDS* Web site, http://www. virusmyth.com, May 1998.
36. Ibid.; *Marketletter*, 16 June 1997.
37."'New' Protease Inhibitor Effects."
38. Ibid.; The Associated Press, 11 June 1997.
39. IAJR Web site, http://refuse-resist.com/iajr/; Merck and Hoffman-La Roche package inserts for Crixivan and Invirase.
40. Koliadin, personal communication with the author.
41. Mark Konlee, "The Worst of the Drug Combination Therapies," *Positive Health News* 16 (Spring 1998): 4.
42. Ibid.
43. "Dangerous Drug Cocktails," *Rethinking AIDS* Web site, http://www.virusmyth. com, May 1998.
44. Ibid.; *POZ*, July 1997.
45. Ostrum, "Apocalypse Now."
46. Ibid.
47. Ibid., 27.
48. Lauritsen, "Protease Inhibitors in Provincetown."
49. Christie, "From Hype."
50. Ibid.
51. "Protease Hero Dies," *Continuum* 5(1): 3 (Autumn 1997).
52. Ibid.
53. R. Lacayo, "Hope With an Asterisk," *Time*, 30 December 1996, 82.
54. "Protease Inhibitors," *Continuum* 5(2): 4 (Winter 1997/1998).
55. Ibid.: *AIDS* 15 (1997).
56. "Combos Fail," *Continuum* 5(1): 2 (Autumn 1997).
57. Christie, "From Hype," 12.
58. Ibid.
59. Ibid.
60. *AIDS* 1 (7): 127-128 (January 1993). See also Mark Konlee, "AZT Suppresses Immune Response," *Positive Health News* 16 (Spring 1998): 4-5; L. Mercure et al., *Immunology* 1 (4): 482-85 (July 1994); Michael Oldstone, *New England Journal of Medicine* (30 November 1997).
61. Konlee, "AZT Suppresses."
62. Ibid.
63. Ibid.; B. Tindall et al., *AIDS* 7(1): 127-28 (January 1993); Mecure et al., op cit.; Oldstone, op cit.; E. Benbrik et al., *J Neurol Sci* 149 (1): 19-25 (July 1997).
64. Gorman, "Disease Detective."
65. N. Ostrum, "Will San Francisco Turn into Auschwitz II?" *New York Native*, 9 September 1996.
66. Ibid.
67. C. Farber, "AIDS: Words from the Front."
68. Ian Young, "Prescription for Suicide: Gays, AZT and Mind Control," *New York Native* 12 September 1988.
69. Elmer-Dewitt, "Turning the Tide."
70. A. Purvis, "The Global Epidemic," *Time*, 30 December 1996, 76-77.
71. "UN's Chaper Combos," *Continuum* 5(2): 3 (Winter 1997/1998); *Financial Times*, 6 November 1997.
72. Ostrum, "Apocalypse Now."
73. Ibid.
74. Ibid.'
75. Ostrum, "Will San Francisco Turn."

76. Ibid.
77. Ibid.
78. Ostrum, "Apocalypse Now," 29-31.
79. Ibid.
80. "Adult Day-Care Centers?" IAJR Web site, http://refuse-resist.com/iajr/, 2 March 1997.
81. Ibid.
82. Ibid.
83. Johnson, "Viral Load," 36.
84. Samuel Taylor Coleridge, "Fears in Solitude," in *The Selected Poetry and Prose of Samuel Taylor Coleridge*, ed. Donald A. Stauffer (New York: The Modern Library, 1951), 68.

PART SIX: ALTERNATIVE TREATMENTS
The New Atlantis

1. Lord Bacon, *New Atlantis*, in *The Moral and Historical Works of Lord Bacon*, ed. Joseph Devey (London: George Bell and Sons, 1890), 286.

Nutritional and Lifestyle Changes

1. S. A. Cohen and D. P. Kotler, "Malnutrition and Acquired Immunodeficiency Syndrome," *Current Opinions in Gastrenterology* 7 (1991):, 284-89; C. McCorkindale, K. Dybevik, A.M. Coulston, and K.P. Sucher, "Nutritional Status of HIV-Infected Patients During the Early Disease Stages," *Journal of American Dietetic Association* 90 (9): 1236 (1990). See also D. R. Hoover, A.J. Saah, H. Bacellar et al., "Signs and Symptoms of 'Asymptomatic' HIV-1 Infection in Homosexual Men. Multicenter AIDS Cohort Study," *Journal of Acquired Immune Deficiency Syndrome* 6 (1):66-71 (January 1993).
2. E. D. Ehrenpreis, D. R. Ganger, G. T. Kochvar et al., "D-xylose Malabsorption: Characteristic Finding in Patients with the AIDS Wasting Syndrome and Chronic Diarrhea," *Journal of Acquired Immune Deficiency Syndrome* 5 (10): 1047-50 (October 1992).
3. M. K. Baum, G. Shor-Poster, P. I. Bonvehi, et al., "Influence of HIV Infection on Vitamin Status and Requirement," *Annals of the New York Academy of Science* 669 (1992): 165-74.
4. G. Lake-Bakaar et al., "Gastric Secretory Failure in Patients with the Acquired Immunodeficiency Syndrome (AIDS)," *Annals of Internal Medicine* (1988): 502-04.
5. D. J. Noon, "Chinese Medicinal Herbs (Directory of Antiviral and Immunomodulatory Therapies for AIDS, Part 1)," *CDC AIDS Weekly*, 45 (1990): 2; J. C. Melchoir, D. Salmon et al., "Resting Energy Expenditure is Increased in Stable, Malnourished HIV-Infected Patients," *American Journal of Clinical Nutrition* 53 (2): 437-41 (1991).
6. Melchior, "Resting Energy Expenditure."
7. F. B. Cerra et al., "Improvement in Immune Function in ICU Patients by Enteral Nutrition Supplemented with Arginine, RNA, and Menhaden Oil is Independent of Nitrogen Balance," *Nutrition* 7 (3): 193 (May-June 1991; F. B. Cerra et al., "Nutrient Modulation of Inflammatory and Immune Function," *American Journal of Surgery* 161 (2): 230-34 (1991).
8. G. R. Harriman, P. D. Smith, M. K. Horne et al., "Vitamin B_{12} Malabasorption in Patients with Acquired Immunodeficiency Syndrome," *Archives of Internal Med icine* 149 (9): 2039-41 (1989); R. Beach et al., "Correlation Between Vitamin B_{12} Levels and Cognitive Ability Noted in Patients with HIV Infection," *CDC AIDS Weekly* 11 (2) (1992); R. S. Beach, R. Morgan et al., "Plasma Vitamin B_{12} Level

as a Potential Cofactor in Studies of Human Immunodeficiency Virus Type 1-Related Cognitive Changes," *Arch Neurol* 49 (5): 501-06 (May 1992).

9. Harriman, XMiaht, Horne et al. "Vitamin B12 Malabsorption": Mantero-Atienza, M.K. Baum, R. Morgan, F. Wilkie et al., "Vitamin B12 in Early Human Immunodeficiency Virus-1 Infection," *Archives of Internal Medicine* 151 (May 1991): 1019-20.
10. F. L. Wilkie, C. Eisdorfer, R. Morgan et al., "Cognition in Early Human Immunodeficiency Virus Infection," *Arch Neurol* 47 (April 1990): 433-40.
11. H. Heseker, W. Kubler, V. Pudel, and J. Westenhoffer, "Psychological Disorders as Early Symptoms of Mild to Moderate Vitamin Deficiency," *Annals of the New York Academy of Science* 669 (30 September 1992): 352-57.
12. D. J DeNoon, "Chinese Medicinal Herbs"; Beach et al., "Correlation Between Vitamin B12 Levels and Cognitive Ability Noted in Patients with HIV Infection," *CDC AIDS Weekly* 11 (2) (13 July 1992); Beach, et al.,"Specific Nutrient Abnormalities in Asymptomatic HIV-1 Infection," *AIDS* 6 (1992): 701-08.
13. P. Boudes, E. Henrion-Geant, L. Mandelbrot et al., "Folate, Vitamin B12 and HIV Infection," *Lancet* 335 (1990): 1401-02.
14. Cerra et al., "Nutrient Modulation of Inflammatory."
15. Ibid.
16. N. D. Penn et al., "The Effects of Dietary Supplementation with Vitamins A, C and E on Cell-Mediated Immune Function in Elderly Long-Stay Patients: A Randomized Control Trial," *Age and Aging* 20 (3): 169-174 (1991). (Summarized as "Vitamins and Immune Function in the Elderly," in *Nutrition Research Newsletter*, 10 (7-8): 83 (July-August 1991).
17. Ibid.
18. S. N. Meydani, "Dietary Modulation of Immune Response in the Aged," *Age* 14 (1991): 108-15.
19. S. N. Meydani, M. Hayek, and L. Coleman, "Influence of Vitamin E and B6 on Immune Response," *Annals of the New York Academy of Science* n 669 (1992): 125-37.
20. K. Schmidt, "Antioxidant Vitamins and Beta-Carotene: Effects on Immunocompetence," *American Journal of Clinical Nutrition* 53 (1 suppl.): 383S-5S. (1991).
21. H. S. Garewal, N. M. Ampel, R. R. Watson et al., "A Preliminary Trial of Beta-Carotene in Subjects Infected with the Human Immunodeficiency Virus," *Journal of Nutrition* 122 (3) (suppl.): 728-32 (1992).
22. L. A. McKeown, "Beta Carotene Lifts CD4 Counts," *Medical Tribune* 1 (1993).
23. S. Harakeh, R. J. Jariwalla, and L. Pauling, "Suppression of Human Immunodeficiency Virus Replication by Ascorbate in Chronically and Acutely Infected Cells," *Proceedings of the National Academy of Sciences* 87 (September 1990): 7245-49.
24. S. Harakeh and R.J. Jariwalla, "Comparative Study of the Anti-HIV Activities of Ascorbate and Thiol-Containing Reducing Agents in Chronically HIV-Infected Cells," *American Journal of Clinical Nutrition* 54 (6) (supplement): 1231S-1335S (1991).
25. Ibid.
26. Ibid.
27. S. R. Gogu, J. J. Lertora et al., "Protection of Zidovudine-Induced Toxicity Against Murine Erythroid Progenitor Cells by Vitamin E," *Exp. Hematol* 19 (7): 649-52 (August 1991); S. R. Gogu, B. S. Beckman, and Agrawal, K. C., "Amelioration of Zidovudine-Induced Fetal Toxicity in Pregnant Mice," *Antimicrob Agents Chemother* 36 (11): 2370-74 (November 1992).

28. A. Hind, "Nutrients as Modulators of Immune Function," *Canadian Medical Association Journal* 145 (1): 35 (1July 1991).

Nutrional Depletion and Augementation

1. C. De Simone et al., "High Dose L-carnitine Improves Immunologic and Metabolic Parameters in AIDS Patients," *Immunopharmacol Immunotoxicol* 15 (1): 1-12 (January 1993).
2. E. Sinforiani et al., "Neuropsychological Changes in Demented Patients Treated with Acetyl-L-Carnitine," *International Journal of Clinical Pharmacology Research* 10 (1-2): 69-74 (1990).
3. 1851. E. Bonavita, "Study of the Efficacy and Tolerability of L-acetylcarnitine Therapy in the Senile Brain," *International Journal of Clinical and Pharmacol Ther Toxicol* 24 (9): 511-16 (September 1986).
4. M. F. Bernet et al., "Lactobacillus Acidophilus LA 1 Binds to Cultured Human Intestinal Cell Lines and Inhibits Cell Attachment and Cell Invasion by Enterovirulent Bacteria," *Gut* 35 (4): 483-89 (April 1994).
5. Ibid. See also H. Link-Amster et al., "Modulation of a Specific Humoral Immune Response and Changes in Intestinal Flora Mediated Through Fermented Milk Intake," *FEMS Immunol Med Microbiol* 10 (1): 53-61 (November 1994).
6. S. Salminen and M. Deighton, "Lactic Acid Bacteria in the Gut in Normal and Disordered States," *Digestive Disorders* 10 (4): 227-238 (1992).
7. M. R. Gismondo et al., "Competitive Activity of a Bacterial Preparation on Colonization and Pathogenicity of C. pylori. A Clinical Study," *Clin Ter*, 34 (1): 41-46 (15 July 1990).
8. G. Pecorella et al., "The Effect of Lactobacillus Acidophilus and Bifidobacterium Bifidum on the Intestinal Ecosystem of the Elderly Patient," *Clin Ter* 140 (1): 3-10 (January 1992).
9. L. Motta, et al., "Study on the Activity of a Therapeutic Bacterial Combination in Intestinal Motility Disorders in the Aged," *Clin Ter* 38 (1): 27-35 (15 July 1991).
10. D. L. Witsell et al., "Effect of Lactobacillus Acidophilus on Antibiotic-associated Gastrointestinal Morbidity: A Prospective Randomized Trial," *Journal of Otolaryngol* 24 (4): 230-33 (August 1995).
11. L. Alm, "Acidophilus Milk for Therapy in Gastrointestinal Disorders," *Nahrung*, 28 (6-7): 683-84 (1984).
12. V. Gotz et al., "Prophylaxis Against Ampicillin-associated Diarrhea with a Lactobacillus Preparation, " *American Journal of Hosp Pharm* 36 (6): 754-57 (June 1979; I. Contardi, "Oral Bacteria Therapy in Prevention of Antibiotic-Induced Diarrhea in Childhood," *Clin Ter* 136 (6): 409-413 (13 March 1991).
13. G. Perdigon et al., "Systemic Augmentation of the Immune Response in Mice by Feeding Fermented Milks with Lactobacillus Casei and Lactobacillus Acidophilus," *Immunology* 63 (1): 17-23 (January 1988).
14. G. Perdigon et al., "Enhancement of Immune Response in Mice Fed with Streptococcus Thermophilus and Lactobacillus Acidophilus," *Journal of Dairy Science* 70(5): 919-26 (May 1987).
15. S. J. Klebanoff and R. W. Coombs, "Viricidal Effect of Lactobacillus Acidophilus on Human Immunodeficiency Virus Type 1: Possible Role in Heterosexual Transmission," *Journal of Exp Medicine* 174 (1): 289-92 (1 July 1991).
16. A. Baur et al., "Inhibition of HIV-infectivity and Replication by Alpha-lipoic Acid," *International Conference on AIDS* 7 (1): 110 (16-21 June 1991); A. Baur et al., "Alpha-lipoic Acid Is an Effective Inhibitor of Human Immuno-deficiency Virus (HIV-1) Replication," *Klin Wochenschr* 69 (15): , 722-724 (2 October 1991).
17. Percorella et al., "The Effects of Lactobacillus"; Gismundo et al., "Competitive Activity"; V. F. Kuznetsov et al., "Intestinal Dysbacteriosis in Yersiniosis Patients

and the Possibility of its Correction with Biopreparations," *Ter Arkh* 66 (11): 17-18 (1994).

18. J. M. Saavedra et al., "Feeding of Bifidobacterium Bifidum and Streptococcus Thermophilus to Infants in Hospital for Prevention of Diarrhoea and Shedding of Rotavirus," *Lancet* 344 (8929): 1046-49 (15 October, 1994).
19. L. C. Duffy et al., "Effectiveness of Bifidobacterium Bifidum in Mediating the Clinical Course of Murine Rotavirus Diarrhea," *Pediatric Research* 35 (6): 690-95 (June 1994).
20. N. N. Mal'tseva et al., "Immunomodulating Properties of Various Microbes— representatives of Normal Intestinal Microflora," *Antibiot Khimioter* 37 (12): 41-43 (December 1992). For effects on immune system, see K. Sekine et al., "Adjuvant Activity of the Cell Wall of Bifidobacterium Infantis for In Vivo Immune Responses in Mice," *Immunopharmacol Immunotoxicol* 16 (4): 589-609 (November 1994).
21. C De Simone et al., "Effect of Bifidobacterium Bifidum and Lactobacillus Acidophilus on Gut Mucosa and Peripheral Blood B Lymphocytes," *Immunopharmacol Immunotoxicol* 14 (1-2): 331-40 (1992). See also, for immunity, Sekine, "Adjuvant Activity."
22. G.W. Elmer et al., "Biotherapeutic Agents. A Neglected Modality for the Treatment and Prevention of Selected Intestinal and Vaginal Infections," *JAMA* 275 (11): 870-76 (20 March 1996).
23. A. D. Pivazyan, "Boron Modified Peptides as Inhibitors of HIV-1 Protease," *Int Conf AIDS* 9 (1): 230 (6-11 June 1993).
24. L. P. Hale and B. F. Haynes, "Bromelain Treatment of Human T Cells Removes CD44, CD45RA, E2/MIC2, CD6, CD7, CD8, and Leu 8/LAM1 Surface Molecules and Markedly Enhances CD2-Mediated T Cell Activation," *Journal of Immunology* 149 (12): 3809-16 (15 December 1992).
25. See the review article, K. Folkers and A. Wolaniuk, 'Research on Coenzyme Q10 in Clinical Medicine and in Immunomodulation," *Drugs Exp Clin Res* 11 (8): 539-45 (1985).
26. K. Folkers et al., "Biochemical Deficiencies of Coenzyme Q10 in HIV-infection and Exploratory treatment," *Biochem Biophys Res Commun* 153 (2): 888-96 (16 June 1988).
27. K. Folkers et al., "Coenzyme Q10 Increases T4/T8 Ratios of Lymphocytes in Ordinary Subjects and Relevance to Patients Having the AIDS Related Complex," *Biochem Biophys Res Commun* 176 (2): 786-791 (30 April 1991).
28. K. Folkers et al., "The Activities of Coenzyme Q10 and Vitamin B6 for Immune Responses," *Biochem Biophys Res Commun* 1931): 88-92 (28 May 1993).
29. M. A. Jacobson et al., "Possible Protective Effects of Dehydroepiandrosterone (DHEA) and/or DHEA-sulfate (DHEA-S) in HIV Infection," *International Conference on AIDS* 5: 560 (4-9 June 1989).
30. E. Henderson et al., "Dehydroepiandrosterone (DHEA) and Synthetic DHEA Analogs are Modest Inhibitors of HIV-1 IIIB Replication," *AIDS Res Hum Retroviruses* 8 (5): 625-31 (May 1992). See also Jacobson et al., "Possible Protective Effect of Dehydroepiandrosterone (DHEA)."
31. J. S. James, "DHEA: Modest Viral Load Reduction in Patients," *AIDS Treatment News* 252 7-8 (2 August 1996). See also P. Salvato et al., "Viral Load Response to Augmentation of Natural Dehydroepiandrosterone (DHEA)," *International Conference on AIDS* 11 (2): 124 (7-12 July 1996).
32. C. Thompson et al., "Effects of Plasma Dehydroepiandrosterone (DHEA) Levels on HIV RNA Plasma Levels as Measured by PCR," *International Conference onAIDS* 11 (2): 311 (7-12 July 1996).

33. D. Hasheeve et al., "DHEA: A Potential Treatment for HIV Disease," *International Conference on AIDS* 10 (1): 223 (7-12 August 1994).

34. A. T. Palamara et al., "Evidence for Antiviral Activity of Glutathione: In Vitro Inhibition of Herpes Simplex Virus Type 1 Replication," *Antiviral Research* 27 (3): 237-53 (June 1995).

35. D. H. Baker, "Cellular Antioxidant Status and Human Immunodeficiency Virus Replication," *Nutrition Review* 50 (11): 15-18.

36. S. M. Liang et al., "Regulation by Glutathione of Interleukin-4 Activity on Cytotoxic T Cells," *Immunology* 75 (3): 435-40 (March 1992).

37. T. Kalebic et al., "Suppression of Human Immunodeficiency Virus Expression in Chronically Infected Monocytic Cells by Glutathione Ester & N-acetyl-cysteine," *Proceedings of the National Academy of Sciences* 88 (February 1991): 986-90.

38. Ibid.

39. David Ho and S. D. Douglas, "Glutathione and N-acetylcysteine Suppression of Human Immunodeficiency Virus Replication in Human Monocyte/macrophages in Vitro," *AIDS Research Human Retroviruses* 8 (7): 1249-53 (July 1992).

40. A. T. Palamara et al., "Glutathione Directly Inhibits Late Stages of the Replication Cycle of HIV and Other Viruses," *International Conference on AIDS* 9 (1): 231 (6-11 June 1993).

41. Valenzuela et al., *Planta Medica* 55 (1989): 42.

42. Dr. Joan Priestey, interview with the author, in *The Pain, Profit and Politics of AIDS*.

43. Dr. Chistopher Calapai, lecture at AIDS Symposium, 11 December 1991.

44. N. A. Darwish and M. Shaarawy, "Effect of Treatment with Povidone-iodine Vaginal Pessaries on Thyroid Function," *Postgraduate Medical Journal* 69 (suppl 3): S39-42 (1993). See also H. Yu and M. Tak-Yin, "The Efficacy of Povidone-iodine Pessaries in a Short, Low-dose Treatment Regime on Candidal, Trichomonal and Non-specific Vaginitis," *Postgraduate Medical Journal* 69 (suppl. 3): S58-S61 (1993): R. Grio et al., "Effectiveness of Povidone-iodine in the Treatment of Non-specific Vaginitis," *Minerva Ginecol* 4 (24): 129-31 (April 1990).

45. S. J. Isenberg et al., "Povidone-iodine for Pphthalmia Neonatorum Prophylaxis," *American Journal of Ophthalmology* 118 (6): 701-06 (15 December 1994). For effects on children, S. J. Isenberg et al., "A Controlled Trial of Povidone-iodine as Prophylaxis Against Ophthalmia Neonatorum," *New England Journal of Medicine* 332 (9): 562-66 (2 March 1995). For effects on adults, G. Schuhman and B. Vidic, "Clinical Experience with Povidone-iodine Eye Drops in Patients with Conjunctivitis and Keratoconjunctivitis," *Journal of Hospital Infection* 6 (suppl. A): 173-75 (March 1995).

46. J. M. Hay et al., "Povidone-iodine Enema as a Preoperative Bowel Preparation for Colorectal Surgery. A Bacteriologic Study," *Dis Colon Rectum* 32 (1): 9-13 (January 1989).

47. R. Rahn, "Review Presentation on Povidone-iodine Antisepsis in the Oral Cavity," *Postgraduate Medical Journal* 69 (suppl 3): S4-S9 (1993).

48. For nasal, see H. Masano et al., "Efficacy of Intranasal Application of Povidone-iodine Cream in Eradicating Nasal Methicillin-resistant Staphylococcus Aureus in Neonatal Intensive Care Unit (NICU) Staff," *Postgraduate Medical Journal* 69 (suppl. 3): S122-25 (1993). For mouth, see R .F. Muller et al., "Efficacy of a PVP-iodine Compound on Selected Pathogens of the Oral Cavity in Vitro," *Dtsch Zahnarztl Z* 44 (5): 366-69 May 1989).

49. V. K. Manna et al., "The Effect of Povidone-iodine Paint on Fungal Infection," *Journal of Intern Med Research* 12 (2): 121-23 (1984).

50. E. G. Friedrich Jr and T. Masukawa, "Effect of Povidone-iodine on Herpes Genitalis," *Obstet Gynecol* 45 (3): 337-39 (March 1975).

51. M. S. Amstey and S. Metcalf, "Effect of Povidone-iodine on Herpesvirus Type 2, in Vitro," *Obstet Gynecol* 46 (5): 528-29 (November 1975).
52. P. D. Goldenheim, "In Vitro Efficacy of Povidone-iodine Solution and Cream Against Methicillin-resistant Staphylococcus Aureus," *Postgraduate Medical Journal* (69) (suppl. 3): S62-65 (1993). See also A. R. Kohlhaas and J. L. Sandrik, "The Antimicrobial Properties of Povidone-iodine Methylmethacrylate Complex. A Preliminary Report," *Clinical Orthop* 113 (November-December 1975): 184-86.
53. On rats, P. W. Caufield et al., "Effect of Topically-applied Solutions of Iodine, Sodium Fluoride, or Chlorhexidine on Oral Bacteria and Caries in Rats," *Journal of Dental Research* 60 (5): 927-32 (May 1981). On rabbits, L. G. Sharma et al., "Evaluation of Topical Povidone-iodine Versus Gentamycin in Staphylococcus Coagulase Positive Corneal Ulcers—An Experimental Study," *Indian Journal of Ophthalmology* 38 (1) 30-32 (January-March 1990).
54. Y. Yura et al., "Inhibition of Herpes Simplex Virus Replication by Genistein, an Inhibitor of Protein-tyrosine Kinase," *Arch Virol* 132 (3-4): 451-61 (1993).
55. M. Mintz, et al., "Levodopa Therapy Improves Motor Function in HIV-infected Children with Extrapyramidal Syndromes," *Neurology* 47 (6): 1583-88 (December 1996).
56. M. W. Wichmann et al., "Melatonin Administration Attenuates Depressed Immune Functions Trauma-Hemorrhage," *Journal of Surgical Research* 63 (1): 256-62 (June 1996).
57. M. C. Caroleo et al., "Melatonin Restores Immunodepression in Aged and Cyclophosphamide-treated Mice," *Annals of the New York Academy of Science* 719 (31 May 1994): 343-52: M. C. Caroleo et al., "Melatonin as Immunomodulator in Immunodeficient Mice," *Immunopharmacology* 23 (2): (March-April 1992) 81-89.
58. G. J. Maestroni et al., "Role of the Pineal Gland in Immunity. III. Melatonin Antagonizes the Immunosuppressive Effect of Acute Stress via an Opiatergic Mechanism," *Immunology* 63 (3): (March 1988): 465-69.
59. B. de Quay et al., "Glutathione Depletion in HIV-infected Patients: Role of Cysteine Deficiency and Effect of Oral N-acetylcysteine," *AIDS* 6 (8): 815-19 (August 1992).
60. J. S. James, "NAC: First Controlled Trial, Positive Results," *AIDS Treatment News* 250 (5 July 1996): 1-3.
61. R. Kinscherf et al., "Effect of Glutathione Depletion and Oral N-acetyl-cysteine Treatment on CD4+ and CD8+ Cells," *FASEB Journal* 8 (6): 448-51 (1 Apri 1994). Also R. L. Roberts et al., "N-acetylcysteine Enhances Antibody-dependent Cellular Cytotoxicity in Neutrophils and Mononuclear Cells from Healthy Adults and Human Immunodeficiency Virus-infected Patients," *Journal of Infectious Disease* 172 (6): 1492-1502 (December 1995).
62. David Ho and S. D. Douglas, "Glutathione and N-acetylcysteine Suppression of Human Immunodeficiency Virus Replication in Human Monocyte/Macrophages in Vitro," *AIDS Research Hum Retroviruses* 8 (7): 1249-53 (July 19920.
63. M. Roederer et al., "N-acetylcysteine: A New Approach to Anti-HIV Therapy," *AIDS Research and Human Retroviruses* 8 (2): 209-17 (February 1992).
64. P. A. Raju et al., "Glutathione Precursor and Antioxidant Activities of N-acetylcysteine and Oxothiazolidine Carboxylate Compared in Vitro Studies of HIV Replication," *AIDS Research and Human Retroviruses* 10 (8): 961-67 (August 1994). Also of interest, M. Roederer et al., "N-acetylcysteine: Potential for AIDS Therapy,"*Pharmacology* 46 (3): 121-29 (1993); M. Roederer et al., "Cytokine-stimulated Human Immunodeficiency Virus Replication is Inhibited by N-acetyl-L-cysteine," *Proceedings of the National Academy of Sciences* 87 (12): 4884-88 (June 1990).

65. R. E. Walker et al., "The Safety, Pharmacokinetics, and Antiviral Activity of N-acetylcysteine in HIV-Infected Individuals," *International Conference on AIDS* 8(1): Mo8 (19-24 July 1992).

66. E. Ylar, et al., "N-acetylcysteine Enhances T Cell Functions and T Cell Growth in Culture," *Int Immunol* 5 (1): 97-101 (January 1993).

67. R. Schlegel et al., "Inhibition of VSV Binding and Infectivity by Phosphatidylserine: Is Phosphatidylserine a VSV-binding Site?," *Cell* 32 (2): 639-46 (February 1983).

68. V. Guarcello et al., "Phosphatidylserine Enhances the Ability of Epidermal Pharmacological Suppression of Humoral Response," *Immunopharmacology* 19 (3): 185-95 (May-June 1990).

69. Mark Konlee, "Selenium Increases Glutathione Levels, Lowers Beta 2 Microglobulin Levels, Fights Cancer & Improves Survival in AIDS," *Positive Health News* 16 (spring 1998): 11.

70. Ibid.

71. Ibid.

72. L. Kiremidjian-Schumacher et al., "Supplementation with Selenium Augments the Functions of Natural Killer and Lymphokine-activated Killer Cells," *Biological Trace Element Research* 52 (3): 227-39 (June 1996).

73. M. Bonomini et al., "Effects of Selenium Supplementation on Immune Parameters in Chronic Uraemic Patients on Haemodialysis," *Nephrol Dial Transplant* 10 (9): 1654-61 (1995).

74. A. Peretz et al., "Effects of Selenium Supplementation on Immune Parameters in Gut Failure Patients on Home Parenteral Nutrition," *Nutrition* 7 (3): 215-21 (May-June 1991).

75. L. Santamaria and A. Bianchi-Santamaria, "Carotenoids and Vitamin A in Prevention, Adjuvant Cancer Therapy, Mastalgia Treatment and AIDS-Related Complex," *Annals of the New York Academy of Sciences* 691 (31 December 1993): 255-58; D. L. Karter, A.J. Karter et al., "Vitamin A Deficiency in Non-Vitamin-Supplemented Patients with AIDS: A Cross-sectional Study," *Journal of Acquired Immune Deficiency Syndrome and Human Retrovirology* 8 (2): 199-203 (1 February 1995).

76. G. Coodley, "Vitamin B6 to Increased NK Cell Activity and Beta Carotene for Increasing NK Cell Counts," *Positive Health News* (Fall 1997): 11.

77. Ibid.

78. L. K. Altman, "Vitamin A Deficiency Linked to Transmission of AIDS Virus from Mothers to Infants," *New York Times* 3 February 1995.

79. Ibid. See also, R. D. Semba et al., "Vitamin A Deficiency, Infant Mortality, and Mother-to-Child Transmission of HIV," *Lancet* 343 (8913): 1593-97 (25 June 1994): R. W. Nduati et al., "Human Immunodeficiency Virus Type 1-infected Cells in Breast Milk: Association with Immunosuppression and Vitamin A Deficiency," *Journal of Infectious Disease* 172 (6), 1461-81 (December 1995).

80. R. D. Semba et al., "Abnormal T-cell Subset Proportions in Vitamin-A-deficient Children," *Lancet* 341 (8836): 5-8 (2 January 1993).

81. P. P. Glasziou and D. E. Mackerras, "Vitamin A Supplementation in Infectious Diseases: A Meta-Analysis," *British Medical Journal* 306 (6874); 366-70 (6 February 1993). See also R. D. Semba, "Vitamin A, Immunity, and Infection," *Clinical Infectious Disease* 193): 489-99 (September 1994).

82. R. D. Semba, et al. "Vitamin A Deficiency and Wasting as Predictors of Mortality in Human Immunodeficiency Virus-infected Injection Drug Users," *Journal of Infectious Disease* 171 (5): 1196-1202 (May 1995).

83. R. D. Semba et al., "Increased Mortality Associated with Vitamin A Deficiency During Human Immunodeficiency Virus Type 1 Infection," *Archives of Internal*

Medicine 153 (18): 2149-54 (27 September 1993). For vitamin A deficiency in rats, see K. M. Nauss et al., "Ocular Infection with Herpes Simplex Virus (HSV-1) in Vitamin A-deficient and Control Rats," *Journal of Nutrition* 115 (10): 1300-15 (October 1985).

84. R. R. Watson et al., "Immunostimulatory Effects of Beta-carotene on T-cell Activation Markers and NK Cells in HIV Infected Patients," *International Conference on AIDS* 5 (4-9 June 1989): 663.

85. G.O. Coodley et al., "Beta-carotene in HIV Infection," *Journal of Acquired Immune Deficiency Syndrome* 6(3): 272-76 (March 1993).

86. H. S. Garewal et al., "A Preliminary Trial of Beta-carotene in Subjects Infected with the Human Immunodeficiency Virus," *Journal of Nutrition* 122 (3 suppl.): 728-32 (March 1992). See also "Vitamin B6 to Increased NK Cell Activity."

87. G. van Poppel et al., "Effect of Beta-carotene on Immunological Indexes in Healthy Male Smokers," *American Journal of Clinical Nutrition* 57 (3): 402-07 (March 1993).

88. C.J. Fuller et al., "Effect of Beta-carotene Supplementation on Photosuppression of Delayed-type Hypersensitivity in Normal young Men," *American Journal of Clinical Nutrition* 56 (4): 684-90 (October 1992).

89. For pneumonia, see M. E. Rupp et al., "Measles Pneumonia: Treatment of a Near-Fatal Case with Corticosteroids and Vitamin A," *Chest* 103 (5): 1625-26 (May 1993); K. D. Pletsitnyi et al., "Effect of Vitamin A on Immunological Status of Patients with Chronic Pneumonia," *Vopr Med Khim* 28 (5): 119-22 (September-October 1982). For surgery, see B. E. Cohen et al., "Reversal of Postoperative Immunosuppression in Man by Vitamin A," *Surg Gynecol Obstet* 149 (5): 658-62 (November 1979). For leukemia, R. H. Prabhala et al., "Influence of Beta-carotene on Immune Functions," *Annals of the New York Academy of Sciences*, 691 (31 December 1993): 262-63. Also relevant, K. D. Pletsityi et al., "Further Study of the Role of Vitamin A in Immunologic Reactions," *Vopr Pitan* 1 (January-February 1984): 26-29: R. H. Prabhala et al., "The Effects of 13-cis-retinoic Acid and Beta-carotene on Cellular Immunity in Humans," *Cancer* 67 (6): 1556-60 (15 March 1991).

90. R. R. Watson, M. D. Yahya, H. R. Darban, R. H. Prabhala, "Enhanced Survival by Vitamin A Supplementation During a Retrovirus Infection Causing Murine AIDS," *Life Sciences* 43 (6): xiii-xviii (1988); A. Bendich and S. S. Shapiro, "Effect of Beta-carotene and Canthaxanthin on the Immune Responses of the Rat," *Journal of Nutrition* 116 (11): 2254-62 (November 1986).

91. J. L. Taylor and W. J. O'Brien, "The Effects of Retinoids on the Replication of Herpes Simplex Virus Type 1," *Current Eye Research* 3(3): 481-88 (March 1984); I. F. Abronina et al., "The Beta-carotene Stimulation of Cellular Immunity Reactions in Mice," *Biull Eksp Biol Med* 116 (9): 295-98 (September 1993); N. Nuwayri-Salti and T. Murad, "Immunologic and Anti-immunosuppressive Effects of Vitamin A," *Pharmacology* 30 (4): 181-87 (1985); T. V. Davydova et al., "Further Study of Immuno-correcting Properties of Vitamins A and E in Experimental Chronic Alcoholic Intoxication," *Vopr Pitan* (3): 45-48 (May-June 1988); T. V. Davydova et al., "The Use of Vitamins A and E for the Correction of Immunologic Disorders in Chronic Alcoholic Intoxication,"*Vopr Pitan* (4): 50-52 (July-August 1987).

92. A. Bianchi-Santamaria et al., "Short Communication: Possible Activity of Beta-Carotene in Patients with the AIDS Related Complex: A Pilot Study," *Medical Oncology Tumor Pharmacotherapy* 9 (3): 151-53 (1992).

93. P. Pontiggia et al., "Whole Body Hyperthermia Associated with Beta-carotene Supplementation in Patients with AIDS," *Biomed Pharmacotherapy* 49 (5): 263-65 (1995).

94. A. Coutsoudis et al., "The Effects of Vitamin A Supplementation on the Morbidity of Children Born to HIV-infected Women," *American Journal of Public Health* 85 (8 Pt. 1): 1076-81 (August 1995).

95. D. A. Fryburg et al., "The Immunostimulatory Effects and Safety of Beta-carotene in Patients with AIDS," *International Conference on AIDS*, 8(2): B163 (19-24 July 1992).

96. S. Shoji et al., "Thiamine Disulfide as a Potent Inhibitor of Human Immunodeficiency Virus (type-1) Production," *Biochem Biophys Res Commun* 205 (1): 967-75 (30 November 1994).

97. S. Araki et al., "Enhancement of Resistance to Bacterial Infection in Mice by Vitamin B2," *Journal of Veterinary Medical Science* 57 (4): 599-602 (August 1995).

98. E. Mantero-Atienza et al., "Vitamin B6 and Immune Function in HIV Infection," *International Conference on AIDS* 6(2): 432 (20-23 June 1990).

99. B. C. Herzlich and T. D. Schiano, "Reversal of Apparent AIDS Dementia Complex Following Treatment with Vitamin B12," *Journal of Internal Medicine* 233 (6): 495-507 (June 1993).

100. B. C. Herzlich, T. D. Schiano, Z. Moussa et al., "Decreased Intrinsic Factor Secretion in AIDS: Relation to Parietal Cell Acid Secretory Capacity and Vitamin B12 Malabsorption," *American Journal of Gastroenterology* 87 (12): 1781-88 (December 1992).

101. For an in vitro study, see T. Sakane et al., "Effects of Methyl-B12 on the in Vitro Immune Functions of Human T Lymphocytes," *Journal of Clinical Immunology* 2 (2): 101-09 (April 1982).

102. B. C. Herzlich, T. D. Schiano, Z. Moussa et al., "Decreased Intrinsic Factor Secretion."

103. M. K. Baum et al., "Vitamin B12 and Cognitive Function in HIV Infection," *International Conference on AIDS* 6 (2): 97 (20-23 June 1990).

104. B. Regland et al., "Vitamin B12-induced Reduction of Platelet Monoamine Oxidase Activity in Patients with Dementia and Pernicious Anaemia," *Eur Arch Psychiatry Clin Neurosci* 240 (4-5) 288-291 (1991).

105. R. F. Cathcart, "Vitamin C in the Treatment of AIDS," *Medical Hypotheses* 14 (1984): 423-33.

106. Robert Cathcart, interview by the author, in *The Pain, Profit and Politics of AIDS*.

107. Ibid.

108. Ibid.

109. Ibid.

110. Dr. Christopher Calapai, interview by the author, in *The Pain, Profit and Politics of AIDS* . See also Raxit Jariwalla, interview by the author in *The Pain, Profit and Politics of AIDS*.

111. Joan Priestley, interview by the author in *The Pain, Profit and Politics of AIDS*.

112. Dolores Perri, interview by the author, in *AIDS, The Untold Story*.

113. Joy DeVincenzo, interview by the author, in AIDS, The Untold Story.

114. Robert Cathcart, interview by the author.

115. Ibid.

116. Ibid.

117. Robert Cathcart, interview by the author, in *The Pain, Profit and Politics of AIDS*.

118. J. Fitzherbert, "Genital Herpes and Zinc," *Medical Journal Aust* 1 (1979): 399.

119. G. T. Terezhalmy et. al., "The Use of Water-Soluble Bioflavanoid-Ascorbic Acid Complex in the Treatment of Recurrent Herpes Labialis,"*Oral Surgery* 45 (1978): 56-62.

120. E. M. Peters et al., "Vitamin C Supplementation Reduces the Incidence of Postrace Symptoms of Upper-Respiratory-Tract Infection in Ultramarathon Runners," *American Journal of Clinical Nutrition* (2): 170-74 (February 1993).

121. R. J. Jariwalla and S. Harakeh, "Ascorbic Acid and AIDS: Strategic Functions and Therapeutic Possibilities," in *Nutrition and AIDS* , ed. R. R. Watson (Boca Raton, Fl.: CRC Press, 1994), 125.

122. Ibid.; I. Stone, *The Healing Factor: Vitamin C Against Disease* (New York: Grosset and Dunlap, 1972), 258; C. W. Jungeblut, "Inactivation of Poliomyelitis Virus by Crystalline Vitamin C (ascorbic acid)," *Journal of Experimental Medicine* 62 (1935): 517; M. Holden et al., "Further Experiments on Inactivation of Herpes Virus by Vitamin C (L-ascorbic acid)," *Journal of Experimental Medicine*, 33 (1937): 251; I. J. Kliger et al., "Inactivation of Vaccinia Virus by Ascorbic Acid and Glutathione," *Nature* 139 (1937): 965; W. Langenbusch et al., "Einfluss der Vitamine Auf Das Virus Der Maul-Und Klavenseuch," *Zentralblatt fur Bakteriologie*, 140 (1937): 112; G. Amato, "Azione Dell'Acido Ascorbico Sul Virus Fisso Della Rabbia e Sulla Tossina Tetanica, Giornale de Batteriologia," *Virologia et Immunologia* 19 (1937): 843; M. Lojkin, "Contributions of the Boyce Thompson Institute," *Proceedings Third International Congress of Microbiology* 8 (4) (New York: L.F. Martin, 1940), 281.

123. C. Bucca et al., "Effect of Vitamin C on Histamine Bronchial Responsiveness of Patients with Allergic Rhinitis," *Annals of Allergy* 65 (4): 311-14 (October 1990).

124. C. Bucca et al., "Effects of Vitamin C on Airway Responsiveness to Inhaled Histamine in Heavy Smokers," *European Respir Journal* 2 (3): 229-33 (March 1989).

125. J. Brajtburg et al., "Effects of Ascorbic Acid on the Antifungal Action of Amphotericin B," *Journal of Antimicrobial Chemotherapy* 24 (3): 333-37 (September 1989).

126. S. Banic, "Immunostimulation by Vitamin C," *International Journal of Vitamin and Nutrition Research* 23 (suppl.) 49-52 (1982).

127. J. C. Delafuente et al., "Immunologic Modulation by Vitamin C in the Elderly," *International Journal Immunopharmacology* 8(2): 205-11 (1986). See also R. Anderson et al., "The Effects of Increasing Weekly Doses of Ascorbate on Certain Cellular and Humoral Immune Functions in Normal Volunteers," *American Journal of Clinical Nutrition* 33 (1): 71-76 (January 1980): Penn, "The Effect of Dietary Supplementation with Vitamins A, C and E"; M. Nicol, "Vitamins and Immunity," *Allerg Immunol* 25 (2): 70-73 (February 1993); T. Huwyler et al., "Effect of Ascorbic Acid on Human Natural Killer Cells," *Immunology Letters* 10 (3-4): 173-76 (1985); R. F. Cathcart III, "The Vitamin C Treatment of Allergy and the Normally Unprimed State of Antibodies," *Medical Hypotheses* 21(3): 307-21 (November 1986); J. Schwartz and S. T Weiss, "Relationship Between Dietary Vitamin C Intake and Pulmonary Function in the First National Health and Nutrition Examination Survey (NHANES I)," *American Journal of Clinical Nutrition* 59 (1): `110-114 (January 1994); V. Mohsenin, "Effect of Vitamin C on NO2-induced Airway Hyperresponsiveness in Normal Subjects: A Randomized Double-blind Experiment," *American Review of Respiratory Disease* 136 (6): 1408-11 (December 1987).

128. S. Harakeh, R. J. Jariwalla, and L. Pauling, "Suppression of Human Immunodeficiency Virus Replication by Ascorbate in Chronically and Acutely Infected Cells," *Proceedings of the Natural Academy of Sciences* 87 (September 1990): 7245-49.

129. Ibid.

130. S. Harakeh and R. J. Jariwalla, "Comparative Study of the Anti-HIV Activities of Ascorbate and Thiol-Containing Reducing Agents in Chronically HIV-Infected Cells," *American Journal of Clinical Nutrition* 54 (1991): 1231S-5S. See also R. J. Jariwalla and S. Harakeh, "HIV Suppression by Ascorbate and its Enhancement by a Glutathione Precursor," *International Conference on AIDS* 8 (2): B207 (abstract no. PoB 3697) (19-24 July 1992); S. Harakeh et al., "Suppression of Human

Immunodeficiency Virus Replication by Ascorbate in Chronically and Acutely Infected Cells," *Proceedings of the National Academy of Science* 87 (18): 7245-49 (September 1990); S. Davies et al., "Suppression of Human Immunodeficiency Virus Replication by Ascorbate in Chronically and Acutely Infected Cells," *Journal of Nutritional Medicine* 1 (1990): 345-46.

131. B. D. Rawal et al., "In Vitro Inactivation of Human Immunodeficiency Virus by Ascorbic Acid," *Biologicals* 23 (1): 75-81 (March 1995).

132. S. Harakeh et al., "Mechanistic Aspects of Ascorbate Inhibition of Human Immunodeficiency Virus," *Chem Biol Interact* 91 (2-3): 207-15 (June 1994). See also S. Harakeh and R. J. Jariwalla, "Mechanistic Aspects of Ascorbate Inhibition of Human Immunodeficiency Virus," *Chemico-Biological Interactions* (1994): 207-15.

133. R. J. Jariwalla and S. Harakeh, "Role of an Antioxidant Vitamin in Suppression of Human Immunodeficiency Virus (HIV)," American Society for Virology—12th Annual Meeting, University of California, Davis, CA10-14 July, 1993.

134. See R. J. Jariwalla and S. Harakeh, "Ascorbic Acid and AIDS: Strategic Functions "; J. R. Blakeslee et al., "Human T-Cell Leukemia Virus 1 Induction by 5-Iodo-2'-Deoxyuridine and N-methyl-N'-Nitro-N- Nitrosoguanidine: Inhibition by Retinoids, L-Ascorbic Acid and Dl-Alpha Tocopherol," *Cancer Research* 45 (1985): 3471; S. Harakeh et al., "Suppression of Human Immunodeficiency Virus Replication by Ascorbate in Chronically and Acutely Infected Cells," *Proceedings of the National Academy of Sciences* 87 (1990) 7245; S. Harakeh and R. J. Jariwalla, "Comparative Study of the Anti-HIV Activities of Ascorbate and Thiol-Containing Reducing Agents in Chronically HIV-Infected Cells," *American Journal of Clinical Nutrition* 54 (1991): 1231S; R. J. Jariwalla and S. Harakeh, "HIV Suppression by Ascorbate and Its Enhancement by a Glutathione Precursor," Eighth International Conference on AIDS, Amsterdam, July 1992, B207.

135. Jariwalla & Harakeh, "Ascorbic Acid and AIDS."

136. Ibid.

137. Ibid. and R. F. Cathcart, "A Unique Function for Ascorbate," *Medical Hypotheses* 35 (1991): 32.

138. Jariwalla and Harakeh, "Ascorbic Acid and AIDS."

139. Ibid. R. F. Cathcart, "Vitamin C in the Treatment of Acquired Immune Deficiency Syndrome (AIDS)," *Medical Hypotheses* 14 (1984): 423.

140. Harakeh and R. J. Jariwalla, "Comparative Analysis of Ascorbate and AZT Effects on HIV Production in Persistently Infected Cell Lines," *Journal of Nutritional Medicine* 4 (1994): 393-401.

141. Ibid.

142. Cathcart, "Vitamin C in the Treatment of Acquired Immune Deficiency Syndrome."

143. Ibid.

144. R. J. Jariwalla, "Micro-nutrient Imbalance in HIV Infection and AIDS: Relevance to Pathogenesis and Therapy," *Journal of Nutritional Medicine* 5 (1995). Also R. F. Cathcart, "Vitamin C in the Treatment of Acquired Immune Deficiency Syndrome. "

145. Jariwalla, "Micro-nutrient Imbalance."

146. A. M. Tang et al., "Dietary Micronutrient Intake and Risk of Progression to Acquired Immunodeficiency Syndrome (AIDS) in Human Immunodeficiency Virus Type 1 (HIV-1)-Infected Homosexual Men," *American Journal of Epidemiology* 138 (11): 937-51 (1993).

147. M. Kodama et al., "Autoimmune Disease and Allergy are Controlled by Vitamin C Treatment," *In Vivo* 8 (2): 251-57 (March-April 1994).

148. C. Huag, F. Muller, S. S. Frolant, H. Rollag, P. Aukrust, M. Degre, "Vitamin D3 Improves Subnormal in Vitro Maturation in Monocytes from AIDS Patients," *International Conference on AIDS* 9 (1): 225 (6-11 June 1993).

149. M. J. Stampfer, et al., "Vitamin E Consumption and the Risk of Coronary Disease in Women," *New England Journal of Medicine* 328 (1993): 1449.

150. A. Bendich and L. J. Machlin, "Safety or Oral Intake of Vitamin E," *American Journal of Clinical Nutrition* 48 (1988): 612-619.

151. Y. Wang and R. R. Watson, "Vitamin E Supplementation at Various Levels Alters Cytokine Production by Thymocytes During Retrovirus Infection Causing Murine AIDS," *Thymus* 22 (3): 153-64 (1994).

152. Y. Wang et al., "Vitamin E Supplementation Modulates Cytokine Production by Thymocytes During Murine AIDS," *Immunol Res* 12 (4): 358-66 (1993).

153. Y. Wang et al., "Nutritional Status and Immune Responses in Mice with Murine AIDS are Normalized by Vitamin E Supplementation," *Journal of Nutrition* 124 (10): 2024-32 (October 1994). Also see Y. Wang et al., "Long-term Dietary Vitamin E Retards Development of Retrovirus-induced Disregulation in Cytokine Production," *Clin Immunol Immunopathol* 72 (1): 70-75 (July 1994).

154. K. D. Pletsityi et al., "The Immunocorrecting Effect of Vitamin E During Ethanol Intoxication,"*Vopr Med Khim* 40 (3): 51-53 (May-June 1994).

155. S. N. Meydani et al., "Vitamin E Supplementation Suppresses Prostaglandin E1(2) Synthesis and Enhances the Immune Response of Aged Mice," *Mech Ageing Dev* 34 (2): 191-201 (April 1986). Also, S. Moriguchi et al., "High Dietary Intakes of Vitamin E and Cellular Immune Functions in Rats," *Journal of Nutrition* 120 (9): 1096-1102 (September 1990).

156. A. Bendich et al., "Effect of Dietary Level of Vitamin E on the Immune System of the Spontaneously Hypertensive (SHR) and Normotensive Wistar Kyoto (WKY) Rat," *Journal of Nutrition* 113 (10): 1920-26 (October 1983).

157. S. R. Gogu et al., "Protection of Zidovudine (AZT)-Induced Bone Marrow Toxicity and Potentiation of Anti-HIV Activity with Vitamin E," 90th Annual Meeting of the American Society for Microbiology, Anaheim, Calif., 13-17 May 1990.

158. For calves M. Hidiroglou, et al., "Possible Roles of Vitamin E in Immune Response of Calves," *International Journal of Vitam Nutr Research* 62 (4): 308-11 (1992). For guinea pigs, A. Bendich et al., "Interaction of Dietary Vitamin C and Vitamin E on Guinea Pig Immune Responses to Mitogens," *Journal of Nutrition* 114 (9): 1588-93 (September 1984). For rabbits, K. D. Pletsityi et al., "Effect of Vitamin E on T and B Lymphocyte Numbers in the Peripheral Blood and Various Indicators of Nonspecific Immunity," *Vopr Pitan* 4 (July-August 1984): 42-44. For sheep, R. P. Tengerdy, "The Role of Vitamin E in Immune Response and Disease Resistance," *Annals of the New York Academy of Sciences* 587 (1990): 24-33. Also see A. Bendich, "Vitamin E and Immune Functions," *Basic Life Sci* 49 (1988): 615-20.

159. Y. J. Suzuki and L. Packer, "Inhibition of NF-kappa B Activation by Vitamin E Derivatives," *Biochem Biophys Res Commun* 193 (1): 277-83 (28 May 1993). Also S. R. Gogu et al., "Increased Therapeutic Efficacy of Zidovudine in Combination with Vitamin E," *Biochem Biophys Res Commun*, 165 (1): 401-07 (30 November 1989).

160. Y. Wang and R.R. Watson, "Is Vitamin E Supplementation a Useful Agent in AIDS Therapy?"*Prog Food Nutr Sci* 17 (4): 351-75 (October-December 1993).

161. O. E. Odeleye and R. R. Watson, "The Potential Role of Vitamin E in the Treatment of Immunologic Abnormalities During Acquired Immune Deficiency Syndrome," *Prog Food Nutr Sci* 15 (1-2): 1-19 (1991). Also R. G. Geissler et al., "In Vitro Improvement of Bone Marrow-Derived Hematopoietic Colony

Formation in HIV-Positive patients by Alpha-D-Tocopherol and Erythropoietin," *European Journal of Haematology* 53 (4): 201-06 (October 1994); Y. Wang and R.R. Watson, "Is Vitamin E Supplementation a Useful Agent in AIDS Therapy?" *Prog Food Nutr Sci* 17 (4): 351-75 (October-December 1993); K. V. Kowdley et al., "Vitamin E Deficiency and Impaired Cellular Immunity Related to Intestinal Fat Malabsorption,"*Gastroenterology* 102 (6): 2139-42 (June 1992).

162. R. J. Sokol et al., "Improvement of Cyclosporin Absorption in Children after Liver Transplantation by Means of Water-soluble Vitamin E," *Lancet* 338 (8761): 212-14 (17 July 1991). On its effect on those on dialysis, M. Taccone-Gallucci et al., "Vitamin E Supplementation in Hemodialysis Patients: Effects on Peripheral Blood Mononuclear Cells Lipid Peroxidation and Immune Response," *Clinical Nephrol* 25 (2): 81-86 (February 1986). For its effect on those who have undergone surgery, O. S. Gaidova et al., "The Immunomodulating Properties of Vitamin E in Surgery Involving Artificial Circulation," *Grud Serdechnososudistaia Khir* 12 (December 1990): 30-33.

163. "Vitamin E Supplementation Enhances Immune Response in the Elderly," *Nutr Rev* 50 (3): 85-87 (March 1992). Also S. N. Meydani et al., "Vitamin E Supplementation Enhances Cell-mediated Immunity in Healthy Elderly Subjects," *American Journal of Clinical Nutrition* 52 (3): 557-63 (September 1990); Penn, "The Effect of Dietary Supplementation with Vitamins A, C and E."

164. L. F. Qualtiere et al., "Menaquinone (bacterial vitamin K) Inhibits HIV-1 Induced Syncytia Formation but Not HIV-1 Replication," *International Conference on AIDS* 5 (4-9 June 1989): 566.

165. S. Sazawal et al., "Zinc Supplementation in Young Children with Acute Diarrhea in India," *New England Journal of Medicine* 333 (13) 839-44 (28 September 1995). Also S. Sazawal et al., "Zinc Supplementation Reduces the Incidence of Persistent Diarrhea and Dysentery among Low Socioeconomic Children in India," *Journal of Nutrition* 126 (2): 443-50 (February 1996).

166. I. Brody, "Topical Treatment of Recurrent Herpes Simplex and Post-herpetic Erythema Multiforme with Low Concentrations of Zinc Sulphate Solution," *British Journal of Dermatology* 104 (2): 191-94 (February 1981). Also A. Wahba, "Topical Application of Zinc-solutions: A New Treatment for Herpes Simplex Infections of the Skin?" *Acta Derm Venereol*, 60(2): 175-77 (1980).

167. W. Kneist et al., "Clinical Double-blind Trial of Topical Zinc Sulfate for Herpes Labialis Recidivans," *Arzneimittelforschung* 45 (5): 624-26 (May 1995).

168. J. Duchateau et al., "Beneficial Effects of Oral Zinc Supplementation on the Immune Response of Old People," *American Journal of Medicine* 70 (5): 1001-04 (May 1981).

169. R.S. Beach et al., "Effect of Zinc Normalization on Immunological Function in Early HIV-1 Infection," *International Conference on AIDS* 7 (1): 330 (16-21 June 1991).

170. F. Ancarani et al., "Zinc Therapy in HIV Infected Subjects," *International Conference on AIDS* 9 (1): 493 (6-11 June 1993).

171. Z. Y. Zhang et al., "Zinc Inhibition of Renin and the Protease from Human Immunodeficiency Virus Type 1," *Biochemistry* 30 (36): 8717-21 (10 September 1991).

172. E. Jacobson, "Chinese Herbs: New Hope for AIDS," *Vegetarian Times* 176 (April 1992): 85.

173. Ibid.

174. Ibid.

175. D. J. DeNoon, "Chinese Medicinal Herbs," 2.

176. Dr. Zhang, interview by the author, 22 March 1995.

177. Ibid.

178. Ibid.
179. D. Brown, lecture at AIDS Symposium, 11 December 1991.
180. N. Farnsworth, *Economic and Medicinal Plant Research, Vol. 1* (Academic Press, 1985).
181. A. Weil, *Natural Health, Natural Medicine* (Boston: Houghton Mifflin, 1990).
182. A. Fine and S. Brown, "Cultivation and Clinical Application of Aloe Vera Leaf," *Radiology* 31 (1938). See also L. J. Lorenzetti et al., "Bacteriostatic Property of Aloe Vera," *Journal of Pharm Sci* 53 (1964).
183. M. Angeles, Verastegui et al., "Antimicrobial Activity of Extracts of Three Major Plants from the Chihuahan Desert," *Journal of Ethnopharmacology* 52 (1996): 175-77.
184. Y. Inamori et al., "Antibacterial Activity of Two Chalcones, Xanthoangelol and 4-Hydroxyderricin, Isolated from the Root of Angelica Keiskie KOIDZUMI," *Chem Pharm Bull* 39 (6): 1604-05 (1991).
185. G. Maglivit, *J Clin Lab Immun* 15 (1988): 112-23.
186. Ibid.
187. Chu Da-Tong et al., "Immunotherapy with Chinese Medicinal Herbs: I. Immune Restoration of Local Xenogenic Graft—vs. host reaction in Cancer Patients by Fractioned Astragalus Membranaceous in Vitro," *J Clinical and Lab Immun.* 25 (1983): 119-23.
188. C. Wong and G. Mavligit, "Immunotherapy with Chinese Medicinal Herbs. II. Reversal of Cyclophosphamide-induced Immune Suppression by Administration of Fractionated Astragalus Membranaceous in Vivo," *J Clin and Lab Immuno* 25 (1988): 125-29.
189. M.J. Humphries et al., *Cancer Research* 48 (15 March 1988): 1410-15.
190. Ibid.
191. Yan Sun et al., "Immune Restoration and/or Augmentation of Local Graft vs. Host Reaction by Traditional Chinese Medicinal Herbs," *Cancer* (1 July 1983): 70-73.
192. Cited in Q. Zhang et al., *Ing J Immunopharm* 9 (5): 539-49 (1987).
193. Ibid.
194. D. T. Chu et al., "Immunotherapy with Chinese Medicinal Herbs. II. Reversal of Cyclophosphamide-Induced Immune Suppression by Administration of Fractionated Astragalus Membranaceus in Vivo," *Journal of Clinical Lab Immunol* 25 (3): 125-29(March 1988). See also K. S. Zhao et al., "Enhancement of the Immune Response in Mice by Astragalus Membranaceus Extracts," *Immunopharmacology* 20 (3): 225-33 (November-December 1990); H. Sugiura et al., "Effects of Exercise in the Growing Stage in Mice and of Astragalus Membranaceus on Immune Functions," *Nippon Eiseigaku Zasshi* 47 (6): 1021-31 (February 1993).
195. Cited in *Herbal Gram* #17, printed by the Herbal Research Foundation in Boulder, Colo.
196. Brown, lecture.
197. H. S. Benjamin, et al., "Macrophage Chemiluminescence Modulated by Chinese Medicinal Herbs: Astragalus Membranaceous and Legustrum Lucidum," *Phytotherapy Research* 3(4): 148 (1989).
198. Cited in paper on Chinese astragalus from Botanicare Natural Products, Haifa, Israel.
199. Godhwani et al., (title unknown). Cited in *Herbal Gram* #17, 6.
200. J. M. Grange and R. W. Davey "Antibacterial Properties of Propolis (Bee Glue)," *Journal of the Royal Society of Medicine* 83 (3 March 1990): 159-60. See also F. Focht et al., "Bactericidal Effect of Propolis in Vitro Against Agents Causing

Upper Respiratory Tract Infections," *Arzneimittelforschung* 43 (8): 921-23 (August 1993).

201. For staphylococcus, T. A. Shub et al., "Effect of Propolis on Staphylococcus Aureus Strains Resistant to Antibiotics," *Antibiotiki* 26 (4): 268-71 (April 1981). For trichomonas, J. Starzyk et al., "Biological Properties and Clinical Application of Propolis. II. Studies on the Antiprotozoan Activity of Ethanol Extract of Propolis," *Arzneimittelforschung* 27 (6) 1198-99 (1977). For herpes, M. Amoros et al., "Comparison of the Anti-herpes Simplex Virus Activities of Propolis and 3-methyl-but-2-enyl Caffeate," *Journal of Natural Products* 57 (5): 644-47 (May 1994). See also V. Bankova et al., "Chemical Composition and Antibacterial Activity of Brazilian Propolis," *Z Naturforsch* 50 (3-4): 167-72 (March-April 1995); V. Bankova et al., "Antibacterial Diterpenic Acids from Brazilian Propolis," *Z Naturforsch* 51 (5-6): 277-80 (May-June 1996).

202. IuF Maichuk et al., "The Use of Ocular Drug Films of Propolis in the Sequelae of Ophthalmic Herpes," *Voen Med Zh* (12) (December 1995): 36-39.

203. *International Journal of Immumopharmacology* 11 (1): 395-403 (1989).

204. Christopher Hobbs, "Adaptogens: All Purpose Herbs," *East West* 2 (7): 54 (July-August 1991).

205. Morton Walker, "The Carnivora Cure for Cancer, AIDS and Other Pathologies—Part II," *Townsend Letter for Doctors* 106 (May 1992): 351.

206. Ibid.

207. Ibid., 352-53.

208. T. Fukada et al., "Photodynamic Antiviral Substance Extracted from Chloerella Cells," *Appl Microbio* 16 (11): 1809-10 (November 1968).

209. Dr. Bruce Halstead, interview with the author.

210. Dr. Alan Sacks, lecture given at AIDS Symposium, 11 December 1991.

211. *AIDS Treatment News*, 3 January 1989; *Time*, 19 September 1989.

212. Arno and Feiden, *Against the Odds*, chapter 17.

213. Information provided by C. L. Foster, World Life Institute, Colton, Calif.

214. Dr. Zhang, interview with the author, 22 March 1995.

215. Ibid.

216. Ibid.

217. R. Chang, "On the Treatment of AIDS WIth Chinese Herbal Medicines," in *Ancient Roots: A Modern Medicine: A Cross Cultural Discussion on AIDS* (1990), 102-04.

218. J. Roesler, *International Journal of Immunopharmacology* 13 (1): 27-37 (1991), [cited in *Herbal Gram* #17, 14].

219. V. R. Bauer et al., "Immunologic in Vivo and in Vitro Studies on Echinacea Extracts," *Arzneimittelforschung* 38 (2): 276-81 (February 1988). See also M. Bukovsky et al., "Immunomodulating Activity of Ethanol-water Extracts of the Roots of Echinacea Gloriosa L., Echinacea angustifolia DC. and Rudbeckia speciosa Wenderoth Tested on the Immune System in C57BL6 Inbred Mice," *Cesk Farm* 42 (4): 184-87 (August 1993); C. Steinmuller et al., "Polysaccharides Isolated from Plant Cell Cultures of Echinacea Purpurea Enhance the Resistance of Immunosuppressed Mice Against Systemic Infections with Candida Albicans and Listeria Monocytogenes," *International Journal of Immunopharmacol* 15 (5): 605-14 (July 1993).

220. Gary Null and Barbara Seaman, *For Women Only* (New York: Seven Stories, 1999), 178.

221. Chandra et al., *Fitoterapia* 58 (1): 23-32 (1987).

222. Null, *For Women Only*, 406.

223 L. M. Williamson et al., "Effect of Feverfew on Phagocytosis and Killing of Candida by Neutrophils," *Inflammation* 12 (1): 11-16 (February 1988).

224. J. Abdullah et al., *Journal of the National Medical Association* 80(4): 439-45 (1988). Also S. Yoshida et al., "Antifungal Activity of Ajoene Derived from Garlic," *Appl Environ Microbiol* 53 (3): 615-17 (March 1987): M. Adetumbi et al., "Allium Sativum (garlic) Inhibits Lipid Synthesis by Candida Albicans," *Antimicrob Agents Chemother* 30 (3): 499-501 (September 1986).

225. N. L. Guo et al., "Demonstration of the Anti-viral Activity of Garlic Extract Against Human Cytomegalovirus in Vitro," *Chinese Medical Journal* 106 (2): 93-96 (February 1993).

226. K. V. Singh and N. P. Shukla, *Fitoterapia* 55 (5) 313-15 (1984).

227. S. R. Gogu, J. J. Lertora et al., "Protection of Zidovudine-Induced Toxicity."

228. Ibid.

229. Dr. Zhang, interview with the author, 22 March 1995.

230. Cited in "Garlic Increases NK Cell Function," *Positive Health News* (Fall 1997): 10.

231. T. Abdullah et al., "Garlic as an Antimicrobial and Immune Modulator in AIDS," *International Conference on AIDS* 5 (4-9 June 1989): 466.

232. C. Atzori et al., "Activity of Bilobalide, A Sesquiterpene from Ginkgo Biloba, on Pneumocystis Carinii," *Antimicrob Agents Chemothe* 37 (7): 1492-96 (July 1993).

233. Farnsworth, *Economic & Medicinal Plant Research*.

234. F. Scaglione et al., "Immunomodulatory Effects of Two Extracts of Panax Ginseng," *Drugs Exp Clin Research* 16 (10): 537-42 (1990).

235. S. K. Chong et al., "In Vitro Effect of Panax Ginseng on Phytohaemagglutinin-induced Lymphocyte Transformation," *International Arch Allergy Appl Immunol* 73 (3): 216-20 (1984). See also G. Yang and Y. Yu, "Immunopotentiating Effect of Traditional Chinese Drugs—Ginsenoside and Glycyrrhiza Polysaccharide," *Proceedings of the Chinese Academic Medical Society: Peking Union Medical College* 5 (4): 188-93 (1990).

236. J. Y. Kim et al., "Panax Ginseng as a Potential Immunomodulator: Studies in Mice," *Immunopharmacol Immunotoxicol* 12 (2): 257-76 (1990). See also K. Ohtani et al., (title unknown), *Chem Pharm Bulletin* 37 (10): 2587-91 (1989), B. Kenarova et al., "Immunomodulating Activity of Ginsenoside Rg1 from Panax Ginseng," *Japanese Journal of Pharmacology* 54 (4): 447-54 (December 1990).

237. Y. K. Cho et al., "The Effect of Red Ginseng and Zidovudine on HIV Patients," *International Conference on AIDS* 10 (1): 215 (7-12 August 1994).

3238. Leonard Horowitz, interview with the author.

239. Gary Null, *Get Healthy Now* (New York: Seven Stories, 1999), 193.

240. J. A. Pizzorno, *A Textbook of Natural Medicine* (Seattle: John Bastyr College Publications, 1985). See also G. H. Rabbani et al., "Randomized Controlled Trial of Berberine Sulfate Therapy for Diarrhea due to Enterotoxigenic Escherichia coli and Vibrio cholerae," *Journal of Infectious Diseases* 155 (5): 979-84 (May 1987).

241. D. Sun et al., "Berberine Sulfate Blocks Adherence of Streptococcus Pyogens to Epithelial Cells, Fibronectin, and Hexadeceane," *Antimicrobial Agents and Chemotherapy* 32 (9): 1370-74 (September 1988).

242. Pizzorno. *Textbook of Natural Medicine*. See also R. Pompeii et al., "Antiviral Activity of Glycyrrhizic Acid," *Experientia* 36 (1980): 304; T. Utsunomiya et al., "Glycyrrhizin Improves the Resistance of Thermally Injured Mice to Opportunistic Infection of Herpes Simplex Virus Type I," *Immunology Letters* 44 (1995): 59-66.

243. R. Pompei et al., "Glycyrrhizic Acid Inhibits Virus Growth and Inactivates Virus Particles," *Nature* 281 (5733): 689-90 (25 October 1979).

244. Cited in *Herbal Gram* #17.

245. K. More et al., "Anti-tumor Activities of Edible Mushrooms by Oral Administration," in *Cultivating Edible Fungi*, ed. P. J. Wuest et al. (New York: Elsevier, 1987).

246. K. Aduchi et al., "Potentiation of Host-mediated Anti-Tumor Activity in Mice by B-Glucan Obtained from Grifola Frondosa (Maitake)," *Chem Pharm Bull* 36 (1988): 1000.
247. H. Nanba, "Maitake Mushrooms," *Mushroom Sci* XII (1989): 653.
248. Q. Zhang et al., *International Journal of Immunopharmacology* 9 (5): 539-49 (1987).
249. The first study is K. Humprecht et al., *International Journal of Immunopharmacology* 9(2) 199-209 (1987). The second study is R. Gorter et al., "Anti-HIV and Immunomodulating Activities of Viscum Album," *International Conference on AIDS* 8 (3): 84 (19-24 July 1992).
250. Pizzorno, *Textbook of Natural Medicine*.
251. For herpes, D. B. Mowrey, *The Scientific Validation of Herbal Medicine* (New Canaan, Conn.: Keats Publishing, 1986), 73. For bowel, "Treating Irritable Bowel Syndrome with Peppermint Oil," *British Medical Journal* (6 October 1979): 835-36.
252. Foster, World Life Institute.
253. J. Cyong et al., "Anti-Bacteriodes Fragilis Substance from Rhubarb," *Journal of Ethnopharmacology* 19 (3): 279-83 (May 1987).
254. L. Ma, "Experimental Study on the Immunomodulatory Effects of Rhubarb" *Chung Hsi I Chieh Tsa Chih* 11 (7): 418-419 (July 1991).
255. Wang and Watson, 1994.
256. Hobbs,"Adaptogens: All Purpose Herbs."
257. K. Bradley, "'Informal' U.S. Trials of French Herbal Extract Open," *Bay Area Reporter*, 13 April 1995, 20.
258. D. Merueol, et al., "Therapeutic Agents with Dramatic Antiretroviral Activity and Little Toxicity at Effective Doses: Aromatic Polycyclic Diones Hypericin and Pseudohypericin," *Proceedings of the National Academy of Sciences* 85 (July 1988): 5230-34.
259. Ibid. See also J. Serkedjieva et al., "Antiviral Activity of the Infusion (SHS-174) from Flowers of Sambucus Nigra L., Aerial Parts of Hypericum Perforatum, L, and Roots of Saponaria Offinalis L against Influenza and Herpes Simplex Viruses," *Phytotherapy Research* 4 (3): 97-100 (1990).
260. G. T. Tan et al., "Evaluation of Natural Products as Inhibitors of Human Immunodeficiency Type I (HIV-1) Reverse Transcriptase," *Journal of Natural Products* 54 (2 January 1991): 1.
261. R.E. Kilkuskie et al., "Characterization of Tannins Containing Anti-HIV Activity," *Annals of the New York Academy of Sciences* 616 (1990): 542-44.
262. Null, *Get Healthy Now*, 196.
263. A. Mazumder et al., "Inhibition of Human Immunodeficiency Virus type-1 integrase by Curcumin," *Biochem Pharmacol* 49 (8): 1165-70 (18 April 1995).
264. Z. Sui, et al. "Inhibition of the HIV-1 and HIV-2 proteases by Curcumin and Curcumin Boron Complexes," *Bioorg Med Chem* 1 (6): 415-22 (December 1993). For antifungal effects, A. Apisariyakul et al., "Antifungal Activity of Turmeric Oil Extracted from Curcuma Longa (Zingiberaceae)," *Journal of Ethnopharmacology* 49 (3): 163-69 (15 December 1995).
265. 2189. R. Shihman et al., *Fitoterapia* 55 (5) 313-15 (1988).
266. *Journal of the ADA*, April 1995.
267. Ibid.
268. G. James, "The DNCB Files," *Continuum* 4 (3): 30-31 (September/October 1996).
269. K. Hiarabayashi et al., (title unknown), *Chem Pharm Bulletin* 37 (9): 2410-12 (1989).
270. Dr. Donald J. Brown, interview with the author, in *The Pain, Profit and Politics of AIDS*.

271. D. H. Baker, (title unknown), "Cellular Antioxidant Status and Human Immunodeficiency Virus Replication," *Nutrition Review* 50 (11): 15-18.
272. Webster, *The Great Instauration*, 247.
273. Chuck DeMarco, interview with the author, in *AIDS, The Untold Story*.
274. H. Suzuki et al., "Lignosulfonate, a Water-Solublized Lignin from the Water Liquor of the Pulp Process, Inhibits the Ineffectivity and Cytopathic Effects of Human Immunodeficiency Virus in Vitro," *Agric Bio Chem* 53 (12): 3372 (1989).
275. Arno and Feiden, *Against the Odds*, Chapter 8.
276. J. A. Golden et al., "Prevention of Pneumocystic Carinii Pneuomona by Inhaled Pentamidine," (25 March 1989): 654.
277. J. Nesmith, "Castor Bean Extract Tested as AIDS Drug," *American Statesman*, 11 December 1988.
278. Ibid.
279. J. Roka, "The Inhibition of Viral Replication by R-1103 and C-1983," *Biologische Medizen* (April 1986): 181-89.
280. Dr. Morton Walker, interview with the author, 1992.

Lifestyle Changes
 1. Patrick Donnelly, interview with the author, in *AIDS, The Untold Story*.
 2. Richard Pierce, interview with the author, in *AIDS, The Untold Story*.
 3. L. M. Hecker and D. P. Kotler, "Malnutrition in Patients with AIDS," *Nutrit Rev* 48 (11): 393-401 (November 1990).
 4. For weight increase or decrease, S. Johnston, M. Gomez., A. Jou et al., "Long Term Efficacy of Nutritional Intervention in AIDS Patients," *International Conference on AIDS* 9 (1): 529 (6-11 June 1993). For restoration of tissues, J. E. Tuttle-Newhall, M. P. Veerabagu, E. Mascioli, G. L. Blackburn, "Nutrition and Metabolic Management of AIDS During Acute Illness," *Nutrition* 9 (3): 240-44 (May-June 1993). For lessening effects of AIDS, see L. Boykin, "Nutrition and HIV/AIDS," *International Conference on AIDS* 8 (3): 59 (19-24 July 1992).
 5. B. Abbruzzese, E.W. Richards, R. Clinton, A. Pelfini, F.O. Cope, M.K. Hellerstein, "Enteral Nutrition Supplementation Improves Bowel Function in the HIV+/AIDS Patient," *International Conference on AIDS* (6-11 June 1993): abstract PoB36-2369.
 6. Marjorie Siebert, interview with the author, in *The Pain, Profit and Politics of AIDS*.
 7. Joan Priestley, interview with the author, in *The Pain, Profit and Politics of AIDS*.
 8. Robert Cathcart, interview with the author, in *The Pain, Profit and Politics of AIDS*.
 9. Joan Priestley, interview with the author, in *The Pain, Profit and Politics of AIDS*.
10. C. Schlenzig, H. Jaegar, M. Wehrenberg, J. Poppinger, and H. Rieder, "Physical Exercise Favorably Influences the Course of Illness in Patients with HIV and AIDS," *International Conference on AIDS* 8 (2): B153 (19-22 July 1992).
11. Joan Priestley, interview with the author, *The Pain, Profit and Politics of AIDS*.
12. Paul Epstein, interview with the author, in *AIDS, The Untold Story*.
13. Eric Schneider, interview with the author, in *AIDS, The Untold Story*.
14. Ibid.
15. Abigail Rist-Podreca, interview with the author, in *AIDS, The Untold Story*.
16. Ibid.
17. P. Scherer, "Acupuncture," *Gay Men's Health Crisis Treatment News* 5 (4) (15 May 1991).
18. Ibid.; M. O. Smith, "Results of Chinese Medical Treatment Show Frequent Symptom Relief and Some Apparent Long-Term Remissions," World Congress of Acupuncture and Natural Medicine, Beijing, November 1987.

19. M. Goh and M. Smith, "The Development of Acupuncture Treatment for AIDS in the USA," *American Journal of Acupuncture* 20 (4): 361 (1992); Smith, "Results of Chinese Medical Treatment."

20. M. Goh and M. Smith, "The Development of Acupuncture for AIDS Treatment in the United States," *International Conference on AIDS* 8 (2): B152 (19-24 July 1992).

21. Majid Ali, interview with the author, in *AIDS, The Untold Story.*

22. Ibid.

23. Null, *For Women Only*, 596.

24. M. Naluyange, M. Ssemukasa, and E. Brehony, *International Conference on AIDS* 10 (2): 236 (7-12 August 1994).

25. P. Courmont, A. Morel and I. Bay, "Sur le Pouvoir Infertilisant de quel Ques Essences Vegetables vis-a-vis du Bacille Tuberculosis Humain," *C R Soc Biol*, 1927.

26. J. Valnet, *The Practice of Aromatherapy* (Rochester, Vt.: Healing Arts Press, 1980), 52.

27. N. Teopista, "The Use of Aromatherapy in the Management of People with AIDS: The AIDS Support Organization, Ugandan Experience," *International Conference on AIDS* 9 (1): 497 (6-11 June 1993).

Sir Francis Bacon and Ancient Wisdom

1. Francis Bacon, "The Wisdom of the Ancients," in Francis Bacon, *The Moral and Historical Works of Lord Bacon*, 201.

2. Ibid., 202.

3. Ibid., 227.

4. Ibid, 228.

5. Paolo Rossi, *Francis Bacon: From Magic to Science*, trans. Sacha Rabinovitch (Chicago: University of Chicago Press, 1968), 89.

Dr. Christopher Calapai

1. Dr Christopher Calapai, interview by the author.

2. P. A. Maki and P. R. Newberne, "Dietary Lipids and Immune Function," *Journal of Nutrition* 122 (3): 610 (1992).

3. R. B. Nieman, J. Fleming, R. J, Coker, J. R. W. Harris, and D. M. Mitchell, "The Effect of Cigarette Smoking on the Development of AIDS in HIV-1 Seropositive Individuals," *AIDS* 7 (1993): 705-10.

Dr. Emmanuel Revici

1. E. Revici, "Research in Physiopathology as Basis of Guided Chemotherapy with Special Application to Cancer," 1961, 42.

2. Ibid., 36-37.

3. E. Revici, "Research and Theoretical Background for Treatment of the Acquired Immunodeficiency Syndrome (AIDS)," *Townsend Letter for Doctors* 45 (February-March 1987): 11.

4. Ibid.

5. Ibid.

Oxygen, Ozone, and AIDS

1. H. Kief, "Die Behandlung von Viruserkrankungen Mit Ozon," *Erfahrungsheilkunde* 37 (1988): 3-11.

2. Ibid.

3. In serum, M. T. F. Carpendale and J.K. Freeberg, "Ozone Inactivates HIV at Noncytotoxic concentrations," *Antiviral Research* 16 (1991):, 281-92. In whole blood, K. F. Wagner, D. L. Mayers, G. P. Linette and R. Sheppard, "Effect of

Ozone on HIV in Experimentally Infected Human Blood," 28th Interscience Conferences on Antimicrobial Agents and Chemotherapy, Los Angeles, Calif, 1988.

4. Carpendale and Freeberg, "Ozone Inactivates HIV," 290.

5. Kief, "Die Behandlung." See also M. T. F. Carpendale, "Resolution of AIDS Diarrhea," International Conference on AIDS, 1989.

6. C. R. Steinhart et al., "The Effect of Hyperbaric Oxygenation (HBO) on HIV-Associated Chronic Fatigue and Peripheral Neuropathy," *Alternative & Complementary Therapies* (July/August 1996): 236-40.

7. Kief, "Die Behandlung."

8. Ibid.

9. K. F. Wagner, D. L. Mayers, G. P. Linette and R. Sheppard, "Effect of Ozone on HIV."

10. G. E. Garber, D. W. Cameron, N. Hawley-Foss et al., "The Use of Ozone-Treated Blood in the Therapy of HIV Infection and Immune Disease: A Pilot Study of Safety and Efficacy," *AIDS* 5(8): 981-94 (August 1991). See also M. H. Hooker and B. G. Gazzard, "Ozone-Treated Blood in the Treatment of HIV Infection," *AIDS* 6 (1): 131 (1992).

11. K. H. Wells, J. Latino, J. Gavalchin, and B. J. Poiesz, "Inactivation of Human Immunodeficiency Virus Type 1 by Ozone in Vitro," *Blood* 78 (7): 1882 (1991).

12. F. J. T. Staal, S. W. Ela, M. Roederer, M. T. Anderson et al., "Glutathione Deficiency and Human Immunodeficiency Virus Infection," *Lancet* 339 (8798): 909-912 (11 April 1992). Cited in *CDC AIDS Weekly* , 11 May 1992, 21.

13. Ibid.

14. Bukowski and Welsh, 1985. See also, V. Bocci, "Ozonization of Blood for the Therapy of Viral Diseases and Immunodeficincies: A Hypothesis," *Medical Hypotheses* 39 (1992): 30-34; V. Bocci and L. Paulesu, "Studies on the Biological Effects of Ozone 1. Induction of Interferon Gamma on Human Luekocytes," *Haematologica* 75 (6): 510-15 (1990).

15. John Pittman, interview with the author, 14 March 1995.

16. Ibid.

17. Ibid.

PART SEVEN: A NEW MEDICAL PARADIGM
Alternative Health-Disease-Therapy Concepts

1. P. Duesberg, "Human Immunodeficiency Virus and Acquired Immunodeficiency Syndrome: Correlation but Not Causation," *Proceedings of the National Academy of Sciences* 86 (1989): 755-64.

2. Harakeh, "Comparative Study of the Anti-HIV"; Penn, "The Effects of Dietary Supplementation with Vitamins A, C and E."

3. Carpendale and Freeberg, "Ozone Inactivates HIV"; Kief, "Die Behandlung"; and K. H. Wells, J. Latino, J. Gavalchin, and B. J. Poiesz, "Inactivation of Human Immunodeficiency Virus Type I by Ozone in Vitro," *Blood* 78 (7): 1882-90 (1991).

4. Is there something paradoxical about these similarities of effect? Ozone is a powerful oxidant; vitamin C is a very effective antioxidant. These parallel effects of seeming opposites may deserve more looking into.

5. V. B. Carson, "Prayer, Meditation, Exercise and Special Diets: Behaviors of the Hardy Person with HIV/AIDS," *Journal of the Association of Nurses in AIDS Care* 4 (3): 18-28 (July-September 1993); I. Wolffers and S. de Moree, "Use of Alternative Treatments by HIV-Positive and AIDS Patients in the Netherlands," *Nederlands Tijdschrift Voor Geneeskunde* 138 (6): 307-10 (5 February 1994); I. Wolffers and S. de Moree, "Alternative Treatment as Contribution to Care of pwHIV/AIDS," *International Conference on AIDS* 10 (2): 66 (7-12 August 1994); and W. Anderson, B. B. O'Connor, R. R. MacGregor, and J .S. Schwartz,

"Patients Use and Assessment of Conventional and Alternative Therapies for HIV Infection and AIDS," *AIDS* 7 (4): 561-65 (April 1993).

6. Fred Bingham, interview with the author, in *The Pain, Profits and Politics of AIDS*.
7. Ibid.
8. Ibid.
9. Ibid.
10. Cliff Goodman, interview with the author, in *The Pain, Profits and Politics of AIDS*.
11. Celia Farber, "AIDS Words from the Front," *Spin*, May 1991.
12. Anderson et al., "Patients Use and Assessment of Conventional and Alternative Therapies."
13. Nick Siano, interview with the author, in *AIDS, The Untold Story*.
14. Keith D. Cylar, interview with the author, in *The Pain, Profit, and Politics of AIDS*.
15. Gary Null interview with Bill Thompson, in The Pain, Profits and Politics of AIDS.
16. Peter Hendrickson, interview with the author, in AIDS, The Untold Story.
17. Jifunza Wright, interview with the author, in *The Pain, Profits and Politics of AIDS*.
18. Cliff Goodman, interview with the author, in *The Pain, Profits and Politics of AIDS*.
19. Ibid.

AIDS Conference in Durban, South Africa, in July 2000

1. This is reported in "Opening Ceremony Hoopla," ACT UP New York Web site, http://www.actupny.org/reports/durban, 1 November 2000.
2. Thabo Mbeki, "Speech of Thabo Mbeki, President of South Africa, at the Opening Ceremony of the 13th International AIDS Conference, Durban, July 9, 2000,"ACT UP New York Web site.
3. Ibid.
4. Nelson Mandela, "Closing Address by Former President Nelson Mandela," 13th AnnualAIDS Conference, 14 July 2000, Durban, South Africa." From the ACT UP New York Website, op cit.
5. Ibid.
6. Patrice Lumumba, "A Rejection of European Rule with a Demand for Independence," in *The African Reader: Independent Africa*, eds. Wilfred Cartey and Martin Kilson (New York: Vintage Books, 1970), 88.
7. Anita Allen, personal communication, 5 March 2001.
8. Mbeki, "Speech of Thabo Mbeki."
9. Ibid.
10. Rian Malan, "AIDS in Africa: In Search of the Truth," *Rolling Stone*, 22 November 2001, 100.
11. Ibid.
12. Ibid.
13. Bob Roehr, "AIDS Conference Concludes in Durban," *Bay Area Reporter*, 20 July 2000, available on ACT UP New York Web site.
14. Ibid.

The Community's Role in the Fight Against AIDS

1. Arno and Feiden, *Against the Odds*, 129.
2. Ibid., 131-37.
3. Ibid., 112-13.
4. Ibid., 202.

The New Paradigm and The South African AIDS Conference

1. Leopold Sedar Senghor, "Negritude: A Humanism of the Twentieth Century," in *The African Reader: Independent Africa*, 182.

2. Ibid., 183.
3. Ibid.
4. Ibid., 189.
5. Kwame Nkrumah, "African Socialism Revisited," *inThe African Reader: Independent Africa*, 204.
6. Anita Allen, personal communication, 5 March 2001.
7. Joan Shenton, "Interview with Professor Sam Mhlongo," *New African*, July-August 2000. Availabe on Web site, http://www.virusmyth.com.aids/data/pdlecture.htm.
8. Ibid.
9. Ibid.
10. Roberto Geraldo, "'Co-Factors' Cause AIDS," an article written in June 2000 and posted during the Internet Discussion of the South African Presidential AIDS Advisory Panel. Web address: www.Roberto.Geraldo.com.

The New Paradigm and Good and Evil

1. Rossi, *Francis Bacon*.
2. Ibid., 214.
3. Bacon quoted in Ibid., 159.
4. Eric Hobsbawn, *The Age of Capital 1848-1875* (New York: Mentor, 1975), 296.
5. Robert W. McChesney, *Corporate Media and the Threat to Democracy* (New York: Seven Stories, 1997), 16.
6. Ibid., 14.
7. Eric Bentley, *The Playwright as Thinker: A Study of the Drama in Modern Times* (New York: Meridian, 1955), 235.
8. Rossi, *Francis Bacon*, 9, my emphasis.
9. Jurgen Habermas, *Toward a Rational Society: Student Protest, Science, and Politics* , trans. Jeremy J. Shapiro (Boston: Beacon Press, 1970).
10. Rollo Whitehead, *Poems and Fragments* (New York: P.O.N. Press, 1992).

Appendix I: Chinese Herbs for the Immune System

1. Chang, "On the Treatment of AIDS With Chinese Herbal Medicines."

Appendix XI: Anti-Chronic Fatigue Protocol

1. Weil, *Natural Health*, 271-272.
2. D. A. Vandenbergh et al., "Plant Products as Potential Antivirus Agents," *Bulletin Institute Pasteur* 84 (1986): 101-47.
3. M. Abou-Karem and T. Shier, "A Simplified Plague Reduction Assay for Antiviral Agents from Plants," *Journal of Natural Products* 53 (2): 340-44 (4 March 1990).
4. L. Van Hoof et al., "Plant Antiviral Agents, VI. Isolation of Antiviral Phenolic Glucosides from Populus Cultivir Beaupre by Droplet Counter-Current Chromatography," *Journal of Natural Products* 52 (4): 875-78 (July-August 1989).
5. Y. Tsuchiya et al., "Antiviral Activity of Natural Occurring Flavinoids in Vitro," *Chem Pharm Bull* 33 (9): 3886-91 (29 October 1984).

Appendix XII: Paul M. Holt, B.S., D.C.

1. *Newsletter of People with HIV/AIDS Coalition* , February 1992.

Index

AZT
  adverse effects, 25, 59, 409, 412–413,
    457, 563–564
  in Africa, 170
  and African-Americans, 145–146
  anecdotal report, 589–590
  and anemia, 409, 411, 424
  and bone marrow, 430–431, 474
  and cancer, 406–407, 423–424
  and cancer drugs, 401
  and children, 424
  Concorde study, 418–419
  and ddI, 474
  development, 405–407
  early use, 411–414
  economics, 59, 355, 603–604
  and hemophiliacs, 39, 268, 426
  and HIV-positive pregnant women,
    171
  and longevity, 424–426
  long-term effects, 475
  and long-term survivors, 431–432
  and lymphocytes, 422–423
  manufacturer, 196
  marketing, 453–456
  and media, 448
  and mitochondrial function, 421–422,
    478
  and nutrients, 648
  prescribed wrongfully, 426–428
  and protease inhibitors, 476–477
  rationale for, 429–431
  research, 90
  results, 177
  and T-cell counts, 189
  toxicity, xiv, 216, 402, 406, 411, 418,
    421–424, 654, 655
  trials, 221, 402–403, 407–411,
    415–421, 469, 604–605

babies, 209, 267, 269, 276, 417, 424, 427,
  500
Back, Kurt, 223
Bacon, Sir Francis, xv–xvi, 7, 8–9, 29, 30,
  36, 73–74, 88, 97, 164, 173, 178–179,
  182, 189, 202, 232–233, 254–255,
  286– 287, 321–322, 347, 381–382,
  489–490, 557–559, 602, 611, 612–613,
  616
bactericide phytochemicals, 540
Bactrim, 562

Bakhtin, Mikhail, 374
Baltimore, David, 17–18, 332, 333, 353,
  393, 395
Baltimore (Md.), 396
Barr, David, 599
Barre, Francoise, 31
basil, 525
Bastyr, John, 555, 629–630
Bauman, Zygmunt, 75–76
Baumgarten, Michael, 272
BdZ (Barclays de Zoete Wedd), 437
Beatles, 291
Beauchamp, Dan, 292
Beckstrom, Stephen M., 539
bee propolis, 525
Beirn, Terry, 116–117
Belgian Congo, 314
Belgium, 330
Beltz, Richard, 406
Benbrik, E., 478
Bentley, Eric, 615
Bergalis, Kimberly, 125, 449
Bergman, Jeanne, 483
beta-carotene, 495, 507–509, 643, 652
Bethel, Tom, 200, 418, 445–446, 454
Bey, Hakim, 292–293
BHV-1 (bovine herpes virus), 311
Bialy, Harvey, 18, 108, 466
bifidus, 500
Biggar, Robert, 358
Bigger, Robert, 147
Bihari, Bernard, 604, 653
Bingham, Fred, 589–590
Biocine, 393
biofeedback, 220
biological weapons, 337, 338–346
Biostim, 653–654
biotechnology, 329–333
Birmingham (Ala.), 396
Birmingham Centre for Cultural Studies,
  290–291
bitter melon, 529, 656
Black, Dean, 178, 388
blacks' immunity to AIDS, 332
  See also African-Americans
blood banks, 300
blood plasma, 320
blood pressure, 647
blood tests, 628, 645
  See also testing for AIDS
blood transfusions, 168, 267–269, 431

BMA Foundation for AIDS, 412
Bolognesi, Dani, 369, 394
Bomgardener, Michael, 41, 431
bordering instance, 202
boron, 500
botanicals, 630
Botswana, 318
botulism vaccine, 344
bovine herpes virus (BHV-1), 311
bovine leukemia, 297–298, 299–301, 304,
    327, 334–335
bovine visna virus, 297–299, 304, 311,
    312, 335
brain tumors, 304, 313, 314–315
    See also cancer
Brazil, 302, 320
"breakthroughs," xi, 156
Brenner, Seymour, 440
Bristol-Myers Squibb, 393, 443, 456
Britain, 362, 419, 455
British Columbia, 246
British Health Services, 110
British Medical Association, 412
Broder, Sam, 407, 408
Brody, Stuart, 122, 124–125
bromelain, 501
Brown, Donald, 525
Brown, Donald J., 543
Brown, Tony, 305
Buanoukas, Frank, 367
Buchanan, Patrick, 170
Buianouckas, Frank, 10, 351
bupleurum falcatum, 526
burdock, 526
Burroughs, William, 436
Burroughs-Wellcome, 196–197, 405, 406,
    407, 408, 409, 411–412, 414, 417,
    418–419, 421, 422, 435–444, 446, 447,
    449, 603
    See also Glaxo-Wellcome;
    Wellcome
Burundi, 161, 302, 314
Bush, George H. W., 225, 449
butyl nitrite, 197
Byrnes, S., 261

Calapai, Christopher, 253, 494, 496, 497,
    503, 511–512, 561–566, 585, 588,
    627–628
California, 392
California Primate Research Center, 315

Callen, Michael, 129–130, 131, 427, 432
Campaign Against Health Fraud, 388
Canada, 213–214, 246
cancer
    and AIDS, 97, 370
    and amyl nitrite, 197, 198, 199
    and biological weapons, 340–342,
        345–346, 370–371
    chemotherapy, 430–431, 443, 475
    and cortisol, 245
    and DHEA, 245
    economics, 357
    and garlic, 531
    and maitake mushrooms, 534
    and phytochemicals, 539
    and retroviruses, 18, 86–87
    treatment, 390
    and viruses, 29, 30, 337–338
    See also specific types of cancer
candidiasis, 102, 104, 183
candidicide phytochemicals, 540
canine distemper parovirus (CPV), 310
Cantwell, A., 203, 307, 318
Capra, Fritjof, 175, 294, 364, 601, 602,
    608
Carciviren-V (C-1983), 546
Carey, Jeanne, 483
carnitine, 498, 648, 655
carnivora (dronaea muscipula), 526
Carolina Center for Bio-oxidative
    Medicine, 579
Carr, Andrew, 470
Carrigan, Donald R., 252
Carswell, J. W., 318
Carswell, Wilson, 153
castanospermine, 527
Castells, Manuel, 162–163, 171, 598
castor oil packs, 633
cat AIDS virus, 299
caterpillar fungus, 527
Cathcart, Robert, 441, 511, 512, 513, 515,
    516, 549
Catholic Church, 9, 214
cat leukemia sarcoma, 337–338
Caton, Hiram, 7, 284, 362
CD4 cells, 20, 41, 49, 65, 99, 185–189,
    243–245, 253, 422–423
CD4 count assay, 366
CD8 cells, 65, 187, 189, 253, 422–423,
    477–478
celebrities, 354, 359, 361, 449, 450

cell cultivation techniques, 29
Centers for Disease Control (CDC), ix,
    16, 34–35, 38, 41, 57, 95, 97, 98,
    99–100, 102, 104, 105, 107, 108, 111,
    112–113, 117–118, 120, 129, 164, 168,
    182, 196, 199–200, 201, 206–207, 209,
    211–213, 245, 259, 269, 274, 299, 305,
    335, 358, 365, 545
Central Intelligence Agency (CIA), 339,
    340
cervical cancer, 97, 100, 102, 221
    See also cancer
Chad, 320
Chaitow, Leon, 276
Chandra, T., 530
Chang, R. S., 521, 621–623
Charles Pfizer and Company, 341
chemical weapons, 342
Chermann, Jean-Claude, 31, 35, 64
Chernov, Harvey I., 411
Chiasson, Mary Ann, 477
Chicago, 305
chicken sarcoma, 338
children, 102, 225, 226–227, 248, 417,
    424, 436, 508, 509
chimps, 39, 40, 301
    See also monkeys
China, 329
Chinatown (San Francisco), 231, 232
Chinese herbs, 621–626, 648
Chirac, Jacques, 27–28
Chirimuuta, Richard, 143, 147–148,
    151–152, 169
Chiron, 366
chlorella, 528
Christian Voice, 224–225
Christie, H., 69, 475, 476–477
Christie, Hue, 430
Christoffersen, Ralph, 440
Chung, Nhi, 552–553
CIA (Central Intelligence Agency), 339,
    340
Circle K company, 324
citrus seed extract, 528
Clemenson, Jay, 296
Clerici, Mario, 244
Cline, F., 127
Clinton, Bill, 112, 323, 344, 438
clones. See "fast-track" lifestyle
cloning, 79–82
CMV (cytomegalovirus), 131, 183, 280, 471

Coalition for Lesbian and Gay Rights,
    369
cocaine, 206, 207–208, 212
Coenzyme Q10 (CoQ10), 501, 647, 648,
    652
cofactor concept, 15–16
Cohen, J., 354, 355, 366
Cohen, Stanley, 290–291
cold fusion, 5
Coleridge, Samuel Taylor, 485, 611, 613
colonics, 631
Colter, Sy, 329
community-based research, 90–91, 175,
    529, 604
community-based treatment, 94
Community Research Initiative, 604
community role, 601–605
compound Q, 528–529
Conant, Marcus, 424–425, 426
Concorde trials, 188, 415, 418–419, 424,
    426, 437, 475, 477
condoms, 132
Conlan, Mark, 4–5, 104, 128, 193,
    209–210, 365, 447
Conlan, Passante, 360–361
conservatism, 224–225, 226, 614
conspiracy theorists, 326, 371
consumer culture, 266
contamination, 70–73
Coodley, Gregg, 507, 508–509
Cooke, Blanche Wiesen, 335
Cookson, J., 327
Cooper, L., 242
cooperation, 88–91, 94
Corey, Lawrence, 425–426
correlation-only evidence, 37–39
cortisol, 187, 240, 242–245, 651
Cournoyer, Cynthia, 392
Cox, David, 109
Cox, P., 266
CPV (canine distemper parovirus), 310
crack, 206, 207
    See also cocaine
Craven, Barry, 114, 116
Crisp, Quentin, 231
Crixivan, 472–473
cross-reactions, 646
cruciferous extract, 528
cryptococcosis, 182–183
cumulative burden thesis, 205206
Curran, James, ix–x, xi, 200

Currant, James, 279, 360, 413, 428, 432
Currin, Sean, 127
Curtis, T., 296, 297–298, 303, 304, 305, 308, 309, 310, 311–312, 314, 327, 334, 335
Cylar, Keith, 590
cytokines, 194
cytomegalovirus (CMV), 131, 183, 280, 471
Cytomel, 651

D4T (drug), 476–477
DAAIR (Direct AIDS Alternative Information Resources), 589, 643–648
Dalton, Harlon, 215
DASM (dehydrozndrographolide succinic acid monoester), 529–530
Davis, Karen, 324
Davis (Calif.), 329
ddC (drug), 59, 460, 478
ddI (drug), 59, 170, 443, 460, 474, 478, 648
DEA (Drug Enforcement Administration), 207
Dean, James, 231
death wish, 230–231
Defense Department, 344, 396
De Fries, F., 244
Delaney, Martin, 419–420
Delta I study, 460
DeMarco, Chuck, 544
deMause, Lloyd, 232
Denmark, 125
Denver, 264, 305
Department of Agriculture, 298, 299, 337, 538, 539
Department of Defense, 344, 396
Department of Health and Human Services, xi, 29, 33, 34, 105–106, 168, 207, 367–368
Department of Justice, 305
depression, 241, 273
Detroit Institute for Cancer Research, 406
DeVincenzo, Joy, 512
Dharmananda, Subhuti, 521
DHEA, 243, 501–502, 650
DHEA-S, 243
diabetes, 471
dialogue as part of writing, 374, 382–383
diaries, 645–646

die-off effect, 655
dietary modifications, 548–550, 638
diethyl dithiocarboliate, 653
DiFerdinando, Tom, 187
Diflucan, 549
digestive enzymes, 642, 656
digestive remedy, 641–642
Dingell, John, 29, 33
dioxychlor, 542
Direct AIDS Alternative Information Resources (DAAIR), 589, 643–648
"direct observation" treatment model, 482–484
Disabato, Joe, 447
discordant couples, 119
diseconomies of scale, 364, 365, 366
dissenters, 8–11, 367–372, 596, 609–610
DNA, 17–18, 19, 74, 75, 76, 77, 79, 81–82
DNA chain terminator. *See* AZT
DNCB (dinitrochlorobenzene), 541–542
doctors
    and AZT, 412–414
    misinformation, 245–246
    training, 388–389
dogmatism, 387–390
Donnelly, Patrick, 548
Douglas, S., 503
Douglas, William Campbell, 334–335
dreamtime interpretation, 323, 326, 413, 414
Drotman, Peter, 299
drug cocktails, 438, 457, 460–461, 462–464, 468, 469–474, 475–480, 482–484
    See also protease inhibitors; reverse transcriptase (RT) inhibitors
Drug Enforcement Administration (DEA), 207
drugs
    approval process, 468–470
    deleterious effects, 204–206
    illegal, combined with protease inhibitors, 474
    prescription, 635–636, 653–654
    research, 89–90
    toxicity, 216
    trials, 170–171, 178, 373, 398
        See also AZT; protease inhibitors; reverse transcriptase (RT) inhibitors
drug subculture, 287–290

drug users, xii, xiv, 39, 120, 201–202, 206– 207, 225, 274–275, 279, 288–290, 550
Dubos, Rene, 181, 184, 192, 440
Duesberg, Peter
    AIDS Inc., 353, 354, 355, 362, 363, 365, 367–369
    alternative causes of AIDS, 202, 205–206, 207, 208–209, 210, 211–212, 214, 215–216, 236, 264, 268, 279, 303
    AZT, 402, 403, 429, 430, 439, 466
    evidence against current AIDS hypotheses, 7, 9, 10, 16, 18, 19, 20, 21, 23, 25, 37–38, 39, 41–42, 45, 48, 59, 60, 70, 71, 79–83, 84, 85, 86, 94, 101, 102–103, 104, 107, 113, 115, 119, 120, 132, 164
    new medical paradigm, 610–611
Dugway Proving Ground, 339, 343
Duke, J., 624–626
Duke, James A., 539
Dunlap, W., 455, 456
Duperval, R., 182
DuPont Co., 330
dysidia, 542

echinacea, 530
eclipta alba, 530
economies of scale, 364
ecstasy (MDMA), 474
Edison, Thomas, 376
educators, 109
Einsteinian physics, 8
Eisenhower, Dwight, 238
elderly, 494–495, 498–499, 519
Eleopulos, Eleni, 51, 53, 60, 64, 65–66, 67–68, 69–70, 70–71, 72, 73, 77, 78–79, 81–82, 87, 146, 199, 274, 280–281, 402, 426
    See also Papadopulos-Eleopulos, Eleni
ELISA test, 44–46, 48, 49, 50, 52, 53, 55, 56, 57, 58, 148, 149, 154–155
    See also testing for AIDS
Elliot, Michael Verney, 20, 74–75, 76, 102–103, 122, 209–210, 272–273, 274, 275, 320–321, 391–392, 397, 398, 426, 447–448
Ellison, Bryan J., 8, 20, 25, 202–203, 357–358, 371
Ellner, Michael, 187, 246

emotions, 223–224, 225, 226, 227, 246–248, 247, 550–551
endorphins, 242, 245
England, 174, 355, 412
enteral diet, 496, 549
Environmental Protection Agency (EPA), 496
environmental toxins, 203
epidemic hysteria, 222–225
Epidemic Intelligence Service (EIS), 357–358, 371
Epstein, Paul, 551
Epstein, Steve, 4, 82, 130, 376
Epstein, Steven, 62, 90, 102
Epstein-Barr virus, 253, 269, 280
equine infectious anemia virus (EIAV), 336–337
Essex, Max, 55
ethics, 24
Eulenspiegel Society, 263
Europe, 111–112, 300, 329, 332, 333, 352, 366, 419
European-Australian Collaborative Group trial, 425
Evans-Pritchard, Ambrose, 135
Evatt, Bruce, 168
exercise, 550, 638

Factor VIII concentrate, 268, 281, 449
false test results, 51, 52, 53–58, 59
Farber, Celia, 24, 58–59, 99, 114, 135, 147, 149, 158–159, 164, 165–166, 171, 236, 283, 357, 362, 371, 418, 454, 456, 479
Farnsworth, Norman, 532
Farr's law, 115
"fast lane hypothesis," 128–132
"fast-track" lifestyle, 193, 199, 209, 211, 251, 260, 265–267, 272–275, 279, 280, 290, 292, 306–307, 378–381
fatty acids, short-chain, 496
Fauci, Anthony, 170, 187, 362, 369, 396, 476, 648
FBI (Federal Bureau of Investigation), 369
FDA. *See* Food and Drug Administration (FDA)
Feast, James, xv, 88, 222–223, 237, 374, 375–382, 617
Federal Bureau of Investigation (FBI), 369
Federal Interagency Committee on Recombinant DNA Research, 152

Federici, Silvia, 170–171
Feiden, Karyn L., 420, 436, 545, 604
Feldman, Martin, 423
feline lymphotropic lentivirus (FTLV),
    299
feline panleukopenia virus (FPLV), 310
female-to-male transmission, 122–123, 132
feverfew, 530–531
Fields, Bernard, 24, 28
Fisher, L. M., 455, 456
fisting, 261, 262, 263
Fleming, Thomas R., 425–426
flu shots, 54, 59
Food and Drug Administration (FDA),
    91, 189, 197, 200, 265, 354, 361, 389,
    395, 407, 409, 411, 427, 436, 437, 441,
    465, 469, 471, 475, 529, 544, 545, 644,
    653
Forest, Waves, 305
Forest Pharmaceuticals, 651
Forscher, Bernard, 6
Fort Detrick (Md.), 340, 341, 342, 357
Foss, Laurence, 220, 284, 286
Foucault, Michel, 377
FPLV (feline panleukopenia virus), 310
fragments, 76–78
France, 112, 330, 419
France, David, 449–451
Franchi, Fabio, 15
Franchini, Genoveffa, 316
Francis, Donald, 358, 395
Frankenstein, Dr., 83, 91
Frederick Cancer Research Center, 341
free radicals, 511, 531, 649
Freire, Paulo, 130–131
Friedman, David B., 380
FTLV (feline lymphotropic lentivirus), 299
functional foods, 538–539
Fundamentalists, 285, 339
funding, 83, 116–117, 118, 135–136, 157,
    356–357, 360, 367
fungicide phytochemicals, 540

Gallagher, Maggie, 117
Gallagher, T., 109
Gallo, Robert
    alternative causes of AIDS, 189,
        205–206, 253, 268, 277, 294, 304,
        307, 316, 319–320, 327, 328, 331,
        332, 337–338, 340, 341–342
    alternative treatments, 541

AZT, 451, 475
conventional treatment programs, 383,
    395, 399
evidence against current AIDS
    hypotheses, 3, 4, 7, 9–10, 15,
    21–22, 23, 28–38, 47, 48, 51, 52,
    59, 64, 66–68, 70–71, 73, 78–79,
    83, 84, 87, 147
introduction, ix–xii
new medical paradigm, 593–596, 601
Gambia, 399
Gannett, Lynn, 415–416, 417
garlic, 531–532, 638, 642, 643, 647
Garrett, L., 356, 357, 438
Garrett, Laurie, 309, 599
Gates, Bill, 140, 191, 598
Gay Activist Alliance, 380, 595
gay bathhouses, 131, 200, 202–203, 210,
    231, 264, 292, 378, 379–380
gay discotheques, 210–211, 267, 378–379
gay liberation movement, 130–131, 193,
    210, 211, 272, 292, 381
gay lifestyle vs. practices, 272–275
gay men
    and AIDS in Africa, 169–170
    and amyl nitrite, 197, 199, 201, 206,
        211, 215, 236, 264, 265
    death wish theory, 230–231
    and funding for AIDS, 117–118
    and hepatitis B, 54
    latency period in, 276
    as minority group, xii, xiii
    mortality of, 39
    New Right's views of, 224–225
    and safer sex, 236
    subculture of, 380–381
Gay Men Fighting AIDS (GMFA), 116
Gay Men's Health Crisis (GMHC), xiii,
    293–294, 333, 362, 363, 369, 454–455,
    470
Gazi, Davis, 171
GC protein, 332
GE-3-S (lichen compound), 542
Geisler, W., 111, 121–122, 127, 153, 204,
    213, 336
Gelderblom, Hans, 71–72, 73
GenBank group, 309
genetic predictability, 16
Geng, Y., 311
genistein, 504
Geraldo, Roberto, 610

Public Health Laboratory Service, 109, 245
Public Health Service, ix, 360
Pullman (Wash.), 336–337
Puritans, 381–382, 543

qigong, 630
Quan Yin Herbal support program, 553
Quayle, Dan, 449
Quinn, Matilda, 321

rabbits, 67
rabies vaccine, 330, 334
racism, 138–146, 150, 220, 317, 614
Raeymaekers, Luc, 466
Ramus, Peter, 613
Rand Corporation, 339
Rasmussen, Collin, 526–527
Rasnick, David, 5, 23, 39, 70, 71, 74, 105,
    214, 215–216, 268, 353–354, 355, 367,
    369, 426, 467–468
Ratcliffe, Molly, 99–100
Reagan, Ronald, xii, 27–28, 226
Reagan administration, x–xi, 33, 66, 225,
    248, 365
recipe for whole lemon drink, 640
recreational drugs, 201–202, 210–211,
    216
    See also amyl nitrite (poppers)
Rector-Page, Linda, 633–634
Red Cross, 168
reflexology, 554–555, 631
Regional Smut Research Laboratory, 337
Reinisch, June Machover, 123
reishi mushrooms, 535
Reiter, Russell, 654
Religious Roundtable, 224–225
Rent, 43–44
researchers, medical, 6–8
resiliency, 287
"Respect Yourself, Protect Yourself,"
    117–118
resting energy expenditure (REE),
    493–494
retrovirologists, 353–354
retrovirology, 91–94
retrovirology limitations, 82–91
retroviruses, 17–20, 69, 74
reverse transcriptase (RT) enzyme, 18, 63,
    64, 84–85
reverse transcriptase (RT) inhibitors, 461,
    476–477, 478, 647–648

See also drug cocktails; specific RT in-
    hibitors
Revia, 636, 653
Revici, Emmanuel, 567–573, 585, 588
rhubarb, 535–536
Richardson, Lynda, 483
Richey, Mike, 366
ricin, 545–546
Rifkin, Jeremy, 298–299, 299
Rist-Podrecca, Abigail, 553
Ritonavir, 476, 646
RNA, 17–18, 19, 74, 75–76, 77, 79, 81
Robert Koch Institute, 71–72, 73, 78–79
Roberts, Seth, 31
Robinson, Jo, 654
Roche Laboratories, 456, 480
Rochester (N.Y.), 396
Rockefeller, Jay, 344
Roka, J., 546
Root-Bernstein, Robert, 15, 16, 23,
    38–39, 42–43, 98, 99, 103, 107, 108,
    119–120, 121, 182, 183–184, 253,
    257–258, 262, 263–264, 267, 269,
    272–274, 274–275, 276, 278, 282,
    283–284
Ross, Andrew, 291
Ross, Graham, 421
Rossi, Paolo, 559, 612–613, 616
Roswell Park Cancer Institute, 331
Rothenberg, Kenneth, 220, 284, 286
Rovital-V (R-1103), 546
Rowland, Diane, 324
royal bee jelly, 536
Royal College of Nursing, 109
RT. See reverse transcriptase (RT)
    enzyme; reverse transcriptase (RT)
    inhibitors
Ruanda, 302
Ruanda-Urundi, 314
Rubin, Harry, 16
Ruprecht, Ruth, 395
Russia, 57, 301
Rwanda, 314

Sabin, Albert, 398
Sabin vaccine, 312, 314
Sacks, Alan, 528
S-Adenosyl-Methionine (SAMe), 650,
    651–652
SAGE (organization), 369
Salk, Jonas, 308–309, 312, 315

SAMe (S-Adenosyl-Methionine), 650, 651–652
San bush people, 318
San Diego, 360–361
SANE (Stop AIDS Now or Else), 293
San Francisco, 231, 232, 264, 266, 293, 305, 341
San Francisco AIDS Foundation, 293
San Francisco Chronicle, xvi
San Francisco General Hospital, 89, 175, 425, 476
Sanofi Anthrop, 366
saquinavir (Invirase), 366, 470–471, 472, 476
Scaglione, F., 532
scapegoating, 219, 225, 231–234
schizandra, 536
Schmidt, Casper G.
   AIDS as distraction, 228–230
   AIDS crisis as morality play, 287–288
   AIDS victims' character, 225–228
   biases, 220
   death wish theory, 230–231
   emotional reactions to AIDS crisis, 235–237
   epidemic hysteria, 222–225, 291–292
   initiation rituals, 234–235
   non-debatable ideas, 219
   origin of analysis, 222
   psychological aspects of AIDS, 217, 246–248
   stress and immune system, 237–241, 550
   theories viewed by others, 242–245
   victim roles, 231–233
   weaknesses in gay community, 248–249
Schneider, Eric, 551–552, 553
Schoofs, M., 393, 395, 399
Schoonmaker, Craig, 449
Schoub, Barry, 43, 45
Schuklent, U., 356, 357
Science, x, 31, 63, 66, 67, 68
science as business, 351–352
scientific method, xi, 3–4, 63, 69, 86, 173
Scotland, 355
Scott, Andrew, 19
Seale, J., 328–329
Seale, John, 297, 319–320
Seattle, 223, 396
seaweeds, 542

secondary transmission, 125
Segal, Jacob, 342
selenium, 258, 506–507
self-indulgence, 562
Selik, R. M., 206
Semba, Richard D., 507–508
semen, 262–263, 277–278, 281
Senate Church Commission Hearings, 339
Sencer, David, 335
Senegal, 148, 163, 320
Senghor, Leopold, 607–608
sensory overload, 266–267
sexism, 220–221
sexually transmitted diseases (STDs), 121, 122, 123, 129, 131, 155, 202–203, 208, 210, 231, 252–253, 281
   See also gonorrhea; herpes viruses; syphilis
sexual transmission of AIDS
   anal intercourse
      not spread by, 126–132
      spread by, 123–126
   and family values, 6
   misinformation from media, 133–136
   regional differences, 54
   vaginal intercourse
      not spread by, 119–123
Shakespeare, William, 615
Shanti Project, 293
shark cartilage, 546
Shearer, Gene M., 244, 399
sheep visna virus, 297–298, 327, 334–335
Shenton, Joan, 55–56, 80, 82, 160, 166
Shihman, R., 537–538
Shilts, Randy, x–xi, xii, 30, 33, 35, 89, 143, 144–145, 166, 168, 193, 195–196, 200, 246, 292, 300, 354, 361
Shinitsky, Meier, 541
Shoft, Andrew, 205
Siano, Nick, 246, 590
Siebert, Marjorie, 252, 253, 549
silymarin, 656
simian immunodeficiency virus (SIV), 40, 142, 152, 299, 315–316, 395–396
simian T-cell leukemia viruses, 316
simian virus, 141, 142
Simian Virus 40. *See* SV40 virus
Singapore, 224
SIV. *See* simian immunodeficiency virus (SIV)

vitamin B6, 495, 501, 510, 643, 648
vitamin B12, 494, 510, 643, 652
vitamin C, 280, 495, 503, 510–517,
   557–558, 588, 643, 649, 650, 652
vitamin D, 517
vitamin E, 495, 517–519
vitamin K, 519
vitamins, 628, 637
   See also specific vitamins
Voeller, Bruce, 123, 125, 127
Volberding, Paul, 425
volume displacement, xi

Waitzkin, Howard, 192
Waldholz, M., 440, 469
Walker, M., 242
Walker, Martin, 388, 407, 409, 411, 412,
   414, 436, 437, 442–443, 444, 446, 453,
   455–456
Walker, Morton, 526, 546
Wang, Wuan Xi, 630, 631
Wang, Y., 518
Warhol, Andy, 291
War of the Worlds, 237
War on AIDS, 229–230
War on Drugs, 225
Washington, D.C., 324
Washington National Airport, 343
Washington State University, 336–337
Watney, Simon, 238–239, 339, 362
Watson, R., 254
Weber, Paul F., 456
Webster, Charles, 381–382
Wedemeyer, George, 630
Weger, Nikolaus, 543
Weil, Andrew, 523, 638–639
Weiss, Robin, 412
Wellcome, 388, 412–413, 419, 420, 436,
   443, 453–456
   See also Burroughs-Wellcome;
   Glaxo-Wellcome
Wellcome, Sir Henry, 435–436
Wellcome Diagnostics, 412
Wellcome Foundation, 407–408, 435, 444
Wellcome Trust, 435–436
Weller, Ian, 419, 475–476
Welles, Orson, 237
Wells, Jody, 128
Welsh, Irvine, 195, 288–290

Western Blot test, 46, 47, 49, 50, 52, 53,
   55, 56, 57, 58, 59
   See also testing for AIDS
White, Edmund, 238
White, Ryan, 449
Whitehead, Alfred North, 602
Whitehead, James, 59, 428, 431
Whitehead, Rollo, 618
White House AIDS Commission, 449
whites' immunity to AIDS, 332
Whitleu, Richard, 294
Whitney, Charles, 321–322
Williams, Enoch, 215
Williams, Rev. Reginald, 215
Winklestein, Warren, 82
Wisconsin, 301
Wishard, Ann Marie, 632
Wisniewski, T., 243
Wistar Institute, 330
Wolinsky, Steven, 396
women, 99–100, 102, 221
Women Organized to Respond to Life-
   Threatening Diseases, 455
Woolgar, Steve, 35–36, 38
World Health Organization (WHO),
   54–55, 110–111, 121–122, 144, 147,
   148, 158–159, 164, 166, 302, 304–305,
   316, 317, 331, 333–334, 358, 597
"worried well," 240
Wright, Jifunza, 590–591
Wright, M. P., 57
Wright, Pierce, 304–305, 317

Xao, Q. S., 524

yeast in diet, 549–550
yoga, 554
Young, I., 260, 267
Young, Ian, 479

Z, Dr., 578–579, 580, 585–586, 588
Zaire, 55, 143, 154–155, 171, 302, 320
Zambia, 154, 302
zero sum game, 356–357
Zhang, Quingcai, 522
Zhang, Z. L., 524
Zhang, Z. Y., 520
Zimbabwe, 154
zinc, 257–258, 519–520